Progress in Mathematical Physics
Volume 21

Arlie O. Petters
Harold Levine
Joachim Wambsganss

Singularity Theory and Gravitational Lensing

with 184 illustrations

Birkhäuser
Boston • Basel • Berlin

Arlie O. Petters
Department of Mathematics
Duke University
Durham, NC 27708-0320
U.S.A.

Harold Levine (Professor Emeritus)
Department of Mathematics
Brandeis University
Waltham, MA 02454-9110
U.S.A.

Joachim Wambsganss
Astrophysikalisches Institut Potsdam
Universität Potsdam
14482 Potsdam
Germany

Library of Congress Cataloging-in-Publication Data

Petters, Arlie O., 1964-
Singularity theory and gravitational lensing / Arlie O. Petters, Harold Levine, Joachim Wambsganss.
p. cm.– (Progress in mathematical physics ; v 21)
Includes bibliographical references and indexes.
ISBN 0-8176-3668-4 (acid-free paper) – ISBN 3-7643-3668-4 (acid-free paper)
1. Gravitational lenses–Mathematics. 2. Singularities (Mathematics) I. Levine, Harold, 1928- II. Wambsganss, Joachim, 1961- III. Title. IV. Series.

QN857.5.G7P48 2001
522'.6–dc21

2001025643
CIP

AMS Subject Classifications: 32Sxx. 32S05, 32S20, 34Dxx, 34Dxx, 34D23, 34D30, 55N10, 57Rxx, 57R17, 57R35, 57R45, 57R50, 58Kxx, 58K05, 58K25, 58K30, 58K35, 58K40, 58K50, 58Z05, 78A05, 83Cxx, 83F05, 85A40

Printed on acid-free paper

Birkhäuser

ISBN 0-8176-3668-4 SPIN 10464426
ISBN 3-7643-3668-4

Reformatted from authors' files in LaTeX2ε by TEXniques, Inc., Cambridge, MA
Cover design by Gary Palmatier, Ideas to Images, Santa Rosa, CA
Cover design based on HST/NASA observation of the galaxy cluster CL0024+1654 by W. Colley, J. Tyson, and E. Turner
Printed and bound by Hamilton Printing, Rensselaer, NY
Printed in the United States of America

9 8 7 6 5 4 3 2 1

To my beloved wife Marcia, parents Velina and Cecil Petters, and grandparents Bernice and Gilbert Waight, whose love and unwavering support I am blessed to have.

A. O. Petters

To Renée with gratitude for her loving support of my incomprehensible mathematical life.

H. Levine

For Hanna, Lisa and Annika.

J. Wambsganss

Contents

Foreword

Astronomers do not do experiments. They observe the universe primarily through detecting light emitted by stars and other luminous objects. Since this light must travel through space to reach us, variations in the metric of space affects the appearance of astronomical objects. These variations lead to dramatic changes in the shape and brightness of astronomical sources. Because these variations are sensitive to mass rather than to light, observations of gravitational lensing enable astronomers to probe the mass distribution of the universe.

With gravitational lensing observations, astronomers are addressing many of the most important scientific questions in astronomy and physics:

• *What is the universe made of?* Most of the energy and mass in the universe is not in the form of luminous objects. Stars account for less than 1% of the energy density of the universe. Perhaps, as much as another 3% of the energy density of the universe is in the form of warm gas that fills the space between galaxies. The remaining 96% of the energy density is in some yet unidentified form. Roughly one third of this energy density of the universe is "dark matter," matter that clusters gravitationally but does not emit light. Most cosmologists suspect that this dark matter is composed of weakly interacting subatomic particles. However, most of the energy density of the universe appears to be in an even stranger form: energy associated with empty space. Recent astronomical data suggest that the universe is accelerating. This implies that there is either a cosmological constant or that there is some mysterious "dark energy" that appears to fill empty space. Since mass and energy curve space, observations of distant gravitationally lensed galaxies are our best astronomical tool for determining the distribution of mass and energy in space.

Observations of gravitational lensing in clusters directly measure their mass distribution. They offer the cleanest astronomical test of numerical simulations of the formation of structure in the early universe and some of the best constraints on the properties of dark matter.

Observations of weak gravitational lensing will enable astronomers to trace the evolution of large scale structure. Since this evolution is sensitive to the properties of both the dark matter and the dark energy, these observations promise important insights.

• *How big is the universe?* and *How old is the universe?* By measuring time delays in multiply imaged systems (see Section 4.4), astronomers can infer the size and the age of the universe. While this inference rests upon the uniqueness of the mass distribution inferred for the intervening matter, it does not have the many systematic errors associated with the traditional distance ladder.

• *What makes up most of the mass of the galaxy?* Gravitational lensing observations of stars probe the distribution of mass within our galaxy. These observations suggest that very low luminosity objects, perhaps white dwarfs, are a substantial fraction of the mass in our galaxy. These observations have also placed important constraints on the shape and composition of our galaxy.

• *Are we alone?* When a binary system passes in front of a distant star, it can produce a rich pattern of light variation as caustics move across the star's surface. These observations can reveal the existence of low mass companions and will perhaps yield the first detection of an Earth-like planet around another star.

• *What does the surface of a distant star look like?* Observations of gravitationally lensed systems can also reveal the structure of luminous objects. Astronomers have watched caustics of the gravitational lensing maps move across the surface of a star. These observations have yielded measurements of the surface structure of a star thousands of light years away and test our understanding of stellar atmospheres in novel ways.

• *What are the properties of the super-massive black holes in the centers of galaxies?* Measurements of variations in the luminosity of distant quasars due to gravitational lensing by stars in an intervening galaxy are a sensitive probe of the properties of the quasar. Quasars are thought to be powered by the accretion of gas onto a billion solar mass black hole. If this accretion disk were large, then the effects of lensing would be diluted by the large size of the light source. The rapid variability seen in these quasars implies that the quasar light is emitted from a narrow region around the central black hole.

While gravitational lensing was an important part of astronomy during the 1990's, it will likely be even more essential in this new millennium. The Sloan Digital Sky Survey

http://www.sdss.org

is beginning to report new results on weak gravitational lensing and the deep lens survey

http://dls.bell-labs.com/dls/index.html

is producing its first images. British astronomers have begun construction on the VISTA telescope

http://www.vista.ac.uk

a 4 meter telescope dedicated to gravitational lensing surveys. The National Academy of Sciences' decadal survey has ranked the Large Synoptic Survey Telescope

http://www.dmtelescope.org

an 8 meter telescope dedicated to measure weak gravitational lensing, as one of its top priorities for the coming decade. NASA's Space Interferometric Mission SIM,

http://sim.jpl.nasa.gov

will have a key project focused on gravitational lensing. There are plans to build GEST, a dedicated satellite that aims at detecting planets around distant stars through lensing effects http://bustard.phys.nd.edu/MPS

While observations of gravitationally lensed objects are a very powerful tool for addressing these questions, interpretation of these observations is challenging. Because mass curves space, the mass between us and a distant object distorts the path of light rays. These distortions map the emitted light pattern into an observed light pattern. The properties of the observed light pattern are very sensitive to the existence and location of singularities of the lensing map.

Astronomers need to infer the properties of the lensing map and the properties of the emitting source from the observed light pattern: These two related challenges are the two basic problems of Chapter 2. If astronomers want to address the central inversion problems of gravitational lensing and obtain the full scientific value of these data sets, they will need to understand the mathematics of the lensing map. Many astronomical observables, such as the number of images or the magnification patterns around cusps, can be better understood through a deeper understanding of singularity theory and its implications for gravitational lensing, the focus of this monograph.

For astronomers reading this book, I would encourage them to skip Chapters 7–10 on the first reading as this contains very challenging mathematical material. The latter chapters, particularly Chapters 11, 12, 13 and 15 contain many results that will be useful for theoretical astrophysicists who are interested in more deeply understanding gravitational lensing. These chapters contain many results that will likely be useful in the interpretation of lensing magnifications in both micro and macro lensed systems. After reading these later chapters, astrophysicists will want to return to the middle chapters of the book for a more rigorous understanding of the underlying mathematics.

For mathematicians, gravitational lensing offers an opportunity to see observable implications of singularity theory. Often, physically motivated questions can yield important mathematical insights. The study of gravitational lensing in the weak-field limit has already created an interplay between the singularity theories of Morse, Whitney, Thom, and Arnold. The applications of these theories to gravitational lensing in the weak field limit naturally lead to questions of how to extend their applications to a full general relativistic setting (Section 3.1.3).

This book grew out of a fortunate conversation a decade ago. Arlie Petters, who was then an MIT graduate student, wandered into my office to ask for comments on a new metric solution that he had found for a superconducting cosmic string. When he inquired about whether this was an appropriate topic for a thesis, I encouraged him to look into a topic that had more astronomical relevance. Having just read Roger Blandford and Ramesh Narayan's insightful article on caustics and gravitational lensing, I encouraged him to read the article and perhaps look more deeply into this area. Over the past decade, Arlie Petters and his collaborators have begun the exploration of the rich interface between singularity theory and gravitational lensing. Harold Levine, a leading mathematician working in singularity theory, and Joachim Wambsganss, a leading researcher in the astrophysics of gravitational lensing, have joined forces with Arlie Petters, a leading mathematical physicist, to write an exciting monograph that should be useful to both mathematicians and astronomers interested in this area.

Princeton, NJ, USA
December, 2000

David Spergel

Preface

If there is anything that can bind the heavenly mind of man to this dreary exile of our earthly home and can reconcile us with our fate so that one can enjoy living — then it is verily the enjoyment of the mathematical sciences and astronomy.

JOHANNES KEPLER

What is Gravitational Lensing?

Gravitational lensing is the deflection of light by a gravitational field. In most astronomical or cosmological situations, the light originates from a distant source (e.g., star, galaxy, quasar) and undergoes deflection by the gravitational field of an intervening matter distribution (e.g., star, galaxy, cluster of galaxies) before arriving on earth. This situation will be assumed throughout the book.[1]

The first example of gravitational lensing was observed in 1919 when Sir Arthur Eddington confirmed Einstein's 1915 prediction as to how much starlight is bent by the sun's gravity. Though several theoretical papers appeared on the subject in the 1930s and 1960s, only the relatively recent discoveries of gravitationally lensed quasars (1979) and galaxies (1986) have led to a vigorous interest in lensing among astronomers.

Why is Gravitational Lensing Important?

One reason astronomers and astrophysicists are excited about gravitational lensing is because it provides a powerful probe of the distribution of gravitating matter — including the elusive dark matter — in galaxies, clusters of galaxies, and the universe at large (whereas only a small fraction of the total matter content of the universe is directly observable by its electromagnetic radiation). In addition, lensing provides an independent measure of the age of the universe (Hubble's constant) and constrains the cosmological constant. Furthermore, gravitational lenses act as natural magnifying "telescopes" that reveal the structure of very faint distant objects which could not otherwise be resolved. Lensing may also reveal planetary systems and exotic objects like black holes and cosmic strings.

[1] An example of a different gravitational lensing situation is where the gravitational field of a sufficiently dense body bends the very light it emits — see page 4.

Aim and Scope of the Book

Astrophysicists infer the geometry, composition, and expansion rate of the universe from gravitational lens observations. These inferences rest on a combination of heuristic arguments, intuition gained from numerical simulations, and models based on simple analytical forms for the distribution of dark matter in galaxies and clusters of galaxies. In the process, some theorists oversimplify the richness and complexity of real systems. For this reason generic and stable results in gravitational lensing are important. Genericity implies that certain physical properties are essentially universal among lens models and stability yields the constancy of form of the models under small perturbations.

Singularity theory provides the mathematical machinery needed to address the classification and genericity of the stable properties of gravitational lens systems. It allows us to draw conclusions that are essentially independent of the details of a chosen lens system model.

Along with studying the generic, stable features of general gravitational lens systems, we also analyze the properties of some specific lens models that have predictive power accessible to current or near-future observations. Furthermore, we employ the ideas and results of singularity theory to put gravitational lensing on a rigorous mathematical foundation, and show how the foundation's basic building blocks can be used to solve certain key lensing problems that were impenetrable using previous methods.

Below is a sample of the lensing issues addressed (see Chapter 2 for more):

Astrophysical Aspects (Part II):

- Explore the nature of (dark) compact objects — possibly planets — in our galaxy by interpreting their lensing effect on background stars. Determine the dark matter mass distribution in clusters of galaxies.
- Estimate Hubble's constant or age of the universe from time delays between lensed images.
- Examine the statistical lensing properties of cosmological models and determine the lensing effects due to such exotic matter as cosmic strings.
- Investigate the structure of distant magnified sources using lensing.

Mathematical Aspects (Part III):

- Determine the number of images produced by a generic gravitational lens system.
- Determine lower bounds on the total magnification of a lensed light source and find its magnification cross section, i.e., the area of regions where a light source has magnification greater than some given amount.

- Characterize the generic local and global properties of critical curves, i.e., the set of all infinitely magnified images of all light sources and caustics, i.e., the set of all positions from which a light source is infinitely magnified.

- Link the generic properties of caustics to those of the lens. In particular, given a caustic network with certain generic geometric properties, determine whether this forces the lens to have certain physical features. Conversely, given a lens with certain generic physical features, determine whether this forces the caustics to have a certain geometry.

The above physical and mathematical problems are treated in the context of thin-screen (deflecting mass distribution lies on a plane), weak-field (deflector causes small bending angles) lensing. These approximations work extremely well for the vast majority of the currently observed gravitational lensing events.

Within the previous approximations, most gravitational lensing research focuses only on the situation of a single deflector along the line of sight (single-plane lensing), and so very little is known about the multiple lens plane case. We present (Part III) a detailed, rigorous theory of thin-screen, weak-field lensing that applies to any finite number of lens planes, and highlight the distinctions between single- and multiple-plane lensing. This book also serves as an introduction to the field and an up-to-date, comprehensive exposition of the most important aspects of the subject.

Relation to other Books on Gravitational Lensing

To the best of our knowledge, there are currently three technical books dealing entirely with gravitational lensing: *Gravitational Lenses* (1989, in Russian) by Bliokh and Minokov [Bli-M]; *Gravitational Lenses* (1992) by Schneider, Ehlers and Falco [Schn-EF]; and *Gravitational Lenses and Microlenses* (1997, in Russian) by Zakharov [Zakh97]. These books are oriented towards astrophysicists. In addition, there are at least a dozen conference proceedings on lensing, all of which deal with the astrophysical aspects of the subject. The book by Schneider, Ehlers and Falco broke new ground with its systematic treatment from first principles of the astrophysics of thin-screen, weak-field gravitational lensing.

Our book is the first to develop a mathematical theory of thin-screen, weak-field gravitational lensing. Its mathematical emphasis complements the physical treatments in [Bli-M], [Schn-EF], and [Zakh97], and covers in Part III many mathematical lensing topics that are outside the scope of those books. In addition, *we establish new results that are being published here for the first time.* For example: mathematical formulation of core lensing concepts for any finite number of lens planes (Chapter 6); classification and genericity of stable lens systems (Chapter 8); qualitative local lensing geometry of a generic k-plane lens system, normal form of a double-plane lensing map near folds, general formulation of local convexity and its violation in double-plane lensing (Chapter 9); lower bounds on total magnification, a magnification cross section formula and its high magnification scalings for k-plane lensing (Chapter 13).

The techniques and theorems we present also provide a platform from which to generalize thin-screen, weak-field lensing to a general relativistic setting. Moreover, they can be used as consistency checks on certain geometrical optics results for general spacetimes by going to the thin-screen, weak-field limit. This aspect of our work then complements the book *Ray Optics, Fermat's Principle, and Applications to General Relativity* by Perlick [Per00b], which has some applications to gravitational lensing.

Audience

This book is directed primarily towards mathematicians, mathematical physicists, and theoretical astrophysicists and physicists.

The introductory astrophysical aspects (Part II) of the book aim to be practicable for physical scientists and should be accessible to mathematicians with a basic physics background. The mathematical portion (Part III) is at a level of generality that preserves the essential physical ideas of lensing and facilitates a natural, unified presentation of the subject. Topics from Part III can serve as a basis for a graduate level seminar on the mathematics of gravitational lensing. Part II, along with some selections from Part III, were used by Petters at Duke University and Princeton University for undergraduate courses, seminars, independent study projects, and senior theses on applications of singularity theory to gravitational lensing.

Background

No prior knowledge of gravitational lensing and singularity theory is assumed.

Part II (astrophysical aspects) of the book requires only familiarity with some of the most elementary notions in astronomy and general relativity (e.g., Cohen [Coh], d'Inverno [dIn], and Schutz [Schu]), and a facility with multivariable calculus (e.g., Marsden and Tromba [Mars-T]) and undergraduate linear algebra (e.g., Strang [Str]).

For Part III (mathematical aspects), we assume the fundamentals of topology and manifold theory. Readers unfamiliar with the former may consult Munkres [Mun74], along with Francis [Franc] and Weeks [Wee85], who present the fundamentals of topology using numerous graphic illustrations. Introductions to elementary manifold theory are given by Guillemin and Pollack [Gui-P] and Majthay [Maj].

Notation and Conventions

Abbreviations for the authors' names: A.P. (Petters), H.L. (Levine), and J.W. (Wambsganss).

Astrophysics:

c = speed of light in vacuum $\simeq 3 \times 10^8$ m/sec.

G = gravitational constant $\simeq 6.67 \times 10^{-11}$ m^3/(kg sec^2).

$M_\odot$ = mass of sun $\simeq 1.99 \times 10^{30}$ kg.

$R_\odot$ = radius of sun $\simeq 6.96 \times 10^8$ m.

$1'' = 1$ arcsecond $\simeq 4.85 \times 10^{-6}$ radians (i.e., the angle subtended from the center of a circle by $1/(60^2 \times 360)$ of the circumference).

1 AU = 1 Astronomical Unit = average earth-sun distance $\simeq 1.5 \times 10^{11}$m.

1 pc = 1 parsec = 1 AU/1″ (i.e., the distance at which 1 AU subtends an angle of 1 arcsecond) $\simeq$ 3.26 light years $\simeq 3.09 \times 10^{16}$ m.

H_0 = Hubble's constant $\simeq h \times 50$ km sec^{-1} Mpc^{-1}, where h is a constant (currently believed to be bounded as $1 \lesssim h \lesssim 2$).

The phrase "of order" is commonly used in astronomy to describe the order of magnitude, or a number within a factor of 2 or 3. For example, we can say that there are of order 10,000 astronomers on earth, which could be 8,239 or as many as 23,199.

Mathematics:

Though a detailed Index of Notation is given at the end of the book, the following points are worth noting: The symbol $\approx$ denotes a diffeomorphism rather than an approximation ($\simeq$); a mathematical statement of the form $A \equiv B$ means that A is defined to be B; the expressions $\Delta\psi$ and $\Delta\tau$ denote the Laplacian of ψ and an incremental change in τ, respectively, while Δ_k is a standard k-simplex; and the symbol $\square$ designates the end of a proof. Unless stated to the contrary, every manifold is finite dimensional and without boundary, and all mappings and manifolds are smooth.

Guide to the Book

Part I: Chapter 1 gives a brief overview of the historical development of gravitational lensing. Chapter 2 presents the central lensing problems along which the book develops.

Part II: This is directed at the astrophysics of lensing. Chapter 3 explicitly spells out the physical assumptions and approximations underlying gravitational lensing, and introduces the basics of gravitational-lens optics. This chapter can be followed either by Chapter 4 (Physical Applications) or Chapter 5 (Observations).

Part III: This is aimed at the mathematical aspects of lensing. The core concepts are presented in Chapter 6 (Time Delay and Lensing Maps), while the remaining chapters (Chapters 7–15) cover two broad categories:

Local Aspects: classification and genericity of locally stable lens systems (Chapters 7, 8); local geometry of generic lensing — including caustic metamorphoses (Chapter 9).

Global Aspects: classification and genericity of transverse and globally stable lens systems (Chapters 7, 8); multiple imaging (Chapters 10, 11, 12); magnification (Chapter 13); global geometry of caustics (Chapters 14, 15).

In Part III, Chapter 6 should be read first. After Chapter 6, however, the reader may go directly to any one of the remaining chapters that is of interest, referring back to earlier chapters as needed. *We encourage astrophysicists to look at Chapters 11, 12, 13, and 15, then Chapters 8 and 9.* Roughly speaking, Chapters 8–9, 11–13, and 15 apply to gravitational lensing, the singular-theoretic mathematical tools of Chapters 7, 10, and 14, respectively. Readers should also look for the *Requisite Mathematics* at the start of some of the gravitational lensing sections in Part III.

On a first reading of Part III, physical scientists may skim through the examples, definitions, theorems, and propositions, omitting all proofs. Note that the definitions pertaining to gravitational lensing codify our postulates about gravitational lens systems and enable us to transform physical gravitational lensing problems into mathematical form.

Finally, readers are welcome to send their comments to A.P.

Durham, NC, USA — A. O. Petters
Paris, France — H. Levine
Potsdam, Germany — J. Wambsganss
December, 2000

Acknowledgments

To Harold Levine and Joachim Wambsganss, I extend my deepest appreciation for partaking in this wonderful journey. I am especially thankful to Harold for numerous penetrating discussions on singularity theory and Joachim for many insightful exchanges on the astrophysical and numerical aspects of gravitational lensing.

I am forever grateful to David Spergel for introducing me to gravitational lensing and suspecting early on that the subject would invoke beautiful mathematics. Also, I thank him tremendously for taking the time to write the Foreword.

To Demetrios Christodoulou, I am indebted for strongly supporting this research and engaging me in countless riveting discussions on gravitational physics; to John Mather, for attentively listening and responding to my queries, and providing guidance on several aspects of the stability-theory literature. It is also a pleasure to thank Bertram Kostant for exposing me to Arnold's singularity theory, and Victor Guillemin for introducing me to Harold Levine's work and commenting on the manuscript.

Christopher Kochanek deserves special thanks for reviewing the astrophysics part of the book and providing valuable suggestions. I am grateful to Dan Margalit for meticulously working through the entire manuscript, flushing out several errors. Thanks go to Rennan Barkana, Ian Budden, Wynn Evans, Simonetta Frittelli, Scott Gaudi, Benjamin Goetz, Thomas Kling, Daniel Kubas, Matthew O'Dowd, Shude Mao, Virginia Trimble, Volker Perlick, Kumar Virbhadra, and Hans Witt for profitable discussions and/or reading parts of the manuscript.

To the following colleagues, I offer appreciation for fruitful technical exchanges and hosting me during visits to some of their institutions:

Tom Banchoff	Jeremy Goodman	John Nye
Michael Berry	J. Richard Gott III	Roger Penrose
Roger Blandford	Ramesh Narayan	Peter Schneider
Jürgen Ehlers	Ted Newman	J. Anthony Tyson

My gratitude extends to Gerree Pecht for her diligence and resourcefulness while typesetting parts of the manuscript. I am thankful to Marcia Petters for serving as a tireless sounding board, Bill Massey for being a steadfast moral supporter, and Patricia Anahory Silva for encouragement during the conceptual stages of this work.

I wrote several parts of the book while at the following institutions: Duke University, Max-Planck-Institut für Astrophysik (Garching), MIT, Oxford University, and Princeton University. At Duke University, Yunliang Yu and Andrew Schretter provided excellent computer support that often went far beyond the call of duty. My research was partly funded by the generous support of an Alfred P. Sloan Research Fellowship and a National Science Foundation Faculty Early Career grant. Lastly, this book would not have been possible without the strong support and bounteous patience of the Birkhäuser team of Ann Kostant and Tom Grasso. Special thanks to Elizabeth Loew of TEXniques for the final transformation from manuscript into a beautifully typeset book.

A. O. Petters

This is to thank Arlie Petters for introducing me to the subject of gravitational lensing and for enlisting me as a collaborator. It has been a pleasure for me to work with him and to learn from him something of a new domain of application of the mathematics that occupied me so long ago. I also want to make it clear that all the credit for the enormous task of melding my contribution into the final exposition is due to him. It was his conception, drive and enormous effort that produced this book.

H. Levine

It is a great pleasure to thank Arlie Petters for his continuous and always cheerful support while working on the draft. It has been a wonderful experience to see and feel the similarities and the differences between mathematicians and astrophysicists when dealing with their respective distorted views of the world. Various parts of the writing were done at Princeton University, Max-Planck-Institut für Astrophysik at Garching, Astrophysikalisches Institut Potsdam, Universität Potsdam, and the University of Melbourne. There are many colleagues and friends whom it is a pleasure to thank for help and advice during the preparation of the manuscript. I particularly benefitted from discussions with Bohdan Paczyński, Ed Turner, Jerry Ostriker, Tomislav Kundić, Peter Schneider, Matthias Bartelmann, Stella Seitz, Hans Witt, Paul Schechter, Robert Schmidt, and Rachel Webster. Those colleagues who gave permission for reproduction of their figures and images are thanked as well.

J. Wambsganss

Singularity Theory and Gravitational Lensing

Part I

INTRODUCTION

This part of the book historically highlights some of the key works (before 1979) that predicted the effect of gravitational lensing, established some of the fundamental physical ideas, or applied the theory in physically important ways. It also includes successes and vain attempts at observing the lensing phenomenon. We then overview some central astrophysical and mathematical issues arising from the subject. These issues will drive the astrophysical and mathematical developments in Parts II and III, respectively.

Chapter 1

Historical Highlights

Of course, there is not much hope of observing this phenomenon directly.

ALBERT EINSTEIN, 1936

The probability that nebulae which act as gravitational lenses will be found becomes practically a certainty.

FRITZ ZWICKY, 1937

This chapter historically traces certain important astrophysical aspects of gravitational lensing. The first section deals with the predictions of various gravitational lens phenomena, while the second briefly describes some of their discoveries. A more detailed account of the history of gravitational lensing is given by Trimble [Tri] and Schneider, Ehlers and Falco [Schn-EF].

1.1 Predicting Gravitational Lens Effects

We shall present some of the major astrophysical contributions to gravitational lensing prior to the 1979 discovery of the first lensing effect outside our solar system. For this reason the literature referred to here is essentially of a theoretical nature. The papers predicted/anticipated a lensing phenomenon that later was observed, worked out important physical applications of lensing, or contributed to establishing fundamental physical ideas or principles of the lens theory. Short historical overviews on other aspects of lensing are given in appropriate chapters (e.g., on the discoveries and observations of lens systems — Section 4.1).

Light Deflection

One of the earliest calculations of the deflection of light from a distant source by a foreground gravitational field is due to Henry Cavendish around 1784. Unfortunately, the only evidence that he did such a calculation is on an "isolated scrap" of paper. Will [Wil] offers some details of this story and attempts to reproduce Cavendish's calculation of the bending angle of light passing near the surface of an attracting mass.

The next closest calculations seems to have been in an 1801 paper by Johann von Soldner: "Ueber die Ablenkung eines Lichtstrals von seiner geradlinigen Bewegung, durch die Attraktion eines Weltkörpers, an welchem er nahe vorbeigeht" ["On the Deviation of a Light Ray from its Motion along a Straight Line through the attraction of a Celestial Body which it Passes Closeby"]. This article appeared in "Berliner Astronomisches Jahrbuch für das Jahr 1804" [Sol]; see Jaki's English translation [Jak]. We note that initially von Soldner's paper calculates — using Newtonian mechanics — the bending angle formula for light emitted tangentially from the surface of a gravitating body and travels off to infinity. When applying his formula, however, he reverses the light ray's path, so it comes in from infinity and "grazes" the attracting body [Jak, pp. 945-947] (cf. [Wil, p. 414, Sec. III]). Von Soldner, who later became director of the Munich observatory and a foreign member of the Royal Astronomical Society, figured out: "Thus when a light ray passes by a celestial body, it will, instead of going on in a straight direction, be forced by its attraction to describe a hyperbola whose concave side is directed against the attracting body." He showed that a light ray grazing the sun would be deflected by 0.84 seconds of arc and concluded: "If one could observe the fixed stars very close to the sun, then one would have to take this very much into account." Von Soldner realized that he was far ahead of his time when he apologized at the end: "Hopefully, no one would find it objectionable that I treat a light ray as a heavy body." As a consequence, "Soldner's work on the deflection of light had no influence either on practice or on theory" [Jak].

Even earlier, John Michell in 1783 [Miche] and Pierre-Simon Laplace in 1796 [Lap] independently investigated light being subjected to gravity, but in the context of a body acting on the very light it emits. They both basically arrived at a notion resembling a "black hole," that is, a sufficiently massive star can capture its own light, leaving the star invisible. Laplace stated his result initially without proof, but provided the proof in 1798; see Hawking and Ellis [Haw-E, p. 365] for an English translation. It seems, however, that Michell and Laplace did not write about the situation of light coming from a distant source and undergoing deflection as it passes near a gravitating body. We note, though, that it was Michell and Laplace who inspired the works of Cavendish and von Soldner, respectively (e.g., [Wil, p. 413] and [Jak, pp. 935–936]).

In hindsight, it is slightly surprising that Newton himself did not seem to have studied light deflection by gravity. All the ingredients were there with his gravitational dynamics and the corpuscular theory of light, and the mathematics involved was straightforward compared to other problems he solved successfully. His famous Query 1 of the *Opticks* (1704):

"Do not Bodies act upon Light at a distance, and by their action bend its Rays; and is not this action strongest at the least distance?" [New] is phrased in such a general way that today we tend to assume that it includes gravitational bending as well. This is not the case, though, according to Jaki, who comments [Jak]: "He [Newton], of course, had reflection, refraction and diffraction in mind."

Maybe it is equally surprising that it seems to have taken more than 100 years after von Soldner until somebody else dealt again quantitatively with the problem of light deflection. Even more so, when one learns that a famous contemporary of von Soldner, the physicist-philosopher Georg Christoph Lichtenberg wrote in the year 1796, "Wenn das Licht trotz seiner Geschwindigkeit noch eine Schwere hat, so würde doch so etwas wie Refraktion am Horizonte erscheinen müssen, weil es von der Erde stark gezogen wird" ["If light despite its velocity also had gravity, then something like refraction would have to occur at the horizon, because it would be attracted heavily by the earth"] [Lic].

We know today that the value of the deflection angle that Cavendish and von Soldner calculated is too small by a factor of two.[1] About a century after von Soldner's work, Albert Einstein arrived at a similar numerical value in his first attempt to determine the deflection angle — based on the equivalence principle. Einstein directly addressed the influence of gravity on light in a letter entitled: "Über den Einfluß der Schwerkraft auf die Ausbreitung des Lichtes" ["On the Influence of Gravitation on the Propagation of Light"] [Ein11]; see [Lor-EMW, p. 99] for an English translation of [Ein11]. He predicted that a light ray passing close to a "celestial body" might suffer a deflection towards the side of decreasing gravitational potential. Einstein found a numerical value of 0.83 seconds of arc for the deflection angle of light grazing the sun. In this paper, Einstein also stated that he wished astronomers to investigate this question: "Es wäre dringend zu wünschen, daß sich Astronomen der hier aufgerollten Frage annähmen, auch wenn die im vorigen gegebenen Überlegungen ungenügend fundiert oder gar abenteuerlich erscheinen sollten" ["It would be highly desirable that astronomers take up the question raised here, even if the considerations should seem to be insufficiently founded or entirely speculative"].

By 1915 Einstein succeeded in deriving the correct expression for the light deflection angle with the presentation of the General Theory of Relativity [Ein16, Ein22]. For a light ray passing at a distance R from an object of mass M he found the following deflection angle[2] — see Figure 1.1:

$$\hat{\alpha} = \frac{4GM}{c^2 R}, \tag{1.1}$$

where G is the constant of gravity and c is the speed of light. With the solar values for

[1] Will [Wil] compares Cavendish and von Soldner's determinations of the light deflection angle, and includes Einstein's correction to twice the value.

[2] Notice that the bending angle $\hat{\alpha}$ is independent of the light's wavelength and depends on the impact distance and deflector's mass. See Section 3.1.5.

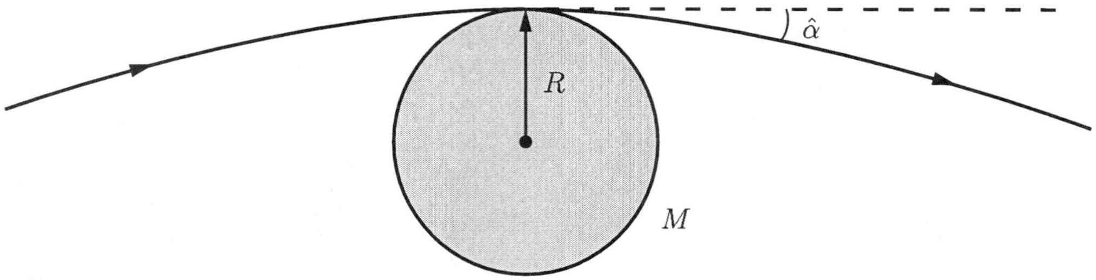

Figure 1.1. Light from a distant source experiences a deflection through angle $\hat{\alpha}$ by a mass M. The angle $\hat{\alpha}$ was calculated incorrectly by Cavendish (1784) and von Soldner (1801) since they used Newton's theory of gravity (see [Wil] for details). It was not until the formulation of General Relativity in 1915 that the correct formula for the bending angle was established by Einstein — see equation (1.1).

radius and mass he obtained

$$\hat{\alpha} = \frac{4GM_\odot}{c^2 R_\odot} \simeq 1.''74.$$

Note that since the light ray grazes the sun, we assume $R \simeq R_\odot$. In Einstein's own words: "Ein an der Sonne vorbeigehender Lichtstrahl erfährt demnach eine Biegung von 1.7 Bogensekunden" ["A light ray grazing the sun therefore experiences a deflection of 1.7 arcseconds"]. We refer readers to [Ray-BB, pp. 59–60] for a derivation of (1.1).

In the following years there were several papers dealing theoretically with light deflection, among them a short letter by Oliver Lodge about the "Einstein deflection" in which he made the point that "it is not permissible to say that the solar gravitational field acts as a lens, for it has no focal length" [Lod]. He pointed out that one would get a focal line starting at "a place seventeen times the distance of Neptune."

Multiple Imaging

While investigating the birth of general relativity, Renn, Sauer and Stachel [Ren-SS] discovered that in notes dated to the spring of 1912 Einstein explored gravitational lensing by a star. He showed that a distant star lying close to the line of sight to a foreground star will be imaged twice, with a faint secondary image increasing the total measurable intensity of the background star [Ren-SS, pp. 185, 186].[3] Einstein did not publish these results until 1936 [Ein36]. The possibility of double images was also mentioned in 1923 by Sir Arthur Eddington in his book *Space, Time and Gravitation* [Edd, p. 133–134]. This effect was also quantitatively studied by Chwolson[Chw] in 1924. Chwolson's article, "Über eine mögliche Form fiktiver Doppelsterne" ["On a possible form of fictitious double stars"], even mentions the "reversed mirror image" (i.e., the negative parity of the second image)

[3] In the same notes, Einstein also determined the 'lens equation' for the system, that is, the equation whose solutions give the image positions [Ren-SS, p. 185].

and cautioned at the end: “Ob der hier angegebene Fall eines fiktiven Doppelsterns auch wirklich vorkommt, kann ich nicht beurteilen” [“Whether the here mentioned case of a fictitious double star really exists I cannot judge”].

Einstein Rings

In 1936 a Czech engineer named Rudi Mandl approached various scientists (including Einstein) with his idea that a foreground star may act as a “gravitational lens” for light coming from a background star. After Mandl’s urging, Einstein wrote a short article for *Science* on the matter, which was published in 1936 [Ein36]. This article was clearly based on Einstein’s 1912 notes on the subject [Ren-SS, p. 186]. The paper included the new result that at perfect alignment a background star undergoing lensing by a foreground one will appear as a very bright ring (hence the name “Einstein ring”). However, Einstein concluded that the size of the ring would be far too small to be resolvable, and that the probability of such a configuration was extremely small. He was pessimistic about a possible detection: “Of course, there is not much hope of observing this phenomenon directly.” According to [Ren-SS, p. 186], Einstein was so skeptical that in his letter to the editor of *Science* he wrote: “Let me also thank you for your cooperation with the little publication, which Mister Mandl squeezed out of me. It is of little value, but it makes the poor guy happy.” Fortunately, history would favor Mandl.

Giant Luminous Arcs, Microlensing, and Blending of Light

Einstein’s article inspired (prior to its publication) his Princeton colleague Henry Norris Russell to write the first popular article on gravitational lenses: In *Scientific American*, he presents lensing as a perfect test of general relativity. Russell invented a hypothetical observer on a planet around the White Dwarf companion of the bright star Sirius to illustrate the relativistic effect of bending of light and to make it “conspicuous to the immediate gaze” [Rus]. In this scenario, Russell seems to be the first to study the lensing effect of extended sources, and his figures even show the typical and now well-known images of giant arcs — magnified and tangentially extended images of a background source that look like crescents.

Russell also noticed that star-star lensing (today we call this galactic microlensing) would produce a change of the observable brightness of the background star due to the relative motion between source, lens and observer, and that such an event would typically last for about a day. He mentioned the problem of blending of light due to the brightness of the lensing star[4] as well as the desirability of an early warning system[5] — all basic

[4]“Blending” here relates to the fact that we do not measure the light of the lensed star alone; but that there is additional light, in this case from the lensing star itself. As a consequence, the “baseline” of the light curve is higher than in the case in which we see the background star alone; thus some properties in a “blended” system may be different from those in an isolated one, e.g., the magnification as a function of time is changed (the amplitude is smaller), and the light curve can be chromatic if the stars involved have different colors.

[5]The idea is to observe a star very often in order to be able to notice a possible change due to gravita-

issues that today's observers have to face (see Section 5.5 on galactic microlensing). In the following year, Tikhov [Tik] determined the intensities of the two images of a star lensed by another star (cf. [Ein36]).

Galaxies as Lenses

Fritz Zwicky — indirectly influenced by Mandl as well — point out in 1937 that "extra-galactic *nebulae* offer a much better chance than *stars* for the observation of gravitational lens effects" [Zwi37a]. So he was the first to recognize the potential of galaxies as gravitational lenses. Zwicky reiterated and emphasized these points about lensing in his 1957 book *Morphological Astronomy* [Zwi57]. Aside from this book, there seems to be a 25-year silence in the field of gravitational lensing following the short, productive period in the mid-1930s.

Lens Renaissance in the 1960s

It was not until the early 1960s when there was another sudden outburst of papers that considered the theoretical aspects of gravitational lensing. In particular, Liebes and Refsdal derived in detail some of the basic equations of gravitational lensing with suggestions for applications.

Liebes. In 1964, Liebes [Lie] gave many applications, including discussions of galactic star-star lensing, lensing of stars in the Andromeda galaxy, lensing by "unobservable bodies," and lensing of galaxies. Liebes critically discussed the probability of detecting these effects, and he even showed figures of "Giant Luminous Arcs."

Refsdal's Method: Time Delay and Hubble's Constant. Refsdal derived in 1964 the relations among the distances between observer, lens, and source, and the angles involved [Ref64a]. He showed how one can determine the mass of the lensing object, and he pointed out that "from theory it is reasonable to believe that the deflection of light is independent of frequency." He cautions though that this "... has not been tested by experiment with any accuracy."

In Refsdal's second lensing paper [Ref64b], he analyzed the lens effect by a galaxy on a light source lying far behind. He mainly had supernovae in mind as distant point-like sources. But he even notes that "star-like objects with intense emission ... have recently been discovered" — today we call them quasars. Refsdal realized the potential usefulness of quasars immediately, and he noted that with these objects "observations at greater distances than with supernova will be possible."

Refsdal found that for a double image there is a difference in arrival time between the two paths connecting source and observer. And since this time delay is inversely propor-

tional lens magnification early enough so that one is able to notify other observers and make more frequent observations possible.

tional to the Hubble constant, he was the first to note that measuring time delay in a lensed system is a way to determine the Hubble constant.

In two more papers, Refsdal suggested the use of gravitational lensing as a test for cosmological theories [Ref66]. A few years later with one of the first "numerical" projects on lensing [Ref70], he employed the ray-tracing technique to study the propagation of light in universes with inhomogeneous mass distributions. He investigated the influence of lensing on "standard candles" and emphasized the importance of magnification bias.[6]

Lensing, Cosmology, and Quasars

Two consequences of lensing with possible cosmological interest were investigated in the late sixties to early seventies: statistical effects of a lens population on background sources, and the possibility that the newly discovered quasars were just active galaxies, strongly magnified by lenses.

Statistical lensing due to local matter inhomogeneities can change the position, shape, and brightness of background sources, compared to a corresponding homogeneous universe [Gun67a, Gun67b, Noo]. If the universe is filled with a population of compact objects, identifying and counting "double images" on different angular scales could help us detect these objects. Press and Gunn showed that this method can work on mass scales from 10^4 to about 10^{17} solar masses [Pre-G].

The once popular hypothesis that quasars are highly magnified Seyfert galaxies was looked at by different authors. The basic conclusion was that although moderate magnification could be involved, lensing alone could not explain the whole quasar phenomenon. The Barnothys proposed a variety of lensing ideas and scenarios. Although some of their ideas are quite bizarre, they are often not given credit for pursuing the idea of gravitational lensing during a period in which the majority of astrophysicists were rather skeptical about this phenomenon [deS, San, Barn, Barn-B].

Specific Lens Models

The uniform transparent lens was explored in 1972 by Clark [Cla]. He gave an elementary discussion of the gravitational deflection of light due to radially and cylindrically symmetric masses.

Bourassa and colleagues developed a complex formalism to describe lensing situations and investigated a spheroidal gravitational lens, calculating bending angles and determining formulae for intensification, distortion and orientation of the images [Bou-KN]. Using this formalism, the theory of transparent gravitational lenses was fully developed. General

[6] Magnification bias is the effect whereby a fraction of all light sources are highly magnified due to gravitational lensing. Although the absolute fraction of them is small, there is usually a large reservoir of faint sources which can be magnified. So, in any sample of sources with a certain cut-off in apparent brightness, the fraction of sources that are affected by lensing is higher than in a "volume-limited" sample.

expressions for the bending angle of a light ray around a bounded elliptical mass distribution were given, and the locations, magnifications, distortions and orientations of images could be calculated [Bou-K]. Cooke and Kantowski calculated time delays for multiple-imaged quasars. These delays are due to the differences in geometrical path lengths and gravitational potential along the light rays [Cook-K].

1.2 Detecting Gravitational Lens Effects

Here we list a number of first observations on various aspects of light deflection. Discoveries as well as failures in the search for certain lensing signatures are described, starting in the early years of the twentieth century. Observational phenomena are described in more detail in Chapter 5.

Measuring Light Deflection at Solar Limb

There were various plans to test Einstein's (wrong!) 1911 prediction for the deflection angle at the solar limb. In 1912 an Argentinian expedition was to observe the solar eclipse in Brazil, but was rained out. Another attempt was made in 1914, when Erwin Freundlich led a group to measure the positions of stars near the sun during the solar eclipse on the Russian Crimea. However, when the observers were in Russia, World War I broke out and they were captured by the Russian soldiers and released only after the eclipse (see, e.g., [Bre, Pai]). So, fortunately for Einstein, the measurement of the light deflection by the sun was not made before the correct value was published.

With the full theory of general relativity [Ein16], the correct value for the light deflection was published. In 1917 the then Astronomer Royal Frank Dyson suggested using the solar eclipse in May 1919 for the observational confirmation. This eclipse was particularly well suited, since it occurred in the Hyades constellation with many bright stars in the field. In two famous expeditions to Sobral (Brazil) and Principe Island (Gulf of Guinea, Africa), Eddington measured the displacement of stars close to the sun during the solar eclipse of May 29, 1919 to within 30% of the value predicted by Einstein (the deflection angle has been confirmed very recently with an accuracy of better than 0.02% [Leb] !) .

This experiment became the first and most famous test of general relativity. The confirmation of the result of general relativity — announced by Dyson and Eddington at a joint meeting of the Royal Society and the Royal Astronomical Society in London on November 6, 1919 — was the basis of Einstein's huge popularity starting in the 1920s.

Vain Attempts for other Lensing Effects

As early as 1923 there were attempts to measure the effect of star-star lensing. According to a footnote in [Zwi37b], E.B. Frost, then director of Yerkes Observatory, outlined a program to look for the lensing effect. The lack of any subsequent paper on this project leads us to interpret this as an unsuccessful attempt. This is symptomatic of the situation in observational lensing prior to the discovery of the double quasar in 1979 (see below): it is difficult to figure out how many people tried to find lenses since negative results are rarely reported in science.

In 1937, Zwicky [Zwi37b] estimated the probability for galaxy-galaxy lensing to be about one percent, and also searched for it observationally.[7] But aside from the remark, "in searching through actual photographs, a number of nebular objects arose our suspicion," there is no indication that he was successful. Twenty years later he lowered the probability a bit [Zwi57], but was still convinced that lenses must be out there: "... it appears certain that we should find gravitational lens effects among the extragalactic nebulae." And later: "It is therefore rather surprising to me that, although the 200-inch telescope has never had specific time allocated to this problem, no gravitational lens effects have been found accidentally." He concludes with: "Our failure to detect any nebulae which act as gravitational lenses, for the present, remains unexplained."

Weak Lensing of Quasars

As mentioned earlier (page 9), Press and Gunn suggested in 1973 a method to detect a cosmological population of compact objects, and from the lack of observed double images, concluded that there cannot exist a significant population of objects exceeding 10^{12} $M_\odot$ since "pairs of QSOs with identical redshifts, separated by a few seconds of arc would hardly have escaped the attention of observers" [Pre-G]. In 1974, Gott and Gunn explained that a pair of quasars at *different* redshifts with separation of only 5 arcseconds is a chance coincidence rather than a physical association involving non-cosmological redshifts (since the lens effect of the nearby quasar can magnify the background quasar considerably). In fact, it is not that unlikely to find two unrelated quasars so close to each other on the sky [Got-G].

Discovery of the First Double Quasar

The breakthrough in gravitational lensing came with the 1979 discovery by Walsh, Carswell, and Weymann [Wals-CW] of the first multiply imaged quasar Q0957+561. It was an accidental observation: Two quasar images were found with a separation of 6.1 arcseconds and the same redshift of $z = 1.41$. Shortly thereafter the lensing galaxy was identified with a redshift of $z = 0.36$. The very readable story of the discovery by Dennis Walsh not only describes how the lens system was found, but is one of the rare pieces in the astronomical literature where one can also learn how science is being done [Wals].

This discovery obviously transformed the field completely. Gravitational lensing stopped being the playground of a few theoreticians; it became something "real." In Figure 1.2, one

[7] He did not specify what lensing effects he was looking for, but most likely he searched for close pairs of galaxies with the more distant one being distorted and/or magnified.

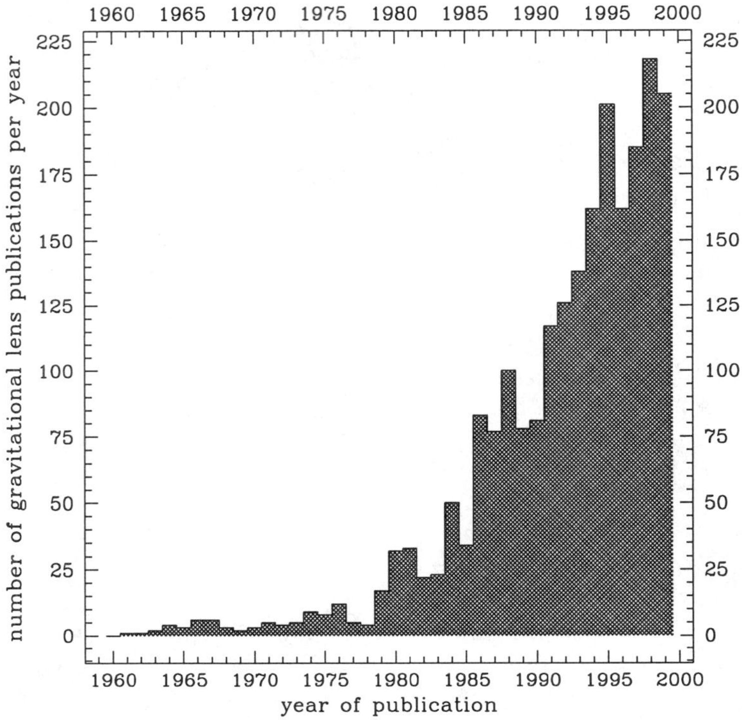

Figure 1.2. Number of gravitational lensing papers per year [Sur]. Number of gravitational lensing papers per year [Sur].

can see that the 1979 double quasar discovery had a significant impact on the number of lens papers published per year. In the following years many more multiple quasars were found (see the current list of accepted and likely lens candidates in Table 5.1, page 145), but a couple of other realizations of lensing still awaited discovery.

Giant Luminous Arcs

When an American and a French team independently announced in 1986 [Lyn-P] and 1987 [Sou-FMP], respectively, their detection of a Giant Luminous Arc in the galaxy cluster Abell 370, there was a great deal of excitement in the astrophysical community. Many possible explanations were published, among them Paczyński's suggestion that they could be extended images of background galaxies [Pac87], lensed by the foreground cluster. By measuring the redshift of the arc, this scenario could be confirmed: The arcs soon turned out to be quite simple things: images of background galaxies gravitationally lensed by a foreground cluster of galaxies. By now more than 30 examples of these Giant Luminous Arcs are known (see Table 5.1). In hindsight one could be surprised about the fact that this phenomenon was not explicitly predicted, and wonder why there was no program to systematically search for such giant arcs prior to their serendipitous discovery. In fact, there are even a couple of arcs seen on photographic plates published before 1986, but the respective authors usually did not comment on them. This is a good example of the tendency to recognize only things one knows.

Arclets and Weak Lensing

The existence of giant arcs requires an almost perfect alignment between a background galaxy and the center of the foreground galaxy cluster. This seems improbable. However, there are quite a lot of background galaxies. Those background galaxies that are not exactly "in line" with the cluster center but are slightly off are affected by lensing as well, although not as dramatically as the "in-line" galaxies. Galaxies near the line of sight and behind a galaxy cluster are coherently expanded tangentially relative to the center of the galaxy cluster. If their intrinsic shapes were perfectly circular, they would be seen as ellipses with their major axes perpendicular to the line connecting them with the center of the galaxy cluster. In contrast to the Giant Arcs, these distorted galaxy images are called "arclets," with no sharp boundary between the two definitions. Many of these arclets have been found around massive galaxy clusters, with the most striking example so far the case of the cluster CL2218 (see Figure 1.3) which displays more than 120 arclets in one image!

A weaker effect has also been detected whereby the lens-induced distortions of background galaxies can no longer be seen as individual galaxies, but rather only in a statistical sense. The signature of this "weak lensing" is a coherent tangential elongation of galaxies relative to the center of the cluster. This effect was searched for by Anthony Tyson as early as the mid-seventies, but only in 1990 could the first coherent weak lensing signal be measured around the galaxy clusters Abell 1689 and CL1409+52 [Tys-VW]. This finally

Figure 1.3. Striking example of a galaxy cluster acting as a gravitational lens: Cluster CL2218 produces a few giant arcs and over 100 arclets: tangentially distorted images of background galaxies (Credits: W. Couch, University of New South Wales, R. Ellis, Cambridge University, and NASA).

became possible because of considerable improvements in three domains: enhancement of the imaging quality of telescopes; reduction in the image blurring due to atmospheric turbulence; and the enhancement of the sensitivity of CCD technology (compared to photographic plates).

Microlensing: Far and Near

A quite different regime of lensing is so-called "microlensing." The name, originally put forward by Paczyński, indicates that the lenses are of stellar masses, and hence only able to produce splitting angles of about a microarcsecond [Pac86a]. There are two variants of microlensing. The one which was observed first is the effect on the light curve of a background quasar that is lensed by individual stars in a deflector galaxy. This was actually predicted as a "perturbation" of lensing by the galaxy as a whole [Chan-R79]. Here — due to the relative motion of quasar, lensing galaxy and observer — the positions of the stars in front of the quasar change with time, and this causes an observable change in brightness [Got81]. This was first detected in the quadruple quasar Q2237+0305 [Irw].

Currently, the more popular type of microlensing scenario is the lensing of a star in the Milky Way or a nearby galaxy by another (closer) star in the disk or Halo of the Milky Way. This type of microlensing was predicted as a possible test for the existence of faint compact objects in the Halo of the Milky Way [Pac86a]. It seemed almost impossible to detect in practice because the low probability (estimated at about 10^{-6}) of a chance alignment between two such stars requires a collection and processing of a huge amount of data. But only six years after the suggestion, three experiments reported successful detections of the microlensing signature [Alc93, Aub, Uda93]. In particular, the lensing of stars near the center of the Milky Way by intervening stars turns out to be a new and very useful tool with many applications: One can, for example, study the structure of the Milky Way (spiral arms) in some detail. Also, one finds, as a by-product of the search, complete catalogs of all kinds of variable stars.

Chapter 2

Central Problems

When the problems in physics become difficult we may often look to the mathematician, who may already have studied such things and have prepared a line of reasoning for us to follow. On the other hand they may not have, in which case we have to invent our own line of reasoning, which we then pass back to the mathematician.

RICHARD FEYNMAN

The combinations that can be formed with numbers and symbols are an infinite multitude. In this thicket how shall we [mathematicians] choose those that are worthy of our attention? Shall we be guided only by whimsy? ... [This] would undoubtedly carry us far from each other, and we would rapidly cease to understand each other. But that is only the minor side of the problem. Not only will physics perhaps prevent us from getting lost, but it will also protect us from a more fearsome danger ... turning around in circles. History [shows that] physics has not only forced us to choose [from the multitude of problems which arise], but it has also imposed on us directions that would never have been dreamed of otherwise .... What could be more useful!

HENRI POINCARÉ

This chapter gives a brief qualitative presentation of some of the most important astrophysical and mathematical problems in gravitational lensing (Section 2.1). To put these problems in perspective, we also present several quotes highlighting some of the differences in the viewpoints of mathematicians and physicists (Section 2.2).

2.1 Two Basic Problems

The theory of gravitational lensing draws fundamentally on analyses of the following complementary problems:[1]

- **Problem 1:** Given a cosmological setting with a matter distribution gravitationally lensing a light source or sources, determine the lensing properties.
- **Problem 2:** Given the effects of gravitationally lensing a light source or sources, determine the properties of the matter deflector, source(s), and cosmological setting.

Astrophysical Aspects (Part I)

Typical matter distributions to be dealt with in Problem 1 are galaxies, clusters of galaxies, black holes, or individual stars. The most direct lensing effects have observable consequences like the multiple image morphology and the spatial and temporal behavior of the lens-induced magnification.

Problem 1 encompasses the following astrophysical issues:

A1: Determine the effects of lensing pointlike and extended sources by compact bodies, galaxies, clusters of galaxies (Sections 3.2.6, 3.2.7, 3.3, and 3.4; Chapters 4 and 5).

A2: Estimate Hubble's constant from time delays between lensed images (Section 4.4 and 5.1).

A3: Examine the statistical lensing properties of cosmological models (Section 4.5).

A4: Determine the lensing effects due to such exotic matter as cosmic strings (Section 4.6).

---

[1]The spirit of Problems 1 and 2 goes back to antiquity. As early as the second century B.C., the mathematician Diocles (180 B.C.) studied the optical properties of specific "burning mirrors" and investigated the construction of such a mirror from given reflective properties. His treatise *On Burning Mirrors* (as commentated by Eutocius, sixth century A.D.) poses:

"... Pythion the Thasian geometer wrote a letter to Conon in which he asked him how to find a mirror surface such that when it is placed facing the sun, the rays reflected from it meet the circumference of a circle. And when Zenodorus the astronomer came down to Arcadia and was introduced to us, he asked us how to find a mirror surface such that when it is placed facing the sun, the rays reflected from it meet a point and thus cause burning. ..." [Too, p. 34]

The first proposition of *On Burning Mirrors* treats the reflective properties of parabolic mirrors, and uses these results to solve the inversion problems of Pythion and Zenodorus. This is followed by studies of how spherical mirrors reflect light; Diocles argues that such surfaces will focus rays to a line segment, not a point. He then raises the problem of finding a "burning mirror" of given focal length [Too, p. 62]: "How do we shape the curvature of the burning mirror when we want the point at which the burning occurs to be at a given distance from the center of the surface of the mirror?" Diocles' approach leads to proving that given a point (focus) and line (directrix) not containing the point, the locus of points equidistant from the line and point is a parabola. He uncovered the parabola from its focus-directrix property. Interested readers can consult Toomer [Too] and Knorr [Kno] for more details.

On the other hand, Problem 2 includes:

A5: Exploring the nature of (dark) compact objects — possibly planets — in our galaxy by interpreting their lensing effect on background stars — galactic "microlensing" (Sections 4.2.1 and 5.5).

A6: Determining the dark matter mass distribution in clusters of galaxies (Sections 4.2.2, 4.2.3, 5.2, 5.3, and 5.6).

A7: Investigating the structure of distant magnified sources using lensing (Section 4.3, 5.2, and 5.4).

A8: Determining limits on the cosmological setting within which lensing occurs (Section 4.5).

Both problems combine to show that gravitational lensing is a powerful probe of the cosmos (cf. Section 4.1); it is used to learn about the lens, light source, and geometry of the universe.

Mathematical Aspects (Part III)

A rigorous mathematical formulation and resolution of Problems 1 and 2 for a general realistic gravitational field is a daunting task. For example, no metric is known that realistically captures the universe's local inhomogeneities and overall isotropy. Similarly, the problem of light deflection due to two rotating interacting black holes is analytically stagnant; no explicit metric is known for such gravitational fields. Fortunately, the vast majority of gravitational lens systems relevant to observational astronomy occur within a small solid angle of our past light cone and are robustly modeled by weak-field spacetimes with a Friedmann background (see Section 3.1). These simplifying physical approximations introduce "thin-screen," "small-angle" gravitational lens models that are mathematically penetrable, yielding analytical progress on Problems 1 and 2:

Problem 1 touches on the following mathematical issues:

M1: Determining counting formulas and bounds on the number of images produced by a generic gravitational lens system, and locate the images (Chapters 11 and 12).

M2: Determining lower bounds on the total magnification of a lensed light source and find the area of regions where a light source has magnification greater than some given amount (Chapter 13)

M3: Characterizing the generic local and global properties of critical curves (i.e., set of all infinitely magnified images of all light sources) and caustics (i.e., set of all positions from which a light source is infinitely magnified). See Chapters 7, 8, 9, 14, and 15.

Problem 2 includes mathematically linking the generic properties of caustics to those of the lens:

M4: Given a caustic network with certain generic geometric properties, determine whether this forces the lens to have certain physical features (Chapter 15). Conversely, given a lens with certain generic physical features, determine whether this forces the caustics to have a certain geometry (Sections 9.2 and 9.3, Chapter 15).

M5: Given a caustic network evolution with certain generic metamorphoses, determine whether this constrains the properties of the lens (Sections 9.3 and 9.5, Chapter 15).

The previous problems can be formulated in rigorous mathematical terms, and invoke interesting and deep mathematical issues. Problems **M1** and **M2** involve Morse theory (e.g., stability of functions, Morse inequalities A and B) and resultants, while Problems **M3** to **M5** call upon the stability and genericity theories of Thom, Whitney, Mather, and Arnold. Moreover, throughout Part III of the book we shall focus heavily (though not exclusively) on generic and stable properties when considering Problems **M1** to **M5**. In other words, we aim at results that do not rely on the details of the potentials or approximations used in modeling lens systems.

2.2 Mathematicians versus Physicists

Our aim is to have the aforementioned central problems drive a thoughtful unlocking and analysis of gravitational lensing's basic mathematical building blocks. This will allow us to lay a rigorous foundation for lensing by way of singularity theory, and provide natural singular-theoretic solutions to some key mathematical lensing problems.

While carrying out the above process, we have to transform gravitational lensing's core astrophysical notions (Part II) into precise mathematical concepts (Part III). In Part II of the book, we shall use primarily physical and heuristic reasoning to discuss lensing, while Part III is dominated by rigorous mathematical arguments. However, throughout the book our intended approach is essentially along the following lines:

> *"One takes care at the beginning to be sure that 'the mathematics is appropriate for the physics,' that is, that everything in the mathematics has physical meaning and that all of the physics one wishes to talk about is describable in terms of the mathematics . . . .*
>
> *To answer a physical question, one first translates that question into various objects . . . in the mathematics, with various properties describing the physical setup. Then one manipulates these objects within the mathematics and translates results back into the physical terms. (These 'translations' ultimately become automatic.)"*

ROBERT GEROCH [Ger, p. 86]

Given that we are attempting to reach mathematicians and physicists, and factoring in the all too common misunderstandings between their "cultures," it is important to look more closely at the respective merits of mathematical and physical reasoning:

> *"Mathematicians are only dealing with the structure of reasoning, and they do not really care what they are talking about. They do not even need to* know *what they are talking about, or, as they themselves say, whether what they say is true. I will explain that. You state the axioms, such-and-such is so, and such-and-such is so. What then? The logic can be carried out without knowing what the such-and-such words mean. If the statements about the axioms are carefully formulated and complete enough, it is not necessary for the man who is doing the reasoning to have any knowledge of the meaning of the words in order to deduce new conclusions in the same language .... In other words, mathematicians prepare abstract reasoning ready to be used if you have a set of axioms about the real world. But the physicist has meaning to all his phrases. That is a very important thing that a lot of people who come to physics by way of mathematics do not appreciate ....*
>
> *When you know what it is you are talking about, that some symbols represent forces, others masses, inertia, and so on, then you can use a lot of common sense, seat-of-the-pants feeling about the world. You have seen various things, and you know more or less how the phenomenon is going to behave. But the poor mathematician translates it into equations, as the symbols do not mean anything to him he has no guide but precise mathematical rigour and care in the argument. The physicist, knows more or less how the answer is going to come out, can sort of guess part way, and so go along rather rapidly."*
>
> RICHARD FEYNMAN [Fey, pp. 55–56]

The astrophysical aspects of the book basically speed along quickly, somewhat in the sense that Feynman mentions. Indeed, there are a lot of advantages in getting a quick "seat-of-the-pants" feel for the physical issues in lensing. At the same time, we are mindful of Polya's comment:

> *"Heuristic reasoning is good in itself. What is bad is to mix up heuristic reasoning with rigorous proof. What is worse is to sell heuristic reasoning for rigorous proof."*
>
> GEORGE POLYA [Pol, p. 113]

The mathematical aspects of the book treat lensing rigorously and precisely, proceeding much more slowly and cautiously than in Part II. Moreover, though in Part III we present the necessary background machinery from singularity theory as the need arises, the more difficult and intricate aspects are treated in separate chapters:

- The stability and genericity theories of Morse, Whitney, Thom, Mather, and Arnold (Chapter 7).

- Morse inequalities with boundary conditions (Chapter 10).
- Computing the Euler characteristic via projectivized rotation numbers of critical points (Chapter 14).

We close this section with the next two quotes, which fit quite well with the philosophy of Part III:

> *"One sometimes hears expressed the view that some sort of uncertainty principle operates in the interaction between mathematics and physics: the greater the mathematical care used to formulate a concept, the less the physical insight to be gained from that formulation. It is not difficult to imagine how such a viewpoint could come to be popular. It is often the case that essential physical ideas of a discussion are smothered by mathematics through excessive definitions, concern over irrelevant generality, etc. Nonetheless, one can make a case that mathematics as mathematics, if used thoughtfully, is almost always useful — and occasionally essential — to progress in theoretical physics.*
>
> *What one often tries to do in mathematics is to isolate some given structure for concentrated, individual study: what constructions, what results, what definitions, what relationships are available in the presence of a certain mathematical structure? But this is exactly the sort of thing that can be useful in physics, for, in a given physical application, some particular mathematical structure becomes available naturally, namely, that which arises from the physics of the problem. Thus mathematics can serve to provide a framework within which one deals only with quantities of physical significance, ignoring other, irrelevant things. One becomes able to focus on the physics. The idea is to isolate mathematical structures, one at a time, to learn what they are and what they can do. Such a body of knowledge, once established, can then be called upon whenever it makes contact with the physics.*
>
> *One could imagine physicists who do not understand, as mathematics, the notion of derivative and the properties of derivatives. Such physicists could still formulate physical laws, for example, by speaking of the 'rate of change of . . . with . . . ' They could use their physical intuition to obtain, as needed in various applications, particular properties of these 'rates of change.' It would be more convenient, however, to isolate the notion 'derivative' once and for all, without direct reference to later physical applications of this concept. One learns what a derivative is and what its properties are: the geometrical significance of a derivative, the rule for taking the derivative of a product, etc. This established body of knowledge then comes into play automatically when the physics requires the use of derivatives. Having mastered the abstract concept 'rate of change' all by itself, the mind is freed for the important, that is, the physical, issues.*
>
> *The only problem is that it takes a certain amount of effort to learn mathematics. . . ."*
>
> ROBERT GEROCH [Ger, pp. 1–2]

"... the influence of physics on mathematics is well understood. However, the contributions of mathematics to physics are not as well understood. It is a common fallacy to suppose that mathematics is important for physics only because it is a useful tool for making computations. Actually, mathematics plays a more subtle role which in the long run is more important. When a successful model is created for a physical phenomenon, that is, a model which can be used for accurate computations and predictions, the mathematical structure itself provides a new way of thinking about the phenomenon. Put slightly differently, when a model is successful it is natural to think of the physical quantities in terms of the mathematical objects which represent them and to interpret similar or secondary phenomena in terms of the same model. Because of this, an investigation of the internal mathematical structure of the model can alter and enlarge our understanding of the physical phenomenon."

MICHAEL REED AND BARRY SIMON [Ree-S, p. ix]

Part II

ASTROPHYSICAL ASPECTS

The aim of Part II is three-fold: First, we present the astrophysical foundations of lensing; that is, we try to determine explicitly and clearly the many physical approximations and assumptions underlying the theory of gravitational lensing (Chapter 3). Second, we show how lensing comes to bear on some important problems in astrophysics and cosmology (Chapter 4). Third, we conclude with a discussion of recent observations of lensing and their applications to the latter fundamental problems (Chapter 5). Furthermore, an additional goal for Part II is to provide a physical basis and guide for the mathematical developments in Part III.

Chapter 3

Basic Physical Concepts

The minute we begin to talk about this world, however, it somehow becomes transformed into another world, an interpreted world, a world delimited by language. . . . What we say about the world, our theories, are like garments—they fit the world to a greater or lesser degree, but none fit perfectly, and none are right for every occasion. . . . Einstein described the nature of physics in the following way:

'Physical concepts are free creations of the human mind, and are not, however it may seem, uniquely determined by the external world. In our endeavor to understand reality we are somewhat like a man trying to understand the mechanism of a closed watch. He sees the face and the moving hands, even hears it ticking, but he has no way of opening the case. If he is ingenious he may form some picture of the mechanism which could be responsible for all the things he observes, but he may never be quite sure his picture is the only one which could explain his observations. He will never be able to compare his picture with the real mechanism and he cannot even imagine the possibility of the meaning of such a comparison.'

BRUCE GREGORY

The physical ingredients of an astrophysical gravitational lens system are explicitly designated and described in Section 3.1: spacetime geometry (Section 3.1.1 and 3.1.3), cosmic distances (Section 3.1.2), matter deflectors (Section 3.1.4), light rays (Section 3.1.5), and light sources (Section 3.1.6). All relevant assumptions and approximations are spelled out. The lensing ingredients and approximations also provide a physical setting within which time delay and lensing maps can be explicitly defined. These mappings are at the heart of the theory of optics for gravitational lenses (Section 3.2). Section 3.2.1 introduces the concept of time delay using Fermat's principle, while Section 3.2.2 derives an explicit expression for time delay. Section 3.2.3 presents the lens equation, which relates a light source's position to the impact positions of its light rays passing through the lens plane.

In Section 3.2.4, the physical variables of gravitational lensing are converted into dimensionless form. These dimensionless variables are then used throughout the rest of the book. Section 3.2.5 introduces the magnification of light sources and discusses magnification as it relates to flux conservation. Section 3.2.6 treats the locus of all formally infinitely magnified images (critical curves) and corresponding positions (caustics) from which a source has at least one such image. Light curves (i.e., the plot of a moving light source's magnification versus time) due to single and binary point-mass lenses are also discussed in Section 3.2.6. Section 3.2.7 treats the distortion of an extended source by lensing, including how convergence and shear are involved. The chapter ends with a discussion of two important families of lens models (Section 3.3) and observable lensing properties (Section 3.4). In the latter, we not only list the observables, but also discuss a family of transformations — "degeneracies" — of the lens system that leave the observables unchanged. The chapter closes with a discussion of weak lensing (no multiple images, but distortions of background galaxies) by the large scale structure of the universe (Section 3.5).

3.1 Ingredients of Gravitational Lens Systems

Any gravitational lens situation consists of five main ingredients: a physical object that emits light; this is usually called the **source** and it can be a star, (radio) galaxy, or quasar. The second part is a matter concentration. It can consist of any kind of physical matter (clusters of galaxies, individual galaxies, ordinary stars, dark matter, black holes) and acts as the deflector, mostly called the **lens**. The third ingredient is somebody with a telescope/detector at a certain distance relative to the source and the lens, the **observer**. The "transmitter" of the information — the **light** — is necessary for our identification of a gravitational lens system. Throughout the book, "light" is meant very broadly, any kind of electromagnetic radiation will be called light in this general sense (e.g., visible light, infrared or ultraviolet radiation, radio waves, X-rays). Finally, **spacetime** (and its geometry), which is defined by a cosmological model, is an essential ingredient of a gravitational lens scenario as well. It is important to emphasize that spacetime must not be thought of as disconnected from the lens. In fact, spacetime will be split into two components arising from the lens(es) and a background, where the lens is thought of as a weak perturbation of the background (see pages 43 and 47). We shall see later that the concepts of distance depend on the underlying cosmological model, and lensing may even help to determine the latter.

The physical arrangement of the ingredients must be such that the source, the lens and the observer are roughly aligned, since otherwise the lensing effect becomes arbitrarily small. In principle, the light emitted by *any* light source in the universe is affected by *every* matter concentration in the universe because the range of the gravitational force is infinite. But in most practical astrophysical situations it suffices to consider only the matter very close to the light ray. This is because the deflection angle falls off with the inverse of the distance between matter concentration and light ray, and the matter density fluctuations average out to zero for realistic cosmological models (which are not dominated by very large scale density perturbations).

The current section attempts to detail the physical assumptions and approximations lurking behind "weak-field, thin-screen" gravitational lens theory. At the same time, one of the goals in Part III of the book is to present "generic" results that are essentially independent of the said approximations/assumptions.

3.1.1 General Relativity and Friedmann Cosmology

The current section reviews some basic ideas from general relativity — such as Einstein's equation, cosmological constant, geometrical and wave optics approximations, and Friedmann cosmology — and the assumptions we make about these notions in the gravitational lensing context.

Einstein's Equation

General relativity follows from the next two statements (cf. [Wald84, p. 73]):[1]

(1) **Spacetime** *is a four dimensional manifold with Lorentzian metric* **g**. *For physical reasons, spacetime is also required to be connected, oriented, and time-oriented* [Sach-W, pp. 27–29].[2]

Loosely speaking, spacetime results when the three dimensions of space and one of time are webbed together to form a four-dimensional continuum (with a Lorentzian geometry) whose curvature gives rise to gravitational effects. In other words, gravity is an aspect of spacetime. The particular Lorentzian geometry needed to model a gravitational situation is determined by the next statement:

(2) *Spacetime's geometry is linked to the stress-energy content* **T** *of matter in spacetime via a physical law called the* **Einstein equation**:[3]

$$\mathbf{Ric} - \frac{1}{2} S\mathbf{g} = \frac{8\pi G}{c^4}\mathbf{T}, \tag{3.1}$$

where **Ric** and S are, respectively, the spacetime Ricci and scalar curvatures.

---

[1] See Misner, Thorne, and Wheeler [Mis-TW], Schutz [Schu], Wald [Wald84] and Weinberg [Wei72] for introductions to the physical aspects of general relativity. Mathematically oriented readers may also see Frankel [Frank], O'Neill [ONe], and Sachs and Wu [Sach-W].

[2] See [ONe, Chap. 5] for a comparison of Lorentzian and Riemannian geometries, and Beem and Ehrlich [Bee-E] for an extensive treatment of Lorentzian geometry.

[3] In local coordinates, the Einstein equation becomes a hyperbolic system of ten coupled nonlinear second order partial differential equations for the components g_{ab} of the metric $\mathbf{g}$: $R_{ab} - (S/2)g_{ab} = (8\pi G/c^4)T_{ab}$.

Informally speaking, the Einstein equation states:

Spacetime grips mass, telling it how to move; and mass grips spacetime, telling it how to curve.

EDWIN TAYLOR AND JOHN WHEELER

For the Einstein equation, we emphasize that when gravitational sources are present, it does not in general make sense to specify a stress-energy tensor $\mathbf{T}$ first and then solve for $\mathbf{g}$ (e.g., [Wald84, p. 73]). This is because $\mathbf{T}$ may explicitly contain $\mathbf{g}$ (e.g., the stress-energy tensor for a perfect fluid contains the metric — see [ONe, pp. 337–341]). In general, the spacetime metric and matter distribution are determined simultaneously.

The left hand side of the Einstein equation is called the *Einstein tensor*: $\mathbf{G} = \mathbf{Ric} - \frac{1}{2}S\mathbf{g}$. It is a symmetric type (0,2) tensor with zero covariant divergence (e.g., [ONe, p. 336]). Consequently, the Einstein equation yields the equation of motion of matter in spacetime, namely, $\operatorname{div}\mathbf{T} = \mathbf{0}$. The latter implies that a particle moving only under the influence of gravity follows a geodesic in spacetime — photons move along null geodesics and mass particles along timelike geodesics (e.g., [Wald84, p. 73]).

The Cosmological Constant

Soon after Einstein completed his general theory of relativity, Friedmann showed that Einstein's theory predicted a universe that is either expanding or contracting. This greatly disturbed Einstein because the philosophical ideas about the universe at the time did not allow anything but a constant or permanent one. As a consequence, Einstein modified his theory. In 1917 he introduced a new term labeled Λ in his field equation, and called Λ the **cosmological constant**:[4]

$$\mathbf{G} + \Lambda\mathbf{g} = \frac{8\pi G}{c^4}\mathbf{T}, \tag{3.2}$$

where $-\infty < \Lambda < \infty$. Notice that the metric has zero divergence and the equations of motion are still preserved. In fact, the left side of (3.2) is the most general symmetric, divergence free type (0,2) tensor that can be built locally out of the metric and its partials up to order two (see [Lov]).

The Einstein equation with cosmological constant allows a static universe as one specific solution. When Edwin Hubble in the late 1920s discovered that galaxies have a recession velocity proportional to their distance, that is, that the universe is expanding, Einstein retracted the cosmological constant and called it "the biggest blunder of my life."

In the last few years cosmologists regained interest in the cosmological constant. Particle physicists think of the cosmological constant as a stress-energy tensor related to the vacuum, but it is not yet possible within quantum field theory to calculate such a vacuum

---

[4]See [Mis-TW, p. 409–412] for more on the cosmological constant.

polarization from first principles. However, there is a problem in that the natural prediction of particle physics is more than 50 orders of magnitudes off what astronomers would like (for details see Weinberg [Wei97]).

The cosmological constant Λ has units of (length)$^{-2}$ and a value of the constant larger than about 10^{-54}cm^{-2} is inconsistent with observations (e.g., Rindler [Rin, p.180]). Astronomers often use the cosmological constant in a slightly modified form:

$$\Omega_\Lambda = \frac{\Lambda}{3(H_0/c)^2}.$$

This way the cosmological constant becomes dimensionless. It is important for the cosmological evolution of the universe that Ω_Λ be of order unity. Only then does the cosmological constant influence the expansion and geometry of the universe significantly (as explored by [Car-PT]).

Various recent astronomical observations point to a "flat" universe ($\Omega_{\text{tot}} = \Omega_{\text{matter}} + \Omega_\Lambda = 1$) with a low matter density ($\Omega_{\text{matter}} \approx 0.3$) and a relatively large cosmological constant ($\Omega_\Lambda \approx 0.7$). This means that the matter density of the universe is smaller than the "critical density" ($\Omega_{\text{matter}} < 1$). However, due to the cosmological constant, the total density is still in agreement with a universe with no spatial curvature, which is a prediction of (the simplest version of) the theory of "inflation."

There is an observational approach to the question of whether there is a nonzero cosmological constant. The brightness of more and more distant objects decreases differently for a universe with or without a cosmological constant. This can be measured if there is a class of objects with constant brightness. Recently increasing evidence was found that a certain class of supernovae are intrinsically all of equal brightness, that is, they can be used as "standard candles." Subsequently, astronomers succeeded in measuring supernovae type Ia at very large distances (redshifts of $z \simeq 1$). It is then possible to obtain the relation between distance and apparent brightness of these objects. Comparison with models for different values of Λ *seem* to suggest that the cosmological constant is nonzero (Riess et al. [Rie]). Filippenko and Riess [Fil-R] reported that results from their data analysis strongly support an eternally expanding universe with a *positive* cosmological constant and current *acceleration* of the expansion. These results are naturally being followed by further study and observations. Also, consult Section 4.5 for how gravitational lensing may be used to set limits on the dimensionless cosmological constant Ω_Λ.

For simplicity, we shall abide by the following convention:

Unless stated to the contrary, the cosmological constant is assumed to be zero.

Geometrical Optics Approximation in Spacetime

The reader may have noticed that we implicitly employ geometrical optics when discussing light undergoing gravitational lensing. This will be the case hereafter:

> *We shall assume that light, meaning any electromagnetic wave, obeys the geometrical optics approximation: In a typical local Lorentz frame (say, a frame at rest relative to nearby galaxies), the reduced wavelength of light (i.e., wavelength divided by 2π) is significantly less than the scale over which the light's amplitude, polarization, and wavelength vary and much shorter than the radius of curvature*[5] *of the spacetime through which the light travels* [Mis-TW, p. 571].[6]

In other words, the wave nature of light — hence, diffraction effects — can be ignored and, loosely speaking, light does not spread like a wave about "warps" in spacetime (since the warps are much bigger than the reduced wavelength).

The geometrical optics approximation yields (e.g., [Mis-TW, p. 571–583]):

(1) *Light is propagated along null geodesics of spacetime* — called **light rays**. Moreover, *the amplitude, polarization, etc., of different light rays do not influence each other.* In addition, *light rays lie in and are orthogonal to surfaces of constant phase.*

(2) *A light ray's polarization vector is orthogonal to the ray and parallel transported along the ray.*

(3) Light obeys the **Law of Photon Conservation**: *The number of photons is conserved.*

Item (3) implies that gravitational lensing does not change the number of photons, but deflects them. In other words, lensing distorts the cross sectional area of light bundles, which affects an image's flux.

In almost all astrophysical lensing situations, the geometrical optics approximation is justified. Astrophysical sources are very large compared to the wavelength of the emitted light, so that wave effects can be neglected. In addition, in cases of two or more images of a lensed source, these light bundles could in principle show interference as well. But since the time delays between images are of the order of weeks or months, much larger than the "coherence length" of a light wave, interference effects can be ignored in these cases as well.

Wave Optics versus Geometrical Optics

There are a few instances, however, where the geometrical optics approximation breaks down: At "caustic crossing events," the geometrical optics approximation formally leads to infinitely high magnifications. This of course is "unphysical." Consequently, for a detailed modeling of this behavior the wave optical treatment is appropriate. In the case of a multiply

[5] In standard cosmology, the radius of curvature is the Hubble radius c/H_0.

[6] See [Per00b] for further discussion.

imaged source near a "fold caustic," the time delay between images can become as small as the period of the light wave (2×10^{-15} secs for optical light). Such a case with respect to quasar lensing was explored by [Jar-P]. Some of these wave aspects of gravitational lensing are studied in the context of "femto-lensing"; see [Gou],[Ulm-G], and [Kol-Tk]. More general aspects of wave optics and its relation to lensing are treated in [Step-H76] [Step-H78], and [Schn-Sc]. Consult Chapter 7 of [Schn-EF] for a more detailed discussion of wave optics in gravitational lensing. In general, however, the geometrical optics approximation suffices for almost all current astrophysical lensing situations. Despite this, it would not be surprising if the wave optics features of gravitational lensing emerge observationally during the next century.

Friedmann Cosmology

On scales of order a hundred million light years or more, the present universe is dominated by matter that is distributed smoothly and isotropically on average, and has no pressure [Mis-TW, p. 703]. This gives rise to a spacetime geometry that is often modeled by a **Friedmann** (or **Friedmann-Lemaître**) **universe**. In fact, most lensing scenarios are treated as perturbations of a Friedmann universe:

> *Unless stated to the contrary, we suppose that gravitational lenses are small local perturbations of a Friedmann universe (i.e., a universe filled with a smooth, pressure-free perfect fluid).*[7]

Let us now briefly review some basic aspects of the Friedmann model.[8] In a Friedmann universe, space is filled with smooth matter that has no pressure (no interaction), and is homogeneous (the smoothed out matter distribution at any two points is roughly similar) and isotropic at each point of space (the universe looks the same in all directions). Note that isotropy at each point implies homogeneity and prevents the curvature of space from varying from point to point. Homogeneity implies that all observers riding on the matter flow of a Friedmann universe can synchronize their watches (e.g., [ONe, p. 359]). The corresponding time, called *cosmic time*, is denoted by τ.

Due to the isotropy in a Friedmann universe, there are coordinates $(\tau, R, \theta, \varphi)$, where (R, θ, φ) are called *comoving coordinates*, in which the geometry of spacetime takes one of the following three forms, where $\mathsf{K} = 1, 0, -1$:[9]

$$\mathbf{g}_F^{\mathsf{K}} \equiv -c^2 d\tau^2 + a^2(\tau) dS_{\mathsf{K}}^2. \tag{3.3}$$

[7]These perturbations are significantly less than the radius of curvature (i.e., the Hubble radius c/H_0) of the universe (cf.[Mis-TW, p. 571, eq. (22.23b)]).

[8]Readers may consult the excellent physical treatments of Friedmann models in [Mis-TW, p. 703, Chap. 27], [Schu, p. 318, Chap. 12], [Wald84, p. 91, Chap. 5], and [Wei72, p. 469, Chap. 15]. Mathematically oriented readers should also see [ONe, p. 332, Chap. 12], [Frank, p. 139], and [Sach-W, p. 159, Chap. 6].

[9]The comoving coordinates are dimensionless and the scaling function a has dimension of length.

We now describe the term $a^2(\tau)dS_{\mathsf{K}}^2$. The function $a(\tau)$ is called the *scaling function* or *expansion factor* (which describes the expansion of the universe). It is defined by

$$\begin{aligned}
\mathsf{K}=1: &\qquad a=\frac{A}{2}(1-\cos u), \quad \tau=\frac{A}{2}(u-\sin u);\\
\mathsf{K}=0: &\qquad a=\left(\frac{9A}{4}\right)^{1/3}\tau^{2/3};\\
\mathsf{K}=-1: &\qquad a=\frac{A}{2}(\cosh u-1), \quad \tau=\frac{A}{2}(\sinh u-u),
\end{aligned}\tag{3.4}$$

where a and τ are given parametrically for $\mathsf{K}=1,-1$, and A is a positive constant. The term

$$\mathbf{g}^{\tau}_{(\mathsf{K})} \equiv a^2(\tau)dS_{\mathsf{K}}^2 \tag{3.5}$$

in (3.3) is the spatial metric of the universe at cosmic time τ. Here, the factor dS_{K}^2 is a metric with constant curvature $\mathsf{K}=-1, 0$, or 1 that is given as follows:

$$dS_{\mathsf{K}}^2=\frac{dR^2}{1-\mathsf{K}R^2}+R^2(d\theta^2+\sin^2\theta d\varphi^2). \tag{3.6}$$

We can put equation (3.6) and, hence, the metric (3.5) of space in a more transparent form by changing the radial coordinate R as follows:

$$R=R_{\mathsf{K}}(\chi)\equiv\begin{cases}\sin\chi & \text{if } \mathsf{K}=1\\ \chi & \text{if } \mathsf{K}=0\\ \sinh\chi & \text{if } \mathsf{K}=-1.\end{cases}\tag{3.7}$$

Then for $\mathsf{K}=1$ we see that the spatial geometry of the universe is a 3-sphere $\mathbf{S}^3(a)$ of radius $a(\tau)$:

$$\mathbf{g}^{\tau}_{(1)}=a^2(\tau)\left[d\chi^2+\sin^2\chi(d\theta^2+\sin^2\theta d\varphi^2)\right],$$

where:

$$0\le\chi<\pi,\quad 0\le\theta<\pi,\quad 0\le\varphi<2\pi.$$

For this reason, the scale factor is sometimes called the *radius of the universe* when $\mathsf{K}=1$. For $\mathsf{K}=0$, the universe's spatial geometry is that of $\mathbf{R}^3$:

$$\mathbf{g}^{\tau}_{(0)}=a^2(\tau)\left[d\chi^2+\chi^2(d\theta^2+\sin^2\theta d\varphi^2)\right],$$

where:

$$0\le\chi<\infty,\quad 0\le\theta<\pi,\quad 0\le\varphi<2\pi.$$

If $\mathsf{K}=-1$, then the universe has the spatial geometry of hyperbolic 3-space $\mathbf{H}^3(a)$ of radius $a(\tau)$:

$$\mathbf{g}^{\tau}_{(-1)}=a^2(\tau)\left[d\chi^2+\sinh^2\chi(d\theta^2+\sin^2\theta d\varphi^2)\right],$$

where:

$$0 \le \chi < \infty, \quad 0 \le \theta < \pi, \quad 0 \le \varphi < 2\pi.$$

For the cases $\mathsf{K} = 1, 0, -1$, the spatial level surfaces, $\chi =$ constant, are 2-spheres parametrized by the spherical polar coordinates (θ, φ). At cosmic time τ, these 2-spheres have the following areas:

$$\text{Area}_{\mathsf{K}}(\tau) = \begin{cases} 4\pi a^2(\tau)\sin^2\chi & \text{if } \mathsf{K} = 1 \\ 4\pi a^2(\tau)\chi^2 & \text{if } \mathsf{K} = 0 \\ 4\pi a^2(\tau)\sinh^2\chi & \text{if } \mathsf{K} = -1. \end{cases} \tag{3.8}$$

If $\mathsf{K} = 1$, then (3.8) yields that the area of the 2-spheres increases from zero at the "North Pole" ($\chi = 0$) to a maximum at the equator ($\chi = \pi/2$), and decreases to zero at the "South Pole" ($\chi = \pi$). In the Euclidean and hyperbolic cases ($\mathsf{K} = 0, -1$), the area of the 2-spheres increase from zero to infinity as χ varies from zero to infinity, with the rate of increase being faster for the hyperbolic case. Consult [dIn, pp. 317–322] and [Mis-TW, pp. 720–725] for details.

Friedmann models with $\mathsf{K} = 0$ are called *Einstein–deSitter models* or *flat universes*. This is because in such models space has a scaled Euclidean metric. In addition, since in (3.6) the factor $(1 - \mathsf{K}R^2)^{-1}$ is approximately unity for $R \ll 1$ (here R is dimensionless), *each Friedmann model is locally a flat universe and, hence, its spatial curvature is locally negligible.*

Topology of the Universe: We presented a Friedmann spacetime as a manifold of the form $I \times \mathsf{S}$ endowed with the Friedmann metric $\mathbf{g}_F^{\mathsf{K}}$. Here, I is an open interval in $\mathbf{R}$; S is either the unit 3-sphere $\mathbf{S}^3$, Euclidean 3-space $\mathbf{R}^3$, or unit hyperbolic 3-space $\mathbf{H}^3$; and S has the metric dS_{K}^2. At cosmic time τ, space is then modeled by the manifold $\mathsf{S}(\tau) = \tau \times \mathsf{S}$ and endowed with metric $\mathbf{g}_\tau^{\mathsf{K}}$; so, space has constant curvature $\mathsf{K}(\tau) = \mathsf{K}/a^2(\tau)$ (e.g., [ONe, p. 343]). In other words, *we have assumed that at cosmic time τ space is globally a complete, simply connected 3-dimensional Riemannian manifold of constant curvature.*[10] If one relaxes the assumption of simple connectedness, then there are other possibilities for the topology of space (e.g., a 3-torus). Readers are referred to Cornish, Spergel and Starkman [Cor-SS] and Weeks [Wee98, Cor-W] for more. For simplicity, however, we require simple connectedness and, hence, follow the usual convention of assuming that space is modeled by $\mathbf{S}^3(a)$, $\mathbf{R}^3$, or $\mathbf{H}^3(a)$. Finally, the space S with the metric dS_{K}^2 will be called a *comoving three-space* and $\tau \times \mathsf{S}$ with $\mathbf{g}_{(\mathsf{K})}^\tau = a^2(\tau)dS_{\mathsf{K}}^2$ is the *proper three-space* at cosmic time τ. In general, the term "comoving" will apply to quantities in S.

The Einstein equation allows us to relate the geometry $\mathbf{g}_F^{\mathsf{K}}$ of a Friedmann universe to the mass density ρ_F of matter in space. Explicitly, the Einstein equation for a Friedmann

[10]Recall that every complete, simply connected 3-dimensional Riemannian manifold of constant curvature K is isometric to $\mathbf{S}^3(r)$ (resp., $\mathbf{R}^3$, $\mathbf{H}^3(r)$) if $K = 1/r^2$ (resp., $0, -1/r^2$); see O'Neill [ONe, p.228, Cor. 25].

universe reduces to two equations that connect the scaling function a of $\mathbf{g}_F^{\mathsf{K}}$ to ρ_F:[11]

$$\left(\frac{\dot{a}(\tau)}{a(\tau)}\right)^2 = \frac{8\pi G}{3}\rho_F(\tau) - \frac{c^2\mathsf{K}}{a^2(\tau)}, \qquad \frac{\ddot{a}(\tau)}{a(\tau)} = -\frac{4\pi G}{3}\rho_F(\tau), \tag{3.9}$$

where $\dot{a}$ and $\ddot{a}$ are the first and second derivatives of a relative to τ.

In a Friedmann universe, *the spacetime paths of galaxies* (i.e., "particles" of the matter flow) *are given by the timelike geodesics with spacetime trajectories having constant spatial coordinates* — that is, the curves $(\tau, R_0, \theta_0, \varphi_0)$ (e.g., [Wei72, p. 413]). *The spatial paths of light rays* (i.e., null geodesics) *in a Friedmann model are geodesics of the spatial metric* dS_{K}^2 [ONe, p. 356, Cor. 26]. Furthermore, we introduce the "conformal time"

$$\mathtt{t} = c\int \frac{d\tau}{a(\tau)}. \tag{3.10}$$

Notice that $\mathtt{t}$ is dimensionless. Since $d\mathtt{t}/d\tau > 0$, we can express τ and (hence) $a(\tau)$ as a function of $\mathtt{t}$. Abusing notation, we follow the usual convention of writing

$$a(\mathtt{t}) = a(\tau(\mathtt{t})).$$

Setting

$$\mathbf{g}_{\mathsf{K}} = -d\mathtt{t}^2 + dS_{\mathsf{K}}^2$$

(e.g., $\mathbf{g}_0$ is the Minkowski metric), we see that the Friedmann metric $\mathbf{g}_F^{\mathsf{K}}$ is conformally related to $\mathbf{g}_{\mathsf{K}}$:

$$\mathbf{g}_F^{\mathsf{K}} = a^2(\mathtt{t})\,\mathbf{g}_{\mathsf{K}}. \tag{3.11}$$

Conformally related metrics have identical light rays up to affine parameter (see Wald [Wald84, pp. 445–449] for more).

Let us consider the dynamics of a Friedmann universe. In all three cases $\mathsf{K} = -1, 0, 1$, the Friedmann model predicts that the universe is expanding and began at a finite time in the past, which is set at $\tau = 0$ by convention. The model also predicts that the birth of the universe is characterized by a "big bang" (e.g., [ONe, p. 348]). This means that $a(\tau) \to 0$ and $\dot{a}(\tau) \to \infty$ as $\tau \to 0$, where $\dot{a} = da/d\tau$ is the "expansion velocity." For the Euclidean case $\mathsf{K} = 0$, the universe's expansion continues with the "expansion velocity" $\dot{a}$ decreasing to zero as $\tau \to \infty$. In the hyperbolic case $\mathsf{K} = -1$, the expansion continues forever with the "expansion velocity" $\dot{a}$ approaching a finite value ($\dot{a} \to 1$ as $\tau \to \infty$). For the spherical case $\mathsf{K} = 1$, note that a and τ trace out a cycloid. In other words, the expansion of space reaches a maximum $a = A$ at $\tau = \pi A/2$, which is followed by a contraction. The contraction continues until $\tau = \pi A$, when the universe ends catastrophically with a "big crunch." The latter means that $a(\tau) \to 0$ and $\dot{a}(\tau) \to -\infty$ as $\tau \to \pi A$. The universe is

[11] For the case with a cosmological constant Λ, add a term $\Lambda/3$ to the right of the equations in (3.9).

said to be *open* if $\kappa = -1, 0$ and *closed* for $\kappa = 1$. See [ONe, pp. 345–353] for details of the previous calculations.

For a treatment of Friedmann cosmology with a nonzero cosmological constant, interested readers may consult the following references: [Bart-S, Sec. 2.1], [Car-PT], [dIn, Chap. 23], [Rin, Sec. 9.10], and [Mis-TW, pp. 771–774].

Letting $\tau = \tau_0$ be the present time, the following constants are useful:

(1) **Hubble's constant:** $H_0 = \dot{a}(\tau_0)/a(\tau_0)$.

(2) **Cosmological density:** $\Omega_0 = \rho_0/\rho_c$, where ρ_0 is the **present mass density** of the universe and $\rho_c = 3H_0^2/8\pi G$ is the **critical density**.

(3) **Deceleration parameter:** $q_0 = -\ddot{a}(\tau_0)/[a(\tau_0)H_0^2] = \Omega_0/2$.[12]

Note that H_0 has physical dimension of $[\text{time}]^{-1}$, while Ω_0 and q_0 are dimensionless. The reciprocal $1/H_0$ is called the **Hubble time**, while c/H_0 is the **Hubble radius** (see [Pad, p. 73] for more).[13] The Hubble radius is the distance where general relativistic effects appear cosmologically, while for length scales that are significantly less, the Newtonian approximation often suffices [Pad, p. 73].

The **age of the universe** can be computed in terms of H_0 and Ω_0 (or q_0) [Wei72, p. 482]:[14]

$$\tau_0 = \frac{1}{H_0}\int_0^1 \left(1 - \Omega_0 + \Omega_0 x^{-1}\right)^{-1/2} dx.$$

Since $(1 + \Omega_0(x^{-1} - 1))^{1/2} \geq 1$ for $0 \leq x \leq 1$, the age of the universe does not exceed the Hubble time:

$$\tau_0 < \frac{1}{H_0}. \tag{3.12}$$

The fate of the universe is also characterized as follows [Wei72, p. 481]:

(i) $\Omega_0 > 1 \Longleftrightarrow \kappa = 1$ (universe expands and collapses).

(ii) $\Omega_0 = 1 \Longleftrightarrow \kappa = 0$ (universe expands forever with expansion velocity slowing down to zero).

---

[12]The last equality follows from the right equation in (3.9). We have $q_0 = \Omega/2 - \Omega_\Lambda$ for the case with a cosmological constant.

[13]The Hubble radius is is also referred to as the *radius of curvature* of the universe.

[14]If there is a cosmological constant, then

$$\tau_0 = \frac{1}{H_0}\int_0^1 \left[(1 - \Omega_0 - \Omega_\Lambda) + \Omega_0 x^{-1} + \Omega_\Lambda x^2\right]^{-1/2} dx.$$

See [Car-PT, p. 508] and [Bart-S, p. 305] for more.

(iii) $0 < \Omega_0 < 1 \Longleftrightarrow \kappa = -1$ (universe expands forever with expansion velocity slowing down to nonzero value).

For $\kappa = 1, 0$, and -1, the age of the universe can be expressed respectively as follows:

$$\begin{aligned}
\tau_0 &= \frac{1}{2H_0}\,\Omega_0(\Omega_0 - 1)^{-3/2}\left[\cos^{-1}(2\Omega_0^{-1} - 1) - 2\Omega_0^{-1}(\Omega_0 - 1)^{1/2}\right],\\
\tau_0 &= \frac{2}{3H_0},\\
\tau_0 &= \frac{1}{H_0}\left[(1 - \Omega_0)^{-1} - \frac{\Omega_0}{2}(1 - \Omega_0)^{-3/2}\cosh^{-1}(2\Omega_0^{-1} - 1)\right]. \qquad (3.13)
\end{aligned}$$

Consult Weinberg [Wei72, pp. 481–484] for proofs and discussion of the above results.

A light ray undergoing gravitational lensing will propagate through a Friedmann universe during most of its travel (except near the lens).

We suppose that in a gravitational lensing scenario, the scaling function $a(\tau)$ changes negligibly during the time delay between different lensed light rays connecting source and observer, and during the travel time of light across the lens (both times being very short compared to the Hubble time H_0^{-1}).

We now discuss time delays between light rays in terms of "redshifts." Assume that two successive light rays are emitted by a source at proper times τ_S and $\tau_S + \Delta\tau_S$, and received by an observer at τ_O and $\tau_O + \Delta\tau_O$, respectively. *Suppose that the observer is at the origin of (R, θ, φ) and the light rays arrive along the R (or radial) direction.* We can view the light rays as orthogonal trajectories to two wave fronts emanating from the source at the aforementioned times. Then $\Delta\tau_O$ and $\Delta\tau_S$ are, respectively, the proper time intervals between the rays — or their successive wave crests — as measured at the observer and source. These intervals are related by (e.g., [dIn, p. 324] and [Wei72, p. 416]):

$$\frac{\Delta\tau_O}{a(\tau_O)} = \frac{\Delta\tau_S}{a(\tau_S)}. \qquad (3.14)$$

The frequency ν_O as measured at the observer is then related to the frequency ν_S measured at the source by

$$\frac{\nu_O}{\nu_S} = \frac{\Delta\tau_S}{\Delta\tau_O}. \qquad (3.15)$$

The **(cosmological) redshift** of light emitted from a source at proper time τ_S and received at the observer at proper time τ_O is defined as the fractional increase in wavelength

$$z = \frac{\lambda_O - \lambda_S}{\lambda_S}.$$

Here λ_S is the wavelength of the light measured near the source at the time of emission, and λ_O is the wavelength measured by the observer at time τ_O when the light arrives. The emission wavelengths λ_S arise from the spectra of hot gases (e.g., hydrogen, helium, etc.) in the aggregate of stars forming the galaxy. If the spectra of the same hot gases are measured in an earth laboratory, then the same emission wavelengths λ_S are assumed to occur. Using (3.14) and (3.15), and $\nu_O/\nu_S = \lambda_S/\lambda_O$, we obtain

$$z = \frac{a(\tau_O)}{a(\tau_S)} - 1. \tag{3.16}$$

Given that for gravitational lensing the scaling function $a(\tau)$ varies negligibly during the time delay between rays connecting source to observer, *all such light rays emitted at cosmic time τ_S have the same redshift to a very good approximation.*[15] Also notice that in an expanding universe (galaxies moving apart), $a(\tau_O) > a(\tau_S)$ (since $\tau_O > \tau_S$), which implies that $z > 0$, that is, wavelengths are lengthened during travel (objects are "redshifted"). For a contracting universe, the reverse happens ($z < 0$), so wavelengths are shortened (objects are "blueshifted"). It also follows from (3.14) and (3.16) that the time intervals $\Delta\tau_O$ and $\Delta\tau_S$ are related via a redshift factor:

$$\Delta\tau_O = (1+z)\Delta\tau_S, \tag{3.17}$$

where z is the redshift of the emitter.

3.1.2 Cosmic Distances

In the gravitational lens equations there always appear the distances between observer-lens, observer-source, and lens-source. As long as the distances involved are small compared to the size of the universe (as, e.g., for microlensing of stars by stars inside the Milky Way), these distances are the "normal" distances between the objects. For cosmological distances, though, one needs to specify the notion of "distance" more accurately. One can define distances in different ways (see [Wei72, pp. 421–427] and [Schn-EF, pp. 110–117] for details):

1. For a Euclidean space, a light source with intrinsic luminosity L (standard candle), the measured apparent luminosity ℓ decreases with the square of the distance d (i.e., $\ell = L/4\pi d^2$).[16] So, knowing the intrinsic luminosity and measuring the apparent luminosity, one can calculate the distance. Generalizing this to a cosmological context,

[15] We note that lensing does not alter redshift (see page 43).

[16] The *total luminosity* L of a point source is the total radiative energy per unit (proper) time of the source. In Euclidean space, the *apparent luminosity* ℓ of a source at rest at distance d (relative to observer) is the total radiative energy per unit (proper) time per unit area (hence, $\ell = L/4\pi d^2$).

the ("uncorrected"[17]) **luminosity distance** of a cosmic light source is defined by

$$D_{lum} = \left(\frac{L}{4\pi\ell}\right)^{1/2}.$$

2. For a source of diameter d_* in a Euclidean space, the measurable angle ϑ_* under which this size appears is inversely proportional to the distance d (i.e., $\vartheta_* = d_*/d$). Knowing the size and measuring the angular diameter is thus another way to determine distance. Generalizing to a cosmological setting, the **angular diameter distance** of a source is defined by

$$D_A = \frac{d_*}{\vartheta_*}.$$

3. Consider a source at a distance d with proper velocity[18] $V_\perp$ transverse to the line of sight. As seen from the observer, the source will move an angular distance $\Delta\,\vartheta_*$ during the observer time interval $\Delta\,t$. In a Euclidean space, the angular distance the source moves on the celestial sphere of radius d during $\Delta\,t$ is $\Delta\,\vartheta_* = (V_\perp \Delta\,t)/d$. Given the transverse velocity $V_\perp$ and $\Delta\,t$, this formula for the measurable proper motion, $\Delta\,\vartheta_*$, yields the distance. Generalizing to the cosmos, the **proper motion distance** of a source is defined by

$$D_{pmd} = \frac{V_\perp \Delta\,t}{\Delta\,\vartheta_*}.$$

4. Suppose that at cosmic time τ_*, we are at the origin of the comoving coordinate system (R, θ, ϕ) and a light source is located at $p_* = (R_*, \theta_*, \varphi_*)$, where R_* is the radial distance to the source from the origin. Then the **proper distance** to the source at cosmic time τ_* is defined by [Wei72, p. 415]:

$$D_{pro}(\tau_*) = a(\tau_*) \int_0^{R_*} \frac{dR}{\sqrt{1 - \kappa R^2}}. \tag{3.18}$$

This is the Riemannian distance from the origin to p_* in the spacelike hypersurface $\tau_* \times \mathsf{S}$ (see page 33).

The following simple relations hold with respect to the redshift z of a source (e.g., [Wei72, p. 423]):

$$D_{lum} = (1+z) D_{pmd} = (1+z)^2 D_A. \tag{3.19}$$

[17] See [Schn-EF, p. 116, eq. (3.79)]

[18] The *proper velocity* of a source is the distance covered by the source (as measured by the observer) per unit proper time of the source.

In our cosmic neighborhood out to several hundred megaparsecs ($z \ll 1$), the four distances agree with each other to a very high degree due to (3.19):

$$D_{lum} \simeq D_A \simeq D_{pmd} \simeq D_{pro}(\tau_*).$$

For example, cosmic sources close ($z \ll 1$) to our galaxy follow **Hubble's law** ([dIn, pp. 325–327] and [Schn-EF, p. 129]):[19]

$$z_S \simeq \frac{v}{c} \simeq \frac{H_0 D_{lum}}{c}, \qquad z \ll 1, \tag{3.20}$$

where z_S, v, and D_{lum} are, respectively, the redshift, radial receding velocity, and luminosity distance of the source. Consequently, we may replace D_{lum} in (3.20) by either D_A, D_{pmd}, or D_{pro}. However, once one goes to large cosmic distances, corresponding to redshifts of $z \simeq 1$ or beyond (i.e., at least of order a billion light years away), then the situation with distances gets more complicated. The reason is that physical space is not Euclidean. As a consequence, the definitions above diverge, and when one talks about distances, one has to specify exactly what distance one has in mind. Fortunately, the distances are related by the above simple formulas.

> *In gravitational lensing,* angular diameter distance *is commonly used. This is not too surprising, since lensing deals with deflection* angles. *Subsequently, we shall always use this distance measurement, unless otherwise noted.*[20]

Readers interested in a discussion of distances in a cosmology with nonzero cosmological constant should see Carroll, Press and Turner [Car-PT, pp. 510–512] and Bartelmann and Schneider [Bart-S, pp. 306–309].

Angular Diameter Distance in Friedmann Models

Suppose that the observer is located at the origin of the spatial coordinates (R, θ, φ) and light rays arrive along the R-direction. Then for a Friedmann universe, the angular diameter distance measured at the observer of an object at radial coordinate distance R_K and whose light left at proper time τ is given (e.g., [Bart-S, p. 307]):

$$D_A = a(\tau) R_K(\chi). \tag{3.21}$$

In general, let $D_A(z_1, z_2)$ be the angular diameter distance of an emitter at redshift z_2 or cosmic time τ_2, from a receiver at z_1 or cosmic time τ_1. Then

$$D_A(z_1, z_2) = a(z_2) R_K(\chi(z_1, z_2)),$$

---

[19] Since locally the universe is approximately flat ($k = 0$), we have $D_A \simeq c(\tau_O - \tau_S)$, where τ_O and τ_S are, resp., the cosmic times when light from the nearby source is received and emitted.

[20] Some lensing theorists prefer to use the proper distance. However, since angular diameter distances in gravitational lensing almost always appear in ratios that are identical for the proper distances, this distinction is not dramatically important.

where $a(z_2)$ is the value of the scaling function at τ_2 and $R_K(\chi(z_1, z_2))$ is the comoving radial coordinate distance between the emitter and receiver. Alternatively, the angular diameter distance D_A can be expressed in terms of the Hubble constant H_0, the redshift z, and the cosmological density Ω_0 (e.g., [Schn-EF, p. 137, eq. (4.48)] and [Sas, p. 770]):[21]

$$D_A(z_1, z_2) = \frac{c}{H_0}|r_A(z_1, z_2)|,$$

where $r_A(z_1, z)$ is the solution, subject to the initial conditions

$$\begin{aligned} r_A(z_1, z_1) &= 0, \\ \frac{d}{dz} r_A(z_1, z)\bigg|_{z=z_1} &= \frac{1}{(1+z_1)^2\sqrt{1+\Omega_0 z_1}}, \end{aligned}$$

of the *"clump-free" Dyer–Roeder equation* [Dye-R72]

$$(1+z)(1+\Omega_0 z)\frac{d^2 r_A}{dz^2} + \left(3 + \frac{1}{2}\Omega_0 + \frac{7}{2}\Omega_0 z\right)\frac{dr_A}{dz} + \frac{3}{2}\Omega_0 r_A = 0. \tag{3.22}$$

Explicitly,

$$r_A(z_1, z_2) = \frac{2(1+z_1)}{\Omega_0^2}[R_1(z_2)R_2(z_1) - R_1(z_1)R_2(z_2)] \tag{3.23}$$

where

$$R_1(z) = [\Omega_0 z - \Omega_0 + 2]/(1+z)^2, \qquad R_2(z) = (\Omega_0 z + 1)^{1/2}/(1+z)^2.$$

In particular,

$$r_A(0, z) = \frac{2}{\Omega_0^2(1+z)^2}\left[\Omega_0 z - (2-\Omega_0)\left(\sqrt{1+\Omega_0 z} - 1\right)\right]. \tag{3.24}$$

The solution $r_A(z_1, z_2)$ is derived, for example, in [Sei-S94, pp. 358–359]. A fundamental relation angular diameter distances obey is *Etherington's reciprocity law* (e.g., [Schn-EF, p. 141]):

$$D_A(z_1, z_2) = \frac{1+z_1}{1+z_2} D_A(z_2, z_1). \tag{3.25}$$

Consequently, we have the asymmetry $D_A(z_1, z_2) \neq D_A(z_2, z_1)$ for $z_1 \neq z_2$. Hence, angular diameter distances do not obey the symmetry property of ordinary distances.

[21] See [Car-PT, pp. 510–512] for the angular diameter distance as a function of the cosmological constant.

Angular Diameter Distance in Inhomogeneous Models

The angular diameter distances we have discussed so far are valid only in the case of a universe dominated by smooth isotropically distributed matter. There is no known exact metric for a general realistic universe that is Friedmann on average, but includes mass density inhomogeneities.

One of the early attempts to model matter inhomogeneities in the cosmos is through the so-called *Swiss-Cheese model.* These models assume that a fraction $\mathtt{f}$ $(0 \leq \mathtt{f} \leq 1)$ of matter is distributed smoothly and homogeneously, and a certain fraction $1 - \mathtt{f}$ appears in clumps. For such models with light bundles that pass far away from all clumps, Dyer and Roeder derived a second-order ordinary differential equation for r_A (hence, the angular diameter distance $D_A = cH_0^{-1}|r_A|$) as a function of redshift z, the cosmological density Ω_0, and $\mathtt{f}$. This differential equation — known as the *generalized Dyer–Roeder equation* — extends (3.22) as follows [Dye-R73]:

$$(1+z)(1+\Omega_0\, z)\,\frac{d^2 r_A}{dz^2} + \left(3 + \frac{1}{2}\Omega_0 + \frac{7}{2}\Omega_0\, z\right)\frac{dr_A}{dz} + \frac{3}{2}\,\mathtt{f}\,\Omega_0\, r_A = 0. \tag{3.26}$$

Hypergeometric functions can be used to solve (3.26), but such solutions have so far been of little practical use (e.g., [Schn-EF, p.141]). However, informative analytical solutions are known for some special cases; for example, $\Omega_0 = 1$ with general $\mathtt{f}$, and $\mathtt{f} = 0$ (all matter in clumps) with general Ω_0. We note that the angular diameter distances corresponding to solutions to (3.26) obey Etherington's reciprocity law (e.g., [Schn-EF, p. 141]). Further discussion of these clumpy models appear in [Schn-EF, pp. 137–143],[Sei-SE], [Sei-S94], and [Sas].

Unfortunately, Swiss-Cheese models are oversimplified (e.g., [Sel94, p. 515] and [Bar, p. 21]). A more realistic approach is to treat the large scale structure mass density perturbations of the universe as a random field with positive and negative fluctuations about the average cosmic density; further discussion appears in Section 3.1.4 (see page 49). For such an approach, it was shown in [Bar, p. 21] that the density perturbations plus a strong primary lens like a galaxy or cluster of galaxies affect the angular diameter distance (cf. [Schn97]).

Choice of Angular Diameter Distance in Lensing

In general, angular diameter distance depends on one's choice of cosmological model. Angular diameter distances relative to a Friedmann metric are useful for pedagogical reasons and work well for lens systems with a weak-field, thin-screen, localized lens where the effects on the system due to the density perturbations of the cosmos, including that of matter between the observer and lens, and lens and source, are negligible. Consequently:

> *Unless stated otherwise, all angular diameter distances will hereafter be relative to the Friedmann metric.*

We add that the choice of angular diameter distance is essentially irrelevant for the stability and genericity theorems of Part III of the book.

Angular diameter distances in gravitational lensing situations will be denoted as follows:

Notation:

(i) d_L = angular diameter distance from observer to lens plane.

(ii) d_S = angular diameter distance from observer to source.

(iii) $d_{L,S}$ = angular diameter distance from lens plane to source.

Remark: In scenarios where lens and source are relatively nearby (so space is roughly Euclidean), the distance $d_{L,S}$ can be approximated by $d_{L,S} = d_S - d_L$, which is not true in the case of cosmological distances. (This particular relation would become simpler in a cosmological context if one uses the proper distances, see footnote 20.) We also note that, strictly speaking, the angular diameter distance d_S of a lensed light source should be defined relative to a cosmological model that already includes the perturbation due to the presence of the lens (e.g., [Sas, p. 757]). However, for many realistic cases the lenses are in the weak-field limit and are so thin and localized that it suffices to calculate d_S relative to the Friedmann metric.

Critical Surface Mass Density and the Einstein Radius

Angular diameter distances allow us to introduce two physically important constants:

(1) **Critical surface mass density:** For a lens system with light source and lens at angular diameter distances d_S and d_L, respectively, we define the critical mass density as follows:

$$\sigma_{crt} = \frac{c^2}{4\pi G} \frac{d_S}{d_L d_{L,S}}. \tag{3.27}$$

The numerical value of the critical surface mass density is roughly $\sigma_{crt} \simeq 0.8$ g cm^{-2} for lens and source redshifts of $z_L = 0.5$ and $z_S = 2.0$, respectively.

The quantity σ_{crt} appears often in lensing. One area of application is in finding a sufficient condition for a lens to produce multiple images of a background source. For example, consider a lens that is a smooth nonsingular isolated "thin" mass distribution. Let σ be the surface mass density resulting from integrating the three-dimensional density of the lens along the line of sight. If $\sigma(\mathbf{r}) > \sigma_{crt}$ at a (regular) point $\mathbf{r}$, then the lens produces multiple images — see Theorem 11.6(1) on page 428 for details.

(2) **Einstein radius:** In 1936, Einstein showed that if a star lies exactly behind a star of mass M, then due to the symmetry, a ring-like image occurs (see "Einstein Rings" in

Section 1.1). This ring has an angular radius, called **angular Einstein radius**, that is given by

$$\theta_E = \sqrt{\frac{4GM}{c^2}\frac{d_{L,S}}{d_L d_S}} = \sqrt{\frac{M}{\pi \sigma_{crt} d_L^2}}. \tag{3.28}$$

The radius of the Einstein ring — **linear Einstein radius** — is

$$R_E = d_L \theta_E = \sqrt{\frac{M}{\pi \sigma_{crt}}}. \tag{3.29}$$

N.B. *In the astrophysics literature, the angle θ_E is commonly called the* **Einstein radius** *for simplicity; bear in mind that θ_E is not the radius R_E of the Einstein ring.*

The angular Einstein radius defines the angular scale for a lensing situation. For cosmologically distant galaxies acting as lenses, the mass is roughly $M \simeq 10^{12} M_\odot$ and distances d_L, d_S, and $d_{L,S}$ are of order 1 gigaparsec. Consequently, the angular Einstein radius is of order one arcsecond (while the linear radius is of order 1 kiloparsec). A star within our galaxy having $M \simeq M_\odot$ and the distances d_L, d_S, and $d_{L,S}$ of order 10 kiloparsecs has an angular Einstein radius of order a milli-arcsecond.

3.1.3 Spacetime Geometry for Gravitational Lensing

The spacetime geometry for gravitational lensing is introduced as a perturbation of a Friedmann universe by a "weak field" spacetime.

Spacetime Geometry Near a Gravitational Lens

We suppose that a gravitational lens and its lensing effects (e.g., multiple images, arcs, etc.) are localized in a very small portion of the sky.[22]

Near a gravitational lens, we assume that gravity is "weak" (spacetime is nearly flat) and, hence, the perturbation due to the lens is described by a Newtonian potential ϕ localized near the lens; the lens changes its position slowly (relative speed of components is much less than c, the speed of light); and matter stresses are small (pressure is much smaller than energy density times c^2). The curvature of space near a gravitational lens is then treated as negligible.

We suppose that the lens is so slowly moving that it can usually be treated as "static," that is, the Newtonian potential ϕ varies negligibly little during the time light rays take to traverse the lens. Furthermore, we assume that ϕ does not alter the energies of photons as they cross the lens; so the redshift of a source's image is unaffected by the potential.

[22] See Sections 3.1.4 and 3.1.5 for more.

These approximations yield that the spacetime geometry *very close* to a gravitational lens has the following "weak-field" form, *excluding the universe's expansion*:[23]

$$\mathbf{g}_{WF} = -\left(1 + 2\frac{\phi}{c^2}\right) c^2 d\tau^2 + \left(1 - 2\frac{\phi}{c^2}\right) dS^2, \tag{3.30}$$

where ϕ is the time-independent, three-dimensional Newtonian potential of the lens and dS^2 is the Euclidean metric on $\mathbf{R}^3$. Of course, sufficiently far from the lens there is no guarantee that space is Euclidean. Also, the typical Newtonian potentials of interest in gravitational lensing are such that $|\phi|/c^2$ is at most of order 10^{-4} (e.g., [Nar-B, p. 365], [Wei72, p. 79]). Consequently:

> *We shall assume that $|\phi|/c^2 \ll 1$ and ignore all terms of order $1/c^3$ or higher occurring in any calculations involving ϕ.*

The assumption that a lens varies negligibly during the time light rays take to cross it is a very good approximation for almost all observed gravitational lensing phenomena: Clusters of galaxies have sizes of order a megaparsec, which means that light takes a few million years to cross them. The dynamical time scale for a cluster of galaxies (i.e., the time it takes them to change their positions significantly), though, is of order the Hubble time (i.e., a few billion years, about a thousand times longer). This ratio is similar for individual galaxies: Our galaxy (the Milky Way) takes about 100 million years to complete a rotation, which exceeds the 100,000 years light takes to traverse the galaxy by three orders of magnitude. The overall shape of the galaxy would not change during the passage of a light ray traversing the galaxy. And even for individual stars this is basically valid: A star in the Milky Way at a distance to the galactic center comparable to that of the sun moves at about 200 km/sec, whereas the velocity of light is greater than 10,000 times higher; so the star's positional change during the passage of a light ray is negligibly small. We remark that although individual light rays are not affected by the change in position of the lensing object, if one looks at a whole sequence of observations of a particular lensed object, changes of lens positions can be noticeable in such a "time series" (see light curves, page 91).

Spacetime Geometry near a Gravitational Lens System

Gravitational lenses act as "weak-field" perturbations of a background Friedmann universe. Consequently, we assume (cf. [Schn-EF, p. 149]):

> *The geometry of spacetime near a gravitational lens system (i.e., source, lens(es), observer, and light rays from source to observer) is approximated by*

[23] Consult [Schn-EF, pp. 121–123], [Mis-TW, p. 435–447]), and [Schu, pp. 205–208] for more on "weak-field" spacetimes.

$$\begin{aligned} \mathbf{g}^{\mathsf{K}}_{GL} &= -\left(1+2\frac{\phi}{c^2}\right)c^2 d\tau^2 + a^2(\tau)\left(1-2\frac{\phi}{c^2}\right)dS^2_{\mathsf{K}} \\ &= a^2(\mathtt{t})\left[-\left(1+2\frac{\phi}{c^2}\right)d\mathtt{t}^2 + \left(1-2\frac{\phi}{c^2}\right)dS^2_{\mathsf{K}}\right]. \end{aligned} \quad (3.31)$$

Here τ is cosmic time, $\mathtt{t}$ conformal time, and ϕ the time-independent Newtonian potential of the perturbation caused by the lens; for ϕ, terms of order greater than $1/c^2$ are ignored.

Unless stated to the contrary, we assume that except very near the lens, the gravitational lensing metric $\mathbf{g}^{\mathsf{K}}_{GL}$ is approximated very well by the Friedmann geometry $\mathbf{g}^{\mathsf{K}}_{F}$. Near the lens when the light rays are deflected, we suppose that $\mathbf{g}^{\mathsf{K}}_{GL}$ is approximated by $\mathbf{g}^{0}_{GL}$ (since the universe is locally flat), where the scale factor a attains approximately the value at the cosmic proper time when a light ray reaches the lens.[24]

The above assumption is used in the derivation of the time delay function (Section 3.2.2) and the lens equation (Section 3.2.3) for single-plane lensing by a primary strong lens. The reader may notice that the assumption ignores the contribution of the large scale structure density perturbations of the universe, which would include matter perturbations between the observer and lens, and lens and source. It was shown in [Schn97] that the lens equation due to a primary strong lens along with large scale structure density perturbations is mathematically equivalent to the single-plane lens equation without cosmological density perturbations. We shall also see in Section 3.5 (page 112) that the lens equation for lensing by the large scale structure (without a primary strong lens) can also be treated mathematically as a single-plane lens equation.

Light Deflection in General Relativistic Models

We present some references for gravitational lensing in a complete general relativistic setting, which will not be pursued in the book, and suggest certain research directions.

For light deflection by Schwarzschild black holes see Chandrasekhar [Cha, pp. 123–134], Frittelli, Kling and Newman [Fri-KN], and Virbhadra and Ellis [Vir-E] (and references therein). Rauch and Blandford [Rau-B] present a detailed study of gravitational lensing in Kerr spacetimes with applications to X-ray variability in active galactic nuclei. Further treatments of light deflection in Kerr spacetimes appear in Chandrasekhar [Cha] and O'Neill [ONe], though applications to gravitational lensing is not their primary focus. It would be useful to undertake a detailed, systematic, general relativistic study of the lensing features (multiple images, magnification, caustic surfaces, etc.) of specific strong gravitational fields like rotating black holes. This would provide a concrete setting for analyzing the differences

[24] Recall that $a(\tau)$ varies very little during the time delay between light rays from source to observer (see page 3.1.1).

and similarities between the *observable* lensing signatures of black holes (possibly rotating) and weak-field compact deflectors, and may even identify new lensing features of the former that might be accessible to observation in the not-too-distant future.

Christodoulou [Chr84, Chr94, Chr99] gives a thorough study of the behavior of light rays in the gravitational collapse of an inhomogeneous spherically symmetric dust cloud and a spherically symmetric scalar field, with seminal applications to naked singularities and the Cosmic Censorship Conjecture.[25] Gravitational lensing by a static spherically symmetric scalar field naked singularity with mass was treated by Virbhadra, Narasimha and Chitre [Vir-NC]. They found that for a sufficiently small scalar charge to mass ratio, the lensing due to the naked singularity is similar to that of a Schwarzschild black hole (e.g., both have one Einstein ring and no "radial critical curve"[26]); for large values of the ratio, the lensing is different (e.g., the naked singularity produces one "radial critical curve" and either zero or two Einstein rings).

Finally, it should be emphasized that the study of gravitational lensing in a complete general relativistic setting is in its infancy and surely is an important direction to explore further. For example, it would be useful to generalize some of the theorems in Part III to Lorentzian manifolds (viewing gravitational lensing in the natural context of spacetime diagrams, null geodesics, etc.). Such extensions would lead to a deeper understanding of the approximations used in the standard weak-field thin-screen approach to lensing as well as shed light on the meaning of genericity and stability in a spacetime context. It would also be useful to apply one's general findings to important concrete cases like black-hole lensing. Perlick [Per00a, Per00b] presents an extensive overview of what is currently known about gravitational lensing in rather general Lorentzian manifolds.

3.1.4 Matter Deflectors: Density Perturbations

The astrophysical objects that are usually considered as gravitational lenses are stars (ordinary stars, white dwarfs, neutron stars), galaxies, clusters of galaxies, and the large scale structure of the universe. The lensing properties of some other types of matter (either smaller mass or more "exotic") have been explored as well: brown dwarfs, planets, comets, black holes, and cosmic strings.

The gravitational lensing described by the metric (3.31) (page 45), which is a weak-field perturbation of a Friedmann metric, will be the primary focus of the book. For this reason, it is useful to present the two categories astrophysicists commonly divide such lensing into:

---

[25]The Cosmic Censorship Conjecture (CCC) is due to Penrose. It is one of the central unsolved problems in general relativity; see Penrose [Pen98, p. 103] and Wald [Wald84, pp. 299–308], [Wald97] for more. The CCC *may* require a non-positive cosmological constant (e.g., [Bri], [Pen98, pp. 116, 118]); note that recent observations suggest a positive cosmological constant (see page 29).

[26]A *radial critical curve* is a circle c that is the lensed image of a circular caustic with the property that a small disc source on the caustic is lensed into arc-like images with one of the images intersecting c radially; see Figure 3.13 on page 99.

Strong lensing: This is due to a very massive lens, which could be a galaxy or cluster of galaxies that produces multiple images of a background source and/or strongly distorts an extended source into a giant arc (e.g., Figure 1.3, page 13) or ring-like shape (e.g., Figure 5.5, page 156). Note that the terminology "strong lensing" is used even though the lenses are in the weak-field limit. Another important regime of strong lensing is **microlensing.** For this situation the lens consists of compact objects (e.g., stars, planets) that are massive enough to create multiple images of a compact source (e.g., star). However, the resulting images and the angular spacings between them are so small (e.g., of order a micro-arcsecond) that individual images are hard to detect. The "micro" images then appear as a single "macro" image whose brightness variations are due to the combined action of the "micro" images; hence, the description "microlensing."

Weak lensing: For this case, the lens is not strong enough to generate multiple images, arcs, or rings, but weakly distorts extended sources (e.g., it produces ellipticity in circular sources — see Figure 3.12, page 98). Such lenses could be a cluster of galaxies or the large scale structure of the universe, while the source is usually a collection of galaxies and, for some studies, the cosmic microwave background radiation from the Big Bang. The weak lensing effect cannot be detected in one image alone, but only in a statistical sense from an ensemble of images.

Weak lensing is treated in Sections 3.5 (page 112) and 4.2.3 (page 128). However, the book focuses primarily on strong lensing, where multiple images, caustic singularities, etc. appear. The reader interested in a more extensive treatment of weak lensing should see [Bart-S] and [Mel].

Density Perturbation due to a Gravitational Lens

Sometimes the notion of density perturbation in gravitational theory causes confusion for non-physicists. To illustrate this concept, imagine a sand box where the sand is leveled out. Let ϱ_s be the amount of sand piled up in the column of sand above a unit area on the bottom of the box (so ϱ_s is proportional to the height of the sand above the bottom). Suppose that the sand in the box is leveled out. Then the value of ϱ_s all around the sand box is constant, say, $\varrho_s = \varrho_0$. Now, create a small hill by uniformly collecting sand from around the box. Adding more sand in a region to create the hill reduces the amount of sand elsewhere in the box by a small amount. Let ϱ_h be the values of ϱ_s restricted to the hill. Then ϱ_h exceeds ϱ_0 by $\varrho_h - \varrho_0$. Outside the hill, the sand is assumed to be roughly leveled out and ϱ_s has approximately the value $\varrho_0 - \epsilon$, where ϵ accounts for the slight height reduction in the sand needed to create the hill. If we had created a hole in the sand, then $\varrho_h - \varrho_0$ would be negative and the value of ϱ_s outside the hole would be slightly increased to $\varrho_s + \epsilon$.

Loosely speaking, the hill or hole is a two-dimensional analog of our gravitational lens. The leveled out sand in the sand box before the hill (or hole) is formed is like the smooth Friedmann universe we use to model the universe on the largest scales. Let $\varrho_L(\tau, \mathsf{x})$ and

$\varrho^{\mathrm{av}}_{\mathrm{unv}}(\tau)$ be, respectively, the density of the lens (e.g., galaxy or cluster of galaxies) at (τ, x) in spacetime and the average density of matter in the universe at cosmic time τ. We think of $\varrho^{\mathrm{av}}_{\mathrm{unv}}$ as approximating the ideal Friedmann density ρ_F. Analogous to the sand box, the density of the lens exceeds (or lags) the average density $\varrho^{\mathrm{av}}_{\mathrm{unv}}$ by[27]

$$\delta^L_{\mathrm{dp}}(\tau, \mathsf{x}) \equiv \varrho^{\mathrm{av}}_{\mathrm{unv}}(\tau)\, \delta^L(\tau, \mathsf{x}) \;=\; \varrho_L(\tau, \mathsf{x}) - \varrho^{\mathrm{av}}_{\mathrm{unv}}(\tau), \tag{3.32}$$

where

$$\delta^L(\tau, \mathsf{x}) \equiv \frac{\varrho_L(\tau, \mathsf{x}) - \varrho^{\mathrm{av}}_{\mathrm{unv}}(\tau)}{\varrho^{\mathrm{av}}_{\mathrm{unv}}(\tau)}. \tag{3.33}$$

The function δ^L_{dp} is called the **density perturbation** due to the lens (relative to the mean density $\varrho^{\mathrm{av}}_{\mathrm{unv}}$), while δ^L is the (dimensionless) **density contrast** of the lens.

The density perturbation δ^L_{dp} is positive for an over-dense lens ("hill") and negative for an under-dense one ("hole" or "void"). For the former situation, light rays passing near the lens tend to converge, while away from the lens they are inclined to diverge. The reverse happens for an under-dense lens (cf. Section 3.2.5, page 87).

It is very difficult to determine the average density of the universe. One of the problems is that the *visible* matter which can be seen directly comprises only a (small) fraction of the total matter. So indirect reasoning is required, namely, estimates from the dynamical state of galaxies in groups or clusters. Consequently, the value of the average matter density of the universe is not well determined. Most astronomers would agree that the average density of the universe is

$$\varrho^{\mathrm{av}}_{\mathrm{unv}} \;\simeq\; \text{(of order) } 3 \times 10^{-30}\,\mathrm{g\,cm^{-3}}.$$

This value correspond to a cosmological density parameter satisfying $0.2 \lesssim \Omega_0 \lesssim 0.4$. In contrast to the case for the universe, the average matter density within galaxy clusters is roughly

$$\varrho^{\mathrm{av}}_{\text{galaxy cluster}} \simeq \text{(few)} \times 10^{-27}\,\mathrm{g\,cm^{-3}},$$

and that inside a spiral galaxy is about

$$\varrho^{\mathrm{av}}_{\mathrm{galaxy}} \simeq 3 \times 10^{-23}\,\mathrm{g\,cm^{-3}},$$

very much higher than the average cosmic density $\varrho^{\mathrm{av}}_{\mathrm{unv}}$. In fact, the majority of known gravitational lenses have densities far exceeding $\varrho^{\mathrm{av}}_{\mathrm{unv}}$. For this reason, we shall always abide by the following:

> *Unless stated to the contrary, the density perturbation due to a gravitational lens is assumed to obey* $\delta^L_{\mathrm{dp}} \geq 0$.

[27]We are not abiding by the standard practice of denoting the average cosmic density and density contrast by $\bar{\rho}$ and δ, respectively (since the overbar will be used largely for complex conjugation and δ will denote the Dirac delta).

For a *local* situation where the lens is a star or collection of stars in our galaxy, and the source is another star in our galaxy, it is not relevant to employ the average density $\varrho_{\rm unv}^{\rm av}$ of the universe. In this case, the average density of our galaxy is used instead of $\varrho_{\rm unv}^{\rm av}$. If the source is a star in a nearby galaxy, say, the Large Magellanic Cloud, then $\varrho_{\rm unv}^{\rm av}$ is replaced by the average density of our Local Group of galaxies. In either case, we have $\delta_{\rm dp}^{L} \geq 0$ (in fact, $\delta_{\rm dp}^{L}(\tau, \mathsf{x}) \gg 0$).

The positivity condition on $\delta_{\rm dp}^{L}$ is consistent with our goal of studying mainly lenses strong enough to produce multiple images and caustics; such lenses always satisfy $\delta_{L} \gg 1$ — **strongly nonlinear lenses**. However, some aspects of weak lensing, where it is possible to have $\delta_{\rm dp}^{L} < 0$ in certain regions of space, will be treated. Note that weak lensing by galaxy clusters involves **nonlinear perturbations** ($\delta_{L} > 1$), which are not strong lenses.

Density Perturbations due to the Large Scale Structure

The matter distribution in the real universe is neither completely homogeneous and smooth — as assumed in a Friedmann universe — nor in a number of discrete mass clumps. The matter density of the universe fluctuates from place to place, and it is not even constant when averaged on scales of about 100 megaparsecs or so. This situation is applicable to weak lensing on cosmological scales. In the standard picture of cosmology, one assumes that the current distribution of matter on the scales of galaxies or larger formed out of gravity acting on small inhomogeneities (e.g., curvature and pressure fluctuations) in the early universe. No one knows the origin of these inhomogeneities, but a favored explanation is that they came from random quantum fluctuations that grew during an **inflationary epoch** in the *very* early universe. **This will be assumed hereafter.** For introductions to the physics and mathematics of cosmic structure formation, see the books by Coles and Lucchin [Col-L, Chap. 14], Kolb and Turner [Kol-Tu, Chap. 9], Padmanabhan [Pad] and Peebles [Pee]. An overview of the current state of the subject is given by Turner and Tyson [Tur-T].

The present matter distribution of the cosmos is usually referred to — in decreasing size — as follows: *large scale structures*, covering scales of hundreds of megaparsecs; *super-clusters* and *clusters of galaxies*, spreading over tens of megaparsecs; and *groups of galaxies* and *individual galaxies* (including gas and dust, and stars) for scales of 100 kiloparsecs and smaller. The matter of large scale structures is concentrated in sheets and filaments that surround large voids with roughly spherical or boxy shapes. These voids are under-dense regions, compared to the average density of the universe and correspond to negative density perturbations. However, the voids are not really "empty" and their sizes can be of order 100 megaparsecs. When strong lensing situations are considered, typically a quasar at a redshift $z > 1$ is involved, which means the quasar is at distances of order a gigaparsec. Consequently, light rays have to cross distances that are still much larger than the dimensions of the large scale structure or the voids. A light ray to such a quasar always traverses many regions that are slightly under-dense and slightly over-dense. Hence, a situation like "viewing a *very* distant object through a void" does not really occur.

We now consider the density contrast δ for the universe. Denoting the mass density of the universe by ϱ, we define

$$\delta(\tau, \mathsf{x}) \equiv \frac{\varrho(\tau, \mathsf{x}) - \varrho^{\mathrm{av}}_{\mathrm{unv}}(\tau)}{\varrho^{\mathrm{av}}_{\mathrm{unv}}(\tau)}. \tag{3.34}$$

Note that $\delta(\tau, \mathsf{x}) \geq -1$ (since $\varrho(\tau, \mathsf{x}) \geq 0$). Regimes where $|\delta(\tau, \mathsf{x})| \ll 1$ are called **linear**, while those with $|\delta(\tau, \mathsf{x})| \gtrsim 1$ are called **nonlinear**. Regimes with $|\delta(\tau, \mathsf{x})|$ of order one are usually called **weakly nonlinear**. The current large scale structures typically satisfy $|\delta_{\mathrm{LSS}}| < 1$, while galaxies are **strongly nonlinear**, $|\delta_{\mathrm{galaxy}}| \gg 1$. **We shall assume that the density perturbations are linear and consider only those occurring over scales significantly less than the Hubble radius** (e.g., [Kol-Tu, p. 325], [Bart-S, p. 313]).

Given that the exact matter distribution of the universe is unknown, we do not expect our current theory of structure formation to predict the exact density contrast of matter at any specific position in space. For this reason, the density contrast δ is treated as a random field and its statistical properties are investigated; see Adler [Adl] for an introduction to random fields. By construction of δ, its expected value (or mean) vanishes:[28] $\mathrm{E}[\delta(\tau, \mathsf{x})] = 0$. Moreover, since the universe is spatially homogeneous and isotropic on average, the random field δ is as well. At cosmic time τ, **homogeneity** implies that $\mathrm{E}[\delta(\tau, \mathsf{x}+\mathsf{x}_0)] = \mathrm{E}[\delta(\tau, \mathsf{x})]$ for any point x_0 in space, while **isotropy** implies $\mathrm{E}[\delta(\tau, \mathrm{Rot}(\mathsf{x}))] = \mathrm{E}[\delta(\tau, \mathsf{x})]$ for any rotation $\mathrm{Rot}(\mathsf{x})$ of x (cf. [Bart-S, p. 323]); the converse is not guaranteed.

Let us now formally Fourier transform the density contrast, that is, we are treating δ as a continuous superposition of independent random waves of all wavelengths:

$$\delta(\tau, \mathsf{x}) = \frac{1}{(2\pi)^3} \int_{\mathbf{R}^3} d\mathbf{k}\, \hat{\delta}(\tau, \mathbf{k})\, e^{-i\mathbf{k}\cdot\mathsf{x}}, \qquad \hat{\delta}(\tau, \mathbf{k}) = \int_{\mathbf{R}^3} d\mathbf{k}\, \delta(\tau, \mathsf{x})\, e^{i\mathbf{k}\cdot\mathsf{x}},$$

where $\hat{\delta}$ is the Fourier transform of δ. Since δ is homogeneous and isotropic on average, the statistics of $\hat{\delta}$ depend only on $|\mathbf{k}|$. We shall assume that each mode $\hat{\delta}(\tau, \mathbf{k}) = \hat{\delta}(\tau, |\mathbf{k}|)$, where τ is fixed, is a Gaussian with variance $\mathrm{Var}_{\hat{\delta}}(\tau, |\mathbf{k}|)$. This implies that the phases of $\hat{\delta}$ are independent and randomly distributed over the interval $[0, 2\pi]$, while the squared amplitudes $|\hat{\delta}|^2$ of $\hat{\delta}$ follow an exponential probability density with expected value $\mathrm{E}[|\hat{\delta}(\tau, |\mathbf{k}|)|^2] = 2\,\mathrm{Var}_{\hat{\delta}}(\tau, |\mathbf{k}|)$ (e.g., [Pad, p. 191]). The quantity

$$P_{\hat{\delta}}(\tau, |\mathbf{k}|) \equiv \mathrm{E}[|\hat{\delta}(\tau, |\mathbf{k}|)|^2]$$

is called the **power spectrum** of δ. An important fact which follows is that δ is **Gaussian** (e.g., [Col-L, p. 284–285]).

[28]The expected value is usually denote by $< \;>$, which will not be used to avoid confusion with the mathematical orderings "$<$" and "$>$."

Since δ is Gaussian with zero mean, its probability density is

$$f_\delta(\mathrm{u}) = \frac{1}{\sqrt{2\pi \,\mathrm{Var}_\delta}} \exp\left[-\frac{\mathrm{u}^2}{2\,\mathrm{Var}_\delta} \right], \tag{3.35}$$

where $\mathrm{u} = \delta(\tau, \mathsf{x})$ with τ fixed and Var_δ is the variance of δ. The variance Var_δ is determined by the power spectrum P_δ (e.g., [Pad, p. 193]):

$$\mathrm{Var}_\delta(\tau) = \mathrm{E}[\delta^2(\tau, \mathsf{x})] = \frac{1}{2\pi^2} \int_0^\infty k^2 dk \; P_\delta(\tau, k),$$

where $k = |\mathbf{k}|$. Note that the statistic Var_δ depends on cosmic time, not on space.

A natural statistic to use for measuring spatial mass-density perturbations is the **two-point correlation function** C_δ of the density contrast δ at different spatial points x and x'. In fact, the average homogeneity and isotropy of δ yields that C_δ depends spatially only on the separation between x and x':

$$C_\delta(\tau, |\mathsf{x} - \mathsf{x}'|) \equiv \mathrm{E}[\delta(\tau, \mathsf{x})\,\delta(\tau, \mathsf{x}')].$$

There is a simple illustration of how C_δ measures spatial mass-density perturbations (e.g., [Col-L, p. 288], [Pad, p. 194]). Suppose that at cosmic time τ, the mass distribution is divided into regions of mean size ℓ with $\delta(\tau, \mathsf{x}) > 0$ (over-dense) and $\delta(\tau, \mathsf{x}) < 0$ (under-dense). On average, points x and x' with separation $|\mathsf{x}-\mathsf{x}'| \ll \ell$ will both be in an over-dense region or an under-dense one; consequently, $\delta(\tau, \mathsf{x})\,\delta(\tau, \mathsf{x}') > 0$ and, hence, $C_\delta(\tau, |\mathsf{x} - \mathsf{x}'|) > 0$. For $|\mathsf{x} - \mathsf{x}'| \gtrsim \ell$, the points x and x' will on average be in different regions, one over-dense and the other under-dense. In this case, we get $\delta(\tau, \mathsf{x})\,\delta(\tau, \mathsf{x}') < 0$, so $C_\delta(\tau, |\mathsf{x} - \mathsf{x}'|) < 0$. The correlation C_δ then attains its first zero at roughly the separation ℓ. Thus, this zero of C_δ characterizes the average size, called the **coherence length**, of the over- and under-dense regions.

The correlation function C_δ is also characterized by the power spectrum P_δ. They form a Fourier transform pair, a result known as the **Wiener–Khintchine theorem** (e.g., [Col-L, p. 287]):

$$C_\delta(\tau, |\mathsf{x}|) = \frac{1}{(2\pi)^3} \int_{\mathbf{R}^3} d\mathbf{k} \; P_\delta(\tau, |\mathbf{k}|) e^{i\mathbf{k}\cdot\mathsf{x}}, \tag{3.36}$$

$$P_\delta(\tau, |\mathbf{k}|) = \int_{\mathbf{R}^3} d\mathsf{x} \; C_\delta(\tau, |\mathsf{x}|) e^{-i\mathbf{k}\cdot\mathsf{x}}.$$

In addition, since δ is a zero mean Gaussian, the joint probability density of δ at any finite number of distinct spatial points $\mathsf{x}_1, \ldots, \mathsf{x}_n$ is a multivariate Gaussian density of the following form:

$$f_\delta(\mathbf{u}) = \frac{1}{(2\pi)^{n/2}\sqrt{\det \mathbf{C}}} \exp\left[-\frac{1}{2} \mathbf{u}^T \mathbf{C}^{-1} \mathbf{u} \right], \tag{3.37}$$

where $\mathbf{u}^T = [u_1 \ \cdots \ u_n]$ with $u_i = \mathcal{S}(\tau, \mathtt{x}_i)$ for τ fixed and $\mathbf{C}$ is the (positive definite) correlation matrix with entries $C_{ij} = C_{\mathcal{S}}(\tau, |\mathtt{x}_i - \mathtt{x}_j|)$. By (3.36), the matrix $\mathbf{C}$ is determined by $P_{\mathcal{S}}$. Hence, equation (3.37) yields that **the power spectrum completely characterizes the statistical properties of** $\mathcal{S}$ (see [Col-L, pp. 285-286] for more). We shall see in Section 3.5 (page 112) that the power spectrum is related to the ellipticity correlation of galaxies distorted by large scale structure weak lensing. Since the ellipticities of galaxies can be determined from observations, weak lensing can be used to constrain models of the power spectrum.

We end this subsection with the following convention aimed at avoiding confusion over whether large scale structure effects are being included when we discuss gravitational lens systems:

Unless stated to the contrary, large scale structure perturbations effects on a given gravitational lens system will be assumed negligible.

The Cosmological Poisson Equation

We now return to the standard situation of gravitational lensing by a strong lens without large scale structure effects.

Consider a gravitational lens system where an observer is at the origin of the spatial comoving coordinates (R, θ, φ) and the center of mass of the lens and a very small light source are at $p_L = (R_L, \theta_L, \varphi_L)$ and $p_S = (R_S, \theta_S, \varphi_S)$, respectively. Denote by τ_L the cosmic proper time when light that left the source (at $\tau = \tau_S$) reaches the lens. Also, let $\mathtt{t}_L$ be the value of the conformal time $\mathtt{t}$ corresponding to τ_L. Let

$$a_L = a(\mathtt{t}(\tau_L)).$$

In the vicinity of the gravitational lens, the spatial metric dS_K^2 is approximated by the Euclidean metric dS^2 and the scale function $a(\mathtt{t})$ varies very little during the time delay between light rays. Consequently, the spacetime geometry near the lens when the light rays arrive is approximated by (cf. Section 3.1.3):

$$\tilde{\mathbf{g}}^0_{GL} = a_L^2 \left[-\left(1 + 2\frac{\phi}{c^2}\right) d\mathtt{t}^2 + \left(1 - 2\frac{\phi}{c^2}\right) dS^2 \right]. \tag{3.38}$$

Here $d\mathtt{t}$ is approximated by

$$d\mathtt{t} = \frac{c}{a_L} d\tau \tag{3.39}$$

and ϕ is the *time-independent* Newtonian potential of the density perturbation due to the lens ($|\phi|/c^2 \ll 1$). By (3.38) and the fact that $|\phi|/c^2 \ll 1$, space is approximately Euclidean close to the lens. Choose dimensionless Cartesian coordinates (x_1, x_2, x_3) (i.e., comoving coordinates[29] near the lens such that the origin is at p_L). We shall always extend the x_3-axis along the line of sight passing through p_L and on to a point near (possibly at) p_S. If space is $\mathbf{R}^3$ outside the lens, then the x_3-axis is a linear path (and (x_1, x_2, x_3) serves as Cartesian coordinates over all of space). However, if space is a 3-sphere or hyperbolic 3-space, then the x_3-axis is not linear when the observer, lens, and source are sufficiently far apart.

The density perturbation δ^L due to the lens is determined by the Einstein equation. In the "static" weak-field limit $\mathbf{g}_{WF}$ (page 44), the Einstein equation reduces to the time-independent Poisson equation (e.g., [dIn, pp. 165–168] and [Ray-BB, pp. 36–37]). For the metric $\tilde{\mathbf{g}}^0_{GL}$ describing spacetime near the lens, we obtain the **cosmological Poisson equation** (e.g., [Fut-S, 2505]) at cosmic time τ_L:

$$\Delta^{(3D)}\phi(\mathsf{x}) = 4\pi a_L^2 G\, \rho(\mathsf{x}), \tag{3.40}$$

where $\Delta^{(3D)}$ is the Laplacian relative to the comoving coordinates $\mathsf{x} = (x_1, x_2, x_3)$ and

$$\rho(\mathsf{x}) \equiv \delta^L_{\text{dp}}(\tau_L, \mathsf{x})$$

with δ^L_{pd} given by (3.32) (page 48). Note that the effects of ρ are assumed nontrivial in the vicinity of the lens and negligible elsewhere, unless explicitly stated to the contrary; moreover, the condition $|\phi|/c^2 \ll 1$ is always enforced. Poisson's equation is solved formally by

$$\phi(\mathsf{x}) = -a_L^2 G \int_{\mathbf{R}^3} \frac{\rho(\mathsf{x}')}{|\mathsf{x} - \mathsf{x}'|} d\mathsf{x}'.$$

Note that ϕ has physical dimensions of $[\text{length}]^2/[\text{time}]^2$.

Refractive Index of Gravitational Lens

It is also useful to view the gravitational field of a matter deflector as having a refractive index. This approach will be utilized in Section 3.1.5 to investigate the bending angles of lensed light rays.

By drawing parallels with an ordinary lens, in 1920 Eddington [Edd, p. 109] interpreted a spacetime with the "weak-field" metric $\mathbf{g}_{WF}$ in (3.30) as "flat," but endowed with a **refractive index**

$$n = \frac{c}{v}, \tag{3.41}$$

where v is the effective spatial speed of a light ray in the gravitational field:

$$v = \left| a(\tau) \frac{dS}{d\tau} \right|. \tag{3.42}$$

[29] To obtain *proper coordinates* r at cosmic time τ_L, we must scale the comoving coordinates $\mathsf{x} = (x_1, x_2, x_3)$ by a_L, i.e., $r = a_L \mathsf{x}$.

More generally, consider a spacetime metric of the form

$$\tilde{\mathbf{g}} = -c^2 A(X) d\tau^2 + a^2(\tau) B(X) dX^2.$$

Here A and B are dimensionless and defined on space, the metric dX^2 is spatial (i.e., no $d\tau$ terms) and $a^2(\tau)dX^2$ has dimension of [length]^2. This metric $\tilde{\mathbf{g}}$ yields the effective speed:

$$v = \left| a(\tau) \frac{dX}{d\tau} \right| = c\sqrt{\frac{A}{B}}$$

and refractive index

$$n = \sqrt{\frac{B}{A}}.$$

Applying this to the metric $\tilde{\mathbf{g}}^0_{GL}$ in (3.38), we obtain the effective speed of light rays near the lens:

$$v_L = c\sqrt{\frac{1 + (2\phi/c^2)}{1 - (2\phi/c^2)}} = c\left(1 + \frac{2}{c^2}\phi\right), \tag{3.43}$$

where $|\phi|/c^2 \ll 1$ and terms of order $1/c^3$ or higher are ignored. The corresponding refractive index (to first order in ϕ/c^2) is[30]

$$n_L = \frac{c}{v_L} = 1 - \frac{2}{c^2}\phi, \qquad \frac{|\phi|}{c^2} \ll 1. \tag{3.44}$$

For the full gravitational lensing metric $\mathbf{g}^{\mathsf{K}}_{GL}$ at cosmic time τ, the refractive index is the same:

$$n(\tau, \mathsf{x}) = 1 - \frac{2}{c^2}\phi(\mathsf{x}), \qquad \frac{|\phi|}{c^2} \ll 1. \tag{3.45}$$

If no lens is present, then by (3.43) the effective speed of a light ray at cosmic time τ_L is $v_L = c$. In the presence of a lens we have $v_L < c$, that is, a light ray passing through a gravitational potential slows down relative to its speed c in the absence of the lens (i.e., when the lens is smoothed out back into the Friedmann background). For example, the density of a particle of mass m at the origin is given by

$$\rho(\mathsf{x}) = \mathrm{m}\delta(\mathsf{x}),$$

where $\delta(\mathsf{x})$ is the three-dimensional Dirac delta at the origin. The Newtonian potential is

$$\phi(\mathsf{x}) = -a_L^2 \frac{G\mathrm{m}}{|\mathsf{x}|}.$$

[30]Note that since ϕ has physical dimensions of [length]^2/[time]^2, the ratio ϕ/c^2 is dimensionless and, so, the refractive index n is dimensionless.

The graph of this function has an infinitely deep "well" at the origin (i.e., $\phi(\mathsf{x}) \to -\infty$ as $|\mathsf{x}| \to \mathbf{0}$). The deeper a ray penetrates this potential "well" (i.e., the closer the ray gets to the origin), the more it is slowed down relative to v_0. However, as light rays get very close to the point mass, the gravitational lensing approximation starts to break down (since our weak-field assumption — namely, $|\phi|/c^2 \ll 1$ — fails).

Single Lens Plane

The gravitational lens (and lensing effects) we consider is assumed to have a "weak" gravitational field and occurs within a very small "cone" about the line of sight to the lens (Section 3.1.3). In addition:

We treat a gravitational lens as "thin."

The physical size of a lensing matter distribution along the line of sight is small compared to the distances between lens and observer, or, lens and source. In practically all cases of known quasar lensing, the observer-lens and lens-source separations are of order thousands of megaparsecs, which exceeds the diameter of a deflector galaxy (roughly 30 kiloparsecs) or cluster of galaxies (about 5 megaparsecs) by a large factor ($\geq$ 1000). For the case of microlensing of stars in our galactic bulge (i.e., nucleus of our galaxy) or towards the Large Magellanic Cloud by foreground stars, the observer-lens and lens-source distances (respectively, about 5 to 25 kiloparsecs) dominate the separation of binary-star gravitational lens systems (order 10^{-2} parsecs).

Consider the local Cartesian coordinates (x_1, x_2, x_3) introduced about the lens at position p_L, where x_3 extends along the line of sight passing through p_L. Since the lens is "thin" and occupies a very small portion of the sky, we can treat the lens as lying in the tangent plane at p_L to the "celestial sphere" passing through p_L and centered at the observer. This tangent plane is called the **lens plane** of the system.

Each gravitational lens is approximated as lying in a lens plane.

Taking into account the expansion of the universe, we shall employ the following proper coordinates at cosmic proper time τ_L:

$$(r_1, r_2, \zeta) = (a_L x_1, a_L x_2, a_L x_3). \tag{3.46}$$

Approximating the lens plane by $\zeta = 0$, we parametrize it with the Cartesian coordinates

$$\mathbf{r} = (r_1, r_2).$$

The vector $\mathbf{r}$ *determines proper distance as measured at the lens,* namely, a light ray impacting the lens plane at $\mathbf{r}$ is at proper distance $|\mathbf{r}|$ in the lens plane. In fact, choose the origin of the spatial comoving coordinates (R, θ, φ) to be at the origin of the system (x_1, x_2, x_3)

(i.e., at p_L). At cosmic time τ_L, suppose that an object at position $\mathbf{r}$ on the lens plane has comoving coordinates given by $(R_1, \theta_1, \varphi_1)$, where R_1 is the radial distance to the object from the origin. Then the proper distance to the object as measured at the lens at cosmic time τ_L is (see (3.18), page 38):

$$D_{prop}(\tau_L) = a_L \int_0^{R_1} \frac{dR}{\sqrt{1 - \kappa R^2}}.$$

Assume that the object is near the origin. Since in the vicinity of the lens, space is approximately Euclidean ($\kappa = 0$), we have

$$D_{prop}(\tau_L) = a_L R_1 = |\mathbf{r}|. \tag{3.47}$$

By integrating the three dimensional density perturbation ρ due to the lens along the ζ-axis, we obtain a projection of ρ to a two-dimensional density perturbation σ on the lens plane. Astrophysicists commonly call the function σ the **surface mass density** or **convergence**[31] of the lens (at cosmic proper time τ_L). Bear in mind, however, that σ is actually the **surface mass density perturbation** due to the lens. Now, assume that the observer and source have ζ-coordinates $\zeta = \zeta_{ob} > 0$ and $\zeta = \zeta_{sr} < 0$, respectively.[32] The surface mass density of the lens at cosmic proper time τ_L is then given by

$$\sigma(\mathbf{r}) = \int_{\zeta_{sr}}^{\zeta_{ob}} \rho(\mathbf{r}, \zeta) d\zeta. \tag{3.48}$$

We remind the reader about our assumption of nonnegative density perturbations (see page 48):

Unless otherwise stated, we are assuming that $\sigma \geq 0$.

The **total mass**, relative to σ, in the lens plane at proper time τ_L is defined by:

$$M = \int_{\mathbf{R}^2} \sigma(\mathbf{r}) d\mathbf{r},$$

where the integration is over the entire lens plane $\mathbf{R}^2$. This, of course, is only formal because the lens occupies a very small portion of the lens plane.

Using the proper coordinates (r_1, r_2, ζ), the cosmological Poisson equation (3.40) becomes

$$\frac{\partial^2 \phi}{\partial r_1^2} + \frac{\partial^2 \phi}{\partial r_2^2} + \frac{\partial^2 \phi}{\partial \zeta^2} = 4\pi G \rho.$$

[31]The reason for the term *convergence* will become apparent on page 97.

[32]Since the source and observer are far apart, the limits $\zeta_{ob} \to \infty$ and $\zeta_{sr} \to -\infty$ are used often.

Here ϕ and ρ are localized near the lens, and $|\phi|/c^2 \ll 1$. Integrating along the ζ-axis, we obtain[33]

$$\Delta^{(2D)} \int_{\zeta_{sr}}^{\zeta_{ob}} \phi(\mathbf{r}, \zeta) d\zeta = 4\pi G \sigma(\mathbf{r}).$$

Here $\Delta^{(2D)}$ is the Laplacian relative to coordinates (r_1, r_2) and we used

$$\begin{aligned} \int_{\zeta_{sr}}^{\zeta_{ob}} \frac{\partial^2 \phi(\mathbf{r}, \zeta)}{\partial \zeta^2} d\zeta &= \frac{\partial}{\partial \zeta} \int_{\zeta_{sr}}^{\zeta_{ob}} \frac{\partial \phi(\mathbf{r}, \zeta)}{\partial \zeta} d\zeta \\ &= \frac{\partial}{\partial \zeta} [\phi(\mathbf{r}, \zeta_{ob}) - \phi(\mathbf{r}, \zeta_{sr})] = 0. \end{aligned}$$

Defining

$$\Psi(\mathbf{r}) \equiv \frac{2}{c^2} \int_{\zeta_{sr}}^{\zeta_{ob}} \phi(\mathbf{r}, \zeta) d\zeta, \tag{3.49}$$

we see that Ψ obeys the two-dimensional Poisson equation:

$$\Delta^{(2D)} \Psi = \frac{8\pi G}{c^2} \sigma. \tag{3.50}$$

Equation (3.50) is solved formally by the following two-dimensional Newtonian potential:

$$\Psi(\mathbf{r}) = \frac{4G}{c^2} \int_{\mathbf{R}^2} d\mathbf{r}' \sigma(\mathbf{r}') \ln \left| \frac{\mathbf{r} - \mathbf{r}'}{d_0} \right|, \tag{3.51}$$

where the integral is only formally over the entire lens plane and d_0 is a constant with physical dimension of [length]. The constant d_0 is usually chosen to be the angular diameter distance d_L to the lens. *We shall see that adding a constant term to Ψ will not change any of the resulting gravitational lensing effects. Consequently, we shall identify potentials that differ by a constant.* Note that since G/c^2 has physical dimensions of [length]/[mass], the potential has $[\Psi]$ = [length].

Multiple Lens Plane

The most dramatic current gravitational lens observations — for example, giant luminous arcs, Einstein rings, multiple quasars — are mainly examples of strong lensing due to a single primary lens at a certain distance. So it makes sense to use the concept of a single lens plane for purposes of modeling and understanding these observations.

The other extreme — namely, the weak lensing regime — is much better described by a "random walk" approach to the deflections of the individual light bundles: All objects

[33] We are assuming that ϕ is twice continuously differentiable, so the second-order partials can be moved to outside the integral.

close to the line of sight between the source and the observer contribute partially to the deflection and distortion of the light bundle. These situations should be properly described by a three-dimensional distribution of lenses, an approximation which would consist ideally of hundreds or thousands of lens planes at varying redshifts. For example, J.W. et al. [Wam-COT] employed 200–300 lens planes to numerically trace light rays through various cosmological models with given power spectra of mass density fluctuations. There is obviously an intermediate regime in which a few lenses along the line of sight at various distances contribute most of the deflection. An approximate description of such a situation would also include two or more lens planes between source and observer.

Depending on what problem one wants to model or analyze, techniques using a single lens plane, a few lens planes, or a large number of lens planes are employed. The geometry and the approximate intensity ratios of most of the multi-quasar systems are well reproduced by the single lens plane approach. However, the failure to model a system with a single lens, or observational indications that the lens consists of two parts at different redshifts, require a two-lens-plane approach. Theoretically, such situations were investigated by, for example, [Koc-A], [Erd-S], and [Pet-Wic95].

A useful general description incorporating the strong, weak, and intermediate lensing regimes is the multiple-lens-plane approach, that is, lensing by at least one lens plane. It is worth noting that weak lensing by large scale structures (Section 3.5, page 112) or clusters of galaxies (Section 4.2.3, page 128), and lensing by a strong lens plus large scale structures (see pages 74, 77) can be viewed as single-plane lensing. Moreover, weak lensing by large scale structures can also be approximated by multiplane lensing [Sei-SE, p. 2357], [Wam-COT]. Various astrophysical aspects of the multiple-lens-plane approach were investigated numerically in [Jar], [Rau], [Wam-COT], [Wam-COX], and [Wam-CO]. Up to a few hundred lens planes were used by the authors. Part III of the book develops a rigorous mathematical theory of gravitational lensing by deflectors on any finite number of lens planes.

When there are two or more gravitational lenses along the line of sight, they are assumed to lie on independent lens planes approximately orthogonal to the line of sight.

3.1.5 Deflected Light Rays and Bending Angles

We saw that the redshift of an image of a lensed source is unaffected by the potential ϕ due to lensing (see page 43). In addition, the time delay between images is so small that the scale factor $a(\tau)$ is practically unchanged during the time delay (page 37). Hence, *all images of a lensed source have approximately the same redshift.*

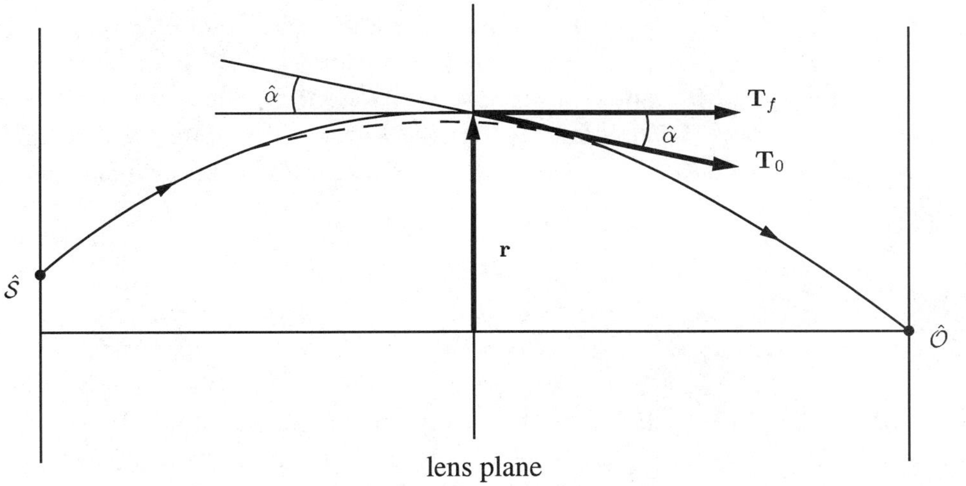

Figure 3.1. At some fixed cosmic time, the light ray's actual spatial path near the lens follows the smooth dashed curve. Since the deflection angles are very small, the ray is approximated by the indicated piecewise-smooth geodesic path, namely, a spatial geodesic from source $\hat{\mathcal{S}}$ to $\mathbf{r}$ on the lens plane L followed by one from $\mathbf{r}$ to the observer $\hat{\mathcal{O}}$. The vectors $\mathbf{T}_f$ and $\mathbf{T}_0$ are, respectively, the final and initial unit tangent vectors to the geodesic from $\hat{\mathcal{S}}$ to $\mathbf{r}$ and from $\mathbf{r}$ to $\hat{\mathcal{O}}$. These vectors determine the angle $\hat{\alpha}$, which approximates how much the smooth ray turns near the lens plane.

Approximating Deflected Light Rays

For the majority of known realistic gravitational lensing scenarios, light rays will experience deflections only by "weak-field" lenses along the line of sight (Section 3.1.4). Consequently:

> *In most realistic lensing situations, the deflection angle of a light ray is very small compared to unity.*

Furthermore, matter between observer and first lens plane, between adjacent lens planes, and between last lens plane and light source is either negligible, or homogeneously distributed. Consequently, the gravitational lens is the only agent of nontrivial perturbations to the ray paths. Each portion of a light ray going from light source to last lens plane, between the lens planes, and from the first lens plane to the observer, is then treated as a null geodesic of the background Friedmann metric. Near the lens, a light ray follows a smooth curved trajectory. Since the deflection angle is very small, we approximate this trajectory by a piecewise-smooth null geodesic path, which projects in space to a piecewise-smooth geodesic — Figure 3.1. Notice that for a spatially Euclidean Friedmann metric ($\kappa = 0$), the light rays are assumed piecewise linear. Of course, this does not apply when space is a 3-sphere or hyperbolic 3-space. We summarize:

Light rays undergoing gravitational lensing are approximated by piecewise-smooth null geodesic curves of the Friedmann metric $\mathbf{g}_F$. *A light ray from a source to the observer then consists of a null geodesic from the source to lens plane and one from the lens plane to observer. At some fixed cosmic time* τ, *the spatial paths of lensed light rays are also approximated by piecewise-smooth geodesics of the spatial metric* $\mathbf{g}^{\tau}_{(\mathsf{K})}$.[34]

The Bending Angle

Consider the piecewise-smooth spatial geodesic path in Figure 3.1 that approximates the spatial trajectory of a light ray (i.e., the path traversing the lens plane along the dashed curve). Denote this path by $\hat{\upsilon}_{\mathbf{r}}$, where $\mathbf{r}$ is the proper impact position of the ray in the lens plane. Let $\mathbf{T}_f(\mathbf{r})$ (resp., $\mathbf{T}_0(\mathbf{r})$) be the final (resp., initial) unit tangent vector of the geodesic segment of $\hat{\upsilon}_{\mathbf{r}}$ starting from the source and terminating at $\mathbf{r}$ (resp., starting from $\mathbf{r}$ and terminating at the observer). Set

$$\hat{\boldsymbol{\alpha}}(\mathbf{r}) = \mathbf{T}_f(\mathbf{r}) - \mathbf{T}_0(\mathbf{r}). \tag{3.52}$$

Then the bending angle of $\upsilon_{\mathbf{r}}$ at $\mathbf{r}$ is: $\hat{\alpha}(\mathbf{r}) = |\hat{\boldsymbol{\alpha}}|$. This approximates the amount by which the actual spatial path of the light ray turns in the vicinity of the lens. Hence, we call $\hat{\boldsymbol{\alpha}}(\mathbf{r})$ and $\hat{\alpha}(\mathbf{r})$, respectively, the **bending angle vector** and **bending angle**[35] at $\mathbf{r}$ of the light ray.

Our goal is to obtain a formula for the bending angle $\hat{\alpha}$. For simplicity, *assume that the spatial geometry of the universe is Euclidean* ($\mathsf{K} = 0$) and employ the proper coordinates (r_1, r_2, ζ) defined in (3.46). Note that these coordinates are now Cartesian and cover all of space at the cosmic proper time τ_L.

Working analogously to regular geometrical optics, view the spatial path of light rays emitted from a (point-like) light source as smooth orthogonal trajectories to the wavefront emanating from the source.[36] At cosmic proper time τ_L, parametrize each light ray's spatial path by arc length s and denote the position vector of the ray by R. Then the light rays are characterized by the following differential equation (see Born and Wolf [Bor-W, p. 122, eq. (2)]):[37]

$$\frac{d}{ds}\left(n\frac{dR}{ds}\right) = \nabla n. \tag{3.53}$$

---

[34]Recall that the spatial path of a null geodesic in a Friedmann universe is a geodesic of the spatial metric dS^2_{K} and, hence, of $\mathbf{g}^{\tau}_{(\mathsf{K})} = a(\tau)dS^2_{\mathsf{K}}$ [ONe, p. 356, Cor. 26; p. 92].

[35]We remind mathematicians that the bending angle $\hat{\alpha}(\mathbf{r})$ is a dimensionless physical quantity and, unless otherwise specified, is assumed to be in terms of the dimensionless entity *radians*.

[36]We are not yet employing the piecewise-smooth geodesic approximation to light ray paths.

[37]Equation (3.53) can also be derived by applying Fermat's principle to the spatial path of light rays, i.e., using the variational principle $\delta \int n ds = 0$; see [Bor-W, pp. 732–734] for details. In Section 3.2.1, we shall use a spacetime version of Fermat's principle to characterize light rays.

Here n is the refractive index of the lens at proper time τ_L (see (3.44)):

$$n = \frac{c}{v_L} = 1 - \frac{2}{c^2}\phi, \qquad \frac{|\phi|}{c^2} \ll 1.$$

The gradient ∇ is defined relative to the proper coordinates (r_1, r_2, ζ).

We shall not use (3.53) to solve for the light ray's position vector R, but rather to determine the amount of turning of the ray's spatial path when it passes the lens plane (i.e., at cosmic time τ_L). First, consider the *curvature vector* to the ray's spatial path. This is the vector defined by

$$\mathbf{K} = \frac{d\mathbf{T}}{ds},$$

where $\mathbf{T}$ is the unit tangent vector to the ray

$$\mathbf{T} = \frac{dR}{ds}.$$

Recall that in arc length parametrization $|dR/ds| = 1$. The curvature vector measures the rate at which the unit tangent $\mathbf{T}$ is turning. By (3.53), we get

$$n\mathbf{K} = \nabla_{\perp} n. \tag{3.54}$$

Here

$$\nabla_{\perp} n = \nabla n - (\nabla n \cdot \mathbf{T})\mathbf{T}.$$

Notice that the vector $\nabla_{\perp} n$ is the orthogonal projection of the vector ∇n into the plane perpendicular to the direction vector $\mathbf{T}$ of the light ray.

Denote the initial (at source) and final (at observer) values of arclength along the light ray by s_{sr} and s_{ob}, respectively. Let $\mathbf{T}_{sr} = \mathbf{T}(s_{sr})$ and $\mathbf{T}_{ob} = \mathbf{T}(s_{ob})$, which are the initial and final unit tangents to the light ray. Now, the piecewise-smooth geodesic path $\hat{v}_{\mathbf{r}}$ in Figure 3.1 approximates the light ray. Moreover, since the geometry of space is assumed to be Euclidean, the path $\hat{v}_{\mathbf{r}}$ will not consist of two curved arcs connecting source-lens and lens-observer as in Figure 3.1 (which depicts the general case), but two linear segments. In addition, it is assumed that $\hat{\alpha}(\mathbf{r}) \ll 1$, so $\mathbf{T}_f(\mathbf{r}) \simeq \mathbf{T}_{sr}$ and $\mathbf{T}_0(\mathbf{r}) \simeq \mathbf{T}_{ob}$. By (3.52), we get

$$\hat{\boldsymbol{\alpha}}(\mathbf{r}) = \mathbf{T}_{sr} - \mathbf{T}_{ob} = -\int_{s_{sr}}^{s_{ob}} \mathbf{K}(s)ds,$$

where the integral is along the piecewise-linear ray $\hat{v}_{\mathbf{r}}$. The vector $\hat{\boldsymbol{\alpha}}$ then approximately measures the total turning of the unit tangent vector of the light ray. In particular, the unit tangent vector $\mathbf{T}_{sr}$ has turned by the angle

$$\hat{\alpha} = |\mathbf{T}_{sr} - \mathbf{T}_{ob}|.$$

Calculating the curvature vector using (3.54) yields

$$\mathbf{K} = \frac{\nabla_\perp n}{n} = \left(-\frac{2}{c^2}\nabla_\perp \phi\right)\left(\frac{1}{1-\frac{2}{c^2}\phi}\right).$$

Ignoring terms of order $1/c^3$ or higher, we get

$$\mathbf{K} = -\frac{2}{c^2}\nabla_\perp \phi.$$

Hence,

$$\hat{\boldsymbol{\alpha}} = \frac{2}{c^2}\int_{s_{sr}}^{s_{ob}} \nabla_\perp \phi \, ds.$$

Because the bending angle of the light ray is assumed to be very small, *we approximate these integrals over the ray by integration along the ζ-axis.* Also, approximate the plane perpendicular to the ray by the (r_1, r_2)-plane (i.e., the lens plane). Then

$$\hat{\boldsymbol{\alpha}}(\mathbf{r}) = \frac{2}{c^2}\int_{\zeta_{sr}}^{\zeta_{ob}} \nabla_{\mathbf{r}}\phi(\mathbf{r}, \zeta) d\zeta, \tag{3.55}$$

where $\nabla_\perp$ is approximated by the gradient operator on the lens plane, namely, $\nabla_{\mathbf{r}} = \mathbf{i}\frac{\partial}{\partial r_1} + \mathbf{j}\frac{\partial}{\partial r_2}$.

Equation (3.55) implies that

$$\hat{\boldsymbol{\alpha}}(\mathbf{r}) = \nabla_{\mathbf{r}}\Psi(\mathbf{r}), \tag{3.56}$$

where Ψ is the two-dimensional Newtonian potential defined by (3.49). The potential Ψ can be expressed as in (3.51), which yields

$$\hat{\boldsymbol{\alpha}}(\mathbf{r}) = \frac{4G}{c^2}\int_{\mathbf{R}^2} d\mathbf{r}'\sigma(\mathbf{r}')\frac{\mathbf{r}-\mathbf{r}'}{|\mathbf{r}-\mathbf{r}'|^2}. \tag{3.57}$$

The integration is only formally over the entire lens plane (since the lens actually occupies a very small portion of the lens plane).

During the derivation of the equation (3.57) for the bending vector $\hat{\boldsymbol{\alpha}}(\mathbf{r})$, we assumed that the spatial geometry of the universe is that of Euclidean space $\mathbf{R}^3$. If space is a 3-sphere or hyperbolic 3-space, then it is still approximately Euclidean in the vicinity of the lens. Moreover, equation (3.57) applies to the Euclidean environment near the lens (because bending angles are small and so $|\mathbf{r}|$ is not very large). Furthermore, the only dependence that (3.57) has on the spatial geometry of the universe is through the scale factor a_L appearing in $\mathbf{r}$ and σ (see (3.46) for $\mathsf{k} = 0$). Hence, it is reasonable to expect that:

> *The bending vector equation (3.57) also holds approximately when space is either a 3-sphere or hyperbolic 3-space.*

In the spherical and hyperbolic cases, the scale factor a_L in (3.57) is calculated via (3.4) with $\kappa = 1$ or $\kappa = -1$. We shall not attempt to justify the above statement. Readers are referred to [Sei-SE, pp. 2365–2366] for a derivation that applies even to multiple plane lensing in "on average" Friedmann models with isolated weak inhomogeneities.

Equation (3.57) implies that the bending angle, $\hat{\alpha} = |\hat{\boldsymbol{\alpha}}|$, does not depend on the light's wavelength:

> *Under our approximations, the deflection of light by gravitational lenses is independent of light's wavelength (or frequency).*

Gravitational lensing is "achromatic" (e.g., X-rays are deflected by exactly the same amount as visible light).[38] Hence, essentially the same lensing effect (e.g., multiple images) should appear when observed at different wavelengths.

Example (Point Masses): A point mass $\mathtt{m}$ at the origin of the lens plane has surface mass density

$$\sigma(\mathbf{r}) = \mathtt{m}\delta(\mathbf{r}),$$

where $\delta(\mathbf{r})$ is the Dirac delta at the origin. The bending angle vector can then be obtained from (3.57):

$$\hat{\boldsymbol{\alpha}}(\mathbf{r}) = \frac{4G\mathtt{m}}{c^2}\frac{\mathbf{r}}{r^2},$$

where $r = |\mathbf{r}|$. Hence, the bending angle is

$$\hat{\alpha}(\mathbf{r}) = \frac{4G\mathtt{m}}{c^2 r}, \tag{3.58}$$

which is manifestly independent of wavelength and coincides with (1.1). A collection of n point masses $\mathtt{m}_i$ at positions $\mathbf{r}_i$ on the xy-plane has surface mass density

$$\sigma(\mathbf{r}) = \sum_{i=1}^{n} \mathtt{m}_i\, \delta(\mathbf{r} - \mathbf{r}_i),$$

where $\delta(\mathbf{r} - \mathbf{r}_i)$ is the Dirac delta at $\mathbf{r}_i$. Consequently,

$$\hat{\boldsymbol{\alpha}}(\mathbf{r}) = \frac{4G}{c^2}\sum_{i=1}^{n} \mathtt{m}_i \frac{\mathbf{r} - \mathbf{r}_i}{|\mathbf{r} - \mathbf{r}_i|^2}.$$

For typical lensing situations, the deflection angle in (3.58) with an impact parameter corresponding roughly to the radius of the lens (grazing) is not larger than one arcminute. This is true for lensing by a galaxy ($\mathtt{m} \simeq 10^{12} M_\odot$, $r \simeq 10$ kpc) and solar type stars ($\mathtt{m} \simeq$

[38] We use the word "light" to refer to any kind of electromagnetic radiation: ranging from radio waves, through infrared emission, visible light, ultraviolet radiation, X-rays to the highest energy gamma rays. There are many sources known in the universe that emit radiation in a few or all of these wavelength ranges.

$1M_\odot$, $r \simeq 7 \times 10^8$ m). Even for white dwarfs (dense stars about the size of the earth, but with an upper mass around $1.25M_\odot$), the deflection angle is much smaller than a degree.

If one deals with neutron stars (stars of such extremely high density and small size that they are composed mainly of tightly packed neutrons), then the deflection angles of light rays just grazing the surface of the star can become large. This is even more extreme for black holes as gravitational lenses. Here the deflection angle can be greater than 360 degrees (e.g., the impact parameter of a light ray from a background source can be such that the ray circles around the black hole many times before escaping to arrive at the observer). However, the intensities of images that correspond to these very large deflection angles are extremely small. See [Mis-TW, pp. 672–679], [Cha, pp. 128–129], and [Vir-E] for more on light deflection by black holes.

3.1.6 Cosmic Light Sources

An essential ingredient of a gravitational lens system is a physical source that emits light.

The tangent plane to the "sphere" centered at the observer and passing through the source is called the **light source plane**.

We suppose that each light source (like a lens) is "thin" and localized on a light source plane that is approximately orthogonal to the line of sight passing through the lens plane(s). In addition, we assume that with or without an intervening lens(es), light rays connecting the source and observer are negligibly affected by the gravitational fields of the source and observer.

Depending on a light source's angular size relative to the angular Einstein radius, sources are divided into two groups: **point-like** *and* **extended**, *which give rise to different observable phenomena.*

The physical sizes of sources in the universe cover a broad range: some sources are quite small (a white dwarf star has a size of only a few thousand kilometers), others are pretty big (a giant galaxy can have a diameter of a million light years), but all are finite in size. What matters for gravitational lensing is the angular size of the source — as seen by the observer — relative to the angular Einstein radius. Using this ratio, we put the physical sources into two categories that give rise to different observational phenomena.

Sources that are (very) small compared to the angular Einstein radius of the lens are usually treated as (and called) **point-like sources.** Gravitational lensing strongly affects their fluxes; they can reach very high magnifications and sometimes multiple images are produced. Examples for this group of sources are quasars lensed by individual galaxies and stars lensed by other stars or compact dark objects.

The exact size of a quasar — more accurately, the region that emits the continuum radiation — is not known, but from a number of observations the estimates are of the order of 0.003 parsec $\simeq 10^{16}$cm or smaller. Seen from distances of a few gigaparsecs, this translates into angular sizes of less than 10^{-6} arcseconds, very small compared to the deflection angles of about an arcsecond or larger for galaxies and clusters. Similarly for stars: diameters of solar type stars are about 1.5×10^{11}cm. At typical distances of 50 kpc (for the Large Magellanic Cloud) this corresponds to angles of order 10^{-6} arcseconds. For big red giant stars with radii of 100 times the solar radius, the angular diameter of 10^{-4} arcseconds is still smaller than the angular Einstein radius of a solar type star at half its distance (which is about 1 milliarcsec), but not by far any more.

The second class of sources — **extended sources** — contains objects that are comparable in size or larger than the angular Einstein radius of the lensing mass, and resolved by the observer. Good examples here are ordinary galaxies that are lensed by clusters of galaxies. Typically the lensing action deforms the sources, that is, the shape of the source appears distorted — often extended in a direction tangential to the center of the lens (cf. Figure 3.13, page 99). In practice this means that the images of background galaxies around foreground clusters are deformed into ellipses that align around the cluster; sometimes they are distorted into small arclets, and occasionally the spectacular luminous arcs are produced.

Other examples of extended sources are radio galaxies. They are sometimes mapped into almost perfect Einstein rings by foreground galaxies acting as lenses. Galaxies at cosmological redshifts, corresponding to distances of 1,000 megaparsecs or more, have angular sizes of one arcsecond or larger. The Einstein radii of foreground galaxies acting as lenses are of similar size; those of galaxy clusters are about a factor of ten larger. The fact that these two sizes are comparable results in the observational phenomena described above, which are often even more spectacular than the multiple images of point sources. In Chapter 5, we will go into more detail on the observational situation.

Unless stated to the contrary, all light sources will be assumed point-like.

3.2 Gravitational Lens Optics

In this section we shall introduce the concepts of time delay and Fermat's principle for both single and multiple lens plane cases. We shall derive the lens equation from the geometry of the deflected rays and comment on positions and magnifications of images for the simplest lens configuration. Finally we shall present the light curve of a star that is microlensed by a moving compact object.

The importance of time delay functions lies in the fact that their critical points characterize the "images" of a gravitationally lensed light source. This is a consequence of the "Principle of Fermat." After qualitatively presenting Fermat's principle and discussing its link with the time delay of gravitationally lensed images, we heuristically compute the time delay function. The section closes with an explicit mathematical expression for the latter.

3.2.1 Fermat's Principle: Time Delays

In a 1657 letter to Cureau de La Chambre, the attorney and great amateur mathematician Pierre de Fermat announced his highly acclaimed *Principle of Least Time* (or *Fermat's Principle*)[Fer]. When expressed in a spacetime context, it has the following intuitive statement:

> **Fermat's Principle:** *A light ray from a source $\mathcal{S}$ (spacetime event) to an observer $\mathcal{O}$ (timelike curve) follows a trajectory that is a stationary value of the arrival times τ, measured relative to the observer's proper time, of all paths from $\mathcal{S}$ to $\mathcal{O}$.*

In Fermat's principle, the observer can also be treated as located at a spacetime event with past-pointing null curves followed from the observer to a source that is a timelike curve (e.g., [Schn-EF, p. 101], [Per00a]).

For each path υ, plot the time measured on the observer's clock when υ arrives; the result is a "surface" of arrival times graphed over the space of paths. The minima, maxima, and (generalized) saddle points of the "arrival time surface" characterize those paths followed by the actual light rays.[39]

The arrival time surface of interest in gravitational lensing is constructed as follows under the approximations of Section 3.1. Consider all paths υ that leave the light source $\mathcal{S}$ at cosmic time τ_S, traveling along null geodesics of the Friedmann metric $\mathbf{g}_F$ from $\mathcal{S}$ to the lens plane, and then along null geodesics from the lens plane to the observer $\mathcal{O}$ — see Figures 3.2 and 3.3. Suppose that at approximately cosmic time τ_L, the impact points on the lens plane of these piecewise-smooth null geodesics υ are parametrized by $\mathbf{r}$, where $\mathbf{r}$ is measured relative to an arbitrary origin (e.g., the center of mass of the lens). Note that since the time delays between paths are assumed to be extremely small, we can choose an approximate cosmic time when the null geodesics arrive near the lens. Assume that $\mathcal{S}$ is at point $\mathbf{s}$ on the light source plane at cosmic time τ_S. Denote by $\tau_{\mathbf{s}}(\mathbf{r})$ the arrival time, measured by the observer $\mathcal{O}$, of the piecewise-smooth null geodesic υ from $\mathcal{S}$ to $\mathcal{O}$ passing through $\mathbf{r}$. In the absence of a lens, that is, the lens is smoothed out back into the Friedmann background,[40] there will be a unique null geodesic l_0 from $\mathcal{S}$ to $\mathcal{O}$ passing through the lens plane at some point $\mathbf{r}_0$. This corresponds to the dashed path in Figures 3.2 and 3.3. Denote by $\tau_{\mathbf{s}}(\mathbf{r}_0)$ its arrival time at the observer.

---

[39] There are actually no maximum light rays for the stated general form of Fermat's principle. This is roughly because the space of all null paths from $\mathcal{S}$ to $\mathcal{O}$ is used, allowing one to always increase the arrival time via a detour (e.g., [Schn-EF, p. 163]). Consult Perlick [Per90a, pp. 1324–1327] [Per00b, pp. 156–164] for a mathematically precise treatment. In gravitational lensing, however, not all null paths connecting source and observer are allowed; in fact, the space of paths is finite dimensional ($2k$-dimensional for k lens planes) and permits maxima light rays.

[40] See the "sand box" analogy on page 47.

The **time delay** of the path υ through $\mathbf{r}$ relative to the unlensed ray υ_0 is defined by

$$\mathcal{T}_s(\mathbf{r}) = \tau_s(\mathbf{r}) - \tau_s(\mathbf{r}_0), \tag{3.59}$$

where the constant term $\tau_s(\mathbf{r}_0)$ is subtracted off for convenience (see computation of $\mathcal{T}_s(\mathbf{r})$ in Section 3.2.2). Fermat's principle implies:

Light rays from the source at $\mathbf{s}$ *to the observer are characterized by critical points of the* **time delay function** $\mathcal{T}_s(\mathbf{r})$ *relative to* $\mathbf{r}$*, that is, the solutions* $\mathbf{r}$ *of the equation*

$$\operatorname{grad} \mathcal{T}_s(\mathbf{r}) = \mathbf{0}, \tag{3.60}$$

where grad *is with respect to* $\mathbf{r}$.

In other words, the gravitationally lensed images of $\mathcal{S}$ are determined by minima, maxima, and saddle points of the function $\mathcal{T}_s$. The graph of $\mathcal{T}_s$ is the relevant arrival time or **time delay surface** for lensing.

References (Fermat's Principle): The importance of using Fermat's principle as a way to understand image formation in gravitational lensing was emphasized by Nityananda (1984, unpublished), Schneider [Schn85], Blandford and Narayan [Bla-N86]), and Nityananda and Samuel [Nit-S]. See [Schn-EF, p. 100] for a physical treatment of Kovner's formulation [Kov90] of Fermat's principle in arbitrary spacetimes. Perlick [Per90a, Per90b, Per95, Per00a, Per00b] gives a mathematically rigorous discussion and proof of the latter — cf. Frankel [Frank, pp. 90–98] for the stationary spacetime case. Also, consult the variational theory of Giannoni, Masiello and Piccione [Gia-MP].

3.2.2 Computing the Time Delay Function

We shall heuristically calculate an explicit expression for the time delay function $\mathcal{T}_s(\mathbf{r})$ in (3.59) using the approximations of Section 3.1.

Consider the gravitational lensing scenario in Figure 3.2. Each deflected ray path has two contributions to its total time delay relative to the unlensed path (dashed path):

$$\mathcal{T}_s(\mathbf{r}) = \{\textit{geometric time delay}\} + \{\textit{potential time delay}\}\,.$$

The geometric time delay arises from the extra path length traveled relative to the unperturbed ray, while the potential term is due to the delay or slowing down of the ray by the gravitational potential well of the lens (cf. Section 3.1.4). The latter effect is also called the *Shapiro time delay* [Sha] (see [dIn, p. 204]).

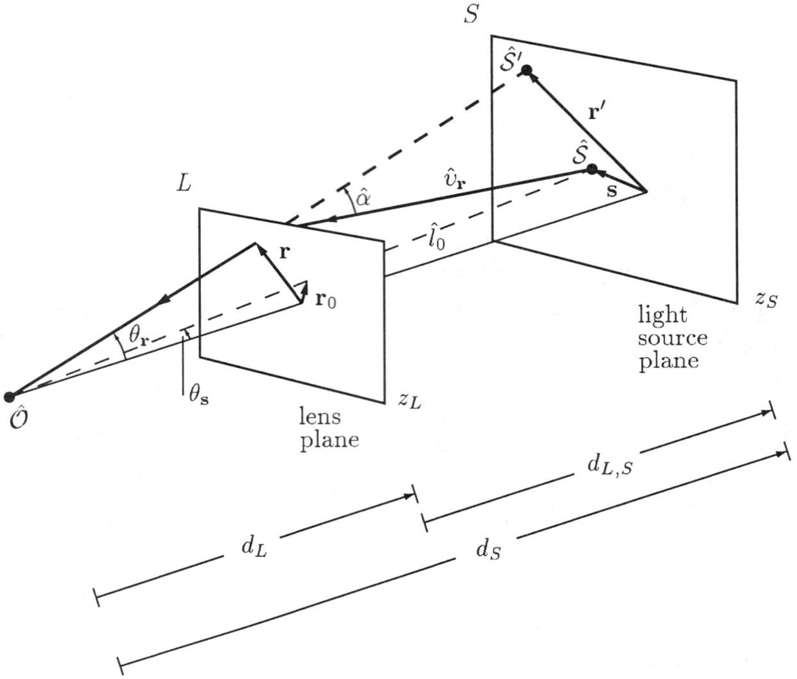

Figure 3.2. Schematic of a gravitational lens situation. At cosmic time τ_S, a light ray is emitted from a source at comoving position $\mathcal{S}$. The ray impacts the lens plane at cosmic time τ_L and arrives at cosmic time τ at an observer with comoving position $\mathcal{O}$. Depicted for cosmic time τ_L are the spatial trajectories $\hat{\upsilon}_{\mathbf{r}}$ (solid piecewise-smooth path) and $\hat{l}_0$ (dashed path) of rays impacting, respectively, at $\mathbf{r}$ when there is a lens and $\mathbf{r}_0$ in the absence of lensing. The basic ingredients of this lensing situation are: observer $\hat{\mathcal{O}}$, lens L (redshift z_L), source $\hat{\mathcal{S}}$ (redshift z_S), and image $\hat{\mathcal{S}}'$; angular diameter distances d_L between observer and lens, d_S between observer and source, and $d_{L,S}$ between lens and source; deflection angle $\hat{\alpha}$, unperturbed source position angle $\theta_{\mathbf{s}}$, position angle $\theta_{\mathbf{r}}$ of image $\hat{\mathcal{S}}'$, and impact parameter $\mathbf{r}$.

Time Delay at the Lens Plane

At roughly cosmic time τ_L, the piecewise-smooth light ray υ from the source passes the lens. Recall that υ's spacetime path traces a null geodesic from the source to lens followed by a null geodesic from the lens to observer. To measure at the lens its time delay relative to the null geodesic path l_0 in the absence of lensing, employ the proper coordinates (r_1, r_2, ζ) defined by (3.46). These coordinates are centered on the lens such that the ζ-axis is along the observer's line of sight and the lens plane corresponds to $\zeta = 0$, with $\mathbf{r} = (r_1, r_2)$ proper coordinates on the lens plane. Recall that since space is approximately Euclidean near the lens, the length $|\mathbf{r}|$ is the proper distance (as measured from the lens) to an object at $\mathbf{r}$ — see (3.47). The rays υ and l_0 impact the lens plane at proper vectors $\mathbf{r}$ and $\mathbf{r}_0$, respectively. Now, the spacelike hypersurface corresponding to cosmic time $\tau = \tau_L$ is of the form $\tau_L \times \mathsf{S}$,

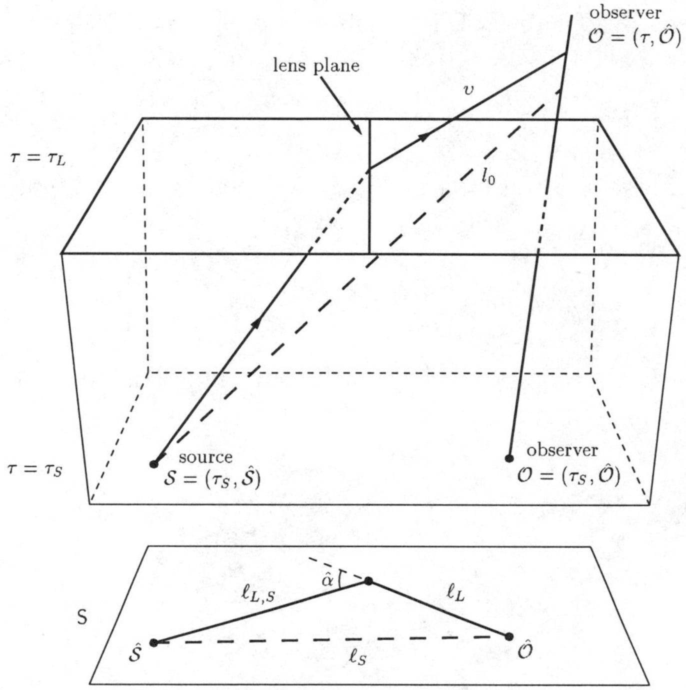

Figure 3.3. The top box is a spacetime region containing the worldlines of a lensed light ray υ and the ray's trajectory l_0 in the absence of lensing. The ray υ departs the source at cosmic time τ_S and arrives at the lens plane at τ_L. The bottom part of the figure depicts the standard comoving space S (i.e., $\mathsf{S} = \mathbf{S}^3$, $\mathbf{R}^3$, or $\mathbf{H}^3$) and the geodesic triangle formed from the the projections of υ and l_0 into S. The lengths ℓ_S, ℓ_L, and $\ell_{L,S}$ are relative to the metric dS^2_{K}.

where S is either the unit 3-sphere $\mathbf{S}^3$, Euclidean space $\mathbf{R}^3$, or the unit hyperbolic 3-space $\mathbf{H}^3$ (see page 33). The spatial path $\hat{\upsilon}_{\mathbf{r}}$ of υ is a piecewise-smooth geodesic in $\tau_L \times \mathsf{S}$ relative to the metric $\mathbf{g}^L_{(\mathsf{K})} = a_L dS^2_{\mathsf{K}}$ (page 60), while the spatial path $\hat{l}_0$ of l_0 is a smooth geodesic $\tau_L \times \mathsf{S}$. The projections of $\hat{\upsilon}_{\mathbf{r}}$ and $\hat{l}_0$ into the comoving space S are, respectively, piecewise-smooth and smooth geodesics of dS^2_{K}. The projections form a geodesic triangle in S — Figure 3.3.

The space $\tau_L \times \mathsf{S}$ can be viewed as having a medium with refractive index n_L given by (3.41). This index is induced by the gravitational lensing metric $\mathbf{g}^{\mathsf{K}}_{GL}$. The length of $\hat{\upsilon}_{\mathbf{r}}$ in $\tau_L \times \mathsf{S}$ is then given by its optical length relative to the refractive medium (e.g., [Bor-W, p. 115]):

$$[\text{length of } \hat{\upsilon}_{\mathbf{r}} \text{ (lens present)}] = \int_{\upsilon_{\mathbf{r}}} n_L a_L d\ell_{\mathsf{K}}.$$

The increment of length $d\ell_{\mathsf{K}}$ is relative to the spatial metric dS^2_{K}. If no lens is present, then

$n_L = 1$; so the length of $\hat{l}_0$ in $\tau_L \times \mathsf{S}$ in the absence of lensing is its length relative to $\mathbf{g}^L_{(\mathsf{K})}$:

$$[\text{length of } \hat{l}_0 \text{ (lens absent)}] = a_L \int_{\hat{l}_0} d\ell_{\mathsf{K}}.$$

Consequently, the time delay (as measured in the lens plane at cosmic time τ_L) of $\hat{\upsilon}_{\mathbf{r}}$ relative to $\hat{l}_0$ is

$$\Delta\,\tau_L = \frac{1}{c}\left[\int_{\hat{\upsilon}_{\mathbf{r}}} n_L a_L d\ell_{\mathsf{K}} - a_L \int_{\hat{l}_0} d\ell_{\mathsf{K}}\right]. \tag{3.61}$$

Writing $\Delta\,\tau_L$ as a sum of geometrical and potential time delays, and using (3.44), we see that $\Delta\,\tau_L = \Delta\,\tau^L_{geom} + \Delta\,\tau^L_{pot}$, where

$$\Delta\,\tau^L_{geom} = \frac{a_L}{c}\left[\int_{\hat{\upsilon}_{\mathbf{r}}} d\ell_{\mathsf{K}} - \int_{\hat{l}_0} d\ell_{\mathsf{K}}\right] = \frac{a_L}{c} L_{\hat{l}_0}(\hat{\upsilon}_{\mathbf{r}}) \tag{3.62}$$

and

$$\Delta\,\tau^L_{pot} = -\frac{2a_L}{c^3}\int_{\hat{\upsilon}_{\mathbf{r}}} \phi\, d\ell_{\mathsf{K}}. \tag{3.63}$$

Notice that

$$L_{\hat{l}_0}(\hat{\upsilon}_{\mathbf{r}}) = \mathrm{Length}_{\mathsf{K}}(\hat{\upsilon}_{\mathbf{r}}) - \mathrm{Length}_{\mathsf{K}}(\hat{l}_0),$$

where $\mathrm{Length}_{\mathsf{K}}(\hat{\upsilon}_{\mathbf{r}})$ and $\mathrm{Length}_{\mathsf{K}}(\hat{l}_0)$ are, respectively, the lengths of $\hat{\upsilon}_{\mathbf{r}}$ and $\hat{l}_0$ (or, equivalently, their projections into S) relative to dS^2_{K}.

Time Delay at the Observer

Using (3.17), the time delay $\Delta\,\tau_L$ measured at the lens is related as follows to the desired *time delay* $\mathcal{T}_{\mathrm{s}}(\mathbf{r})$ *at the observer*:

$$\mathcal{T}_{\mathrm{s}}(\mathbf{r}) = (1 + z_L)\Delta\,\tau_L. \tag{3.64}$$

This can be expressed in terms of the geometrical and potential time delays:

$$\mathcal{T}_{\mathrm{s}}(\mathbf{r}) = \Delta\,\tau_{geom} + \Delta\,\tau_{pot}, \tag{3.65}$$

where

$$\Delta\,\tau_{geom} = (1 + z_L)\Delta\,\tau^L_{geom} \tag{3.66}$$

and

$$\Delta\,\tau_{pot} = (1 + z_L)\Delta\,\tau^L_{pot}. \tag{3.67}$$

Alternatively, the time delay measured at the observer can also be expressed in terms of the present cosmic proper τ_0. In fact, by (3.45) the index of refraction becomes

$$n_0 = 1 - \frac{2}{c^2}\phi. \tag{3.68}$$

Setting $a_0 = a(\mathtt{t}(\tau_0))$, we see that

$$\mathcal{T}_s(\mathbf{r}) = \frac{1}{c}\left[\int_{\hat{\upsilon}_\mathbf{r}} n_0 a_0 \, d\ell_\mathsf{K} - a_0 \int_{\hat{l}_0} d\ell_\mathsf{K}.\right] \tag{3.69}$$

$$= \frac{a_0}{c}\left[L_{\hat{l}_0}(\hat{\upsilon}_\mathbf{r}) - \frac{2}{c^2}\int_{\hat{\upsilon}_\mathbf{r}} \phi \, d\ell_\mathsf{K}.\right], \tag{3.70}$$

which also follows from (3.61), (3.62), and (3.64).

We shall now calculate $\vartriangle \tau_{pot}$ and $\vartriangle \tau_{geom}$.

Potential Time Delay

Equations (3.63) and (3.67) yield

$$\vartriangle \tau_{pot} = -\frac{2(1+z_L)}{c^3}\int_{\hat{\upsilon}_\mathbf{r}} \phi \, a_L d\ell_\mathsf{K}.$$

Since bending angles are very small, we approximate the integral along the ray by integration along the line of sight ζ (see (3.46)):

$$\vartriangle \tau_{pot} = -\frac{2(1+z_L)}{c^3}\int_{\zeta_{sr}}^{\zeta_{ob}} \phi \, d\zeta.$$

Applying (3.49), we then obtain the *potential time delay* at the observer:

$$\vartriangle \tau_{pot} = -\frac{1+z_L}{c}\Psi(\mathbf{r}). \tag{3.71}$$

Geometrical Time Delay

By (3.62) and (3.66), the geometrical time delay measured at the observer is

$$\vartriangle \tau_{geom} = (1+z_L)\frac{a_L}{c}L_{\hat{l}_0}(\hat{\upsilon}_\mathbf{r}), \tag{3.72}$$

where $L_{\hat{l}_0}(\upsilon_\mathbf{r})$ is the (very small) amount by which the geometrical length of $\hat{\upsilon}_\mathbf{r}$ exceeds that of $\hat{l}_0$ relative to the metric dS^2_K.

We shall determine $L_{\hat{l}_0}(\hat{\upsilon}_\mathbf{r})$ for the unit 3-sphere $\mathbf{S}^3$ ($\mathsf{K} = 1$), Euclidean 3-space $\mathbf{R}^3$ ($\mathsf{K} = 0$), and the unit hyperbolic 3-space $\mathbf{H}^3$ ($\mathsf{K} = -1$). To aid in our calculation, let us collect a few results. First, Figure 3.3 implies

$$L_{\hat{l}_0}(\hat{\upsilon}_\mathbf{r}) = \ell_L + \ell_{L,S} - \ell_S. \tag{3.73}$$

Also, since $\hat{\alpha} \ll 1$, we have

$$\ell_S \simeq \ell_L + \ell_{L,S} \tag{3.74}$$

and

$$1 - \cos\hat{\alpha} = 2\sin^2\frac{\hat{\alpha}}{2} \simeq \frac{\hat{\alpha}^2}{2}. \tag{3.75}$$

Second, the geodesic triangle in Figure 3.3 lies in a two-dimensional (complete simply connected totally geodesic) submanifold S^2_{tg} of the standard space S; and S^2_{tg} is isometric to the 2-sphere $\mathbf{S}^2$ for $\mathsf{S} = \mathbf{S}^3$, Euclidean plane $\mathbf{R}^2$ for $\mathsf{S} = \mathbf{R}^3$, and the hyperbolic plane $\mathbf{H}^2$ for $\mathsf{S} = \mathbf{H}^3$ (e.g., [Sak, p. 137] and [ONe, p.105]). Moreover, the geodesic triangle obeys the *Law of Cosines* on S^2_{tg} (e.g., [Sak, p. 138] and [Thu, pp. 75,81]):

$$\mathsf{K} = 0: \quad \ell_S^2 = \ell_L^2 + \ell_{L,S}^2 - 2\ell_L\ell_{L,S}\cos(\pi - \hat{\alpha}), \tag{3.76}$$

$$\mathsf{K} = 1: \quad \cos\ell_S = \cos\ell_L\cos\ell_{L,S} + \sin\ell_L\sin\ell_{L,S}\cos(\pi - \hat{\alpha}), \tag{3.77}$$

$$\mathsf{K} = -1: \quad \cosh\ell_S = \cosh\ell_L\cosh\ell_{L,S} - \sinh\ell_L\sinh\ell_{L,S}\cos(\pi - \hat{\alpha}). \tag{3.78}$$

Third, let $a_S = a(\mathsf{t}(\tau_S))$ with τ_S the cosmic proper time when the ray departs from the source. Equation (3.21) (page 39) implies that the sides of the geodesic triangle in Figure 3.3 relate to Friedmann angular diameter distances as follows:

$$d_L = a_L R_{\mathsf{K}}(\ell_L), \quad d_{L,S} = a_S R_{\mathsf{K}}(\ell_{L,S}), \quad d_S = a_S R_{\mathsf{K}}(\ell_S). \tag{3.79}$$

Here

$$R_{\mathsf{K}}(\ell) = \begin{cases} \sin\ell & \text{if } \mathsf{K} = 1 \\ \ell & \text{if } \mathsf{K} = 0 \\ \sinh\ell & \text{if } \mathsf{K} = -1. \end{cases}$$

Finally, since the position vector $\mathbf{r}'$ of the image $\hat{\mathcal{S}}'$ in Figure 3.2 satisfies $\mathbf{r}'/d_S \simeq \mathbf{r}/d_L$ and the bending angle $\hat{\alpha}$ is the same in $\tau_L \times \mathsf{S}$ and S, we obtain

$$\hat{\alpha}\, d_{L,S} \simeq |\mathbf{r}' - \mathbf{s}| \simeq \left|\frac{\mathbf{r}}{d_L} - \frac{\mathbf{s}}{d_S}\right| d_S. \tag{3.80}$$

Spatially Euclidean Case ($\mathsf{K} = 0$): The law of cosines (3.76) and (3.75) imply that

$$\begin{aligned} \ell_S^2 &= \ell_L^2 + \ell_{L,S}^2 + 2\ell_L\ell_{L,S}\cos\hat{\alpha} \\ &= (\ell_L + \ell_{L,S})^2 - 2\ell_L\ell_{L,S}(1 - \cos\hat{\alpha}) \\ &\simeq (\ell_L + \ell_{L,S})^2 - \ell_L\ell_{L,S}\,\hat{\alpha}^2. \end{aligned} \tag{3.81}$$

Equation (3.81) yields

$$(\ell_L + \ell_{L,S} - \ell_S)(\ell_L + \ell_{L,S} + \ell_S) = (\ell_L + \ell_{L,S})^2 - \ell_S^2 \simeq \ell_L \ell_{L,S}\,\hat{\alpha}^2. \tag{3.82}$$

Consequently, equations (3.73), (3.74), and (3.82) give

$$L_{\hat{\imath}_0}(\hat{\upsilon}_{\mathbf{r}}) \simeq \frac{\ell_L \ell_{L,S}}{\ell_L + \ell_{L,S} + \ell_S}\hat{\alpha}^2 \simeq \frac{\ell_L \ell_{L,S}}{2\ell_S}\hat{\alpha}^2. \tag{3.83}$$

By equations (3.79) with $\kappa = 0$, (3.80), and (3.83), we get

$$L_{\hat{\imath}_0}(\hat{\upsilon}_{\mathbf{r}}) \simeq \frac{1}{a_L}\frac{d_L d_S}{2d_{L,S}}\left|\frac{\mathbf{r}}{d_L} - \frac{\mathbf{s}}{d_S}\right|^2.$$

Hence, if $\kappa = 0$, then the *geometrical time delay* (3.72) becomes

$$\Delta\,\tau_{geom} = \frac{(1+z_L)}{2c}\frac{d_L d_S}{d_{L,S}}\left|\frac{\mathbf{r}}{d_L} - \frac{\mathbf{s}}{d_S}\right|^2. \tag{3.84}$$

Spatially Spherical and Hyperbolic Cases ($\kappa = 1, -1$)**:** We shall apply basically the same methods used in computing $L_{\hat{\imath}_0}(\hat{\upsilon}_{\mathbf{r}})$ for $\kappa = 0$. Combine the cosine laws (3.77) and (3.78) as follows:

$$\cos_\kappa \ell_S = \cos_\kappa \ell_L \cos_\kappa \ell_{L,S} - \kappa \sin_\kappa \ell_L \sin_\kappa \ell_{L,S} \cos\hat{\alpha}, \tag{3.85}$$

where

$$\cos_\kappa x = \begin{cases} \cos x & \text{if } \kappa = 1 \\ \cosh x & \text{if } \kappa = -1, \end{cases} \qquad \sin_\kappa x = \begin{cases} \sin x & \text{if } \kappa = 1 \\ \sinh x & \text{if } \kappa = -1. \end{cases}$$

Using the standard formulas for the cosine and hyperbolic cosine of a sum of angles, equation (3.85) becomes

$$\begin{aligned} \cos_\kappa \ell_S &= \cos_\kappa(\ell_L + \ell_{L,S}) + \kappa(\sin_\kappa \ell_L)(\sin_\kappa \ell_{L,S})(1 - \cos\hat{\alpha}) \\ &\simeq \cos_\kappa(\ell_L + \ell_{L,S}) + \kappa(\sin_\kappa \ell_L)(\sin_\kappa \ell_{L,S})\frac{\hat{\alpha}^2}{2}, \end{aligned} \tag{3.86}$$

where the last step follows from (3.75). Recall that

$$\sin\left(\frac{x+y}{2}\right)\sin\left(\frac{x-y}{2}\right) = -\frac{\cos x - \cos y}{2}$$

and

$$\sinh\left(\frac{x+y}{2}\right)\sinh\left(\frac{x-y}{2}\right) = \frac{\cosh x - \cosh y}{2}.$$

It follows

$$\begin{aligned}(\sin_K \ell_+)\,(\sin_K \ell_-) &= -K\frac{\cos_K(\ell_L+\ell_{L,S})-\cos_K \ell_S}{2} \\ &\simeq (\sin_K \ell_L)(\sin_K \ell_{L,S})\,\frac{\hat{\alpha}^2}{4}.\end{aligned} \tag{3.87}$$

where

$$\ell_+ = \frac{\ell_L+\ell_{L,S}+\ell_S}{2}, \qquad \ell_- = \frac{\ell_L+\ell_{L,S}-\ell_S}{2}.$$

The approximation in (3.87) follows from (3.86). But since $\sin_K x \simeq x$ for $|x|$ sufficiently small, equations (3.73), (3.74), and (3.87) imply that

$$L_{\hat{l}_0}(\hat{\upsilon}_\mathbf{r}) \simeq \frac{(\sin_K \ell_L)(\sin_K \ell_{L,S})}{2\sin_K \ell_+}\,\hat{\alpha}^2 \simeq \frac{(\sin_K \ell_L)(\sin_K \ell_{L,S})}{2\sin_K \ell_S}\,\hat{\alpha}^2. \tag{3.88}$$

Applying (3.79) with $K = 1, -1$ and (3.80) to (3.88) yields

$$L_{\hat{l}_0}(\hat{\upsilon}_\mathbf{r}) \simeq \frac{1}{a_L}\frac{d_L d_S}{2d_{L,S}}\left|\frac{\mathbf{r}}{d_L}-\frac{\mathbf{s}}{d_S}\right|^2.$$

Thus, for $K = 1, -1$ the geometrical time delay at the observer also has the form in (3.84).

Time Delay Function: Single Lens Plane

Adding (3.71) and (3.84), we get the *total time delay measured at the observer* (cf. [Schn-EF, p. 146, eq. (4.67)]):

$$\mathcal{T}_\mathbf{s}(\mathbf{r}) = \frac{(1+z_L)}{c}\frac{d_L d_S}{d_{L,S}}\left[\frac{1}{2}\left|\frac{\mathbf{r}}{d_L}-\frac{\mathbf{s}}{d_S}\right|^2 - \frac{d_{L,S}}{d_L d_S}\Psi(\mathbf{r})\right]. \tag{3.89}$$

We shall call (3.89) the **time delay function** for single plane lensing.

Though $\mathcal{T}_\mathbf{s}(\mathbf{r})$ is not an observable, the time delay, $\mathcal{T}_\mathbf{s}(\mathbf{r}_i) - \mathcal{T}_\mathbf{s}(\mathbf{r}_j)$, between two images $\mathbf{r}_i$ and $\mathbf{r}_j$ can be measured and is related to the Hubble constant; see Sections 4.4 (page 134) and 5.1.1 (page 147).

Remark (Strong Lens plus Large Scale Structures): For the case of gravitational lensing by a primary strong lens along with density fluctuations arising from the large scale structure, the form of the time delay function is modified. Readers are referred to [Bar, p. 21, eq. (20)] and [Schn97, p. 675, eq. (3.7)] for explicit expressions. However, the modified time delay function is analogous mathematically to that of a single-plane time delay function without large scale structure density perturbations [Schn97, p. 676, eq. (3.7)]. Moreover, the corresponding surface mass density perturbation of the potential for this single-plane time delay function is not guaranteed to be nonnegative.

Time Delay Function: Multiple Lens Planes

We begin by calculating the time delay function for gravitational lensing by two lens planes and then extend to k lens planes.

At the present cosmic time τ_0 (i.e., approximately when the light rays arrive at us — the observer), the index of refraction due to the lenses is (cf. (3.45)):

$$n_0 = 1 - \frac{2}{c^2}(\phi_1 + \phi_2),$$

where the potential ϕ_i is localized at the lens plane L_i for $i = 1, 2$. Referring to Figure 3.4, the time delay measured at the observer of the lensed ray $\hat{\upsilon}$, which travels from $\mathbf{r}_3$ to the observer, relative to the unperturbed ray $\hat{l}_3$ is (cf. (3.69)):

$$\mathcal{T}_{\mathbf{s}}(\mathbf{r}_1, \mathbf{r}_2) = \frac{1}{c}\left[\int_{\hat{\upsilon}} n_0 a_0 d\ell_{\mathsf{K}} - a_0 \int_{\hat{l}_3} d\ell_{\mathsf{K}}\right]. \tag{3.90}$$

Here $\mathbf{s}$ is the source's position ($\mathbf{s} = \mathbf{r}_3$). The first (resp., second) integral in (3.90) is the length of $\hat{\upsilon}$ (resp., $\hat{l}_3$) in the presence (resp., absence) of the lenses, where the length is relative to the metric $a_0 dS^2_{\mathsf{K}}$. It follows that

$$\mathcal{T}_{\mathbf{s}}(\mathbf{r}_1, \mathbf{r}_2) = \left[\frac{a_0}{c} L_{\hat{l}_3}(\hat{\upsilon}) - \frac{2a_0}{c^3}\left(\int_{\hat{\upsilon}} \phi_1 \, d\ell_{\mathsf{K}} + \int_{\hat{\upsilon}} \phi_2 \, d\ell_{\mathsf{K}}\right)\right]. \tag{3.91}$$

Recall that

$$L_{\hat{l}_3}(\hat{\upsilon}) = \mathrm{Length}_{\mathsf{K}}(\hat{\upsilon}) - \mathrm{Length}_{\mathsf{K}}(\hat{l}_3),$$

where the lengths are now relative to dS^2_{K}. The first and second terms on the right of (3.91) are the geometrical and potential time delays, respectively.

In Figure 3.4, let $\hat{\upsilon}_{21}$ be the ray from $\mathbf{r}_2$ to $\mathbf{r}_1$ to the observer, and $\hat{\upsilon}_{32}$ the ray from $\mathbf{r}_3$ to $\mathbf{r}_2$ to the observer. The figure shows that

$$L_{\hat{l}_3}(\hat{\upsilon}) = \left[\mathrm{Length}_{\mathsf{K}}(\hat{\upsilon}_{21}) - \mathrm{Length}_{\mathsf{K}}(\hat{l}_2)\right] + \left[\mathrm{Length}_{\mathsf{K}}(\hat{\upsilon}_{32}) - \mathrm{Length}_{\mathsf{K}}(\hat{l}_3)\right].$$

Consequently,

$$\begin{aligned}
\mathcal{T}_{\mathbf{s}}(\mathbf{r}_1, \mathbf{r}_2) &= \frac{a_0}{c}\left[L_{\hat{l}_2}(\hat{\upsilon}_{21}) - \frac{2}{c^2}\int_{\hat{\upsilon}_{21}} \phi_1 \, d\ell_{\mathsf{K}}\right] \\
&\quad + \frac{a_0}{c}\left[L_{\hat{l}_3}(\hat{\upsilon}_{32}) - \frac{2}{c^2}\int_{\hat{\upsilon}_{21}} \phi_2 \, d\ell_{\mathsf{K}}\right] \\
&= \mathcal{T}_{\mathbf{s}}(\mathbf{r}_1) + \mathcal{T}_{\mathbf{s}}(\mathbf{r}_2),
\end{aligned}$$

where the last equality follows from (3.70). In other words, the function $\mathcal{T}_{\mathbf{r}_{i+1}}(\mathbf{r}_i)$ is the time delay measured at the observer of a ray from $\mathbf{r}_{i+1}$ ("source") to $\mathbf{r}_i$ to the observer (deflection occurs only at $\mathbf{r}_i$ on L_i).

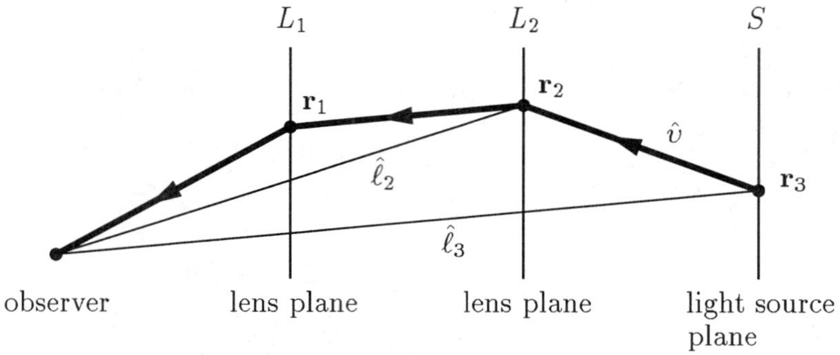

Figure 3.4. Double plane gravitational lensing. For the present cosmic time τ_0, the spatial trajectory of a lensed light ray traveling from a source at $\mathbf{r}_3$ to the observer experiences deflections at $\mathbf{r}_2$ and $\mathbf{r}_1$ on lens planes L_1 and L_2, respectively. If no lenses are present on L_1 and L_2, then the ray follows $\hat{l}_3$. A light ray from $\mathbf{r}_2$ to the observer propagates along $\hat{l}_2$ in the absence of a lens on L_1. All the paths are in $\tau_0 \times \mathsf{S}$.

Extending to k lens planes, we obtain the time delay measured at the observer (relative to the unperturbed path) of a ray connecting a source at $\mathbf{s}$ to the observer and experiencing deflections at impact points $(\mathbf{r}_1, \ldots, \mathbf{r}_k)$:

$$\mathcal{T}_{\mathbf{s}}(\mathbf{r}_1, \ldots, \mathbf{r}_k) = \sum_{i=1}^{k} \mathcal{T}_{\mathbf{r}_{i+1}}(\mathbf{r}_i),$$

where $\mathbf{r}_{k+1} = \mathbf{s}$. By (3.89) we obtain the k**-plane time delay function**

$$\mathcal{T}_{\mathbf{s}}(\mathbf{r}_1, \ldots, \mathbf{r}_k) = \sum_{i=1}^{k} \frac{(1+z_i)}{c} \frac{d_i d_{i+1}}{d_{i,i+1}} \left[\frac{1}{2} \left| \frac{\mathbf{r}_i}{d_i} - \frac{\mathbf{r}_{i+1}}{d_{i+1}} \right|^2 - \frac{d_{i,i+1}}{d_i d_{i+1}} \Psi_i(\mathbf{r}_i) \right]. \qquad (3.92)$$

Here, the redshift z_i is that of the ith lens plane; d_i is the angular diameter distance from the observer to the ith lens plane, where d_{k+1} is to the light source plane; $d_{i,i+1}$ is the angular diameter distance separating the ith and $(i+1)$st lens planes; $\mathbf{r}_i$ is the impact position of the ray on the ith lens plane, with $\mathbf{r}_{k+1} = \mathbf{s}$; and $\Psi_i(\mathbf{r}_i)$ is the two-dimensional Newtonian potential on the ith lens plane. The k-plane time delay function was introduced by Blandford and Narayan [Bla-N86, p. 579].

Applying Fermat's principle now yields that the lensed light rays connecting the source and observer correspond to solutions $(\mathbf{r}_1, \ldots, \mathbf{r}_k)$ of the $2k$ equations:

$$\operatorname{grad}_{\mathbf{r}_i} \mathcal{T}_{\mathbf{s}}(\mathbf{r}_1, \ldots, \mathbf{r}_k) = 0 \quad i = 1, \ldots, k. \qquad (3.93)$$

In this case, the gravitationally lensed images are determined by minima, maxima, and higher order saddles of $\mathcal{T}_{\mathbf{s}}$.

3.2.3 The Lens Equation: Ray Tracing

Single-Plane Lens Equation

Taking the gradient of the single-plane time delay function (3.89) and plugging the result into (3.60) yields the **lens equation**

$$\mathbf{s} = \frac{d_S}{d_L}\mathbf{r} - d_{L,S}\hat{\boldsymbol{\alpha}}(\mathbf{r}). \tag{3.94}$$

For a lensed light ray from a source to the observer, this equation relates the impact position $\mathbf{r}$ of the ray to the source's position $\mathbf{s}$, taking into account the deflection $\hat{\boldsymbol{\alpha}}(\mathbf{r})$ by the lens at $\mathbf{r}$.

Identifying a **lensed image** of the source with the impact vector $\mathbf{r}$, notice that for fixed $\mathbf{s}$ the lensed images are then solutions $\mathbf{r}$ of the lens equation. In general, these solutions are extremely difficult to determine analytically. For this reason, the lens equation is often attacked numerically. However, in Part III (e.g., Chapters 11 and 12) we shall obtain (among other things) counting information on the number of solutions to the lens equation for a given source position.

Now, Fermat's principle still holds if the light rays are reversed (so they travel from observer to source — e.g., [Schn-EF, p. 101], [Per00a]). Tracing a light ray backwards from the observer to the impact point $\mathbf{r}$, we then see that the ray hits the light source plane at the position $\mathbf{s}$ determined by (3.94). In other words, the lens equation can be viewed as defining **ray tracing** from the observer to the light source plane. In fact, it induces a mapping s — called a **lensing map** — from a subset of the lens plane L into the light source plane S:

$$s(\mathbf{r}) = \frac{d_S}{d_L}\mathbf{r} - d_{L,S}\hat{\boldsymbol{\alpha}}(\mathbf{r}). \tag{3.95}$$

Note that the Jacobian matrix of the lensing map s is symmetric.

Remark (Strong Lens plus Large Scale Structures): The lens equation due to a primary strong lens plus large scale structure density fluctuations can be compacted into a modified (and more complicated) single-plane lens equation without large scale structure density fluctuations [Schn97, p. 676] (cf. [Koc-A, p. 1075, eq. (2.6)] and [Kov87]). In addition, the Jacobian matrix of the modified, single-plane lensing map remains symmetric (e.g., [Schn97, p. 675],[Bart-S, p. 390]). However, for multiple-plane gravitational lensing (see page 57), the Jacobian matrix of the associated lensing map need not be symmetric.

Example (Circularly Symmetric Lens): Consider a light ray that impacts the lens plane at distance r from the origin. Suppose that the deflector is circularly symmetric with center of symmetry at the origin of the lens plane. Using (3.57) with the origin at the center of symmetry of the lens, direct calculation yields the bending vector (cf. [Schn-EF, pp. 230–231]):

$$\hat{\boldsymbol{\alpha}}(\mathbf{r}) = \frac{4GM(r)}{c^2 r^2}\mathbf{r}, \tag{3.96}$$

where $M(r)$ is the total mass of the lens inside the disc of radius $r = |\mathbf{r}|$:

$$M(r) = 2\pi \int_0^r \sigma(r')r'dr'.$$

Notice that the bending angle

$$\hat{\alpha}(r) = \frac{4GM(r)}{c^2 r}$$

does not depend on the matter outside radius r. Plugging (3.96) into (3.94), we get the lens equation for a circularly symmetric lens:

$$\mathbf{s} = \left(\frac{d_S}{d_L} - \frac{4Gd_{L,S}M(r)}{c^2r^2}\right)\mathbf{r}.$$

If the source is at the origin, then

$$r^2 = \frac{M(r)}{\pi\sigma_{crt}},$$

where σ_{crt} is the critical density in (3.27). In other words, the source is lensed into a ring of radius r. We shall call this the **Einstein ring** of a circularly symmetric lens and denote its radius by

$$R_E = \sqrt{\frac{M(R_E)}{\pi\sigma_{crt}}}. \tag{3.97}$$

This immediately reduces to the linear Einstein radius (3.29) due to a point lens of mass M.

Multiplane Lens Equation

Applying equation (3.60) to the k-plane time delay function (3.92) leads to the **k-plane lens equation** (Blandford and Narayan [Bla-N86, p. 579]):

$$\mathbf{s} = \frac{d_S}{d_1}\mathbf{r}_1 - \sum_{i=1}^{k} d_{i,S}\hat{\boldsymbol{\alpha}}_i(\mathbf{r}_i), \tag{3.98}$$

where $\hat{\boldsymbol{\alpha}}_i(\mathbf{r}_i) = \mathrm{grad}_{\mathbf{r}_i}\Psi_i(\mathbf{r}_i)$ and the impact vectors $\mathbf{r}_i$ are given recursively:

$$\mathbf{r}_i = \frac{d_i}{d_1}\mathbf{r}_1 - \sum_{\ell=1}^{i-1} d_{\ell,i}\hat{\boldsymbol{\alpha}}_i(\mathbf{r}_i), \qquad (2 \le i \le k). \tag{3.99}$$

As in the single-plane case, we identify the lensed image associated with the light ray having impact vectors $(\mathbf{r}_1, \dots, \mathbf{r}_k)$ with the first impact vector $\mathbf{r}_1$ — see Figure 3.5. Hence, the k-plane lens equation recursively relates the lensed image $\mathbf{r}_1$ to source positive $\mathbf{s}$. It also

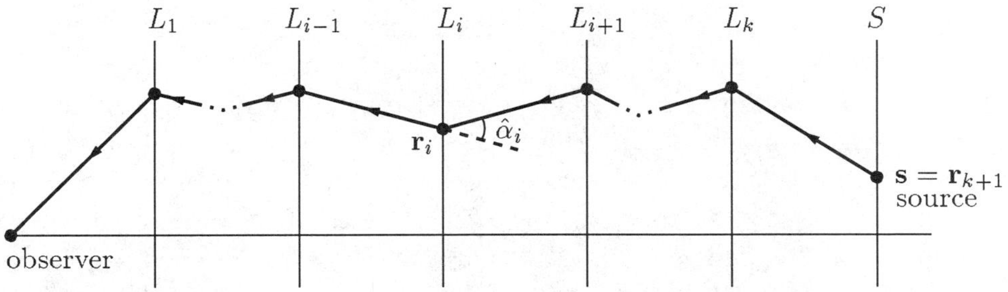

Figure 3.5. Gravitational lensing by matter deflectors on k lens planes. When the ray impacts on the ith lens plane L_i at $\mathbf{r}_i$, it is deflected by angle $\hat{\alpha}_i$. Tracing light rays backwards from observer to the light source plane induces a lensing map from a subset of L_1 into S.

can be used to trace light rays from the observer to the light source plane. A derivation of the k-plane lens equation is given in Section 6.4.2 (cf. [Schn-EF, pp. 282–285]).

Remark (Weak Lensing by Large Scale Structures): Weak lensing by large scale structure density perturbations can be approximated using a multiplane lens equation. The basic idea is to divide the inhomogeneous matter distribution into cubes at varying redshifts and project the matter in each cube to the mid-plane (i.e., "lens plane") of the cube [Sei-SE, p. 2357], [Wam-COT]. However, in a multiplane lensing model of weak lensing, the surface mass density perturbations on the lens planes can be positive or negative. Moreover, for a given light source position $\mathbf{s}$ the multiplane lens equation (3.98) has only one solution, that is, there are no multiple images.

3.2.4 Dimensionless Variables

This section presents the dimensionless variables to be used throughout the rest of the book. The conversion equations should allow the reader to readily transform the dimensionless mathematical variables and formulas used hereafter into ones involving quantities with physical dimensions.

Dimensionless Variables for Single-Plane Lensing

Henceforth, the following dimensionless quantities will be used:

(1) $\mathbf{x} = \mathbf{r}/d_L$ and $\mathbf{y} = \mathbf{s}/d_S$.

(2) $\psi(\mathbf{x}) = (d_{L,S}/d_L d_S)\Psi(d_L\mathbf{x})$.

(3) $\kappa(\mathbf{x}) = \sigma(d_L\mathbf{x})/\sigma_{crt}$, where $\sigma_{crt} = c^2 d_S/4\pi G d_L d_{L,S}$ with $[\sigma_{crt}] = [\text{mass}]/[\text{length}]^2$; this is also expressed as κ *is* σ *in units of* σ_{crt}.

(4) $\boldsymbol{\alpha}(\mathbf{x}) = \text{grad}\,\psi = (d_{L,S}/d_S)\hat{\boldsymbol{\alpha}}(d_L\mathbf{x})$.

Remarks:

(1) If θ is the angle subtended by the impact vector $\mathbf{r}$ at angular diameter distance d_L, then the small angle approximation yields $\theta \simeq |\mathbf{r}|/d_L = |\mathbf{x}|$; similarly, the magnitude $|\mathbf{y}| = |\mathbf{s}|/d_S$ is approximately the angle spanned by $\mathbf{s}$ at angular diameter distance d_S. For this reason, both $\mathbf{x}$ and $\mathbf{y}$ are referred to as **angular vectors**, and $\mathbf{x}$ can be interpreted as inducing a direction from the observer to the impact point $\mathbf{r}$ relative to the line-of-sight. Some authors use the suggestive notation $\boldsymbol{\theta} = \mathbf{r}/d_L$ and $\boldsymbol{\beta} = \mathbf{s}/d_S$. However, we shall abide by the simpler $\mathbf{x}$ and $\mathbf{y}$ notation, leaving $\boldsymbol{\theta}$ and $\boldsymbol{\beta}$ for the context of weak lensing by large scale structures.

(2) A different set of length scales used by certain authors is $\mathbf{x} = \mathbf{r}/(d_L\theta_E) = \mathbf{r}/R_E$ and $\mathbf{y} = \mathbf{s}/(d_S\theta_E)$. We have $\kappa(\mathbf{x}) = \sigma(R_E\,\mathbf{x})/\sigma_{crt}$ and $\boldsymbol{\alpha}(\mathbf{x}) = (d_{L,S}/R_E d_S)\hat{\boldsymbol{\alpha}}(R_E\,\mathbf{x})$ for such scalings. Unless explicitly stated, we shall not assume these conventions.

Poisson's equation (3.50) and the potential (3.51) now take the dimensionless form

$$\Delta\psi(\mathbf{x}) = 2\kappa(\mathbf{x}),$$

where Δ is the Laplace operator on $\mathbf{R}^2$ relative to coordinates $\mathbf{x}$, and

$$\psi(\mathbf{x}) = \frac{1}{\pi}\int_{\mathbf{R}^2} d\mathbf{x}'\kappa(\mathbf{x}')\ln\left|\mathbf{x}' - \mathbf{x}\right|. \tag{3.100}$$

We also obtain

$$\kappa(\mathbf{x}) = \frac{1}{2}\operatorname{div}(\boldsymbol{\alpha})(\mathbf{x}), \tag{3.101}$$

where div is the divergence operator relative $\mathbf{x}$.

Example: A point mass m at the origin of a lens plane has physical surface mass density $\sigma(\mathbf{r}) = \mathrm{m}\delta(\mathbf{r})$, where $\delta(\mathbf{r})$ is the two-dimensional Dirac delta at the origin with physical dimensions $[\delta(\mathbf{r})] = 1/[\text{length}]^2$. This converts to the dimensionless density $\kappa(\mathbf{x}) = (\mathrm{m}/d_L^2\sigma_{crt})\delta(\mathbf{x})$, where $\delta(\mathbf{x}) = d_L^2\delta(d_L\mathbf{x}) = d_L^2\delta(\mathbf{r})$. Let

$$m \equiv \frac{\mathrm{m}}{\pi d_L^2\sigma_{crt}} = \theta_E^2, \tag{3.102}$$

where θ_E is the angular Einstein radius of the mass m. Note that m and $\delta(\mathbf{x})$ are dimensionless. The density κ simplifies as follows: $\kappa(\mathbf{x}) = \pi m\delta(\mathbf{x})$. Inserting this into (3.100), we obtain the dimensionless potential $\psi(\mathbf{x}) = m\ln|\mathbf{x}|$.

The time delay function (3.89) transforms to

$$\mathcal{T}_{d_S\mathbf{y}}(d_L\mathbf{x}) = \vartheta_{L,S}\left[\frac{|\mathbf{x}-\mathbf{y}|^2}{2} - \psi(\mathbf{x})\right], \tag{3.103}$$

where

$$\vartheta_{L,S} = \frac{(1+z_L)}{c}\frac{d_L d_S}{d_{L,S}}.$$

Hereafter, we shall be interested primarily in the critical point properties of $\mathcal{T}_{d_S \mathbf{y}}$. Multiplying $\mathcal{T}_{d_S \mathbf{y}}$ by $\vartheta_{L,S}^{-1}$ does not alter the critical point structure of $\mathcal{T}_{d_S \mathbf{y}}$. Moreover, it simplifies transforms $\mathcal{T}_{d_S \mathbf{y}}$ into a dimensionless form

$$T_{\mathbf{y}}(\mathbf{x}) = \frac{|\mathbf{x} - \mathbf{y}|^2}{2} - \psi(\mathbf{x}).$$

The lens equation (3.94) also converts into a dimensionless form

$$\mathbf{y} = \mathbf{x} - \boldsymbol{\alpha}(\mathbf{x}),$$

where $\boldsymbol{\alpha}(\mathbf{x}) = \text{grad}\, \psi(\mathbf{x})$. This is the bending angle vector associated with ψ. Note that the gradient operator grad is relative to the dimensionless rectangular coordinates $\mathbf{x}$. The corresponding dimensionless lensing map is defined by

$$\boldsymbol{\eta}(\mathbf{x}) = \mathbf{x} - \boldsymbol{\alpha}(\mathbf{x}). \tag{3.104}$$

In other words, $\boldsymbol{\eta} = s/d_S$, where s is the (dimensional) lensing map in equation (3.95), that is, $s(d_L \mathbf{x}) = d_S \mathbf{x} - d_{L,S} \hat{\boldsymbol{\alpha}}(d_L \mathbf{x})$.

Dimensionless Variables for multiplane Lensing

Henceforth, the following dimensionless quantities will be used:

(1) $\mathbf{x}_i = \mathbf{r}_i/d_i$ and $\mathbf{y} = \mathbf{s}_i/d_S = \mathbf{r}_{k+1}/d_S$.

(2) $\psi_i(\mathbf{x}_i) = (d_{i,S}/d_i d_S)\Psi(d_i \mathbf{x}_i)$, where Ψ has physical dimension of length.

(3) $\kappa_i(\mathbf{x}_i) = \sigma_i(d_i \mathbf{x}_i)/\sigma^i_{crt}$, where $\sigma^i_{crt} = c^2 d_i/4\pi G d_i d_{i,S}$ with $[\sigma^i_{crt}] = [\text{mass}]/[\text{length}]^2$.

(4) $\boldsymbol{\alpha}_i(\mathbf{x}_i) = (d_{i,S}/d_S)\hat{\boldsymbol{\alpha}}_i(d_i \mathbf{x}_i) = \text{grad}_i\, \psi_i(\mathbf{x}_i)$, the gradient operator grad_i is defined relative to the dimensionless rectangular coordinates $\mathbf{x}_i$ on the ith lens plane.

On the ith lens, we then have the dimensionless potential

$$\psi_i(\mathbf{x}_i) = \frac{1}{\pi} \int_{\mathbf{R}^2} d\mathbf{x}\, \kappa_i(\mathbf{x}) \ln |\mathbf{x}_i - \mathbf{x}|.$$

This potential satisfies the dimensionless Poisson equation

$$\Delta_{\mathbf{x}_i} \psi_i(\mathbf{x}_i) = 2\kappa_i(\mathbf{x}_i),$$

where $\Delta_{\mathbf{s}_i}$ is the Laplacian relative to $\mathbf{x}_i$. Set

$$\vartheta_i = \frac{(1 + z_i)}{c} \frac{d_i d_{i+1}}{d_{i,i+1}}, \quad \beta_{i,j} = \frac{d_{i,j} d_S}{d_j d_{i,S}}, \quad \beta_i = \beta_{i,i+1}.$$

Then equation (3.92) becomes

$$T_{\mathbf{y}}(\mathbf{x}_1, \ldots, \mathbf{x}_k) = \sum_{i=1}^{k} \vartheta_i \left[\frac{|\mathbf{x}_{i+1} - \mathbf{x}_i|^2}{2} - \beta_i \psi_i(\mathbf{x}_i) \right].$$

N.B. Unlike the single plane case, the time delay function $T_{\mathbf{y}}(\mathbf{x}_1, \ldots, \mathbf{x}_k)$ is not dimensionless, but has physical dimension $[T_{\mathbf{y}}(\mathbf{x}_1, \ldots, \mathbf{x}_k)] = [\text{time}]$. Furthermore, if $k = 1$, then the k-plane time delay function reduces to $T_{\mathbf{y}}(\mathbf{x}_1)$, which actually corresponds notationally to $\vartheta_{L,S}^{-1} T_{\mathbf{y}}(\mathbf{x})$, not the dimensionless function $T_{\mathbf{y}}(\mathbf{x})$.

The lens equation (3.98) transforms into the dimensionless form

$$\mathbf{y} = \mathbf{x}_1 - \sum_{i=1}^{k} \boldsymbol{\alpha}_i(\mathbf{x}_i).$$

The points $\mathbf{x}_i$ are given recursively (cf. (3.99)):

$$\mathbf{x}_i = \mathbf{x}_1 - \sum_{\ell=1}^{i-1} \beta_{\ell,i} \boldsymbol{\alpha}_i(\mathbf{r}_i), \qquad (2 \leq i \leq k).$$

3.2.5 Magnification and Flux Conservation

Definition of Magnification

The gravitational lens effect does not produce any photons (Law of Photon Conservation, page 30); it only deflects existing photons. As a result, it is possible that from some lensed sources in the universe there are more photons reaching the earth than would get here if all the matter in the universe (including the lens) were smoothed out, while from other sources fewer photons can reach the earth. This is called magnification or de-magnification of a source (usually meant relative to a smoothed out universe — see discussion below on magnification and flux conservation). In fact, what actually happens is that we see the source under a bigger or smaller solid angle, which translates into a larger or smaller surface area of the light emitting source. Sometimes this effect is called "amplification." However, since there is not really an amplifier involved (in the sense that there is energy put into the system), Burke suggested at a 1989 Toulouse lensing conference that the terminology "(de-)magnification" should be used instead.

Let us now look more closely at the meaning of magnification. Since the deflection angle of a light ray depends on the ray's impact parameter, two neighboring rays are deflected by different amounts. As a consequence, a whole bundle of rays with a circular cross section gets deformed and distorted. In particular, the cross sectional areas of the distorted bundle and the undistorted one (in the absence of lensing) are, in general, different from each other. This illustrates the effect of "magnification" (or "de-magnification") mentioned above. The ratio between the solid angle $\Delta\,\Theta$ subtended by an image of a small extended lensed source and the solid angle $\triangle \Theta_S$ the source subtends in the absence of lensing defines **magnification**:

$$\text{magnification of source's image} \equiv \mu = \frac{\triangle \Theta}{\triangle \Theta_S}. \tag{3.105}$$

We can also express the magnification of an image of a lensed source as a ratio of fluxes. In fact, recall that at observed frequency ν the **monochromatic flux** of a light source is defined by

$$F_\nu^S = \frac{\text{radiative energy}}{\text{unit area} \cdot \text{unit time} \cdot \text{unit frequency interval}}$$

and **surface brightness** is (e.g., Peebles[Pee, p. 157]):

$$\mathrm{I}_\nu = \frac{\text{radiative energy}}{\text{unit area} \cdot \text{unit time} \cdot \text{unit solid angle} \cdot \text{unit frequency interval}}.$$

By construction, these quantities are related:

$$F_\nu^S = \mathrm{I}_\nu \triangle \Theta_S.$$

The surface brightness of a source obeys the following fundamental result:

> **The Brightness Theorem**: *Suppose that non-interacting light rays with frequency ν_s leave a source with redshift z_s and surface brightness I_{ν_s}, as measured by an observer at rest at the source. Then an observer on earth (redshift $z = 0$) will measure the surface brightness of the source to be as follows, where the frequency ν_s has shifted to $\nu = \nu_s/(1 + z_s)$:*
>
> $$\mathrm{I}_\nu = \left(\frac{\nu}{\nu_s}\right)^3 \mathrm{I}_{\nu_s} = \frac{\mathrm{I}_{\nu_s}}{(1 + z_s)^3}.$$

This result yields that for a non-interacting family of photons, the quantity I_ν/ν^3 is the same for observers along each photon's world line. The proof of the Brightness Theorem employs Liouville's theorem (e.g., Arnold [Arn89b, p. 69]). Readers are referred to [Mis-TW, pp. 583–588], [Pee, pp. 253–254], and [Schn-EF, pp. 115–117] for details.

Denote the observed surface brightness of the source in the absence of lensing by I_ν, where ν is the frequency measured on earth of the received light. Let $\mathrm{I}'_{\nu'}$ be the observed surface brightness of an image of the source in the presence of lensing. When there is a lens, the cosmic scaling function changes negligibly during the travel time of light across the lens (page 36). Moreover, deflection by the lens leaves the frequency of the light unchanged (page 63). Consequently, in the presence of lensing an observer on earth will measure the source's redshift z'_s and frequency ν' of the received light to be the unlensed values: $z'_s = z_s$ and $\nu' = \nu$. Applying the Brightness Theorem, we get

$$\mathrm{I}'_{\nu'} = \left(\frac{\nu'}{\nu_s}\right)^3 \mathrm{I}_{\nu_s} = \left(\frac{\nu}{\nu_s}\right)^3 \mathrm{I}_{\nu_s} = \mathrm{I}_\nu,$$

which expresses the following:

> **Conservation of Surface Brightness**: *The surface brightness of an image of a lensed source coincides with the surface brightness of the source in the absence of lensing.*

This implies that the monochromatic flux $F'_{\nu'}$ of the source's lensed image subtending a solid angle $\Delta\,\Theta$ satisfies

$$F'_{\nu'} = \mathrm{I}'_{\nu'}\,\Delta\,\Theta = \mathrm{I}_\nu\,\Delta\,\Theta \equiv F_\nu,$$

which is measurable. Hence, the magnification is a ratio of monochromatic fluxes:

$$\mu = \frac{F_\nu}{F_\nu^S},$$

which is not directly measurable (since F_ν^S is unknown). However, when there are multiple images, the magnification ratios of the images is measurable:

$$\frac{\mu_i}{\mu_j} = \frac{F_i}{F_j},$$

where F_i is the monochromatic flux of the ith image.

We now express magnification in terms of the lensing map $\boldsymbol{\eta}$ of a gravitational lens system. Suppose that in the absence of lensing a small source about the point $\mathbf{s} = (s_1, s_2)$ on a light source plane at angular diameter distance d_S spans the following area:

$$\Delta\,A_{\mathbf{s}} = \Delta\,s_1 \Delta\,s_2.$$

By definition of solid angle, we have

$$\Delta\,\Theta_S = \frac{\Delta\,A_{\mathbf{s}}}{d_S^2}.$$

If the small source is lensed and has a lensed image about the point $\mathbf{r} = (r_1, r_2)$ on a lens plane at angular diameter distance d_L, then the image spans a solid angle

$$\Delta\,\Theta = \frac{\Delta\,A_{\mathbf{r}}}{d_L^2},$$

where

$$\Delta\,A_{\mathbf{r}} = \Delta\,r_1 \Delta\,r_2$$

is the area of the lensed image. Since $\mathbf{x} = \mathbf{r}/d_L$ and $\mathbf{y} = \mathbf{s}/d_S$, the magnification of the lensed image is

$$\mu = \frac{\triangle A_{\mathbf{r}}}{d_L^2} \frac{d_S^2}{\triangle \mathbf{s}} = \frac{\triangle A_{\mathbf{x}}}{\triangle A_{\mathbf{y}}},$$

where

$$\triangle A_{\mathbf{x}} = \triangle \mathrm{u} \triangle \mathrm{v}, \qquad \triangle A_{\mathbf{y}} = \triangle y_1 \triangle y_2$$

with $\mathbf{x} \equiv (\mathrm{u}, \mathrm{v})$ and $\mathbf{y} \equiv (y_1, y_2)$.

Consider the lensing map $\boldsymbol{\eta} : L \to S$, where L and S are the lens and light source planes, respectively. The absolute value of the determinant of the Jacobian matrix of $\boldsymbol{\eta}$ is the ratio of an infinitesimal area in the light source plane to its corresponding area in the lens plane. The magnification of a lensed image is then approximately the absolute value of the inverse of the Jacobian determinant at the lensed image. Given a lensing map $\boldsymbol{\eta}$, we then define **the magnification of a lensed image at x of a point-like light source at y** $= \boldsymbol{\eta}(\mathbf{x})$ as follows:

$$\mu_{\mathbf{y}}(\mathbf{x}) = \frac{1}{|\det[\mathrm{Jac}\,\boldsymbol{\eta}](\mathbf{x})|}, \tag{3.106}$$

where the Jacobian matrix $(\mathrm{Jac}\,\boldsymbol{\eta})(\mathbf{x})$ is called the **magnification matrix**.[41] Sometimes we shall denote $\mu_{\mathbf{y}}(\mathbf{x})$ more simply by $\mu(\mathbf{x})$ or even μ. For certain lenses, the following holds (Schneider [Schn84, p.123]):

> **The Magnification Theorem**: *If an isolated, gravitational lens has a positive density perturbation,*[42] *then its action on any light source will produce a lensed image with magnification of at least one.*

Theorem 13.1 (page 469) gives a mathematically precise formulation and extension of the previous result.

Dropping the absolute value in (3.106), one obtains the **signed magnification**:

$$\mu_{\mathbf{y}}^{sgn}(\mathbf{x}) = \frac{1}{\det[\mathrm{Jac}\,\boldsymbol{\eta}](\mathbf{x})}.$$

The **total magnification of a point-like light source at y** is

$$\mu_{tot}(\mathbf{y}) = \sum_{\mathbf{x}} \mu_{\mathbf{y}}(\mathbf{x}),$$

where the sum runs over all lensed images $\mathbf{x}$ of $\mathbf{y}$. The **magnification of an extended light source** with surface brightness $\mathrm{I}(\mathbf{y})$ is defined by (e.g., [Schn-EF, p. 215]):

$$\mu_{Ex} = \frac{\int d\mathbf{y}\ \mathrm{I}(\mathbf{y})\ \mu_{tot}(\mathbf{y})}{\int d\mathbf{y}\ \mathrm{I}(\mathbf{y})}, \tag{3.107}$$

---

[41] Some authors denote the magnification matrix $(\mathrm{Jac}\,\boldsymbol{\eta})(\mathbf{x})$ by $A(\mathbf{x})$. We shall not employ this notation.

[42] See page 47 for a discussion of the density perturbation due to a gravitational lens. Note that the positivity condition on the density perturbation is always assumed and is repeated in the theorem for emphasis.

where the integrals are over the source and $\mu_{tot}(\mathbf{y})$ is the total magnification of a point source at $\mathbf{y}$. The light source is often chosen to be a disc with uniform surface brightness. However, at times a limb darkening model is used for the brightness profile (e.g., Figure 5.10, page 164).

Example (Point Mass): The lens equation for a particle of mass M is given by

$$\mathbf{y} = \mathbf{x}\left(1 - \frac{m}{x^2}\right). \tag{3.108}$$

Here m is given by (3.102) and $x = |\mathbf{x}|$. We quote several basic facts about this lens system; see Section 6.3.3 (page 187) for a detailed discussion. If the source is not at the origin, then the source has two images (i.e., the lens equation has two solutions):

$$\mathbf{x}_+ = \frac{1}{2}\left(y + \sqrt{y^2 + 4\theta_E^2}\right) \mathbf{e}_{\vartheta_s}$$

and

$$\mathbf{x}_- = \frac{1}{2}\left(y - \sqrt{y^2 + 4\theta_E^2}\right) \mathbf{e}_{\vartheta_s},$$

where $y = |\mathbf{y}|$, $\theta_E^2 = m$ via (3.102), and

$$\mathbf{y} = y(\cos\vartheta_s, \sin\vartheta_s) \equiv y\, \mathbf{e}_{\vartheta_s}.$$

The magnifications of the images $\mathbf{x}_\pm$ are

$$\mu(\mathbf{x}_+) = \frac{1}{1 - (\theta_E/x_+)^4}, \qquad \mu(\mathbf{x}_-) = \frac{1}{(\theta_E/x_-)^4 - 1},$$

where $x_\pm = |\mathbf{x}_\pm|$. The total magnification can then be expresed as

$$\mu = \mu(\mathbf{x}_+) + \mu(\mathbf{x}_-) = \frac{y^2 + 2\theta_E^2}{y\sqrt{y^2 + 4\theta_E^2}}.$$

Define u to be the angular separation between the point lens at the origin and source in units of the angular Einstein radius:

$$u = \frac{y}{\theta_E}.$$

Then the total magnification of a point mass lens simplifies as follows:

$$\mu = \frac{u^2 + 2}{u\sqrt{u^2 + 4}}. \tag{3.109}$$

Note that this value is always larger than one!

If the light source is a radius R disc with uniform surface brightness, then using (3.109) it can be shown that the source's magnification has the following maximum (e.g., [Schn-EF, pp. 38, 240]):

$$\mu_{Ex}^{max} = \frac{\sqrt{4 + R^2}}{R}.$$

Magnification, De-Magnification and Flux Conservation

In earlier years there was some confusion about the use of "magnification" in gravitational lensing. This originated from two different ways of viewing isolated lens situations. One could compare the lensing effects of various matter clumps with a universe in which all matter is completely homogeneously smeared out ("filled-beam approximation"), or with a completely empty universe ("empty-beam approximation"). In the latter case, even with only a single lens in the universe, every source would appear magnified compared to the "empty" situation. But because of the Law of Photon Conservation (page 30), this scenario is unphysical.

In the filled-beam approximation, a light bundle gets slightly focused if there is more matter inside the beam than "on average" in the universe. This results in a net magnification. On the other hand, a light bundle that encounter less than an average amount of matter inside its beam gets slightly defocussed with respect to the smoothed out universe; such sources are seen slightly de-magnified. One could describe the latter scenario as if under-dense regions in the universe act effectively as "negative" matter (in the sense that they effectively "repel/diverge" light rays instead of attracting them); but this is only the consequence of the situation in which the attraction of the matter outside the beam is larger than the attraction of the matter inside the beam. In summary, the net average magnification of all objects in the universe due to lensing is one. **Lensing does not create any photons; it can only redistribute them.**

An easy way to visualize this is as follows: Imagine a sphere around an isotropically radiating source. Each surface element of equal size receives exactly the same number of photons, that is, observers at any position on this sphere would measure the same apparent brightness for this source. If now there is a matter clump somewhere inside the sphere, this clump will attract photons and deflect their paths by various amounts, depending on the impact parameters. Hence, some surface elements of that sphere will receive more photons than in the scenario without the clump, and others – those on "the other side" – will receive fewer photons. Observers at the latter positions would see the source slightly de-magnified, whereas observers on the former positions would get a magnified view of the source. The (occasionally large) "magnification" for a few observers is compensated by the small "de-magnification" for most observers. (For the real universe, a clean description of this issue is given, e.g., by [Sel94, Sel96], or [Bar], using the weak lensing formalism based on the density power spectrum.)

The above (de-)magnification issue was explored by J.W. et al. [Wam-COX] using the lensing effects of the large scale matter distribution of the universe on "supernovae as standard candles." They showed with numerical simulations that most high redshift supernovae are slightly de-magnified with respect to a smoothed out matter distribution, whereas some are moderately magnified, and a small fraction is highly de-magnified.

3.2.6 Critical Curves, Caustics, and Light Curves

Critical Curves and Caustics

For a lens system with lensing map $\boldsymbol{\eta}$, the positions $\mathbf{x}$ in the lens plane where $\det[\mathrm{Jac}\,\boldsymbol{\eta}](\mathbf{x}) = 0$, formally have infinite magnification. They are called **critical curves** — the locus of all images with formally infinite magnification. The corresponding locations in the light source plane are the **caustics**. In other words, the caustics due to a gravitational lens are the critical values of the associated lensing map. When caustics are curves, the smooth arcs are called **folds**, while **cusps** are the points where two abutting fold arcs have coincident tangents with the fold arcs on opposite sides of the double tangent. For example, the caustics in the top left panel of Figure 3.6 have six cusps and eight fold arcs. See Section 7.5.1 (page 244) for a precise definition of folds and cusps.

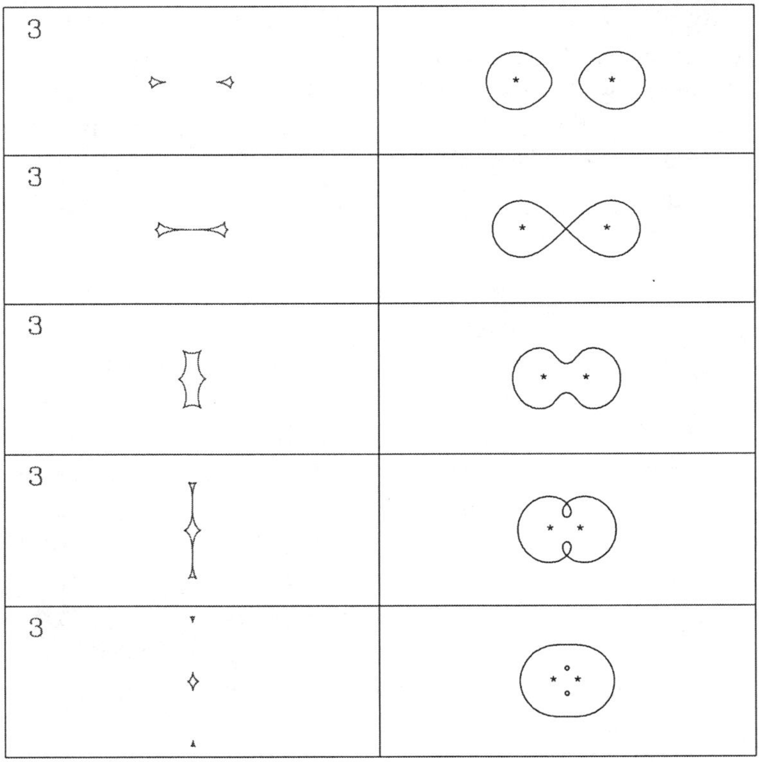

Figure 3.6. Caustics (left column) and critical curves (right column) due to two point masses (asterisks): equal mass, decreasing distances from top to bottom. The number "3" indicates how many lensed images are produced for a source outside the caustics; five lensed images are produced of a source inside any of the caustics.

Example (Point Mass): In the simple case of a point mass lens, equation (3.108) yields $\det[\text{Jac}\,\boldsymbol{\eta}](\mathbf{x}) = 1 - (\theta_E/x)^4$, where $x = |\mathbf{x}|$. The critical curve is given by the condition $x = \theta_E$, which is a ring. By the lens equation (3.108), this ring is mapped to the point $\mathbf{y} = 0$ in the light source plane. In other words, a point mass lens produces a point caustic ($\mathbf{y} = 0$). A source at the origin is then lensed into an infinitely bright ring — an Einstein ring with angular radius θ_E. This coincides with our earlier result for a general circularly symmetric lens. Section 6.3.3 (page 187) gives a detailed treatment of a point mass lens.

Example (Two Point Masses — Single Lens Plane): If one breaks the circular symmetry by adding another point mass, then the lens equation takes the form

$$\mathbf{y} = \mathbf{x} - m_1 \frac{\mathbf{x} - \boldsymbol{\xi}_1}{|\mathbf{x} - \boldsymbol{\xi}_1|^2} - m_2 \frac{\mathbf{x} - \boldsymbol{\xi}_2}{|\mathbf{x} - \boldsymbol{\xi}_2|^2}, \tag{3.110}$$

where $\boldsymbol{\xi}_1$ and $\boldsymbol{\xi}_2$ are the positions of the masses m_1 and m_2, respectively. The caustic structure in this case depends on the distance between the two lenses. Figure 3.6 illustrates examples of the caustics and critical curves as decreasing distances $|\boldsymbol{\xi}_1 - \boldsymbol{\xi}_2|$ between the lenses when $m_1 = m_2$. Three images are seen of a source outside the caustic, while sources inside have five images (Theorem 11.11). Readers are referred to Section 15.2.2 (page 514) for more on binary point mass lenses; also, see Theorem 11.11 (page 439) and page 468 for the number of lensed images and total magnification. Schneider and Weiss [Schn-W86] treat the equal-mass case in detail.

Caustics and Wavefronts

It is also instructive to view gravitational lensing — in particular, caustics — from a wavefront perspective. Within our geometrical optics approximation, light rays emitted at a certain instant from an isotropically radiating (point-like) source can be treated as orthogonal trajectories to a surface of constant phase — a **wavefront** — spreading out in all directions from the source. In fact, one obtains a continuous family of wavefronts since real sources emit continuously. Shortly after emission, a wavefront is diffeomorphic to a 2-sphere, but upon passing through a lens plane, the front gets distorted and can develop singularities such as self-intersections and cusp ridges where two sheets abut — see Figures 3.7. and 3.8. Those singularities corresponding to points through which the cross section of an infinitesimal bundle of light rays of the wavefront degenerates into a line or point are *caustics*. In general, caustics on a wavefront are points that are light-ray conjugate to the observer. In other words, light rays pile up in the vicinity of caustic points, which is why an observer near a caustic sees the source as highly magnified.

As a wavefront moves, its caustics trace out a surface in space called a *caustic surface*. Points on such a surface are where the observer's past light cone is no longer an immersed submanifold. In fact, caustic surfaces consist of several sheets typically joined along edges and vertices. These surfaces are extremely complicated in general. However, the stable ones are locally either fold sheets, cusped ridges, swallowtails, elliptic umbilics, or hyperbolic

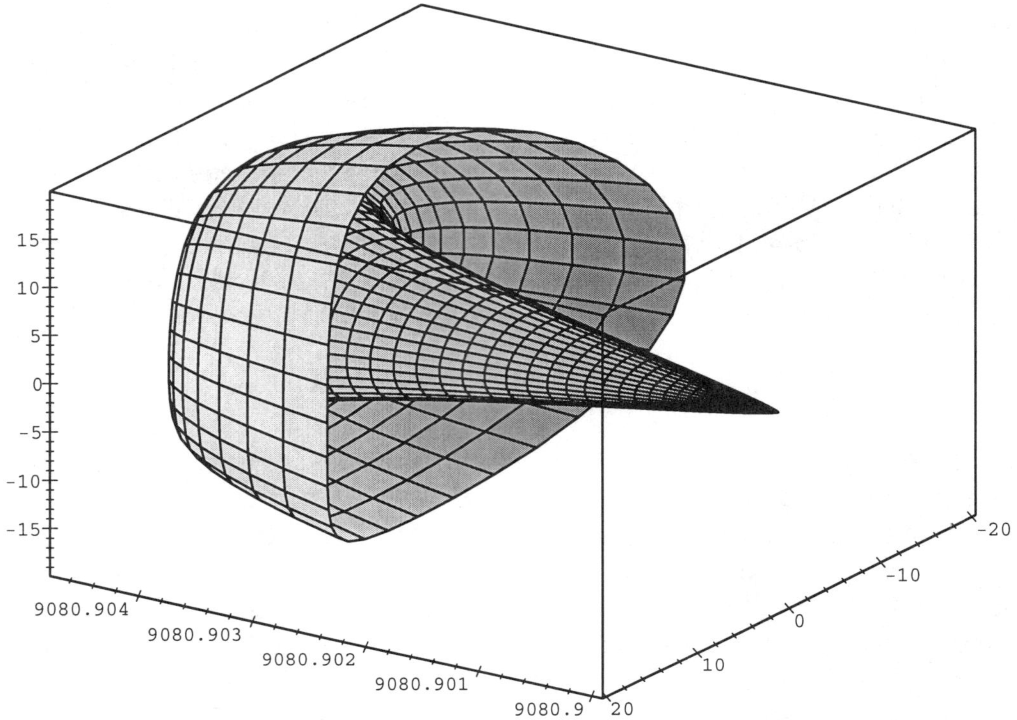

Figure 3.7. Portion of a single wavefront spreading out from an observer at the origin. The front passed through an elliptical potential lens on a lens plane at position 90 along the axis shown with 9000-range values. The lens caused part of the wavefront to lag behind, creating the spike at the center. This spike is along the optical axis and represents the endpoints of light rays that are severely retarted by the gravitational lens. (Courtesy of Frittelli.)

umbilics (cf. Figure 9.15, page 376). See Arnold [Arn92][Arn91], Ehlers and Newman [Ehl-N] for more on wavefront singularities and caustic surfaces; also, consult Friedrich and Stewart [Fri-S], Stewart [Stew], and Perlick [Per00a] (and references therein) for a general relativistic perspective.

Remark: As different sheets of a wavefront pass through an observer, several lensed images are seen of the light source. In addition, the time delays between lensed-image pairs arise from sheets on the front lagging behind each other. This approach was used by Refsdal [Ref64b] in 1964 to relate such time delays to Hubble's constant (also, see [Ref94, p. 168]). Furthermore, the magnification of these lensed images are determined by the principal curvatures of the wavefront at the points of impact (Berry and Upstill [Berr-U], Nityananda [Nit]).

Caustic curves were defined earlier as critical values of a lensing map η, where η is interpreted as tracing light rays from an observer to a light source plane S. In order to relate the caustics of η to wavefronts, imagine that the observer emits light continuously and isotropically. This produces a continuous train of wavefronts spreading out from the observer (i.e., the wavefronts have an ordering in time, each wavefront occurs before and after some other wavefront). Suppose that after the wave-train passes through a gravitational lens, the train is sliced at cosmic time τ_s by the light source plane S. Then S intersects a collection of caustics arising from a continuous family of different wavefronts in the train. This intersection produces a set of caustics on S that forms the caustics of the lensing map $\boldsymbol{\eta}$. Equivalently, the caustics of $\boldsymbol{\eta}$ come from S slicing through, at cosmic time τ_s, the caustic surface created by the wave-train. Notice that the caustic points of $\boldsymbol{\eta}$ arising from different wavefronts have lag times relative to each other. Also, since the lens system is assumed "static" (unless stated otherwise) and the observer is an isotropic continuous emitter, the caustics of $\boldsymbol{\eta}$ do not vary as the wave-train moves through S. However, in Section 9.5 (page 375) and Section 15.3.2 (page 535), we shall investigate non-static lens systems, where caustics evolve and undergo metamorphoses.

This subsection concludes with an illustration (Figures 3.7 and 3.8) created by Simonetta Frittelli of the caustics, lagging sheets, self-intersection, and overall distortion of a wavefront by gravitational lensing due to an elliptical potential.

Light Curves for a Single Point Mass Lens

The case of an isolated stellar mass lens moving relative to the observer's line of sight is currently of interest. We shall only briefly summarize the most important results of this system. For further study, Paczyński [Pac96], gives a comprehensive review of the theoretical and observational aspects of this type of microlensing. Also, see Section 4.3 for applications to the structure of quasars.

The total magnification μ of a point source being lensed by a point mass depends entirely on the "impact parameter" u (by equation (3.109), page 86). If u is a function of time (at the observer), then the magnification is also $\mu = \mu(t)$. So if the lens and source move relative to each other — which in general is the case for any two objects in space — we measure different total magnifications with time and get a microlensed "light curve." In general, a **light curve** is the plot of a source's total magnification as a function of observer time as the source moves along a trajectory. Note that the impact parameter u controls how close the source passes the lens (for transverse motion cross the sky). For example, if $u \leq 1$ (so $y \leq \theta_E$), then the source passes within the Einstein ring — see the left part of Figure 3.9.

The time scale of such a "microlensing event" is defined as the time it takes to cross a distance equal to the linear Einstein radius:

$$t_E = \frac{R_E}{v_\perp}, \tag{3.111}$$

where $v_\perp$ is the (relative) transverse velocity of the lens. Applied to the lensing of stars in the central part of the Milky Way by intervening stars, this amounts to a few months for

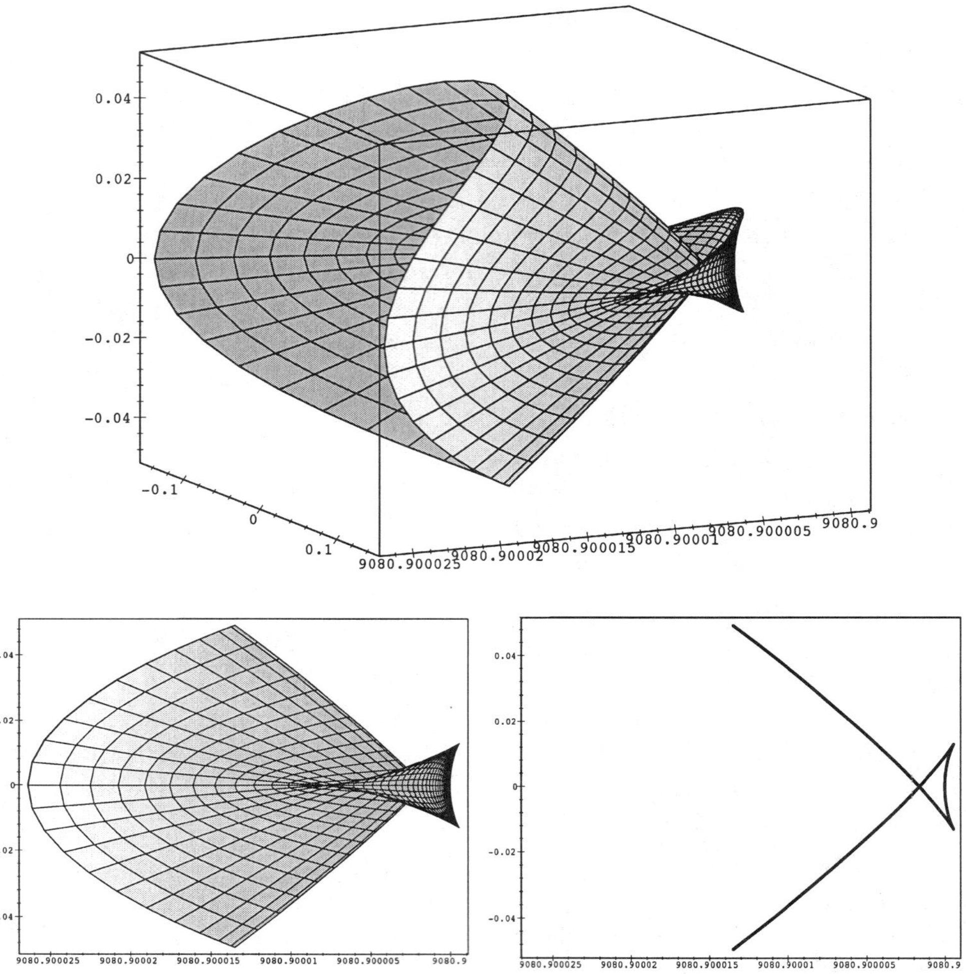

Figure 3.8. *Top:* Zooming into the tip of the spike in Figure 3.7, we see cusp ridges (which are caustic curves on the wavefront) terminating at "swallowtail" points (cf. the swallowtail surface in Figure 9.15, page 376). Note that the perspective is slightly rotated to show the singularities more clearly. *Bottom left:* A rotated view of the top surface. This shows a profile of the singularities at the tip of the spike (so the grid does *not* depict a family of wavefronts). The whole wavefront is moving from right to left in the panel. *Bottom right*: A curve resulting from slicing through the surface in the bottom left panel along the optical axis. A self-intersection in the wavefront is revealed. The two cusp points lie on cusp ridges of the wavefront. When the caustics on the wavefront first intersect a light source plane S, the intersection consists of two caustic points on S, namely, the points where the cusp ridges vanish. As the front moves through S, the two caustic points split into four points that trace out a lips-shaped caustic (similar to the one in the bottom right panel in Figure 3.16, page 107). (Courtesy of Frittelli.)

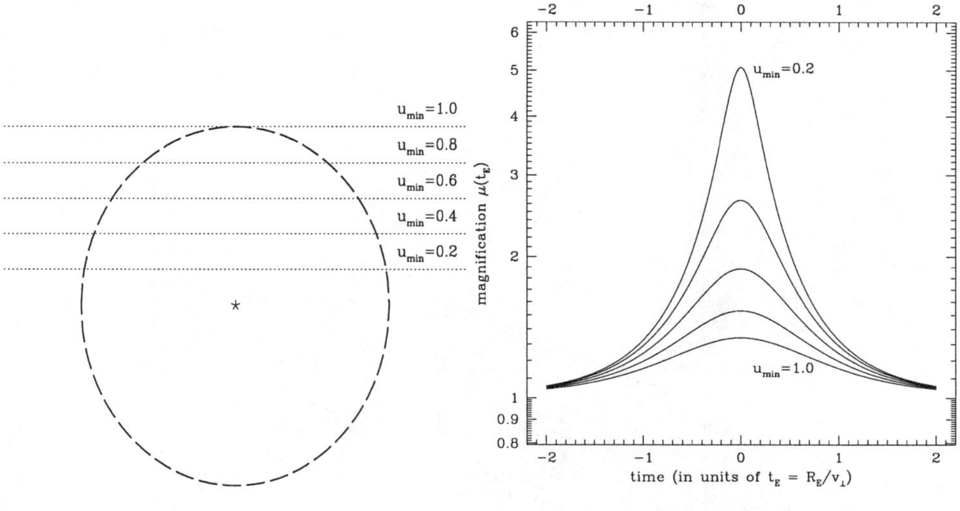

Figure 3.9. Light curves due to a point mass lens. The left panel shows five different tracks for relative motion between lens and source, parametrized by the minimum impact parameter $u_{min} = 0.2, 0.4, ..., 1.0$. In the right panel the corresponding light curves are displayed.

realistic values of distances, transverse velocity and lensing mass

$$t_E \simeq 0.2yr \left(\frac{M}{M_\odot}\right)^{0.5} \left(\frac{d_L}{10kpc}\right)^{0.5} \left(1 - \frac{d_L}{d_S}\right)^{0.5} \left(\frac{v_\perp}{200\text{km/sec}}\right)^{-1}. \tag{3.112}$$

The impact parameter as a function of time is given as

$$u(t) = \left[u_{min}^2 + \left(\frac{t - t_{max}}{t_E}\right)^2\right]^{\frac{1}{2}}. \tag{3.113}$$

Here u_{min} is the closest approach of the point lens and source, and t_{max} is the corresponding time (at the observer). The time t_{max} is when the total magnification reaches its maximum (since the caustic is at the origin). Examples of light curves due to a point mass lens with various minimum impact parameters are displayed in Figure 3.9.

The cross section for microlensing is often defined such that the dimensionless impact parameter is at most unity, $u_{\min} \leq 1$. In other words, the approach of the source is inside or on the Einstein ring of the lens ($y \leq \theta_E$). Applying (3.109), this translates into a minimum for the peak of the total magnification:[43]

$$\mu_{max} \geq \frac{1}{\sqrt{5}}\left[\left(u + \frac{1}{u}\right) + \frac{1}{u}\right] \geq 3/\sqrt{5} \simeq 1.3416. \qquad (u \leq 1).$$

[43]Note that $f(u) = u + 1/u$ has the global minimum value $f(1) = 2$ on $(0, 1]$.

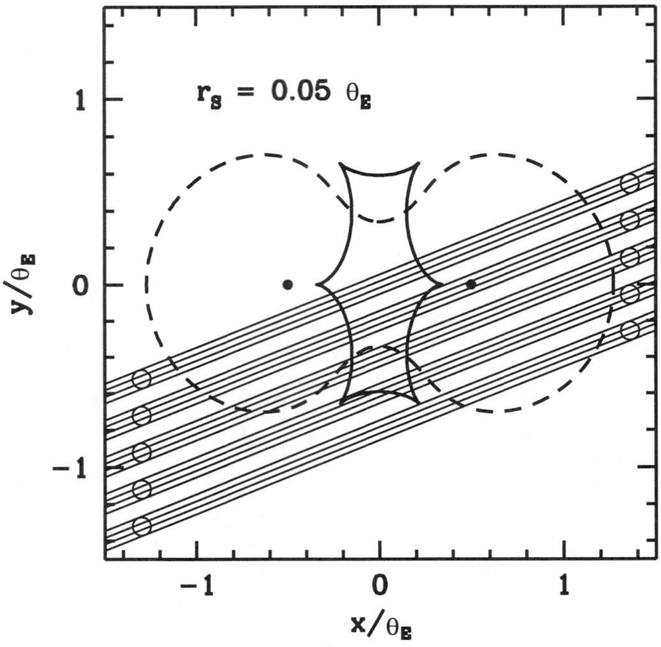

Figure 3.10. Caustic (solid curve) and critical curve (dashed curve) due to two point masses m_1 and m_2 (bold dots). The mass fraction is $m_2/m_1 = 1$ and each light source has angular radius $r_s = 0.05\theta_E$, where θ_E is the angular Einstein radius for the mass $M = m_1 + m_2$. (Courtesy of Paczyński.)

Here $\mu_{max} = \mu(t_{max})$. This can also be seen from the peaks in Figure 3.9. Note that the lower bound is attained when the closest approach is at the Einstein ring. Readers are also referred to Section 5.5 (e.g., page 159) for a discussion of how to identify observationally lensing by moving point mass.

Light Curves for a Binary Point Mass Lens

A caustic, critical curve, and sample light curves due to two point masses are shown in Figure 3.10. Five tracks are shown along which a disc-shaped light source moves. See Section 15.2.2 (page 514) for more on this system.

As each disc-shaped light source in Figure 3.10 moves along the indicated trajectory, the source's magnification μ_e varies. This is plotted in Figure 3.11 as a function of time, where the magnification is given in terms of stellar magnitudes Δ mag $\equiv 2.5 \log \mu_e$ and time is normalized by t_0, which is the time it takes to traverse the linear Einstein for the system — see (3.111) on page 91; the Einstein ring is defined relative to the total mass $M = m_1 + m_2$.

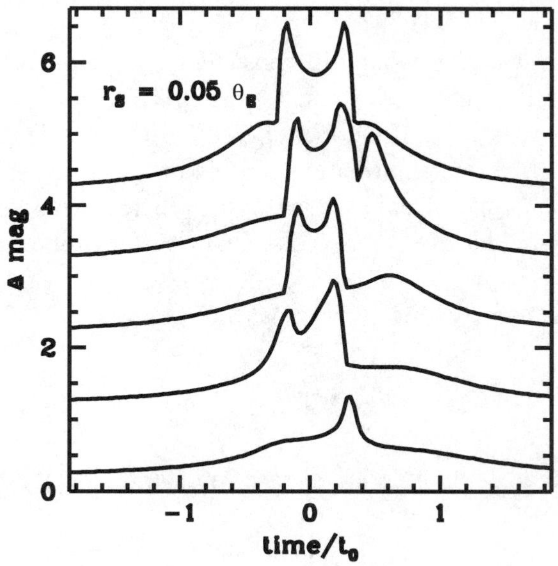

Figure 3.11. Light curves corresponding to the tracks in Figure 3.10, where the top light curve is due to the top track in Figure 3.10, etc. (Courtesy of Paczyński [Pac96].)

In Figure 3.11, notice the high peaks due to caustic crossings and the corners that result from fold crossings. In the bottom light curve, there is a high magnification peak that results from the source passing just outside the bottom right cusp in Figure 3.10. A similar peak is shown in the second light curve from the top (see the rightmost peak); here, the source passes outside the middle cusp on the right in Figure 3.10. Such increases in magnification outside cusps are due to "lobes" of high magnification. These "lobes" occur generically outside cusps due to lensing by any finite number of lens planes (see Figure 9.5 on page 334).

3.2.7 Shear, Convergence, and Extended Sources

Having treated the magnification of pointlike sources, we now turn to how the image of an extended source is distorted by gravitational lensing. This requires introducing the notions of shear and convergence. We then discuss how shear and convergence appear in the lensing of a small extended source not on a caustic; see Section 4.2.3 (page 128) for applications to weak lensing. The section ends with some examples of lensing of extended sources on caustics.

Shear and Convergence

In gravitational lensing, it is not enough to consider only the matter components near a light bundle from a source. Matter components far away from the light bundle can cause an additional light deflection as well. If the matter distribution outside the bundle is roughly cylindrically symmetric, then its contribution cancels out. But if the mass distribution is asymmetric (e.g., as in an edge-on spiral galaxy), then there is an additional "pull" on the light rays along one particular direction. This has to be taken into account and is often called "external shear."

In general, shear has two components defined by

$$\Gamma_1(\mathbf{x}) = \frac{1}{2}[\psi_{\mathrm{uu}}(\mathbf{x}) - \psi_{\mathrm{vv}}(\mathbf{x})], \qquad \Gamma_2(\mathbf{x}) = \psi_{\mathrm{uv}}(\mathbf{x}),$$

where $\mathbf{x} = (\mathrm{u}, \mathrm{v})$. Together they define a tracefree symmetric matrix called the **shear matrix** at $\mathbf{x}$:

$$\mathbf{\Gamma}(\mathbf{x}) = \begin{bmatrix} \Gamma_1(\mathbf{x}) & \Gamma_2(\mathbf{x}) \\ \Gamma_2(\mathbf{x}) & -\Gamma_1(\mathbf{x}) \end{bmatrix}.$$

Choose the following parametrization:

$$\theta(\mathbf{x}) = \frac{1}{2}\tan^{-1}\left[\frac{\Gamma_2(\mathbf{x})}{\Gamma_1(\mathbf{x})}\right], \qquad \Gamma(\mathbf{x}) = \sqrt{\Gamma_1^2(\mathbf{x}) + \Gamma_2^2(\mathbf{x})}. \tag{3.114}$$

We call $\theta(\mathbf{x})$ and $\Gamma(\mathbf{x})$, respectively, the **angle of shear** and **magnitude of shear at x**. It follows that

$$\Gamma_1(\mathbf{x}) = \Gamma(\mathbf{x})\cos[2\theta(\mathbf{x})],$$

$$\Gamma_2(\mathbf{x}) = \Gamma(\mathbf{x})\sin[2\theta(\mathbf{x})].$$

In this parametrization, we see that the shear at $\mathbf{x}$ produced along angle $\theta(\mathbf{x})$ is identical to that produced along $\theta(\mathbf{x}) + \pi$. Consequently, it suffices to assume that $0 \le \theta(\mathbf{x}) < \pi$.

In studying the lensing of small extended sources, we need to consider the geometric action of the inverse lensing map $\boldsymbol{\eta}^{-1}$ in a sufficiently small neighborhood of a noncaustic point $\mathbf{y}$ in the light source plane. For such a case, we can approximate $\boldsymbol{\eta}^{-1}$ (resp., $\boldsymbol{\eta}$) by its Jacobian matrix $[(\mathrm{Jac}\,\boldsymbol{\eta})(\mathbf{x})]^{-1}$ (resp., $(\mathrm{Jac}\,\boldsymbol{\eta})(\mathbf{x})$).

Let us now express $[(\mathrm{Jac}\,\boldsymbol{\eta})(\mathbf{x})]^{-1}$, as well as $(\mathrm{Jac}\,\boldsymbol{\eta})(\mathbf{x})$, in terms of the shear matrix $\mathbf{\Gamma}(\mathbf{x})$ and surface mass density $\kappa(\mathbf{x})$. The Jacobian matrix of $\boldsymbol{\eta}$ at $\mathbf{x}$ is

$$(\mathrm{Jac}\,\boldsymbol{\eta})(\mathbf{x}) = [1 - \kappa(\mathbf{x})]\,\mathbf{I} - \mathbf{\Gamma}(\mathbf{x}), \tag{3.115}$$

where $\mathbf{I}$ is the 2×2 identity matrix. Let $R[\theta(\mathbf{x})]$ be counterclockwise rotation through angle $\theta(\mathbf{x})$ relative to the u-axis:

$$R[\theta(\mathbf{x})] = \begin{bmatrix} \cos[\theta(\mathbf{x})] & -\sin[\theta(\mathbf{x})] \\ \sin[\theta(\mathbf{x})] & \cos[\theta(\mathbf{x})] \end{bmatrix}.$$

Then

$$(\mathrm{Jac}\,\boldsymbol{\eta})(\mathbf{x}) \quad = \quad R[\theta(\mathbf{x})] \begin{bmatrix} 1-\kappa(\mathbf{x})-\Gamma(\mathbf{x}) & 0 \\ 0 & 1-\kappa(\mathbf{x})+\Gamma(\mathbf{x}) \end{bmatrix} R[-\theta(\mathbf{x})]. \tag{3.116}$$

Since

$$\det[\mathrm{Jac}\,\boldsymbol{\eta}](\mathbf{x}) = [1-\kappa(\mathbf{x})]^2 - \Gamma^2(\mathbf{x}), \tag{3.117}$$

we have the following at points $\mathbf{x}$ in the lens plane that the lensing map $\boldsymbol{\eta}$ sends to $\mathbf{y}$:

$$[(\mathrm{Jac}\,\boldsymbol{\eta})(\mathbf{x})]^{-1} = \frac{[1-\kappa(\mathbf{x})]\,\mathbf{I} + \boldsymbol{\Gamma}(\mathbf{x})}{[1-\kappa(\mathbf{x})]^2 - \Gamma^2(\mathbf{x})}.$$

Hence

$$[(\mathrm{Jac}\,\boldsymbol{\eta})(\mathbf{x})]^{-1} = R[\theta(\mathbf{x})] \begin{bmatrix} \frac{1}{1-\kappa(\mathbf{x})-\Gamma(\mathbf{x})} & 0 \\ 0 & \frac{1}{1-\kappa(\mathbf{x})+\Gamma(\mathbf{x})} \end{bmatrix} R[-\theta(\mathbf{x})]. \tag{3.118}$$

Notice that if there is no shear at $\mathbf{x}$ (i.e., $\boldsymbol{\Gamma}(\mathbf{x}) = \mathbf{0}$), but $\kappa(\mathbf{x})$ is nonzero, then (3.118) shows approximately that $\boldsymbol{\eta}^{-1}$ maps a small radius r disc-shaped source, which is not on a caustic, into a disc of radius $r/|1-\kappa(\mathbf{x})|$ on the lens plane. The term $\kappa(\mathbf{x})$ causes isotropic focusing (or diverging) of light rays — see Figure 3.12. For this reason, the surface mass density $\kappa(\mathbf{x})$ is also called the **convergence** (**expansion**, or **Ricci focusing**) due to the lens.

References: Further treatments of shear and convergence in gravitational lensing appear in Blandford et al. [Bla-SBV] and Seitz et al. [Sei-SE]. Readers are referred to Sachs [Sach], Penrose [Pen66], and O'Neill [ONe, Sec. 5.7] for more on shear and convergence in general relativity.

Lensing of Small Extended Sources not on Caustics

Suppose that there is an extended source whose projection into the light source plane is a disc of radius r centered at $\mathbf{y}$. If the disc is sufficiently small and does not lie on a caustic, and if a lensed image of the source is about $\mathbf{x}$, then the inverse Jacobian matrix

$[(\text{Jac}\,\boldsymbol{\eta})(\mathbf{x})]^{-1}$ approximates the action of $\boldsymbol{\eta}^{-1}$ in mapping the disc source into the lensed image around $\mathbf{x}$. We saw above that in the absence of shear at $\mathbf{x}$, the disc-shaped source is lensed into a disc image, but with radius $r/|1-\kappa(\mathbf{x})|$. There is an isotropic magnification of the source — Figure 3.12(a).

When shear is present, there is a tidal force on an infinitesimal light bundle. Specifically, the shear $\boldsymbol{\Gamma}(\mathbf{x})$ is responsible for an anisotropic stretching of light rays (astigmatism). The magnitude $\Gamma(\mathbf{x})$ of shear affects the amount of distortion and the angle of shear $\theta(\mathbf{x})$ determines the angle (relative to the u-axis) along which the distortion occurs. For example, the matrix in (3.118) will map a disc-shaped source of radius r centered at $\mathbf{y}$ into an ellipse shaped image centered at $\mathbf{x}$ — see Figure 3.12(b). The major axis of the ellipse is at angle $\theta(\mathbf{x})$ and has length $2r/|1-\kappa(\mathbf{x})-\Gamma(\mathbf{x})|$. Note that the distortion is then along the line through the major axis. The minor axis has length $2r/|1-\kappa(\mathbf{x})+\Gamma(\mathbf{x})|$. The area of the elliptical image equals the area of disc source times $[(1-\kappa(\mathbf{x}))^2-\Gamma^2(\mathbf{x})]^{-1}$.

Let us reverse the situation above. If a small disc centered at $\mathbf{x}$ in the lens plane is now traced via the lensing map $\boldsymbol{\eta}$ back to the light source plane, and if the disc does not lie on a critical curve, then the action of $\boldsymbol{\eta}$ can be approximated by the Jacobian matrix of $\boldsymbol{\eta}$ in (3.116). This time the disc is mapped into an ellipse shaped region with major axis along angle $\theta(\mathbf{x})+\pi/2$. The shearing action is now rotated by angle $\pi/2$ and the resulting caustics will be distorted along the line at angle $\theta(\mathbf{x})+\pi/2$.

Convention: When tracing light rays from the light source plane to the observer, we shall call $\theta_{SO}(\mathbf{x})=\theta(\mathbf{x})$ the angle of shear. On the other hand, if light rays are traced from the observer to the light source plane, then $\theta_{OS}(\mathbf{x})=\theta_{SO}(\mathbf{x})+\pi/2$ is the angle of shear.

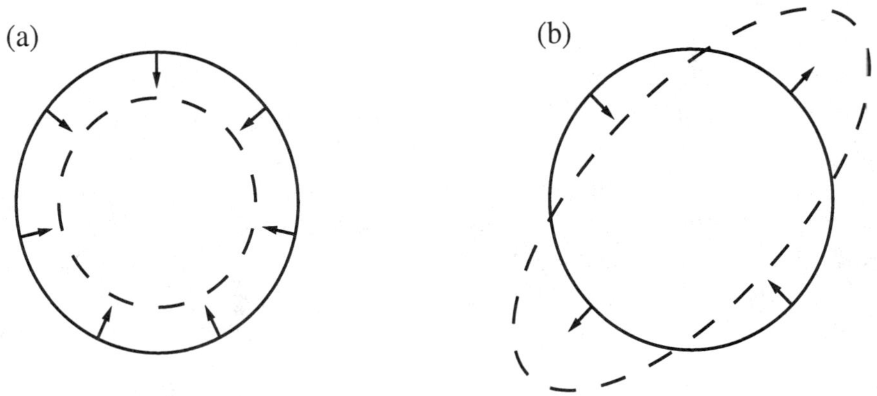

Figure 3.12. Isotropic and anisotropic effects due to lensing of a small extended disc-shaped source not on a caustic. *(a)* Light from the source isotropically converges to a disc-shaped image (dashed curve); shear is absent. *(b)* In the presence of shear and convergence, the image (dashed curve) of the source is distorted into an elliptical shape.

Lensing of Extended Sources On Caustics

Example (Circularly Symmetric Lenses): A disc-shaped extended source centered on the point caustic due to a point mass lens will appear as an annulus. For the case of a nonsingular circularly symmetric lens, there are two caustics: a circle and the point at the center of the circle. The corresponding critical curves are two concentric circles, where the caustic of the outer one corresponds via the lensing map to the point caustic. Point sources inside the caustic circle have three images, while a source outside the circle has one image. Figure 3.13 illustrates how this lens acts on an extended source. Notice that the arc-like images near the outer critical curve are tangentially elongated and the image intersecting the inner critical curve is radially stretched. For this reason, the outer and inner critical curves are called, respectively, **tangential** and **radial**. Consult Narayan and Bartelmann [Nar-B, p. 391] for further discussion of circularly symmetric lenses.

Example (Elliptical Potentials): These non-circularly symmetric potentials have a very simple analytical form and provide an instructive rough model of elliptical galaxy lenses (Blandford and Kochanek [Bla-K87b]).

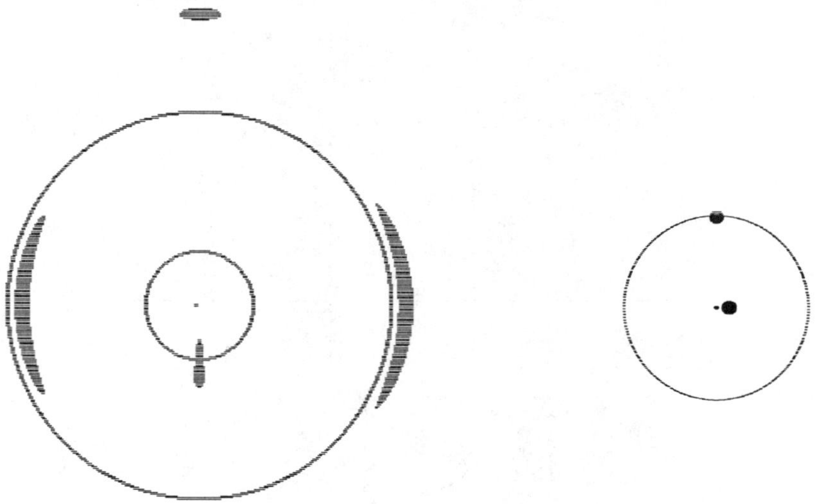

Figure 3.13. Lensing of extended sources (shown by the discs on the right) by a nonsingular circularly symmetric lens. The dot and circle on the right are the caustics, while the two concentric circles on the left are the critical curves. The source near the point caustic is lensed into a faint image corresponding to the central dot on the left and two arcs-like images *tangential* to the outer critical curve. For the source on the caustic circle, there is a *radially* stretched image intersecting the inner critical curve and a tangential image at the top, outside the outer critical curve. (Courtesy of Bartelmann.)

Elliptical potentials are of the following form (see Section 3.3.2, page 105):

$$\Psi_{ep}(\mathbf{r}) = 4\pi \frac{\sigma_v^2}{c^2} \sqrt{r_{cr}^2 + (1-\epsilon)x^2 + (1+\epsilon)y^2},$$

where $\mathbf{r} = (x, y)$, the constant r_{cr} represents the core radius of the galaxy, and the ellipticity ϵ is constrained to $0 \leq \epsilon \lesssim 0.2$ for physical reasons. The lensing action of an elliptic potential on an extended light source is shown in Figure 3.14. The partial ring in the bottom left panel is analogous to the gravitational lensing of the radio source MG1131+0456. We note that a point source inside the astroid caustic will have five images, while one in the region bounded by the astroid and elliptically shaped caustic has three images. A point source in the exterior of the latter caustic has one image.

References: For more on lensing of extended sources see Kochanek, Keeton and McLeod [Koc-KM], Kochanek [Koc92], Kassiola, Kovner and Blandford [Kas-KB], and Nye [Nye].

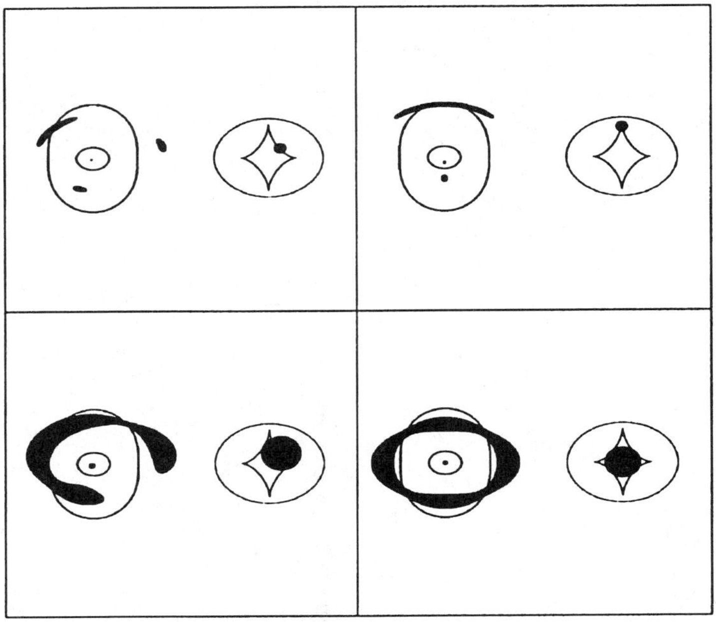

Figure 3.14. The right part of each panel shows an extended disc-shaped light source (bold circle) against the caustics due to an elliptical gravitational lens potential. On the left of every panel are the resulting arcs and rings, shown against the critical curves, into which the extended source is deformed. (Courtesy of Bartelmann.)

3.3 Two Important Families of Lens Models

This section unifies the simple lens models we have seen so far into two important classes, and concludes with a comparison of the models' regimes of applicability.

3.3.1 Point Masses with Continuous Matter and Shear

We shall describe an effective simple lens model used for a **microlensing scenario**. The latter means that the action of a deflector galaxy on light rays (e.g., from a quasar) is divided into contributions from stars and continuous matter in a local region near the rays, and an external shear from infinity due the action of the galaxy as a whole. It will be supposed that the density of continuous matter and the shear from infinity are constant across the local region.

Point Masses

Most astrophysical objects are "big," compared to the sizes of things we know from everyday life. Nevertheless, in some contexts the spatial extent of certain objects can be neglected. They can be treated as "point-like" if the size is very small compared to the relevant length scale.[44]

In gravitational lensing the relevant length scale is defined by the linear Einstein radius. This gives the typical positional offset of a lensed image due to the deflection of light rays in the potential of the lensing object. For a sun-like star acting as a lens at a "cosmological" distance (i.e., at "redshift" of about $z \simeq 0.2$, corresponding to a distance of about 650Mpc $\simeq 2 \times 10^{27}$cm), the radius of 8×10^{12}cm is more than three orders of magnitude smaller than the linear Einstein radius $R_E \simeq 3 \times 10^{16}$cm, which justifies the **point lens** approximation. On the other hand, a galaxy with a mass of 10^{11} solar masses and a radius of about 10 kpc ($\simeq 3 \times 10^{22}$ cm) acting as a lens has a linear Einstein ring radius of comparable size (9×10^{21} cm). Hence in this case the mass distribution of the galaxy has to be taken into account and the point mass treatment is not justified.

An ensemble of g point masses m_i at positions $\mathbf{r}_i$ on a lens plane has surface mass density

$$\sigma(\mathbf{r}) = \sum_{i=1}^{g} \mathrm{m}_i \delta(\mathbf{r} - \mathbf{r}_i).$$

[44] A prototypical example of such a situation is the motion of the planets around the sun. Although the earth has a radius of more than 6,000 kilometers (and the sun more than 100 times that), the dynamics of the planetary system is described very accurately with Kepler's laws, which use the "point mass approximation": the sun and the planets are treated as if all their mass were concentrated in a point at their center. This is justified since the distance between the sun and earth (150 million kilometers) corresponds to more than 100 times the combined radii of the sun and earth.

Here $\delta(\mathbf{r}-\mathbf{r}_i)$ is the Dirac delta at $\mathbf{r}_i$. By (3.51), the corresponding two-dimensional Newtonian potential is

$$\Psi_{pt}(\mathbf{r}) = \frac{4G}{c^2}\sum_{i=1}^{g} \mathrm{m}_i \ln\left|\frac{\mathbf{r}-\mathbf{r}_i}{d_0}\right|.$$

In dimensionless form,

$$\psi_{pt}(\mathbf{x}) = \sum_{i=1}^{g} m_i \ln\left|\frac{\mathbf{x}-\mathbf{x}_i}{d_0}\right|, \qquad m_i = \frac{\mathrm{m}_i}{\pi d_L^2 \sigma_{crt}}. \tag{3.119}$$

Letting $\mathbf{x} = (\mathrm{u}, \mathrm{v})$ and $\mathbf{x}_i = (\mathrm{u}_i, \mathrm{v}_i)$, the components of shear at $\mathbf{x}$ due to this potential are

$$\begin{aligned}\Gamma_1(\mathbf{x}) &= \sum_{i=1}^{g} m_i \frac{(\mathrm{u}-\mathrm{u}_i)^2 - (\mathrm{v}-\mathrm{v}_i)^2}{|\mathbf{x}-\mathbf{x}_i|^2}\\ \Gamma_2(\mathbf{x}) &= \sum_{i=1}^{g} m_i \frac{2(\mathrm{u}-\mathrm{u}_i)(\mathrm{u}-\mathrm{v}_i)}{|\mathbf{x}-\mathbf{x}_i|^2}.\end{aligned}$$

It follows that the shear $\Gamma = (\Gamma_1, \Gamma_2)$ vanishes in the limit $|\mathbf{x}| \to \infty$. Employing polar coordinates, we write $\mathbf{x}_i = |\mathbf{x}_i|(\cos\vartheta_i, \sin\vartheta_i)$ to obtain the shear at the origin:

$$\begin{aligned}\Gamma_1(\mathbf{0}) &= \sum_{i=1}^{g} m_i \frac{\cos 2\vartheta_i}{\mathrm{u}_i^2}\\ \Gamma_2(\mathbf{0}) &= \sum_{i=1}^{g} m_i \frac{\sin 2\vartheta_i}{\mathrm{v}_i^2}.\end{aligned}$$

Appearance of the angular variable $2\vartheta_i$ (rather than ϑ_i) reflects the fact that a point lens at a polar coordinate position $(|\mathbf{x}_i|, \vartheta_i)$ produces exactly the same tidal force at the origin as one at position $(|\mathbf{x}_i|, \vartheta_i + \pi)$.

Continuous Matter

The concept of **continuously distributed matter** relates to the length scales/angular scales as well. What is important in this respect is the following: If the angular Einstein radius of a lensing object is much smaller than the angular size of the source object, and if there are "many" lensing objects of this kind, then the surface mass distribution can be considered "continuous" in this particular situation. Examples for such continuous matter distributions are clouds of gas or dust inside a lensing galaxy. The gas and dust "particles" are much smaller than any luminous astrophysical object that can serve as a "source" and hence this continuous-matter-concept is always justified in these circumstances. Stars inside a galaxy lensing a background quasar are **not** continuously distributed matter: The angular Einstein radius of a star is comparable to the size of a quasar. These same stars, however, can be treated as "continuous matter" if the lensed object is a background galaxy, whose angular size is orders of magnitude larger than the stellar angular Einstein radius. The potential due a continuous matter distribution with constant surface mass density $\sigma(\mathbf{r}) = \sigma_c$ is

$$\Psi_{cts}(\mathbf{r}) = \frac{4G}{c^2}\frac{\sigma_c}{2}|\mathbf{r}|^2, \tag{3.120}$$

or, in dimensionless form,

$$\psi_{cts}(\mathbf{x}) = \frac{\kappa_c}{2}|\mathbf{x}|^2,$$

where $\kappa_c = \sigma_c/\pi\sigma_{crt}$. Notice that as is expected intuitively, this potential produces no shear: $\Gamma_1(\mathbf{x}) = 0 = \Gamma_2(\mathbf{x})$.

Shear From Infinity

In a microlensing scenario, we suppose that there is a constant **shear** γ **from infinity** due to the action of the deflector galaxy as whole. This originates from an asymmetric distribution of matter very far away from the light rays (e.g., spiral galaxies would produce a tug along a particular direction). Explicitly, the potential describing the shear from infinity due to this distant matter is

$$\Psi_{sh}(\mathbf{r}) = -\frac{4\pi G\sigma_{crt}}{c^2}\frac{\gamma}{2}[(x^2 - y^2)\cos 2\theta + 2xy\sin 2\theta],$$

where $\mathbf{r} = (x, y)$ and $0 \leq \theta < \pi$ is a constant. In dimensionless form,

$$\psi_{sh}(\mathbf{x}) = -\frac{\gamma}{2}[(\mathrm{u}^2 - \mathrm{v}^2)\cos 2\theta + 2\mathrm{uv}\sin 2\theta]. \tag{3.121}$$

It follows that the shear due to this potential at any point $\mathbf{x}$ has components,

$$\Gamma_1(\mathbf{x}) = \gamma\cos 2(\theta + \pi/2)$$

and

$$\Gamma_2(\mathbf{x}) = \gamma\sin 2(\theta + \pi/2).$$

Consequently, the shear relative to tracing light rays from the light source plane to the observer is along the line at angle $\theta_{SO} \equiv \theta(\mathbf{x}) = \theta + \pi/2$ (see page 98). Reversing the direction, we obtain the shear with respect to ray tracing from the observer to the light source plane. This shear is along the line at angle $\theta_{OS} \equiv \theta_{SO}(\mathbf{x}) + \pi/2 = \theta + \pi$, or, $\theta_{OS} = \theta$. In single-plane lensing, it is enough to choose the angle of shear θ_{OS} along the u-axis (i.e., $\theta = 0$). Of course, we shall not do this for multiplane lensing since shear can act at different angles on various lens planes.

Point Masses Plus Continuous Matter Plus Shear

A general simple "local" description of a gravitational microlensing scenario (of, e.g., a background quasar) is given by a surface mass density consisting of two matter components — point mass distribution and continuously distributed matter — and a constant shear γ from infinity. The potential model for this system is the superposition of potentials (3.119), (3.120), and (3.121):

$$\psi_{ptcs}(\mathbf{x}) = \frac{\kappa_c}{2}|\mathbf{x}|^2 - \frac{\gamma}{2}\left[(\mathrm{u}^2 - \mathrm{v}^2)\cos 2\theta + 2\mathrm{uv}\sin 2\theta\right] + \sum_{i=1}^{g} m_i \ln|\mathbf{x} - \boldsymbol{\xi}_i|. \tag{3.122}$$

The potential ψ_{ptcs} yields a surface mass density

$$\kappa_{ptcs}(\mathbf{x}) = \kappa_c + \sum_{i=1}^{g} m_i \delta(\mathbf{x} - \boldsymbol{\xi}_i).$$

The shear components due to ψ_{ptcs} satisfy $\lim_{|\mathbf{x}|\to\infty} \Gamma_1(\mathbf{x}) = \gamma\cos 2(\theta + \pi/2)$ and $\lim_{|\mathbf{x}|\to\infty} \Gamma_2(\mathbf{x}) = \gamma\sin 2(\theta + \pi/2)$. Sufficiently far away from the point masses, the magnitude of shear is dominated by the shear from infinity. The latter acts along the line at angle θ (or, equivalently, $\theta + \pi$) relative to tracing light rays from observer to the light source plane. Consequently, the caustics due to ψ_{ptcs} experience an overall stretching along the line at angle θ.

The lens equation due to ψ_{ptcs} is as follows, where the shear from infinity is chosen along the u-axis, relative to ray tracing from observer to light source plane:

$$\mathbf{y} = \begin{bmatrix} 1 - \kappa_c + \gamma & 0 \\ 0 & 1 - \kappa_c - \gamma \end{bmatrix}\mathbf{x} - \sum_{i=1}^{g} m_i \frac{\mathbf{x} - \boldsymbol{\xi}_i}{|\mathbf{x} - \boldsymbol{\xi}_i|^2}. \tag{3.123}$$

The solutions $\mathbf{x}$ of this equation yield the number of lensed images and their positions. See Sections 11.4, 11.5, and 11.6 for a detailed study of these issues.

Example (Point Mass with Shear): For a point mass at the origin with shear γ from infinity, the lens equation (3.123) becomes

$$\mathbf{y} = \begin{bmatrix} 1 + \gamma & 1 \\ 1 & 1 - \gamma \end{bmatrix}\mathbf{x} - m\frac{\mathbf{x}}{|\mathbf{x}|^2}. \tag{3.124}$$

Samples of the corresponding caustics and critical curves are shown in Figure 3.15. Four images are produced of sources inside the caustics and two images for those outside. This lens was first investigated by Chang and Refsdal [Chan-R79] and is commonly called a **Chang–Refsdal lens**. See Section 15.2.1 (page 509) for more.

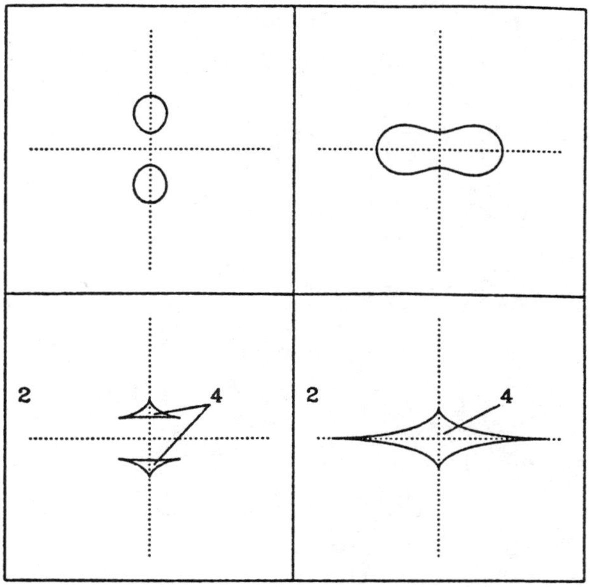

Figure 3.15. Caustics *(bottom curves)* and critical curves *(top curves)* due to a point lens with shear from infinity. The numbers indicate how many images are seen of a source located in the given region.

3.3.2 Elliptical Lens Models

Galaxies are often modeled by elliptical mass distributions. Suppose that a galaxy is at angular diameter distance d_L, has proper core radius r_c, and its stars have line-of-sight velocity dispersion σ_v. Assume that the lens is centered around the origin with the (dimensionless) coordinates (u, v) lying, respectively, along the major and minor axes of the lens. Kassiola and Kovner [Kas-K] have studied in detail the elliptical surface mass densities that model this lens. They even calculated the corresponding two-dimensional potential, though it is analytically complicated [Kas-K, p. 457].

By treating elliptical mass distributions as projections of oblate spheroids, Keeton and Kochanek [Kee-K98, p. 158] found simpler analytical expressions for the potential, which are equivalent to those in [Kas-K]. They considered an **isothermal ellipsoid lens**, that is, a lens with surface mass density

$$\kappa_{ie}(\mathbf{x}) = \frac{b_I}{2\sqrt{q^2(s_c^2 + \mathrm{u}^2) + \mathrm{v}^2}}.$$

Here $s_c = r_c/d_L$, b_I is a constant depending on the eccentricity and velocity dispersion σ_v of the mass distribution, and q is the length of the minor axis divided by the major axis (see

[Kee-K98, pp. 157–158]). The associated potential ψ_{ie} and bending angle vector $\boldsymbol{\alpha}_{ie}$ are

$$\psi_{ie}(\mathbf{x}) = \mathrm{u}\alpha_1 + \mathrm{v}\alpha_2 - b_I s_c \ \ln\sqrt{(h(\mathbf{x}) + s_c)^2 + (1-q^2)\mathrm{u}^2}, \qquad \boldsymbol{\alpha}_{ie} = (\alpha_1, \alpha_2),$$

where

$$\begin{aligned} \alpha_1 &= \frac{b_I}{\sqrt{1-q^2}} \tan^{-1}\left[\frac{\mathrm{u}\sqrt{1-q^2}}{h(\mathbf{x}) + s_c}\right], \\ \alpha_2 &= \frac{b_I}{\sqrt{1-q^2}} \tanh^{-1}\left[\frac{\mathrm{v}\sqrt{1-q^2}}{h(\mathbf{x}) + q^2 s_c}\right], \end{aligned}$$

with $h(\mathbf{x}) = q^2(s_c^2 + \mathrm{u}^2) + \mathrm{v}^2$. The signed magnification is given by

$$\mu_{ie}^{sgn}(\mathbf{x}) = 1 - \frac{b_I}{h(\mathbf{x})} + \frac{b_I^2 \, s_c}{h(\mathbf{x})[(h(\mathbf{x}) + s_c)^2 + (1-q^2)\mathrm{u}^2]}.$$

If $s_c = 0$, then the lens is called a **singular isothermal ellipsoid** and has a singularity at the origin. This is consistent with most galaxy models having three-dimensional mass densities with cuspidated centers. Notice that for a singular isothermal ellipsoid lens, the level curves, $b_I/h(\mathbf{x}) =$ constant, determine the contours for *both* the signed magnification and the surface mass density. IF $s_c > 0$, then the model κ_{ie} is called a **softened isothermal ellipsoid** or **isothermal ellipsoid with finite core**. Readers should see [Kee-K98] for more on these models and others, along with applications to lensing by spiral galaxies.

From a pedagogical point of view, it is more convenient to work with the following analytically simpler semi-realistic potential — **elliptical potential**:

$$\psi_{ep}(\mathbf{x}) = C\sqrt{x_{cr}^2 + (1-\epsilon)\mathrm{u}^2 + (1+\epsilon)\mathrm{v}^2},$$

where $C = (4\pi\sigma_v^2 d_{L,S})/(c^2 d_S) = \sigma_v^2/G\sigma_{crt}d_L$ with σ_v the line-of-sight velocity dispersion and $x_{cr} = r_{cr}/d_L$ with r_{cr} the proper core radius. For ellipticity ϵ in the range $0 \le \epsilon \lesssim 0.2$, elliptical potentials provide a reasonable approximation to an elliptical mass distribution [Kas-K, p. 460]. The corresponding surface mass density is

$$\kappa_{ep}(\mathbf{x}) = C\frac{\left(2x_{cr}^2 + (1-\epsilon^2)^2|\mathbf{x}|^2\right)}{2\left(x_{cr}^2 + (1-\epsilon)\mathrm{u}^2 + (1+\epsilon)\mathrm{v}^2\right)^{3/2}}$$

and the lens equation becomes

$$\mathbf{y} = \begin{bmatrix} 1 - \mathrm{E}(\mathbf{x})(1-\epsilon) & 1 \\ 1 & 1 - \mathrm{E}(\mathbf{x})(1+\epsilon) \end{bmatrix}\mathbf{x},$$

where

$$\mathrm{E}(\mathbf{x}) = \frac{C}{\sqrt{x_{cr}^2 + (1-\epsilon)\mathrm{u}^2 + (1+\epsilon)\mathrm{v}^2}}.$$

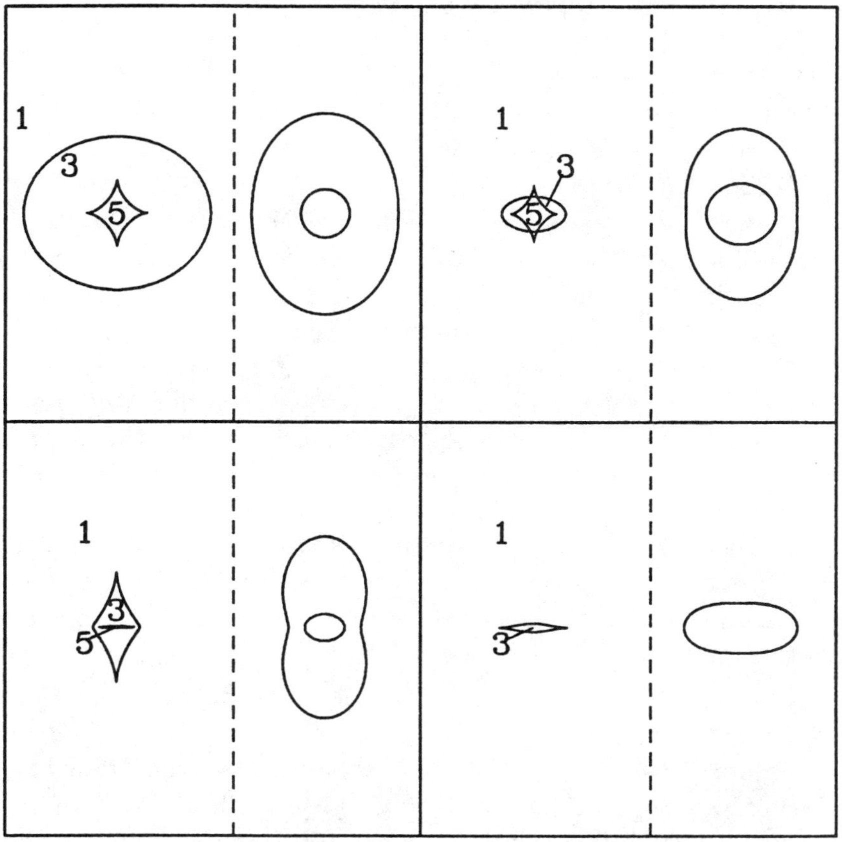

Figure 3.16. Caustics (solid curves) and critical curves (dotted curves) due to an elliptical potential. The numbers indicate how many images are see of a source situated in the given region. Note that there is a single lips-shaped caustic in the bottom right panel.

Figure 3.16 shows some of the caustics due to ψ_{ep} and the multiple images it produces; see Section 11.3 (page 426) for more on multiple imaging due to this potential, as well as general nonsingular isolated lenses.

If the core radius and ellipticity vanish (i.e., $r_c = 0$ and $\epsilon = 0$), then the elliptical potential describes a **singular isothermal sphere**

$$\psi_{sis}(\mathbf{x}) = C|\mathbf{x}|.$$

The associated surface mass density has a singularity at the origin:

$$\kappa_{sis} = \frac{C}{2}\frac{1}{|\mathbf{x}|}.$$

3.3.3 Point Mass versus Elliptical Lens Models

In gravitational lensing, there are two commonly used models of lenses: point and extended mass distributions. The former way is not just for simplicity, but is also justified in many astrophysical scenarios. The "extended" mass distributions can be treated in various ways. A relatively manageable model of an "extended lens" is the isothermal ellipsoid (or the semi-realistic elliptical potential for a quick first approximation).

There is a quantitative criterion which can be used to estimate whether the simple point lens treatment is justified. This is the ratio between the physical angular (resp., linear) size of the lens and the size of its angular (resp., linear) Einstein radius. If this ratio is much smaller than unity, then a point lens treatment is justified. Astrophysical scenarios in which this is the case are, for example, the searches for "MACHOS" (faint MAssive Compact Halo Objects with Jupiter-like masses) in the outer regions of the Milky Way. There the lenses are stars or more compact objects. Their Einstein radii are two or more orders of magnitude bigger than their physical sizes. Similarly, in the treatment of "microlensing" in distant quasars, point lenses provide a valid approximation. Again, ordinary stars inside distant galaxies act as lenses on an even more distant object, a quasar. And again the ratio between physical angular size and angular Einstein radius of these micro-lenses is very small compared to unity.

The astrophysical case in which this approximation is not valid and in which elliptical mass models are usually applied are to distant galaxies acting as lenses. Here the angular Einstein radius is usually of order one arcsecond, whereas the physical extent of the galaxy is of this order or even larger. Since most (projected) mass distributions of galaxies are roughly elliptical, an elliptical mass model is appropriate and indeed seems to reproduce many of the observed lensing features relatively well. The inner regions of distant massive clusters of galaxies can also be modeled to first order as elliptical mass distributions (because most of the mass in a galaxy cluster is smoothly distributed, with the visible galaxies making up only a few percent of the mass). The Einstein radii/physical sizes of these inner regions are of order 20 arcseconds. Such models reproduce relatively well the "Giant Luminous Arcs." However, for a detailed modeling and understanding of the lens properties of galaxy clusters, in particular of the "weak lensing," more detailed mass distributions need to be considered.

3.4 Observable Lensing Properties

To put the theoretical point of view into perspective, we briefly present those quantities that currently can be observed from a gravitational lens system. A more extended treatment appears in Chapter 5. This is followed by a discussion of a family of transformations of a lens system that leave the observables invariant.

Current Observables

Almost all the observational properties of a lens systems are dimensionless (e.g., angles and ratios of intensities). An exception is the time delay (i.e., the arrival time difference for a pair of light rays traveling from source and to observer). There are two types of gravitational lens observations/detections: *static* (i.e., an image configuration, a spectrum) and *dynamic* ones (i.e., light curves).

Lensed point sources are usually multiply imaged. The apparent brightnesses and intensity ratios — hence, relative magnification — of the images are measurable. For the intensity ratios, the image separation must be large enough to be resolved, which is a function of our technology. [In fact, what is "large enough" in this context depends on the wavelength and instrument used. Before the launch of the Hubble Space Telescope, in optical astronomy the lower boundary was a little under one arcsec. This has been pushed by almost a factor of ten now. In radio astronomy with the powerful VLBI technique, the resolution can be higher than a milliarcsec.] In some cases, an image shows variations in apparent brightness over time ("microlensing"). These fluctuations can be either variations intrinsic to the source, or changes induced by small gravitational lenses along the line of sight. For unresolved point sources (stars), the lens effect can be measured only by their characteristic brightness changes in time.

Lensed extended sources (optical/radio galaxies) can show up in a variety of ways. Strongly affected sources are usually highly distorted, in some cases multiply imaged as well. The most conspicuous examples are the Einstein rings and the Giant Luminous Arcs. Furthermore, the arclets or the statistical alignment of weakly distorted images (coherent shape distortions) are clear signatures of lensing of extended sources.

Images of multiple quasars, giant luminous arcs, Einstein rings, arclets, and weak lensing are examples of *static* observations. Quasar microlensing and galactic microlensing on the other hand are *dynamic* observations: one needs a large number of individual exposures over an extended period of time in order to measure a change in the apparent brightness of a source. The same is true for a measurement of the time delay between the two or more images of a multiple quasar, which can be used to determine the Hubble constant.

Below are some of the current observables associated with gravitational lensing:

- Source and lens redshifts (i.e., z_S and z_L).
- Number of images, their relative positions and angular separations.
- Apparent brightnesses of images and their relative magnifications; light spectra of the lens.
- Time variation in apparent brightness of individual images.
- Time delay between images.
- Image shape (for lensed extended sources).

Of course, the observable properties of a gravitational lens system depend on the size of the source, "strength" of the lens (i.e., its mass), and distances involved. Consequently, the above list of observables does not have to be measurable or applicable to a given lens system. By now a whole list of various observed phenomena related to lensing is known. In addition to the original discovery of a multiply imaged quasar, a large number of other realizations of the lens effect has been found (Chapter 5).

Degeneracies in Observables

Observations of a gravitational lens system with, for example, multiple images of a background quasar provide a number of constraints for a theoretical model of this system: relative image positions, magnification ratios, and time delays. However, as Falco, Gorenstein and Shapiro [Fal-GS, Goren-FS] first showed, there are transformations between the model parameters that leave these observables invariant. In other words, observations of relative image positions, magnification ratios, and time delays can never uniquely determine the model parameters. In [Fal-GS], it was shown that adding a uniform sheet of matter, reducing the value of the Hubble constant, and changing the (unobservable) source position leaves the positions, the relative magnifications, and the time delays between pairs of images unchanged. When applied to the double quasar Q0957+561, this implies that a measurement of the time delay cannot be used uniquely to determine the Hubble constant. However, coupled with additional data (e.g., a measurement of velocity dispersion of of the lensing galaxy), the degeneracy can be broken [Fal-SMD] (see page 149 of Section 5.1.1).

In their systematic and detailed discussion of various parameter degeneracies, the authors [Goren-FS] distinguished four different invariant transformations: prismatic, similarity, magnification, and combined transformations. For example, a prismatic transformation translates the source position, by a constant vector,

$$\mathbf{y} \to \mathbf{y}' = \mathbf{y} - \boldsymbol{\alpha}_0.$$

and adds a constant contribution to the bending vector:

$$\boldsymbol{\alpha}(\mathbf{x}) \to \boldsymbol{\alpha}'(\mathbf{x}) = \boldsymbol{\alpha}(\mathbf{x}) + \boldsymbol{\alpha}_0.$$

The latter means that the potential transforms as (up to a constant)

$$\psi(\mathbf{x}) \to \psi'(\mathbf{x}) = \psi(\mathbf{x}) + \boldsymbol{\alpha}_0 \cdot \mathbf{x}.$$

A prismatic transformation leaves the image positions, magnification ratios, and time delays invariant [Goren-FS, p. 698]. In fact, the position $\mathbf{x}$ of an image is determined from the lens equation (3.94),

$$\mathbf{x} = \mathbf{y} + \boldsymbol{\alpha}(\mathbf{x}).$$

If $\mathbf{y}$ and $\boldsymbol{\alpha}(\mathbf{x})$ are replaced by $\mathbf{y}'$ and $\boldsymbol{\alpha}'(\mathbf{x})$, respectively, then the lensed image position (hence, relative positions $\mathbf{x}_i - \mathbf{x}_j$) is unchanged: $\mathbf{x} \to \mathbf{x}' = \mathbf{x}$. From the potential ψ', it is also clear that the magnification matrix (hence, image magnifications and magnification ratios) is unaffected.

Turning to the time delay $\Delta\tau_{ij}$ between lensed images $\mathbf{x}_i$ and $\mathbf{x}_j$, we have from (3.103) that

$$\Delta\tau_{ij} = \mathcal{T}_{d_S\mathbf{y}}(d_L\mathbf{x}_i) - \mathcal{T}_{d_S\mathbf{y}}(d_L\mathbf{x}_j).$$

Explicitly,

$$\Delta\tau_{ij} = \frac{(1+z_L)d_Ld_S}{cd_{L,S}}\left\{\frac{|\mathbf{x}_i - \mathbf{y}|^2 - |\mathbf{x}_i - \mathbf{y}|^2}{2} - [\psi(\mathbf{x}_i) - \psi(\mathbf{x}_j)]\right\}.$$

The lens equation yields

$$\mathbf{x}_i - \mathbf{y} = \boldsymbol{\alpha}(\mathbf{x}_i).$$

Hence,

$$\begin{aligned} |\mathbf{x}_i - \mathbf{y}|^2 - |\mathbf{x}_j - \mathbf{y}|^2 &= [(\mathbf{x}_i - \mathbf{y}) - (\mathbf{x}_j - \mathbf{y})] \cdot [(\mathbf{x}_i - \mathbf{y}) + (\mathbf{x}_j - \mathbf{y})] \\ &= [\mathbf{x}_i - \mathbf{x}_j] \cdot [\boldsymbol{\alpha}(\mathbf{x}_i) - \boldsymbol{\alpha}(\mathbf{x}_j)]. \end{aligned}$$

Thus,

$$\Delta\tau_{ij} = \frac{(1+z_L)d_Ld_S}{cd_{L,S}}\left\{\frac{[\mathbf{x}_i - \mathbf{x}_j] \cdot [\boldsymbol{\alpha}(\mathbf{x}_i) + \boldsymbol{\alpha}(\mathbf{x}_j)]}{2} - [\psi(\mathbf{x}_i) - \psi(\mathbf{x}_j)]\right\}.$$

Let us apply a prismatic transformations to $\Delta\tau_{ij}$. We already know that the relative image positions $\mathbf{x}_i - \mathbf{x}_j$ are invariant under such a transformation:

$$\mathbf{x}_i - \mathbf{x}_j \to \mathbf{x}'_i - \mathbf{x}'_j = \mathbf{x}_i - \mathbf{x}_j.$$

Now note that

$$\frac{[\mathbf{x}'_i - \mathbf{x}'_j] \cdot [\boldsymbol{\alpha}'(\mathbf{x}_i) + \boldsymbol{\alpha}'(\mathbf{x}_j)]}{2} = \frac{[\mathbf{x}_i - \mathbf{x}_j] \cdot [\boldsymbol{\alpha}(\mathbf{x}_i) + \boldsymbol{\alpha}(\mathbf{x}_j)]}{2} + \boldsymbol{\alpha}_0 \cdot (\mathbf{x}_i - \mathbf{x}_j)$$

and

$$\psi'(\mathbf{x}_i) - \psi'(\mathbf{x}_j) = \psi(\mathbf{x}_i) - \psi(\mathbf{x}_j) + \boldsymbol{\alpha}_0 \cdot (\mathbf{x}_i - \mathbf{x}_j).$$

Hence, replacing $\boldsymbol{\alpha}$ and ψ in the time delay $\Delta\tau_{ij}$ by $\boldsymbol{\alpha}'$ and ψ', respectively, leaves $\Delta\tau_{ij}$ unchanged. Thus, the time delay between lensed images is invariant under a prismatic transformation. Readers are referred to [Goren-FS] for a detailed treatment of the other invariant transformations.

3.5 Weak Lensing by Large Scale Structures

Weak lensing by the large scale structure of the universe deals with a gravitational lensing regime where no multiple images or caustics occur. Rather it touches on the statistics of the weak distortions[45] (namely, magnification and shearing) of background galaxies by the large scale structure density perturbations of the universe. These statistics can then be used, for example, to constrain models of the large scale structure density perturbations. The idea of using weak lensing to probe the large scale structure was suggested by Gunn [Gun67a, Gun67b] in 1967. The inclusion of power-spectrum methods was initiated by Blandford [Bla90], Blandford et al. [Bla-SBV], and Miralda–Escudé [Mir], and further developed by several authors (Kaiser [Kai92, Kai98], Seljak [Sel94, Sel96], Bar-Kana [Bar], Villumsen [Vill], Jain and Seljak [Jain-Sel], etc.). Readers are referred to the previous papers and the recent weak-lensing reviews by Bartelmann and Schneider [Bart-S] and Mellier [Mel] for more.

Assumptions

In weak lensing by the large scale structure of the universe, we follow a thin bundle of light rays rather than an individual light ray (as was done previously). To get a handle on the cross sectional distortions of a light ray bundle, the following physically reasonable assumptions in **WL1-2** are made (e.g., [Bart-S, p. 391]):

WL1: The spacetime geometry for weak lensing is approximated by the following first-order perturbation of the Friedmann metric:[46]

$$\mathbf{g}_{lss}^{\mathrm{K}} = \tilde{a}^2(\mathrm{t})\left[-\left(1+2\frac{\phi}{c^2}\right)d\mathrm{t}^2+\left(1-2\frac{\phi}{c^2}\right)\left(d\chi^2+R_{\mathrm{K}}^2(\chi)d\omega^2\right)\right], \tag{3.125}$$

where $\tilde{a}$ is the scaling function, t is conformal time, $d\omega^2 = d\theta^2 + \sin^2\theta d\varphi^2$, and

$$R_{\mathrm{K}}(\chi) = \begin{cases} \mathrm{K}^{-1/2}\sin(\mathrm{K}^{1/2}\chi) & \text{if } \mathrm{K} > 0 \\ \chi & \text{if } \mathrm{K} = 0 \\ (-\mathrm{K})^{-1/2}\sinh[(-\mathrm{K})^{1/2}\chi] & \text{if } \mathrm{K} < 0. \end{cases} \tag{3.126}$$

Here $R_{\mathrm{K}}(x)$ has dimension of length, while $\tilde{a}(\mathrm{t})$ is dimensionless. It is assumed in (3.125) that (i) mass density perturbations have velocities much less than c (e.g., galaxies in a cluster have velocity of order 1,000 km/s); (ii) Newtonian potentials ϕ obey $|\phi|/c^2 \ll 1$ (in fact, $|\phi|/c^2$ is at most of order 10^{-4} for typical lensing situations); and (iii) the scale of

[45] Without lensing, the orientations of background galaxies is random, while with lensing the orientations tend to be tangential relative to the density perturbations (e.g., [Bla-SBV, p. 601]).

[46] Note that in (3.125) we are *not* assuming that $\mathrm{K} = 1, 0, -1$ as was done earlier for K in (3.31) (page 45); however, this can always be accomplished via coordinate changes — see [dIn, p. 318]. Also, the key differences between (3.125) and (3.31) lie in how ϕ is affected by the assumptions for the two lensing scenarios.

each mass density perturbation of interest is much smaller than the Hubble radius (e.g., a cluster of galaxies is of order one megaparsec, while the Hubble radius is of order 3,000 megaparsecs).

As in strong lensing,[47] the potential ϕ obeys the cosmological Poisson equation (e.g., Futamase and Sasaki [Fut-S, 2505]):

$$\Delta^{(3D)}\phi(\tau, \mathrm{x}) = 4\pi G\, \tilde{a}^2(\mathrm{t}(\tau))\, \varrho_{\mathrm{unv}}^{\mathrm{av}}(\tau)\, \delta(\tau, \mathrm{x}),$$

where $\Delta^{(3D)}$ is the Laplacian relative to the comoving coordinates, $\varrho_{\mathrm{unv}}^{\mathrm{av}}$ is the average mass density of the universe, and δ is the density contrast (see page 50).

WL2: It is assumed that the *change* in the comoving separation between any two rays in a thin light ray bundle is so much smaller than the comoving separation of their unperturbed paths,[48] that the difference of the potential's perpendicular gradients[49] at the two rays can be calculated along the unperturbed paths and for their unperturbed separation [Bart-S, p. 390].[50] This is called the **Born approximation** (terminology originating from the first Born approximation in scattering theory; e.g., [Arf-W, p. 521]).

Lensing Map for Large Scale Structures

Consider a thin light ray bundle emanating from an observer at $\chi = 0$. The weak lensing due to the large scale structures of the universe will cause the bundle's cross section to distort slightly. To investigate the distortion, we first define a two-dimensional "angular vector" $\boldsymbol{\theta}$ that will parametrize all the light rays in the bundle relative to a fixed ray in the bundle called a *fiducial ray*. Denote by $\mathbf{T}_{fd}$ the unit tangent vector in the direction of motion to a fiducial ray's spatial path at the cosmic time when the bundle leaves the observer. Suppose that another light ray in the thin bundle has spatial unit tangent $\mathbf{T}$ in the direction of motion. In the spatial (three-dimensional) tangent space at the observer, we have $\mathbf{T} = \mathrm{proj}_{\mathbf{T}_{fd}} \mathbf{T} + \mathrm{proj}_{\mathbf{T}_{fd}^{\perp}} \mathbf{T}$, where $\mathbf{T}_{fd}^{\perp}$ is the plane orthogonal to $\mathbf{T}_{fd}$ and the notation $\mathrm{proj}_{\mathbf{b}}\, \mathbf{a}$ refers to the orthogonal projection of $\mathbf{a}$ to $\mathbf{b}$. The angle θ_0 between $\mathbf{T}$ and $\mathbf{T}_{fd}$ satisfies $|\theta_0| \ll 1$ (since the bundle is thin), so $\mathrm{proj}_{\mathbf{T}_{fd}} \mathbf{T} = (\cos\theta_0)\mathbf{T}_{fd} \simeq \mathbf{T}_{fd}$. In other words, $\mathbf{T} \simeq \mathbf{T}_{fd} + \boldsymbol{\theta}$, where $\boldsymbol{\theta} \equiv \mathrm{proj}_{\mathbf{T}_{fd}^{\perp}} \mathbf{T}$ and $|\boldsymbol{\theta}| \simeq \theta_0$. We then think of $\boldsymbol{\theta}$ as an *angular vector* determining a point on the unit sphere $\mathbb{S}^2$ about the observer, namely, the direction unit vector $\mathbf{T}$ of a light ray emanating from the observer at an angle $|\boldsymbol{\theta}|$ relative to the fiducial ray. Note that the fiducial ray is determined by $\boldsymbol{\theta} = \mathbf{0}$. Hence, the angular vector $\boldsymbol{\theta}$ parametrizes the thin bundle of light rays emanating from the observer.

---

[47]See equation (3.40), page 53.

[48]That is, the rays' paths in the absence of lensing, which are null geodesics of the Friedmann metric.

[49]These gradients are essentially with respect to directions transverse to the χ-direction.

[50]A similar approximation was used on page 62 for single ray lensing, where the integral of the transverse potential gradient along a light ray was approximated by one over a straight line (the ζ-axis).

Next we specify a useful parametrization along the light rays. For the background Friedmann metric (i.e., $\mathbf{g}^{K}_{lss}$ with $\phi \equiv 0$), the null geodesics emanating from the observer at $\chi = 0$ move radially with $\chi = \mathtt{t}_0 - \mathtt{t}$, where $\mathtt{t}_0$ is the present conformal time and $\mathtt{t}_0 > \mathtt{t}$. In light of the Born approximation, it is convenient to parametrize the light rays of $\mathbf{g}^{K}_{lss}$ by χ as well (e.g., [Jain-Sel, p. 571]). In fact, null geodesics of $\mathbf{g}^{K}_{lss}$ obey $d\mathtt{t} = (1 - 2\phi/c^2)d\ell$, where $d\ell$ is the comoving path length of the geodesic; use of the Friedmann parametrization then relates $d\ell$ to $d\chi$ via $d\chi = (1 - 2\phi/c^2)d\ell$.

The cross sectional distortion of the light ray bundle will now be considered using the separation between two arbitrary rays in the bundle. We shall follow the light ray $l_{\boldsymbol{\theta}}$ determined by $\boldsymbol{\theta}$ and the fiducial light ray $l_{\mathbf{0}}$ (given by $\boldsymbol{\theta} = \mathbf{0}$) from the observer ($\chi = 0$) to some coordinate location χ across the cosmos. Let $\mathbb{R}(\boldsymbol{\theta}, \chi)$ be the two-vector giving the proper transverse separation between $l_{\boldsymbol{\theta}}$ and $l_{\mathbf{0}}$ at χ. Here the separation is transverse relative to the radial direction χ (which is the overall direction of the light bundle). Let $l'_{\boldsymbol{\theta}}$ and $l'_{\mathbf{0}}$ be $l_{\boldsymbol{\theta}}$ and $l_{\mathbf{0}}$, respectively, in the absence of lensing (i.e., $l'_{\boldsymbol{\theta}}$ and $l'_{\mathbf{0}}$ are Friemann null geodesics). The proper transverse separation between $l'_{\boldsymbol{\theta}}$ and $l'_{\mathbf{0}}$ at χ is $\mathbb{R}'(\boldsymbol{\theta}, \chi) \simeq d_A(\chi)\boldsymbol{\theta}$, where $d_A(\chi) = \tilde{a}(\chi)R_{\mathrm{K}}(\chi)$ with $\tilde{a}(\chi)$ the value of the scaling function at $\mathtt{t} = \mathtt{t}_0 - \chi$. Note that $d_A(\chi)$ is the Friedmann angular diameter distance from the observer to χ (see (3.21), page 39). Furthermore, the Friedmann null geodesics $l'_{\boldsymbol{\theta}}$ and $l'_{\mathbf{0}}$ can be expressed as

$$l'_{\boldsymbol{\theta}}(\chi) = (\mathtt{t}(\chi), R_{\mathrm{K}}(\chi)\boldsymbol{\theta}, \chi), \qquad l'_{\mathbf{0}}(\chi) = (\mathtt{t}(\chi), \mathbf{0}, \chi),$$

where $\mathtt{t}(\chi) = \mathtt{t}_0 - \chi$ and $R_{\mathrm{K}}(\chi)\boldsymbol{\theta} = \mathbb{R}'(\theta, \chi)/\tilde{a}(\chi)$ is the comoving separation at χ between $l'_{\boldsymbol{\theta}}$ and $l'_{\mathbf{0}}$.

We are interested in the angular vector difference between the proper transverse separation for lensing and without lensing, that is, the two-vector

$$\boldsymbol{\alpha}_{lss}(\boldsymbol{\theta}, \chi) \equiv \frac{\mathbb{R}'(\boldsymbol{\theta}, \chi) \; - \; \mathbb{R}(\boldsymbol{\theta}, \chi)}{d_A(\chi)}.$$

The vector $\boldsymbol{\alpha}_{lss}(\boldsymbol{\theta}, \chi)$ is the **angular relative deflection vector** at χ of $l_{\boldsymbol{\theta}}$ relative to $l_{\mathbf{0}}$ due to weak lensing. Using the assumptions in **WL1-2**, it can be shown that (e.g., [Bart-S, pp. 389–390]):

$$\boldsymbol{\alpha}_{lss}(\boldsymbol{\theta}, \chi) = \frac{2}{c^2}\int_0^{\chi} d\chi \; \frac{R_{\mathrm{K}}(\chi - \chi')}{R_{\mathrm{K}}(\chi)} \; \left[(\nabla_{\perp}\phi)(l'_{\boldsymbol{\theta}}(\chi')) \; - \; (\nabla_{\perp}\phi)(l'_{\mathbf{0}}(\chi'))\right]. \tag{3.127}$$

Here the transverse gradient $\nabla_{\perp}$ is the two-dimensional gradient with respect to coordinates $(x_1, x_2) = (R_{\mathrm{K}}(\chi')\theta_1, R_{\mathrm{K}}(\chi')\theta_2)$, where $\boldsymbol{\theta} = (\theta_1, \theta_2)$.

For fixed χ', Taylor expands ϕ to the second-order in the transverse direction $\mathrm{x}_{tr} \equiv (x_1, x_2)$ relative to the fiducial ray

$$\begin{aligned}\phi(\mathtt{t}(\chi'), \mathrm{x}_{tr}, \chi) \;&=\; \phi(\mathtt{t}(\chi'), \mathbf{0}, \chi') + \mathrm{x}_{tr} \cdot [(\mathrm{grad}_{\mathrm{x}_{tr}}\, \phi)(\mathtt{t}(\chi'), \mathbf{0}, \chi')] \\ &\quad + \frac{1}{2}\, \mathrm{x}_{tr}\, [(\mathrm{Hess}_{\mathrm{x}_{tr}}\, \phi)(\mathtt{t}(\chi'), \mathbf{0}, \chi')]\, (\mathrm{x}_{tr})^T,\end{aligned} \tag{3.128}$$

where $\mathbf{x}_{tr}^T$ is the transpose of $\mathbf{x}_{tr}$, while $\mathrm{grad}_{\mathbf{x}_{tr}}\,\phi$ and $\mathrm{Hess}_{\mathbf{x}_{tr}}\,\phi$ are, respectively, the gradient and Hessian matrix of ϕ relative to the coordinates $\mathbf{x}_{tr}$. Note that $\mathrm{grad}_{\mathbf{x}_{tr}} = \nabla_\perp$. Approximate ϕ using (3.128) and take the transverse gradient $\nabla_\perp$, where $\nabla_\perp = \mathrm{grad}_{\mathbf{x}_{tr}}$, we obtain

$$(\nabla_\perp\phi)(l'_\theta(\chi')) - (\nabla_\perp\phi)(l'_0(\chi')) = [(\mathrm{Hess}_{\mathbf{x}_{tr}}\,\phi)(l'_0(\chi')]\,\mathbf{x}_{tr}^T. \tag{3.129}$$

The **shear tensor** ([Kai92, p. 273]) associated with (3.127) is defined to be the Jacobian matrix of $\boldsymbol{\alpha}_{lss}$ relative to $\boldsymbol{\theta}$:

$$\mathbf{S}_{lss}(\boldsymbol{\theta},\chi) \equiv (\mathrm{Jac}_{\boldsymbol{\theta}}\ \boldsymbol{\alpha}_{lss})(\boldsymbol{\theta},\chi).$$

Using (3.129), the components of $\mathbf{S}_{lss}$ are

$$\mathbf{S}_{lss}^{ij}(\boldsymbol{\theta},\chi) \equiv \frac{2}{c^2}\int_0^\chi d\chi\ \frac{R_K(\chi')\,R_K(\chi-\chi')}{R_K(\chi)}\ \frac{\partial^2\phi}{\partial x_i\partial x_j}(l'_0(\chi')). \tag{3.130}$$

Readers are referred to Bar-Kana [Bar] for a detailed treatment, including a scenario that incorporates a strong lens.

Now, defining

$$\boldsymbol{\beta}_{lss}(\boldsymbol{\theta},\chi) \equiv \frac{\mathbb{R}(\boldsymbol{\theta},\chi)}{d_A(\chi)},$$

we obtain that

$$\boldsymbol{\beta}_{lss}(\boldsymbol{\theta},\chi) = \boldsymbol{\theta} - \boldsymbol{\alpha}_{lss}(\boldsymbol{\theta},\chi), \tag{3.131}$$

where $\boldsymbol{\alpha}_{lss}$ is given by (3.127). *Unless stated otherwise, we shall assume that χ is fixed and notationally suppress its appearance.* Equation (3.131) then becomes

$$\boldsymbol{\beta}_{lss}(\boldsymbol{\theta}) = \boldsymbol{\theta} - \boldsymbol{\alpha}_{lss}(\boldsymbol{\theta}). \tag{3.132}$$

For fixed χ, we can then view $\boldsymbol{\beta}_{lss}$ as defining a mapping from a patch (parametrized by $\boldsymbol{\theta}$) on the unit celestial sphere $\mathbb{S}^2$ about the observer to a patch on the celestial sphere at χ.[51] The large scale structure lensing map $\boldsymbol{\beta}_{lss}$ has the same mathematical form as the usual single-plane lensing map (see (3.104), page 81):

$$\boldsymbol{\eta}(\mathbf{x}) = \mathbf{x} - \boldsymbol{\alpha}(\mathbf{x}).$$

This allows certain mathematical facts about $\boldsymbol{\eta}$ to be applied to $\boldsymbol{\beta}_{lss}$.

[51] The reader is reminded that $\boldsymbol{\beta}_{lss}$ is not unique (since it is defined relative to an arbitrarily chosen fiducial ray in the bundle).

Convergence, Shear, Magnification, and the Power Spectrum

With a lensing map $\boldsymbol{\beta}_{lss}$ available, we can proceed as in Section 3.2.7 (page 95) to investigate how a small extended source at χ, seen from the observer along the direction $\boldsymbol{\theta}$, is distorted by the lensing action of $\boldsymbol{\beta}_{lss}^{-1}$. Notice that since there are no multiple images or caustics, the mapping $\boldsymbol{\beta}_{lss}$ is a 1-1 correspondence and its Jacobian matrix Jac $\boldsymbol{\beta}_{lss}$ is invertible. Moreover, since the shear tensor $\mathbf{S}_{lss}$ is symmetric (by (3.130)) and because Jac $\boldsymbol{\beta}_{lss} = \mathbf{I} - \mathbf{S}_{lss}$, we see that Jac $\boldsymbol{\beta}_{lss}$ is *symmetric* (which also holds for Jac $\boldsymbol{\eta}$).

The lensing action of $\boldsymbol{\beta}_{lss}^{-1}$ can be approximated by $[\text{Jac}\,\boldsymbol{\beta}_{lss}]^{-1}$. As in the single-plane case, a disc-shape source of radius r is lensed into an ellipse shaped image with major and minor axes having the following lengths, respectively (see page 98):

$$a_\theta = \frac{2r}{|1-\kappa_e(\boldsymbol{\theta}) - |\boldsymbol{\gamma}_e(\boldsymbol{\theta})||}, \qquad b_\theta = \frac{2r}{|1-\kappa_e(\boldsymbol{\theta}) + |\boldsymbol{\gamma}_e(\boldsymbol{\theta})||}. \tag{3.133}$$

Let us explain the notation in (3.133). The *effective convergence* κ_e is given by (cf. (3.101), page 80):[52]

$$\kappa_e(\boldsymbol{\theta}) \equiv \frac{1}{2}\,\text{div}_\theta(\boldsymbol{\alpha}_{lss})(\boldsymbol{\theta}) = \frac{1}{c^2}\int_0^\chi d\chi\,\frac{R_K(\chi')\,R_K(\chi-\chi')}{R_K(\chi)}\,(\Delta_{\mathrm{x}_{tr}}\phi)(l_0'(\chi')).$$

Here div_θ is the divergence operator relative to $\boldsymbol{\theta}$, while $\Delta_{\mathrm{x}_{tr}}$ is the Laplace operator with respect to x_{tr}. Now introduce an *effective potential* ψ_e formally as follows (cf. (3.100), page 80):

$$\psi_e(\boldsymbol{\theta}) \equiv \frac{1}{\pi}\int_{\mathbf{R}^2} d\theta'\,\kappa_e(\boldsymbol{\theta}')\ln\left|\boldsymbol{\theta}'-\boldsymbol{\theta}\right|.$$

Note that $\Delta_{\boldsymbol{\theta}}\psi_e = 2\kappa_e$, where $\Delta_{\boldsymbol{\theta}}$ is the Laplace operator relative to $\boldsymbol{\theta}$. Then the *effective complex shear* $\boldsymbol{\gamma}_e$ is defined by (cf. (3.114), page 96):

$$\boldsymbol{\gamma}_e(\boldsymbol{\theta}) \equiv \frac{1}{2}\left[\frac{\partial^2\psi_e}{\partial\theta_1^2}(\boldsymbol{\theta}) - \frac{\partial^2\psi_e}{\partial\theta_2^2}(\boldsymbol{\theta})\right] + i\,\frac{\partial^2\psi_e}{\partial\theta_1\partial\theta_2}(\boldsymbol{\theta}),$$

where $\boldsymbol{\theta} = (\theta_1, \theta_2)$. Also, notice that the *effective signed magnification* μ_e^{sgn} obeys (cf. (3.117), page 97):

$$\mu_e^{sgn}(\boldsymbol{\theta}) \equiv \frac{1}{\det[\text{Jac}\,\boldsymbol{\beta}_{lss}](\boldsymbol{\theta})} = \frac{1}{[1-\kappa_e(\boldsymbol{\theta})]^2 - |\boldsymbol{\gamma}_e|^2(\boldsymbol{\theta})}.$$

One way of describing the distortion of the aforementioned ellipse shaped image is to use a *complex ellipticity* defined as follows:

$$\boldsymbol{\epsilon}_e(\boldsymbol{\theta}) = \frac{1-b_\theta/a_\theta}{1+b_\theta/a_\theta}\,e^{2i\varphi_e},$$

[52] Note that κ_e may take on positive and negative values.

where φ_e is the angle made by the major axis with the θ_1-axis.[53] Under the weak lensing approximation — that is, $|\kappa_e| \ll 1$ and $|\boldsymbol{\gamma}_e| \ll 1$ — the ellipticity, shear, convergence, and signed magnification are related (e.g., [Bart-S, p. 339, eq. (4.12); p. 396]):

$$\boldsymbol{\epsilon}_e(\boldsymbol{\theta}) = \frac{\boldsymbol{\gamma}_e(\boldsymbol{\theta})}{1 - \kappa_e(\boldsymbol{\theta})}, \qquad \mu_e^{sgn}(\boldsymbol{\theta}) \simeq 1 + 2\kappa_e(\boldsymbol{\theta}).$$

Assuming that the large scale structure of the universe is described by density fluctuations that are homogeneous, isotropic, and Gaussian (see the subsection on page 49), the weak lensing approximation yields that the auto-correlations for ellipticity, shear, convergence, and signed magnification are the same except for a factor of 4:

$$\begin{aligned} A_{\epsilon}(|\boldsymbol{\zeta}|) &\equiv \mathrm{E}[\boldsymbol{\epsilon}_e(\boldsymbol{\theta})\bar{\boldsymbol{\epsilon}}_e(\boldsymbol{\theta}+\boldsymbol{\zeta})] = \mathrm{E}[\boldsymbol{\gamma}_e(\boldsymbol{\theta})\bar{\boldsymbol{\gamma}}_e(\boldsymbol{\theta}+\boldsymbol{\zeta})] \\ &= \mathrm{E}[\kappa_e(\boldsymbol{\theta})\kappa_e(\boldsymbol{\theta}+\boldsymbol{\zeta})] = \frac{1}{4}\mathrm{E}[\mu_e^{sgn}(\boldsymbol{\theta})\mu_e^{sgn}(\boldsymbol{\theta}+\boldsymbol{\zeta})]. \end{aligned} \tag{3.134}$$

Here $\boldsymbol{\zeta}$ is an angular vector direction, E[.] is the expected value, and the over-bar denotes complex conjugate. In addition, the auto-correlation function A_ϵ is related to the power-spectrum of the density perturbations:

$$\begin{aligned} A_{\epsilon}(|\boldsymbol{\zeta}|) &= \frac{9H_0^4\Omega_0^2}{8\pi c^4}\int_0^{\chi_H} d\chi\, \frac{R_{\mathrm{K}}^2(\chi)\, \mathrm{W}^2(\chi,\chi)}{a^2(\chi)} \\ &\quad \times \int_0^{\infty} kdk\; P_{\delta}(\chi,k)\, \mathrm{J}_0(R_{\mathrm{K}}(\chi)k|\boldsymbol{\zeta}|), \end{aligned} \tag{3.135}$$

where χ_H is the horizon distance (i.e., the value of χ for infinite redshift), W is a weighting function involving the distance distribution of the sources (e.g., [Bart-S, p. 392]), P_δ is the mass-density power spectrum (page 50),[54] and J_0 is the zeroth Bessel function. Readers are referred, for example, to the review [Bart-S, p. 397] for proofs of (3.134) and (3.135).

Equation (3.135) relates the ellipticity auto-correlation function A_ϵ to the power spectrum P_δ of the density contrast δ of the universe. Models of the formation of large scale structures in the cosmos determine a certain form for P_δ that in turn affects the form of A_ϵ (e.g., [Bart-S, p. 398]). Since A_ϵ can be calculated from observations (e.g., [Bac], [Kai-WL], [van00], [Wittm00]), large scale structure weak lensing provides a way of constraining models of how those structures were formed. Readers should see [Bart-S] and references therein for further applications of weak lensing by large scale structures.

[53] The complex ellipticity $\boldsymbol{\epsilon}_e$ can be generalized to images of lensed non-disc-shaped sources by using the second brightness moments of the image (e.g., [p. 339]Bar-Schneider).

[54] Note that the χ coordinate in $P_\delta(\chi,k)$ determines the dependence of P_δ on cosmic time (cf.[Bart-S, p. 325]).

Chapter 4

Physical Applications

If we want to know whether a particular way of talking about the world is valuable, we look for the predictions the theory makes and compare these with observations.

BRUCE GREGORY

Gravitational lensing has changed in the past two decades from an interesting curiosity into a useful astrophysical tool. In fact, it has a lot of important applications, ranging from studies in the solar system, to galactic structure, to probing cosmological objects. Some of the practical consequences in lensing have been known for a long time and have been predicted early on, whereas other applications popped up only recently. In fact, gravitational lensing is one of the few branches in astronomy/astrophysics, which is conceptually simple enough that one can use it to make quantitative predictions. This feature of being able to calculate or prove a result and then go out and confirm/detect it observationally — generally hailed as one of the principles of good science — has produced some particularly nice examples in lensing, with the Einstein light deflection at the solar limb being the most famous one.

In Section 4.1, we overview Zwicky's seminal work on how lensing can be used to probe the universe. The remaining sections address how to employ lensing to determine the nature of dark matter (Section 4.2), the structure of quasars (Section 4.3), Hubble's constant (Section 4.4), limits on cosmology (Section 4.5), and exotic objects like cosmic strings (Section 4.6).

4.1 Probing the Universe with Gravitational Lensing

Gravitational lenses are very useful astrophysical tools. In Section 3.1 the three ingredients of gravitational lens systems were introduced. Through astronomical observations, one can learn about and investigate in detail all three "participants" in a gravitational lens scenario: the source, the lens, and the geometry of the universe.

The matter along the line of sight to distant objects can act like a magnifying glass on all sources behind: This use of lensing as a "telescope" helps us to see faint sources bright, and to detect and study sources which would be too faint for detection without being lensed. The unique signatures of gravitational lensing events allow us to identify the lens systems, to measure their individual properties, and to analyze the occurrence of lensing statistically. Ultimately, lensing should even give us answers to the questions on size, age and geometry of the universe.

The Lens According to Zwicky

Fritz Zwicky led the way to direct applications of the gravitational lens effect [Zwi37a]. He pointed out three reasons why lensing might turn out to be of interest:

- Lensing would provide an additional test for general relativity.
- Lensing would enable us to see nebulae at distances farther than ordinarily achievable (with all its consequences concerning cosmology).
- Lensing can be used to directly measure the masses of galaxies.

These statements have been repeated quite often since. The second point is certainly the safest of these statements: Due to the magnification effect of lensing, an object that would be too faint to be detected can be brightened enough to be visible in a flux-limited sample. For example, this has been exploited in some of the giant luminous arcs. These are galaxies at high redshifts, and only because of the lens-induced magnification (and even then requiring great effort), was it possible to take spectra and study galactic and stellar properties at high redshifts. However, since the exact magnification rate is usually not known, and since the magnification furthermore is not necessarily constant across the whole source (parts of the galaxy could get a higher magnification than others), it is not straightforward to draw conclusions about the underlying source population.

Zwicky's first and third items are a little bit more delicate. In fact, they are intertwined: If one believes general relativity, then one can get an estimate of the mass of the lensing objects. Lens systems with four (or more) images of the background quasar allow a very precise determination of the lens mass. The first lens for which the mass could be determined with high accuracy was the quadruple quasar Q2237+0305. The mass inside the circle determined by the four images was found to better than 2% [Rix], [Wam-P].[1] At the time, this result was probably the best mass determination for any galaxy. More recently, lensing has been used for mass determinations and even reconstructions of mass profiles of

[1] This assumes a known value for the Hubble constant.

clusters of galaxies [Kai-S, Fah-KSW]. As the issue of cluster masses is far from settled, lensing is considered a very useful independent tool here.

Zwicky's first point — using lensing as an additional test for general relativity — worked so far only for the light deflection at the solar limb (which had been done when he wrote his paper). In this situation we know both the positions of the stars with and without deflection, and we know the mass of the lensing object — the sun — very accurately. In basically all the other cases of lensing, we do not have an independent accurate measurement of the lensing mass. Consequently, we currently cannot confirm general relativity independently[2] for such lensing effects. However, we are able to measure the velocity dispersions of galaxies, which are a rough measure of the galaxy mass. It is fair to say that these masses and the measured deflection angles are consistent with the general theory of relativity.

4.2 Dark Matter

One of the most pressing problems in astronomy/cosmology is: What is the nature of the dark matter? It has been known for decades that some observed phenomena ("flat" rotation curves of galaxies, apparently bound clusters of galaxies) can be explained only if there is more matter than we "see" directly there. More recently, the inflationary model of the early universe predicts the average density of the universe to be equal to the critical density (i.e., the cosmological density $\Omega_0 \simeq 1$). This means that the universe is flat (spatially Euclidean), just at the balance between an open universe which expands forever and a closed one which re-collapses some time in the future. This critical density is much higher than the sum of all the matter density in known objects, measured by a large variety of observations. Early on this was called the "missing mass problem;" by now people mostly refer to it as the "dark matter problem".

There are basically *two* dark matter problems: The first one concerns the normal or "baryonic" matter: Are there more baryons[3] in the universe than we can detect in stars/galaxies/clusters of galaxies via their emission of photons? Such dark baryonic matter could be hidden in a non-luminous or weakly luminous compact form (i.e., white dwarf stars, brown dwarfs, neutron stars). The second dark matter problem states: Is there any non-baryonic, "exotic" form of dark matter in the universe? Particle physicists have come up with a big list of possible candidates for non-baryonic dark matter, none of which has been confirmed experimentally yet. Indeed, if the density of the universe is equal to the critical one, then there *must* be non-baryonic dark matter. The reason is that the primordial nucleosynthesis in the very early universe produces the light elements (hydrogen, helium, lithium) in very

---

[2] In other words, if the deflection angle were only half the Einstein value, doubling the masses of the galaxies would result in the same configuration at the sky.

[3] Baryons are a certain class of elementary particles – comprising, e.g., protons and neutrons – that are subject to the strong interaction; baryons consist of three quarks.

characteristic mass ratios, which strongly depend on the total mass of the universe (Kolb and Turner [Kol-Tu]). From the observed relative abundances of these elements, it is obvious that the total amount of baryonic matter in the universe can comprise only a few percent of the critical density, the exact fraction depending on the value of the Hubble constant.

Gravitational lensing becomes an important tool in attacking the problem of dark matter, because it can detect a large variety of compact mass distributions. Lensing does not distinguish between luminous and dark, baryonic or exotic matter. Whenever there is a certain amount of spatially concentrated matter, light rays are attracted and deflected by its gravity. The concentrated mass distribution leaves its signature on the background light sources, and this can be measured.

There is emphasis though on the word "concentrated." If the elusive dark matter exists in the form of elementary particles, and is completely smoothly distributed, then its lensing effects (if any) are hard to detect, basically negligible. In the next section, we list several types of dark matter candidates that can be detected by lensing. Due to the different detection techniques, we split them into two groups: (compact) objects of stellar mass or smaller, and (extended) dark matter in galaxy clusters.

4.2.1 Detecting Compact Dark Objects and Planets

We shall describe several types of compact dark matter objects that can be found via gravitational lensing. These objects are known to exist. What is not known is whether they make up a significant fraction of the (baryonic) dark matter.

MACHOs in Halos of Galaxies

Stars and gas in the outer parts of galaxies move with a much higher velocity in their orbits than one would expect if they felt only the gravity of the visible matter. This requires, in the outskirts of these galaxies, the existence of a certain amount of matter that does not emit any visible light — hence called "dark matter." It is believed that the dark matter forms a spherical mass distribution — called a "halo" — that surrounds the visible galaxy. "Conservative" candidates[4] for this dark matter are low mass stars or "brown dwarfs," objects more massive than a planet, but not as massive as a star (a commonly assumed upper mass limit for brown dwarfs is 8% of the mass of the sun; objects with masses below that value cannot start hydrogen fusion in their central stellar core). As early as 1981, J. Richard Gott III suggested that these "heavy halos" around galaxies could be made of low mass stars and that gravitational lensing can be used as a test of this hypothesis [Got81].

The closest place to look for dark matter is in our own Milky Way. In 1986, Bohdan Paczyński realized that it can be tested observationally whether the dark halo of our own galaxy is made of such brown dwarfs or other compact dark objects with masses roughly in the range between planets and stars [Pac86a]. His idea is very simple and elegant: If

[4] "Conservative" in the sense that we know these candidates exist elsewhere in the universe.

our galaxy's halo is made of compact dark matter, then every now and then one of these objects (later labeled MACHOs by [Grie], for Massive Compact Halo Objects) would pass in front of a background star outside the Milky Way, and thus would magnify the apparent brightness of the background star in a very simple and predictable way, which depends only on the impact parameter between lens and source. The probability of such an event is quite small, only about one out of a million stars is significantly affected at any time by gravitational lensing. That means one has to measure the brightness of many million stars very often, ideally several times per night, to be able to pick this glowing needle in the cosmic haystack. A convenient part of the sky for such an experiment is the Large Magellanic Cloud (LMC), a small neighbor galaxy to the Milky Way, with its many individually measurable stars.

This idea, which sounded like science fiction originally, attracted the attention of a number of people. By 1991, three different groups set out to observe the MACHOs. They set up very effective telescope and computer facilities, and started to regularly monitor a large number of stars in the LMC, of order 10^7 stars. In 1993, the first detections of microlensing events were reported, and by now over 500 such events have been detected. The observations are discussed in detail in Section 5.5.

Binary Systems and Planets

Probably 50% or more of all stars are in binary or multiple stellar systems, that is, they have at least one nearby stellar companion. If such a system acts as a gravitational lens, it can – depending on the projected distance between the two objects – produce a very characteristic multi-peaked light curve of the background star (for very small separations, the two stars act effectively as one lens with the sum of their masses; for very large separations they act essentially as two isolated lenses). Such a light curve of a background star is generated because the double lens produces a characteristic magnification distribution, consisting of one or more closed caustic curves. A "binary lens event" looks very different from the smooth light curve produced by a single lens: It has one, two or even more caustic crossings, which result in light curves with multiple peaks, occasionally very high magnification events. It is not trivial, though, to reproduce the exact parameters of the double lens involved in such an event from an observed light curve. The reason is that quite a number of physical parameters are involved: mass ratio, separation, impact parameter, and "impact angle."

If the mass ratio between the two lenses deviates significantly from one, we reach the regime of planetary systems. A mass ratio of 10^{-3} is already characteristic of the relative masses of Jupiter and the sun, and 10^{-5} is not too far from the earth-sun mass ratio. For such extreme mass ratios, the effect of the stellar companion on the light curve is only a slight perturbation compared with the case of an isolated microlensing event (Mao and Paczyński [Mao-P]). But with a good time coverage and an accurate determination of brightness, it is feasible to detect galactic planets in this way in the near future (J.W. [Wam97]). Thus, the search for planets is a natural extension of a variant of the MACHO-type experiments that look towards stars at the center of the Milky Way. This planet detection method is complementary to other searches for planets. It has a few advantages as well:

- Planets out to distances of many kiloparsecs can be detected this way.
- It is possible to detect earth-mass planets from ground based telescopes with the microlensing technique.
- The microlensing technique is most sensitive to planet-star separations that correspond roughly to 1 to 3 times the earth-sun distance, which overlaps with the "habitable zone" of solar type stars.
- This planet search technique does not suffer from the selection effects of other methods (e.g., it is not preferentially sensitive to massive planets in close orbits around nearby stars); this means, in the long run, that the microlensing planet search technique is well suited for statistical investigations of how frequently planets of certain masses occur.

A disadvantage of this lensing method compared to other methods for the detection of planets is that the signature of a planet will be there only once for a short time, and it cannot be confirmed later. Furthermore, as the "signal" — the perturbation of the light curve — can be quite complicated, it is not clear how unique a planet microlensing event will be.

We now describe in more detail the geometry and the technique of the microlensing planet hunt.

Numerical Simulations of Microlensing by Planets

To illustrate the effect of a planetary companion on the microlensed light curve of a background star, we use two-dimensional magnification patterns and their corresponding light curves.

The first magnification pattern is for an isolated lens — top left panel of Figure 4.1. This is the magnification in the light source plane. The distribution is spherically symmetric around the (projected) position of the lens, with the magnification increasing monotonically in the direction towards the lens. The center is a degenerate point caustic (see Sections 6.3.3, page 187, for details).

In the top right panel of Figure 4.1, there is a magnification pattern similar to that in the left panel. After a close inspection, the reader may recognize that the right panel has a small deviation in the symmetry and monotony: Close to the center there is a caustic curve with several cusps. This is the magnification pattern of a system consisting of two objects m_1 and m_2 at positions -0.52 and 0.52, respectively, on the x-axis, which is in units of the angular Einstein radius for $M = m_1 + m_2 = 1$. The x-axis is in units of the angular Einstein

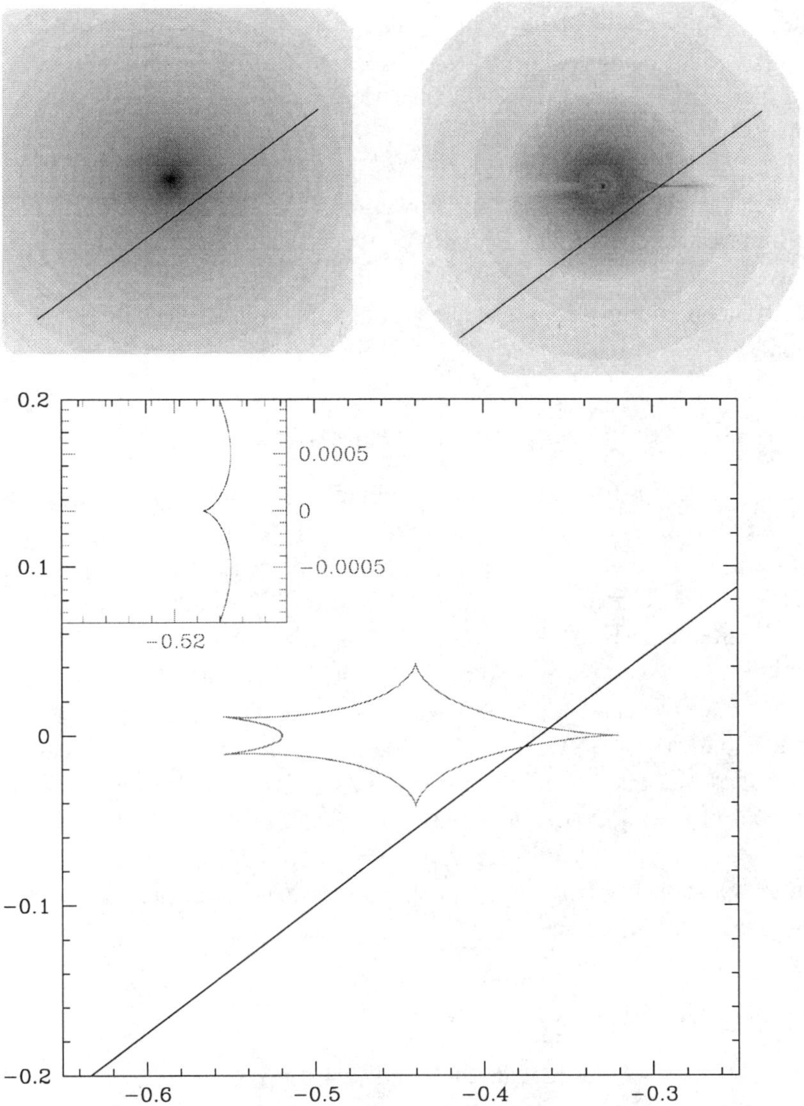

Figure 4.1. *Top panels*: Magnification patterns due to point mass or star *(top left)* and the star with a small planetary companion *(top right)*. Darker regions have higher magnification. The caustic in the top left panel is a point located at the dark spot. *Bottom panel*: The caustic associated with the magnification pattern in the top right panel. Notice the cusp in the close up of the leftmost part of the caustic. The straight lines indicate the relative motion of a background source.

radius. One object (star) dominates the mass, that is, $m_1 = 0.999$, while the companion's mass is only $m_2 = 0.001$.

The bottom panel of Figure 4.1, depicts the caustic associated with the magnification pattern in the top right panel of the figure. The caustic curve with cusps is now clearly visible.

All three panels in Figure 4.1 contain a straight line. These lines reflect arbitrarily positioned tracks along which the light curve of a background star is evaluated. The corresponding light curves are shown in Figure 4.2. These light curves reflect the microlensing action of a single stellar lens and binary lens consisting of a star and a planet. For both systems, the transverse velocity is 220 km/sec. Observed examples of lensing by a single stellar lens have durations between several days and a few hundred days.

A more extended numerical treatment of detecting planets by their effects on light curves is given by J.W. [Wam97].

Cosmologically Distributed Compact Dark Objects

If there are "dark" galaxies (or isolated massive black holes, for that matter) in the universe in large enough numbers to make up a fair fraction of the total matter density, gravitational lensing can detect them. For such large masses, the time scale for changing image configurations is quite large, compared to a human life time. This can be seen by plugging in typical values for the linear Einstein radius and transverse velocity of a galaxy — 5 kiloparsecs and 600 km/sec, respectively — into equation (3.111) (page 91). The resulting time scale is of order 10,000 years! So, we cannot measure the change of magnification as a function of time, but rather "static" experiments are applicable. Such objects would produce double images of background sources, namely, quasars, galaxies, and radio sources. Current telescopes are sensitive to a lens mass range between roughly $10^6 M_\odot$ and $10^{15} M_\odot$. Images would be detectable by different methods, ranging from interferometric imaging with joint radio telescopes (VLBI) for the small masses/separations to standard radio or optical techniques searching for multiple images.

4.2.2 Dark Matter in Galaxy Clusters

As early as 1933, Fritz Zwicky [Zwi33] noticed that the velocities of galaxies in clusters correspond to a total cluster mass that exceeds the luminous matter by more than an order of magnitude, if one assumes the clusters to be bound systems. This situation has not changed dramatically since then. However, gravitational lensing can now be used to actually detect and measure the dark matter in galaxy clusters, and even map its projected two-dimensional distribution on the lens.

Massive clusters of galaxies unavoidably act as lenses on background sources. The shapes and brightnesses of galaxies that are located *behind* clusters of galaxies must be affected by the gravitational distortion of the massive galaxy cluster in the foreground.

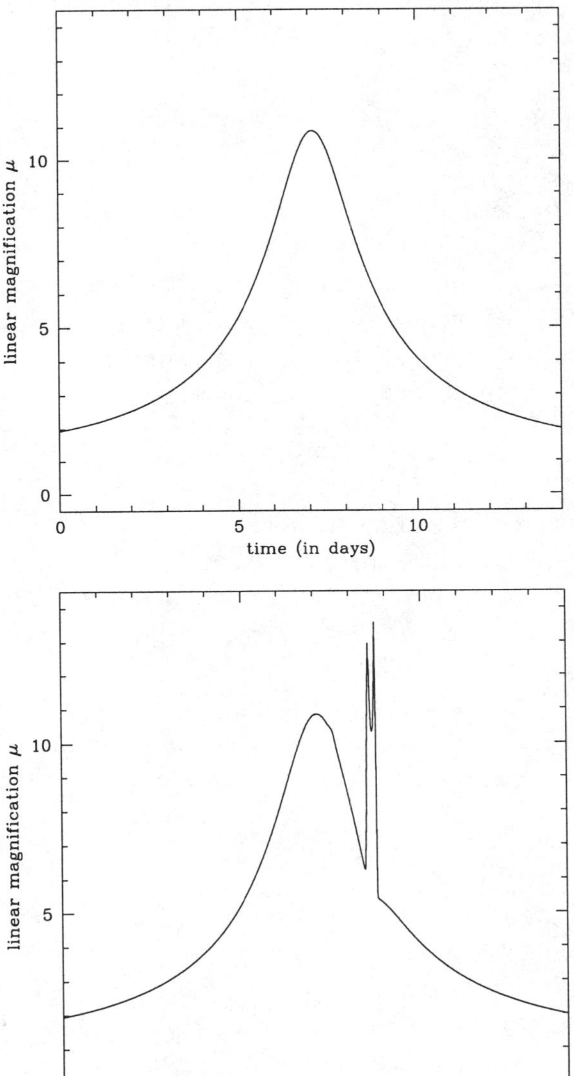

Figure 4.2. Light curves corresponding to the tracks starting from the lower left in Figure 4.1. The top panel shows the light curve for the track in the top left panel of Figure 4.1. The bottom panel depicts the light curve for the track in the top right or bottom panel of Figure 4.1. Notice that the spikes occur over less than a day, while the overall lensing event is shown takes about 15 days.

There are two "modes" of cluster lensing: the "strong" regime, which produces "Giant Luminous Arcs," and the "weak" regime, which deforms the images of background galaxies very slightly into arclets — tangentially aligned ellipses. Both these effects make it possible to determine the mass of the clusters.

Strong Lensing Regime: Cluster Mass via Giant Luminous Arcs

The giant arcs produced in the strong lensing case are located very close to the Einstein ring radius of the cluster. In the simplest model of a circularly symmetric mass distribution for the cluster, the large arcs will form approximately along the critical line, which marks the Einstein ring. So one can compute a rough estimate of the lensing mass by determining the radius of curvature and interpreting it as the linear Einstein radius R_E of the lens system. Explicitly, for a circularly symmetric matter distribution with Einstein ring of radius R_E, equation (3.97) shows that the total mass inside the angular radius $\theta_E = R_E/d_L$ is

$$M(\theta) = \pi (d_L \theta_E)^2 \sigma_{crt}, \tag{4.1}$$

where d_L is the angular diameter distance to the lens.

More detailed modeling of the lensing clusters which allows for asymmetry in the mass distribution via the visible galaxies plus an unknown dark matter component gives us more accurate determinations of the total cluster mass and its distribution. These elaborated models use a number of constraints, like the position of the different (up to five) images of one source, their magnification, orientation and parity. From the local perturbations of the images (or the lack thereof), one can estimate the influence of the mass in visible galaxies versus that in the more equally distributed dark matter. See Kneib et al. [Kne] for more.

Weak Lensing Regime: Cluster Mass via Arclets

Weak lensing, on the other hand, can be detected in a statistical sense, and used to reproduce the relative mass distribution in the galaxy cluster. Given the wide interests in this topic, we reserve Section 4.2.3 for its discussion. We add that weak and strong lensing methods consistently result in cluster masses that are much higher than the luminous masses, and thus confirm Zwicky's original hypothesis that there is an unknown form of dark matter in clusters of galaxies. In Sections 5.2 and 5.6 some exciting observed examples of both types of cluster lensing are presented.

4.2.3 Cluster Mass Reconstruction via Weak Lensing

Based on an elegant idea first proposed by Kaiser and Squires in 1993 [Kai-S], a number of mathematical methods have been developed in the last several years for the reconstruction of the mass distribution in galaxy clusters from the small distortion effects of the cluster on background galaxies. The goal is to obtain a two-dimensional map of the projected mass distribution (i.e., the normalized surface mass density or convergence $\kappa = \sigma/\sigma_{crit}$) of the galaxy cluster by using the information about the cluster potential hidden in the distorted images of background galaxies.

The Kaiser and Squires technique uses the fact that the shear (Γ_1, Γ_2) as well as the (dimensionless) surface mass density κ can be expressed as second partial derivatives of the lensing potential (see Section 3.2.7):

$$\kappa(\mathbf{x}) = \frac{\psi_{uu}(\mathbf{x}) + \psi_{vv}(\mathbf{x})}{2},$$

$$\Gamma_1(\mathbf{x}) = \frac{\psi_{uu}(\mathbf{x}) - \psi_{vv}(\mathbf{x})}{2},$$

$$\Gamma_2(\mathbf{x}) = \psi_{uv}(\mathbf{x}),$$

where $\mathbf{x} = (u, v)$. Taking the Fourier transform of κ, Γ_1, and Γ_2 yields

$$\hat{\kappa}(\mathbf{k}) = -\frac{(k_1^2 + k_2^2)\hat{\psi}(\mathbf{k})}{2},$$

$$\hat{\Gamma}_1(\mathbf{k}) = -\frac{(k_1^2 - k_2^2)\hat{\psi}(\mathbf{k})}{2},$$

$$\hat{\Gamma}_2(\mathbf{k}) = -k_1 k_2\, \hat{\psi}(\mathbf{k}).$$

The following relation between $\hat{\kappa}$, $\hat{\Gamma}_1$, and $\hat{\Gamma}_2$ immediately follows:

$$\hat{\kappa} = \frac{k_1^2 - k_2^2}{k_1^2 + k_2^2}\, \hat{\Gamma}_1 + \frac{2k_1 k_2}{k_1^2 + k_2^2}\, \hat{\Gamma}_2, \tag{4.2}$$

where $\mathbf{k} = (k_1, k_2)$.

We shall see that in weak lensing the components Γ_1 and Γ_2 are measurable, so their Fourier transforms $\hat{\Gamma}_1$ and $\hat{\Gamma}_2$ can be determined. Equation (4.2) then allows us to compute $\hat{\kappa}$. Taking the inverse Fourier transform of $\hat{\kappa}$ yields the desired surface mass density κ [Kai-S, p. 442]:

$$\kappa(\mathbf{x}) = \frac{1}{\pi} \int d\mathbf{x}'\, \mathrm{Re}[\mathcal{D}^*(\mathbf{x} - \mathbf{x}')\gamma(\mathbf{x}')], \tag{4.3}$$

where γ is the complex shear

$$\gamma(\mathbf{x}) = \Gamma_1(\mathbf{x}) + i\Gamma_2(\mathbf{x})$$

and $\mathcal{D}^*$ the complex conjugate of

$$\mathcal{D}(\mathbf{x}) = \frac{(v^2 - u^2) - 2iuv}{|\mathbf{x}|^4}, \qquad \mathbf{x} = (u, v).$$

Equation (4.3) expresses the surface mass density (or convergence) κ as a convolution with the complex kernel $\mathcal{D}$. The method works in practice because the shear field $\boldsymbol{\gamma}(\mathbf{x})$ can be observationally determined. To see this, we need some preliminary concepts and results (that will lead to (4.5) below).

The *complex ellipticity* of an elliptically shaped image is defined as

$$\boldsymbol{\epsilon}^I = \epsilon_1 + i\epsilon_2 = \frac{1-r^I}{1+r^I}\,e^{2i\theta}, \qquad r^I \equiv \frac{b^I}{a^I}.$$

The quantities a^I and b^I are the lengths of the major and minor axes, respectively, of the elliptical shaped image, while θ is the angle the major axis makes with the u-axis.

For a source that is a small disc of radius r not lying on a caustic, we have (see page 98):

$$a^I = \frac{2r}{|1-\kappa-|\boldsymbol{\gamma}||}, \qquad b^I = \frac{2r}{|1-\kappa+|\boldsymbol{\gamma}||}.$$

The eigenvalues of $\det[\operatorname{Jac}\boldsymbol{\eta}](\mathbf{x})$ are

$$\lambda_1 = 1-\kappa+\Gamma, \qquad \lambda_2 = 1-\kappa-\Gamma,$$

where $\Gamma = |\boldsymbol{\gamma}|$. For the majority of cases of interest, we suppose that $\lambda_2 > 0$, which implies $\lambda_1 > 0$ and, hence, $\det[\operatorname{Jac}\boldsymbol{\eta}] > 0$. Under such an assumption, we have the following complex ellipticity for a disc-shaped source (cf. 4.4)):

$$\boldsymbol{\epsilon}^I = \frac{\boldsymbol{\gamma}}{1-\kappa} \equiv \boldsymbol{\gamma}_{rd}.$$

The quantity $\boldsymbol{\gamma}_{rd}$ is called the *reduced shear.*

If the source is non-circular and $\lambda_2 > 0$ in the region of interest, then the ellipticity $\boldsymbol{\epsilon}^I$ of the image is related to the ellipticity $\boldsymbol{\epsilon}^S$ of the source as follows (Seitz and Schneider [Sei-S96, p. 390]):

$$\boldsymbol{\epsilon}^I = \frac{\boldsymbol{\epsilon}^S + \boldsymbol{\gamma}_{rd}}{1-\boldsymbol{\epsilon}^S\,\boldsymbol{\gamma}^*_{rd}}. \tag{4.4}$$

Applying (4.4) to sources with randomly distributed ellipticity distributions, we have

$$\left\langle |\boldsymbol{\epsilon}^S| \right\rangle = 0.$$

The average (over a finite area on the sky) of the magnitude of the ellipticity becomes (Schramm and Kayser [Schr-K, p. 7]):

$$\langle |\boldsymbol{\epsilon}| \rangle = |\boldsymbol{\gamma}_{rd}| = \frac{|\boldsymbol{\gamma}|}{1-\kappa}.$$

For weak lensing, that is,

$$\kappa \ll 1 \quad \text{and} \quad |\boldsymbol{\gamma}| \ll 1,$$

this results directly in the shear values

$$|\gamma| \simeq \langle|\epsilon|\rangle\,, \qquad \Gamma_1 \simeq \langle\epsilon_1\rangle\,, \qquad \Gamma_2 \simeq \langle\epsilon_2\rangle\,. \tag{4.5}$$

The determination of the shear is based on a measurement of the ellipticity of a large number of galaxies (of order thousands). In practice a number of difficulties arise. One is the so-called *mass-sheet degeneracy* (cf. Section 3.4, page 110). This relates to the fact that we obtain an identical ellipticity pattern for a lensing situation in which the convergence κ and magnitude of shear γ undergo the transformations

$$\gamma \rightarrow s\gamma$$

and

$$(1-\kappa) \rightarrow s(1-\kappa),$$

where s is an arbitrary constant. In other words, the reduced shear γ_{rd} is unchanged.

Another problem with determining the shear is that in the original description, the reconstruction method can be applied only in the weak lensing regime. In addition, the convolution described in (4.3) formally extends over the entire sky, but all astronomical data sets are only finite in extend. Furthermore, to obtain a physical surface mass density σ from the convergence $\kappa = \sigma/\sigma_{crt}$, one needs to know the value of the critical surface mass density σ_{crit} in (3.27) (page 42). This requires knowledge of the redshifts/distances of the lensed galaxies which is often not available, because they are usually very faint.

These problems have attracted a lot of attention in recent years, which made "cluster mass reconstruction" one of the most active branches in the field of gravitational lensing. One way to deal with the mass sheet degeneracy, for example, is to determine the magnification of a background source in addition to its shear [Broa-TP] [Bart-N]. The difficulty with the finite field size of the observational data has been treated in many different ways, see [Kai95], [Squ-K], [Schn95], [Bart95], and [Sei-S96]. Various generalizations of the reconstruction method into the strong lensing regime were provided (e.g., [Sei-S95] and [Squ-K]). Examples of reconstructed surface mass densities appear in Figures 5.4 (page 154) and 5.12 (page 168).

For lack of space we refer readers to more detailed accounts of cluster mass reconstruction: A good review on weak lensing can be found in [Nar-B]. The latest observational results on cluster mass reconstruction are summarized in [Mel]. In [Bart-S] a thorough, comprehensive treatment of the theory of weak lensing is given with a discussion of many different mass reconstruction methods.

4.3 Structure of Quasars via Microlensing

Quasars are assumed to consist of a hot accretion disk of matter around a massive black hole, and lie in the centers of certain (possibly all) galaxies. Quasars emit a huge amount of light in all wavelengths through the release of gravitational energy when this matter falls into the central black hole. The size of such an accretion disk around the central black hole, which supposedly emits most of the "continuum" light (in contrast to the broad or narrow emission lines which are produced much farther out), is roughly of order 10^{15} cm (it depends on the luminosity and black hole's mass). At cosmological distances of order 10^{28} cm, the disk subtends an angle of 2×10^{-8} arcsecs, which seems far beyond the resolving capabilities of even the most optimistic proposed future telescopes in space.

On the other hand, gravitational lensing provides a possibility to study the structure of quasars on such small scales. It works as follows: a massive galaxy acts as a gravitational lens on a moving background quasar and produces two (or more) "macro" images of the quasar. The galaxy consists of individual stars; each star acts as a "micro-lens." So the quasar's light is not deflected by a smooth mass distribution, but rather by a large number of (almost point) lenses. This ensemble of point lenses produces a complicated magnification distribution in the quasar plane (light source plane) — Figures 4.3.

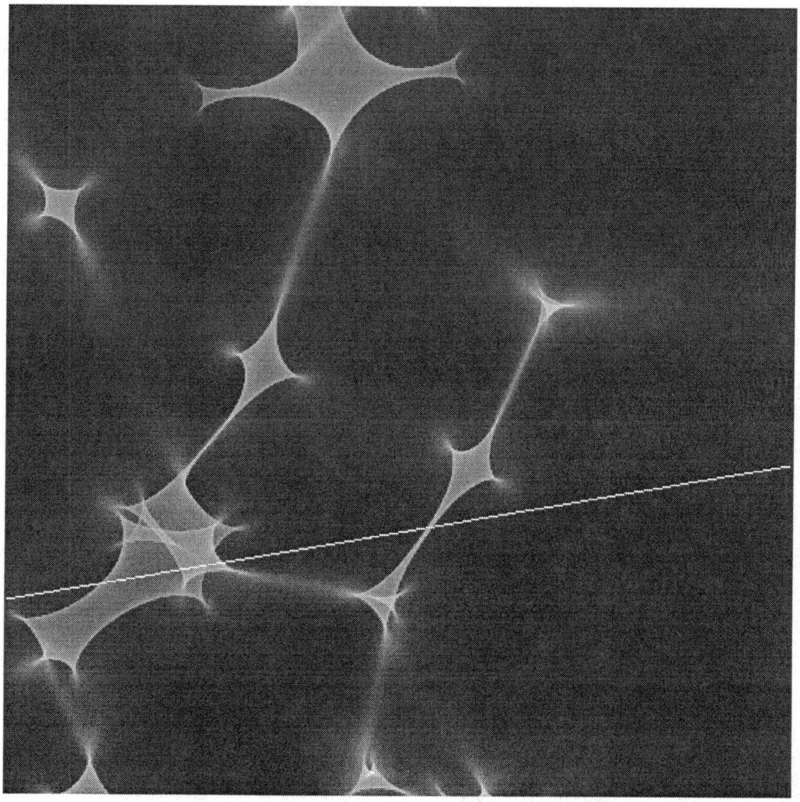

Figure 4.3. The straight white line represents the path of a quasar through a magnification pattern due to stars in a deflector galaxy. The grey scale ranges from low (darker) to high magnification (lighter).

Since observer, lens and quasar move relative to each other, the quasar encounters different parts of this magnification pattern at different times; occasionally the quasar even crosses one or more of the very sharp caustics — see the light curve in Figures 4.4. Locally these caustics are almost straight lines, and the magnification on one side of them is very high. So when the quasar crosses such a (fold) caustic, its finite size becomes important: The outer parts get magnified first, then the central parts.

Microlensing effects, that is, the lensing effects due to stars in a galaxy lens on a background moving light source, were first observed by Irwin et al. [Irw] in Image A of the quadruple-imaged quasar Q2237+0305. The signature of a microlensing caustic crossing event depends on the physical size of the quasar accretion disk, its brightness profile, and on the local geometry of the micro-caustic. If one can observe the brightness variations of a quasar for a couple of such caustic crossing events (see the light curve in Figure 4.4), presumably under various relative angles between the fold caustic and direction of motion, then in principle one can reconstruct the quasar brightness profile [Gri-KS]. This is similar to a "tomographic technique." Furthermore, J.W., Paczyński and Schneider [Wam-PS] also

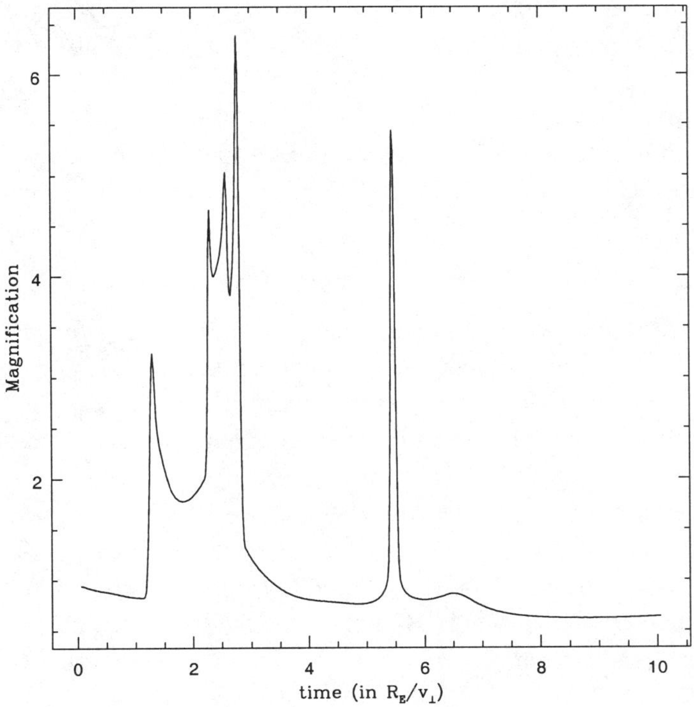

Figure 4.4. This light curve corresponds to the quasar's path through the caustics (beginning at the bottom left part) in Figure 4.3. Notice the effects of the dense caustics in the lower left corner of the magnification pattern — they produce four spikes in the left part of the light curve.

used the microlensing-induced flux variations in Image A of Q2237+0305 to show that the (the optical continuum region of the) source Q2237+0305 has a radius not larger than about 2×10^{15} cm. In [Wam90], J.W. gives a detailed discussion, including numerous illustrations, of microlensing magnification patterns with applications to the lensed quasar Q2237+0305.

4.4 Time Delay and Hubble's Constant

Sjur Refsdal [Ref64b] was the first to point out in 1964 that the lens effect can be used to determine the Hubble constant H_0. This is an important value in astrophysics and cosmology since it determines both the age and the size of the universe. The Hubble constant H_0 was originally defined as the constant of proportionality between the receding velocity v (redshift) of nearby objects and their luminosity distance d — Hubble's law (3.20):

$$v = H_0 d.$$

Note that for nearby sources ($z << 1$), the luminosity, angular diameter, and proper distances are approximately the same — see (3.19). A measurement of the Hubble constant establishes the extragalactic distance scale: Knowing H_0 we can determine the distance of an object from the (measurable) receding velocity. The age of the universe is then roughly $\tau_0 = d/v = 1/H_0$.

Refsdal suggested using a doubly imaged supernova at high redshift — he even mentioned the just newly discovered quasars (!) — and showed how the value of H_0 depends on the time delay between the two images. The method works in the following way:

Relating Time Delay to the Age of the Universe

The observation of a static gravitational lensing scenario with two quasar images seen around a lensing galaxy provides only *dimensionless* numbers: the angular separation between the images, the positions of the images relative to the lensing galaxy and the magnification ratio between the images are the measurable quantities. If one imagines now a situation with a universe that is twice as big and a galaxy with twice the mass, then the observational situation would be identical — see Figure 4.5.

So it appears impossible to say anything about the size of the universe (i.e., c/H_0) from this kind of observation. However, as Refsdal found out in 1964, there is a way to determine the scale of the universe via lensing. He noted that the total time delay $\Delta\, \tau_{\text{delay}}$ between the two quasar images is a direct measure of the size of the universe: It is a *dimensional* quantity! The two parts of the time delay — the geometrical and the gravitational delays — turn out to scale similarly. As can be seen from Figure 4.5, the geometrical time delay is different for the two situations described above: It is roughly 1.5 times as big for the larger scenario. In other words, the geometrical time delay scales as $1/H_0$ since it is the same *fraction* of the total path length in both scenarios. Similarly, the figure shows that the

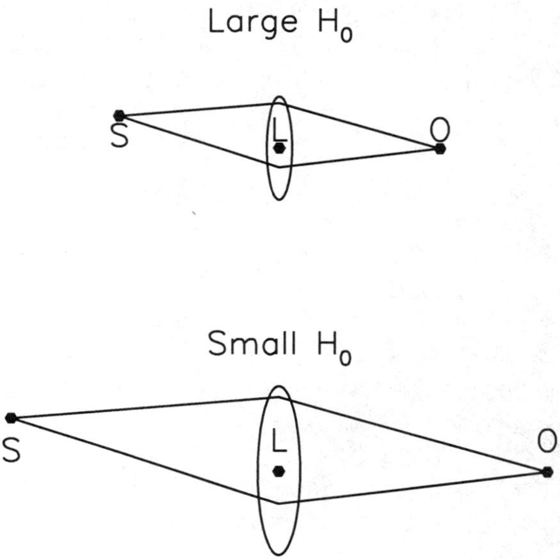

Figure 4.5. Two gravitational lensing scenarios that result in exactly the same observing situation in terms of image positions, image separation, and magnification ratio. Since the size of each system scales like $d = c/H_0$, the larger one (with smaller Hubble constant H_0) represents a universe that is slightly over 1.5 times as big as the smaller (and the lensing galaxy is slightly over 1.5 times as massive). Furthermore, the geometrical and potential time delays both scale inversely proportional to Hubble's constant H_0 (with proper adjustments of the mass). This implies that a measurement of the total time delay (geometric + potential) between images can be used to measure the physical scale of the universe. (Courtesy of Bartelmann.)

linear size and mass of the lens scales as $1/H_0$, so the potential time delay also scales in the same way. Hence, the total time delay $\Delta\, \tau_{\text{delay}}$ is proportional to $1/H_0$, where the constant of proportionality depends on the lens and the geometry of the lens system. This implies that a measurement of the total time delay means the determination of the physical "size" or scale of the universe. More explicitly, Refsdal [Ref64b] found (see [Nar-B, pp. 403–406] and [Bla-N92, pp. 330–334], and [Schn-EF, pp. 147, 171, 481] for more):

$$H_0 = K_1 \frac{(\Delta\, \theta)^2}{\Delta\, \tau_{\text{delay}}},$$

where $\Delta\, \theta$ is the angular separation between the images, $\Delta\, \tau_{\text{delay}}$ is the time delay, and K_1 is a constant whose exact value depends on the mass distribution of the lens. In turn, the *age of the universe* τ_0 is inversely proportional to the Hubble constant by (3.12) on page 35: $\tau_0 = K_2/H_0$, where the constant K_2 is bounded above one. Thus

$$\tau_0 = \frac{K_2}{K_1} \frac{\Delta\, \tau_{\text{delay}}}{(\Delta\, \theta)^2}.$$

For a very rough estimate of the age of the universe τ_0, one might plug in numbers from the double quasar Q0957+561: $\Delta\,\tau_{\text{delay}} \simeq 1.1$ years and $\Delta\,\theta \simeq 6$ arcseconds $\simeq 2.9 \times 10^{-5}$ radians. Assuming $K_1 \simeq K_2 \simeq 1$, this results in an age of the universe of roughly 13 billion years.

Requirements for a Determination of the Hubble Constant

It is not easy to determine the time delay in a gravitational lens system—hence, the Hubble constant—via this method. The reasons are many-fold:

- A time delay can be measured only if the **quasar** is **variable**. A certain change in the quasar brightness detected in one image can be measured in the other image(s) as well, after a certain delay.
- The angular positions of the **quasar images** have to be **well separated** so that it is easily possible to record their brightnesses individually to high accuracy.
- One needs a **large number of brightness measurements** of the quasar images to compile reliable light curves. The total light curve needs to be a few times as long as the value of the time delay; it needs to be well sampled in time (ideally one data point every night) and highly accurate. This mode of operation (a few short observations of many nights) is still relatively unusual for telescopes. It is much more common to use a telescope for several full nights in a row. (Hence, often telescope time allocation committees do not approve this kind of proposal.)
- The step from the time delay to the Hubble constant requires a unique **good model of the lensing system**; often one can get good "fits" to the observed data with different lensing potentials.

For these reasons, it took almost 20 years to establish the first measured time delay for the double quasar Q0957+561. There were early reports that the time delay should be somewhere around 400–420 days ("short delay"). Later, other groups reported values of about 530–540 days ("long delay"). This resulted in quite different values for the Hubble constant and the value of the "age of the universe" as well. These values were mutually exclusive, and in the mid 1990s it became a matter of faith what value people believed.

Recent Determination of Hubble's Constant

This controversial issue about the value of the time delay in the double quasar Q0957+561 was put to rest only after a dedicated monitoring campaign in the years 1994–1996. A measurement of a significant drop in the light curve of the supposedly "leading" Image A in December 1995 was used to "predict" that the same feature should appear in the trailing Image B either in February 1996 or in June 1996, depending on what the real value of the time delay is. The "shorter" of the two values — as originally proposed by Schild [Schi90] — was confirmed at (417 ± 3) days, resulting in a Hubble constant value around $H_0 = 64$ km sec^{-1} Mpc^{-1} with an accuracy of about 10% (Kundić et al. [Kun97]). Figure 5.2 in Section 5.1.1 shows the data supporting the shorter delay. The reader may consult Section 5.1.1 and Keeton and Kochanek [Kee-K97] for more on recent determinations of the Hubble constant via time delays.

Remark: For a given gravitational lens situation with two (or more) images and measured redshifts of lens and source, we have seen that the time delay between the images can be used to determine the Hubble constant. A necessary requirement for such an analysis is a good model for the mass distribution of the lens. Of course, an important part of this is the breaking of the degeneracies in the lens model (Section 3.4), which can be accomplished for the quasar Q0957+561 (page 149). Conversely, if the value of the Hubble constant is accurately known, then a measurement of the time delay could be used for the mass determination of the lens. In practice this would work well only in the case of a very simple lens, that is, a point lens (see equations (6.17) and (3.102)). But all known lenses are galaxies or galaxy clusters, which have extended mass distributions. Consequently, determining the mass is not as simple as in the point mass case. It is easily imaginable that more than one mass distribution would be consistent with the measured time delays. So although this method might look like an interesting possibility for the (medium distant) future to determine masses of galaxies, in practice it seems unlikely to work better than other methods.

4.5 Statistical Lensing and Limits on Cosmology

There are various cosmogonic models that try to explain the evolution of the universe. Almost all are based on a standard "hot big bang" scenario, but they differ in the amount and type of matter they contain. The matter content of a cosmological model is defined as $\Omega_0 = \rho_0/\rho_c$, where ρ_0 is the actual density of the universe and ρ_c is the density that is required to ultimately slow down the expansion completely. Some cosmological models invoke a "vacuum density" or cosmological constant Λ. We shall briefly name just three currently popular cosmological models here:

- A "flat" model that has exactly the critical matter density, $\Omega_{matter} = 1$, e.g., the standard cold or hot dark matter models.

- A "flat" model with a low density value $\Omega_{matter} < 1$, but with a cosmological constant Λ, so that $\Omega_{matter} + \Omega_\Lambda = 1$, e.g., Lambda-dominated cold dark matter [here $\Omega_\Lambda = \Lambda/(3H_0^2)$].

- An "open" model with a low density value $\Omega_{matter} < 1$, e.g., open cold dark matter.

It is an important task in cosmology to find out which of these models describes the observed universe best. Many methods and tests have been proposed and used in attempts to answer this question, with no final word yet. Various independent observations in the last years though point to a matter content of $\Omega_{\text{matter}} = 0.2 - 0.4$, with or without cosmological constant.

Gravitational lensing provides an additional and independent test to check which of these cosmological models describes the real universe best. The underlying idea is the following: The different cosmological models are "constructed" such that they describe the universe "today" appropriately (e.g., capture the observed frequency of galaxy clusters, or, the separation distribution of galaxies in our cosmic neighborhood). However, the "histories" or the evolution of these toy universes are different from each other. Qualitatively the behavior is similar; all these models describe an expanding universe in which matter started smoothly distributed and formed more compact structure with time due to gravitational attraction. But quantitatively these models make different predictions for properties in the past: Some cosmological models predict that galaxies and galaxy clusters form relatively early after the Big Bang, and hence could act as lenses at high redshift, whereas other models prefer relatively recent galaxy formation.

A model dominated by a cosmological constant Λ would provide much longer baselines or larger volumes for lensing, making it more probable for high redshift objects to be lensed [Car-PT] — see discussion in next subsection. The larger backwards in time one gets, the bigger the differences between the cosmological models. Loosely speaking, certain cosmological models produce many more multiple quasars and giant luminous arcs than others. This aspect of lensing has been explored in particular to find limits on the cosmological constant Λ through the statistical analysis of multiple images in various lens surveys, and comparison with theoretical models. Statistical gravitational lensing — the frequency of multiple lensing — can also be used as a tool to determine the cosmological model that "correctly" describes the universe. Currently, comparison of the observed frequency of the lensed quasars, makes it unlikely that the observed universe is Λ-dominated [Maoz-R, Rix-MTF]: a scenario with a large value of the cosmological constant would predict more lensed quasars than what is observed. Similar studies also apparently exclude the standard Cold Dark Matter Scenario ($\Omega_{\text{matter}} = 1$) as a viable model for structure formation [Wam-COT, Koc95]. The investigation of the statistics of giant arcs – lensed images of distant galaxies around clusters of galaxies – point in a similar direction [Bart98].

It has also been suggested early on that statistical lensing can detect (or exclude) any significant population of compact objects in currently accessible mass ranges, that is, resolvable angular splittings [Pre-G, Can]. This has been applied to quite a range of masses by now (cf. [Dal, Nem]), basically excluding that a large fraction of matter in the universe can be in compact objects between 10^6 to 10^{15} $M_\odot$. A number of surveys are underway to extend this range to the lower end. This application in particular rules out any significant number of "dark galaxies," or super-massive black holes.

On the Cosmological Constant

The way the cosmological constant affects the statistics of gravitational lenses is manyfold. The most obvious effect of the cosmological constant is its influence on volume in space. More accurately, the volume per unit redshift at large redshift is higher for universes with a large cosmological constant, as was first pointed out by Turner [Tur] (also, see [Car-PT, pp. 512–514]. Since the angular diameter distances enter in the value of the critical surface mass density, a large value of the cosmological constant reduces the latter, which means for a given physical surface mass density, a galaxy is more likely to act as a lens in a cosmological constant dominated universe. More subtle effects enter as well, e.g., a cosmological constant affects both the formation of structure in the universe and the evolution of individual galaxies and galaxy clusters.

To be able to conclude anything about the value of the cosmological constant from the number of observed multiple quasars, it is therefore necessary to have a good understanding of its effects on the frequency of multiple images and also to have a large enough observational sample for comparison with these theoretical predictions. Since the total number of lensed quasars found in "well-defined" surveys is still small, the conclusions from such analyses are still a bit uncertain. So far only upper limits on the value of the cosmological constant could be obtained, and even these have recently been challenged. Many groups have investigated this issue (see [Fuk-FK, Fuk-FKT, Koc96, Maoz-R, Fal-KM]). The most recent and most thorough one so far concludes that the (in-)frequency of gravitationally lensed quasars in well-defined samples does not allow a value of the cosmological constant larger than $\Omega_\Lambda \leq 0.62$ (at the 95% likelihood level for flat cosmological models), if data on multiple quasars in the radio regime and in the optical regime are combined [Fal-KM].

4.6 Exotic Matter: Cosmic Strings

One of the important concepts of modern particle physics is the spontaneous symmetry breaking in the early universe. This can result in topological defects of various dimensions, like domain walls (two-dimensional), cosmic strings (one-dimensional) and magnetic monopoles (zero-dimensional). Originally, cosmic strings seemed an appealing concept for astrophysicists and cosmologists: cosmic strings were considered as possible "seeds" for galaxy formation.

What makes cosmic strings interesting in relation to lensing is the fact that the space near a string has a "conical nature." In other words, space around a cosmic string is flat, but there is a "deficit angle," a wedge with a certain opening angle is removed. Points on the two faces of the wedge are identified, which means that objects lying within a certain angle behind a straight cosmic string are seen twice by a distant observer (e.g., Vilenkin [Vil] and Kolb and Turner [Kol-Tu, p. 226]). All sources that are not behind the string within this angle are seen only once.

The "strength" of a cosmic string is determined by its linear mass density ς_{cs}, which has "reasonable" values of order (cf. [Vil]):

$$\frac{G\,\varsigma_{cs}}{c^2} \simeq 4 \times 10^{-6}(\Omega_0\,h_{100})^{-1},$$

where G is the gravitational constant, Ω_0 is the cosmological density, and $h_{100} = H_0/(100$ km sec^{-1} Mpc$^{-1})$ with H_0 the Hubble constant. The deflection angle of a straight cosmic string is given by

$$\vartheta_{cs} = \frac{4\pi G\,\varsigma_{cs}}{c^2} \simeq 10(\Omega_0\,h_{100})^{-1}\text{arcsec}, \tag{4.6}$$

(assuming the value of $G\,\varsigma_{cs}/c^2$ from above). The deflection angle is independent of the impact parameter. For an infinitely long straight string — the simplest imaginable configuration — the double images are all unmagnified, undistorted and all image pairs have the same separation, which is proportional to the mass per unit length of the string [Got85]. The angular region that is multiply imaged covers objects with impact "angles" smaller than $2\vartheta_{cs}$ [Vil].

More complex lensing configurations are produced by curved strings, in particular, string loops have a complicated lensing signature. The lensing properties of various string configurations were recently summarized by [And].

4.7 Outlook

Every now and then some skepticism can be heard of the kind that gravitational lensing has not fulfilled its promises to measure the size and the age of the universe, or the masses and the spatial distribution of its components. Well, considering that the field was born as an observational science only in 1979, it is not too surprising that lensing is only beginning to mature. Conversely, would it not be all too suspicious if such a newcomer could instantly produce results superior to those obtained in more than half a century of hard work (involving many thousand man and woman years and numerous telescope nights) in trying to determine the Hubble constant or the correct cosmological model?

Without much doubt, in the coming years lensing will help to establish a more accurate value of the Hubble constant and will provide very significant results in the fields discussed here. This is not blind unjustified optimism, but rather it is based on the latest successes, the momentum and attention lensing gets, the next generation of telescopes that go on line fairly soon (see the Foreword), and the big surveys, like the Sloan Digital Sky Survey, which will find hundreds of quasar lenses with well-defined selection criteria.

In addition to the astrophysical applications of gravitational lensing that have been thought of by now, many more will be proposed and employed in the coming years. In an attempt to prognosticate the future of gravitational lensing, Press [Pre] not just predicts a continuing rise of scientists and papers dealing with lensing: "Gravitational lens studies are rapidly becoming mainstream astronomy." He continues: "Gravitational lens effects

. . . are present along virtually every line of sight out to . . . [cosmological] distances," and consequently predicts that: "Gravitational lensing may well disappear as a sub-specialty in astronomy, and instead become simply a ubiquitous observational technique." Press lists more than a dozen future applications, which will quantitatively and qualitatively change our knowledge about the universe.

Chapter 5

Observations of Gravitational Lensing

For better or for worse, there is little evidence that we have any idea of what reality looks like from an absolute point of view. We only know what the world looks like from our *point of view....*

In Heisenberg's words, 'What we observe is not nature itself, but nature exposed to our method of questioning. Our scientific work in physics consists in asking questions about nature in the language we possess and trying to get an answer from experiment by the means that are at our disposal.' And Bohr's, 'It is wrong to think that the task of physics is to find out how nature is. Physics concerns only what we can say *about nature.'*

... Physics has a clear criterion for the choice of a language—physicists choose the language that allows their predictions most closely to fit their observations.

BRUCE GREGORY

The first detection of a gravitational lens phenomenon happened in 1979 [Wals-CW], when an impressive number of theoretical studies of the lens effect had already been published (cf. Chapter 1). By now the majority of scientists working in gravitational lensing and most papers appearing in journals deal with observations. Basically, there are two types of observational projects on lensing: the individual object approach, where the (often serendipitous) discovery/confirmation of a gravitational lens system or a refined new observation is reported. A complementary approach deals with gravitational lens surveys: Scientists systematically search for lenses among candidates that fulfill certain criteria (defined by source redshift, brightness, position on the sky, detectability in the radio/X-ray/infrared regime, morphology, or others). The first approach is more frequent, although the second one is more liked by theoreticians, who prefer well-defined samples in their attempts to make (statistical) sense of the observations.

In this chapter, we review the current observational situation for gravitational lensing. The reader may be cautioned, though, that in such a rapidly evolving field "current" can be outdated tomorrow. The observations divide into different categories by various criteria, for example, by sources: observations of unresolved or point sources (quasars, stars) versus those of resolved or extended sources (optical galaxies, radio jets). Equally justified, one can use the type of lens as the criterion of subdivision: cluster lenses, galaxy lenses, and stellar lenses. Yet another criterion could be the mass of the lens: macro-lensing, in which case the various images can be seen individually, versus microlensing, where only the combined magnification of all images can be seen. Related would be the concepts of "static" lensing, for which one can detect only the image configuration, and "dynamic" lensing, where one can see a change in configuration or magnification with time. Finally, the strength of the lensing can be used as a division: strong lensing with high magnification and possibly multiple images versus weak lensing, visible only as a statistical effect. We decided to use a pragmatic (rather than a dogmatic) approach, choosing the categories as they emerged over time from various observations. In other words, we employ a mixture of the above criteria that is most often used in praxis.

In Table 5.1, we list several different lensing phenomena that have been discovered and/or predicted in the last two decades. In the second column, the number of observed examples in this category is given as known at the time this book went into print. These numbers certainly have gone up since, so the ones here may serve as "snapshots" or lower limits. In the third column, year and reference are listed for the cases in which this particular phenomenon was predicted.[1] In the last column the year of discovery or first detection is given together with the corresponding reference. The phenomena are ordered roughly by year of first discovery. We distinguish — somewhat arbitrarily — eight sub-fields. There is no reason to believe that the discovery of "new" lensing phenomena has stopped now.

Due to the enormous amount of observational material that has been compiled on lensed and lensing objects, it is impossible to list or comment on all the cases individually. Therefore, we decided to give a general overview about certain particular phenomena and the scientific goals that have been achieved so far. Subsequently, we discuss only one or two prototype examples in each group, so that one can get an impression of what the observational properties and problems are for a given class of lenses. Section 5.1 treats multiply imaged quasars with emphasis on the double quasar Q0957+561. Lensing by galaxy clusters appears in Section 5.2. The arcs in the cluster CL0024+1654 are used to illustrate this phenomenon. Section 5.3 deals with Einstein rings using the ring 1938+166 as a key example. Quasar microlensing is taken up in Section 5.4 and the quadruple quasar Q2237+0305 is discussed. Microlensing in our galaxy (e.g., the galactic bulge and halo) is treated in Section 5.5. Section 5.6 investigates weak lensing of background galaxies by galaxy clusters, and the reconstruction of the galaxy cluster mass profile. The latter is illustrated for the galaxy cluster CL1358+62.

[1] It is rare in astrophysics that there are predictions for discoveries that are confirmed later. This is one of the attractive features of gravitational lensing; its principles are so simple that one really can make firm and quantitative predictions, one of the classical features desired for a scientific discipline.

Phenomenon	**Known Cases**	**Discovered/ First Detected**	**Predicted**
Multiple Images (quasars/radio sources)	≥60	1979 [Wals-CW]	1912 [Ren-SS] 1923 [Edd] 1924 [Chw]
Giant Luminous Arcs	≥30	1986/87 [Lyn-P], [Sou-FMP]	(1937 [Rus])
Einstein Rings	≥12	1988 [Hew]	1924 [Chw] 1936 [Ein36]
Quasar Microlensing	>5	1989 [Irw]	1979 [Chan-R79] 1981 [Got81]
Arclets Weak Cluster Lensing	> 100	1990s (various)	–
Galactic Microlensing	> 500	1993 [Aub], [Alc93] [Uda93]	1986 [Pac86b]
Galaxy-Galaxy Lensing	≥3	1983 [Val-TJ] 1996/7 [Del-T]	1937 [Zwi37a] [Zwi37b]
Lensing by Large-Scale Structure	>4	2000 [Bac], [Kai-WL] [van00], [Wittm00]	(various)

Table 5.1. Various types of gravitational lens observations. Listed are: number of known cases per category, year and reference of discovery of this particular lensing phenomenon, and year and reference of its prediction.

5.1 Multiple Quasars

Quasars[2] were the first objects found to be gravitationally lensed by a deflector outside our solar system. In 1979, two images of the quasar Q0957+561 were discovered serendipidously by Dennis Walsh and his collaborators in a survey of radio sources [Wals-CW] (see Section 1.2). Although it was not entirely clear initially, very soon the objects were confirmed to be two images of the same quasar rather than a physical pair of quasars.

This illustrates one of the problems of identifying multiple images of quasars as real gravitational lenses: When is a two/three/four image configuration accepted as a bona fide lens system? Various research groups used to apply different criteria, but over the years a sort of consensus list of criteria has emerged. The ideal quasar lens system fulfills all the criteria listed here:

- There are two or more point-like images with the same color.
- The spectra of the various images are identical or *very* similar to each other.
- The redshifts (or distances) of the quasar images are identical or *very* similar.
- There is a lens (most likely a galaxy) found between the images, with a measured redshift much smaller than the quasar redshift.
- Over time, the fluxes measured from the various images follow an identical light curve, except for certain lags — the time delays.

In the rare cases where all these criteria are matched, there is general agreement that the multiple quasar system is a "real" gravitational lens. If one or more are not fulfilled, the quasar system is somewhere on the line between being an "almost safe lens" and a "lens candidate." There are more and less conclusive criteria: If a time delay is measured between the light curves of two or more quasar images, then this is almost a "proof," whereas two point images with the same color are still a relatively weak argument for lensing. Since not all of the above criteria are established for most of the known multiple quasar systems, it is not easy to say how many gravitationally lensed quasar systems exist. A recent compilation by Keeton and Kochanek [Kee-K96] put together 29 quasars as lenses or lens candidates in three probability "classes," leaving out the "less promising" candidates.

Gravitationally lensed quasar systems are used individually to investigate the lens and the source. They are also analyzed statistically to get information about the population of lenses in the universe, their distribution in distance (i.e., cosmic time) and mass, and hence the cosmological model. It is too extensive to give a detailed account here, including these applications, of all the known multiple quasar systems. Instead, we decided to describe

[2]Most theorists believe that quasars are short-term active states of galaxies, fueled by (super-)massive black holes in their centers.

the double quasar Q0957+561A,B in some detail. It is the best studied and longest known quasar lens system, and it exhibits a large variety of features. With the measurement of the time delay and the first determination of a value for the Hubble constant H_0 from a gravitational lens, this system is indeed a record holder in many aspects.

In the future, it is very likely that a simpler or "cleaner" lens system will turn out to be a better candidate for determining H_0 than Q0957+561A,B (the radio rings with compact components look like promising candidates). Ultimately, a whole ensemble of lens systems with measured time delays will give the best lensing value for H_0 (cf. [Sche]).

5.1.1 A Case Study: The Double Quasar Q0957+561

The story of this famous lens system can be found in the discoverers' own account [Wals]. The double quasar was originally found in a radio survey and subsequently identified in the optical light as well. To date, more than 100 scientific papers have appeared on this gravitationally lensed quasar, many more than on any other gravitational lens. We shall summarize what is known about the system.

Optical Observations

The optical appearance in Figure 5.1 (left panel) shows two relatively bright point-like sources that are separated by 6.1 arcseconds. The redshifts of the two images are $z_S = 1.41$. Slightly off the connecting line, close to the component B at the bottom, a fuzzy patch can be seen — the lensing galaxy. Its redshift is $z = 0.36$. In fact, this galaxy is part of a whole cluster of galaxies at about the same redshift. This is the reason for the relatively large separation for a galaxy-type lens (typically galaxies produce splitting angles of only about one arcsecond). In this system, the mass in the galaxy cluster helps to increase the deflection angles to create such a large image separation.

Radio Observations: VLA, MERLIN, VLBI, and Jets

The system Q0957+561 was originally discovered in radio wavelengths. A recent image by the MERLIN radio telescope is shown in Figure 5.1, right panel. The positions of the two point-like objects coincide with the optical sources. There is no radio emission at the position of the galaxy: The lensing galaxy is (as almost all galaxies) radio-quiet. But it also tells us that a possible third image of the quasar must be very faint, below the detection limit of all radio observations. In the right panel of Figure 5.1, a "jet" can be seen emerging from image A (at the top). It is not unusual for radio quasars to have such a "jet" feature, which is most likely matter being emitted from the central engine of the quasar with very high speed. The jet is not seen in image B (lower point source), which means it must lie outside the region that is multiply imaged. Only the central part of the quasar happens to be inside this region. With the VLBI technique (Very Long Baseline Interferometry), it is possible to map the two quasar images with milli-arcsecond resolution. On these scales, two jets are

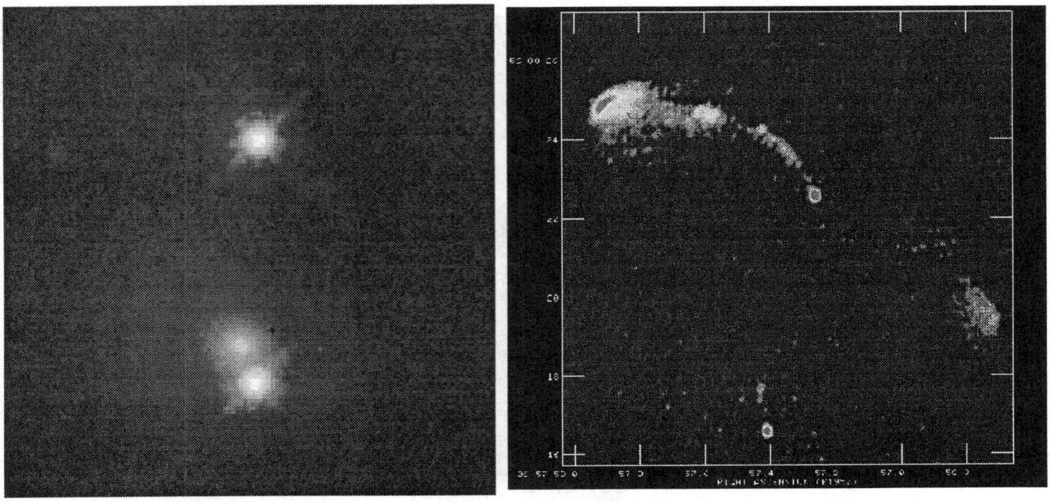

Figure 5.1. The double quasar Q0957+561A,B: Images in optical and radio light. *Left:* Optical image from a Hubble Space Telescope I-band exposure, image B at the bottom with the slightly blurred lensing galaxy nearby is separated by 6.1 arcseconds from the image A at the top (image courtesy of E.E. Falco and the CASTLE collaboration, NASA/STScI). *Right:* Radio observation taken with the MERLIN array of telescopes. The quasar images are point-like, a radio jet is emerging from image A only (image courtesy of N. Jackson, MERLIN).

ejected from both images A and B. Within these jets individual knots are resolved, which can be used for detailed modeling of the lens potential.

Light Curves – Time Delay – Hubble's Constant – Microlensing

As stated above, a virtual "proof" of a gravitational lens system is a measurement of the "time delay" $\Delta\tau$, that is, the relative shift of the light curves of the two or more images, $I_A(\tau)$ and $I_B(\tau)$, so that $I_B(\tau) = \text{const} \cdot I_A(\tau + \Delta\tau)$. In addition, a measured value of the time delay — in conjunction with a good theoretical model of this lens system — can be used to determine the Hubble constant H_0. The inverse of the Hubble constant directly relates to the age of the universe (equation (3.12)). A large amount of astronomers' time and telescope time has been used to determine H_0, and still there is not yet general agreement on its value. Lensing offers a new, additional, and independent method to determine this hard-to-get value (Section 4.4), and it is comparatively inexpensive.

Shortly after the discovery of the double quasar Q0957+561, a number of groups started to "monitor" this lens system with the goal of measuring the time delay $\Delta\tau$ between its two images. The two quasar images were observed as often as possible, with the goal to record their brightnesses accurately as a function of time, and correlate the two light curves to find the value of $\Delta\tau$. It turned out to be more difficult than originally thought. A necessary requirement for this method to be successful is that the quasar varies intrinsically.

Otherwise it is not possible to find a unique value of $\Delta\tau$. The quasar seemed to cooperate: Its brightness changed with time. So this important condition was fulfilled.

Soon some researchers claimed to have found a value for the time delay. However, different groups arrived at different values. After more than a decade, the measured values of the time delay clustered around two intervals: at about 415 days and roughly 530 days. This discrepancy was solved only recently. It had become possible through (almost) online data reduction and (almost) nightly monitoring to identify a change in the light curve in the leading image A and predict the dates of the occurrence of this feature in the trailing light curve B for the two fashionable values of the time delays [Kun95]. The predicted feature was measured in image B after about 417 days, and no noticeable change was seen in image B at the time corresponding to the formerly claimed longer time delay (Kundić [Kun97]). Hence, the shorter of the two popular time delay values, originally proposed by Schild [Schi90], was confirmed — see Figure 5.2.

The second requirement for a determination of the Hubble constant from gravitational lensing is a good model of the lens. In the system Q0957+561, this is complicated due to the fact that the lens consists of a galaxy embedded within a galaxy cluster, which produces a degeneracy in modeling: From the information of the lensed quasar images alone, one can construct a whole family of models with the mass divided differently between the cluster and the galaxy, and still fulfill most of the observational constraints [Goren-FS]. This is an example of the degeneracies cited in Section 3.4 (page 110). On the other hand, there exists a huge amount of data from optical, radio and X-ray observations that provide many additional constraints for the lens models. That is, with the measurement of the velocity dispersion in the lensing galaxy [Fal-SMD] and the availability of weak lensing

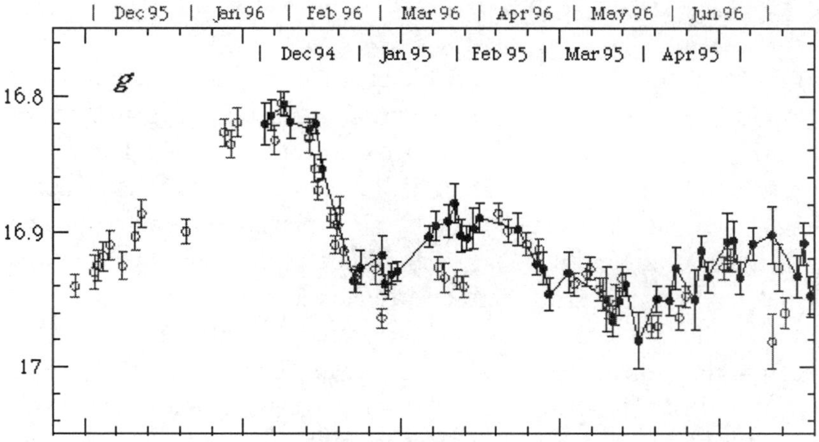

Figure 5.2. The double quasar Q0957+561A,B: Optical light curves of images A (filled symbols) and B (open symbols). The latter is shifted by a time delay of 417 days (image courtesy of Tomislav Kundić).

data to constrain the contribution of the cluster [Fis-BRT], it became possible to get some handle on this degeneracy. The resulting values for the Hubble constant are in the range (65 ± 8) km/sec/Mpc [Kun97],[Fal-SMD]. Using the *Hubble Space Telescope*, Keeton et al. [Kee-00] recently succeeded in detecting the host galaxy of this quasar. Some parts of the host galaxy are seen twice, other parts are even imaged four times. Analyzing this wealth of data, it was shown in [Kee-00] that essentially all previously published models of this lens system are inconsistent with these newly found distortions. The lensing galaxy's mass distribution follows very closely its light distribution. These new observations finally break the degeneracy in the mass distribution between galaxy and galaxy cluster, the shear contribution of the cluster is relatively small. With these new observations, there are very good prospects to reduce the remaining uncertainties and produce a much more accurate value of the Hubble constant.

The fact that the light curves of components A and B are so similar for the roughly 160 days of "overlap" displayed in Figure 5.2 puts some constraints on the action of microlensing in this system. The light bundles pass the galaxy with angular separations of one and five arcseconds, respectively (Figure 5.1). The one closer to the center of the galaxy passes through regions with visible stars, the other one (image B) passes through the dark halo of the galaxy and hence probes possible compact objects there. From comparison with simulations, it can be concluded that the halo of the lensing galaxy cannot be made entirely of Jupiter-type objects, because then we should see larger differences between the two light curves [Schm-W]. There seems to be some fluctuations in the difference light curve on longer time scales, which may be explored to get constraints on objects with larger masses [Schi96].

5.2 Galaxy Clusters as Strong Lenses: Giant Luminous Arcs

In 1986, two groups discovered a new gravitational lensing phenomenon: magnified, distorted and strongly elongated images of background galaxies lying behind a foreground cluster of galaxies [Lyn-P, Sou-FMP]. Soon more examples were found. These Giant Luminous Arcs were not immediately identified as lensed images of extended sources, although lensing was discussed as one of the possible explanations. These thin and elongated features (up to about 20 arcseconds long) were found in the central parts of very massive clusters of galaxies. Most of them were clearly curved towards the center of the cluster looking like pieces of a circle with radius of curvature of about 20 arcseconds. Their color was different from the typical cluster galaxies; they tended to be bluer. Paczyński [Pac87] pointed out in 1987 that it is likely that such an arc is a piece of an Einstein ring forming an elongated image of a background galaxy. This suggestion was confirmed later in 1987, when redshifts of arcs were measured which showed that the sources are far beyond the location of the galaxy cluster.

There are some interesting aspects related to the "Giant Arcs." One is that they had been around before 1986, but no one noticed. There are published images of clusters on which one can clearly identify these arcs. But they were not recognized or commented on. The second interesting point is that these arcs were not predicted. At the time of their discovery, gravitational lensing was well established as an observational science; the only known lensed objects, however, were quasars, unresolved (i.e., very small) sources. Although there were published simulations of extended sources which clearly showed highly magnified curved images, no one predicted explicitly that such arc-like images must be seen, if there are galaxies behind clusters and if clusters have centrally concentrated mass distributions. By now there are dozens of examples of Giant Arcs (Table 5.1). Not all of them are as long, nicely curved, and striking as the first ones; there is a transition to (more abundant) smaller arcs and arclets.

Giant Arcs can be exploited in two aspects, as is typical for many lensing phenomena. First, they provide us with highly magnified galaxies at very high redshifts, which are too faint to be detected or analyzed in their unlensed state. Hence, we can study these infant or proto-galaxies in their early evolutionary stages, relatively shortly after the big bang. The other useful application of the arcs is to take them as tools for studying the potential and mass distribution of the lensing galaxy cluster (Section 4.2.2).

Some general results from the analysis of giant arcs in galaxy clusters are: Clusters of galaxies are dominated by dark matter. The typical mass-to-light ratios[3] are $M/L \geq 100 M_\odot/L_\odot$. The distribution of this dark matter follows roughly the distribution of the light in the galaxies, in particular, in the central part of the cluster. Lensing shows that the core radius of a galaxy cluster (i.e., the radius at which the mass profile of the cluster flattens towards the center) must be smaller than the distance between arcs and cluster center (between 20 and 30 arcsec); otherwise no arcs would form. This is in contrast to what was expected from the distribution of the X-ray emitting gases.

Giant Arcs in clusters can be analyzed statistically as well. Gioia and Luppino [Gio-L] investigated 41 galaxy clusters for arcs, which had been selected as bright in X-rays from the Einstein Medium Sensitivity Survey. They found that 14 of these clusters contain bright gravitationally lensed arcs, clearly indicating a strong relation between having high X-ray emission and being very massive and compact. The statistics of arcs can even be used as a tool to distinguish between different cosmological models (Wu and Mao [Wu-M], Bartelmann et al. [Bart98]), but today there is not enough data for conclusive statements.

Recently, the new record holder for the most distant individual object in the universe turned out to be a gravitationally lensed arc. Franx et al. [Franx] found the distorted image of a galaxy at a redshift of $z = 4.92$ in the field of the galaxy cluster CL1358+62. In contrast to most other arcs, this one is relatively red. The reason is that due to this very high redshift, the Lyman Alpha emission of the galaxy (which is emitted in the ultra-violet part of the electromagnetic spectrum at a wavelength of 1216Å) is shifted by a factor of $z + 1 \simeq 6$

[3]The mass-to-light ratio is the relative amount of matter that is there for each "unit" of light. It is typically expressed in solar units, where $M_\odot$ is the mass of the sun (2×10^{33} grams) and $L_\odot$ its luminosity (4×10^{33} ergs/sec).

to the red part at a wavelength of 7204Å. So once again, the gravitational telescope helps us uncover otherwise inaccessibly distant cosmic objects.

In the next section, we discuss an example of a recently analyzed lens system consisting of a galaxy cluster producing many striking arcs. For more information on observations and analyzes of giant arcs, see the recent review article by Fort and Mellier [For-M].

5.2.1 Fantastic Arcs in Galaxy Cluster CL0024+1654

In an impressive image taken with the refurbished Hubble Space Telescope, Colley, Tyson and Turner [Col] obtained deep exposures of the cluster of galaxies CL0024+1654 (redshift $z = 0.39$) in a blue and a red filter. The combined image (Figure 5.3) shows very nicely the reddish images of cluster galaxies — the brightest of them concentrated around the center — and the bluish arcs. There are four blue images which all have a similar shape: similar to the Greek letter Θ. Four of them lie roughly on a circle and are tangentially elongated, whereas the fifth image is relatively close to the center of the cluster.

The image is displayed in Figure 5.3. All the images are resolved and show structure. The substructure is very similar in all of the images (e.g., the bright fish hook-like feature at one end of the arcs), except that two of them are mirror inverted! It turns out to be a five-image configuration produced by the massive galaxy cluster. All five arcs are images of the same galaxy, lying far behind the cluster at a much higher redshift and most likely a galaxy in formation. This is a spectacular example of the power of the universe as a gravitational telescope. In [Col], the authors also showed that the lensing effect in CL0024+1654 — which produces a magnification of roughly a factor of ten — combined with the resolution of the HST of 0.1 arcsec can be used to yield a resolution that effectively corresponds to 0.01 arcsec, unprecedented in direct optical imaging. They show in particular, that mapping the five images "backward" with their model for the cluster lens, they get basically identical source morphology for all five of them, hence confirming that the five arcs are images of one source. The striking figure below depicts the reconstructed surface mass density for the dark matter in the cluster CL0024+1654 shown in Figure 5.3 [Tys-KD].

5.3 Einstein Rings

If a point-like light source in the background and a point-like gravitational lens in the foreground are perfectly aligned (as seen from an observer), then the resulting image configuration is a circle with infinitely high magnification (page 89). This was first pointed out by [Chw] in 1926. and independently by Einstein [Ein36] in 1936. These images are called "Chwolson–Einstein rings" or, more commonly, "Einstein rings." Of course, in nature there are neither point sources nor point lenses, nor does it ever occur that three points are exactly aligned. Fortunately, a ring-like image configuration is possible even for not quite ideal conditions: extended lens, finite source, and almost perfect alignment.

Figure 5.3. Hubble Space Telescope image of the strong-lens galaxy cluster CL0024+1654 with highly magnified giant luminous arcs (Courtesy of Wes Colley and NASA/STScI.)

Any (transparent) spherically-symmetric centrally-condensed lens produces a degenerate point-caustic, resulting in a ring image of a source that is (partly) covering the caustic (page 99). Realistic lenses (like galaxies) that are not perfectly symmetric produce little astroid caustics. Moreover, as long as the source covers the whole caustic, there is always a ring-like image, though with structure (e.g., Figure 3.14). For larger sources, it is more likely that they would cover the whole caustic. Since the region of a quasar that emits radio emission is presumably much larger than the optical emission region, it is much more likely that Einstein rings are found in the radio regime rather than in the optical.

The sources of most Einstein rings have both an extended and a compact component. The latter is often seen as a double image, separated by roughly the diameter of the Einstein ring; occasionally even four point images are detected. In some cases, monitoring of the radio flux showed that the compact source is variable. This gives an opportunity to measure the time delay in these systems, and hence, possibly determine the Hubble constant.

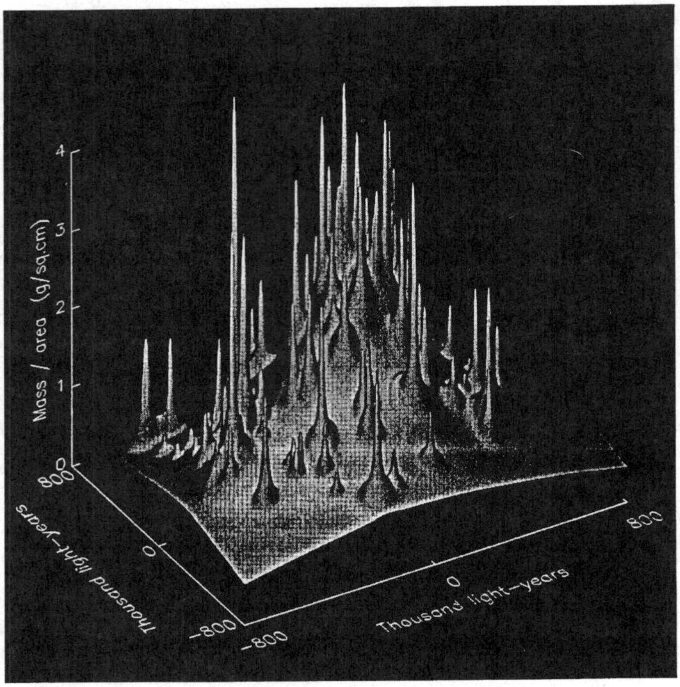

Figure 5.4. Reconstruction of the surface mass density of the galaxy-cluster lens system CL0024+1654 from a strong lensing analysis. The graph of the surface mass density is shown. The reconstruction is due to Tyson, Kochanski and Dell'Antonio [Tys-KD]. They estimated the cluster's mass to be of order $10^{16} M_{\odot}$. (Courtesy of Tony Tyson.)

Einstein ring systems provide three advantages over the multiple quasar systems for the ultimate goal of determining the Hubble constant. First of all, the extended image structure provides many constraints on the lens model, so that it can be much better determined than in the case of just two, three, or four point-like quasar images. Second, since the separation of the images (or the diameter of the rings) is of order one or two arcseconds, the expected time delay is likely to be much shorter than the one in the double quasar (in fact, it can be arbitrarily short, if the source happens to be very close to the point caustic). Consequently, one does not have to wait very long to establish a time delay.[4]

The third advantage of Einstein ring systems is that since the emitting region of the radio flux is presumably much larger than that of the optical continuum flux, the radio light curves of the different images are not affected by microlensing. Hence, the radio light curves between the images should agree quite well with each other. It is important to add that Einstein rings also help us learn about the mass distribution of galaxies at large

[4]In general, one has to monitor the light curves of both sources for a duration that corresponds to at least three or five times the time delay — ideally more than ten times — in order to establish a reliable time delay.

distances (redshift range $z_S \simeq 0.5$ to 1.0). The many constraints that the images of Einstein rings provide, allow us to determine the slope of the mass distribution in the lensing galaxy.

In 1988, the first example of an "Einstein ring" was discovered by Jackie Hewitt [Hew]. With high resolution radio observations, the extended radio source MG1131+0456 turned out to be a ring with a diameter of about 1.75 arcseconds. The source is a radio lobe at a redshift of $z_S = 1.13$, whereas the lens is a galaxy at $z_L = 0.85$. In the meantime, about a dozen Einstein rings have been discovered. For the Einstein ring in MG 1654+561 and with the assumption that the radially averaged surface mass density of the lens behaves like $\sigma(r) \propto r^s$, it was found that the value of s lies between $-1.1 \leq s \leq -0.9$ (an isothermal sphere would have exactly $s = -1$); see Kochanek [Koc95].

Another interesting application is the non-detection of a central image in the Einstein rings. For singular lenses, there should be no central image (since the image's magnification tends to zero as it approaches the singularity — Chapter 13). However, many galaxy models predict a transparent finite core in the mass distribution of a galaxy. The non-detection of the central de-magnified image puts strong constraints on the size of the core radii — they must be smaller than $r_c \leq 200$ parsecs.

The observed Einstein rings are not perfectly circular symmetric. Sometimes there are one or two interruptions along the circle (but there is always a clear lack of radio emission in the inner/central part of the ring). This reflects the fact that most galaxies are not really cylindrical symmetric, as viewed from the earth. Consequently, instead of the degenerate point caustic, the lens produces a small asteroid caustic. The appearance of the image can nevertheless be very similar to an Einstein ring, when the source size is (much) larger than the extent of the caustic.

Recently, a nice theory of Einstein rings was developed by Kochanek, Keeton and McLeod [Koc-KM]. They show that the exact structure of the Einstein ring is determined in equal parts by the shape of the lens potential and by the shape of the lensed galaxy. Applying this to the three examples PG 1115+080, B 1608+656, and B 1938+666, they determined the axis ratios of the lensed galaxies. Furthermore, they showed that Einstein rings are by far the best lens systems for the determination of the Hubble constant.

5.3.1 The Einstein Ring 1938+666

We now present a relatively new example of an Einstein ring. The gravitational lens system 1938+666 was discovered in 1992 during a systematic search for lenses with radio telescopes (JVAS: **J**odrell Bank - **V**ery Large Array - **A**strometric **S**urvey). A radio "image" of this object showed a dominant arc-like structure plus two pairs of point-like objects within a region of about one arcsecond (King et al. [Kin]). Optical and infrared observations detected a very red object at the position of the radio lens. This system was mapped with the new infrared camera (NICMOS) aboard the Hubble Space Telescope in August 1997, and identified a ring image with a diameter of 0.95 arcsecond plus some emission from the center of the ring (Figure 5.5). The straightforward interpretation, which is supported by a

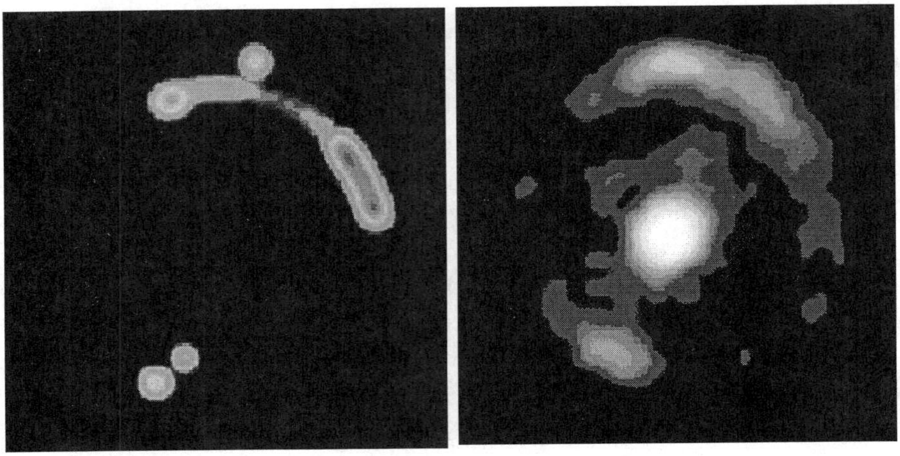

Figure 5.5. *Left Panel:* Image of the radio source 1938+666 with the MERLIN array of radio antennas. The arc (or incomplete ring) shows that the radio source is strongly lensed by an intervening galaxy which happens to be almost perfectly aligned between the radio source and earth. The lensing galaxy is not visible because it does not contain a radio source. *Right Panel:* Infrared Image of the same system 1938+666 taken with the Hubble Space Telescope [Kin]. Here one sees both the almost perfect Einstein ring image of the background galaxy and the lensing galaxy as the bright spot in the center of the ring (images courtesy of Neal Jackson and NASA STScI).

simple model, is that we are seeing a lensing galaxy surrounded by the Einstein ring image of a background source. The total brightness of the ring plus the central condensation is 18.0 ± 0.3 magnitudes in the H-band, with about 40% of this flux due to the ring image. The redshifts of the source and the lens are not known yet.

A model for this system consisting of an isolated elliptical potential (the major axis of the galaxy has a position angle of 150 degrees) can easily reproduce the observed configuration in the infrared and in the radio at the same time.

5.4 Quasar Microlensing

There are currently a lot of examples of multiply imaged quasars undergoing lensing by galaxies or galaxy clusters. These lenses contain stars, and because each of these quasar images is seen through a galaxy, many stars are directly in front of the quasar. The surface mass density of the stars is of order the critical surface mass density, because the quasar is "macro-lensed" (i.e., multiply imaged).

For such a high surface mass density, the stars produce a quite complicated magnification pattern in the plane of the quasar. The pattern consists of many caustics and varying magnification as a function of quasar position. Due to the relative motion of quasar, galaxy and observer, this arrangement of caustics changes its position relative to the quasar. Consequently, at different times the quasar is magnified differently (cf. Figures 4.3 and 4.4).

Therefore, the apparent brightness of the quasar image changes with time, which should occur for all images of a particular quasar, but in an uncorrelated way.

Such fluctuations contain information about the lensing objects, the transverse velocity, the size of the emission region in the quasar, and even the brightness profile of the quasar. In the case of galactic microlensing, it is quite difficult to disentangle the information about the lens' ma.ss and velocity. Here, this is much more complicated because we are not dealing with an individual lens any more, but rather with an ensemble of lenses, combining their action in a nonlinear way. As a result the mass determinations of an individual lens with quasar microlensing seems intractable. It may be possible, however, to determine statistically the typical mass of the microlenses from the durations and amplitudes of microlensing events, and comparison with simulations. Observationally, this requires frequent high resolution observations of quasars with high accuracy over many years.

The problem of disentangling intrinsic variations of the quasar from microlens induced fluctuations can be dealt with in two ways: If the time delay is measured[5] (e.g., as is the case in Q0957+561), one has to shift the light curves by that amount. Then deviations between the light curves can be attributed to microlensing. The other case is that of four quasar images with a high degree of symmetry (e.g., Q2237+0305). Here the expected time delay is of the order a day or so. Hence, any intrinsic variation of the quasar should show up in all four images at about the same time. Each relative intensity change between the images on time scales of weeks or months must therefore be attributed to microlensing. So far there are indications of microlensing at least in the multiple quasars Q2237+0305, Q0957+561, and Q1115+080.

5.4.1 Microlensing in Quadruple Quasar Q2237+0305

So far, the best example of a microlensed quasar is the quadruple quasar Q2237+0305 (see Figure 5.6). The first evidence of microlensing was found in this system in 1989 by Irwin et al. [Irw]. They found that image A was variable relative to the other images. Many groups have observed this lens, first only once or twice per year. More recently, a number of "monitoring campaigns" were started to get well-sampled light curves of the four images with high photometric accuracy.

Figure 5.6 displays light curves (covering a period of about eight years) for the four images of Q2237+0305.

The figure shows quite a bit of rise and fall in the light curves for the images A and C (top curves), whereas images B and D (bottom curves) are relatively steady at their faint level. The difference between the maximum and minimum of the image A light curve is about 0.7 magnitudes. Non-astrophysicists should note that the horizontal axis in Figure 5.6 is in *Julian day numbers* or JD.[6]

---

[5]This already implies that microlensing cannot be dominant, because otherwise it would be difficult to identify the two or more light curves with each other.

[6]A JD counts the total number of days since 12 hours Universal Time (Greenwich noon) on January 1 in

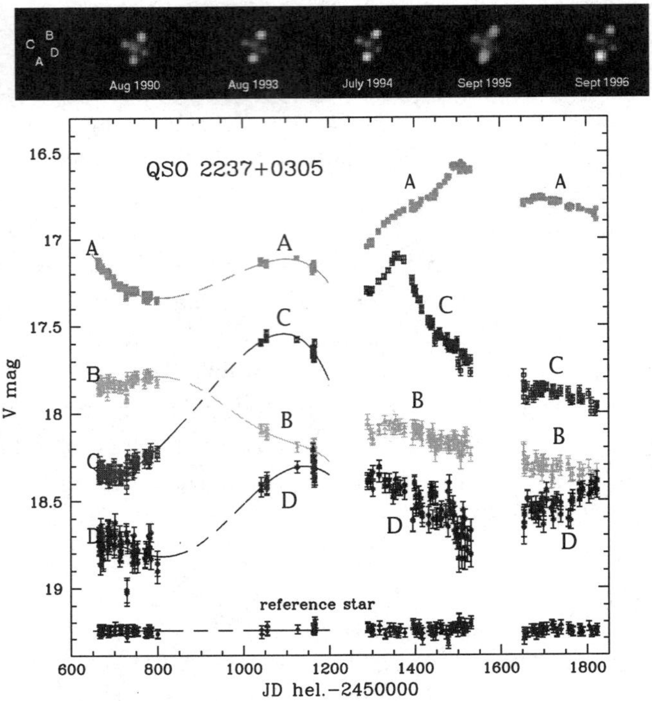

Figure 5.6. The top panel shows the quadruple quasar system Q2237+0305 at five instants of time from 1990 to 1996; it can be seen that the relative brightness of the four images changes with time. In particular, the relative brightness between images A and B (cf. labels at left) is variable. The bottom panel shows recent fluctuations of the four images as light curves over a period of about three years. The light curves of images A to D correspond, respectively, to the curves with the given labels (images courtesy of Roy Ostenson, Przemek Wozniak, and the OGLE team).

The most likely interpretation of the uncorrelated fluctuations in Figure 5.6 is microlensing by ordinary low mass stars in the central part of the lensing galaxy. Due to the relative motion of the quasar, the lensing galaxy, and the Milky Way, the quasar is magnified differently at different times. The high amplitude "events" are probably caustic crossings. They are not arbitrarily sharp spikes, because the quasar is not a point source, but an extended object. In fact, when comparing the observed fluctuations in this quasar with simulations, J.W. [Wam90] showed that the "size of the quasar" (i.e., the continuum emitting region) cannot be larger than about 2×10^{15}cm. Otherwise, we would not be able to see such pronounced individual peaks in the light curves.

We add that usually only the relative transverse motion of the lensing galaxy is considered when simulating the microlensing effect. But since the individual stars in the galaxy

the year 4713 B.C.; see http://www.astronomy.ohio-state.edu/~pogge/Ast350/timesys.html for more on time systems, including the HJD (Heliocentric Julian Date).

are moving as well, this introduces additional fluctuations, due to the change of the caustic network with time. A nice visualization of the moving and changing caustics, based on an analysis of microlensing by random motion of individual stars in the lensing galaxy (J.W. and Kundić [Wam-K]), can be obtained as a movie from the Astrophysical Journal Video Series (1995, Volume 450, Number 1, Part 1).

5.5 Galactic Microlensing: From MACHOS to Planets

The suggestion [Pac86b] to search for dark compact objects in the halo of the Milky Way caused three big experiments to start monitoring the brightnesses of millions of stars: The MACHO, EROS, and OGLE collaborations. Their first results were reported almost simultaneously in September of 1993 [Alc93, Aub, Uda93]. The various groups followed different strategies. To cover different mass scales for the (unknown) lensing objects, the monitoring frequency varied from about two observations per month up to about 10 times per night. The groups had also different emphases. OGLE originally monitored only the galactic bulge, whereas MACHO and EROS looked at the Large Magellanic Cloud as well. Readers should also see Paczyński's article [Pac96] for a comprehensive, recent review of microlensing in our galaxy and galaxies of the Local Group.

Criteria for Identifying Microlensing Events

It had been obvious from the beginning that a massive monitoring campaign would produce a very large number of variable stars, and that the few microlensing events per 10 million stars would rather be like a needle in a haystack of other variables. However, the microlensing light curves had several characteristics that make it possible to distinguish them from all kinds of (known) variable stars. For an individual microlensing event the following statements must hold:

- The light curve is achromatic, symmetric, and follows equation (3.109) (see page 86).
- The event does not repeat.
- The lensed star does not have a peculiar spectrum (which would make it suspicious for a new kind of variable star).

For a large number of microlensing light curves produced by many isolated "Machos," the following statistical arguments must hold:

- The positions of the lensed stars in the sky must be distributed proportionally to the star density of the underlying population.
- The positions of the lensed stars in the color-magnitude diagram must be distributed proportionally to the density of the stars there (otherwise, they could indicate a new kind of variable stars).

- The distribution of maximum magnifications μ_{max} must follow a predetermined relation (see (3.109), page 86): $\mu_{max} = (u_{min}^2 + 2)/(u_{min}\sqrt{u_{min}^2 + 4})$, where u_{min} is the minimum impact parameter u_{min}. Now consider only isolated microlensing events where a source's trajectory (projected in the lens plane) at least touches the Einstein ring of the Macho (i.e., $0 \leq u_{min} \leq 1$) and assume that the impact parameters u_{min} are equally distributed on the interval [0, 1] (since the tracks are random relative to the background stars). This means that each value of u_{min} in the interval is equally likely (i.e., the probability distribution is constant). Then for u_{min} sufficiently small, we have $\mu_{max} \propto 1/u_{min}$ and so 10% of the cases have $u_{min} \leq 0.1$ or $\mu_{max} \geq 10$, while 1% have $u_{min} \leq 0.01$ or $\mu_{max} \geq 100$. Hence, this predicts that for small minimum impact parameters it is approximately 10 times more likely to find a maximum magnification of at least 10 rather than at least 100.

Complications

Many observed microlensing light curves are neither perfectly achromatic, symmetric, nor exactly described by the shape shown in Figure 3.9 (cf. equation (3.109)). There are various reasons for this, and we can list here some of them:

- *Binary Lenses:* More than half the stars exist in double or multiple star arrangements. If their projected distances are roughly comparable to the angular Einstein radius of the total mass, these stars act as a binary lens which produce light curves that are different from the smooth ones shown in Figure 3.9. Light curves due to a binary lens can have "spiky" high amplitude features. Figures 5.7 to 5.11 show, respectively, crossings of fold and cusp caustics produced by a binary lens.

- *Repeating Events:* Even if the double star acting as a lens has a wide separation, so that each lens acts as a single rather than binary lens, it is imaginable (though seemingly unlikely) that the separation of the two stars is just along the direction of the relative motion. This means that there could be two individual microlensing events for the same source star.

- *Parallax:* If the duration of the microlensing event (cf. equation (3.112) is of order a year, then the motion of the earth around the sun introduces a second (vector) component to the relative velocity vector between source, lens and observer. This results in a (small) deviation from the smooth curves. At least one such event has been observed [Alc95].

- *Extended Source:* The idealized scenario for microlensed light curves mentioned above assumed point sources. Since in the real world all physical sources are finite, at some point there are deviations between ideal and real behavior. The finite size of the source can lead to higher or lower amplitude events (Witt [Wit95]).

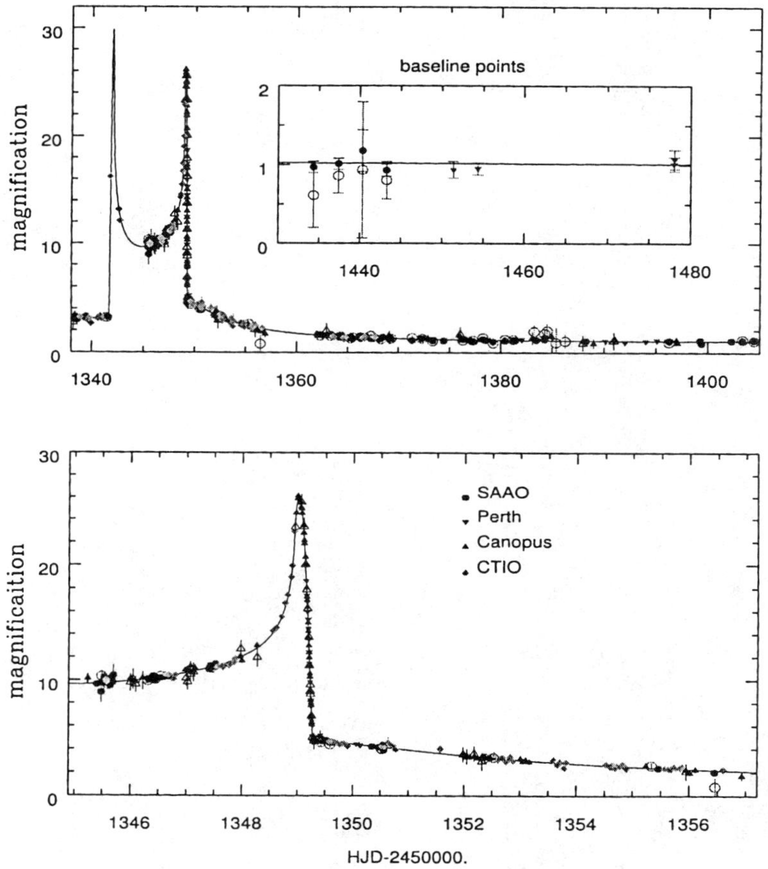

Figure 5.7. Microlensing light curve of a light source crossing fold caustics due to a binary lens in the direction of the Galactic Bulge. The data is from the PLANET collaboration's observations of the event OGLE-1999-BUL-23. It shows data from four participating telescopes: SAAO, Perth, Canopus, and CTIO. The closed symbols are for the I-band, while the open symbols are in the V-band. The "U" shape displays the typical form produced by a light source crossing two nearby fold arcs (see Figure 9.3 on page 331). The bottom panel shows a close-up of the light curve's right peak. The horizontal axis denotes time using the Heliocentric Julian Date. See Footnote 6, p. 158. (Courtesy of the PLANET collaboration [Alb00b].)

- *Binary Source/Blending/Achromaticity:* If an unresolved double star is gravitationally lensed, then only one of the two partners is magnified. Should the two stars have different colors, the color of the combined light curve changes during the event due to the differential magnification. The same is true if there is no physical pair of stars, but their projected positions happen to be near each other in the sky plane.

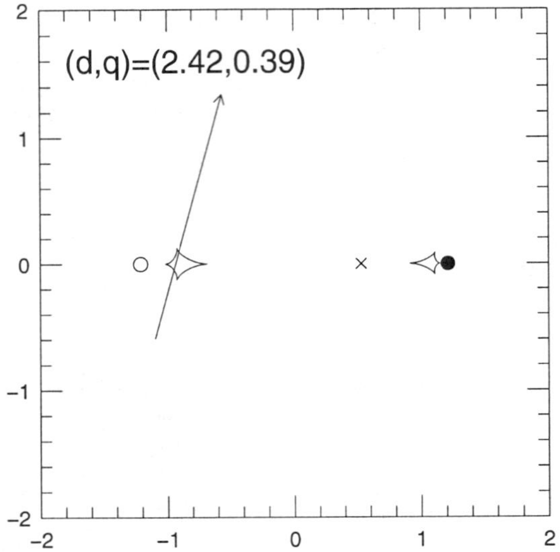

Figure 5.8. Caustics due to a binary point-mass lens with separation $d = 2.42$ (wide binary — cf. page 517) and mass ratio $q = m_2/m_1 = 0.39$. The more massive component is m_1 located at $\bullet$, while m_2 is at $\circ$. The center of mass of the lenses is shown by $\times$ and the origin is at the geometric center of the mass positions. A light source following the indicated trajectory has a light curve given by the one fitting the data in Figure 5.7. It is the PLANET collaboration's best-fit model for the system [Alb00b]. (Courtesy of the PLANET collaboration.)

These effects make the whole field more complicated, but more interesting as well. In fact, *the "binary lens" scenario allows us to observe double stars with high mass ratio. This offers the exciting possibility of using microlensing to detect planetary systems around distant stars!*[7]

In Figure 5.9, the indicated light curve fitting the data requires a model with a two point mass lens, light source blending, extended light source size, and limb darkening of the source — see [Alb99] for detailed treatment of this model. The shape of the light curve shows features typical of "outside" crossings through an isolated cusp. Figure 9.5 on page 334 illustrates the qualitative generic behavior of light curves near cusp crossings in the case of lensing by any finite number of lens planes. Other examples for the binary lens case are given in Figure 3.11, page 95.

The cusp crossing of Figure 5.9 is shown in more detail in Figure 5.10. It also depicts the PLANET collaboration's best-fit light curves for when the source has a light intensity profile modeled with limb darkening versus uniform brightness across the stellar disc. Limb darkening clearly presents a better fit. In the figure, there is a discontinuous change in the

[7]The subsection *Binary Systems and Planets* on page 123 describes a technique for planet detection using microlensing. Section 15.2.2 (page 514) and Section 15.2.3 (page 524) further discuss binaries, with application to the star-planet regime.

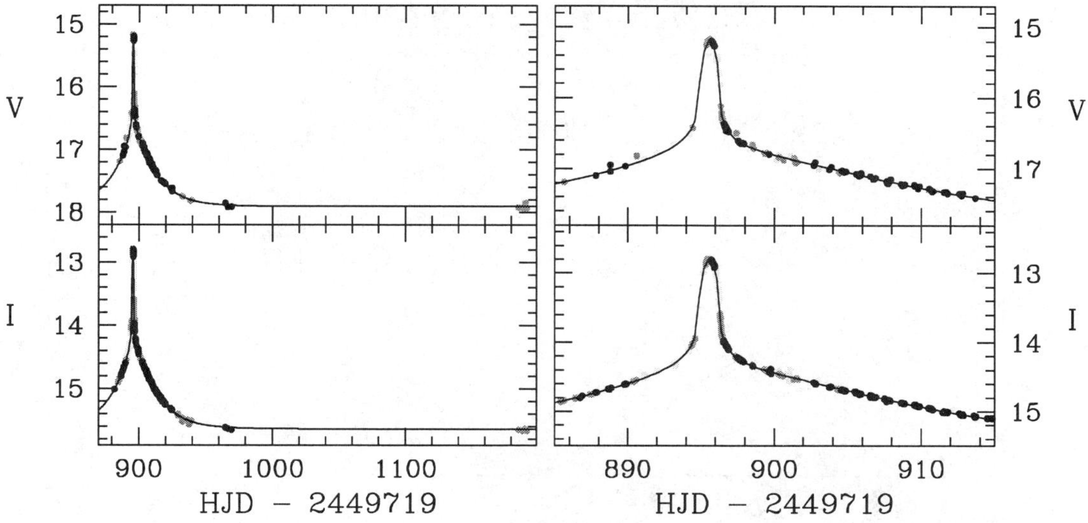

Figure 5.9. *Left:* Microlensing light curve of a light source passing outside a cusp point produced by a binary lens. This event (MACHO-1997-BUL-28) is shown in the I- and V-bands, and spans a 300 day period. It was observed by the PLANET collaboration [Alb99], which also provided a best-fit model indicated by the solid curve — see Figures 5.10 and 5.11. *Right:* An enlargement showing the 30-day period near the peak. (Courtesy of the PLANET collaboration.)

slope at HJD-2449719 = 896.4 in the light curves of the bottom left panels. This occurs when part of the source's limb (outer edge) passes through one of the fold caustic curves abutting the cusp — see Figure 5.11.

The Current State

By now more than 500 microlensing events have been detected, with the vast majority towards the bulge of our galaxy. Recently, the MACHO collaboration published 14 events towards the Large Magellanic Cloud. The durations of the events range from nine to 200 days. The current situation is nicely summarized in Paczyński's review article [Pac96]. The impact of the microlensing searches can easily be seen when looking at the following numbers. According to a compilation by the Surdejs [Sur], in 1995 a total of about 200 papers were published in the general field of gravitational lensing (Figure 1.2, page 12). As many as 64 of them — almost a third — deal with galactic microlensing alone!

The color distribution of the "lensed" stars in the center of the Milky Way and the distribution of their maximum magnifications show no peculiarities. There is no evidence, though, that these events towards the Milky Way's center are related to dark matter; all can be explained by ordinary low mass stars. The quantitative picture is such that the number of events towards the center is about three times higher than expected! This clearly shows that the Milky Way's structure was not understood well enough when the experiments started. Hence, lensing is also a new tool for the study of the structure of our galaxy. The number

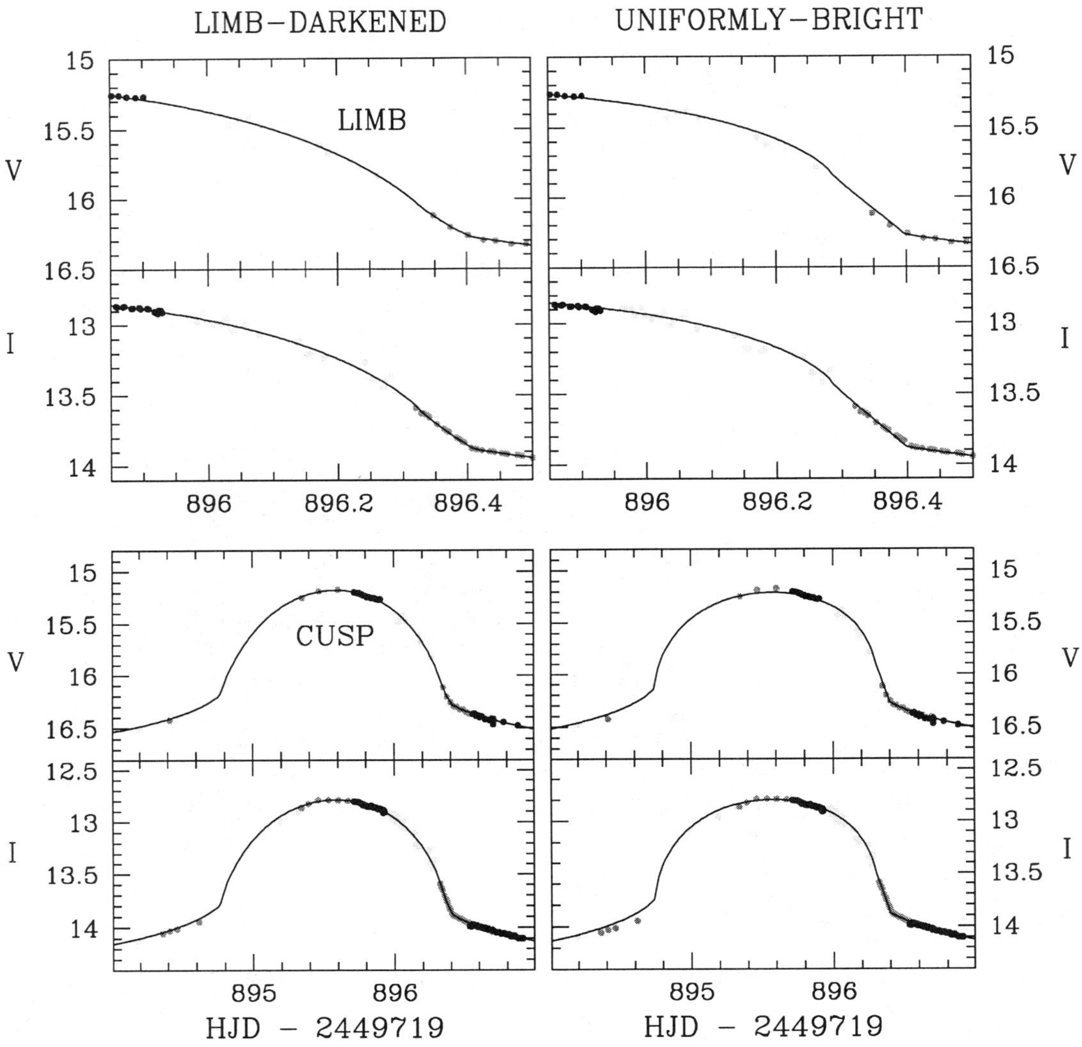

Figure 5.10. Cusp crossing event MACHO-1997-BUL-28 (cf. Figure 5.9). These light curves are for the PLANET collaboration's best fitting limb-darkening model superimposed on the V and I data sets for a 3-day period centered on the cusp-crossing (bottom) and a 16-hour period during which the stellar limb swept over the cusp (top)[Alb99]. Note the discontinuous change in the slope at 896.4. *Right:* Same for the best uniform bright source model. (Courtesy of the PLANET collaboration.)

of events towards the Large Magellanic Cloud is lower than one would have expected if the halo were made entirely of compact objects. The most recent analysis estimates that less than 50% of the matter in the halo could be made out of machos. It could be much less than that as well. However, the most likely mass of the lenses is of order half a solar mass, which is higher than what one would naively expect.

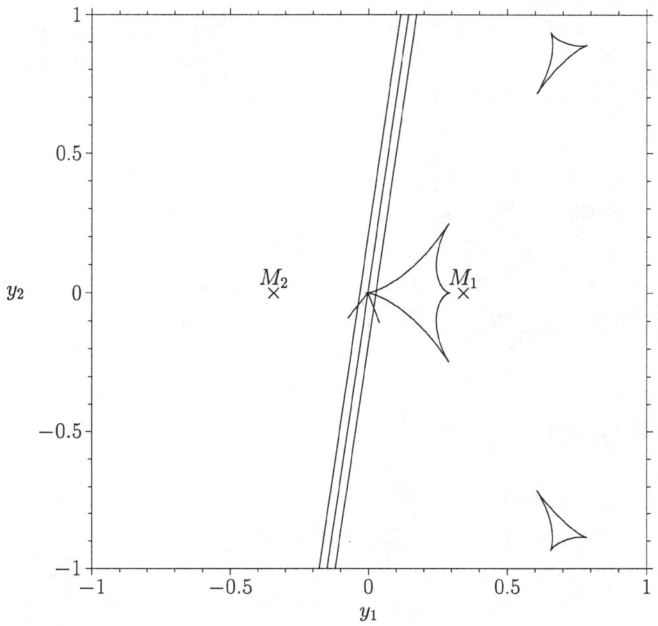

Figure 5.11. A disc shaped light source is centered on the inner parallel line and its diameter coincides with the distance between the two outer lines. As the source traverses the indicated path, it passes through an isolated cusp on the central caustic. Note that part of the source's limb passes through the fold arcs abutting the cusp. The caustics are due to two point masses M_1 and M_2 ($M_1 > M_2$), whose projected positions in the light source plane are shown by the symbol $\times$. The light curve for this system fits the data in Figure 5.9 [Alb99]. (Courtesy of the PLANET collaboration.)

Observations of Planets via Microlensing

There are three collaborations that monitor about 10 million stars in the galactic bulge, the central part of the Milky Way (i.e., they determine the brightness of these stars regularly in short intervals). This is done to identify those stars that are magnified by an intervening star acting as a gravitational lens. The teams (MACHO, EROS, and OGLE) have automated their software to the point that they are able to identify a background star that starts to show an increase in brightness characteristic of a microlensing event. These rare cases are identified immediately and interested astronomers are "alerted," that is, they get e-mail about position and brightness of each candidate microlens event.

Now, a different set of teams, named PLANET (acronym means: "Probing Lensing Anomalies NETwork"), GMAN ("Global Microlensing Alert Network"), and MPS ("Microlensing Planet Search"), is also involved in microlensing monitoring. However, they follow only the alerted candidates, with a different set of telescopes and much higher monitoring frequency, ideally a few times per hour. The goal is to measure a very well covered light curve for all the microlensing events. This is in order to detect any deviations from the smooth single-lens light curve ("anomalies"). So far, more than several hundred microlensing events towards the direction of the galactic bulge have been found. A fair fraction of them display deviations indicative of nearly equal mass binary lenses, rather than star-planet mass ratios. It is to be expected that these microlensing searches will soon detect planetary companions as well. Their signature would be a short term deviation (of order a day or shorter!) from the smooth point-lens light curve. This requires working on a "24 hour night shift" [Sack]. Gaudi and Sackett [Gau-S] developed an algorithm for determining detection efficiencies, that is, the probability of detecting a binary point mass lens from a given light curve data, where the light source's trajectory is assumed to be random. See Albrow et al.[Alb00a] for an application of this algorithm to real data.

5.6 Weak Lensing

The most well-known features of gravitational lensing are connected with spectacular astrophysical phenomena: double quasars, giant luminous arcs, Einstein rings. These phenomena are all related to the presence of caustics, involving very high magnification and strong deformation of the sources, and/or multiple imaging. This regime is often called "strong lensing." Only about one out of 500 far away (high redshifted) objects are known to be affected by strong lensing. In contrast, "weak lensing" deals with all measurable effects of light deflection outside the regime of high magnification and multiple imaging. All fluctuations in the matter distribution between us and distant light sources affect the measurable properties of the sources in three ways: The angular size and shape of extended objects is modified; the apparent brightnesses of the objects are affected; and their positions can be changed (cf. Sections 3.2.5 and 3.2.7). This was emphasized by James Gunn more than 30 years ago [Gun67a, Gun67b]. Since small fluctuations in matter density are ubiquitous in the universe, weak gravitational lensing should be observable "everywhere" as well. In principle, weak lensing acts along each line of sight in the universe, since each photon's path is affected by matter inhomogeneities along or near its path. It is just a matter of how accurately we can measure (Press [Pre]).

The weak lensing signature can be detected on an *individual* object only if the unlensed shape, brightness, and/or position are known. This is rarely the case. There are instances, however, where it is possible. Light deflection at the solar limb observable during a solar eclipse is an example of positional weak lensing. The Macho-type light curves for large impact parameters illustrate the weak lensing effect on brightness. Both phenomena can be detected because they lead to temporal changes in the observed objects due to "moving" lenses, and the relevant un-lensed parameters can be determined before or after the lensing event occurs. But since weak lensing usually acts on a large region of the sky, it can also be established by measuring a small effect on a large number of sources. In this *statistical* sense the weak lensing effect can, for example, be seen as the coherent deformation of the shape of (extended) background sources around foreground galaxy clusters. (Even the 1919 light deflection at the solar limb was measured only in a statistical sense, namely, for several dozen stellar positions, because it was too difficult to do for an individual star.)

We shall see that weak lensing can be both a blessing and a curse for the astronomer. The statistically coherent small deformation of background galaxies seen around galaxy clusters can be used to determine the cluster mass distribution. On the other hand, the statistical incoherent lens-induced change of the apparent brightness of "standard candles" decreases the accuracy of the determination of cosmological parameters. The idea to use the distortion and tangential alignment of faint background galaxies to map the mass distribution of foreground galaxies/clusters is not new. The first observational attempts go back to 1978/79, when J. Anthony Tyson and John Jarvis were trying to measure the positions and orientations of the then newly discovered faint blue galaxies, which were suspected to be at large distances. Due to the not quite adequate techniques at the time (photographic plates), these efforts ended unsuccessfully [Val-TJ]. Only with the advent of the new technology of CCD cameras, with much improved imaging properties of telescope optics, with a successful reduction in the "seeing" at the telescopes sites, and an accurate determination of galaxy shapes, could the weak lensing effect ultimately be found.

Weak lensing is one of two sub-disciplines within the field of gravitational lensing with the highest rate of growth in the last couple of years (along with galactic microlensing). The reason is that the two necessary requirements for detailed exploration of "small" shape distortions became available a few years ago: Technical equipment needs to be up to the job (large format high resolution cameras with excellent imaging properties) and statistical methods to analyze the data need to be available. We shall present only one type of weak lensing observations: The reconstruction of the mass distribution of galaxy clusters from weak lensing. Other applications of weak lensing that are currently under hot pursuit by many groups are, for example, constraining the redshift distribution of the faint galaxies from weak lensing, studying the effects of weak lensing on the determination of cosmological parameters from the use of "standard candles," galaxy-galaxy weak lensing, and the weak lensing effects of the large scale structure of the universe. Readers are referred to [Bart-S] and [Mel], and [Bac], [Kai-WL], [van00], [Wittm00] for recently observed large scale structure weak lensing.

Cluster Mass Reconstruction

The first real coherent weak lensing signal of distorted background galaxies was measured by Tyson, Valdes and Wenk [Tys-VW] around the galaxy clusters Abell 1689 and CL1409+52. They showed that the orientation of background galaxies — the angle of the semi-major axes of the elliptical isophotes relative to the center of the cluster — was more likely to be tangentially oriented rather than radially relative to the cluster. For an unaffected population of background galaxies one would expect no preferential direction. This analysis is based on the assumption that the major axes of the background galaxies are intrinsically randomly oriented.

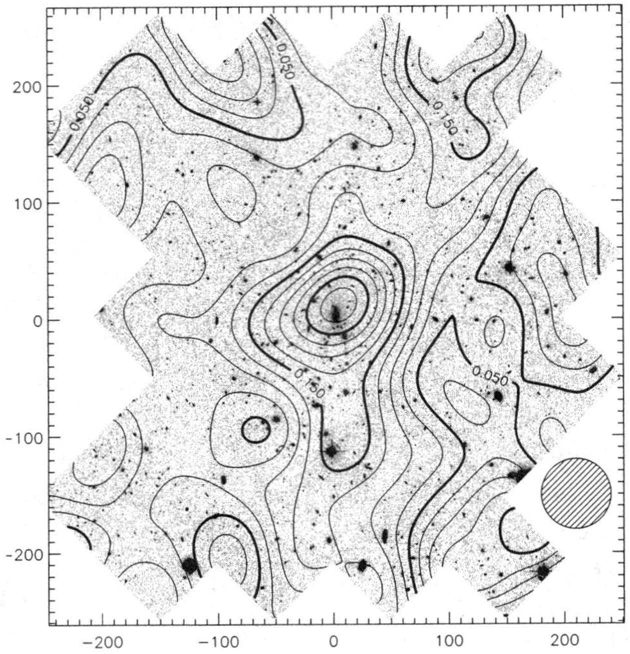

Figure 5.12. Reconstructed mass distribution of galaxy cluster CL1358+62 from a weak lensing analysis is shown as contour lines superposed on the image taken with the Hubble Space Telescope [Hoe-FKS]. The map is smoothed with a Gaussian of size 24 arcsec (see shaded circle). The center of the mass distribution agrees with the central elliptical galaxy. The numbers indicate the reconstructed surface mass density in units of the critical density (Credits: Henk Hoekstra).

With the powerful method developed in 1993 by Kayser and Squires, such a signal could be used to reconstruct the surface mass distribution (which includes the dark matter) of the cluster [Kai-S]; see Section 4.2.3. As mentioned in that section, the original reconstruction method, though very elegant, was still not perfect. The reconstruction technique was refined, extended, and generalized by several authors (e.g., [Bart-NSS, Broa-TP, Kai95, Schn-Se, Squ-K, Sei-KSS]). Maximum likelihood techniques, non-linear reconstructions, and methods using the magnification effect rather than the distortion effect, complement each other.

In Figure 5.12, an example is shown of a reconstruction by Hoekstra et al. [Hoe-FKS] of the mass profile of the galaxy cluster CL1358+62.

Part III

MATHEMATICAL ASPECTS

The goal of Part III is to take first steps towards a general mathematical theory of gravitational lensing. Specifically, we shall attempt to develop a mathematical infrastructure for optics in weak gravitational fields due to thin matter deflectors. This will be accomplished by addressing certain key mathematical problems in gravitational lensing. In the process, we shall present lensing results that are "universal," that is, they do not rely on the many approximations and assumptions used in thin-screen, weak-field lensing. We illustrate these "universal" results with specific physically important lens models, and even analyze these models for additional quantitative results.

Chapter 6

Time Delay and Lensing Maps

Philosophy is written in that great book which ever lies before our gaze — I mean the universe — but we cannot understand if we do not first learn the language and grasp the symbols in which it is written. The book is written in the mathematical language, and the symbols are triangles, circles and other geometrical figures, without the help of which it is impossible to conceive a single word of it, and without which one wanders in vain through a dark labyrinth.

GALILEO GALILEI

We shall develop two useful perspectives that lie at the heart of the mathematical theory of gravitational lensing. One viewpoint uses light ray arrival times, that is, *time delay functions*, and the other employs ray tracing, which involves *lensing maps*. The goal of this chapter is to precisely define these mappings (including subsidiary concepts like light rays, lensed images, magnification, etc.) for single and multiplane lensing, relate them mathematically, and discuss their relative advantages. In the process, we shall also extend part of earlier work by A.P. [Pet91, Pet96b], where first steps were taken towards expressing the core concepts of weak-field gravitational lensing in precise mathematical form.

A major challenge we faced with the current chapter was to abstract gravitational lensing mathematically without losing sight of the essential physical ideas. Too often the latter falls victim to exorbitant definitions and inconsequential (self-indulgent) generalities. We tried our best to chart a middle course. For this reason, definitions are constructed at a level of generality that strongly couples with the most physically important gravitational lens models, while allowing nontrivial progress on the problems posed in Chapter 2. On the other hand, we shall also explore mathematical features that add to our understanding of gravitational lensing, even if they seem beyond the scope of current physical observations.

Outline: Section 6.1 formalizes the notion of a gravitational lens potential, while Section 6.2 deals with its single-plane lensing features. The latter includes precise treatments of the following concepts: time delay functions, light rays, lensing maps, lensed images, magnification, critical points, and caustics. We discuss the interrelationships of these notions, highlighting how one concept yields an alternative view of another. For example, by using Fermat's Principle to identify a lensed image with its associated light ray, we can assign the light ray's "index" to the lensed image. This allows a classification of lensed images according to "index" (i.e., critical point type: minima, saddles, or maxima). The ideas of Section 6.2 are illustrated in Section 6.3 through the analysis of three simple lens models: uniform sheet, singular isothermal sphere, and point mass lens. At the end of Section 6.3, we present a table of gravitational lens models. Section 6.4 generalizes the concepts of Section 6.2 to multiple lens planes. This section starts with the notions of a lens plane and angular diameter distance. The section then explicitly expresses the domains of multiplane time delay and lensing maps, where the latter is introduced recursively and leads to the notion of "obstruction points." At the end of Section 6.4, we also show that the ray-tracing perspective naturally expresses a multiplane lensing map as a composition of maps. Section 6.5 details the connections between multiplane time delay and lensing maps (as well as light rays, lensed images, etc.). It entails a discussion of the relative advantages of time delay and lensing maps in proving certain key facts about gravitational lensing.

6.1 Gravitational Lens Potentials

We shall mathematically abstract the notion of a thin-screen weak-field gravitational lens; see Section 3.1.4 (page 46) for a physical treatment.

Let A be a finite set of points in $\mathbf{R}^2$. We think of A as corresponding to the "singularities" of the lens, that is, points where the surface mass density is unbounded. Let $L = \mathbf{R}^2 - A$ and denote by Δ the Laplace operator on $\mathbf{R}^2$ relative to the dimensionless rectangular coordinates $\mathbf{x}$ (cf. Section 3.2.4, page 79).

Definition 6.1 [1] A *gravitational lens potential with singularities* A is a smooth function $\psi : L \longrightarrow \mathbf{R}$ satisfying the following:
(1) The function ψ is subharmonic, that is, $\Delta\psi(\mathbf{x}) \geq 0$ for all $\mathbf{x} \in L$.
(2) For each $\mathbf{a} \in A$, as $\mathbf{x} \to \mathbf{a}$ either $\psi(\mathbf{x}) \to -\infty$ or $\Delta\psi(\mathbf{x}) \to \infty$.[2]

If $A = \emptyset$, then ψ is called *nonsingular* (or, *transparent*); otherwise, ψ is *singular*. A singularity $\mathbf{a} \in A$ such that $\psi(\mathbf{x}) \to -\infty$ as $\mathbf{x} \to \mathbf{a}$ will be called an *infinite singularity* of ψ. Hereafter denote grad ψ by $\boldsymbol{\alpha}$, which we shall term the *bending angle vector field* due to ψ. Note that the domain of $\boldsymbol{\alpha}$ is L. The magnitude $|\boldsymbol{\alpha}|$ is the *bending angle* due to

[1] Definition 6.1 readily extends to potentials with weaker differentiability and an infinite number of singularities (cf. cosmic strings – Section 4.6). However, our assumptions suffice for most physically relevant gravitational lens models. Also, note that the condition $\Delta\psi \geq 0$ does not hold everywhere for weak lensing by large scale structures — see Section 3.1.4. But strong lensing will be our main focus.

[2] Note: If $\psi(\mathbf{x}) = \ln|\mathbf{x} - \mathbf{a}|$, then $\psi(\mathbf{x}) \to -\infty$, but $\Delta\psi(\mathbf{x}) = 0$, as $\mathbf{x} \to \mathbf{a}$. For $\psi(\mathbf{x}) = |\mathbf{x}|$, as $\mathbf{x} \to \mathbf{0}$ we get $\Delta\psi(\mathbf{x}) = |\mathbf{x}|^{-1} \to \infty$, while $\psi(\mathbf{x}) \to 0$.

ψ. The (dimensionless) *surface mass density perturbation* (cf. Section 3.1.4, page 46) or *convergence* (cf. page 97) associated with ψ is defined by

$$\kappa = \frac{\Delta\psi}{2},$$

which is a smooth function on L. Following common practice, we shall refer to κ simply as a *surface mass density*. If ψ is nonsingular (resp., singular), then κ is also termed *nonsingular* (resp., *singular*). Call κ *subcritical* (resp., *supercritical, critical*) *at* $\mathbf{x}$ if $\kappa(\mathbf{x}) < 1$ (resp., $\kappa(\mathbf{x}) > 1$, $\kappa(\mathbf{x}) = 1$).

The *shear matrix* at $\mathbf{x}$ due to ψ is defined by (Section 3.2.7, page 95):

$$\boldsymbol{\Gamma}(\mathbf{x}) = \begin{bmatrix} \Gamma_1(\mathbf{x}) & \Gamma_2(\mathbf{x}) \\ \Gamma_2(\mathbf{x}) & -\Gamma_1(\mathbf{x}) \end{bmatrix}, \tag{6.1}$$

where

$$\Gamma_1(\mathbf{x}) = \frac{1}{2}[\psi_{\mathrm{uu}}(\mathbf{x}) - \psi_{\mathrm{vv}}(\mathbf{x})], \qquad \Gamma_2(\mathbf{x}) = \psi_{\mathrm{uv}}(\mathbf{x}),$$

with $\mathbf{x} = (\mathrm{u}, \mathrm{v})$. Employ a polar coordinate parametrization of the components of $\boldsymbol{\Gamma}$ as follows:

$$\Gamma_1(\mathbf{x}) = \Gamma(\mathbf{x}) \cos 2\theta(\mathbf{x})$$

and

$$\Gamma_2(\mathbf{x}) = \Gamma(\mathbf{x}) \sin 2\theta(\mathbf{x}).$$

Here

$$\Gamma(\mathbf{x}) = \sqrt{\Gamma_1^2(\mathbf{x}) + \Gamma_2^2(\mathbf{x})}, \qquad 0 \leq \theta(\mathbf{x}) < \pi.$$

The angle $\theta(\mathbf{x})$ is restricted to the latter interval since the shear at $\mathbf{x}$ along angle $\theta(\mathbf{x})$ coincides with that along $\theta(\mathbf{x}) + \pi$. We then call $\Gamma(\mathbf{x})$ the *shear function* or *magnitude of shear*, and $\theta(\mathbf{x})$ the *angle of shear*. Note that since the shear at $\mathbf{x}$ along angle $\theta(\mathbf{x})$ coincides with that along $\theta(\mathbf{x}) + \pi$, we restrict $\theta(\mathbf{x})$ to the interval $[0, \pi)$.

Remark: The shear matrix contributes an anisotropic cross sectional stretching of a light beam due to matter outside the beam. In particular, a small disc light source not on a caustic is mapped to an elliptically shaped lensed image. See Section 3.2.7 (page 95) for more.

We conclude this section with two very important families of gravitational lens potentials (see Section 3.3).

Isothermal Ellipsoids and Elliptical Potentials

Isothermal ellipsoidal gravitational lens potentials are defined as follows (cf. Section 3.3.2, page 105):

$$\psi_{ie}(\mathbf{x}) = \mathrm{u}\alpha_1 + \mathrm{v}\alpha_2 - b_I s_c \ \ln \sqrt{(h + s_c)^2 + (1 - q^2)\mathrm{u}^2},$$

where

$$\alpha_1 = \frac{b_I}{\sqrt{1-q^2}} \tan^{-1}\left[\frac{\mathrm{u}\sqrt{1-q^2}}{h(\mathbf{x})+s_c}\right],$$

$$\alpha_2 = \frac{b_I}{\sqrt{1-q^2}} \tanh^{-1}\left[\frac{\mathrm{v}\sqrt{1-q^2}}{h(\mathbf{x})+q^2 s_c}\right]$$

with $\mathbf{x} = (\mathrm{u}, \mathrm{v})$, $b_I > 0$, $q > 0$, $s_c \geq 0$, and $h(\mathbf{x}) = q^2(s_c^2 + \mathrm{u}^2) + \mathrm{v}^2$. These potentials are realistic models of galaxies acting as gravitational lenses (see Section 3.3.2).

We shall call s_c the *(dimensionless) core radius* of ψ_{ie}. If $s_c = 0$, then ψ_{ie} is called a *singular isothermal ellipsoid*, while for $s_c > 0$ the model is a **softened isothermal ellipsoid** (or *isothermal ellipsoid with finite core*). The surface mass density and bending angle vector field associated with ψ_{ie} are, respectively:

$$\kappa_{ie}(\mathbf{x}) = \frac{b_I}{2\sqrt{q^2(s_c^2+\mathrm{u}^2)+\mathrm{v}^2}}, \qquad \boldsymbol{\alpha}_{ie}(\mathbf{x}) = (\alpha_1, \alpha_2).$$

A very simple class of potentials that are used as semi-realistic models of galaxies are the *elliptical type potentials*:

$$\psi_{ep}(\mathbf{x}) = C\sqrt{x_{cr}^2 + (1-\epsilon)\mathrm{u}^2 + (1+\epsilon)\mathrm{v}^2},$$

where $C > 0$, $x_{cr} \geq 0$, and the *ellipticity* ϵ is nonnegative. **Due to physical reasons, the ellipticity will henceforth be assumed to obey** $0 \leq \epsilon \lesssim 0.2$; for larger values, the potential has unphysical features (e.g., dumbbell-shaped mass density contours [Kas-K, p. 460]). The term x_{cr} is called the *(dimensionless) core radius* of ψ_{ep}.

The mass density and bending angle vector field of ψ_{ep} are given, respectively, by

$$\kappa_{ep}(\mathbf{x}) = \frac{C^4}{2\psi_{ep}^3(\mathbf{x})}\left[2x_{cr}^2 + (1-\epsilon^2)^2|\mathbf{x}|^2\right]$$

and

$$\boldsymbol{\alpha}_{ep}(\mathbf{x}) = \frac{C^2}{\psi_{ep}(\mathbf{x})}((1-\epsilon)\mathrm{u}, (1+\epsilon)\mathrm{v}).$$

For $x_{cr} > 0$, the potential ψ_{ep} is nonsingular. If $x_{cr} = 0$, then the bending angle vector field $\boldsymbol{\alpha}_{ep}$ is discontinuous at $\mathbf{x} = \mathbf{0}$ and $\kappa_{ep}(\mathbf{x}) \to \infty$ as $\mathbf{x} \to \mathbf{0}$, that is, a singularity occurs at the origin $\mathbf{0}$ (though $\psi_{ep}(\mathbf{0})$ is finite).

If $x_{cr} = 0$ and $\epsilon = 0$, then ψ_{ep} is called a *singular isothermal sphere* and denoted by ψ_{sis}. In this case, the bending angle vector field has constant magnitude:

$$|\boldsymbol{\alpha}_{sis}(\mathbf{x})| = C.$$

For $x_{cr} > 0$ and $\epsilon = 0$, the potential ψ_{ep} is called a *softened isothermal sphere* (or *isothermal sphere with finite core*) and denoted by ψ_{fis}. Finally, the case with $x_{cr} > 0$ and $0 < \epsilon \lesssim 0.2$ is referred to as an *an elliptical potential with singular core* if $x_{cr} = 0$ and *softened* (or *finite*) *core* if $x_{cr} > 0$.

Point Mass Potentials with Continuous Matter and Shear

A gravitational lens potential due to g point masses (e.g., stars) on a plane with continuous matter having constant density κ_c, and experiencing a constant shear γ from infinity along angle θ, is of the form (cf. Section 3.3.1, page 101):

$$\psi_{ptcs}(\mathbf{x}) = \frac{\kappa_c}{2}|\mathbf{x}|^2 - \frac{\gamma}{2}\left[(\mathrm{u}^2 - \mathrm{v}^2)\cos 2\theta + 2\mathrm{uv}\sin 2\theta\right] + \sum_{\ell=1}^{g} m_\ell \ln|\mathbf{x} - \boldsymbol{\xi}_\ell|, \qquad (6.2)$$

where $m_\ell \geq 0$, $\kappa_c \geq 0$, $\gamma \geq 0$, and $0 \leq \theta < \pi$. Physically, the constant γ is the magnitude of shear from infinity along angle θ relative to ray tracing from observer to the light source plane.[3]

Without loss of generality, we shall choose the angle θ along the u-axis (i.e., $\theta = 0$) in the case of single-plane lensing:[4] Then

$$\psi_{ptcs}(\mathbf{x}) = \frac{\kappa_c}{2}|\mathbf{x}|^2 - \frac{\gamma}{2}(\mathrm{u}^2 - \mathrm{v}^2) + \sum_{\ell=1}^{g} m_\ell \ln|\mathbf{x} - \boldsymbol{\xi}_\ell|.$$

This yields a surface mass density that is discontinuous at the point mass positions $\boldsymbol{\xi}_1, \ldots, \boldsymbol{\xi}_g$, but is smooth otherwise: $\kappa_{ptcs}(\mathbf{x}) = \kappa_c$, where $\mathbf{x} \in \mathbf{R}^2 - \{\boldsymbol{\xi}_1, \ldots, \boldsymbol{\xi}_g\}$. We may interpret κ_{ptcs} as a distribution and write

$$\kappa_{ptcs} = \pi(m_1\delta_{\boldsymbol{\xi}_1} + \cdots + m_g\delta_{\boldsymbol{\xi}_g}) + \kappa_c,$$

where $\delta_{\boldsymbol{\xi}_\ell}$ is the Dirac-delta distribution centered at $\boldsymbol{\xi}_\ell$. The bending angle vector field of ψ_{ptcs} is given by

$$\boldsymbol{\alpha}_{ptcs}(\mathbf{x}) = ((\kappa_c - \gamma)\mathrm{u}, (\kappa_c + \gamma)\mathrm{v}) + \sum_{\ell=1}^{g} m_\ell \frac{\mathbf{x} - \boldsymbol{\xi}_\ell}{|\mathbf{x} - \boldsymbol{\xi}_\ell|^2}.$$

As $\mathbf{x} \to \boldsymbol{\xi}_\ell$, where $1 \leq \ell \leq g$, we have $|\boldsymbol{\alpha}_{ptcs}(\mathbf{x})| \to \infty$. Hence, the potential ψ_{ptcs} has g singularities corresponding to the point mass positions.

Remark: For a point mass potential with continuous matter and shear we have that $|\boldsymbol{\alpha}_{ptcs}(\mathbf{x})| \to \infty$ as $|\mathbf{x}|$ approaches a singularity $\boldsymbol{\xi}_\ell$, and, if $\kappa_c \neq \gamma$, then $|\boldsymbol{\alpha}_{ptcs}(\mathbf{x})| \to \infty$ as $|\mathbf{x}| \to \infty$. However, a singular isothermal sphere, which has the origin as its singularity, behaves differently: $|\boldsymbol{\alpha}_{sis}(\mathbf{x})| = C$ as $|\mathbf{x}| \to 0$ and $|\mathbf{x}| \to \infty$.

[3] See the subsection "Shear From Infinity" on page 103.

[4] In multiplane lensing, the condition $\theta_0 = 0$ is too restrictive since shear need not act along the same direction on every lens plane.

6.2 Single-Plane Lensing

Time delays (i.e., arrival time differences) and ray tracing naturally determine two mappings: "time delay functions" and "lensing maps." Our goal is to define these maps precisely for gravitational lenses on a single lens plane. The reader should consult Sections 3.2.2 (page 67) and 3.2.3 (page 77) for the physical basis of the current section.

6.2.1 Definition of Critical Point

The precise notion of a "critical point" is fundamental to our mathematical treatment.

Definition 6.2 Let $f : X \to Y$ be smooth map between manifolds, where $\dim X = n$ and $\dim Y = m$. Then:

(1) A point $x \in X$ is a *critical point* of f if $\operatorname{rank}[d_x f] < \min\{n, m\}$. If x is a critical point of f, then $f(x)$ is a *critical value* of f. Denote the set of critical points of f by $\operatorname{Crit}(f)$.

(2) If $x \notin \operatorname{Crit}(f)$, then x is a *regular point* of f. If $y \in Y$ is not a critical value of f, then y is a *regular value* of f (even if $y \notin f[X]$).

Remark: Some authors refer to a point x in X as a *singular point* of $f : X \to Y$ if $\operatorname{rank}[d_x f] < \min\{n, m\}$, and a *critical point* if $\operatorname{rank}[d_x f] < m$. We avoid this usage since the term "singularity" is reserved for a point where the potential or surface mass density of a gravitational lens "blows up" (Definition 6.1). Moreover, the latter is consistent with the use of "singularity" in general relativity — see Wald [Wald84, pp. 212–16].

The following is a basic fact about critical points (see [Gol-G, pp. 30–36] for detailed proof):

Theorem 6.3 (Sard's Theorem) *The set of critical values of a smooth map $f : X \to Y$ has measure zero in Y.*

If the set of regular values of f is not dense in Y, then f has a nonempty open set of critical values. This contradicts Sard's theorem since open subsets of a manifold do not have zero measure. Thus:

Corollary 6.4 (Brown's Theorem) *The set of regular values of a smooth map $f : X \to Y$ is dense in Y.*

Under certain conditions, the regular values also form an open set

Proposition 6.5 *If $f : X \to Y$ is smooth proper map,*[5] *where* $\dim X \geq \dim Y$, *then the set of regular values of f is open in Y.*

[5] Recall that a map $f : X \to Y$ is *proper* if the f preimage of each compact subset of Y is compact in X.

Proof: We need the following elementary lemma:

Lemma. If $f : X \to Y$ is a smooth proper map, then $f(X)$ is closed.

Proof: Assume that $f(X)$ is not closed. Let y be a point in the closure of $f(X)$ such that $y \notin f(X)$. Then there is a sequence of points $\{y_i\}$ in $f(X)$ converging to y. The set $C = \{y, y_1, y_2, y_3, \dots\}$ is compact. Since f is proper, the preimage $f^{-1}(C)$ is also compact. Because X is metrizable (e.g., [Gol-G, p. 24]), compactness of $f^{-1}(C)$ is equivalent to sequentially compactness. Let $\{x_j\}$ be a sequence of points in $f^{-1}(C)$ such that $f(x_j) = (y_i)_j$ is a subsequence of $\{y_i\}$ converging to y. Continuity of f then yields that $f(x_j)$ converges to $f(x)$. Hence, $y = f(x) \in f(X)$. This is a contradiction. □ **(Lemma)**

Let Reg(f) be the set of regular points of f and denote the set of regular values of f by Regv(f). Note that Regv(f) includes points not in $f(X)$. Since f restricted to Reg(f) is a submersion and, hence, an open map, we see that $f(\mathrm{Reg}(f))$ is open. Because

$$\mathrm{Regv}(f) = [Y - f(X)] \bigcup [f(\mathrm{Reg}(f))],$$

the preceding lemma yields that Regv(f) is open in Y. □

Let U be open in $\mathbf{R}^n$ and suppose that $f : U \to \mathbf{R}$ is smooth. Then $x \in U$ is a critical point of f if and only if grad $f(x) = 0$. A critical point x of f is called *nondegenerate* (resp., *degenerate*) if $\det[\mathrm{Hess}\, f](x) \neq 0$ (resp., $\det[\mathrm{Hess}\, f](x) = 0$), where Hess f is the Hessian of f. Moreover, when all critical points are nondegenerate, f is termed *nondegenerate* or *Morse*; otherwise, f is *degenerate*.

Finally, let $f : U \subseteq \mathbf{R}^2 \to \mathbf{R}^2$, where U is open, be smooth. Then a point x in U is a critical point of f if and only if $\det[\mathrm{Jac}\, f](\mathbf{x}) = 0$, where Jac f is the Jacobian matrix of f.

6.2.2 Time Delay Functions and Light Rays

Single-Plane Time Delay Functions

The following mathematically abstracts the physical time delay function of Section 3.2.2 (page 67):

Definition 6.6 Let $\psi : L \to \mathbf{R}$ be a gravitational lens potential and set $S = \mathbf{R}^2$. For $\mathbf{y} \in S$, the smooth function $T_\mathbf{y} : L \longrightarrow \mathbf{R}$ defined by

$$T_\mathbf{y}(\mathbf{x}) = \frac{|\mathbf{x} - \mathbf{y}|^2}{2} - \psi(\mathbf{x}) \tag{6.3}$$

is a *single-plane time delay function* induced at *light source position* $\mathbf{y}$ by ψ. The function $T : L \times S \longrightarrow \mathbf{R}$ given by

$$T(\mathbf{x}, \mathbf{y}) = T_\mathbf{y}(\mathbf{x}),$$

is a *single-plane time delay family* parametrized by the plane S and induced by ψ.

We call L and S, respectively, the *lens plane* and *light source plane* of $T_{\mathbf{y}}$. The graph of $T_{\mathbf{y}}$ is termed a *time delay surface.* Moreover, if ψ is nonsingular (resp., singular), then call $T_{\mathbf{y}}$ *nonsingular* (resp., *singular*).

Note that the time delay function in Definition 6.3 is dimensionless — see Section 3.2.4, page 79.

Convention: *Unless stated to the contrary, we suppose that in the expression for* $T_{\mathbf{y}}$, *the light source parameters* $\mathbf{y} = (y_1, y_2)$ *and all parameters associated with* ψ *are fixed.*

Light Rays and Critical Points

Motivated by Fermat's principle (Section 3.2.1, page 66), we refer to each critical point of $T_{\mathbf{y}}$ as a *light ray from* $\mathbf{y}$ *that is lensed by* ψ (or, simply, a *light ray* of $T_{\mathbf{y}}$). In other words, we identify a light ray from $\mathbf{y}$ to the observer with the ray's impact point on the lens plane. If $\mathbf{x}_0$ is a light ray of $T_{\mathbf{y}}$, then $T_{\mathbf{y}}(\mathbf{x}_0)$ is termed the *time delay* of $\mathbf{x}_0$.

Define a light ray $\mathbf{x}_0$ to be *nondegenerate* (resp., *degenerate*) if $\mathbf{x}_0$ is nondegenerate (resp., degenerate). The *index* of a nondegenerate light ray $\mathbf{x}_0$ is the number of negative eigenvalues of $[\text{Hess}\, T_{\mathbf{y}}](\mathbf{x}_0)$. By Sylvester's Law of Inertia (e.g., [Art, p. 245]), the index is independent of coordinates. Morse's lemma (e.g., [Maj, p. 224]) yields that for each nondegenerate light ray $\mathbf{x}_0$ of index i, there is a (curvilinear) coordinate system $\lambda(\mathrm{u}, \mathrm{v}) = (\nu_1, \nu_2)$ centered at $\mathbf{x}_0$ such that the time delay function is given as follows relative to λ:

$$T_{\mathbf{y}} \circ \lambda^{-1}(\nu_1, \nu_2) = T_{\mathbf{y}}(\mathbf{x}_0) + \begin{cases} \nu_1^2 + \nu_2^2 & \text{if } i = 0 \\ -\nu_1^2 + \nu_2^2 & \text{if } i = 1 \\ -\nu_1^2 - \nu_2^2 & \text{if } i = 2. \end{cases} \tag{6.4}$$

A more detailed discussion of Morse's lemma will appear in Section 7.4.1 (page 238). Equation (6.4) classifies a nondegenerate light ray by its index i: *The light ray* $\mathbf{x}_0$ *is either a local minimum* ($i = 0$), *saddle* ($i = 1$), *or local maximum* ($i = 2$). *Hence, all nondegenerate light rays are isolated.*

The critical point type of a light ray is readily determined by the quadratic form of $T_{\mathbf{y}}$ at $\mathbf{x}_0$:

$$q(t_1, t_2) = T_{\mathrm{uu}}(\mathbf{x}_0)\, t_1^2 + 2T_{\mathrm{uv}}(\mathbf{x}_0)\, t_1 t_2 + T_{\mathrm{vv}}(\mathbf{x}_0)\, t_2^2,$$

where $(t_1, t_2) \in \mathbf{R}^2$. For notational simplicity, we set $T_{\mathbf{y}}(\mathbf{x}) = T(\mathrm{u}, \mathrm{v})$ when taking partial derivatives relative to u and v. Then (e.g., [Str, pp. 322–28]):

(i) A light ray $\mathbf{x}_0$ is a *local minimum* if and only if $q(t_1, t_2) > 0$ for all (t_1, t_2) in $\mathbf{R}^2 - \{\mathbf{0}\}$, if and only if

$$T_{\mathrm{uu}}(\mathbf{x}_0) > 0 \quad \text{and} \quad \det[\text{Hess}\, T_{\mathbf{y}}](\mathbf{x}_0) = (T_{\mathrm{uu}} T_{\mathrm{vv}} - T_{\mathrm{uv}}^2)(\mathbf{x}_0) > 0;$$

(ii) a *local maximum* if and only if $q(t_1, t_2) < 0$ for all (t_1, t_2) in $\mathbf{R}^2 - \{\mathbf{0}\}$, if and only if

$$T_{\mathrm{uu}}(\mathbf{x}_0) < 0 \quad \text{and} \quad \det[\text{Hess}\, T_{\mathbf{y}}](\mathbf{x}_0) > 0; \ \text{and}$$

(iii) a *saddle* if and only if $q(t_1, t_2)$ attains both positive and negative values. A nondegenerate light ray $\mathbf{x}_0$ is a saddle if and only if $\det[\text{Hess } T_{\mathbf{y}}](\mathbf{x}_0) < 0$.

Note that *if* $\mathbf{x}_0$ *is a minimum light ray (resp., maximum), then* $\Delta T(\mathbf{x}_0) > 0$ (resp., $\Delta T(\mathbf{x}_0) < 0$). In fact, if $\mathbf{x}_0$ is a minimum, then $T_{\mathrm{uu}}(\mathbf{x}_0) = q(1,0) > 0$ and $T_{\mathrm{vv}}(\mathbf{x}_0) = q(0,1) > 0$; hence, $\Delta T(\mathbf{x}_0) > 0$. (resp., $\Delta T(\mathbf{x}_0) < 0$).

6.2.3 Local Geometry of Time Delay Surfaces

This section looks more closely at the geometry of time delay surfaces in a neighborhood of its critical points, that is, points corresponding to light rays (cf. [Bla-N86, p. 570]).[6]

Let $\mathbf{x}$ be a light ray of a source at $\mathbf{y}$ and consider the corresponding point

$$X = (\mathbf{x}, T_{\mathbf{y}}(\mathbf{x}))$$

in the time delay surface of $T_{\mathbf{y}}$. The *principal curvatures* at X coincide with the eigenvalues of $[\text{Hess } T_{\mathbf{y}}](\mathbf{x})$, namely,

$$\lambda_1(\mathbf{x}; \mathbf{y}) = 1 - \kappa(\mathbf{x}) + \Gamma(\mathbf{x}), \qquad \lambda_2(\mathbf{x}; \mathbf{y}) = 1 - \kappa(\mathbf{x}) - \Gamma(\mathbf{x}).$$

Hence, the light ray $\mathbf{x}$ is a minimum (resp., maximum) if and only if the principal curvatures at X are both positive (resp., negative). The point $\mathbf{x}$ is a saddle light ray if and only if the principal curvatures at X have opposite signs.

Unlike the principal curvatures, the *mean curvature* at X does not involve the shear. It is related to the surface mass density or convergence due to ψ:

$$H(X) = \frac{\lambda_1(\mathbf{x}; \mathbf{y}) + \lambda_2(\mathbf{x}; \mathbf{y})}{2} = 1 - \kappa(\mathbf{x}).$$

Note that $2[1 - \kappa(\mathbf{x})] = \Delta T(\mathbf{x})$. In addition, the magnitude of shear obeys

$$\Gamma(\mathbf{x}) = \frac{\lambda_1(\mathbf{x}; \mathbf{y}) - \lambda_2(\mathbf{x}; \mathbf{y})}{2}.$$

Consequently, shear occurs at a light ray $\mathbf{x}$ if and only if the principal curvatures at X are distinct. Equivalently, there is no shear at $\mathbf{x}$ if and only if X is an *umbilic point* (i.e., the normal curvature is constant at X). Moreover, if all points of a connected open subset of a time delay surface are umbilic points, then the open set is a subset of either a sphere or plane (e.g., [doC, p. 147]).

The *Gaussian curvature* at $X = (\mathbf{x}, T_{\mathbf{y}}(\mathbf{x}))$, where $\mathbf{x}$ is a light ray, is given by

$$G(X) = \lambda_1(\mathbf{x}; \mathbf{y})\lambda_2(\mathbf{x}; \mathbf{y}) = \det[\text{Hess } T_{\mathbf{y}}](\mathbf{x})$$

[6] We refer readers to [Ber-G], [doC], and [Dub-FN] for excellent introductions to the elementary differential geometry employed in this subsection.

(since grad $T_{\mathbf{y}}(\mathbf{x}) = 0$ for $\mathbf{x}$ a light ray).[7] We have the following possibilities for the Gaussian curvature at the light ray point X (cf. [Dub-FN, pp. 80–85] and [Lip, pp. 177–179]):

(1) $G(X) > 0$, where both principal curvatures positive at X (*elliptic point*): The time delay surface bends upwards in a neighborhood of X, lying locally on one side of the tangent plane through X. *The principal curvatures at X are both positive if and only if the light ray* $\mathbf{x}$ *is a minimum.* Figure 6.1(a).

(2) $G(X) > 0$, where both principal curvatures negative at X (*elliptic point*): The time delay surface bends downwards in a neighborhood of X, lying locally on one side of the tangent plane through X. *The principal curvatures at X are both negative if and only if the light ray* $\mathbf{x}$ *is a maximum.* Figure 6.1 (a).

(3) $G(X) < 0$, where the principal curvatures having opposite signs at X (*hyperbolic point*): The time delay surface at X has a saddle shape in a neighborhood of X, lying locally on both sides of the tangent plane through X. *The principal curvatures at X have opposite signs if and only if the light ray* $\mathbf{x}$ *is a saddle.* Figure 6.1(b).

(4) $G(X) = 0$, where either one principal curvature is nonzero at X (*parabolic point*) or both vanish at X (*planar point*): The time delay surface flattens in a neighborhood of X (and it will be shown that the corresponding lensed image $\mathbf{x}$ has infinite "magnification" — see (6.7) on page 183). In the parabolic case, the surface can lie locally on one side (e.g., parabolic cylinder) or both sides (e.g., surface with cubic sections $(\mathrm{u}, \mathrm{u}^3)$) of the tangent plane through X. For the planar case, the surface has a higher degree of contact at X with the tangent plane than in the parabolic case. The local form of the surface about a planar point varies wildly (e.g., [Ber-G, pp. 138, 391]). Note that planar points are umbilic points and $\kappa(\mathbf{x}) = 1$ (critical) at planar points. Overall, *at least one principal curvature at X vanishes if and only if the light ray* $\mathbf{x}$ *is degenerate.* Figure 6.1(c,d).

6.2.4 Lensing Maps, Lensed Images, and Magnification

Lens Equation, Lensed Images, and Lensing Maps

Critical points of the time delay function $T_{\mathbf{y}} : L \to \mathbf{R}$ are identified by the Principle of Fermat with light rays from a light source at $\mathbf{y}$. Consequently, the vanishing of the gradient of $T_{\mathbf{y}}$ yields an equation,

$$\mathbf{y} = \mathbf{x} - \boldsymbol{\alpha}(\mathbf{x}),$$

called the *lens equation* (cf. Section 3.2.3, page 77). This equation links the light source position $\mathbf{y}$ with the point $\mathbf{x} \in L$ where a light ray impacts on the lens plane L. We then define a *lensed image* (relative to ψ) of a light source at $\mathbf{y}$ to be a solution $\mathbf{x}$ of the lens equation.

The equivalence of the lens equation and

$$\operatorname{grad} T_{\mathbf{y}}(\mathbf{x}) = \mathbf{0}$$

[7] Recall: At a general point X, we have $G(X) = \det[\text{Hess } T_{\mathbf{y}}](\mathbf{x})/(1 + |\operatorname{grad} T_{\mathbf{y}}(\mathbf{x})|^2)^2$.

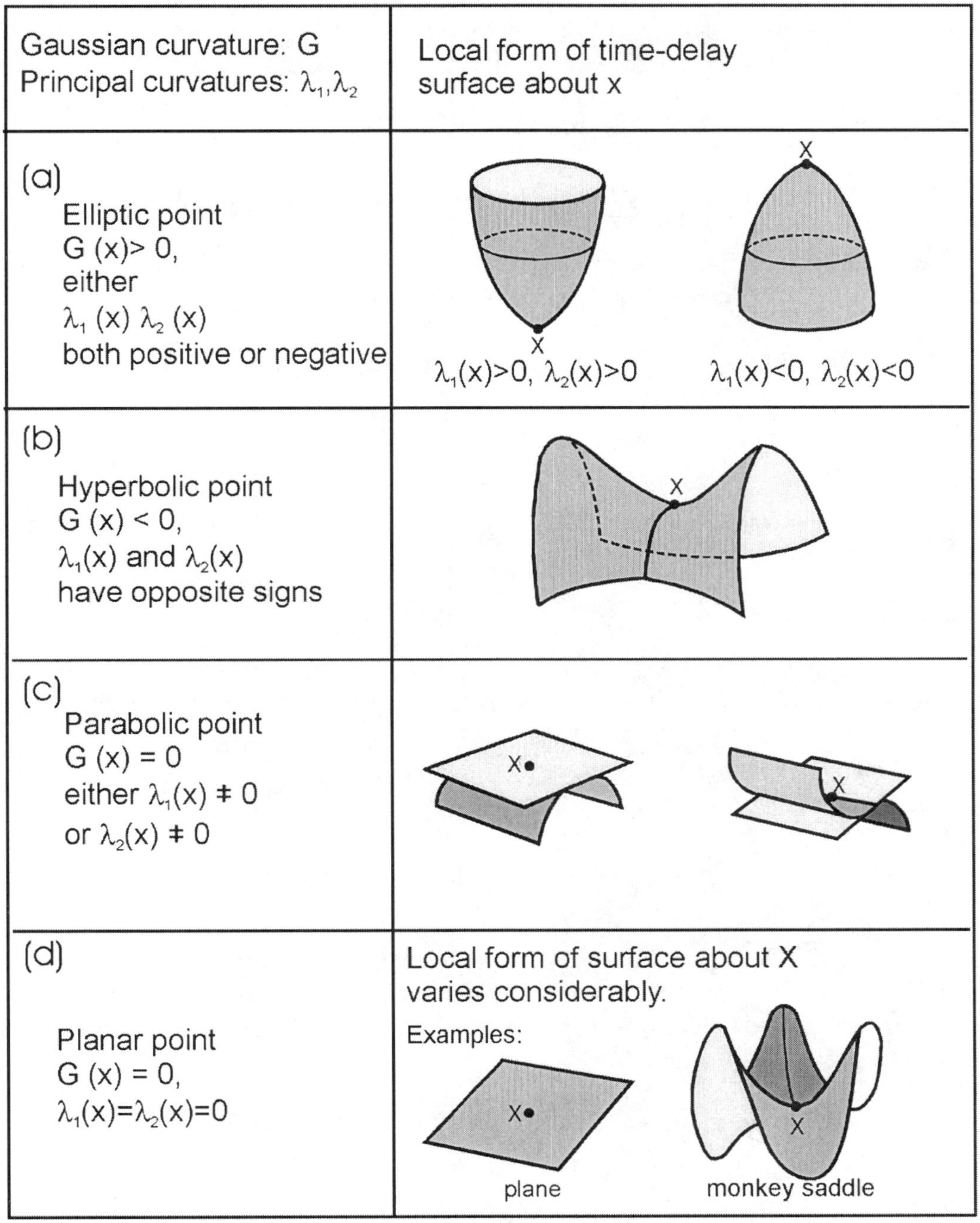

Figure 6.1. Local geometry of a time delay surface in a neighborhood of a point $X = (\mathbf{x}, T_{\mathbf{y}}(\mathbf{x}))$, where $\mathbf{x}$ is a light ray of $T_{\mathbf{y}}$. In (a,b), the light ray $\mathbf{x}$ corresponds to a local minimum, maximum, and saddle, respectively. In (c,d), the light ray is a degenerate critical point.

means that our notion of "lensed image" coincides with that of a "light ray." We can now speak of the *nondegeneracy* and *index* of lensed images. However, the concepts of "lensed image" and "light ray" do differ in the multiplane case. This is because we shall identify a light ray with the points on all lens planes where the ray impacts, while a lensed image corresponds to the point on the first lens plane (counting in the direction from observer to light source) where a light ray crosses.

The lens equation naturally yields the following map (cf. Section 3.2.3, page 77):

Definition 6.7 Let $\psi : L \to \mathbf{R}$ be a gravitational lens potential. A *single-plane lensing map* induced by ψ is a (smooth) map $\boldsymbol{\eta} : L \to S$ defined by $\boldsymbol{\eta}(\mathbf{x}) = \mathbf{x} - \boldsymbol{\alpha}(\mathbf{x})$.[8]

The fibre $\boldsymbol{\eta}^{-1}(\mathbf{y})$ *consists of all lensed images of a light source at* $\mathbf{y}$.

Magnification of Lensed Images

The *magnification* of a lensed image $\mathbf{x} \in L$ of a light source at $\mathbf{y} \in S$, where $\boldsymbol{\eta}(\mathbf{x}) = \mathbf{y}$, is defined as follows (cf. Section 3.2.5, page 82):

$$\mathrm{Mag}(\mathbf{x}; \mathbf{y}) = \frac{1}{|\det[\mathrm{Jac}\, \boldsymbol{\eta}](\mathbf{x})|}, \qquad \boldsymbol{\eta}(\mathbf{x}) = \mathbf{y}.$$

Roughly speaking, the function $\mathrm{Mag}(\mathbf{x}; \mathbf{y})$ is the ratio of the area of a small region about $\mathbf{x}$ to the area of a small region about its $\boldsymbol{\eta}$-image (which contains $\mathbf{y}$). In other words, magnification measures area distortion. If $\mathrm{Mag}(\mathbf{x}; \mathbf{y}) > 1$ (resp., $\mathrm{Mag}(\mathbf{x}; \mathbf{y}) < 1$), then $\mathbf{x}$ is *magnified* (resp., *de-magnified*). When $\mathrm{Mag}(\mathbf{x}; \mathbf{y}) = 1$, call $\mathbf{x}$ *unmagnified.* It is also common to introduce a *signed magnification*:

$$\mu_{sgn}(\mathbf{x}; \mathbf{y}) = \frac{1}{\det[\mathrm{Jac}\, \boldsymbol{\eta}](\mathbf{x})}, \qquad \boldsymbol{\eta}(\mathbf{x}) = \mathbf{y}.$$

Note that $\mu_{sgn}(\mathbf{x}; \mathbf{y}) = (-1)^i \mathrm{Mag}(\mathbf{x}; \mathbf{y})$, where i is the index of the lensed image $\mathbf{x}$, and $\mathrm{Jac}\, \boldsymbol{\eta}$ is *symmetric* (which need not hold in multiplane gravitational lensing).

The magnification of lensed images will now be related to the gravitational lens potential and Gaussian curvature. We have

$$\begin{aligned} \det[\mathrm{Jac}\, \boldsymbol{\eta}](\mathbf{x}) &= 1 - \mathrm{trace}[\mathrm{Hess}\, \psi](\mathbf{x}) + \det[\mathrm{Hess}\, \psi](\mathbf{x}) \\ &= [1 - \kappa(\mathbf{x})]^2 - \Gamma^2(\mathbf{x}), \qquad \mathbf{x} \in L, \end{aligned} \tag{6.5}$$

where Γ is the shear function due to ψ. Furthermore,

$$\det[\mathrm{Jac}\, \boldsymbol{\eta}](\mathbf{x}) = \lambda_1(\mathbf{x}; \mathbf{y})\lambda_2(\mathbf{x}; \mathbf{y}) = \det[\mathrm{Hess}\, T_{\mathbf{y}}](\mathbf{x}), \qquad \mathbf{x} \in L. \tag{6.6}$$

[8]Single-plane lensing maps are dimensionless (cf. Section 3.2.4, page 79).

Hence, if $\mathbf{x}$ is a lensed image of $\mathbf{y}$, then we obtain a link between magnification and the Gaussian curvature

$$\mathrm{Mag}(\mathbf{x}; \mathbf{y}) = \frac{1}{|G(\mathbf{x}, T_{\mathbf{y}}(\mathbf{x}))|}. \tag{6.7}$$

Thus, *the magnification of a lensed magnification increases if and only if the Gaussian curvature of the associated light ray decreases.* In particular, the Gaussian curvature vanishes at degenerate light rays.

The *total magnification* of a light source at $\mathbf{y}$ is[9]

$$\mathrm{Mag}_{tot}(\mathbf{y}) = \sum_{\mathbf{x}\in\boldsymbol{\eta}^{-1}(\mathbf{y})} \mathrm{Mag}(\mathbf{x}; \mathbf{y}). \tag{6.8}$$

Since the cardinality of the fibre $\boldsymbol{\eta}^{-1}(\mathbf{y})$ is in general unknown,[10] the meaning of the above sum must be clarified. It is given precisely as follows:

$$\sum_{\mathbf{x}\in\boldsymbol{\eta}^{-1}(\mathbf{y})} \mathrm{Mag}(\mathbf{x}; \mathbf{y}) = \begin{cases} \textit{the supremum of the set of all finite sums } \mathrm{Mag}(\mathbf{c}_1; \mathbf{y})+ \\ \cdots + \mathrm{Mag}(\mathbf{c}_n; \mathbf{y}), \textit{ where } \mathbf{c}_1, \ldots, \mathbf{c}_n \textit{ are distinct lensed} \\ \textit{images in } \boldsymbol{\eta}^{-1}(\mathbf{y}). \end{cases}$$

If the supremum does not exist, then the total magnification is said to be *infinite*. When $\boldsymbol{\eta}^{-1}(\mathbf{y})$ has finitely many elements, equation (6.8) reduces to a usual sum.

Critical Curves, Caustics, and Light Curves

The set $\mathrm{Crit}(\boldsymbol{\eta})$ of critical points of $\boldsymbol{\eta} : L \to S$ consists of all points $\mathbf{x}$ in L where $\mathrm{rank}[\mathrm{Jac}\,\boldsymbol{\eta}](\mathbf{x}) < 2$, that is, $\mathbf{x}$ satisfies $\det[\mathrm{Jac}\,\boldsymbol{\eta}](\mathbf{x}) = 0$. In other words, $\mathrm{Crit}(\boldsymbol{\eta})$ is the collection of all infinitely magnified lensed images (relative to ψ) of a light source at $\mathbf{y}$ for all $\mathbf{y}$ in the light source plane S. We shall see that "generically," the set $\mathrm{Crit}(\boldsymbol{\eta})$ consists of disjoint simple closed curves (see Sections 7.5, 8.2, and 8.4). Whenever a collection of critical points forms a curve, we call such curves *critical curves*. The *caustics* of $\boldsymbol{\eta}$, denoted $\mathrm{Caustic}(\boldsymbol{\eta})$, is the set of all critical values of $\boldsymbol{\eta}$. Equivalently, the set $\mathrm{Caustic}(\boldsymbol{\eta})$ consists of all points $\mathbf{y}$ in S where a light source has at least one infinitely magnified lensed image.[11] Figure 6.2 shows caustics due to many point masses with continuous matter and shear. Finally, equation (6.6) implies that the set of degenerate light rays of $T_{\mathbf{y}}$ for all $\mathbf{y}$ in S coincides with $\mathrm{Crit}(\boldsymbol{\eta})$. For this reason, we also refer to elements of $\mathrm{Crit}(\boldsymbol{\eta})$ as *degenerate* lensed images.

---

[9]See (3.107) on page 85 for the magnification of an extended source.

[10]It can be uncountable! For a circularly symmetric lens with a light source on the line of sight through the origin, the fibre $\boldsymbol{\eta}^{-1}(\mathbf{y})$ is a circle; see Sections 3.2.3 (page 77).

[11]See page 340 for a discussion of lensed images of light sources on caustics; also, consult Figure 11.1, page 427.

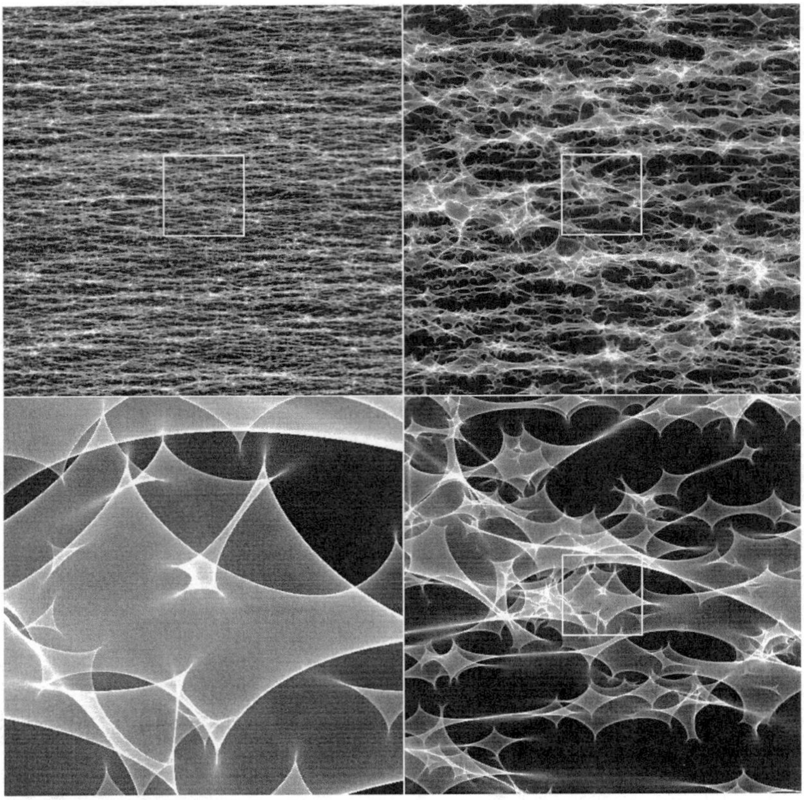

Figure 6.2. A caustic network and its magnification pattern due to several hundred thousand point masses with continuous matter and shear along the horizontal. The grey scale depicts regions of high (lighter) and low (darker) magnification. The white curves are caustics (infinite magnification). (Courtesy of J.W. [Wam90, p. 127].)

If a light source moves along a straight line through a caustic network, then the graph of the total magnification of the light source as a function of position along the path (or, function of observer time as the source moves along the path) is called a *light curve* (cf. Section 3.2.6, page 88). Light curves are physically important. In microlensing, they provide an important technique for detecting planets (see Figure 4.2, page 127).

Magnification Patterns

The shape of a light curve is affected by the magnification pattern a lens produces on a light source plane S. Such a pattern can be represented by assigning to each point of S a shade indicating the total magnification of a light source at the given point — see Figure 6.2. Magnification patterns are important in the study of microlensing (see Section 4.3 and 5.4 and 5.5) and lead naturally to magnification cross section problems (Section 13.3).

6.3 Simple Models

Our goal is to determine the gravitational lensing signature of the uniform sheet, singular isothermal sphere, and point mass. In particular, we shall consider the following problems (cf. Section 2.1):

- Count the number of lensed images and find their locations.
- Determine the total magnification.
- Characterize the local and global properties of all critical points and caustics.

Addressing these issues for the aforementioned simple potentials is accessible via calculation, though somewhat tedious. For more complicated lenses (e.g., many point mass lenses), a *direct* analytical treatment of the above problems is practically impossible. These lenses require the powerful methods of singularity theory, which will be developed and applied in the forthcoming chapters.[12] The section ends with a table summarizing several gravitational lens models.

6.3.1 Uniform Sheet

The gravitational lens potential of a uniform sheet of matter with constant density is given by

$$\psi_c(\mathbf{x}) = \frac{\kappa_c}{2}|\mathbf{x}|^2,$$

where κ_c is a positive constant. Since ψ_c is nonsingular, the lens plane L coincides with $\mathbf{R}^2$. The corresponding time delay function and lensing map are

$$T^c_{\mathbf{y}}(\mathbf{x}) = \frac{|\mathbf{x}-\mathbf{y}|^2}{2} - \frac{\kappa_c}{2}|\mathbf{x}|^2, \quad \text{and} \quad \boldsymbol{\eta}_c(\mathbf{x}) = \mathbf{x}(1-\kappa_c),$$

respectively. Note: $\det[\text{Jac}\,\boldsymbol{\eta}_c](\mathbf{x}) = (1-\kappa_c)^2$ and $(\partial^2 T^c_{\mathbf{y}}/\partial \mathrm{u}^2)(\mathbf{x}) = 1-\kappa_c$.

Non-critical case ($\kappa_c \neq 1$): *Only one lensed image* $\mathbf{x}_0$ *of a light source at* $\mathbf{y}$ *occurs if* $\kappa_c \neq 1$. In fact, the lens equation is $\mathbf{y} = \mathbf{x}(1-\kappa_c)$. This yields a nondegenerate lensed image

$$\mathbf{x}_0 = \frac{\mathbf{y}}{1-\kappa_c}.$$

The magnification of $\mathbf{x}_0$ is

$$\text{Mag}(\mathbf{x}_0;\mathbf{y}) = \frac{1}{(1-\kappa_c)^2},$$

which is independent of the light source position $\mathbf{y}$. *If* $\kappa_c < 1$, *then* $(\partial^2 T^c_{\mathbf{y}}/\partial \mathrm{u}^2)(\mathbf{x}_0) > 0$ *and* $\det[\text{Jac}\,\boldsymbol{\eta}_c](\mathbf{x}_0) > 0$. *Consequently, the lensed image* $\mathbf{x}_0$ *is a minimum and magnified; for* $\kappa_c > 1$, *it is a maximum, which is de-magnified for* $\kappa_c > 2$. Moreover, since $\det[\text{Jac}\,\boldsymbol{\eta}_c]$ never vanishes, *no critical points (hence, caustics) occur.*

[12] Some sophisticated numerical methods are available for investigating many point mass lenses (e.g., J.W. [Wam90, Wam-WS]). However, our aim is to provide mathematical results about gravitational lensing.

Critical case ($\kappa_c = 1$): *If* $\mathbf{y} \neq \mathbf{0}$ *and* $\kappa_c = 1$, *then no lensed image exists* (since $\text{grad}\, T^c_{\mathbf{y}}(\mathbf{x}) = -\mathbf{y} \neq \mathbf{0}$ for all $\mathbf{x} \in \mathbf{R}^2$). Note that for the noncritical case with $\mathbf{y} \neq \mathbf{0}$, the lensed image $\mathbf{x}_0$ "runs off to infinity" (getting infinitely bright) as $\kappa_c \to 1$. This is why no lensed image occurs for $\kappa_c = 1$ and $\mathbf{y} \neq \mathbf{0}$. If $\kappa_c = 1$ and $\mathbf{y} = \mathbf{0}$, then the time delay function is constant: $T_{\mathbf{y}}(\mathbf{x}) = 0$. In this situation, every point on the lens plane L is a degenerate lensed image. Hence, *when* $\kappa_c = 1$, *a light source at* $\mathbf{y} = \mathbf{0}$ *appears as an infinitely bright plane.*[13] *The set of caustics consists of the point* $\mathbf{y} = \mathbf{0}$.

6.3.2 Singular Isothermal Sphere

A singular-isothermal-sphere has a gravitational lens potential of the form

$$\psi_{sis}(\mathbf{x}) = C|\mathbf{x}|,$$

which induces the time delay function

$$T^{sis}_{\mathbf{y}}(\mathbf{x}) = \frac{|\mathbf{x} - \mathbf{y}|^2}{2} - C|\mathbf{x}|,$$

and lens equation

$$\mathbf{y} = \mathbf{x}\left(1 - \frac{C}{|\mathbf{x}|}\right), \tag{6.9}$$

or, lensing map $\boldsymbol{\eta}_{sis}(\mathbf{x}) = \mathbf{x}(1 - C|\mathbf{x}|^{-1})$.

Exploiting the circular symmetry of ψ_{sis}, write $\mathbf{y}$ and $\mathbf{x}$ in polar coordinates, say, $\mathbf{y} = r_s \mathbf{e}_{\theta_s}$ and $\mathbf{x} = r\mathbf{e}_\theta$, where $\mathbf{e}_\beta = (\cos\beta, \sin\beta)$. Also, notice that $\det[\text{Jac}\, \boldsymbol{\eta}_{sis}](\mathbf{x}) = 1 - C/r$ and $(\partial^2 T^{sis}_{\mathbf{y}}/\partial \mathbf{u}^2)(\mathbf{x}) = 1 - (C/r)\sin^2\theta$.

Case I : Light source not at origin ($\mathbf{y} \neq \mathbf{0}$). Take the dot product of (6.9) with itself to obtain

$$r^2 - 2Cr + (C^2 - r_s^2) = 0, \tag{6.10}$$

where $r_s > 0$.

Sub-case 1 ($0 < r_s < C$): *Two lensed images* $\mathbf{x}_\pm$ *occur.* In fact, the lensed images are $\mathbf{x}_\pm = (r_\pm \cos\theta_\pm, r_\pm \sin\theta_\pm)$, where $r_\pm = C \pm r_s$ are the solutions of equation (6.10). To determine $\theta_\pm$ note that by (6.9) the lensed images $\mathbf{x}_\pm$ are proportional to $\mathbf{y}$:

$$\mathbf{x}_\pm = \frac{\mathbf{y}}{1 - C/r_\pm^2}.$$

[13] This may remind the reader of focusing by a parabolic mirror. Incoming light rays parallel to the axis of such a mirror impact at each point of the mirror and converge to the focus. An observer at the focus sees the portion of the sky reflected in the mirror as (formally) infinitely bright. Moreover, if the light rays converging at the focus originated from a finite distance, then they all have the same path length (which is analogous to our $T_{\mathbf{y}}(\mathbf{x}) = 0$).

Since $r_+ > C$ and $r_- < C$, we see that $\mathbf{x}_+$ (resp., $\mathbf{x}_-$) is in the same (resp., opposite) direction as $\mathbf{y}$. Consequently, $\theta_+ = \theta_s$ and $\theta_- = \theta_s + \pi$. Hence

$$\mathbf{x}_\pm = (r_s \pm C)\, \mathbf{e}_{\theta_s},$$

where $\mathbf{e}_{\theta_s} = (\cos\theta_s, \sin\theta_s)$. The lensed image $\mathbf{x}_+$ lies outside the ring $r = C$, while $\mathbf{x}_-$ is inside — see Figure 6.3. Moreover, the following relation is immediate:

$$\mathbf{y} = \frac{\mathbf{x}_+ + \mathbf{x}_-}{2}.$$

Since $\det[\mathrm{Jac}\,\boldsymbol{\eta}_{sis}](\mathbf{x}_+) > 0$, $\det[\mathrm{Jac}\,\boldsymbol{\eta}_{sis}](\mathbf{x}_-) < 0$, and $(\partial^2 T_{\mathbf{y}}^{sis}/\partial \mathrm{u}^2)(\mathbf{x}_+) > 0$, *the lensed image* $\mathbf{x}_+$ *is a minimum and* $\mathbf{x}_-$ *a saddle.*

The *magnifications* of the lensed images are

$$\mathrm{Mag}(\mathbf{x}_+;\mathbf{y}) = 1 + \frac{C}{r_s}, \qquad \mathrm{Mag}(\mathbf{x}_-;\mathbf{y}) = \frac{C}{r_s} - 1.$$

Note that the minimum $\mathbf{x}_+$ is magnified. In addition, we get the relation

$$\frac{\mathrm{Mag}(\mathbf{x}_+;\mathbf{y}) - \mathrm{Mag}(\mathbf{x}_-;\mathbf{y})}{2} = 1. \tag{6.11}$$

The average of the lensed image magnification difference is the magnification of the unlensed light source.

Sub-case 2 ($r_s = C$): If $r_s = C$, then $\mathbf{x}_+ = 2\mathbf{y}$, $\mathbf{x}_- = \mathbf{0}$, $\mathrm{Mag}(\mathbf{x}_+;\mathbf{y}) = 2$, and $\mathrm{Mag}(\mathbf{x}_-;\mathbf{y}) = 0$. Hence, *there is one lensed image* $\mathbf{x}_+$*, which is a magnified minimum. The saddle lensed image* $\mathbf{x}_-$ *disappears at the singularity.*

Sub-case 3 ($r_s > C$): Since $\mathbf{x}_-$ does not exist for $r_s > C$, *there is one lensed image* $\mathbf{x}_+ = (r_s + C)\,\mathbf{e}_{\theta_s}$*, which is a magnified minimum with* $\mathrm{Mag}(\mathbf{x}_+;\mathbf{y}) = 1 + C/r_s$. Note, however, that $1 < \mathrm{Mag}(\mathbf{x}_+;\mathbf{y}) < 2$, which differs from Sub-case 2.

Case II: Light source at origin ($\mathbf{y} = \mathbf{0}$). Equation (6.9) implies that the set of lensed images forms the circle $r = C$. Since $\det[\mathrm{Jac}\,\boldsymbol{\eta}_{sis}](\mathbf{x}) = 1 - C/r = 0$, *the light source appears as an infinitely bright circle.*

The set of critical points is the circle $r = C$ *and the set of caustics consists of the point* $\mathbf{y} = \mathbf{0}$. See Figure 6.3.

6.3.3 Point Mass

A point mass m at the origin has gravitational lens potential

$$\psi_{pt}(\mathbf{x}) = m \ln|\mathbf{x}|,$$

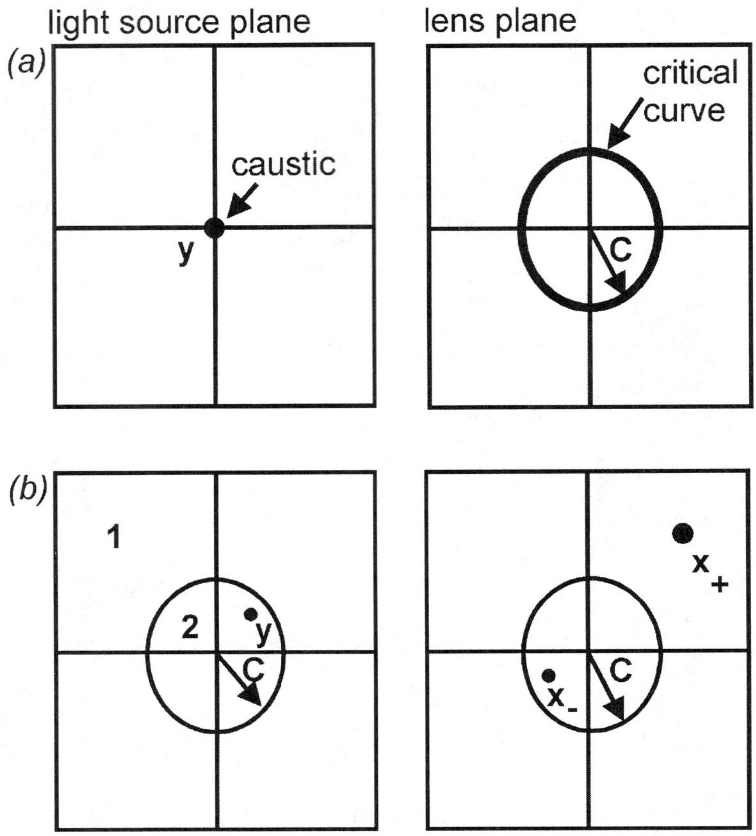

Figure 6.3. Caustic, critical curve and lensed images produced by a singular isothermal sphere. *(a)* Given a light source at $\mathbf{y}$, when $\mathbf{y}$ is at the caustic (origin), the light source appears as a formally infinitely magnified ring (critical curve). *(b)* The numbers indicate how many lensed images are seen of a light source. If $0 < |\mathbf{y}| < C$, then there are two lensed images: $\mathbf{x}_+$ (magnified minimum) and $\mathbf{x}_-$ (saddle). For $|\mathbf{y}| \geq 1$, only one lensed image occurs, namely, $\mathbf{x}_+$.

which induces the time delay function

$$T_{\mathbf{y}}^{pt}(\mathbf{x}) = \frac{|\mathbf{x}-\mathbf{y}|^2}{2} - m \ln|\mathbf{x}|,$$

and lens equation

$$\mathbf{y} = \mathbf{x}\left(1 - \frac{m}{|\mathbf{x}|^2}\right), \tag{6.12}$$

or, lensing map $\boldsymbol{\eta}_{pt}(\mathbf{x}) = \mathbf{x}(1 - m|\mathbf{x}|^{-2})$. We proceed analogously to the case of a singular isothermal sphere. Assuming polar coordinates, call the circle $r = \sqrt{m}$ the

Einstein (or *Chwolson–Einstein*) *ring* of ψ_{pt}. Note: $\det[\text{Jac}\,\boldsymbol{\eta}_{pt}](\mathbf{x}) = 1 - (m^2/r^4)$ and $(\partial^2 T_{\mathbf{y}}^{pt}/\partial \mathrm{u}^2)(\mathbf{x}) = m(r^2 - 2\mathrm{u}^2)/r^4$. Then the following hold:[14]

Case I: Light source not at origin ($\mathbf{y} = \mathbf{0}$). *There are two lensed images* $\mathbf{x}_\pm = (r_\pm \cos\theta_\pm, r_\pm \sin\theta_\pm)$ *for* $\mathbf{y} \neq \mathbf{0}$. In fact, taking the dot product of (6.12) with itself implies that

$$r^4 - (r_s^2 + 2m)r^2 + m^2 = 0, \tag{6.13}$$

which has two solutions:

$$r_\pm^2 = \frac{1}{2}\left[(r_s^2 + 2m) \pm r_s\sqrt{r_s^2 + 4m}\right]. \tag{6.14}$$

Hence,

$$r_\pm = \frac{1}{2}\left(\pm r_s + \sqrt{r_s^2 + 4m}\right) \tag{6.15}$$

(since we are interested only in positive values of r). To find the angles $\theta_\pm$, note that by (6.12) we have

$$\mathbf{x}_\pm = \frac{\mathbf{y}}{1 - m/r_\pm}.$$

Since $r_+ > \sqrt{m}$ and $r_- < \sqrt{m}$, the image $\mathbf{x}_+$ (resp., $\mathbf{x}_-$) is in the same (resp., opposite) direction as $\mathbf{y}$: $\theta_+ = \theta_s$ and $\theta_- = \theta_s + \pi$. In other words, the lensed image $\mathbf{x}_+$ lies outside the Einstein ring (on the same side as the light source), while $\mathbf{x}_-$ is inside (on the side opposite the light source). Consult Figure 6.4. The image positions are given explicitly by

$$\mathbf{x}_+ = \frac{1}{2}\left(r_s + \sqrt{r_s^2 + 4m}\right)\mathbf{e}_{\theta_s}$$

and

$$\mathbf{x}_- = \frac{1}{2}\left(r_s - \sqrt{r_s^2 + 4m}\right)\mathbf{e}_{\theta_s}.$$

The following interesting relation is immediate:

$$\mathbf{x}_+ + \mathbf{x}_- = \mathbf{y}. \tag{6.16}$$

Since

$$\frac{\partial^2 T_{\mathbf{y}}^{pt}}{\partial \mathrm{u}^2}(\mathbf{x}_+) = \frac{m}{r_+^2}\frac{r_+^2 - 2\mathrm{u}_+^2}{r_+^2} > 0, \qquad \det[\text{Jac}\,\boldsymbol{\eta}_{pt}](\mathbf{x}_+) = 1 - (m^2/r_+^4) > 0,$$

the lensed image $\mathbf{x}_+$ *is a minimum. The other image* $\mathbf{x}_-$ *is a saddle.*

[14] Compare our modern treatment with that of Chwolson 1924[Chw] and Einstein 1936 [Ein36].

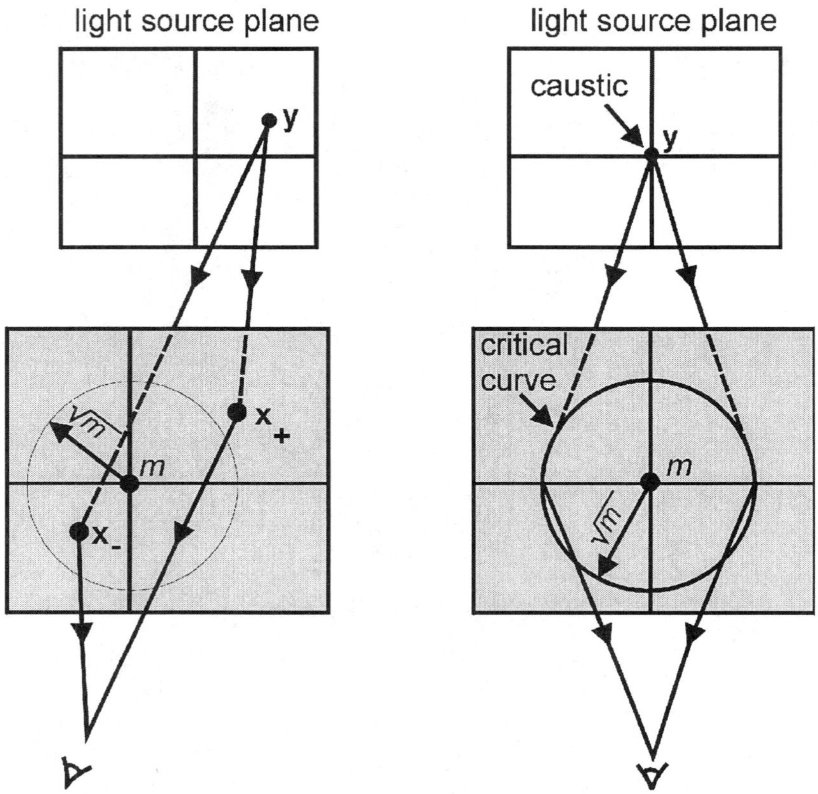

Figure 6.4. Lensed images and caustic due to point mass lens m. The shaded planes are in the foreground, while the light source $\mathbf{y}$ is in the background. *Left:* Two lensed images, $\mathbf{x}_+$ (magnified minimum) and $\mathbf{x}_-$ (saddle), are seen for $\mathbf{y} \neq \mathbf{0}$. *Right:* If the light source position $\mathbf{y}$ is at the caustic (i.e., the origin), then the light source appears as a formally infinitely magnified ring (critical curve) called an *Einstein ring*.

Let us consider the time delay between the images $\mathbf{x}_+$ and $\mathbf{x}_-$. Using the expression for $T_{\mathbf{y}}^{pt}$ and equation (6.16), we obtain the the time delay between the images, namely,

$$T_{\mathbf{y}}(\mathbf{x}_+) - T_{\mathbf{y}}(\mathbf{x}_-) = \frac{r_-^2 - r_+^2}{2} + m \ln \frac{r_-}{r_+}. \tag{6.17}$$

Note that *knowing* $T_{\mathbf{y}}(\mathbf{x}_+) - T_{\mathbf{y}}(\mathbf{x}_-)$ *and* $r_\pm$ *yields a determination of the (dimensionless) mass* m.

The lensed image magnifications are

$$\mathrm{Mag}(\mathbf{x}_+; \mathbf{y}) = \frac{1}{1 - (m^2/r_+^4)}$$

and

$$\mathrm{Mag}(\mathbf{x}_-; \mathbf{y}) = \frac{1}{(m^2/r_-^4) - 1}.$$

Note that *the minimum image* $\mathbf{x}_+$ *is magnified.* Equation (6.15) implies that $r_+ r_- = m$. Direct calculation, along with the latter, yields

$$\text{Mag}(\mathbf{x}_+; \mathbf{y}) - \text{Mag}(\mathbf{x}_-; \mathbf{y}) = 1. \tag{6.18}$$

The difference in lensed image magnifications is the magnification of the unlensed light source.

Now, applications of equations (6.13) and (6.14) imply that

$$r_+^4 + r_-^4 = (r_s^2 + 2m)^2 - 2m^2$$

and

$$r_+^4 - r_-^4 = r_s(r_s^2 + 2m)\sqrt{r_s^2 + 4m}.$$

By employing the above equations, the total magnification can be expressed in terms of the light source position:

$$\text{Mag}_{tot}(\mathbf{y}) = \frac{r_s^2 + 2m}{r_s\sqrt{r_s^2 + 4m}} = \frac{1}{2}\left[\frac{1}{\sqrt{1 + (4m/r_s^2)}} + \sqrt{1 + (4m/r_s^2)}\right]. \tag{6.19}$$

For the magnifications of $\mathbf{x}_\pm$, use equations (6.18) and (6.19) to obtain

$$\text{Mag}(\mathbf{x}_+; \mathbf{y}) = \frac{\text{Mag}_{tot}(\mathbf{y}) + 1)}{2} = \frac{1}{4}\left[\frac{1}{\sqrt{1 + (4m/r_s^2)}} + \sqrt{1 + (4m/r_s^2)} + 2\right]$$

and

$$\text{Mag}(\mathbf{x}_-; \mathbf{y}) = \frac{2\text{Mag}_{tot}(\mathbf{y}) - 1}{2} = \frac{1}{4}\left[\frac{1}{\sqrt{1 + (4m/r_s^2)}} + \sqrt{1 + (4m/r_s^2)} - 2\right].$$

As the light source moves to "infinity" (i.e., $r_s \to \infty$), the positions and magnifications of the lensed images behave as follows:

(i) $\mathbf{x}_+ \to \infty$ and $\text{Mag}(\mathbf{x}_+; \mathbf{y}) \to 1$;

(ii) $\mathbf{x}_- \to \mathbf{0}$ (i.e., $\mathbf{x}_-$ approaches the point mass) and $\text{Mag}(\mathbf{x}_-; \mathbf{y}) \to 0$, that is, $\mathbf{x}_-$ becomes dimmer as it tends towards the point mass.

Case II: Light source at origin ($\mathbf{y} = \mathbf{0}$). If $\mathbf{y} = \mathbf{0}$, then the lens equation is solved by all points on the Einstein ring. In fact, *the set of critical points is the Einstein ring, while the set of caustics consists of the point* $\mathbf{y} = \mathbf{0}$. Hence, *a light source at the origin appears as an infinitely magnified ring.* Examples of the light curves due this systems are shown in Figure 3.9 on page 93.

Figure 6.4 summarizes the basic features of lensing due to a point mass lens: double images, critical curve (i.e., Einstein ring), and caustic point.

Section 6.3 now concludes with a table of gravitational lens models.

6.3.4 A Gallery of Gravitational Lens Models

Uniform sheet
Potential: $\psi_c(\mathbf{x}) = \frac{\kappa_c}{2}|\mathbf{x}|^2$, where $\kappa_c > 0$.
Surface mass density: $\kappa_c(\mathbf{x}) = \kappa_c$.
Bending angle vector: $\boldsymbol{\alpha}_c(\mathbf{x}) = \kappa_c \mathbf{x}$.
Singularities: none.
Applications: sheet of continuous matter with constant surface mass density κ_c; a finite uniform sheet (e.g., disc) approximately models the central parts of clusters of galaxies with large cores.

Point mass(es)
Potential: $\psi_{pt}(\mathbf{x}) = \sum_{\ell=1}^{g} m_\ell \ln|\mathbf{x} - \boldsymbol{\xi}_\ell|$.
Surface mass density: $\kappa_{pt}(\mathbf{x}) = \pi \sum_{\ell=1}^{g} m_\ell \delta_{\boldsymbol{\xi}_\ell}$.
Bending angle vector: $\boldsymbol{\alpha}_{pt}(\mathbf{x}) = \sum_{\ell=1}^{g} m_\ell \frac{\mathbf{x}-\boldsymbol{\xi}_\ell}{|\mathbf{x}-\boldsymbol{\xi}_\ell|^2}$.
Singularities: point mass positions $\mathbf{x}_\ell$.
Applications:
(1) Single point mass lens: simple first approximation to some two-image lens systems (image configuration); accurate model for most galactic microlensing cases (light curve).
(2) Binary point mass lens: rough approximation to lensing galaxy clusters with two dominant galaxies (Luminous Arcs); very good model for some galactic microlensing events (light curve).
(3) Many point mass lens: microlensing of quasars by galaxies; potentially globular clusters as lenses.

Uniform sheet with shear and point mass(es)
Potential: $\psi_{ptcs}(\mathbf{x}) = \frac{\kappa_c}{2}|\mathbf{x}|^2 - \frac{\gamma}{2}(\mathrm{u}^2 - \mathrm{v}^2) + \sum_{\ell=1}^{g} m_\ell \ln|\mathbf{x} - \boldsymbol{\xi}_\ell|$.
Surface mass density: $\kappa_{ptcs}(\mathbf{x}) = \kappa_c + \pi \sum_{\ell=1}^{g} m_\ell \delta_{\boldsymbol{\xi}_\ell}$.
bending angle vector:
$\boldsymbol{\alpha}_{ptcs}(\mathbf{x}) = ((\kappa_c + \gamma)\mathrm{u}, (\kappa_c - \gamma)\mathrm{v}) + \sum_{\ell=1}^{g} m_\ell \frac{\mathbf{x}-\boldsymbol{\xi}_\ell}{|\mathbf{x}-\boldsymbol{\xi}_\ell|^2}$.
Singularities: point mass positions $\mathbf{x}_\ell$.
Applications:
(1) Single/Binary point mass lens with continuous matter and shear: as above, more realistic.
(2) Multiple point mass lens with continuous matter and shear: realistic microlensing of quasars (e.g., quasar 2237+0305).

Isothermal sphere: singular/softened core
Potential: $\psi_{sis}(\mathbf{x}) = C|\mathbf{x}|$; $\psi_{fis}(\mathbf{x}) = C\sqrt{x_{cr}^2 + |\mathbf{x}|^2}$.
C = positive constant depending on velocity dispersion.
Surface mass density: $\kappa_{sis}(\mathbf{x}) = C/2|\mathbf{x}|$
$\kappa_{fis}(\mathbf{x}) = C(x_{cr}^2 + 2|\mathbf{x}|^2)/2(x_{cr}^2 + |\mathbf{x}|^2)^{3/2}$.
Bending angle vector: $\boldsymbol{\alpha}_{sis}(\mathbf{x}) = C\mathbf{x}/|\mathbf{x}|$; $\boldsymbol{\alpha}_{fis}(\mathbf{x}) = C\mathbf{x}/\sqrt{x_{cr}^2 + |\mathbf{x}|^2}$.
Singularities: origin for '*sis*' (i.e., $x_{cr} = 0$) and none for 'fis' (i.e., $x_{cr} > 0$).
Applications: semi-realistic model of galaxies as lenses; useful both for modeling of individual lens systems as well as for statistical studies of lensing cross sections.

Elliptical potential: singular/softened core
Potential: $\psi_{ep}(\mathbf{x}) = C\sqrt{x_{cr}^2 + (1-\epsilon)\mathrm{u}^2 + (1+\epsilon)\mathrm{v}^2}$.
x_{cr} = positive constant: core radius; singular case: $x_{cr} = 0$.
ϵ = ellipticity; restricted to $0 < \epsilon \lesssim 0.2$ for physical reasons [Kas-K, p. 460].
Surface mass density: $\kappa_{ep}(\mathbf{x}) = C^4\left(2x_{cr}^2 + (1-\epsilon^2)^2|\mathbf{x}|^2\right)/2\psi_{ep}^3(\mathbf{x})$.
bending angle vector: $\boldsymbol{\alpha}_{ep}(\mathbf{x}) = C^2\left((1-\epsilon)\mathrm{u}, (1+\epsilon)\mathrm{v}\right)/\psi_{ep}(\mathbf{x})$.
Singularities: origin for $x_{cr} = 0$ (singular core) and none for $x_{cr} > 0$ (softened core).
Applications: semi-realistic models of lensing galaxies or lensing clusters (image configuration).

Isothermal ellipsoids: singular/softened core
Potential: $\psi_{ie}(\mathbf{x}) = \mathrm{u}\alpha_1 + \mathrm{v}\alpha_2 - b_I s_c \ln\sqrt{(h(\mathbf{x}) + s_c)^2 + (1-q^2)\mathrm{u}^2}$,
where $\alpha_1 = (b_I/\sqrt{1-q^2})\ \tan^{-1}\left[(\mathrm{u}\sqrt{1-q^2})/(h(\mathbf{x}) + s_c)\right]$,
$\alpha_2 = (b_I/\sqrt{1-q^2})\ \tanh^{-1}\left[(\mathrm{v}\sqrt{1-q^2})/(h(\mathbf{x}) + q^2 s_c)\right]$,
with $\mathbf{x} = (\mathrm{u}, \mathrm{v})$, $b_I > 0$, $q > 0$, $s_c \geq 0$, and $h(\mathbf{x}) = q^2(s_c^2 + \mathrm{u}^2) + \mathrm{v}^2$.
Surface mass density: $\kappa_{ie}(\mathbf{x}) = b_I/2\sqrt{q^2(s_c^2 + \mathrm{u}^2) + \mathrm{v}^2}$.
Bending angle vector: $\boldsymbol{\alpha}_{ie}(\mathbf{x}) = (\alpha_1, \alpha_2)$.
Singularities: origin for $s_c = 0$ (singular core) and none for $s_c > 0$ (softened core).
Applications: realistic models of lensing galaxies.

6.4 Multiple Plane Lensing

We generalize the mathematical notions of time delay functions and lensing maps to multiple lens planes (cf. Sections 3.2.2 and 3.2.3 on pages 67 and 77, resp.). Unlike the single-plane case of Section 6.2, the form of these maps will now explicitly involve angular diameter distances. Section begins with lens planes and angular diameter distances, which is followed by multiple plane time delay functions and lensing maps.

Lens Planes and Angular Diameter Distances

We have used the notions "lens planes" and (Friedmann) "angular diameter distances" mainly within a physical context. However, such concepts must also be precisely abstracted before they can be used in mathematical theorems. The reader should consult Sections 3.1.2 (page 37) and 3.1.4 (page 46) for the physical basis of these concepts.

Definition 6.8 Let ψ be a gravitational lens potential with domain L. A *lens plane at redshift* z is a triple (L, ψ, z).

By abuse of terminology and notation, we shall often refer to a lens plane at redshift z simply as a *lens plane*. In this context, the potential ψ is said to be a *gravitational lens potential at redshift* z. Also, we shall use the abbreviation L or (L, z) when no confusion is likely to arise by omitting element(s) of the triple (L, ψ, z).

Angular diameter distances are now defined as follows (cf. equations (3.24) and (3.25), page 40):

Definition 6.9 For a lens plane L at redshift z_L, the *Friedmann angular diameter distance of L as measured by an observer at redshift* 0 is the positive real number given by

$$d_L = \frac{c}{H_0}|r_A(z_L)|,$$

where

$$r_A(z) = \frac{2}{\Omega_0^2(1+z)^2}\left[\Omega_0 z - (2-\Omega_0)\left(\sqrt{1+\Omega_0 z} - 1\right)\right].$$

Here H_0 is Hubble's constant and Ω_0 the current cosmological density of the matter dominated Friedmann universe.[15]

Definition 6.10 For any two lens planes L_1 and L_2 at redshifts z_1 and z_2, respectively, the *Friedmann angular diameter distance of* L_2 *as measured from* L_1 is the positive real number

$$d_{1,2} = \frac{c}{H_0}|r_A(z_1, z_2)|.$$

Here

$$r_A(z_1, z_2) = \frac{2(1+z_1)}{\Omega_0^2}\left[R_1(z_2)R_2(z_1) - R_1(z_1)R_2(z_2)\right]$$

[15] See page 35 for the constants H_0 and Ω_0.

with $R_1(z) = [\Omega_0 z - \Omega_0 + 2]/(1+z)^2$ and $R_2(z) = (\Omega_0 z + 1)^{1/2}/(1+z)^2$, and

$$d_{2,1} = \frac{1+z_2}{1+z_1} d_{1,2} \tag{6.20}$$

is the *angular diameter distance of L_1 as measured from L_2.*

Remark: Note that angular diameter distances are not distances in the usual mathematical sense (since $d_{1,2} \neq d_{2,1}$). Furthermore, there are non-Friedmann choices of angular diameter distance. Such choices depend on the assumed cosmology — see Section 3.1.2, page 37 — and are generally more complicated to define. For our purposes, the explicit form of the angular diameter distance will be immaterial. The Friedmann case is presented simply for concreteness.

6.4.1 Multiplane Time Delay Functions

We now extend Definition 6.6 to multiple lens planes.

Let S be a plane at redshift z_S and let $(L_1, \psi_1, z_1), \ldots, (L_k, \psi_k, z_k)$ be lens planes at redshifts $z_1 < \cdots < z_k < z_S$.

The quantities $\vartheta_{i,j}$ and $\beta_{i,j}$

When dealing with several lens planes, the following notation is convenient:

Notation: Given lens planes L_i and L_j at redshifts $z_i < z_j$, set:

(1) $\vartheta_{i,j} = (1+z_i)(d_i d_j / c d_{i,j})$ and $\vartheta_i = \vartheta_{i,i+1}$, where $d_k = d_{L_k}$.

(2) $\beta_{i,j} = (d_{i,j} d_S)/(d_j d_{i,S})$ and $\beta_i = \beta_{i,i+1}$. Here, d_S and $d_{i,S}$ are the angular diameter distances of S as measured from the observer (at redshift 0) and from L_i, respectively.

Note that $\beta_k = 1$, $\vartheta_k = (1+z_k)(d_k d_S / c d_{k,S})$, and $\beta_{i,j} = \vartheta_{i,S}/\vartheta_{i,j}$. In addition, direct application of the definition of $\vartheta_{i,j}$ yields:

Lemma 6.11 *For planar sets L_i, L_j and L_k at redshifts $z_i < z_j < z_k$, we have*

$$\frac{1}{\vartheta_{i,k}} = \frac{1}{\vartheta_{i,j}} + \frac{1}{\vartheta_{j,k}}. \tag{6.21}$$

This result will be used in deriving the multiplane "lens equation" (see proof of Theorem 6.13):

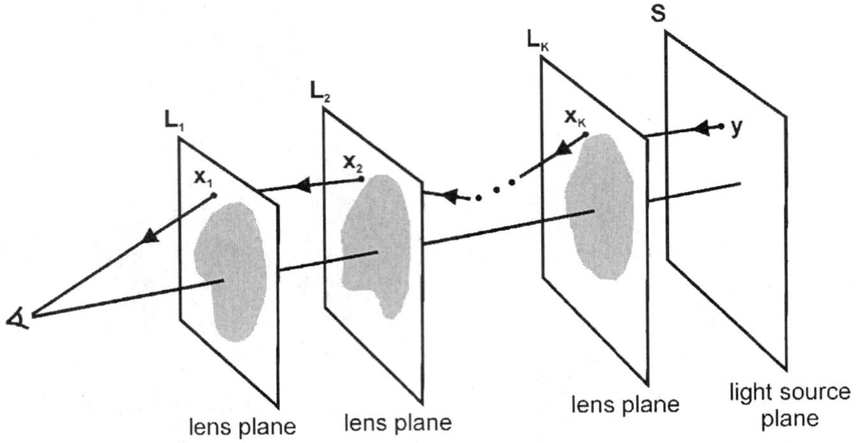

Figure 6.5. Schematic of k plane gravitational lensing. A ray emanating from a light source at $\mathbf{y}$ experiences multiple deflections on its way to the observer.

Time Delay Function for k Planes

Let $L_{(k)} = L_1 \times \cdots \times L_k$. Then (cf. Section 3.2.2, page 67):

Definition 6.12 For $\mathbf{y} \in S$, the smooth function $T_{\mathbf{y}} : L_{(k)} \longrightarrow \mathbf{R}$ defined by

$$T_{\mathbf{y}}(\mathbf{x}_1, \ldots, \mathbf{x}_k) \quad = \quad \sum_{i=1}^{k} \vartheta_i \left[\frac{|\mathbf{x}_i - \mathbf{x}_{i+1}|^2}{2} - \beta_i \psi_i(\mathbf{x}_i) \right], \tag{6.22}$$

where $\mathbf{x}_{k+1} = \mathbf{y}$, is a *$k$-plane time delay function* induced at the *light source position* $\mathbf{y}$ by gravitational lens potentials $\psi_1, \ldots, \psi_k$.[16] The family of time delay functions $T : L_{(k)} \times S \longrightarrow \mathbf{R}$ given by

$$T(\mathbf{X}, \mathbf{y}) = T_{\mathbf{y}}(\mathbf{X})$$

is a called a *k-plane time delay family* parametrized by the light source plane S.

We call L_i and S, respectively, the ith *lens plane* and *light source plane* of $T_{\mathbf{y}}$ — see Figure 6.5. The graph of $T_{\mathbf{y}}$ is termed the *time delay hypersurface* of $T_{\mathbf{y}}$.

Convention: *Unless stated to the contrary, we suppose that in (6.22) the parameters ϑ_i, β_i, $\mathbf{y}$, and all parameters associated with ψ_i $(1 \leq i \leq k)$, are fixed. Only the variables $\mathbf{x}_1, \ldots, \mathbf{x}_k$ vary.*

[16]All variables and parameters in (6.22) are dimensionless except ϑ_i, which has units of *time*. Hence, $T_{\mathbf{y}}$ also has units of time. Also, observe that if $k = 1$, then equation (6.22) coincides with equation (6.3) up to the insignificant factor ϑ_1 (since $\beta_1 = 1$ and $\vartheta_1 = \vartheta_{L,S}$).

Let us take a closer look at the domain of $T_{\mathbf{y}}$. We have $\mathrm{Dom}(T_{\mathbf{y}}) = L_1 \times \cdots \times L_k$, where $L_i = E_i - A_i$ with $E_i = \mathbf{R}^2$ and A_i the set of singularities of ψ_i $(i = 1, \ldots, k)$. Set $E_{(k)} = E_1 \times \cdots \times E_k$. Then the domain of $T_{\mathbf{y}}$ can be expressed as follows for all natural numbers $k \geq 1$:[17]

$$\prod_{i=1}^{k}(E_i - A_i) = E_{(k)} - \bigcup_{i=1}^{k} E_1 \times \cdots \times E_{i-1} \times A_i \times E_{i+1} \times \cdots \times E_k.$$

Denote the elements of A_i by $\mathbf{a}_{i,1}, \ldots, \mathbf{a}_{i,g_i}$ and let

$$\mathbf{A}_i = \bigcup_{j=1}^{g_i} E_1 \times \cdots \times E_{i-1} \times \{\mathbf{a}_{i,j}\} \times E_{i+1} \times \cdots \times E_k,$$

where $i = 1, \ldots, k$. Each term of this union is a $2(k-1)$-dimensional hyperplane in $\mathbf{R}^{2k}$ passing through $\mathbf{a}_{i,j}$ (since $E_{(k)} = \mathbf{R}^{2k}$), while $\mathbf{A}_i$ is a *disjoint* union of these hyperplanes. Moreover, the set $\mathbf{A}_1 \cup \cdots \cup \mathbf{A}_k$ is a $2(k-1)$-dimensional "rectangular grid" of hyperplanes and consists of all points in $\mathbf{R}^{2k}$ where $T_{\mathbf{y}}$ is not smooth. The domain of $T_{\mathbf{y}}$ is then the complement in $\mathbf{R}^{2k}$ of this "grid":

$$\mathrm{Dom}(T_{\mathbf{y}}) = \mathbf{R}^{2k} - \bigcup_{i=1}^{k} \mathbf{A}_i. \tag{6.23}$$

We now transfer some basic notions about single-plane time delay functions to the multiplane setting.

Notation: Set $\mathbf{X} = (\mathbf{x}_1, \ldots, \mathbf{x}_k)$ and $|\mathbf{X}| = (|\mathbf{x}_1|^2 + \cdots + |\mathbf{x}_k|^2)^{1/2}$.

Term a k-plane time delay function $T_{\mathbf{y}}$ *nonsingular* (resp., *singular*) if the potentials $\psi_1, \ldots, \psi_k$ are nonsingular (resp., singular).

The Principle of Fermat allows us to define a *light ray* from $\mathbf{y}$ that is lensed by potentials $\psi_1, \ldots, \psi_k$ at redshifts $z_1 < \cdots < z_k$ (or, simply, *light ray* of $T_{\mathbf{y}}$) to be a critical point $\mathbf{X}_0$ of $T_{\mathbf{y}}$.

We call a light ray $\mathbf{X}_0$ of $T_{\mathbf{y}}$ *nondegenerate* if $\mathbf{X}_0$ is a nondegenerate critical point; otherwise, $\mathbf{X}_0$ is *degenerate*. Given a nondegenerate light ray $\mathbf{X}_0$, the number of negative eigenvalues of $[\mathrm{Hess}\, T_{\mathbf{y}}](\mathbf{X}_0)$ is called the *index* of $\mathbf{X}_0$. In addition, Morse's lemma (e.g., [Gol-G, pp. 65–68]) implies that about each nondegenerate light ray $\mathbf{X}_0$ of index i there is a (curvilinear) coordinate system $\lambda(u_1, \ldots, u_{2k}) = (\nu_1, \ldots, \nu_{2k})$ centered at $\mathbf{X}_0$ such that in these coordinates we have that

$$T_{\mathbf{y}} \circ \lambda^{-1}(\nu_1, \ldots, \nu_{2k}) = T_{\mathbf{y}}(\mathbf{X}_0) - \nu_1^2 - \cdots - \nu_i^2 + \nu_{i+1}^2 + \nu_{i+2}^2 + \cdots + \nu_{2k}^2. \tag{6.24}$$

[17] Inductively employ: (i) $A \times (B - C) = (A \times B) - (A \times C)$, $(A - B) \times C = (A \times C) - (B \times C)$, (ii) $(A - B) - C = A - (B \cup C)$, and (iii) if $C \subseteq A$, then $A \cup (B - C) = A \cup B$.

Hence, equation (6.24) classifies nondegenerate light rays by their indices: $\mathbf{X}_0$ is either a local minimum ($i = 0$), local maximum ($i = 2k$), or, generalized saddle ($1 \leq i \leq 2k - 1$). Thus, all nondegenerate light rays are isolated.

As in the single-plane case, the critical point type of a multiplane light ray can be determined from the quadratic form of $T_{\mathbf{y}}$ at $\mathbf{X}_0$:

$$q(t_1, \ldots, t_{2k}) = \sum_{i=1}^{2k} T_{u_i u_i}(\mathbf{X}_0, \mathbf{y})\, t_i^2 + 2 \sum_{1 \leq i < j \leq k} T_{u_i u_j}(\mathbf{X}_0, \mathbf{y})\, t_i t_j,$$

where $(t_1, \ldots, t_{2k}) \in \mathbf{R}^{2k}$ and we write $T_{\mathbf{y}}(\mathbf{X}) = T(u_1, \ldots, u_{2k}, \mathbf{y})$ when taking partial derivatives relative to u_i. A light ray $\mathbf{X}_0$ is (i) a *local minimum* if and only if $q(t_1, \ldots, t_{2k}) > 0$ for all $(t_1, \ldots, t_{2k})$ in $\mathbf{R}^{2k} - \{\mathbf{0}\}$; (ii) a *local maximum* if and only if $q(t_1, \ldots, t_{2k}) < 0$ for all $(t_1, \ldots, t_{2k})$ in $\mathbf{R}^{2k} - \{\mathbf{0}\}$; and (iii) a *generalized saddle* if and only if $q(t_1, \ldots, t_{2k})$ attains both positive and negative values.

In Section 6.2.3, we treated the local geometry of time delay surfaces. For the case of k-plane lensing, we are dealing with the Riemannian geometry of $2k$-dimensional manifolds, namely, time delay hypersurfaces. Given a k-plane time delay function $T_{\mathbf{y}}$, the eigenvalues $\lambda_1, \ldots, \lambda_{2k}$ of the Hessian matrix Hess $T_{\mathbf{y}}$ are the *principal curvatures* of the associated time delay hypersurface. The arithmetic mean of the principal curvatures define the *mean curvature*:

$$H = \frac{\lambda_1 + \cdots + \lambda_{2k}}{2k}.$$

If $\mathbf{X}_0$ is a light ray of $T_{\mathbf{y}}$, then the *Gauss–Kronecker curvature* of the time delay hypersurface at $\mathbf{X}_0$ is

$$G = \lambda_1 \cdots \lambda_{2k} = \det[\text{Hess}\, T_{\mathbf{y}}]. \tag{6.25}$$

Readers are referred to [doC, p. 129] and [Sas, p. 49] for more on the geometric quantities G and H, and the second fundamental form from which they are constructed. However, note that not all authors use the same signs when defining the second fundamental form.

6.4.2 Multiplane Lensing Maps

Consulting Figure 6.6, trace a light ray backwards from the observer to the point $\mathbf{x}_1$ ("impact vector" on first lens plane), continuing on through the various lens planes, and, after exiting the last lens plane (i.e., L_k) through $\mathbf{x}_k$, conclude at the light source position $\mathbf{y}$. This induces a "lensing map" $\boldsymbol{\eta}$ relating $\mathbf{x}_1$ to $\mathbf{y}$. Our goal is to construct $\boldsymbol{\eta}$.

Lens Equation for k Planes

The light source position $\mathbf{y}$ was expressed recursively by Blandford and Narayan [Bla-N86, eq. (6.3), p. 579] in terms of the impact vector $\mathbf{x}_1$ on the first lens plane L_1. We formulate their result as follows:

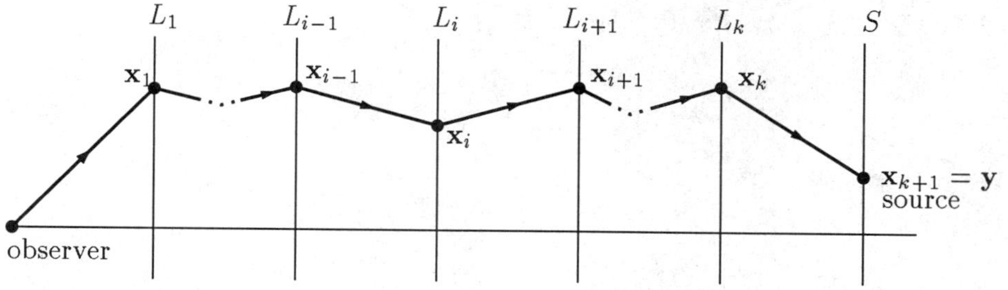

Figure 6.6. Light rays traced backwards from the observer induce a *lensing map* from a subset P of the first lens plane L_1 into S, the light source plane.

Theorem 6.13 *Let $T_{\mathbf{y}} : L_1 \times \cdots \times L_k \longrightarrow \mathbf{R}$ be a k-plane time delay function induced at $\mathbf{y} \in S$ by gravitational lens potentials $\psi_1, \ldots, \psi_k$ at redshifts $z_1 < \cdots < z_k$ with $z_k < z_S$. Then a point $(\mathbf{x}_1, \ldots, \mathbf{x}_k)$ in $L_1 \times \cdots \times L_k$ is a light ray of $T_{\mathbf{y}}$ if and only if*

$$\mathbf{x}_i = \mathbf{x}_1 - \sum_{j=1}^{i-1} \beta_{j,i} \boldsymbol{\alpha}_j(\mathbf{x}_j), \quad (2 \le i \le k+1), \tag{6.26}$$

where $\mathbf{x}_{k+1} = \mathbf{y}$.

Proof: A point $\mathbf{X} = (\mathbf{x}_1, \ldots, \mathbf{x}_k)$ in $L_1 \times \cdots \times L_k$ is a light ray of $T_{\mathbf{y}}$ if and only if $\mathbf{X}$ satisfies the k vector equations

$$\text{grad}_{\mathbf{x}_i} T_{\mathbf{y}}(\mathbf{X}) = \mathbf{0}.$$

These equations are equivalent to

$$\mathbf{x}_2 - \mathbf{x}_1 = -\beta_1 \boldsymbol{\alpha}_1(\mathbf{x}_1) \tag{6.27}$$

$$\mathbf{x}_{i+1} - \mathbf{x}_i = \frac{\vartheta_{i-1}}{\vartheta_i}(\mathbf{x}_i - \mathbf{x}_{i-1}) - \beta_i \boldsymbol{\alpha}_i(\mathbf{x}_i), \quad (2 \le i \le k), \tag{6.28}$$

where $\boldsymbol{\alpha}_i$ is the bending angle vector field of ψ_i. Applying equations (6.27) and (6.28) recursively yields

$$\mathbf{x}_{i+1} - \mathbf{x}_i = -\frac{1}{\vartheta_i} \sum_{\ell=1}^{i} \vartheta_\ell \beta_\ell \boldsymbol{\alpha}_\ell(\mathbf{x}_\ell), \quad (1 \le i \le k),$$

which, upon using the technical result,

$$\sum_{r=1}^{i} \frac{1}{\vartheta_i} \left(\sum_{\ell=1}^{r} \vartheta_\ell \beta_\ell \boldsymbol{\alpha}_\ell(\mathbf{x}_\ell) \right) = \sum_{\ell=1}^{i} \vartheta_\ell \beta_\ell \boldsymbol{\alpha}_\ell(\mathbf{x}_\ell) \left(\sum_{r=\ell}^{i} \frac{1}{\vartheta_r} \right),$$

implies that

$$\mathbf{x}_{i+1} - \mathbf{x}_1 = -\sum_{\ell=1}^{i} \vartheta_\ell \beta_\ell \boldsymbol{\alpha}_\ell(\mathbf{x}_\ell) \left(\sum_{r=\ell}^{i} \frac{1}{\vartheta_r} \right), \quad (1 \le i \le k). \tag{6.29}$$

Lemma 6.11 gives

$$\sum_{r=\ell}^{i} \frac{1}{\vartheta_r} = \frac{1}{\vartheta_{\ell,i+1}}.$$

Note that $(\vartheta_\ell \beta_\ell)/\vartheta_{\ell,i+1} = \beta_{\ell,i+1}$. Hence, equation (6.29) becomes

$$\mathbf{x}_{i+1} = \mathbf{x}_1 - \sum_{j=1}^{i} \beta_{j,i+1} \boldsymbol{\alpha}_j(\mathbf{x}_j), \qquad (1 \leq i \leq k). \ \square$$

Theorem 6.13 yields the light source position $\mathbf{y}$ in terms of the impact vector $\mathbf{x}_1$ by way of the *k-plane lens equation* (cf. Section 3.2.3, page 77):

$$\mathbf{y} = \mathbf{x}_1 - \sum_{\ell=1}^{k} \boldsymbol{\alpha}_\ell(\mathbf{x}_\ell), \tag{6.30}$$

where $\mathbf{x}_{\ell+1}$ is given recursively by (6.26) for $\ell = 1, \ldots, k-1$ and we used $\beta_{j,k+1} = 1$. As in the single-plane case, the k-plane lens equation will induce the "multiplane lensing map" $\boldsymbol{\eta}$.

Partial Lensing Maps and Obstruction Points

Before defining the k-plane lensing map $\boldsymbol{\eta}$, we need to construct $k+1$ partial lensing maps $\boldsymbol{\eta}_1, \ldots, \boldsymbol{\eta}_{k+1}$. Each $\boldsymbol{\eta}_i$ will correspond to tracing light rays from the observer to the ith lens plane L_i, where the light rays experience deflections at $L_1, \ldots L_{i-1}$. In other words, the map $\boldsymbol{\eta}_i$ is from a subset P_{i-1} of L_1 into L_i. The desired k-plane lensing map $\boldsymbol{\eta}$ will be $\boldsymbol{\eta}_{k+1}$.

We now precisely define $\boldsymbol{\eta}_1, \ldots, \boldsymbol{\eta}_{k+1}$ inductively. For $i = 1, \ldots, k$, set $L_i = E_i - A_i$, where $E_i = \mathbf{R}^2$ and A_i are the singularities of ψ_i, and set $E_{k+1} = \mathbf{R}^2$. First, *let* $\boldsymbol{\eta}_1 : E_1 \to E_1$ *be the identity map.* Second, set $\mathsf{B}_1 = A_1$ and $P_1 = E_1 - \mathsf{B}_1$. Using (6.26), *define* $\boldsymbol{\eta}_2$ *to be the smooth map* $\boldsymbol{\eta}_2 : P_1 \longrightarrow E_2$ *given by*[18]

$$\boldsymbol{\eta}_2 = \boldsymbol{\eta}_1 + \beta_{1,2}(\boldsymbol{\alpha}_1 \circ \boldsymbol{\eta}_1).$$

Third, let $\mathsf{B}_2 = \boldsymbol{\eta}_2^{-1}(A_2)$ and $P_2 = E_1 - \mathsf{B}_1 \cup \mathsf{B}_2$. Motivated by (6.26), *let* $\boldsymbol{\eta}_3$ *be the smooth map* $\boldsymbol{\eta}_3 : P_2 \longrightarrow E_3$ *defined by*[19]

$$\boldsymbol{\eta}_3 = \boldsymbol{\eta}_1 - \beta_{1,3}(\boldsymbol{\alpha}_1 \circ \boldsymbol{\eta}_1) - \beta_{2,3}(\boldsymbol{\alpha}_2 \circ \boldsymbol{\eta}_2).$$

Continuing the above process inductively, suppose that $\boldsymbol{\eta}_1, \ldots, \boldsymbol{\eta}_{i-1}$ are already known. Then the sets $P_1, P_2, \ldots, P_{i-2}, \mathsf{B}_1, \mathsf{B}_2, \ldots, \mathsf{B}_{i-3}$, and B_{i-2} are known. In order to define

[18] Note: $\boldsymbol{\alpha}_1 \circ \boldsymbol{\eta}_1 = \boldsymbol{\alpha}_1$, which is smooth on P_1. Also, since $\boldsymbol{\eta}_2(\mathbf{x}_1) = \mathbf{x}_2$, i.e., $\boldsymbol{\eta}$ maps a point of the first lens plane into the plane of the second lens, we choose E_2 as the target space for $\boldsymbol{\eta}_2$.

[19] (1) Because the domain of the bending angle vector field $\boldsymbol{\alpha}_\ell$ is $E_\ell - A_\ell$, where $\ell = 1, \ldots, k$, the map $\boldsymbol{\alpha}_2 \circ \boldsymbol{\eta}_2$ is smooth on P_2. (2) The target E_1 of $\boldsymbol{\eta}_1$ is identified with the target E_3 of $\boldsymbol{\eta}_3$.

η_i, let $\mathsf{B}_{i-1} = \eta_{i-1}^{-1}(A_{i-1})$ and $P_{i-1} = E_1 - \mathsf{B}_1 \cup \cdots \cup \mathsf{B}_{i-2} \cup \mathsf{B}_{i-1}$. Since A_{i-1} is closed, so is $\mathsf{B}_1 \cup \cdots \cup \mathsf{B}_{i-1}$ (by continuity of η_{i-1}). Consequently, P_{i-1} is open (since P_{i-2} is open).[20] We then employ (6.26) to *define* η_i *as the smooth map* $\eta_i : P_{i-1} \longrightarrow E_i$ *given by*

$$\eta_i \;=\; \eta_1 - \sum_{\ell=1}^{i-1} \beta_{\ell,i}(\boldsymbol{\alpha}_\ell \circ \eta_\ell), \quad (2 \le i \le k+1), \tag{6.31}$$

where η_1 is the identity map $E_1 \to E_1$. Since $\eta_i(\mathbf{x}_1) = \mathbf{x}_i$, Figure 6.6 allows us to interpret η_i as tracing a light ray backwards from the observer to the plane E_i containing the ith deflector, with deflection due to gravitational lens potentials $\psi_1, \ldots, \psi_{i-1}$ on the first $i-1$ lens planes $L_1, \ldots, L_{i-1}$.[21] Hence, for $i = 1, \ldots, k+1$ we call η_i an $(i-1)$*-plane partial lensing map*.

Notice that *for* $i = 2, \ldots, k+1$, *the domain* P_{i-1} *of* η_i *is the maximal subset of* E_1 *where* $\boldsymbol{\alpha}_1 \circ \eta_1, \ldots, \boldsymbol{\alpha}_{i-1} \circ \eta_{i-1}$ *are smooth.*[22] In addition, the union $\mathsf{B}_1 \cup \cdots \cup \mathsf{B}_{i-1}$ $(2 \le i \le k+1)$ consists of all points in the plane $E_1 \supseteq L_1$ where a light ray either terminates by hitting a singularity of ψ_1, or, passes through and terminates at a singularity of one of the potentials $\psi_2, \ldots, \psi_{i-1}$, never reaching the plane E_i. For this reason, we define:

Definition 6.14 Elements of the closed set $\mathsf{B}_1 \cup \cdots \cup \mathsf{B}_{i-1}$ are (*light-path*) *obstruction points* of η_i.

Obstruction points were introduced by A.P. [Pet95b, p. 4277] as part of a link between gravitational lenses and the geometry of their caustics (see Section 15.4, page 544).

The domain P_{i-1} of η_i then excises all backward-traced light rays impacting an obstruction point.

Also note that Theorem 6.13 can be written in terms of partial lensing maps: *A point* $(\mathbf{x}_1, \ldots, \mathbf{x}_k)$ *in* $L_1 \times \cdots \times L_k$ *is a light ray of* $T_\mathbf{y}$ *if and only if* $\mathbf{x}_1 \in P_k$ *and* $\mathbf{x}_i = \eta_i(\mathbf{x}_1)$ *for each* $i = 1, \ldots, k+1$, *where* $\mathbf{x}_{k+1} = \mathbf{y} = \eta_{k+1}(\mathbf{x}_1)$.

Lensing Map for k Planes

The desired lensing map is $\eta = \eta_{k+1}$. Let $P = P_k$ and $\mathsf{B} = \mathsf{B}_1 \cup \cdots \cup \mathsf{B}_{k-1}$. Then the following generalizes Definition 6.7 (single-plane case):

Definition 6.15 Let S be a plane at redshift z_S. A *k-plane lensing map* induced by gravitational lens potentials $\psi_1, \ldots, \psi_k$ at redshifts $z_1 < \cdots < z_k$ with $z_k < z_S$, is a (smooth) map

---

[20] Note: $P_{i-1} \subseteq P_{i-2} \subseteq \cdots \subseteq P_1 = L_1$ and $\mathsf{B}_1 \cap \cdots \cap \mathsf{B}_{i-1} = \emptyset$ for $i = 2, \ldots, k+1$.

[21] The map η_1 corresponds to "trivial" lensing, that is, all bending angle vector fields vanish.

[22] In fact, if each $\boldsymbol{\alpha}_\ell \circ \eta_\ell$, where $\ell = 1, \ldots, i-1$, is smooth at some point $\mathbf{b}$ in $\mathsf{B}_1 \cup \cdots \cup \mathsf{B}_{i-1}$, say, $\mathbf{b} \in \mathsf{B}_j = \eta_j^{-1}(A_j)$, then this contradicts the fact that $\boldsymbol{\alpha}_j \circ \eta_j$ is not smooth at $\mathbf{b}$ (since $\boldsymbol{\alpha}_j$ is not smooth at points of A_j).

$\boldsymbol{\eta} : P \longrightarrow S$ defined by[23]

$$\boldsymbol{\eta}(\mathbf{x}_1) = \mathbf{x}_1 - \sum_{i=1}^{k} (\boldsymbol{\alpha}_i \circ \boldsymbol{\eta}_i)(\mathbf{x}_1), \tag{6.32}$$

where $\boldsymbol{\eta}_i$ is given by (6.31). We call S and B, respectively, the *light source plane* and the set of *obstruction points* of $\boldsymbol{\eta}$.

Explicitly:

$$\boldsymbol{\eta}(\mathbf{x}_1) = \mathbf{x}_1 - \sum_{i=1}^{k} \boldsymbol{\alpha}_i(\mathbf{x}_i), \qquad \mathbf{x}_i = \boldsymbol{\eta}_i(\mathbf{x}_1) = \mathbf{x}_1 - \sum_{j=1}^{i-1} \beta_{j,i}\boldsymbol{\alpha}_j(\mathbf{x}_j), \quad (2 \leq i \leq k).$$

Informally speaking, the set B consists of all points on the first deflector plane through which a backward-traced light ray will not reach the light source plane. Notice that if all the gravitational lens potentials are nonsingular, then the lensing map has no obstruction points (and, hence, is defined on all of $\mathbf{R}^2$).

Concepts like "lensed image," "magnification," etc., carry over readily to k-plane lensing maps. A *lensed image* of a light source at $\mathbf{y}$, relative to potentials $\psi_1, \ldots, \psi_k$ at redshifts $z_1 < \cdots < z_k$, is an element of the fibre $\boldsymbol{\eta}^{-1}(\mathbf{y})$. In other words, a lensed image is a solution $\mathbf{x}_1$ of the *lens equation*: $\mathbf{y} = \boldsymbol{\eta}(\mathbf{x}_1)$. The *magnification* of lensed image $\mathbf{x}_1$ is

$$\mathrm{Mag}(\mathbf{x}_1; \mathbf{y}) = \frac{1}{|\det[\mathrm{Jac}\, \boldsymbol{\eta}](\mathbf{x}_1)|},$$

while the *total magnification* of a light source at $\mathbf{y}$ is (cf. page 183):

$$\mathrm{Mag}_{tot}(\mathbf{y}) = \sum_{\mathbf{x} \in \eta^{-1}(\mathbf{y})} \mathrm{Mag}(\mathbf{x}_1; \mathbf{y}).$$

Note that $\mathrm{Jac}\, \boldsymbol{\eta}$ need not be symmetric (though it is in the single-plane case). *Critical points* of $\boldsymbol{\eta}$ are points $\mathbf{x}_1$ satisfying $\det[\mathrm{Jac}\, \boldsymbol{\eta}](\mathbf{x}_1) = 0$. If a collection of critical points of $\boldsymbol{\eta}$ forms a curve, then we call the curve a *critical curve* — see Figure 6.7. The critical values of $\boldsymbol{\eta}$ are called *caustics.* A point $\mathbf{y}$ is a caustic point if and only if a light source at $\mathbf{y}$ has at least one infinitely magnified lensed image.

Multiplane Lensing Maps as Compositions

We introduced the notion of a k-plane lensing map $\boldsymbol{\eta}$ recursively. Equivalently, H.L. observed that $\boldsymbol{\eta}$ can be expressed as a composition of maps. In fact, define maps $\boldsymbol{\lambda}_1 : L_1 \to E_2$ and $\boldsymbol{\lambda}_i : E_{i-1} \times L_i \to E_{i+1}$ using equations (6.27) and (6.28), respectively. These maps

[23] Multiplane lensing maps are dimensionless. In addition, we denote elements of P by $\mathbf{x}_1$ to remind the reader that such points lie on the first lens plane (counting from observer to light source plane).

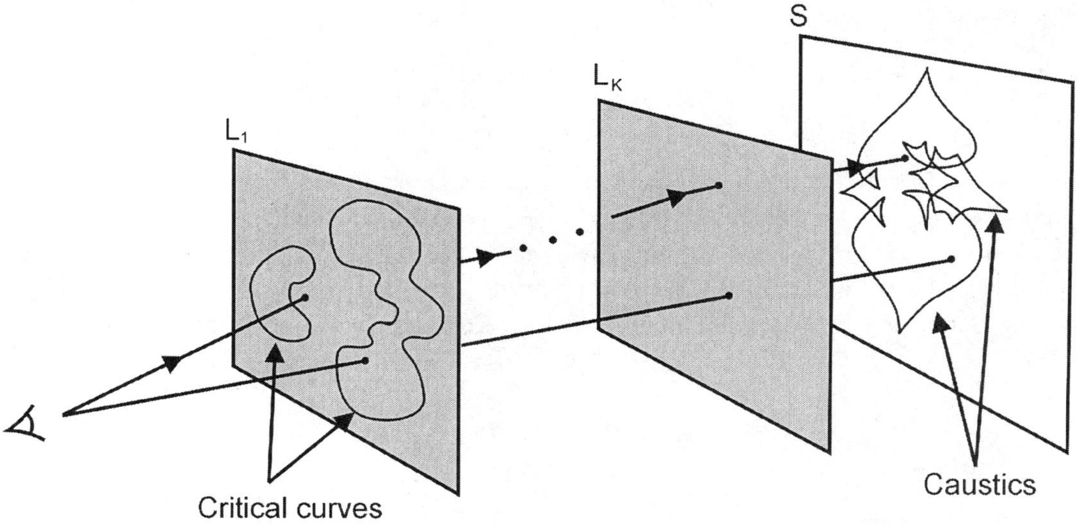

Figure 6.7. Critical curves and caustics associated with a k-plane lensing map defined by tracing light backwards from the observer to S.

describe the path of a backward-traced light ray as it traverses three consecutive lens planes — see Figure 6.6. Let $\hat{\boldsymbol{\lambda}}_1 : L_1 \to L_1 \times E_2$ and $\hat{\boldsymbol{\lambda}}_i : E_{i-1} \times L_i \to L_i \times E_{i+1}$ be respectively, suspensions of $\boldsymbol{\lambda}_1$ and $\boldsymbol{\lambda}_i$, that is,

$$\hat{\boldsymbol{\lambda}}_1(\mathbf{x}_1) = (\mathbf{x}_1, \boldsymbol{\lambda}_1(\mathbf{x}_1)), \qquad \hat{\boldsymbol{\lambda}}_i(\mathbf{x}_{i-1}, \mathbf{x}_i) = (\mathbf{x}_i, \boldsymbol{\lambda}_i(\mathbf{x}_{i-1}, \mathbf{x}_i)).$$

Now, given maps $f : A \to B$ and $h : C \to D$, where $B \subseteq E$ and $C \subseteq E$, the composition $h \circ f$ is defined with domain $f^{-1}(C \cap B)$. Since the target of $\hat{\boldsymbol{\lambda}}_{i-1}$ and domain of $\hat{\boldsymbol{\lambda}}_i$ are contained in $E_{i-1} \times E_i$, we can compose these maps to obtain $\boldsymbol{\eta}$ [Lev-P]:

$$\boldsymbol{\eta} = \boldsymbol{\lambda}_k \circ \hat{\boldsymbol{\lambda}}_{k-1} \circ \cdots \circ \hat{\boldsymbol{\lambda}}_2 \circ \hat{\boldsymbol{\lambda}}_1.$$

The mappings $\boldsymbol{\lambda}_i$ will be used in the next section.

6.5 Relating Time Delay and Lensing Maps

We begin with a lemma extending equation (6.6) to k lens planes. Let $T : L_1 \times \cdots \times L_k \times S \longrightarrow \mathbf{R}$ and $\boldsymbol{\eta} : P \longrightarrow S$ be, respectively, a k-plane time delay family and lensing map induced by gravitational lens potentials $\psi_1, \ldots, \psi_k$ at redshifts $z_1 < \cdots < z_k$.

Notation: Define a map $\boldsymbol{\Upsilon} : P \longrightarrow P \times L_2 \times \cdots \times L_k$ by

$$\boldsymbol{\Upsilon}(\mathbf{x}_1) = (\boldsymbol{\eta}_1(\mathbf{x}_1), \boldsymbol{\eta}_2(\mathbf{x}_1), \ldots, \boldsymbol{\eta}_k(\mathbf{x}_1)).$$

Since $\boldsymbol{\eta}_1(\mathbf{x}_1) = \mathbf{x}_1$, the range $\boldsymbol{\Upsilon}[P]$ is the graph of $(\boldsymbol{\eta}_2, \ldots, \boldsymbol{\eta}_k)$ over P. Hence, $\boldsymbol{\Upsilon}[P]$ is diffeomorphic to P.

Lemma 6.16 *For each* $\mathbf{x}_1$ *in* P, *where* $\boldsymbol{\eta}(\mathbf{x}_1) = \mathbf{y}$, *we have*

$$\det[\mathrm{Jac}\,\boldsymbol{\eta}](\mathbf{x}_1) = \frac{\det[\mathrm{Hess}\,T_{\mathbf{y}}](\boldsymbol{\Upsilon}(\mathbf{x}_1))}{(\vartheta_1 \cdots \vartheta_k)^2}.$$

Lemma 6.16 shows that in the multiplane case, magnification is dimensionless and computable via a time delay function. Also, if $\mathbf{x}_1$ is a lensed image associated with a light ray $\mathbf{X} = (\mathbf{x}_1, \ldots, \mathbf{x}_k)$ of $T_{\mathbf{y}}$, then Lemma 6.16 and equation (6.25) yield that *the magnification of the lensed image* $\mathbf{x}_1$ *is inversely proportional to the Gauss–Kronecker curvature at* $(\mathbf{X}_0, T_{\mathbf{y}}(\mathbf{X}_0))$ *of the associated time delay hypersurface.* Hence, the flatter a time delay hypersurface is about a critical point the higher the magnification for the corresponding lensed images.

A proof of Lemma 6.16 appears at the end of this section.

Notation:

(1) $\mathrm{Light}(T_{\mathbf{y}})$ = set of all light rays of $T_{\mathbf{y}}$.

(2) $\mathrm{Light}(T) = \bigcup_{\mathbf{y}\in S} \mathrm{Light}(T_{\mathbf{y}})$.

(3) $\mathrm{Degen}(T)$ = set of all degenerate light rays of $T_{\mathbf{y}}$ for all $\mathbf{y} \in S$.

(4) $\mathrm{Caustic}(T)$ = set of $\mathbf{y} \in S$ such that $T_{\mathbf{y}}$ has at least one degenerate light ray.

The next theorem links light rays, lensed images, critical curves, and caustics.[24]

Theorem 6.17 *Let* $T : L_1 \times \cdots \times L_k \times S \longrightarrow \mathbf{R}$ *and* $\boldsymbol{\eta} : P \longrightarrow S$ *be, respectively, a k-plane time delay family and lensing map induced by gravitational lens potentials* $\psi_1, \ldots, \psi_k$ *at redshifts* $z_1 < \cdots < z_k$. *Then:*

(1) $\boldsymbol{\eta}^{-1}(\mathbf{y}) = \boldsymbol{\Upsilon}^{-1}\left[\mathrm{Light}(T_{\mathbf{y}})\right]$ *for all* $\mathbf{y} \in S$.

(2) $\mathrm{Light}(T) = \boldsymbol{\Upsilon}[P]$.

(3) $\mathrm{Crit}(\boldsymbol{\eta}) = \boldsymbol{\Upsilon}^{-1}\left[\mathrm{Degen}(T)\right]$.

(4) $\mathrm{Caustic}(\boldsymbol{\eta}) = \mathrm{Caustic}(T)$.

Proof: (1–2) Theorem 6.13 implies that if $\mathbf{x}_1 \in \boldsymbol{\eta}^{-1}(\mathbf{y})$, then $\boldsymbol{\Upsilon}(\mathbf{x}_1)$ is a critical point of $T_{\mathbf{y}}$; conversely, if $(\mathbf{x}_1, \ldots, \mathbf{x}_k)$ is a critical point of $T_{\mathbf{y}}$, then $(\mathbf{x}_1, \mathbf{x}_2, \ldots, \mathbf{x}_k) = \boldsymbol{\Upsilon}(\mathbf{x}_1)$ and $\mathbf{x}_1 \in \boldsymbol{\eta}^{-1}(\mathbf{y})$. Hence, $\mathrm{Light}(T_{\mathbf{y}}) = \boldsymbol{\Upsilon}[\boldsymbol{\eta}^{-1}(\mathbf{y})]$. Because $P = \bigcup_{\mathbf{y}\in S} \boldsymbol{\eta}^{-1}(\mathbf{y})$, we have $\mathrm{Light}(T) = \boldsymbol{\Upsilon}[P]$. (3–4) By Lemma 6.16, $\boldsymbol{\Upsilon}[\mathrm{Crit}(\boldsymbol{\eta})] = \mathrm{Degen}(T)$. In addition, $\boldsymbol{\eta}(\mathbf{x}_1)$ is a caustic of $\boldsymbol{\eta}$ if and only if $T_{\boldsymbol{\eta}(\mathbf{x}_1)}$ has at least one degenerate light ray. □

We list some immediate consequences of the previous theorem. First, Theorem 6.17(1) implies that *the map* $\boldsymbol{\Upsilon}$ *restricts to a bijection from the set* $\boldsymbol{\eta}^{-1}(\mathbf{y})$ *of all lensed images of a*

[24] See [Lev-P] for a generalized nonsingular version.

light source at $\mathbf{y}$ *onto the set* $\text{Light}(T_{\mathbf{y}})$ *of all light rays from* $\mathbf{y}$ *to observer.* Using the identification afforded by Υ, we shall abuse terminology and speak of the *nondegeneracy* and *index* of lensed images, including *minimum, index i saddle,* and *maximum* lensed images. Note that $\text{Light}(T_{\mathbf{y}}) = \boldsymbol{\eta}^{-1}(\mathbf{y})$ for $k = 1$. Second, Theorem 6.17(2,3) shows that $\text{Light}(T)$ *and* $\text{Degen}(T)$ *are homeomorphic, respectively, to* P *and* $\text{Crit}(\boldsymbol{\eta})$. In the single plane case, $\text{Degen}(T) = \text{Crit}(\boldsymbol{\eta})$. Third, Theorem 6.17(4) yields that *a light source on a caustic has at least one degenerate light ray.*

Theorem 6.17 shows that multiple imaging, critical points, and caustics can be investigated through either the time delay or lensing map. In some situations, however, there are certain advantages in choosing one map over the other. For example, Theorem 6.17 (1) yields that the total number of lensed images of a point source at $\mathbf{y}$ is the cardinality of the fibre $\boldsymbol{\eta}^{-1}(\mathbf{y})$. However, to get lower bounds and counting formulas for the total number of lensed images, $\text{Light}(T_{\mathbf{y}})$ is less difficult to investigate than $\boldsymbol{\eta}^{-1}(\mathbf{y})$ (see Chapters 11 and 12). Nevertheless, it is best to investigate $\boldsymbol{\eta}^{-1}(\mathbf{y})$ when seeking upper bounds on the number of lensed images due to point mass lenses (Sections 11.5 and UBNLI). Also, Theorem 6.17(4) shows that caustics of $\boldsymbol{\eta}$ coincide with $\text{Caustic}(T)$. The local geometry of caustics can be found using T or $\boldsymbol{\eta}$ (Chapters 8 and 9). On the other hand, the global geometry is more readily treated by employing the lensing map (Chapter 15).

Proof of Lemma 6.16: Let $\mathfrak{L} = P \times L_2 \times \ldots \times L_k \times S$ and $\text{grad}_{\mathbf{x}_i} = \partial/\partial\mathbf{x}_i$. Define a map $\Upsilon_S : P \longrightarrow \mathfrak{L}$ by $\Upsilon_S(\mathbf{x}_1) = (\boldsymbol{\eta}_1(\mathbf{x}_1), \boldsymbol{\eta}_2(\mathbf{x}_1), \ldots, \boldsymbol{\eta}_k(\mathbf{x}_1), \boldsymbol{\eta}_{k+1}(\mathbf{x}_1))$. Note that $\Upsilon_S[P]$ is diffeomorphic to P (because $\Upsilon_S[P]$ is the graph over P of $(\boldsymbol{\eta}_2, \cdots, \boldsymbol{\eta}_{k+1})$). Define maps $T_i : L_i \times L_{i+1} \longrightarrow \mathbf{R}$ by

$$T_i(\mathbf{x}_i, \mathbf{x}_{i+1}) = \vartheta_i \left[\frac{|\mathbf{x}_{i+1} - \mathbf{x}_i|^2}{2} - \beta_i \psi_i(\mathbf{x}_i) \right].$$

Equations (6.27) and (6.28) imply that the graph of $\boldsymbol{\lambda}_1$ is the solution set of

$$\frac{\partial T}{\partial \mathbf{x}_1} = \frac{\partial T_1}{\partial \mathbf{x}_1} = \mathbf{0}$$

and the graph of each map $\boldsymbol{\lambda}_i$ is the solution set of

$$\frac{\partial T}{\partial \mathbf{x}_i} = \frac{\partial T_{i-1}}{\partial \mathbf{x}_i} + \frac{\partial T_i}{\partial \mathbf{x}_i} = \mathbf{0},$$

where $2 \leq i \leq k$ (see end of Section 6.4.2 for $\boldsymbol{\lambda}_i$).

Write $\mathbf{x}_i = (\mathrm{u}_i, \mathrm{v}_i)$ for $i = 1, \ldots, k$, and let $\mathbf{x}_{k+1} = \mathbf{y} = (\mathrm{u}_{k+1}, \mathrm{v}_{k+1})$. For $1 \leq m \leq k$ and $1 \leq n \leq k+1$, introduce 2×2 (in general, nonsymmetric) matrices as follows:

$$\mathbf{T}_{mn} = \begin{bmatrix} T_{\mathrm{u}_m \mathrm{u}_n} & T_{\mathrm{u}_m \mathrm{v}_n} \\ T_{\mathrm{v}_m \mathrm{u}_n} & T_{\mathrm{v}_m \mathrm{v}_n} \end{bmatrix}.$$

Written formally: $\mathbf{T}_{mn} = \frac{\partial^2 T}{\partial \mathbf{x}_m \partial \mathbf{x}_n}$. We have $\mathbf{T}_{i,i} = (\vartheta_{i-1} + \vartheta_i)\mathbf{I} - \beta_i \vartheta_i (\text{Hess}\, \psi_i)$ and $\mathbf{T}_{i,i+1} = -\vartheta_i \mathbf{I}$ for $i = 1, \ldots, k$, where $\vartheta_0 = 0$ and $\mathbf{I}$ is the 2×2 identity matrix. Also, $\mathbf{T}_{mn} = \mathbf{0}$ if $|m - n| \neq 0, 1$.

Equations (6.27) and (6.28) yield that $\mathbf{T}_{i\,i+1}\,(\partial \boldsymbol{\lambda}_i / \partial \mathbf{x}_i) + \mathbf{T}_{i\,i}$ and $\mathbf{T}_{i\,i+1}\left(\partial \boldsymbol{\eta}_i / \partial \mathbf{x}_{i-1}\right) + \mathbf{T}_{i\,i-1}$ vanish on the graph of $\boldsymbol{\lambda}_i$, and $\mathbf{T}_{12}\,(\partial \boldsymbol{\lambda}_1 / \partial \mathbf{x}_1) + \mathbf{T}_{11}$ vanishes on the graph of $\boldsymbol{\lambda}_1$. Now, note that $\Upsilon_S(P)$ is the intersection of the graphs of $\boldsymbol{\lambda}_i$ ($i = 1, \ldots, k$), that is, the set of all points $(\mathbf{x}_1, \mathbf{x}_2, \ldots, \mathbf{x}_{k+1}) \in \mathfrak{L}$ such that $\mathbf{x}_2 = \boldsymbol{\lambda}_1(\mathbf{x}_1)$, $\mathbf{x}_3 = \boldsymbol{\lambda}_2(\mathbf{x}_1, \mathbf{x}_2), \ldots, \mathbf{x}_{k+1} = \boldsymbol{\lambda}_k(\mathbf{x}_{k-1}, \mathbf{x}_{k+1})$ (or, more simply $\mathbf{x}_{j+1} = \boldsymbol{\eta}_i(\mathbf{x}_1)$, $i = 1, \ldots, k$). Thus, on $\Upsilon_S(P)$ we can use the above equations to replace $\mathbf{T}_{i\,i}$ by $-\mathbf{T}_{i\,i+1}\,(\partial \boldsymbol{\lambda}_i / \partial \mathbf{x}_i)$, $\mathbf{T}_{i\,i-1}$ by $-\mathbf{T}_{i\,i+1}\,(\partial \boldsymbol{\lambda}_i / \partial \mathbf{x}_{i-1})$, $i = 2, \ldots, k$, and $\mathbf{T}_{11}$ by $-\mathbf{T}_{12}\,(\partial \boldsymbol{\lambda}_1 / \partial \mathbf{x}_1)$. All of the preceding replacements are valid on $\Upsilon_S(P)$. *To keep the notation as simple as possible, we shall suppress this restriction to the range of* Υ_S.

Next, for each $1 \leq m \leq k$ and $1 \leq n \leq k+1$, the $2k \times 2(k+1)$ matrix $\mathbf{T} = [\mathbf{T}_{mn}]$ of 2×2 blocks has the form

$$\mathbf{T} = \begin{bmatrix} \mathbf{T}_{11} & \mathbf{T}_{12} & \mathbf{0} & \mathbf{0} & \ldots \mathbf{0} & \mathbf{0} \\ \mathbf{T}_{21} & \mathbf{T}_{22} & \mathbf{T}_{23} & \mathbf{0} & \ldots \mathbf{0} & \mathbf{0} \\ \mathbf{0} & \mathbf{T}_{32} & \mathbf{T}_{33} & \mathbf{T}_{34} & \ldots \mathbf{0} & \mathbf{0} \\ \cdots & \cdots & \cdots & \cdots & \cdots & \cdots \\ \mathbf{0} & \ldots & \mathbf{0} & \mathbf{T}_{k\,k-1} & \mathbf{T}_{k\,k} & \mathbf{T}_{k\,k+1} \end{bmatrix}.$$

The previous paragraph yields that $\mathbf{T} = \mathbf{D}\widehat{\mathbf{D}}$ on $\Upsilon_S[P]$, where

$$\mathbf{D} = \begin{bmatrix} \mathbf{T}_{12} & \mathbf{0} & \mathbf{0} & \mathbf{0} & \ldots & \mathbf{0} & \mathbf{0} \\ \mathbf{0} & \mathbf{T}_{23} & \mathbf{0} & \mathbf{0} & \ldots & \mathbf{0} & \mathbf{0} \\ \mathbf{0} & \mathbf{0} & \mathbf{T}_{34} & \mathbf{0} & \ldots & \mathbf{0} & \mathbf{0} \\ \cdots & \cdots & \cdots & \cdots & \cdots & \mathbf{0} & \mathbf{0} \\ \mathbf{0} & \ldots & \mathbf{0} & \mathbf{0} & \ldots & \mathbf{T}_{k-1\,k} & \mathbf{0} \\ \mathbf{0} & \ldots & \mathbf{0} & \mathbf{0} & \ldots & \mathbf{0} & \mathbf{T}_{k\,k+1} \end{bmatrix}$$

and

$$\widehat{\mathbf{D}} = \begin{bmatrix} -\frac{\partial \boldsymbol{\lambda}_1}{\partial \mathbf{x}_1} & \mathbf{I} & \mathbf{0} & \mathbf{0} & \mathbf{0} & \dots & \mathbf{0} \\ -\frac{\partial \boldsymbol{\lambda}_2}{\partial \mathbf{x}_1} & -\frac{\partial \boldsymbol{\lambda}_2}{\partial \mathbf{x}_2} & \mathbf{I} & \mathbf{0} & \mathbf{0} & \dots & \mathbf{0} \\ \mathbf{0} & -\frac{\partial \boldsymbol{\lambda}_3}{\partial \mathbf{x}_2} & -\frac{\partial \boldsymbol{\lambda}_3}{\partial \mathbf{x}_3} & \mathbf{I} & \mathbf{0} & \dots & \mathbf{0} \\ \dots & \dots & \dots & \dots & \dots & \dots & \dots \\ \mathbf{0} & \dots & \mathbf{0} & -\frac{\partial \boldsymbol{\lambda}_{k-1}}{\partial \mathbf{x}_{k-2}} & -\frac{\partial \boldsymbol{\lambda}_{k-1}}{\partial \mathbf{x}_{k-1}} & \mathbf{I} & \mathbf{0} \\ \mathbf{0} & \dots & \dots & \mathbf{0} & -\frac{\partial \boldsymbol{\lambda}_k}{\partial \mathbf{x}_{k-1}} & -\frac{\partial \boldsymbol{\lambda}_k}{\partial \mathbf{x}_k} & \mathbf{I} \end{bmatrix}.$$

Delete the last column of $\widehat{\mathbf{D}}$, denoting the resulting matrix by $\mathbf{A}$. *Then the product* $\mathbf{DA}$ *is the Hessian matrix of the time delay family* T*, which is written as* Hess T.

We shall refer to the entries and rows in the block matrix $\mathbf{A}$ as if they were single entries rather than 2×2 matrices. Beneath the $\mathbf{I}$ entry in the $(j-1)^{st}$ row and j^{th} column, there are two possibly nonzero entries, namely, $-\frac{\partial \boldsymbol{\lambda}_j}{\partial \mathbf{x}_j}$ and $-\frac{\partial \boldsymbol{\lambda}_{j+1}}{\partial \mathbf{x}_j}$. We proceed to eliminate everything in the jth column, except the $\mathbf{I}$ in the $(j-1)^{st}$ row, by multiplying the $(j-1)^{st}$ row by $\frac{\partial \boldsymbol{\lambda}_j}{\partial \mathbf{x}_j}$ and adding the result to the jth row, for $j = 2, \dots, k$, and by multiplying the $(j-1)^{st}$ row by $\frac{\partial \boldsymbol{\lambda}_{j+1}}{\partial \mathbf{x}_j}$ and adding it to the $(j+1)^{st}$ row, for $j = 2, \dots, k-1$.

The following is the matrix resulting from the above operations on $\mathbf{A}$:

$$\begin{bmatrix} \mathbf{A}_1 & \mathbf{I} & \mathbf{0} & \mathbf{0} & \mathbf{0} & \dots & \mathbf{0} \\ \mathbf{A}_2 & \mathbf{0} & \mathbf{I} & \mathbf{0} & \mathbf{0} & \dots & \mathbf{0} \\ \mathbf{A}_3 & \mathbf{0} & \mathbf{0} & \mathbf{I} & \mathbf{0} & \dots & \mathbf{0} \\ \dots & \dots & \dots & \dots & \dots & \dots & \dots \\ \mathbf{A}_{k-1} & \dots & \mathbf{0} & \mathbf{0} & \mathbf{0} & \mathbf{0} & \mathbf{I} \\ \mathbf{A}_k & \dots & \dots & \mathbf{0} & \mathbf{0} & \mathbf{0} & \mathbf{0} \end{bmatrix} \tag{6.33}$$

We now show that

$$\mathbf{A}_k = -\mathrm{Jac}(\boldsymbol{\eta}) = -\frac{\partial \boldsymbol{\eta}_{k+1}}{\partial \mathbf{x}_1}$$

on $\boldsymbol{\Upsilon}_S[P]$ by proving inductively that

$$\mathbf{A}_j = -\frac{\partial \boldsymbol{\eta}_{j+1}}{\partial \mathbf{x}_1}, \qquad j = 1, \ldots, k.$$

First, note that

$$\mathbf{A}_1 = -\frac{\partial \boldsymbol{\lambda}_2}{\partial \mathbf{x}_1} = -\frac{\partial \boldsymbol{\eta}_2}{\partial \mathbf{x}_1}.$$

Suppose that we have performed all the row operations involving the first j rows of our matrix. We consider the altered $(j-1)^{st}$ and j^{th} rows, and the unaltered $(j+1)^{st}$ row:

$$\begin{bmatrix} -\frac{\partial \boldsymbol{\eta}_j}{\partial \mathbf{x}_1} & \mathbf{0} & \ldots & \mathbf{0} & \mathbf{I} & \mathbf{0} & \ldots & \ldots & \mathbf{0} \\ -\frac{\partial \boldsymbol{\eta}_{j+1}}{\partial \mathbf{x}_1} & \mathbf{0} & \ldots & \ldots & \mathbf{0} & \mathbf{I} & \mathbf{0} & \ldots & \mathbf{0} \\ \mathbf{0} & \ldots & \ldots & \mathbf{0} & -\frac{\partial \boldsymbol{\lambda}_{j+1}}{\partial \mathbf{x}_j} & -\frac{\partial \boldsymbol{\lambda}_{j+1}}{\partial \mathbf{x}_{j+1}} & \ldots & \ldots & \mathbf{0} \end{bmatrix}$$

Performing the row operations described above we obtain

$$\mathbf{A}_{j+1} = -\left\{ \left(\frac{\partial \boldsymbol{\lambda}_{j+1}}{\partial \mathbf{x}_j}\right)\left(\frac{\partial \boldsymbol{\eta}_j}{\partial \mathbf{x}_1}\right) + \left(\frac{\partial \boldsymbol{\lambda}_{j+1}}{\partial \mathbf{x}_{j+1}}\right)\left(\frac{\partial \boldsymbol{\eta}_{j+1}}{\partial \mathbf{x}_1}\right) \right\}.$$

By definition,

$$\boldsymbol{\eta}_{j+2}(\mathbf{x}_1) = \boldsymbol{\lambda}_{j+1}(\boldsymbol{\eta}_j(\mathbf{x}_1), \boldsymbol{\eta}_{j+1}(\mathbf{x}_1)).$$

Recalling that the element $(\boldsymbol{\eta}_j(\mathbf{x}_1), \boldsymbol{\eta}_{j+1}(\mathbf{x}_1))$ lies in $L_j \times L_{j+1}$, the chain rule gives

$$\mathbf{A}_{j+1} = -\frac{\partial \boldsymbol{\eta}_{j+2}}{\partial \mathbf{x}_1},$$

as required. Since the matrix in equation (6.33) resulted from $\mathbf{A}$ by adding scalar multiples of one row to another, these two matrices have the same determinant. Hence

$$\det[\text{Hess}\, T](\boldsymbol{\Upsilon}_S(\mathbf{x}_1)) = \det \mathbf{D} \det \mathbf{A} = \left[\prod_{i=1}^{k} \det[\mathbf{T}_{i\,i+1}](\boldsymbol{\Upsilon}_S(\mathbf{x}_1))\right] \det[\text{Jac}\, \boldsymbol{\eta}](\mathbf{x}_1).$$

But

$$[\text{Hess}\, T](\boldsymbol{\Upsilon}_S(\mathbf{x}_1)) = \begin{bmatrix} [\text{Hess}\, T_{\mathbf{y}}](\boldsymbol{\Upsilon}(\mathbf{x}_1)) & \mathbf{0} \\ * & \mathbf{I} \end{bmatrix},$$

which yields $\det[\text{Hess}\, T](\boldsymbol{\Upsilon}_S(\mathbf{x}_1)) = \det[\text{Hess}\, T_{\mathbf{y}}](\boldsymbol{\Upsilon}(\mathbf{x}_1))$. □

Chapter 7

Critical Points and Stability

Functions, just like living beings, are characterized by their singularities.

PAUL MONTEL

The 1959 Bonn lectures of René Thom and the notes, *Singularities of Differentiable Maps,* by H.L. [Lev59] were a seminal first step towards a general mathematical theory of the critical points and stability of differentiable maps.[1] This theory is commonly called *singularity theory* (not to be confused with "singularities" in general relativity — see the remark on page 176). It grew out of the earlier works of Morse, Whitney, and Thom, and initiated a flurry of activity leading to far reaching theories on stability (Mather), Lagrangian and Legendrian singularities (Arnold), and much more (e.g., Bruce and Mond [Bru-M]).

The current chapter is a collection of several topics in singularity theory.

> *Due to the many hard and technical results in singularity theory, we suggest that (to avoid getting bogged down) readers unfamiliar with the subject should simply skim through (or skip) this chapter on a first reading, referring back to it as needed in later chapters. Indeed the mathematically disinclined reader may want to skip this material entirely, picking up the plot of the story in Section 7.6 (page 266; see Theorem 7.76, page 275), where we explain in the concrete language of elementary calculus what certain key aspects of this singularity theory means.*

This material will be used to analyze the local and global properties of caustics and multiple lensed images in gravitational lensing. In fact, many of the theorems, propositions, and lemmas in this chapter will be needed in Chapters 8 and 9 to study the stability and

[1] Recall that a *critical point* of $f : X \to Y$, where $\dim X = n$ and $\dim Y = m$, is a point $x \in X$ such that $\text{rank}[d_x f] < \min\{n, m\}$; see Section 6.2.1, page 176.

critical-point structure of time delay functions, lensing maps, and time delay families. Two additional topics in singularity theory that are not covered in the current chapter will be treated in Chapters 10 (*Morse Inequalities*) and 14 (*Computing the Euler Characteristic*). Their applications to gravitational lensing will appear in Chapters 11, 12, 13, and 15.

Outline: Section 7.1 introduces the basic notions of jets, Whitney C^∞-topology, and transversality (and multijet transversality). The section ends with Thom's Transversality Theorem and its generalization to multijets. With the language of Section 7.1 available, we summarize Mather stability theory in Section 7.2. This includes the links between local stability, transverse stability, infinitesimal stability, and global stability. Section 7.3 discusses singularity manifolds. These manifolds are crucial in studying the transverse stability of maps. The tools of Sections 7.1, 7.2, and 7.3 are employed to study Morse theory (i.e., stability/critical-point theory of functions), in Section 7.4 and Whitney theory (i.e., stability/critical-point theory of maps between 2-manifolds) in Section 7.5. Thom catastrophe theory (i.e., stability/critical-point theory of families of functions) is treated in Section 7.6. The chapter closes with Arnold's theory of Lagrangian singularities in Section 7.7.

References: Consult Golubitsky and Guillemin[Gol-G] and Martinet [Mart] for a detailed introduction to the fundamentals of singularity theory, and Arnold et al. [Arn-GVI], [Arn-GVII], [Arn93a], [Arn93b] for a comprehensive treatment. We shall closely follow the notation of [Gol-G].

7.1 Jets and Transversality

The concepts and results of this section are needed for Mather stability theory (Section 7.2). Applications to gravitational lensing appear in Chapter 8.

7.1.1 Jets, Multijets, and Whitney C^∞-Topology

Jets

Consider a smooth function $f = (f_1, \ldots, f_m)$ from an open set $U \subseteq \mathbf{R}^n$ into $\mathbf{R}^m$. Let t_i^k be the kth order Taylor polynomial of f_i at p, excluding the constant term. Then the multi-tuple $j^k f(p) = (p, f(p), t_1^k(p), \ldots, t_m^k(p))$ represents a "k-jet" of f at p; here, $j^0 f(p) = (p, f(p))$. For distinct points $p_1, \ldots, p_s \in \mathbf{R}^n$, an "$s$-fold k-jet" or "multijet" of f is $j_s^k f(p_1, \ldots, p_s) = (j^k f(p_1), \ldots, j^k f(p_s))$. We now define these notions precisely and in a coordinate-free way.

Let X and Y be smooth manifolds. Denote the set of all smooth maps from X into Y by $C^\infty(X, Y)$. Let TX and TY be the tangent bundles of X and Y, respectively. Let $f, h \in C^\infty(X, Y)$ and $p \in X$. Then f has *zeroth order contact* with h at p if $f(p) = h(p)$. It has *first order contact* with h at p if $df : TX \to TY$ has zeroth order contact with $dh : TX \to TY$ at p (i.e., $d_p f = d_p h$). For k a positive integer, f has *kth order contact* with h at p if df has $(k-1)$-order contact with dh at p. We shall abbreviate the phrase "f has kth order contact with g at p" by "$f \sim_k g$ at p."

In the Euclidean case, the map f having kth order contact involves Taylor coefficients. Let $U' \subseteq \mathbf{R}^n$ be open and suppose that $f, h \in C^\infty(U', \mathbf{R}^m)$. Set $\tau = (\tau_1, \ldots, \tau_n)$, where each τ_i is a nonnegative integer, and $|\tau| = \tau_1 + \cdots + \tau_n$. Let

$$\frac{\partial^{|\tau|}}{\partial x^\tau} = \frac{\partial^{|\tau|}}{\partial x_1^{\tau_1} \cdots \partial x_n^{\tau_n}},$$

where $x_1, \ldots, x_n$ are coordinates in U'. Write $f = (f_1, \ldots, f_m)$ and $h = (h_1, \ldots, h_m)$. *Then $f \sim_k h$ at $p \in U'$ if and only if*

$$\frac{\partial^{|\tau|} f_i}{\partial x^\tau}(p) = \frac{\partial^{|\tau|} h_i}{\partial x^\tau}(p),$$

whenever $0 \le |\tau| \le k$ *and* $1 \le i \le m$ [Gol-G, p. 37]. In other words, their kth-order Taylor expansions at p coincide.

Let $C^\infty(X, Y)_{(p,q)} = \{f \in C^\infty(X, Y) : f(p) = q\}$. The relation $f \sim_k h$ at p is an equivalence relation on $C^\infty(X, Y)_{(p,q)}$. Denote the resulting set of equivalence classes by $J^k(X, Y)_{(p,q)}$. An element of $J^k(X, Y)_{(p,q)}$ is called a *k-jet of mappings* from X to Y with *source* p and *target* q. If $\sigma \in J^k(X, Y)_{(p,q)}$, then write $p = \text{source}(\sigma)$ and $q = \text{target}(\sigma)$. Form the following disjoint union:

$$J^k(X, Y) = \dot{\bigcup_{(p,q) \in X \times Y}} J^k(X, Y)_{(p,q)}.$$

Notice that $J^0(X, Y) = X \times Y$. Let $\alpha : J^k(X, Y) \to X$ be the *source map* $\sigma \mapsto \text{source}(\sigma)$ and $\beta : J^k(X, Y) \to Y$ the *target map* $\sigma \mapsto \text{target}(\sigma)$.

The space $J^k(X, Y)$ can be endowed with a smooth manifold structure. In fact, let us construct the coordinate systems. Set $\dim X = n$ and $\dim Y = m$. Let $\phi : U \subset X \to \mathbf{R}^n$ and $\psi : V \subset Y \to \mathbf{R}^m$ be charts, where $\phi(U) = U'$ and $\psi(V) = V'$. Set

$$J^k_{U,V}(X, Y) = (\alpha \times \beta)^{-1}(U \times V) = \dot{\bigcup_{(p,q) \in U \times V}} J^k(X, Y)_{(p,q)},$$

that is, $J^k_{U,V}(X, Y)$ is the set of all k-jets of mappings from X to Y with source in U and target in V. Define a map

$$\xi_1 : J^k_{U,V}(X, Y) \to J^k(U', V')$$

by

$$\xi_1(\sigma) = [\psi \circ f \circ \phi^{-1}]^k_{\phi(p)},$$

where f is a representative of σ and $[\psi \circ f \circ \phi^{-1}]^k_{\phi(p)}$ is the equivalence class of $\psi \circ f \circ \phi^{-1}$ in $J^k(U', V')_{(\phi(p),\psi(q))}$ with $q = f(p)$. Let P^k_n be the vector space of n-variable polynomials of degree less than or equal to k, and such that their constant term vanishes. Set $P^k_{n,m} = P^k_n \times \cdots \times P^k_n$ (m-times). Define a second map

$$\xi_2 : J^k(U', V') \to U' \times V' \times P^k_{n,m}$$

by

$$\xi_2(\sigma') = (p', q', t^k h_1, \ldots, t^k h_m).$$

Here, $h = (h_1, \ldots, h_m)$ is a representative of σ', where $h(p') = q'$, and $t^k h_i$ is the k-order Taylor polynomial of h_i without the constant term. The mapping

$$\xi_{U,V} = \xi_2 \circ \xi_1 : J^k_{U,V}(X, Y) \to U' \times V' \times P^k_{n,m}$$

is a bijection. A manifold structure can then be induced on $J^k(X, Y)$ such that the bijections $\xi_{U,V}$ become charts (e.g., [Gol-G, p. 40] and [MatII, p. 258]). Note that ξ_1 and ξ_2 are diffeomorphisms.

Suppose that $\lambda : X \to X'$ and $\nu : Y \to Y'$ are diffeomorphisms. Define a map

$$\xi : J^k(X, Y) \to J^k(X', Y')$$

by

$$\xi(\sigma) = [\nu \circ f \circ \lambda^{-1}]^k_{\lambda(p)},$$

where f is a representative of σ and $[\nu \circ f \circ \lambda^{-1}]^k_{\lambda(p)}$ is the equivalence class of $\nu \circ f \circ \lambda^{-1}$ in $J^k(X', Y')_{(\lambda(p),\nu(q))}$ with $q = f(p)$. *Then $\xi : J^k(X, Y) \to J^k(X', Y')$ is a diffeomorphism.* In particular, we have the diffeomorphism $J^k(U', V') \approx J^k(U, V)$. Hence

$$J^k_{U,V}(X, Y) \approx J^k(U, V).$$

Each point of $J^k(X, Y)$ then has a neighborhood diffeomorphic to some $J^k(U, V)$.

Notation: Set $J^k(n, m) = J^k(\mathbf{R}^n, \mathbf{R}^m)_{(0,0)}$ and $J^k(n) = J^k(n, 1)$.

The bijection between $J^k(n, m)$ and $P^k_{n,m}$ induces a Euclidean-space manifold structure on $J^k(n, m)$. Consequently, if $U \subseteq X$ and $V \subseteq Y$ are coordinate neighborhoods, then

$$J^k(U, V) \approx U \times V \times J^k(n, m),$$

where

$$\begin{aligned} \dim J^k(n, m) &= \dim P^k_{n,m} \\ &= nm + \frac{n(n+1)}{2} m + \cdots + \binom{n+k-1}{n-1} m. \end{aligned}$$

Notice that $\binom{n+k-1}{n-1}$ is the dimension of the vector space of all homogeneous polynomials of degree k in n variables (e.g., [Cas-H, p. 117]).

Define $\pi : J^k(X, Y) \to X \times Y$ by $\pi = \alpha \times \beta$, that is,

$$\pi(\sigma) = (\text{source}(\sigma), \text{target}(\sigma)).$$

Then $J^k(X, Y)$ is a fibre bundle over $X \times Y$ with fibre $J^k(n, m)$ and projection π (e.g., [Gol-G, p. 41] [Micho, p.5]). We obtain a vector bundle when $Y = \mathbf{R}^m$. Call $J^k(X, Y)$ the *k-jet bundle of mappings from X to Y.*

Definition 7.1 Given f in $C^\infty(X, Y)$, the *k-jet of f* is the map $j^k f : X \to J^k(X, Y)$ given by $j^k f(p) = [f]_p^k$, where $[f]_p^k$ is the equivalence class of f in $J^k(X, Y)_{(p,q)}$, where $q = f(p)$.

The k-jet $j^k f$ of $f \in C^\infty(X, Y)$ is a smooth map [Gol-G, p.40]. Note that $(j^0 f)(X) =$ graph f.

Let $U \subseteq \mathbf{R}^n$ be a coordinate neighborhood. For $f \in C^\infty(U, \mathbf{R}^m)$, we identify $j^k f(p)$ with the following multi-tuple of numbers:

$$\left(p, f(p), \left\{\frac{\partial^{|\tau|} f_i}{\partial x^\tau}(p)\right\}\right), \qquad \text{for } 1 \le |\tau| \le k,\ 1 \le i \le m.$$

Here, any representative of $[f]_p^k$ is used and the numbers in curly brackets string out in some fixed (arbitrary) order. For example, if $f \in C^\infty(U, \mathbf{R})$, then we may write

$$j^2 f(p) = (p, f(p), z_1, \ldots, z_m, z_{11}, \ldots, z_{1n}, z_{22}, \ldots, z_{2n}, \ldots, z_{nn}),$$

where

$$z_i = \frac{\partial f}{\partial x_i}(p), \qquad z_{ij} = \frac{\partial^2 f}{\partial x_i \partial x_j}(p),$$

and the upper triangular entries of (Hess f)(p) are listed row-wise (with the symmetry $z_{ij} = z_{ji}$ taken into account). Hence

$$\dim J^2(U, \mathbf{R}) = 1 + 2n + \binom{n+1}{2}.$$

Multijets

Later, we shall need to consider sets of k-jets at distinct source points with the same target points. This requires generalizing k-jets to *multijets*. Sections 7.4.2 and 7.5.3 apply this notion to the stability of functions and maps between 2-manifolds, respectively.

Let $X^s = X \times \cdots \times X$ (s-times) and $X^{(s)} = \{(x_1, \ldots, x_s) \in X^s : x_i \neq x_j \text{ if } i \neq j\}$. Note that $X^{(s)}$ is open in X^s. Let $\alpha : J^k(X, Y) \to X$ be the *source map* $\sigma \mapsto \text{source}(\sigma)$ and $\beta : J^k(X, Y) \to Y$ the *target map* $\sigma \mapsto \text{target}(\sigma)$. Denote the s-fold Cartesian product of α with itself by $\alpha^s : J^k(X, Y)^s \to X^s$. The set

$$J^k_s(X, Y) \equiv (\alpha^s)^{-1}[X^{(s)}]$$

is called an *s-fold k-jet bundle*. It consists of all s-tuples $(w_1, \ldots, w_s)$ with $w_i \in J^k(X, Y)$ and $\alpha(w_i) \neq \alpha(w_j)$ for $i \neq j$. Since $X^{(s)}$ is open, *the space $J^k_s(X, Y)$ is an open submanifold of $J^k(X, Y)^s$. The manifold $J^k_s(X, Y)$ is a fibre bundle over $X^{(s)} \times Y^s$ with fibre $[J^k(n, m)]^s$ and projection $\pi^s : J^k_s(X, Y) \to X^{(s)} \times Y^s$ given by $\pi^s = (\alpha \times \beta)^s$.* A *multijet bundle* is any such fibre bundle.

If $U \subseteq X$ and $V \subseteq Y$ are coordinate neighborhoods, then the following diffeomorphisms hold:

$$(\pi^s)^{-1}(U^{(s)} \times V^s) \approx U^{(s)} \times V^s \times [J^k(n, m)]^s \approx J^k_s(U, V).$$

Consequently, *each point of $J^k_s(X, Y)$ has an open set diffeomorphic to some $J^k_s(U, V)$.* Note that if $(w_1, \ldots, w_s)$ is a point of $J^k_s(X, Y)$, where x_i and y_i are the respective source and target points of w_i for $i = 1, \ldots, s$, then we can choose

$$U = \bigcup_{i=1}^{s} U_i, \qquad V = \bigcup_{i=1}^{s} V_i,$$

where U_i and V_i are coordinate neighborhoods of x_i and y_i, respectively, with $U_i \cap U_j = \emptyset = V_i \cap V_j$ for $i \neq j$.

Definition 7.2 The *s-fold k-jet* of a smooth map $f : X \to Y$ is the map $j^k_s f : X^{(s)} \to J^k_s(X, Y)$ given by

$$j^k_s f(x_1, \ldots, x_s) = (j^k f(x_1), \ldots, j^k f(x_s)).$$

A *multijet* of f is some s-fold k-jet of f.

Notation: Set $J^k_1(X, Y) = J^k(X, Y)$ and $j^k_1 f = j^k f$.

Whitney C^∞-Topology

Intuitively speaking, if f and h are smooth functions on an open subset U of $\mathbf{R}^n$, then f is "C^∞ close" to h if their values and partial derivatives of all orders are close. The Whitney C^∞-topology precisely captures this situation for maps between manifolds.

Definition 7.3 Let X and Y be manifolds, and k a nonnegative integer. For any subset $\mathcal{W} \subset J^k(X,Y)$, let $M(\mathcal{W}) = \{f \in C^\infty(X,Y) : j^k f(X) \subset \mathcal{W}\}$. The family of sets,

$$W_k = \{M(\mathcal{W}) : \mathcal{W} \text{ open in } J^k(X,Y)\},$$

is a basis for the Whitney C^k-topology, while $W = \bigcup_k W_k$ is a basis for the *Whitney C^∞-topology.*

Note that every open set in the Whitney C^k-topology is open in the Whitney C^{k+1}-topology and (hence) in the Whitney C^∞-topology. Open sets relative to the Whitney C^k-topology are called *C^k-open*, while those in the Whitney C^∞-topology are *C^∞-open.*

Convention: *We always assume that $C^\infty(X,Y)$ has the Whitney C^∞-topology.*

It is no restriction to assume that all manifolds under consideration are metrizable [Gol-G, p. 24]. For any space considered, we shall adopt a universal notation for such a metric without making any distinction for different spaces. So if Q is one of our manifolds and s and t points in Q, we denote by $|s-t|$ the Q-metric distance between s and t. We shall use general properties of metrics, but never use the special form suggested by this notation. We then define certain C^k-open "balls" around an element $f \in C^\infty(X,Y)$: For $\mu : X \to \mathbf{R}_+$, where $\mathbf{R}_+ = (0,\infty)$, a continuous function, let

$$V_\mu^k(f) = \{h \in C^\infty(X,Y) : |j^k f(x) - j^k h(x)| < \mu(x) \text{ for all } x \in X\}$$

and for $A \subset X$, set

$$V_{\mu,A}^k(f) = \{h \in C^\infty(X,Y) : |j^k f(x) - j^k h(x)| < \mu(x) \text{ for } x \in A\}.$$

These "balls" are Whitney C^k-open sets, the first being of the form $M(\mathcal{W})$, where

$$\mathcal{W} = \{\sigma \in J^k(X,Y) : |\sigma - j^k f(x)| < \mu(x) \text{ for } \alpha(\sigma) = x\},$$

and the second being the union of all $V_\nu^k(f)$ where the union is over all continuous ν such that $\nu : X \to \mathbf{R}_+$ and $\nu|_A \le \mu|_A$.

A subset $\mathcal{O} \subseteq C^\infty(X,Y)$ is open in the Whitney C^∞-topology if and only if for every $f \in \mathcal{O}$ there is a continuous function $\mu : X \to \mathbf{R}_+$ and an integer $k \ge 0$ such $V_\mu^k(f) \subseteq \mathcal{O}$.

Note. If in the expressions, $V_\mu^k(f)$ or $V_{\mu,A}^k(f)$, the function μ is replaced by a constant, say ϵ, then we interpret that constant as a constant function.

For $C^\infty(U,\mathbf{R}^m)$, where $U \subseteq \mathbf{R}^n$ is open, we have

$$V_\epsilon^k(f) = \left\{h \in C^\infty(U,\mathbf{R}^m) : \left|\frac{\partial^{|\tau|} h_i}{\partial x^\tau}(x) - \frac{\partial^{|\tau|} f_i}{\partial x^\tau}(x)\right| < \epsilon\right\},$$

for $0 \le |\tau| \le k$, $1 \le i \le m$, and $x \in U$. The "ball" $V_\epsilon^k(f)$ consists of all maps in $C^\infty(U,\mathbf{R}^m)$ whose values and partial derivatives up to order k approximate those of f to within ϵ.

Loosely speaking, if $\mathfrak{H}$ is open and dense in $C^\infty(X, Y)$, then a property shared by the elements of $\mathfrak{H}$ not only "permeates" $C^\infty(X, Y)$, but each element is surrounded by an entire neighborhood of points with that property. Indeed, any features shared by members of an open dense subset of $C^\infty(X, Y)$ are "ubiquitous" in $C^\infty(X, Y)$.

Despite the extreme fineness of the Whitney C^∞-topology, it has an important property. In fact, define a subset of a topological space to be *residual* if it is the countable intersection of open dense subsets of the space. Then $C^\infty(X, Y)$ is Baire, that is, each residual set in $C^\infty(X, Y)$ is dense [Gol-G, p. 44] [MatV, p. 309]. Showing that a subset $\mathfrak{F}$ of $C^\infty(X, Y)$ is residual is perhaps the most practical way of establishing the "genericity" of any property shared by elements of $\mathfrak{F}$. Ideally, we would like $\mathfrak{F}$ to be open and dense.

7.1.2 Transversality and Multijet Transversality

Let H and K be submanifolds of a manifold X. Then H and K *intersect transversely*, written $H \pitchfork K$, if either $H \cap K = \emptyset$, or, we have $H \cap K \neq \emptyset$ and $T_x H + T_x K = T_x X$ for all $x \in H \cap K$. More generally:

Definition 7.4 Let $f \in C^\infty(X, Y)$ and suppose that Q is a submanifold of Y. Then f is *transverse to* Q at x, written $f \pitchfork Q$ at x, if either $f(x) \notin Q$, or, $f(x) \in Q$ and

$$d_x f(T_x X) + T_{f(x)} Q = T_{f(x)} Y.$$

We shall say that f is *transverse to* Q, denoted $f \pitchfork Q$, if f is transverse to Q at all x in X.

It immediately follows that *a map* $f \in C^\infty(X, Y)$ *is transverse to every open submanifold of* Y. Also, suppose H and K are submanifolds of X. Let $i : H \to X$ be the inclusion map. *Then* $i \pitchfork K$ *if and only if* $H \pitchfork K$ (since $d_p i(T_p H) + T_p X = T_p H + T_p K$). In general, $f \pitchfork Q$ *if and only if* (graph f) $\pitchfork (X \times Q)$ *in* $X \times Y$ (e.g., [Maj, pp. 184–185]).

Example: Two one-dimensional submanifolds of the plane intersect transversely if and only if they are either disjoint or cross only at nonzero angles. In addition, let $f \in C^\infty(X, Y)$, where $\dim X = \dim Y = 2$, and suppose that at $x \in X$ we have $\text{rank}(d_x f) = 1$. Then f is transverse to a one-manifold $Q \subseteq Y$ at x if and only if either $f(x) \notin Q$ or the line $d_x f(T_x X)$ is not tangent to Q at $f(x)$. Figure 7.1 illustrates these remarks in the plane.

In the remainder of this section, we present in detail several basic results to be employed in later proofs.

Differentiable Equivalence and Transversality

We first point out that transversality is preserved by "differentiable equivalence" of maps. Let x be a point of a manifold X and let Y be another manifold. A *local map of* X *at* x *into* Y is a (smooth) map from a neighborhood of x into Y. Define a local map λ of X at x into Y to be a *local diffeomorphism at* x if there is an open set U about x and open set $V \subseteq Y$ such that $\lambda|_U$ is a diffeomorphism of U onto V.

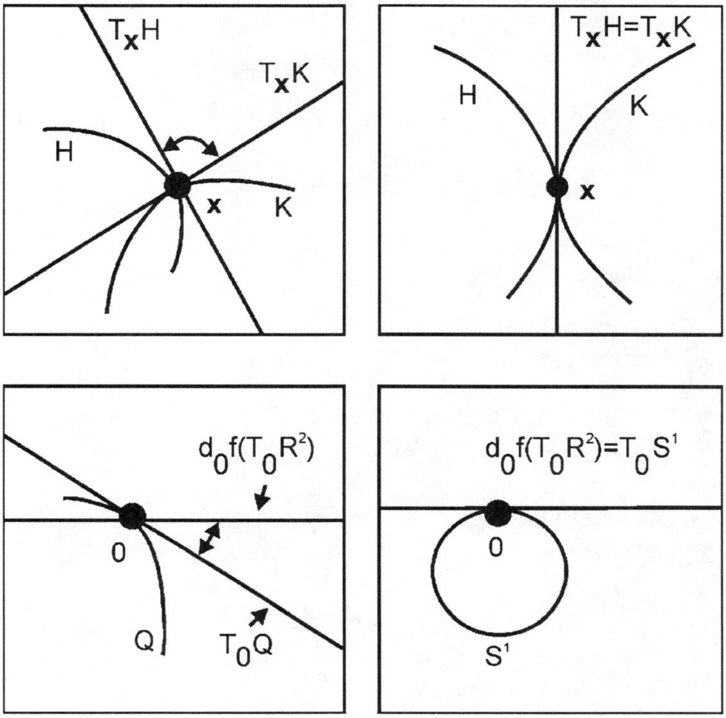

Figure 7.1. *(a)* Two one-dimensional manifolds in $\mathbf{R}^2$ cross transversely at x exactly when their tangent lines at x cross at a nonzero angle. *(b)* The mapping f is transverse to Q (bottom left panel), but not to the circle S^1 (bottom right panel).

Let f be a local map of X at x into Y and h another local map that is from X' at x' into Y'. Then *h at x' is differentiably equivalent to f at x* if there are smooth local diffeomorphisms $\lambda : U \subseteq X \to U' \subseteq X'$ at x with $\lambda(x) = x'$ and $\nu : V \subseteq Y \to V' \subseteq Y'$ at $f(x) \in Y$ with $\nu(f(x)) = h(x')$, such that $h = \nu \circ f \circ \lambda^{-1}$ at all points of some neighborhood of x. Note that differentiable equivalence at a given point is an equivalence relation.

Proposition 7.5 *Using the above notation, suppose that f at x is differentiably equivalent to h at x'. Assume that $Q \subseteq Y$ and $Q' \subseteq Y'$ are submanifolds with $\nu[Q \cap V] = Q' \cap V'$. Then $f \pitchfork Q$ at x if and only if $h \pitchfork Q'$ at x'.*

Proof: First, note that $f(x) \in Q$ if and only if $h(\lambda(x)) \in Q'$. Suppose $f \pitchfork Q$ at x with $f(x) \in Q$, i.e.,

$$d_x f(T_x X) + T_{f(x)} Q = T_{f(x)} Y. \tag{7.1}$$

Since λ, ν, and $\nu|_Q : Q \to Q'$ are diffeomorphisms, we have

$$d_x f(T_x X) = d_x f \circ (d_x \lambda)^{-1}[T_{\lambda(x)} X'], \qquad T_{f(x)} Q = (d_{f(x)} \nu)^{-1}[T_{h \circ \lambda(x)} Q'],$$

and

$$T_{f(x)}Y = (d_{f(x)}\nu)^{-1}[T_{h\circ\lambda(x)}Y'].$$

Multiplying equation (7.1) by $d_{f(x)}\nu$ and using $d_{\lambda(x)}h = d_{f(x)}\nu \circ d_x f \circ (d_x\lambda)^{-1}$, we obtain

$$d_{\lambda(x)}h(T_{\lambda(x)}X') + T_{h\circ\lambda(x)}Q' = T_{h\circ\lambda(x)}Y'.$$

Conversely, if $h \pitchfork Q'$ at $\lambda(x)$ with $h \circ \lambda(x) \in Q'$, then working backwards and multiplying by $(d_{f(x)}\nu)^{-1}$ yields $f \pitchfork Q$ at x. □

Codimension and Transversality

Given a submanifold $Q \subseteq Y$, the integer $\operatorname{codim}_Y Q = \dim Y - \dim Q$ is the *codimension* of Q in Y.

Proposition 7.6 *Let $f \in C^\infty(X, Y)$, $Q \subseteq Y$ a submanifold, and $\dim X < \operatorname{codim}_Y Q$. If $f \pitchfork Q$, then $f(X) \cap Q = \emptyset$.*

Proof: Since $f \pitchfork Q$, we have for $f(x) \in Q$ that

$$\begin{aligned}
\dim Y &= \dim[d_x f(T_x X) + T_{f(x)}Q] \\
&\leq \dim[d_x f(T_x X)] + \dim[T_{f(x)}Q] \\
&= (\dim[T_x X] - \dim[\operatorname{Kernel}(d_x f)]) + \dim Q \\
&\leq \dim X + \dim Q,
\end{aligned}$$

so $\dim X \geq \operatorname{codim}_Y Q$, which is a contradiction. □

The following is a useful criterion for transversality:

Proposition 7.7 *Let X and Y be manifolds with $Q \subset Y$ a submanifold of codimension r and $y \in Q$. Suppose that:*

(1) *There is an open set $V \subseteq Y$ about y and a map $h \in C^\infty(V, \mathbf{R}^r)$ such that $Q \cap V = h^{-1}(0)$ and $\operatorname{rank}[d_y h] = r$.*

(2) *The map f is in $C^\infty(X, Y)$ and $f(x) = y$ for some $x \in X$.*

Then $f \pitchfork Q$ at x if and only if the map $h \circ f : f^{-1}(V) \to \mathbf{R}^r$ satisfies $\operatorname{rank}[d_x(h \circ f)] = r$.

Proof: First note that the assumptions on h yield that $\operatorname{Kernel}(d_y h) = T_y Q$. Assume that f is transverse to Q at $x \in X$: $d_x f(T_x X) + T_{f(x)} Q = T_{f(x)}Y$. Applying dh to this equation gives $d_x(h \circ f)(T_x X) = T_0(\mathbf{R}^r)$. Note that $T_x f^{-1}(V) = T_x X$. The rank condition is now immediate. Conversely, suppose that $d_y h \circ d_x f\,(T_x X) = T_0(\mathbf{R}^r)$. Recall that if $T : Z \to W$ is a linear transformation between finite dimensional vector spaces, and if $T(S) = W$ for some linear subspace $S \subseteq Z$, then $S + \text{Kernel}(T) = Z$.[2] Applying this to $d_y h$ yields $d_x f(T_x X) + T_y Q = T_y Y$, i.e, $f \pitchfork Q$ at x. □

The next result is fundamental:

Theorem 7.8 *Let $f \in C^\infty(X, Y)$ and $Q \subseteq Y$ a submanifold such that $f(X) \cap Q \neq \emptyset$.*

(1) *If $f \pitchfork Q$, then $f^{-1}(Q)$ is a submanifold of X, $\text{codim}_X f^{-1}(Q) = \text{codim}_Y Q$, and $T_x[f^{-1}(Q)] = (d_x f)^{-1}[T_{f(x)} Q]$.*

(2) *If $f \pitchfork Q$ at $x \in X$, then there is an open neighborhood U of x such that $f|_U \pitchfork Q$. Thus, $U \cap f^{-1}(Q)$ is a submanifold with $\text{codim}_X [U \cap f^{-1}(Q)] = \text{codim}_Y Q$.*

Detailed proofs of Theorem 7.8 appear in [Gol-G, Thm. 4.4, p. 52] and [Gib, pp. 41–42].

By Theorem 7.8, if $f \pitchfork \{y\}$ and $y \in f(X)$, then $f^{-1}(y)$ is an $(n - m)$-dimensional manifold, where $\dim X = n$ and $\dim Y = m$, and $T_x[f^{-1}(y)] = \text{Kernel}(d_x f)$ for all $x \in f^{-1}(y)$. Note that if $f \pitchfork \{y\}$ at x and $f(x) = y$, then $m \leq n$ (since $\text{codim}_Y\{y\} \leq \dim X$ by the contrapositive of Proposition 7.6).

Suppose that H, K are submanifolds of X and the inclusion map $i : H \to X$ is transverse to K. If $H \cap K \neq \emptyset$, then Theorem 7.8(1) yields that $H \cap K = i^{-1}[K]$ *is a submanifold of X and* $\text{codim}_X(H \cap K) = \text{codim}_X K + \text{codim}_X H$ (since $\text{codim}_H(H \cap K) = \text{codim}_X K$).

Restrictions, Submersions, and Transversality

Lemma 7.9 *Let $f \in C^\infty(X, Y)$, $Q_1 \subseteq Y$ be a submanifold, and $X_1 = f^{-1}(Q_1)$. Suppose $f(x_1) \in Q_1$ and $f \pitchfork Q_1$ at $x_1 \in X_1$. If $w \in T_{x_1} X$, then $d_{x_1} f(w) \in T_{f(x_1)} Q_1$ if and only if $w \in T_{x_1} X_1$.*

Proof: By Theorem 7.8(2), there is an open set U about x_1 such that $U_1 = U \cap X_1$ is a submanifold of X (so $T_{x_1} X_1 = T_{x_1} U_1$). Denote the restriction of f to U_1 by f_1. If $w \in T_{x_1} U_1$, then $d_{x_1} f(w) \in T_{f(x_1)} Q_1$ (since $d_{x_1} f(w) = d_{x_1} f_1(w)$ for all w tangent to U_1 at x_1). Conversely, let $h : Y \to \mathbf{R}^r$ be a smooth map with $\text{rank}[d_{f(x_1)} h] = \text{codim}\, Q_1$ and $h^{-1}(0) = Q_1$ in a neighborhood of $f(x_1)$. Now, $T_{x_1} X = T_{x_1} U_1 \oplus S$ for some subspace $S \subset T_{x_1} X$. Since $T_{f(x_1)} Q_1 = \text{Kernel}[d_{f(x_1)} h]$ and $f \pitchfork Q_1$, applying dh to $T_{f(x_1)} Y = d_{x_1} f[T_{x_1} X] + T_{f(x_1)} Q_1$ yields

$$T_0 \mathbf{R}^r = d_{f(x_1)} h \circ d_{x_1} f[T_{x_1} X] = d_{f(x_1)} h \circ d_{x_1} f[S].$$

[2] Indeed, for each $z \in Z$ we have $T(z) = w = T(s)$ for some $s \in S$. So, $0 = T(z) - T(s) = T(z - s)$, that is, $z - s \in \ker(T)$.

But $\dim S = \text{codim}_X U_1 = \text{codim}_Y Q_1 = r$. Thus, $d_{f(x_1)}h \circ d_{x_1} f : S \to T_0\mathbf{R}^r$ is an isomorphism. Now, suppose $d_{x_1} f(w) \in T_{f(x_1)}Q_1$ and write $w = w_1 + w_2$, where $w_1 \in T_{x_1}U_1$ and $w_2 \in S$. We have

$$0 = d_{f(x_1)}h(d_{x_1} f(w)) = d_{f(x_1)}h(d_{x_1} f(w_2)),$$

where the first equality is by hypothesis and the second follows from $d_{x_1} f(w_1) = d_{x_1} f_1(w_1)$ lying in $T_{f(x_1)}Q_1 = \text{Kernel}[d_{f(x_1)}h]$. Since $(d_{f(x_1)}h \circ d_{x_1} f)|_S$ is an isomorphism, we get $w_2 = 0$. Thus, $w = w_1 \in T_{x_1}U_1 = T_{x_1}X_1$. □

Proposition 7.10 *Let $f \in C^\infty(X, Y)$ and let Q_1, Q_2 be submanifolds of Y, where $Q_2 \subset Q_1$. Suppose $f \pitchfork Q_1$ and set $X_1 = f^{-1}(Q_1)$. Then $f \pitchfork Q_2$ if and only if the restriction $f|_{X_1} : X_1 \to Q_1$ is transverse to Q_2.*

Proof: Suppose that $f|_{X_1} \pitchfork Q_2$ and $f(x) \in Q_2$. Then

$$\begin{aligned} T_xY &= d_x f(T_xX) + T_{f(x)}(Q_1) \qquad (\text{since } f \pitchfork Q_1) \\ &= d_x f(T_xX) + d_x(f|_{X_1})(T_xX_1) + T_{f(x)}(Q_2) \qquad (\text{since } f|_{X_1} \pitchfork Q_2) \\ &= d_x f(T_xX) + T_{f(x)}(Q_2), \end{aligned}$$

i.e., f is transverse to Q_2.

Conversely, suppose $f \pitchfork Q_2$ and $f(x) \in Q_2$. Let $f_1 = f|_{X_1} : X_1 \to Q_1$. We want to show that $f_1 \pitchfork Q_2$. Since Q_1 and Q_2 are manifolds with Q_2 a submanifold of Q_1, there is a smooth map $h \times k : Y \to \mathbf{R}^r \times \mathbf{R}^s$ such that: (1) in a neighborhood of a point $y \in Y$, we have $(h \times k)^{-1}(0, 0) = Q_2$ and $h^{-1}(0) = Q_1$, (2) $\text{rank}[d_y(h \times k)] = r + s$ and $\text{rank}[d_y h] = r$, and (3) $r + s = \text{codim}_Y Q_2$ and $r = \text{codim}_Y Q_1$. Let $x_1 \in X$ be such that $f(x_1) = y \in Q_2$. Then $x_1 \in X_1$ and Proposition 7.7 yields

$$\begin{aligned} f \pitchfork Q_1 \text{ at } x_1 &\iff d_{x_1}(h \circ f) \text{ is surjective.} \\ f \pitchfork Q_2 \text{ at } x_1 &\iff d_{x_1}[(h \times k) \circ f] \text{ is surjective.} \end{aligned}$$

Now, set $(h(y), k(y)) = (y_1, y_2)$ and consider the mapping $d_{x_1}(k \circ f_1) : T_{x_1}X_1 \to T_{y_2}\mathbf{R}^s$, where $y_2 = (k \circ f_1)(x_1)$. Let v be any element of $T_{y_2}\mathbf{R}^s$. Then $(0, v)$ is an element of $T_{y_1}\mathbf{R}^r \times T_{y_2}\mathbf{R}^s = d_{x_1}[(h \times k) \circ f](T_{x_1}X)$. Consequently, there is a $w \in T_{x_1}X$ such that $d_{x_1}[(h \times k) \circ f](w) = (0, v)$. This yields that $d_{x_1}(k \circ f)(w) = v$ and $d_{x_1}(h \circ f)(w) = 0$. Since $d_{x_1} f(w) \in \text{Kernel}[d_{f(x_1)}h] = T_{f(x_1)}Q_1$, and because Lemma 7.9 gives $d_{x_1} f(w) \in T_{f(x_1)}Q_1$ if and only if $w \in T_{x_1}X_1$, we see that $d_{x_1}(k \circ f)$ maps the element w of $T_{x_1}X$ to v. Thus, the map $d_{x_1}(k \circ f)|_{T_{x_1}X_1} = d_{x_1}(k \circ f_1)$ is surjective. □

Proposition 7.11 *Let $f \in C^\infty(X, Y)$ and $f = \beta \circ h$, where $h : X \to Z$ and $\beta : Z \to Y$ and β is a submersion. Let $Q \subset Y$ be a submanifold. Then $f \pitchfork Q$ if and only if $h \pitchfork \beta^{-1}(Q)$.*

Proof: Suppose that $f \pitchfork Q$ and $f(x) \in Q$, i.e.,

$$d_{h(x)}\beta[d_x h(T_x X)] + T_{\beta(h(x))}Q = T_{\beta(h(x))}Y.$$

Apply $(d_{h(x)}\beta)^{-1}$ to both sides:

$$d_x h(T_x X) + (d_{h(x)}\beta)^{-1}(T_{\beta(h(x))}Q) = (d_{h(x)}\beta)^{-1}(T_{\beta(h(x))}Y).$$

Using Theorem 7.8(1) and the fact that β is a submersion, we obtain

$$d_x h(T_x X) + T_{h(x)}[\beta^{-1}(Q)] = T_{h(x)}Z,$$

i.e., $h \pitchfork \beta^{-1}(Q)$. Conversely, assume that $h \pitchfork \beta^{-1}(Q)$, where β is a submersion. Then

$$d_x h(T_x X) + T_{h(x)}[\beta^{-1}(Q)] = T_{h(x)}Z,$$

which yields (after applying $d_{h(x)}\beta$ to both sides and using Theorem 7.8(1)):

$$d_x f(T_x X) + T_{f(x)}(Q) = d_{h(x)}\beta[T_{h(x)}Z] = T_{\beta(h(x))}Y.$$

Hence, f is transverse to Q. □

General Position and Transversality

Define vector subspaces K_1 and K_2 of a vector space H to be in *general position* in H if

$$\operatorname{codim}_H(K_1 \cap K_2) = \operatorname{codim}_H K_1 + \operatorname{codim}_H K_2,$$

or, equivalently, if $K_1 + K_2 = H$.

Notation: Let $X^2 = X \times X$ and $X^{(2)} = X^2 - \triangleright X$, where $\triangleright X = \{(x, x) \in X^2 : x \in X\}$.

Proposition 7.12 *Let $f \in C^\infty(X, Y)$ and let $f^{(2)} : X^{(2)} \to Y^2$ be the restriction $(f \times f)|_{X^{(2)}}$. Suppose that f is an immersion. Then the following are equivalent:*

(1) $f^{(2)} \pitchfork \triangleright Y$.

(2) *For any pair of distinct points x_1, x_2 in $f^{-1}(y)$, where $y \in Y$, the vector subspaces $d_{x_1} f(T_{x_1} X)$ and $d_{x_2} f(T_{x_2} X)$ are in general position in $T_y Y$.*

Proof: Set $\bar{x} = (x_1, x_2)$, $\bar{y} = f^{(2)}(\bar{x}) = (y, y)$, $H_1 = d_{x_1} f(T_{x_1} X)$, and $H_2 = d_{x_2} f(T_{x_2} X)$. We have

$$\begin{aligned} (2) \iff\ & T_y Y = H_1 + H_2 \\ \iff\ & \dim T_y Y = \dim H_1 + \dim H_2 - \dim(H_1 \cap H_2). \end{aligned}$$

The latter is equivalent to

$$\begin{aligned}
\dim[(T_yY)^2] &= 2\dim T_yY \\
&= \dim H_1 + \dim H_2 + \dim Y - \dim(H_1 \cap H_2) \\
&= \dim(H_1 \oplus H_2) + \dim Y - \dim[(H_1 \oplus H_2) \cap T_{\bar{y}} \rhd Y] \\
&\quad (\text{since } (H_1 \oplus H_2) \cap T_{\bar{y}} \rhd Y \cong H_1 \cap H_2) \\
&= \dim(H_1 \oplus H_2 + T_{\bar{y}} \rhd Y) \\
&\quad (\text{since } \dim Y = \dim(\rhd Y) = \dim(T_{\bar{y}} \rhd Y)).
\end{aligned}$$

Hence:

$$\begin{aligned}
(2) &\iff \dim[(T_yY)^2] = \dim(H_1 \oplus H_2 + T_{\bar{y}} \rhd Y) \\
&\iff (T_yY)^2 = H_1 \oplus H_2 + T_{\bar{y}} \rhd Y \\
&\iff f^{(2)} \pitchfork \rhd Y \text{ at } \bar{x} \\
&\iff (1). \ \square
\end{aligned}$$

Consult [Gol-G, Lem. 3.6, p. 83] for a generalization of Proposition 7.12 to immersions with normal crossings.

Thom's Transversality Theorem

The next result is used in the proof of Thom's transversality theorem:

Lemma 7.13 *Let D be a manifold, $Q \subseteq Y$ a submanifold, and $j : D \to C^\infty(X, Y)$ a map.*[3] *Suppose $\Phi : X \times D \to Y : (x, d) \mapsto j(d)(x)$ is smooth. If $\Phi \pitchfork Q$, then the set $C = \{d \in D : j(d) \pitchfork Q\}$ is dense in D. In fact, the set C is the complement in D of a set of measure zero.*

See [Gol-G, p. 53, Lem. 4.6] for a proof of Lemma 7.13.

Theorem 7.14 (Thom's Transversality Theorem) *Let W be any submanifold of $J^k(X, Y)$. Then*

$$\mathfrak{T}_W = \{f \in C^\infty(X, Y) : j^k f \pitchfork W\}$$

is a residual set. In addition, the set $\mathfrak{T}_W$ is open if W is closed.

A detailed proof of Thom's transversality theorem appears in [Gol-G, Thm. 4.9, p. 54]. Since residual sets in $C^\infty(X, Y)$ are dense, the theorem yields that there is an arbitrarily small perturbation f of any map in $C^\infty(X, Y)$ such that the k-jet of f is transverse to W. We may think of the submanifold W as defining for each point of X a set of constraints on the values and partial derivatives up to order k of maps in $C^\infty(X, Y)$.

[3] We do not assume that j is continuous.

Differentiable Equivalence and Multijet Transversality

Using Proposition 7.5, multijet transversality for maps between manifolds will be manifolds expressed in terms of the case for maps between Euclidean spaces (see Proposition 7.15).

We begin with some preliminary notation and observations before stating the main proposition. *First,* let $f \in C^\infty(X, Y)$ and $x = (x_1, \dots, x_s) \in X^{(s)}$, where $\dim X = n$ and $\dim Y = m$. Choose charts $\lambda_i : U_i \subseteq X \to \mathbf{R}^n$ and $\nu_i : V_i \subseteq Y \to \mathbf{R}^m$ about x_i and $f(x_i)$, respectively, with $\lambda_i(U_i) = U_i'$ and $\nu_i(V_i) = V_i'$ such that $U_i \cap U_j = \emptyset = U_i' \cap U_j'$ for $i \neq j$, where $i, j = 1, \dots, s$. Let $U' = \bigcup_{i=1}^s U_i'$ and $V' = \bigcup_{i=1}^s V_i'$ and set

$$h_i = \nu_i \circ f \circ \lambda_i^{-1} : U_i' \to V_i'.$$

Second, the sets

$$W = J^k_{U_1, V_1}(X, Y) \times \cdots \times J^k_{U_s, V_s}(X, Y)$$

and

$$W' = J^k_{U_1', V_1'}(U', V') \times \cdots \times J^k_{U_s', V_s'}(U', V')$$

are open in $J^k_s(X, Y)$ and $J^k_s(U', V')$, respectively. Note that W contains $j^k_s f(x)$. By construction (see page 211),

$$J^k_{U_i, V_i}(X, Y) \approx J^k(U_i, V_i), \qquad J^k_{U_i', V_i'}(U', V') \approx J^k(U_i', V_i'),$$

and

$$J^k(U_i, V_i) \approx J^k(U_i', V_i').$$

We make the following identifications:

$$W = J^k(U_1, V_1) \times \cdots \times J^k(U_s, V_s), \quad W' = J^k(U_1', V_1') \times \cdots \times J^k(U_s', V_s').$$

Third, consider the diffeomorphism

$$\xi_i : J^k(U_i, V_i) \to J^k(U_i', V_i'), \qquad \xi_i(\sigma_i) = j^k(\nu_i \circ g_i \circ \lambda_i^{-1})(p),$$

where g_i is a representative of σ_i. Here, $\sigma_i \in J^k(U_i, V_i)_{(p_i, q_i)}$, where $q_i = g_i(p_i)$, and $\xi_i(\sigma_i) \in J^k(U_i', V_i')_{(\lambda_i(p_i), \nu_i(q_i))}$. Observe that $\xi_i \circ (j^k f) = (j^k h_i) \circ \lambda_i$ on a neighborhood of $x_i \in U_i$. It follows that the map

$$\xi = \xi_1 \times \cdots \times \xi_s$$

is a diffeomorphism from W onto W' (using the above identifications).

Fourth, set

$$N = U_1 \times \cdots \times U_s, \quad N' = U_1' \times \cdots \times U_s'.$$

Then $N \subset X^{(s)}$ and $N' \subset (U')^{(s)}$. Define a diffeomorphism $\lambda : N \to N'$ by

$$\lambda = \lambda_1 \times \cdots \times \lambda_s.$$

Consider the map

$$h : U' \to V', \quad h|_{U_i'} = h_i.$$

Then h at $\lambda_i(x_i)$ is differentiably equivalent to f at x_i, where $i = 1, \dots, s$. We also have the following at all points $(x_1, \dots, x_s)$ in some open subset of N:

$$\begin{aligned}(\xi \circ j_s^k f)(x_1, \dots, x_s) &= \big((\xi_1 \circ j^k f)(x_1), \dots, (\xi_s \circ j^k f)(x_s)\big)\\ &= \big((j^k h_1 \circ \lambda_1)(x_1), \dots, (j^k h_s \circ \lambda_s)(x_s)\big)\\ &= (j_s^k h \circ \lambda)(x_1, \dots, x_s).\end{aligned}$$

We are now ready to state our proposition. Let $Q \subseteq J_s^k(X, Y)$ be a submanifold containing the element $j_s^k f(x) = (j^k f(x_1), \dots, j^k f(x_s))$. Since W is open in $J_s^k(X, Y)$, the set $Q \cap W$ is an open submanifold of Q. Hence,

$$Q' = \xi[Q \cap W]$$

is a submanifold of $J_s^k(U', V')$. Applying Proposition 7.5, we obtain:

Proposition 7.15 *Using the above notation, we have that $j_s^k f : X^{(s)} \to J_s^k(X, Y)$ is transverse to Q at $x = (x_1, \dots, x_s)$ if and only if $j_s^k h : (U')^{(s)} \to J_s^k(U', V')$ is transverse to Q' at $\lambda(x) = (\lambda(x_1), \dots, \lambda(x_s))$.*

Multijet Transversality via Jet Transversality

We shall express multijet transversality to certain "orbits" in terms of jet transversality.

Each point of $J_s^k(X, Y)$ has an open set diffeomorphic to some $J_s^k(U, V)$, where $U \subseteq X$ and $V \subseteq Y$ are coordinate neighborhoods (see page 214). We treat $J_s^k(U, V)$ as a submanifold of $J_s^k(X, Y)$ under such an identification. Denote the group of diffeomorphisms of $\mathbf{R}^n$ mapping 0 to 0 by Diff(n). For $i = 1, \dots, s$, let $Q_i(n, m)$ be an orbit submanifold of $J^k(n, m)$ that is invariant under left action by Diff$(n) \times$ Diff(m), that is, if $(\lambda_1, \lambda_2) \in$ Diff$(n) \times$ Diff(m) and $j^k f(x) \in Q_i(n, m)$, then:

$$(\lambda_1, \lambda_2)(j^k f(x)) = j^k(\lambda_2 \circ f \circ \lambda_1^{-1})\,(\lambda_1(x)) \in Q_i(n, m).$$

Set $Q_i = U \times V \times Q_i(n, m)$, where $U \subseteq X$ and $V \subseteq Y$ are coordinate neighborhoods, and let $Q = Q_1 \times \cdots \times Q_s$. Since

$$J_s^k(U, V) \approx U^{(s)} \times V^s \times [J^k(n, m)]^s,$$

we treat Q as a submanifold of $J_s^k(U, V)$. Under these identifications, we obtain the orbit:

$$\begin{aligned} Q \cap J_s^k(U, V) &= [(U \times V \times Q_1(n, m)) \times \cdots \times (U \times V \times Q_s(n, m))] \\ &\quad \cap \left[U^{(s)} \times V^s \times (J^k(n, m))^s\right] \\ &= U^{(s)} \times V^s \times [Q_1(n, m) \times \cdots \times Q_s(n, m)]. \end{aligned}$$

Hence, the intersection $Q \cap J_s^k(U, V)$ is a submanifold of $J_s^k(X, Y)$.

Proposition 7.16 *Using the notation and identifications above, the following holds for* $f \in C^\infty(U, V)$*:*

$$j_s^k f \pitchfork [Q \cap J_s^k(U, V)] \Longleftrightarrow j^k f \pitchfork Q_i \quad \text{for } i = 1, \ldots, s.$$

Proof: Let $x = (x_1, \ldots, x_s) \in U^{(s)}$. Since $J_s^k(U, V)$ is an open submanifold of $[J^k(U, V)]^s$, we have that

$$\begin{aligned} T_{j_s^k f(x)} J_s^k(U, V) &= T_{j_s^k f(x)} J^k(U, V)^s \\ &= \left(T_{j^k f(x_1)} J^k(U, V), \ldots, T_{j^k f(x_s)} J^k(U, V)\right). \end{aligned}$$

Moreover,

$$\begin{aligned} d_x(j_s^k f)[T_x U^{(s)}] + T_{j_s^k f(x)}[Q] &= (d_{x_1}(j^k f)[T_{x_1} U], \ldots, d_{x_s}(j^k f)[T_{x_s} U]) + \\ &\qquad \left(T_{j^k f(x_1)}[Q_1], \ldots, T_{j^k f(x_s)}[Q_s]\right) \\ &= \Big(d_{x_1}(j^k f)[T_{x_1} U] + T_{j^k f(x_1)}[Q_1], \ldots, \\ &\qquad d_{x_s}(j^k f)[T_{x_s} U] + T_{j^k f(x_s)}[Q_s]\Big). \end{aligned}$$

Hence, $j_s^k f \pitchfork Q$ if and only if

$$d_{x_i}(j^k f)[T_{x_i} U] + T_{j^k f(x_i)}[Q_i] = T_{j^k f(x_i)} J^k(U, V), \qquad \text{for } i = 1, \ldots, s. \square$$

The Multijet Transversality Theorem

Thom's Transversality Theorem extends to multijets as follows (see [Gol-G, Thm. 4.13, p. 57] and [MatV, Prop. 3.3, p. 310] for proofs):

Theorem 7.17 (Multijet Transversality Theorem) *Let* $f \in C^\infty(X, Y)$ *and* W *a submanifold of* $J_s^k(X, Y)$*. Then*

$$\mathfrak{T}_W = \{f \in C^\infty(X, Y) : j_s^k f \pitchfork W\}$$

is a residual set. In addition, the set $\mathfrak{T}_W$ *is open if* W *is compact.*

In this case, a submanifold W yields for each collection of s distinct points of X, a set of constraints on the values and partial derivatives up to order k of maps in $C^\infty(X, Y)$. We shall see that what is important and new about multijet transversality involves the elements of W whose target s-tuple have some repeated points of Y (e.g., Sections 7.4.2 and 7.5.3).

Note that since any countable intersection of residual sets is a residual set, we have:

Corollary 7.18 *Let* $\{W_i\}_{i\in I}$ *be a family of countably many submanifolds of* $J_s^k(X, Y)$. *Then* $\bigcap_{i\in I} \mathfrak{T}_{W_i}$ *is a residual subset of* $C^\infty(X, Y)$.

7.2 Mather Stability Theory

In a mathematical tour de force, John Mather published a series of six seminal papers[MatI]–[MatVI] from 1968 to 1971 that: (1) established a powerful criterion for deciding the stability of a map, (2) determined the dimension ("nice dimensions") for which stability is a dense (and open) property, and (3) classified the locally stable maps (via certain algebras yielding "normal forms") for all nice dimensions. Our goal is to summarize briefly those aspects of Mather stability theory needed for gravitational lensing. A detailed introduction to his work requires a separate volume. Readers are referred to the excellent exposition by Golubitsky and Guillemin[Gol-G].

7.2.1 Local Stability

Informally speaking, a map f is "stable at x" if for any sufficiently small perturbation $\tilde{f}$ of f, there is a point $\tilde{x}$ near x such that $\tilde{f}$ at $\tilde{x}$ is differentiably equivalent to f at x. In this case, the perturbation $\tilde{f}$ has the same critical point structure, up to coordinate change, as f in sufficiently small neighborhoods of $\tilde{x}$ and x, respectively. We now express this concept precisely.

Stable Map-Germs

The concept of "map-germs" provides a convenient way to express the local behavior of maps. Consider the set of all smooth local maps of X at x into Y (i.e., maps from a neighborhood of x into Y). Two local maps f and $\tilde{f}$ in this set are *germ equivalent at* x if $f = \tilde{f}$ on some neighborhood of x. Germ equivalence at x defines an equivalence relation on the set of all smooth local maps of X at x into Y. Each corresponding equivalence class is called a *map-germ* or, *germ of* f *at* x, denoted $[f]_x$. When confusion is unlikely, we shall also denote the germ $[f]_x$ and its representative f, by f. Note that all representatives of $[f]_x$ assign the same value to x.

Notation: The map-germ $[f]_x$ will also be written as $f : (X, x) \to (Y, y)$, where $y = f(x)$.

Definition 7.19 Two map-germs $[f]_x$ and $[\tilde{f}]_{\tilde{x}}$ are *differentiably equivalent* if for each representative $f \in [f]_x$ and $\tilde{f} \in [\tilde{f}]_{\tilde{x}}$, the local map f at x is differentiably equivalent to $\tilde{f}$ at $\tilde{x}$.

Equivalently, two map-germs $f : (X, x) \to (Y, y)$ and $\tilde{f} : (X, \tilde{x}) \to (Y, \tilde{y})$ are differentiably equivalent if and only if there are germs of diffeomorphisms at x and y, say, $\lambda : (X, x) \to (X, \tilde{x})$ and $\nu : (Y, y) \to (Y, \tilde{y})$, such that $[\tilde{f}]_{\tilde{x}} = [\nu \circ f \circ \lambda^{-1}]_{\tilde{x}}$.

Let f and $\tilde{f}$ be smooth local maps of X at x into Y and X at $\tilde{x}$ into Y, respectively. Then *f at x is differentiably equivalent to $\tilde{f}$ at $\tilde{x}$ if and only if $[f]_x$ and $[\tilde{f}]_{\tilde{x}}$ are differentiably equivalent.*

Definition 7.20 The map-germ $[f]_x$ of a map $f \in C^\infty(X, Y)$ is *stable* if for any neighborhood U of x, there is a neighborhood $\mathfrak{F}$ of f in $C^\infty(X, Y)$ such that for each map h in $\mathfrak{F}$ there is a point x' in U with $[h]_{x'}$ differentiably equivalent to $[f]_x$.

Definition 7.21 A map f in $C^\infty(X, Y)$ is *stable at* $x \in X$ if the germ $[f]_x$ is stable. If f is stable at all points of X, then f is *locally stable.*

Proposition 7.22 *A map $f \in C^\infty(X, Y)$ is stable at all points $x \in X$ with* $\operatorname{rank}[d_x f] = \min\{n, m\}$, *where* $\dim X = n$ *and* $\dim Y = m$.

Proof: Let $B_\epsilon(x)$ be an open ball of radius ϵ centered at x and consider the following open set in $C^\infty(X, Y)$:

$$V^1_{\epsilon, B_\epsilon(x)}(f) = \left\{h \in C^\infty(X, Y) : |j^1 f(x') - j^1 h(x')| < \epsilon \text{ for all } x' \in B_\epsilon(x)\right\}.$$

For $h \in V^1_{\epsilon, B_\epsilon(x)}(f)$ with ϵ sufficiently small, the values and first partials of h in $B_\epsilon(x)$ differ by an arbitrarily small amount from those of $f|_{B_\epsilon(x)}$. Relative to coordinates at x and $f(x)$, let $(\operatorname{Jac} f)(x)$ and $(\operatorname{Jac} h)(x)$ be, resp., the Jacobian matrices of f and h at x. For each nonzero entry in $(\operatorname{Jac} f)(x)$, the corresponding one in $(\operatorname{Jac} h)(x)$ is also nonzero. Consequently, if $n \geq m = \operatorname{rank}[d_x f]$, then $\operatorname{rank}[d_x h] = \operatorname{rank}[d_x f]$. Hence, h at x and f at x are differentiably equivalent to the standard submersion $(x_1, \ldots, x_n) \mapsto (x_1, \ldots, x_m)$ at 0. Thus, h at x is differentiably equivalent to f at x. Similarly, if $m \geq n = \operatorname{rank}[d_x f]$, then $\operatorname{rank}[d_x h] = \operatorname{rank}[d_x f]$. In this case, h at x and f at x are differentiably equivalent to the standard immersion $(x_1, \ldots, x_n) \mapsto (x_1, \ldots, x_n, 0, \ldots, 0)$ at 0. Hence, h at x is differentiably equivalent to f at x, i.e., $[h]_x$ is differentiably equivalent to $[f]_x$. Therefore, f is stable at x. □

Infinitesimally Stable Map–Germs and Determinacy

A useful (computable) notion for investigating stability is "infinitesimal stability."[4]

[4] The original motivations for this concept are recalled in [Gol-G, p. 74–78].

Let $f \in C^\infty(X, Y)$ and let $f^*(TY)$ be the pull-back bundle of Y by f. By construction $f^*(TY)$ is a vector bundle over X. Let $s : X \to f^*(TY)$ be a section of $f^*(TY)$. Since $f^*(TY)_x = T_{f(x)}Y$ for $x \in X$, we have $s(x) \in T_{f(x)}Y$ and can treat s as a vector field along f, that is, a map $X \to TY$ satisfying $\pi_Y \circ s = f$, where π_Y is the projection $TY \to Y$. Working backwards, a vector field along f can also be identified with a section of $f^*(TY)$. Hence, when speaking of germs of vector fields along f, we shall freely employ the identification with germs of sections of $f^*(TY)$.

For a vector bundle E over X, denote by $\Gamma^\infty(E)_p$, the set of germs of smooth sections $X \to E$ of E at p. Then:

Definition 7.23 Let $f \in C^\infty(X, Y)$ and $p \in X$ with $f(p) = q$. The map-germ $[f]_p$ is *infinitesimally stable* if for every germ $[\tau] \in \Gamma^\infty(f^*(TY))_p$ of a vector field along f, there are germs of vector fields on X and Y, say $[\zeta]_p \in \Gamma^\infty(TX)_p$ and $[\omega]_q \in \Gamma^\infty(TY)_q$, such that

$$[\tau]_p = [df \circ \zeta]_p + [\omega \circ f]_p, \tag{7.2}$$

where $df : TX \to TY$ is the differential map of f.

Let $x_1, \ldots, x_n$ and $y_1, \ldots, y_n$ be coordinates centered at p and q, respectively. In these coordinates, equation (7.2) becomes

$$\tau_i = \sum_{\ell=1}^{n} \frac{\partial f_i}{\partial x_\ell} \zeta_\ell + \omega_i(f_1, \ldots, f_m)$$

where $1 \le i \le m,\ f = (f_1, \ldots, f_m)$, and

$$\tau = \sum_{i=1}^{m} \tau_i \frac{\partial}{\partial y_i}, \qquad \zeta = \sum_{\ell=1}^{n} \zeta_\ell \frac{\partial}{\partial x_\ell}, \qquad \omega = \sum_{i=1}^{m} \omega_i \frac{\partial}{\partial y_i}.$$

Consider the map-germ $[f]_x$, where $f \in C^\infty(X, Y)$ and $x \in X$. Since every neighborhood of x contains an open set that is diffeomorphic to $\mathbf{R}^n$ and this diffeomorphism can be chosen so that x maps to $0 \in \mathbf{R}^n$, the germ $[f]_x$ is differentiably equivalent to $[h]_0$, where h is a local map of $\mathbf{R}^n$ at 0 into $\mathbf{R}^m$ with $h(0) = 0$. Hence, it suffices to study the vector space $\mathcal{E}_0(n, m)$ of germs of local maps of $\mathbf{R}^n$ at 0 into $\mathbf{R}^m$ mapping $0 \in \mathbf{R}^n$ to $0 \in \mathbf{R}^m$.

Two germs $[f]$ and $[h] \in \mathcal{E}_0(n, m)$ are *k-equivalent* (or, have *kth-order contact*) if for any representatives $f \in [f]$ and $h \in [h]$, the kth-order Taylor expansions of f and h coincide at $0 \in \mathbf{R}^n$. This is an equivalence relation in $\mathcal{E}_0(n, m)$.

The equivalence class of $[f] \in \mathcal{E}_0(n, m)$ under k-equivalence is called the *k-jet* of $[f]$, denoted $j^k[f]$. Note that $j^k[f] = (j^k[t^k f_1], \ldots, j^k[t^k f_m])$, where $f = (f_1, \ldots, f_m)$ and $t^k f_i$ is the kth-order Taylor expansion at 0 of f_i.

Notation: We shall identify the vector space $J_{\mathcal{E}_0(n,m)}$ of k-jets of germs in $\mathcal{E}_0(n, m)$ with $J^k(n, m) = J^k(\mathbf{R}^n, \mathbf{R}^m)_{(0,0)}$. By abuse of notation, we shall also use $J^k(n, m)$ to denote $J_{\mathcal{E}_0(n,m)}$. Also, recall that we set $J^k(n) = J^k(n, 1)$.

Definition 7.24 A germ $[f] \in \mathcal{E}_0(n, m)$ is *k-determined* if every germ in $\mathcal{E}_0(n, m)$ that is k-equivalent to $[f]$ is differentiably equivalent to $[f]$.

If $[f] \in \mathcal{E}_0(n, m)$ is k-determined, then every representative f of $[f]$ is such that f at $0 \in \mathbf{R}^n$ is differentiably equivalent to the kth-order Taylor expansion of f at $0 \in \mathbf{R}^n$.

The map-germ $[f]_x$ of $f \in C^\infty(X, Y)$ is *k-determined* if $[f]_x$ is differentiably equivalent to a k-determined germ in $\mathcal{E}_0(n, m)$.

Theorem 7.25 *Let $f \in C^\infty(X, Y)$ and $x \in X$. If the map-germ of f at x is infinitesimally stable, then the germ is $(m + 1)$-determined, where $m = \dim Y$.*

Theorem 7.26 *Let $f \in C^\infty(X, Y)$ and $x \in X$. If the map-germ of f at x is infinitesimally stable, then the germ is stable.*

Proofs of Theorems 7.25 and 7.26 appear in [MatIV, p. 234] and [Arn-GVI, Chap. 7], respectively.

7.2.2 Transverse and Global Stability

Transverse Stability

Let Diff(X) and Diff(Y) be, respectively, the groups of diffeomorphisms of X and Y. The group Diff$(X) \times$ Diff(Y) has a natural left action on the s-fold k-jet bundle $J_s^k(X, Y)$. Let $\sigma = (j^k f(x_1), \ldots, j^k f(x_s))$ be an element of $J_s^k(X, Y)_{(p,q)}$, where $p = (x_1, \ldots, x_s) \in X^{(s)}$ and $q = (f(x_1), \ldots, f(x_s)) \in Y^s$. For $(\lambda_1, \lambda_2) \in \text{Diff}(X) \times \text{Diff}(Y)$, let

$$(\lambda_1, \lambda_2)(\sigma) = (j^k(\lambda_2 \circ f \circ \lambda_1^{-1})(\lambda_1(x_1)), \ldots, j^k(\lambda_2 \circ f \circ \lambda_1^{-1})(\lambda_1(x_s))).$$

Denote by $\mathcal{D}_\sigma^s$ the orbit of σ in $J_s^k(X, Y)$ under the above action. *Then $\mathcal{D}_\sigma^s$ is a submanifold of $J_s^k(X, Y)$ for all $\sigma \in J_s^k(X, Y)$* [MatV, p.303] [Gol-G, p.138].

Let $\sigma = (\sigma_1, \ldots, \sigma_s)$ be an s-fold k-jet in $J_s^k(X, Y)$, where $\sigma_i \in J^k(X, Y)_{(x_i, y_i)}$ for $i = 1, \ldots, s$. Then σ is called a *diagonal element* if $y_1 = \cdots = y_s$.

Definition 7.27 Let $f \in C^\infty(X, Y)$ and $m = \dim Y$. Then f is *transverse stable* if for $s = 1, \ldots, m + 1$ and every diagonal element σ in $J_s^m(X, Y)$, we have $j_s^m f \pitchfork \mathcal{D}_\sigma^s$.

If f is transverse stable and $\sigma = j^m f(x)$, where $x \in X$, then $j^m f \pitchfork \mathcal{D}_\sigma$ at x for all $x \in X$. It follows that the map-germ $[f]_x$ is infinitesimally stable for all $x \in X$ (see [Gol-G, Thm. 5.13 on p. 138]). Theorem 7.26 now yields for each $x \in X$ that $[f]_x$ is stable. Hence:

Theorem 7.28 *If a map f in $C^\infty(X, Y)$ is transverse stable, then f is locally stable.*

Global Stability

Local stability globalizes as follows:

Definition 7.29 A map $f \in C^\infty(X, Y)$ is *globally stable* (or, simply, *stable*) if there is a neighborhood $\mathfrak{N}$ of f in $C^\infty(X, Y)$ such that each map h in $\mathfrak{N}$ is *differentiably equivalent* to f, that is, there are smooth diffeomorphisms $\lambda_1 : X \to X$ and $\lambda_2 : Y \to Y$ such that $f = \lambda_2 \circ h \circ \lambda_1^{-1}$.

By definition, a globally stable map f is locally stable and every map h sufficiently close to f has the same critical-point properties up to diffeomorphism, as f. Moreover, the values and partials of h are near those of f for *all* points of X. In addition, the set of stable maps in $C^\infty(X, Y)$ is open (because by Definition 7.29 each map in $\mathfrak{N}$ is stable).

Let us extend the notion of infinitesimally stable map-germs to maps (cf. Definition 7.23).

Definition 7.30 A map $f \in C^\infty(X, Y)$ is *infinitesimally stable* if for every vector field τ along f, there are vector fields ζ and ω on X and Y, respectively, such that

$$\tau = df \circ \zeta + \omega \circ f.$$

A continuous map $f : X \to Y$ is *proper* if for every compact set D in Y, the set $f^{-1}(D)$ is compact in X. The next theorem characterizes global stability:

Theorem 7.31 (Mather's Stability Theorem) *Suppose that a map f in $C^\infty(X, Y)$ is proper. Then the following are equivalent:*

(1) *f is globally stable.*

(2) *f is infinitesimally stable.*

(3) *f is transverse stable.*

Now assume that in Theorem 7.31 the hypothesis of properness is dropped. Then globally stable maps, as well as infinitesimally stable maps, are still transverse stable. However, transverse stable maps are not necessarily either globally stable or infinitesimally stable. In addition, globally stable maps are not necessarily infinitesimally stable. Proofs of all these results, including Theorem 7.31, appear in [MatV, pp. 313, 317] and [Gol-G, Chap.V].

A natural issue that arises is to investigate the genericity and classification of stable maps. More precisely, denote the subspaces of proper and proper stable maps in $C^\infty(X, Y)$ by $C^\infty_{pr}(X, Y)$ and $C^\infty_{prs}(X, Y)$ be, respectively. We note that the subspace $C^\infty_{pr}(X, Y)$ is nonempty and open in $C^\infty(X, Y)$ (see Lemma 8.1, page 309). Then:

- Determine the pairs of natural numbers (n, m), where $n = \dim X$ and $m = \dim Y$, for which $C^{\infty}_{prs}(X, Y)$ is dense in $C^{\infty}_{pr}(X, Y)$. Such pairs are called *nice dimensions.*
- Classify (up to differentiable equivalence) the stable map-germs of elements in $C^{\infty}_{pr}(X, Y)$ for nice dimensions.

Mather settled these issues in [MatVI]. The next theorem summarizes his results on the nice dimensions.

Theorem 7.32 *Let $n = \dim X$ and $m = \dim Y$. The subspace $C^{\infty}_{prs}(X, Y)$ is open and dense in $C^{\infty}_{pr}(X, Y)$ if and only if the pair (n, m) satisfies one of the following, where $q = m - n$:*

(1) $m < 7q + 8$ *and* $q \geq 4$.

(2) $m < 7q + 9$ *and* $3 \geq q \geq 0$.

(3) $m < 8$ *and* $q = -1$.

(4) $m < 6$ *and* $q = -2$.

(5) $m < 7$ *and* $q \leq -3$.

Figure 7.2 depicts the nice dimensions (n, m) of Theorem 7.32. Consult Zakalyukin [Zaka92] for a generalization of nice dimensions.

The nice dimension $(n, 1)$ is where Morse theory applies (Section 7.4), while $(2, 2)$ uses Whitney's singularity theory (Section 7.5). Whitney also investigated the pairs (n, m) for $2 \leq m \leq 5$ and $m = 2n - 2$, $m = 2n - 1$, and $m \geq 2n$ [Whi58].

7.3 Singularity Manifolds

We are interested in decomposing the set $\mathrm{Crit}(f)$ of critical points[5] of maps f in $C^{\infty}(X, Y)$ into disjoint subsets $S_{r_1,\dots,r_k}(f)$, where each captures a specific drop in $\mathrm{rank}(df)$ and $\mathrm{rank}(d(f|))$. Here, $f|$ will indicate the given successive restrictions of f to subsets of $\mathrm{Crit}(f)$.

The subsets $S_{r_1,\dots,r_k}(f)$ will be studied via certain "singularity manifolds" $S_{r_1,\dots,r_k}(X, Y)$ in $J^k(X, Y)$. This will lead to results of the form

$$x \in S_{r_1,\dots,r_k}(f) \text{ if and only if } j^k f(x) \in S_{r_1,\dots,r_k}(X, Y).$$

The singularities $S_r(X, Y)$ and $S_{r_1,r_2}(X, Y)$ were first treated by Thom [Tho] and H.L. [Lev59], respectively, while the general case $S_{r_1,\dots,r_k}(X, Y)$ is due to Boardman [Boa]. Our applications to gravitational lensing will involve primarily $S_r(X, Y)$ and $S_{r_1,r_2}(X, Y)$. The main references for this section are [Lev59], [Gol-G], and [Arn-GVI].

[5] Section 6.2.1, page 176.

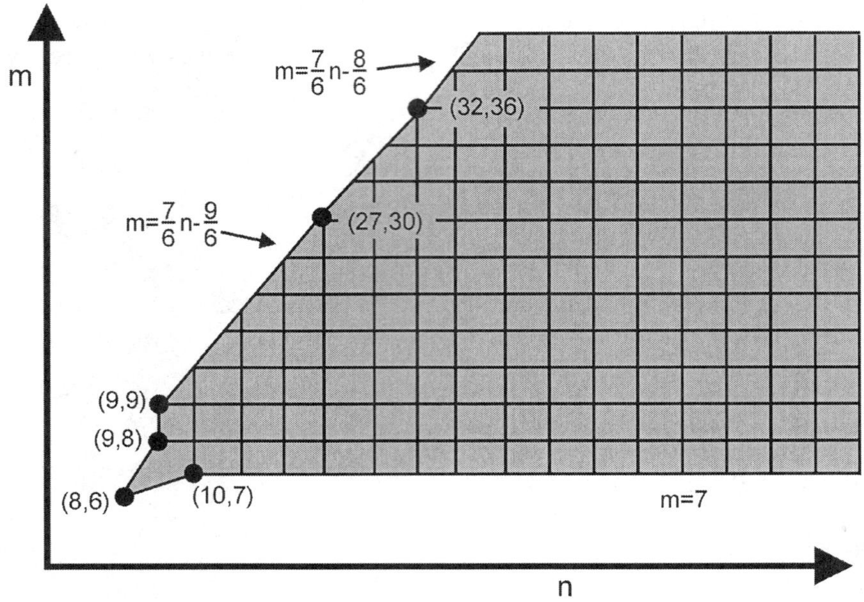

Figure 7.2. Mather's nice dimensions for the space of smooth maps from an n-manifold X into an m-manifold Y. These are all pairs (n, m) except those occurring inside and on the boundary of the shaded region.

7.3.1 Singularities $S_r(X, Y)$ and $S_r(f)$

Singularity manifolds are defined generally as follows:

Definition 7.33 An *s-fold singularity manifold of order k* is a submanifold of $J_s^k(X, Y)$ that is invariant under the group $\mathrm{Diff}(X) \times \mathrm{Diff}(Y)$.

The orbits $\mathcal{D}_\sigma^s$ (see Section 7.2.2, page 229) are s-fold singularity manifolds of order k in $J_s^k(X, Y)$, while *every s-fold singularity manifold of order k in $J_s^k(X, Y)$ is a union of disjoint orbits $\mathcal{D}_\sigma^s$.*

Notation: $\mathcal{D}_\sigma = \mathcal{D}_\sigma^1$.

A 1-fold singularity manifold of order k will be called a *k-singularity* (or *singularity manifold of order k*).

Let σ be in $J^1(X, Y)$, say $\sigma \in J^1(X, Y)_{(x, f(x))}$. Define

$$\mathrm{corank}(\sigma) \equiv \mathrm{corank}(d_x f) = \min\{n, m\} - \mathrm{rank}(d_x f),$$

where $n = \dim X$, $m = \dim Y$, and f is a representative of σ. Note that if $\sigma = j^1 f(x)$ and $\mathrm{corank}(\sigma) > 0$, then x is a critical point of f.

Theorem 7.34 *Let*

$$S_r(X, Y) = \{\sigma \in J^1(X, Y) : \operatorname{corank}(\sigma) = r\}.$$

Then $S_r(X, Y)$ is a submanifold of $J^1(X, Y)$ and

$$\operatorname{codim}_{J^1(X,Y)} S_r(X, Y) = r(|n - m| + r).$$

See [Lev59, p. 11], [Gib, p. 56], and [Gol-G, p. 61] for proofs. Note that since $\operatorname{codim}_{J^1(X,Y)} S_0(X, Y) = 0$, *the submanifold $S_0(X, Y)$ is open in $J^1(X, Y)$.* Hence, we immediately have $j^1 f \pitchfork S_0(X, Y)$.

The manifolds $S_r(X, Y)$ are 1-singularities in $J^1(X, Y)$:

Proposition 7.35 *Let $f \in C^\infty(X, Y)$ and $\sigma = j^1 f(p)$. If* $\operatorname{corank}(\sigma) = r$, *then the orbit $\mathcal{D}_\sigma$ in $J^1(X, Y)$ is $S_r(X, Y)$.*

Proof: By definition $\mathcal{D}_\sigma$ is a subset of $S_r(X, Y)$. Conversely, if $\tau \in S_r(X, Y)$, say $\tau = (j^1 h)(x')$, then since $\operatorname{rank}[d_{x'} h] = \ell - r$, where $\ell = \min\{n, m\}$, there are coordinate changes centered at x' and $h(x')$, such that the Jacobian matrices of h at x' and f at x have the same form, $\begin{Vmatrix} I_{\ell-r} & 0 \\ 0 & 0 \end{Vmatrix}$, where $I_{\ell-r}$ is the $(\ell - r) \times (\ell - r)$ identity matrix. Hence, τ is in the orbit $\mathcal{D}_\sigma$. □

By Proposition 7.35, we get:

Proposition 7.36 *Let $\ell = \min\{n, m\}$ with $n = \dim X$ and $m = \dim Y$. Then the jet space $J^1(X, Y)$ decomposes into a disjoint union of orbits:*

$$J^1(X, Y) = S_0(X, Y) \,\dot{\bigcup}\, S_1(X, Y) \,\dot{\bigcup}\, \cdots \,\dot{\bigcup}\, S_\ell(X, Y).$$

Notation:

(1) $S_r(f) = \{x \in X : \operatorname{corank}(d_x f) = r)\}$.

(2) ${}_{\mathrm{T}}S_r(f) = \{x \in S_r(f) : j^1 f \pitchfork S_r(X, Y) \text{ at } x\}$. The "T" in ${}_{\mathrm{T}}S_r(f)$ is a reminder that a transversality condition is employed.

Note that by construction, $S_0(f) = {}_{\mathrm{T}}S_0(f)$. Also, we have

$$x \in S_r(f) \textit{ if and only if } j^1 f(x) \in S_r(X, Y). \tag{7.3}$$

If $j^1 f \pitchfork S_r(X, Y)$ at $x \in S_r(f)$, then Theorem 7.8 (page 219) implies that there is a neighborhood V of x in X such that $S_r(f) \cap V$ is a submanifold of X with

$$\operatorname{codim}_X [S_r(f) \cap V] = \operatorname{codim}_{J^1(X,Y)} S_r(X, Y) = r(|n - m| + r).$$

Hence, if ${}_{\top}S_r(f) \neq \emptyset$, then ${}_{\top}S_r(f)$ is a submanifold of X and

$$\text{codim}_X[{}_{\top}S_r(f)] = r(|n-m|+r).$$

Notation: For any submanifold $S(X,Y)$ of $J^k(X,Y)$, let

$$\begin{aligned} {}_{\top}S(X,Y) &= \{j^{k+1}f(x) \in J^{k+1}(X,Y) : j^k f(x) \in S(X,Y) \\ &\qquad \text{and } j^k f \pitchfork S(X,Y) \text{ at } x\}. \end{aligned}$$

For example,

$$\begin{aligned} {}_{\top}S_r(X,Y) &= \{j^2 f(x) \in J^2(X,Y) : j^1 f(x) \in S_r(X,Y) \\ &\qquad \text{and } j^1 f \pitchfork S_r(X,Y) \text{ at } x\}. \end{aligned}$$

It follows that ${}_{\top}S_r(f) = (j^2 f)^{-1}({}_{\top}S_r(X,Y))$, namely,

$$x \in {}_{\top}S_r(f) \textit{ if and only if } j^2 f(x) \in {}_{\top}S_r(X,Y). \tag{7.4}$$

The Singularities $\mathbf{S_r(n, m)}$

Denote by Diff(n) the group of diffeomorphisms of $\mathbf{R}^n$ mapping 0 to 0. A *k-singularity* in $J^k(n,m)$ is a submanifold of $J^k(n,m)$ that is invariant under left action by Diff$(n) \times$ Diff(m). Let

$$S_r(n,m) = \{j^1 f(x) \in J^1(n,m) : \text{corank}(d_x f) = r\}.$$

Then $S_r(n,m)$ is a submanifold of $J^1(n,m)$ and

$$\text{codim}_{J^1(n,m)} S_r(n,m) = r(|n-m|+r).$$

Moreover, the manifolds $S_r(n,m)$ are 1-singularities in $J^1(n,m)$. Hence, $J^1(n,m)$ is a union of finitely many disjoint orbits:

$$J^1(n,m) = S_0(n,m) \dot{\bigcup} S_1(n,m) \dot{\bigcup} \cdots \dot{\bigcup} S_\ell(n,m).$$

In addition, if $U \subseteq \mathbf{R}^n$ and $V \subseteq \mathbf{R}^m$ are coordinate neighborhoods, then

$$\begin{aligned} J^1(U,V) &\approx U \times V \times J^1(n,m) \qquad \text{(diffeomorphism)} \\ &= U \times V \times \dot{\bigcup}_{r=0}^{\ell} S_r(n,m) \\ &= \dot{\bigcup}_{r=0}^{\ell} [U \times V \times S_r(n,m)], \end{aligned}$$

where $\ell = \min\{n,m\}$. Note that

$$J^1(U,V) \approx U \times V \times \mathbf{R}^{nm}.$$

We have

$$S_r(U,V) \approx U \times V \times S_r(n,m) \qquad \text{(diffeomorphism)}.$$

Consult [Lev59, pp. 10–16, 43,44] and [Gib, p. 56] for more.

7.3.2 Singularities $S_{r,s}(X, Y)$ and $S_{r,s}(f)$

Suppose that ${}_{\mathsf{T}}S_r(f) \neq \emptyset$. Since ${}_{\mathsf{T}}S_r(f)$ is a manifold, define

$$S_{r,s}(f) = \{x \in {}_{\mathsf{T}}S_r(f) : \operatorname{rank}\left[d_x(f|_{{}_{\mathsf{T}}S_r(f)})\right] = \dim {}_{\mathsf{T}}S_r(f) - s\},$$

where $0 \leq s \leq \dim[{}_{\mathsf{T}}S_r(f)]$; otherwise, $S_{r,s}(f) = \emptyset$. Notice that

$${}_{\mathsf{T}}S_r(f) = \dot{\bigcup_s} S_{r,s}(f),$$

where ${}_{\mathsf{T}}S_0(f) = S_0(f) = S_{0,0}(f)$.

Definition 7.37 Let $f \in C^\infty(X, Y)$. Then f is *1-generic* if $j^1 f \pitchfork S_r(X, Y)$ for $r = 0, \ldots, \ell$, where $\ell = \min\{n, m\}$ with $n = \dim X$ and $m = \dim Y$.

Notation: Set

$$\nu_{r,s}(n, m) = (m - n + r)r + \frac{s}{2}[(m - n + r)(2r - s + 1) - 2r + 2s]. \tag{7.5}$$

For example, if $n = m$, then

$$\nu_{r,s}(n, n) = r^2 + \frac{s}{2}[2r^2 - rs + 2s - r].$$

Theorem 7.38 *Let $f \in C^\infty(X, Y)$ be 1-generic. If integers r, s satisfy $(r + \max\{n - m, 0\}) \geq s \geq 0$, then there are 2-singularities in $J^2(X, Y)$, denoted $S_{r,s}(X, Y)$, such that:*

(1) *$x \in S_{r,s}(f)$ if and only if $j^2 f(x) \in S_{r,s}(X, Y)$.*

(2) *$S_{r,s}(X, Y)$ is a submanifold of the manifold $S'_r(X, Y) = \pi_{2,1}^{-1}[S_r(X, Y)]$, where $\pi_{2,1} : J^2(X, Y) \to J^1(X, Y)$ is the natural projection,[6] and*

$$\operatorname{codim}_{J^2(X,Y)} S_{r,s}(X, Y) = \nu_{r,s}(n, m).$$

Theorem 7.38 is due to H.L. [Lev59, p. 55]. Consult [Lev59, pp. 47–56], [Gol-G, p. 154, Thm 4.7], and [Gib, pp. 185–186] for further details. We add that *the manifolds $S_{r,s}(X, Y)$, as well as $S_r(X, Y)$, are constructed independently of f* (see [Lev64] and [Boa]).

Note that since $S_0(X, Y)$ is open in $J^1(X, Y)$, the manifold $S'_0(X, Y)$ is open in $J^2(X, Y)$. Moreover, the 2-singularity $S_{0,0}(X, Y)$ is also open in $J^2(X, Y)$ because $\operatorname{codim}_{J^2(X,Y)} S_{0,0}(X, Y) = 0$. It follows that $j^2 f \pitchfork S_{0,0}(X, Y)$.

Notation: By construction, ${}_{\mathsf{T}}S_r(X, Y) \subseteq S'_r(X, Y)$. Set
$D_r(X, Y) = S'_r(X, Y) \setminus {}_{\mathsf{T}}S_r(X, Y)$ and write

$$D(X, Y) = \dot{\bigcup}_{r=0}^{\ell} D_r(X, Y).$$

In this notation, $S'_r(X, Y) = {}_{\mathsf{T}}S_r(X, Y) \dot{\bigcup} D_r(X, Y)$.

---

[6] Explicitly, if $r \geq s$ and $[f]^r_p$ the r-jet of f in $J^r(X, Y)_{(p, f(p))}$, then the projection $\pi_{r,s} : J^r(X, Y) \to J^s(X, Y)$ is given by $\pi_{r,s}([f]^r_p) = [f]^s_p$.

Proposition 7.39 *The jet space $J^2(X, Y)$ decomposes as follows:*

$$J^2(X, Y) = \left[\dot{\bigcup}_{r,s} S_{r,s}(X, Y)\right] \dot{\bigcup} D(X, Y),$$

where $0 \leq r \leq \min\{n, m\}$ *and* $0 \leq s \leq (r + \max\{n - m, 0\})$.

Proof: Let $\ell = \min\{n, m\}$. By Proposition 7.36, we have

$$J^2(X, Y) = \dot{\bigcup}_{r=0}^{\ell} S'_r(X, Y).$$

By (7.4), $x \in {}_{\top}S_r(f)$ if and only if $j^2 f(x) \in {}_{\top}S_r(X, Y)$. Since ${}_{\top}S_r(f) = \dot{\bigcup}_s S_{r,s}(f)$, where $0 \leq s \leq (r + \max\{n - m, 0\})$, Theorem 7.38(1) yields that $x \in S_{r,s}(f)$ if and only if $j^2 f(x) \in S_{r,s}(X, Y)$. Consequently,

$${}_{\top}S_r(X, Y) = \dot{\bigcup}_s S_{r,s}(X, Y),$$

where $0 \leq s \leq (r + \max\{n - m, 0\})$. The result now follows. □

Definition 7.40 Let $f \in C^\infty(X, Y)$. If f is 1-generic and $j^2 f \pitchfork S_{r,s}(X, Y)$ for $r = 0, \ldots, \ell$, where $\ell = \min\{n, m\}$, and $0 \leq s \leq (r + \max\{n - m, 0\})$, then f is called *2-generic.*

Theorem 7.41 *Let $f \in C^\infty(X, Y)$ be 2-generic with $n = \dim X$ and $m = \dim Y$. For $r = 0, \ldots, \ell$, where $\ell = \min\{n, m\}$, and $(r + \max\{0, n - m\}) \geq s$, the space $S_{r,s}(f)$ is a submanifold of $S_r(f)$ and*

$$\operatorname{codim}_X S_{r,s}(f) = \operatorname{codim}_{J^2(X,Y)} S_{r,s}(X, Y) = \nu_{r,s}(n, m).$$

See [Lev59, p. 55], [Gol-G, Cor. 4.9, p. 154], and [Arn-GVI, pp. 44–59] for proofs. Theorems 7.38 and 7.41 were generalized by Boardman[Boa] to singularities $S_{r_1,\ldots,r_k}(X, Y)$ — see [Arn-GVI, pp. 43–59] and [Gib, pp. 174–191] for an introduction.

The Singularities $\mathbf{S_{r,s}(n, m)}$

Let $U \subseteq X$ and $V \subseteq Y$ be coordinate neighborhoods. Then

$$J^2(U, V) = \left[\dot{\bigcup}_{r,s} S_{r,s}(U, V)\right] \dot{\bigcup} D(U, V) \approx U \times V \times J^2(n, m). \tag{7.6}$$

The manifold $J^2(n, m)$ has infinitely many orbits for all pairs (n, m) with $n^2 + m^2 < nm(n + 1)/2$ (see [Lev59, p. 16]). This is unlike the case for $J^1(n, m)$, which has finitely many orbits. Notice that

$$J^2(n, m) = J^1(n, m) \times D^2(n, m),$$

where $D^2(n, m)$ is the space of all second-order partial derivatives of the component functions of a map from $\mathbf{R}^n$ into $\mathbf{R}^m$.

Results similar to Theorems 7.38 and 7.41 hold for $J^k(n, m)$. In fact, define $f \in C^\infty(\mathbf{R}^n, \mathbf{R}^m)_{(0,0)}$ to be 1-*generic* if $j^1 f \pitchfork S_r(n, m)$ for $r = 0, \dots, \ell$, where $\ell = \min\{n, m\}$. The map f is called 2-*generic* if f is 1-generic and $j^2 f \pitchfork S_{r,s}(n, m)$ for all $0 \le s \le (r + \max\{n - m, 0\})$.

Suppose that the integers r and s satisfy $0 \le s \le (r + \max\{n - m, 0\})$. *Then there are 2-singularities in $J^2(n, m)$, denoted $S_{r,s}(n, m)$, such that:*

(1) $x \in S_{r,s}(f)$ *if and only if* $j^2 f(x) \in S_{r,s}(n, m)$.

(2) $S_{r,s}(n, m)$ *is a submanifold of the manifold* $S_r'(n, m) = \pi_{2,1}^{-1}[S_r(n, m)]$, *where* $\pi_{2,1} : J^2(n, m) \to J^1(n, m)$ *is the natural projection. Moreover,*

$$\operatorname{codim}_{J^2(n,m)} S_{r,s}(n, m) = \nu_{r,s}(n, m).$$

(3) *If f is 2-generic, then $S_{r,s}(f)$ is a submanifold of $S_r(f)$ and*

$$\operatorname{codim}_{\mathbf{R}^n} S_{r,s}(f) = \operatorname{codim}_{J^2(n,m)} S_{r,s}(n, m) = \nu_{r,s}(n, m).$$

Proofs of these results first appeared in [Lev59, pp. 47-55]. Note that

$$\operatorname{codim}_{J^2(n,m)} S_{r,s}(n, m) = \operatorname{codim}_{J^2(U,V)} S_{r,s}(U, V).$$

Description of $\mathbf{S_{1,0}(n, n)}$: Though the description of the singularities $S_r(n, m)$ and $S_r(X, Y)$ are straightforward, it is unfortunately not simple to describe $S_{r,s}(n, m)$ or $S_{r,s}(X, Y)$. Readers are referred to [Gol-G, pp. 152–154] and [Lev59, pp. 47–55] for the constructions of $S_{r,s}(n, m)$ and $S_{r,s}(X, Y)$, respectively. However, the description of $S_{1,0}(n, n)$ is less involved (cf. [Lev59, p. 50]). In fact, a necessary and sufficient condition for a point $x \in \mathbf{R}^n$ to be in $S_{1,0}(f)$ translates into a necessary and sufficient condition for $(j^2 f)(0)$ to be in $S_{1,0}(n, n)$. Explicitly, the 1-jets of maps in $S_1(n, n)$ are the orbits of the 1-jets at 0 of $\{E(u, x) = u, F(u, x) = 0\}$, where $E = (E_1, \dots, E_{n-1})$ and $u = (u_1, \dots, u_{n-1})$. The set of 2-jets above this set is the orbit in $J^2(n, n)$ of $\{E(u, x) = u, F(u, x) = q(u, x)\}$, where q is a quadratic form in u and x. The equation defining $S_1(n, n)$ is $F_x = 0$ and the transversality of 1-jets to $S_1(n, n)$ is that the map $(u, x) \mapsto F_x(u, x)$ has rank 1 at 0. This is an openness condition on $S_1(n, n) \times D^2(n, n)$, *so* ${}_{\mathsf{T}}S_1(n, n)$ *is open in* $S_1(n, n) \times D^2(n, n)$. Also, note that $\pi_{2,1}^{-1}(S_1(n, n)) = S_1(n, n) \times D^2(n, n)$ and, for $f = (E, F)$, we have $\dim\left[d_0(f|_{{}_{\mathsf{T}}S_1(f)})\right] = n - 1 = \dim[{}_{\mathsf{T}}S_1(f)]$. The condition for the 2-jet to be in $S_{1,0}(n, n)$ is that $F_{xx} = q_{xx}$ is nonzero at 0. In other words, the 2-singularity $S_{1,0}(n, n)$ is given by strict inequalities in second partial derivatives. Hence $S_{1,0}(n, n)$ *is open in* $S_1(n, n) \times D^2(n, n)$.

We conclude this with expressions for $J^2(n,m)$ and $S_{r,s}(U,V)$ in terms of $S'_r(n,m)$. Given a submanifold $S(n,m)$ of $J^k(n,m)$, set

$$\begin{aligned} {}_{\mathrm{T}}S(n,m) \;=\; & \{j^{k+1}f(x) \in J^{k+1}(n,m) : j^k f(x) \in S(n,m) \\ & \text{and } j^k f \pitchfork S(n,m) \text{ at } x\}. \end{aligned}$$

Let $D_r(n,m) = S'_r(n,m) \backslash {}_{\mathrm{T}}S_r(n,m)$ and

$$D(n,m) = \dot{\bigcup}_{r=0}^{\ell} D_r(n,m),$$

where $\ell = \min\{n,m\}$. We then write $S'_r(n,m) = {}_{\mathrm{T}}S_r(n,m) \dot{\bigcup} D_r(n,m)$. Analogous to Proposition 7.39, we obtain

$$J^2(n,m) = \left[\dot{\bigcup}_{r,s} S_{r,s}(n,m)\right] \dot{\bigcup} D(n,m)$$

where $0 \le r \le \ell$ and $0 \le s \le (r + \max\{n-m, 0\})$. It follows that

$$S_{r,s}(U,V) \approx U \times V \times S_{r,s}(n,m). \tag{7.7}$$

7.4 Morse Theory

In 1925, Marston Morse introduced a beautiful and powerful subject now called *Morse Theory*. Broadly speaking, Morse theory studies the critical-point structure of real-valued maps. It gave impetus to the creation of an analogous theory for mappings between manifolds. Using tools from Sections 7.1, 7.2, and 7.3, we shall discuss the stability aspects of Morse theory, leaving the Morse inequalities for Chapter 10. Applications to gravitational lensing will appear in Section 8.2 and Chapters 11 to 13.

7.4.1 Nondegenerate Functions and Morse's Lemma

A critical point of a smooth function $f : U \subseteq \mathbf{R}^n \to \mathbf{R}$, where U is open, is a point $x \in U$ with grad $f(x) = 0$. The critical point x is *nondegenerate* if $\det[\text{Hess } f](x) \neq 0$. If all critical points of f are nondegenerate, then f is also called *nondegenerate* or *Morse*. The *index* of a nondegenerate critical point is i if $(\text{Hess } f)(x)$ has i negative eigenvalues. Morse was the first to emphasize the importance of nondegenerate critical points and their indices. In fact, the Morse Lemma completely characterizes a nondegenerate function in the vicinity of its critical points.

We shall formulate Morse's Lemma for functions on manifolds. Let $f \in C^\infty(X, \mathbf{R})$ and choose a chart $\phi : U \subseteq X \to \mathbf{R}^n$ about a point $x \in X$. The second derivative of $f \circ \phi^{-1}$:

$\phi(U) \subseteq \mathbf{R}^n \to \mathbf{R}$ at $\phi(x)$ is a symmetric bilinear form $d^2_{\phi(x)}(f \circ \phi^{-1}) : \mathbf{R}^n \times \mathbf{R}^n \to \mathbf{R}$ given by

$$d^2_{\phi(x)}(f \circ \phi^{-1})(v, w) = \sum_{i,j=1}^{n} v_i w_j \frac{\partial^2 (f \circ \phi^{-1})}{\partial x_i \partial x_j}(\phi(x)),$$

where $v = (v_1, \ldots, v_n)$ and $w = (w_1, \ldots, w_n)$. Here, $T_{\phi(x)}\mathbf{R}^n$ is identified with $\mathbf{R}^n$.

Let x be a critical point of f (i.e., $d_x f = 0$). Then the symmetric bilinear form $\operatorname{Hess}_x f : T_x X \times T_x X \to \mathbf{R}$ given by

$$\operatorname{Hess}_x f = \left(d^2_{\phi(x)}(f \circ \phi^{-1})\right) \circ (d_x\phi \times d_x\phi)$$

is called the *Hessian* of f at x. Using the formula for the second derivative of a composition of functions and the fact that x is a critical point, this definition can be shown to be independent of the chart ϕ (e.g., [Ber-G, p. 132]). The *Hessian matrix* of f at x relative to a basis $b_1, \ldots, b_n$ of $T_x X$ is the matrix with entries $(\operatorname{Hess}_x f)(b_i, b_j)$, where $1 \le i, j \le n$.

The *null space* of $\operatorname{Hess}_x f$ is the set of all vectors $v \in T_x X$ with $(\operatorname{Hess}_x f)(v, w) = 0$ for all $w \in T_x X$. If the null space of $\operatorname{Hess}_x f$ is the zero subspace, then $\operatorname{Hess}_x f$ is called *nondegenerate*. The critical point x is also termed *nondegenerate* (*Morse*) if $\operatorname{Hess}_x f$ is nondegenerate. We call f *nondegenerate* if all its critical points are nondegenerate.

If $\operatorname{Hess}_x f$ is nondegenerate, then there is an integer i, where $0 \le i \le n$, and a basis $e_1, \ldots, e_n$ of $T_x X$ such that:

(1) $\operatorname{Hess}_x f(e_\ell, e_k) = \varepsilon_\ell \delta_{\ell k}$, where $\varepsilon_\ell = -1$ if $0 \le \ell \le i$ and $\varepsilon_\ell = 1$ if $i < \ell \le n$.

(2) The integer i is independent of the choice of basis $e_1, \ldots, e_n$.

See [Cas-H, p. 10] and [Art, p. 245] for discussion (and proofs) of the above two results. The integer i is called the *index* of the critical point x (or *index* of $\operatorname{Hess}_x f$). Note that the index of x is the number of negative eigenvalues of the Hessian matrix of f at x relative to some basis of $T_x X$.

Morse's Lemma uses the index of a critical point to characterize nondegenerate functions:

Theorem 7.42 (Morse's Lemma) *Let $f \in C^\infty(X, \mathbf{R})$, where $\dim X = n$.*[7] *Then at each point x in X there is a chart $\xi : U \subseteq X \to \mathbf{R}^n$ centered at x such that if $\xi(p) = (\nu_1(p), \ldots, \nu_n(p)) \equiv (\nu_1, \ldots, \nu_n)$ for each $p \in U$, then:*

(1) *The point x is a regular point of f if and only if*

$$(f \circ \xi^{-1})(\nu_1, \ldots, \nu_n) = f(x) + \nu_1$$

in a neighborhood of the origin in $\mathbf{R}^n$.

[7] Unless stated to the contrary, the usual orientation of $\mathbf{R}$ is always assumed.

(2) *The point x is a nondegenerate critical point of f of index i if and only if*

$$(f \circ \xi^{-1})(\nu_1, \ldots, \nu_n) = f(x) - \nu_1^2 - \cdots - \nu_i^2 + \nu_{i+1}^2 + \cdots + \nu_n^2$$

in a neighborhood of the origin in $\mathbf{R}^n$.

Detailed proofs of Morse's Lemma are given in [Gol-G, pp. 65–68] and [Hir, pp. 145–147].

By the Morse lemma, a critical point of a nondegenerate function is one of $n+1$ types: minima ($i = 0$), maxima ($i = n$), or generalized saddle of index i with $1 \leq i \leq n-1$.

Corollary 7.43 *Each nondegenerate critical point of $f \in C^\infty(X, \mathbf{R})$ is isolated.*

Proof: Let x be a nondegenerate critical point of f. By the Morse Lemma, f at $x \in X$ is differentiably equivalent to $h = f \circ \xi^{-1}$ at $0 \in \mathbf{R}^n$. But there is a neighborhood N about 0 with 0 the only critical point of h. Since differentiable equivalence preserves critical points, there is only one critical point in $\phi^{-1}(N)$, namely, x.□

Corollary 7.44 *Let $f \in C^\infty(X, \mathbf{R})$ with C a compact subset of X. Suppose that each critical point of f in C is nondegenerate. Then f has finitely many critical points in C.*

Proof: Let $\mathrm{Crit}_C(f)$ be the set of all critical points of f in C. Since the elements of $\mathrm{Crit}_C(f)$ are isolated, the set $C - \mathrm{Crit}_C(f)$ is open in C. Now, for each critical point x, let U_x be a proper open set in C containing x and no other critical point. Then the collection

$$\left\{U_x \bigcup [C - \mathrm{Crit}_C(f)] : x \in \mathrm{Crit}_C(f)\right\}$$

forms an open cover of C. Compactness of C implies that this cover must have a finite subcover. The latter contains all of the open sets U_x. Hence $\mathrm{Crit}_C(f)$ is finite. □

By Corollary 7.44, *if $f \in C^\infty(X, \mathbf{R})$ is nondegenerate with X compact, then f has finitely many critical points.*

7.4.2 Stability of Functions

This section characterizes locally stable, transverse stable, and globally stable proper functions (see Theorem 7.46).

Before presenting the main theorem, we need a lemma. Let $\beta : J^k(X, \mathbf{R}) \to \mathbf{R}$ be the target map (i.e., $\beta(\sigma) = y$ for $\sigma \in J^k(X, \mathbf{R})_{(x,y)}$). Set

$$S_1^{[2]}(X, \mathbf{R}) = [S_1(X, \mathbf{R}) \times S_1(X, \mathbf{R})] \cap J_2^1(X, \mathbf{R}) \cap [(\beta \times \beta)^{-1}(\triangleright\mathbf{R})].$$

About each point of $J_2^1(X, \mathbf{R})$ there is an open set diffeomorphic to

$$N = U^{(2)} \times (\mathbf{R} \times \mathbf{R}) \times (J^1(n, 1) \times J^1(n, 1))$$

for some coordinate neighborhood $U \subseteq \mathbf{R}^n$, where $U^{(2)} = U \times U - \triangleright U$. Consequently, the space $S_1^{[2]}(X, \mathbf{R})$ is given locally by

$$S = U^{(2)} \times \triangleright\mathbf{R} \times (\{0\} \times \{0\}).$$

Since S is a submanifold of N, we can induce a submanifold structure on $S_1^{[2]}(X, \mathbf{R}) \subset J_2^1(X, \mathbf{R})$.

Lemma 7.45 *Let $f \in C^\infty(X, \mathbf{R})$. Then:*

(1) $j^1 f \pitchfork S_1(X, \mathbf{R})$ *if and only if f is nondegenerate.*

(2) $j_2^1 f \pitchfork S_1^{[2]}(X, \mathbf{R})$ *if and only if the critical points of f have distinct values.*

Proof: Suppose that $\dim X = n$. By Proposition 7.15 (page 224), it suffices to prove the first part of the lemma for $X = U$, where $U \subseteq \mathbf{R}^n$ is a coordinate neighborhood. For the second part of the lemma, we choose X to be a coordinate neighborhood $U \subseteq \mathbf{R}^n$ formed from a disjoint union of coordinate neighborhoods at a pair of distinct source points from X_1 and X_2 (see page 214).

(1) Consider $J^1(U, \mathbf{R})$, where U is a coordinate neighborhood. Since $J^1(n, 1) \approx \mathbf{R}^n$, we have

$$J^1(U, \mathbf{R}) \approx U \times \mathbf{R} \times \mathbf{R}^n.$$

First, we shall apply Proposition 7.7 with $X = U, Y = J^1(U, \mathbf{R})$, and $Q = S_1(U, \mathbf{R}) \approx U \times \mathbf{R} \times \{0\}$. A point of $J^1(U, \mathbf{R})$ is of the form $q = (x, f(x), \text{grad } f(x))$, where $x = (x_1, \ldots, x_n)$. Define $h : J^1(U, \mathbf{R}) \to \mathbf{R}^n$ by $h(x, f(x), w_1, \ldots, w_n) = [w_1 \cdots w_n]^T$, where $w_i = \frac{\partial f}{\partial x_i}$ for $i = 1, \ldots, n$. Notice that $d_q h = [\text{Hess } f \ \mathbf{0}_n \ I_n]$. Hence, $\text{rank}[d_q h] = n$. Since

$$J^1(U, \mathbf{R}) = S_0(U, \mathbf{R}) \dot{\bigcup} S_1(U, \mathbf{R}),$$

it follows that $q \in S_1(U, \mathbf{R})$ if and only if x is a critical point of $f \in C^\infty(U, \mathbf{R})$. Consequently, $h^{-1}(0) = S_1(U, \mathbf{R})$.

All x with $j^1 f(x) \notin S_1(U, \mathbf{R})$ are regular and yield $j^1 f \pitchfork S_1(U, \mathbf{R})$ at x. Suppose x is a critical point of $f \in C^\infty(U, \mathbf{R})$. Note that $j^1 f(x) = (x, f(x), w_1, \ldots, w_n)$. Since by Theorem 7.34, $\text{codim}_{J^1(U,\mathbf{R})} S_1(U, \mathbf{R}) = n$, Proposition 7.7 yields that $j^1 f \pitchfork S_1(U, \mathbf{R})$ at x if and only if $\text{rank}[d_x(h \circ j^1 f)] = n$. But $d_x(h \circ j^1 f) = (\text{Hess } f)(x)$. Hence, $j^1 f \pitchfork S_1(U, \mathbf{R})$ at a critical point x if and only if x is nondegenerate.

(2) For the coordinate neighborhood U mentioned at the start of the proof, we have

$$J_2^1(U, \mathbf{R}) \approx U^{(2)} \times (\mathbf{R} \times \mathbf{R}) \times (\mathbf{R}^n \times \mathbf{R}^n),$$

where $U^{(2)} = U \times U - \triangleright U$ and

$$S_1^{[2]}(U, \mathbf{R}) = U^{(2)} \times \triangleright\mathbf{R} \times \{0\} \times \{0\}.$$

An element of $J_2^1(U, \mathbf{R})$ is of the form

$$\tau = (x_1, x_2, f(x_1), f(x_2), \text{grad}\, f(x_1), \text{grad}\, f(x_2)).$$

Note that $\tau \in S_1^{[2]}(U, \mathbf{R})$ if and only if x_1 and x_2 are distinct critical points of f with the same value, $f(x_1) = f(x_2)$. Now, suppose $j_2^1 f \pitchfork S_1^{[2]}(U, \mathbf{R})$. Since

$$\text{codim}_{J_2^1(U,\mathbf{R})} S_1^{[2]}(U, \mathbf{R}) = 2n + 1 > \dim U = n,$$

Proposition 7.6 implies $j_2^1 f(U^{(2)}) \cap S_1^{[2]}(U, \mathbf{R}) = \emptyset$. Consequently, we have $f(x_1) \neq f(x_2)$ for distinct critical points x_1 and x_2. Conversely, if the latter holds, then $j_2^1 f(x_1, x_2) \notin S_1^{[2]}(U, \mathbf{R})$, i.e., $j_2^1 f \pitchfork S_1^{[2]}(U, \mathbf{R})$ at (x_1, x_2). □

Theorem 7.46 *Let $f \in C^\infty(X, \mathbf{R})$. Then:*

(1) *The function f is locally stable if and only if f is nondegenerate.*

(2) *The function f is transverse stable if and only if f is nondegenerate and has distinct critical values.*

(3) *If f is proper, then f is globally stable if and only if f is transverse stable.*

Proof: (1) Suppose f is stable at $x \in X$. This means that for any open set $U \subseteq X$ about x, there is a neighborhood $\mathfrak{F}$ of f in $C^\infty(X, \mathbf{R})$ such that for all $h \in \mathfrak{F}$, there is a point $x' \in U$ with h at x' differentiably equivalent to f at x. Lemma 7.45(1) yields that $j^1 f \pitchfork S_1(X, \mathbf{R})$ if and only if f is nondegenerate. Application of Thom's Transversality Theorem (Theorem 7.14) then implies that the set of nondegenerate functions in $C^\infty(X, \mathbf{R})$ is dense. Hence, there is a nondegenerate function h in $\mathfrak{F}$. In particular, h at x' is differentiably equivalent to f at x. Thus, x is either a regular point or nondegenerate critical point. Therefore, f is nondegenerate.

Conversely, assume f is nondegenerate. Then f is locally stable at all regular points in X (by Proposition 7.22). Now, suppose $x \in X$ is a critical point of f. Let $B_\epsilon(x)$ be an open ball of radius ϵ centered at x. Set

$$V^k_{\epsilon, B_\epsilon(x)}(f) = \left\{h \in C^\infty(X, \mathbf{R}) : |j^k f(x') - j^k h(x')| < \epsilon \text{ for all } x \in B_\epsilon(x)\right\}.$$

For k sufficiently large and ϵ sufficiently small, the values and partials in $B_\epsilon(x)$ up to order k of $h \in V^k_{\epsilon, B_\epsilon(x)}(f)$ differ by an arbitrarily small amount from those of $f|_{B_\epsilon(x)}$. Since x is a nondegenerate critical point, the latter shows that h has a nondegenerate critical point x' near x with the same index, say i, as x. By the Morse Lemma (Theorem 7.42), there is an open set U' about x' such that $h|_{U'}$ is differentiably equivalent to a quadratic of index i. But $f|_{B_\epsilon(x)}$ is also differentiably equivalent to this quadratic (again by Morse's Lemma). Hence, for k sufficiently large and ϵ sufficient small, h at x' is differentiably equivalent to f at x. Thus, the map-germ, $[f]_x$, of f at x is stable, i.e., f is stable at x.

(2) Corollary 7.15 implies that it is enough to prove part (2) of the theorem for $X = U$, where $U \subseteq \mathbf{R}^n$ is the coordinate neighborhood mentioned at the start of the proof of Lemma 7.45. Since we must apply Definition 7.27 (page 229), suppose that $\sigma = (\sigma_1, \sigma_2)$ is a diagonal element of $J_2^1(U, \mathbf{R})$, that is, $\text{target}(\sigma_1) \neq \text{target}(\sigma_2)$. Since

$$J_2^1(U, \mathbf{R}) \approx U^{(2)} \times (\mathbf{R} \times \mathbf{R}) \times (\mathbf{R}^n \times \mathbf{R}^n),$$

we can express σ in the form $\sigma = (j^1 f(x_1), j^1 f(x_2))$, where $x_1 \neq x_2$ and $f(x_1) = f(x_2)$, that is, $\sigma = j_2^1 f(x_1, x_2)$ with $f(x_1) = f(x_2)$.

Case I: $\text{corank}(\sigma_1) = \text{corank}(\sigma_2) = 0$.

This case is equivalent to the differentials $d_{x_1} f = \text{grad}\, f(x_1)$ and $d_{x_2} f = \text{grad}\, f(x_2)$ having full rank, i.e., x_1 and x_2 are regular points of f. Consequently,

$$\mathcal{D}_\sigma^2 \approx U^{(2)} \times \triangleright\mathbf{R} \times \mathbf{R}^n \times \mathbf{R}^n.$$

Suppose $j_2^1 f \pitchfork \mathcal{D}_\sigma^2$ at (x_1, x_2) with $j_2^1 f(x_1, x_2) \in \mathcal{D}_\sigma^2$. Then x_1 and x_2 are regular points. Conversely, if x_1 and x_2 are regular, then:

$$\begin{aligned} & d(j_2^1 f)[T_{(x_1,x_2)} U^{(2)}] + T_{j_2^1 f(x_1,x_2)}[\mathcal{D}_\sigma^2] \\ &= \mathbf{R}^n \times \mathbf{R}^n \times \mathbf{R} \times \mathbf{R} \times d_{x_1}(df)[\mathbf{R}^n] \times d_{x_2}(df)[\mathbf{R}^n] \\ &\qquad + \mathbf{R}^n \times \mathbf{R}^n \times T_{(f(x_1), f(x_2))}[\triangleright\mathbf{R}] \times \mathbf{R}^n \times \mathbf{R}^n \\ &= \mathbf{R}^n \times \mathbf{R}^n \times \mathbf{R} \times \mathbf{R} \times \mathbf{R}^n \times \mathbf{R}^n \\ &= T_{j_2^1 f(x_1,x_2)}[J_2^1(U, \mathbf{R})]. \end{aligned}$$

Hence, $j_2^1 f \pitchfork \mathcal{D}_\sigma^2$ at (x_1, x_2), where $(j_2^1 f)(x_1, x_2) \in \mathcal{D}_\sigma^2$. Thus, $j_2^1 f \pitchfork \mathcal{D}_\sigma^2$ at (x_1, x_2), where $(j_2^1 f)(x_1, x_2) \in \mathcal{D}_\sigma^2$, if and only if x_1 and x_2 are regular.

Case II: $\text{corank}(\sigma_1) = 1$ and $\text{corank}(\sigma_2) = 0$.

We have x_1 a critical point and x_2 regular. Consequently,

$$\mathcal{D}_\sigma^2 \approx U^{(2)} \times \triangleright\mathbf{R} \times \{0\} \times \mathbf{R}^n.$$

Suppose that $j_2^1 f \pitchfork \mathcal{D}_\sigma^2$ at (x_1, x_2), where $(j_2^1 f)(x_1, x_2) \in \mathcal{D}_\sigma^2$. Then:

$$\begin{aligned} & d(j_2^1 f)[T_{(x_1,x_2)} U^{(2)}] + T_{(j_2^1 f)(x_1,x_2)}[\mathcal{D}_\sigma^2] \\ &= \mathbf{R}^n \times \mathbf{R}^n \times \{0\} \times \mathbf{R} \times d_{x_1}(df)(\mathbf{R}^n) \times d_{x_2}(df)(\mathbf{R}^n) \\ &\quad + \mathbf{R}^n \times \mathbf{R}^n \times T_{(f(x_1), f(x_2))} \triangleright \mathbf{R} \times \{0\} \times \mathbf{R}^n \\ &= T_{(j_2^1 f)(x_1,x_2)} J_2^1(U, \mathbf{R}) \\ &= \mathbf{R}^n \times \mathbf{R}^n \times \mathbf{R} \times \mathbf{R} \times \mathbf{R}^n \times \mathbf{R}^n. \end{aligned}$$

Consequently, $d_{x_1}(df)(\mathbf{R}^n) = \mathbf{R}^n$, i.e., $(\text{Hess}\, f)(x_1)$ is invertible. Hence, the critical point x_1 is nondegenerate and x_2 is regular. Conversely, if x_1 is a nondegenerate critical point and

x_2 regular, then the above equalities show that $j_2^1 f \pitchfork \mathcal{D}_\sigma^2$ at (x_1, x_2), where $(j_2^1 f)(x_1, x_2) \in \mathcal{D}_\sigma^2$. Thus, $j_2^1 f \pitchfork \mathcal{D}_\sigma^2$ is equivalent to f being nondegenerate (i.e., $j^1 f \pitchfork S_1(U, \mathbf{R})$).

Case III: $\operatorname{corank}(\sigma_1) = \operatorname{corank}(\sigma_2) = 1$.

For this case, x_1 and x_2 are critical points, and

$$\mathcal{D}_\sigma^2 \approx S_1^{[2]}(U, \mathbf{R}) = U^{(2)} \times \triangleright\mathbf{R} \times \{\mathbf{0}\} \times \{\mathbf{0}\}.$$

By Lemma 7.45 (2), the condition $j_2^1 f \pitchfork S_1^{[2]}(U, \mathbf{R})$ is equivalent to f assigning distinct values to distinct critical points.

Cases I–III (where Case I is superfluous) show that f is transverse stable if and only if f is nondegenerate and has distinct critical values.

(3) This is a consequence of Mather's stability theorem (Theorem 7.31). □

7.5 Whitney Singularity Theory

In a seminal 1955 paper [Whi55], Hassler Whitney investigated the critical-point structure of maps between planes.[8] We shall present Whitney's theory using the methods of Sections 7.1, 7.2, and 7.3. Applications to gravitational lensing are given in Sections 8.2 and 9.1–9.4.

Throughout Section 7.5, assume that dim X = dim Y = 2.

7.5.1 Definition of Folds and Cusps

Given a general map between two dimensional manifolds, we shall decompose its set of critical points[9] into disjoint subsets with specific critical-point properties.

For $f \in C^\infty(X, Y)$, the set of critical points is

$$\operatorname{Crit}(f) = \{x \in X : \operatorname{rank}(d_x f) = 0 \text{ or } 1\}.$$

By definition,

$$S_r(f) = \{x \in X : \operatorname{corank}(d_x f) = r\},$$

where $\operatorname{corank}(d_x f) = 2 - \operatorname{rank}(d_x f)$. It follows that the set $S_0(f)$ consists of regular points, while $S_1(f)$ and $S_2(f)$ are, respectively, rank 1 and 0 critical points of f. The domain of f then decomposes as follows:

$$X = S_0(f) \dot{\bigcup} \operatorname{Crit}(f),$$

[8] See Arnold, Gusein-Zade and Varchenko[Arn-GVI, pp. 12–26] for a summary of Whitney's singularity theory. We note that Whitney also considered the case for maps $\mathbf{R}^n \to \mathbf{R}^m$, where $2 \le m \le 5$, $m = 2n - 2$, $m = 2n - 1$, and $m \ge 2n$ [Whi58].

[9] Section 6.2.1, page 176.

where

$$\mathrm{Crit}(f) = S_1(f) \dot{\bigcup} S_2(f).$$

The symbol $\dot{\bigcup}$ indicates disjoint union.

Decomposition of $S_1(f)$

Let us now split the rank 1 critical points $S_1(f)$ into a disjoint union of subsets with specific types of critical-point behavior. This will naturally lead to the notions of "folds" and "cusps." In fact, we shall introduce these notions in a coordinate independent way (see Section 9.4, page 367, for characterizations of folds and cusp in the plane via directional derivatives in rectangular coordinates).

Suppose that $S_1(f) \neq \emptyset$. Consider

$$\begin{aligned} {}_{\mathrm{T}}S_1(X, Y) &= \{j^2 f(x) \in J^2(X, Y) : j^1 f(x) \in S_1(X, Y) \\ &\qquad \text{and } j^1 f \pitchfork S_1(X, Y) \text{ at } x\} \end{aligned} \tag{7.8}$$

and

$${}_{\mathrm{T}}S_1(f) = (j^2 f)^{-1}[{}_{\mathrm{T}}S_1(X, Y)].$$

Elements of ${}_{\mathrm{T}}S_1(f)$ are called *transverse rank* 1 *critical points* of f. By (7.3), we have

$${}_{\mathrm{T}}S_1(f) = \{x \in S_1(f) : j^1 f \pitchfork S_1(X, Y) \text{ at } x\}.$$

If $j^1 f \pitchfork S_1(X, Y)$ at $p \in S_1(f)$, then there is a neighborhood V of p in X such that $V \bigcap S_1(f)$ is a 1-dimensional boundaryless submanifold of X (see Theorem 7.8). Consequently, *if* ${}_{\mathrm{T}}S_1(f) \neq \emptyset$, *then* ${}_{\mathrm{T}}S_1(f)$ *is diffeomorphic to a union of disjoint circles and/or real lines.* Write

$$S_1(f) = {}_{\mathrm{T}}S_1(f) \dot{\bigcup} {}_{\mathrm{T}}S_1^c(f),$$

where

$${}_{\mathrm{T}}S_1^c(f) = \{x \in S_1(f) : j^1 f \text{ is not transverse to } S_1(X, Y) \text{ at } x\}.$$

Decomposition of ${}_{\mathrm{T}}S_1(f)$

For ${}_{\mathrm{T}}S_1(f)$ nonempty, we obtain the further decomposition

$${}_{\mathrm{T}}S_1(f) = S_{1,0}(f) \dot{\bigcup} S_{1,1}(f).$$

Here

$$S_{1,0}(f) = \{x \in {}_{\mathrm{T}}S_1(f) : \mathrm{rank}[d_x(f|_{{}_{\mathrm{T}}S_1(f)})] = 1\}$$

$$S_{1,1}(f) = \{x \in {}_{\mathrm{T}}S_1(f) : \mathrm{rank}[d_x(f|_{{}_{\mathrm{T}}S_1(f)})] = 0\}.$$

If $S_{1,1}(f)$ is nonempty, then it decomposes as follows:

$$S_{1,1}(f) = {}_{\mathsf{T}}S_{1,1}(f) \dot{\bigcup} {}_{\mathsf{T}}S^c_{1,1}(f),$$

where

$$\begin{aligned} {}_{\mathsf{T}}S_{1,1}(f) &= \{x \in S_{1,1}(f) : j^2 f \pitchfork S_{1,1}(X, Y) \text{ at } x\}; \\ {}_{\mathsf{T}}S^c_{1,1}(f) &= \{x \in S_{1,1}(f) : j^2 f \text{ is not transverse to } S_{1,1}(X, Y) \text{ at } x\}. \end{aligned}$$

Note that

$$\begin{aligned} {}_{\mathsf{T}}S_{1,1}(X, Y) = \{ & j^3 f(x) \in J^3(X, Y) : j^2 f(x) \in S_{1,1}(X, Y) \\ & \text{and } j^2 f \pitchfork S_{1,1}(X, Y) \text{ at } x\}. \end{aligned} \tag{7.9}$$

It follows that

$${}_{\mathsf{T}}S_{1,1}(f) = (j^3 f)^{-1}[{}_{\mathsf{T}}S_{1,1}(X, Y)].$$

Hence

$${}_{\mathsf{T}}S_1(f) = S_{1,0}(f) \dot{\bigcup} {}_{\mathsf{T}}S_{1,1}(f) \dot{\bigcup} {}_{\mathsf{T}}S^c_{1,1}(f).$$

Folds and Cusps

The critical values of f will be called *caustic points.* The following definition divides critical and caustic points of f into three types:

Definition 7.47 Let $f \in C^\infty(X, Y)$, where $\dim X = \dim Y = 2$.

(1) Elements of $S_{1,0}(f)$ are called *fold critical points* of f. Points in $f[S_{1,0}(f)]$ are termed *fold caustic points.*

(2) Elements of ${}_{\mathsf{T}}S_{1,1}(f)$ are called *cusp critical points* of f. Points in $f[{}_{\mathsf{T}}S_{1,1}(f)]$ are termed *cusp caustic points.*

(3) A point of $\mathrm{Crit}(f)$ that is neither a fold nor cusp critical point is called a *degenerate critical point*, and the corresponding critical value is a *degenerate caustic point.*

Since $S_{1,0}(f) = (j^2 f)^{-1}[S_{1,0}(X, Y)]$ and ${}_{\mathsf{T}}S_{1,1}(f) = (j^3 f)^{-1}[{}_{\mathsf{T}}S_{1,1}(X, Y)]$, *the fold and cusp critical points of f involve conditions on the* 2- *and* 3-*jets of f, respectively* (cf. Theorem 7.25, page 229).

Notation: Set $\mathrm{Folds}(f) = S_{1,0}(f)$ and $\mathrm{Cusps}(f) = {}_{\mathsf{T}}S_{1,1}(f)$. The set of degenerate critical points will be denoted by $\mathrm{Degen}(f)$.[10]

[10] Some authors denote the set of cusp critical points of f by $S_{1,1,0}(f)$ or $S^2_1(f)$.

Remarks on Terminologies for Folds and Cusps

(1) Some authors call elements of $S_{1,1}$ *cusps*, while points of ${}_{\top}S_{1,1}$ are termed *regular* (or, *simple*) *cusps*. We avoid the former usage since elements of $S_{1,1}$ may not look like cusps. For example, consider the map $f : \mathbf{R}^2 \to \mathbf{R}^2$ defined by

$$f(\mathrm{u}, \mathrm{v}) = (\mathrm{u}, \mathrm{uv}).$$

Then

$$S_1(f) = (\text{v-axis}) = {}_{\top}S_1(f).$$

[The rightmost equality follows from Proposition 9.6.] Since $f|_{{}_{\top}S_1(f)} \equiv 0$, we have

$$S_{1,1}(f) = (\text{v-axis}).$$

(2) Mathematicians often refer to fold and cusp critical points simply as *fold* and *cusp points*, respectively. However, astrophysicists often use the terms *fold* and *cusps points* to refer to fold and cusp caustic points. For these reasons, we shall abide by the terminologies in Definition 7.47.

(3) In singularity theory literature, one will also find that fold and cusp critical points are called *transverse folds* and *transverse cusps*, respectively. This refers to the transversality conditions used to define them.

Decomposition of Crit(**f**)

We now have the following decomposition of the set of critical points of f:

$$\mathrm{Crit}(f) = \mathrm{Folds}(f) \dot{\bigcup} \mathrm{Cusps} \dot{\bigcup} \mathrm{Degen}(f),$$

where

$$\mathrm{Degen}(f) = {}_{\top}S^c_{1,1}(f) \dot{\bigcup} {}_{\top}S^c_1(f) \dot{\bigcup} S_2(f)$$

are the degenerate critical points of f. Elements of $\mathrm{Degen}(f)$ will become important when we consider 1-parameter evolutions of caustics of f (see Section 9.5, page 375).

7.5.2 Characterizations of Folds and Cusps

Whitney's Forms for Folds and Cusps

The sections classifies folds and cusps in terms of local curvilinear coordinate expressions known as Whitney's "folds" and "pleats."

Lemma 7.48 *Let f be in $C^\infty(X, Y)$, where* $\dim X = \dim Y = 2$, *and denote the origin of* $\mathbf{R}^2$ *by* $\mathbf{0}$.

(1) *A point x of X is in $S_1(f)$ if and only if there are charts $\lambda : U \subseteq X \to \mathbf{R}^2$ and $\nu : V \subseteq Y \to \mathbf{R}^2$ centered at x and $f(x)$, respectively, such that if $\lambda(p) = (\mathrm{u}, \mathrm{v})$ and $\nu(q) = (E, F)$ for $p \in U$ and $q \in V$, then the map $h = \nu \circ f \circ \lambda^{-1}$ is given by*

$$\begin{cases} E = \mathrm{u} \\ F = C(\mathrm{u}, \mathrm{v}), \end{cases}$$

where $C(\mathrm{u}, 0) = 0$, and obeys $C_{\mathrm{v}}(\mathbf{0}) = 0$. In other words:
$x \in S_1(f) \iff \mathbf{0} \in S_1(h) \iff C_{\mathrm{v}}(\mathbf{0}) = 0$.

(2) $x \in {}_TS_1(f) \iff \mathbf{0} \in {}_TS_1(h) \iff C_{\mathrm{v}}(\mathbf{0}) = 0,\ [C_{\mathrm{vu}}(\mathbf{0}), C_{\mathrm{vv}}(\mathbf{0})] \neq \mathbf{0}$.

(3) **Fold**: *The point x is a fold critical point of f* (i.e., $x \in S_{1,0}(f)$)
$\iff \mathbf{0} \in S_{1,0}(h) \iff C_{\mathrm{v}}(\mathbf{0}) = 0$ *and* $C_{\mathrm{vv}}(\mathbf{0}) \neq 0$.

(4) $x \in S_{1,1}(f) \iff \mathbf{0} \in S_{1,1}(h) \iff C_{\mathrm{v}}(\mathbf{0}) = C_{\mathrm{vv}}(\mathbf{0}) = \mathbf{0}$ *and* $C_{\mathrm{vu}}(\mathbf{0}) \neq 0$.

(5) **Cusp:** *The point x is a cusp critical point of f* (i.e., $x \in {}_TS_{1,1}(f)$)
$\iff \mathbf{0} \in {}_TS_{1,1}(h) \iff C_{\mathrm{v}}(\mathbf{0}) = C_{\mathrm{vv}}(\mathbf{0}) = 0,\ C_{\mathrm{vu}}(\mathbf{0}) \neq 0$, *and* $C_{\mathrm{vvv}}(\mathbf{0}) \neq 0$.

Proof: (1) Let $x \in S_1(f)$. Choose charts $\lambda' : U' \subseteq X \to \mathbf{R}^2$ and $\nu' : V' \subseteq Y \to \mathbf{R}^2$ centered at x and $f(x)$, resp. Write $\lambda'(p) = (\mathrm{u}', \mathrm{v}')$ and $\nu'(q) = (E', F')$ for $p \in U'$ and $q \in V'$. Set $h' = \nu' \circ f \circ (\lambda')^{-1}$ and

$$\begin{cases} E' = A(\mathrm{u}', \mathrm{v}') \\ F' = B(\mathrm{u}', \mathrm{v}'). \end{cases}$$

Then

$$\operatorname{Jac} h' = \begin{bmatrix} A_{\mathrm{u}'} & A_{\mathrm{v}'} \\ B_{\mathrm{u}'} & B_{\mathrm{v}'} \end{bmatrix}$$

has rank 1 at $\mathbf{0}$ (since $x \in S_1(f)$). Since some entry of $\operatorname{Jac} h'(\mathbf{0})$ is nonzero, say, $A_{\mathrm{u}'}(\mathbf{0}) \neq 0$. Introduce new coordinates in the domain of h':

$$\begin{cases} \tilde{\mathrm{u}} = A(\mathrm{u}', \mathrm{v}') \\ \tilde{\mathrm{v}} = \mathrm{v}'. \end{cases}$$

Note that since $A_{\mathrm{u}'}(0, 0) \neq 0$, the Jacobian matrix of the above transformation has full rank (hence, Inverse Function Theorem applies). Also, the above new coordinates are centered at the origin. Expressing u' in terms of $\tilde{\mathrm{u}}$ and v', we get

$$\begin{cases} E' = A(\mathrm{u}', \mathrm{v}') = \tilde{\mathrm{u}} \\ F' = B(\mathrm{u}', \mathrm{v}') = \tilde{B}(\tilde{\mathrm{u}}, \mathrm{v}'). \end{cases}$$

The Jacobian matrix of this transformation relative to $(\tilde{\mathrm{u}}, \mathrm{v}')$ is

$$\begin{bmatrix} 1 & 0 \\ \tilde{B}_{\tilde{\mathrm{u}}} & \tilde{B}_{\mathrm{v}'} \end{bmatrix}.$$

Because $\operatorname{rank}[\operatorname{Jac} h'](\mathbf{0}) = 1$, we obtain $\tilde{B}_{\mathrm{v}'}(\mathbf{0}) = 0$. Change coordinates in the target of h':

$$\begin{cases} \tilde{E} = E' \\ \tilde{F} = F' - \tilde{B}(E', 0). \end{cases}$$

Set

$$C(\tilde{\mathrm{u}}, \mathrm{v}') = F'(\tilde{\mathrm{u}}, \mathrm{v}') - \tilde{B}(E', 0) = \tilde{B}(\tilde{\mathrm{u}}, \mathrm{v}') - \tilde{B}(\tilde{\mathrm{u}}, 0).$$

Dropping the tildes and setting $v = v'$ yield

$$\begin{cases} E = \mathrm{u} \\ F = C(\mathrm{u}, \mathrm{v}), \end{cases} \tag{7.10}$$

where $C(\mathrm{u}, 0) = 0$ and $C_{\mathrm{v}}(\mathbf{0}) = 0$. Notice that since $C(\mathrm{u}, 0) = 0$, each representative of the k-jet of the map germ of C at $(0, 0)$ has no Taylor terms purely in u.

We shall denote the charts that transform f to the form in (7.10) by $\lambda : U \subseteq X \to \mathbf{R}^2$ and $\nu : V \subseteq Y \to \mathbf{R}^2$, where λ and ν are centered at x and $f(x)$, respectively. Hence, $(\nu \circ f \circ \lambda^{-1})(\mathrm{u}, \mathrm{v}) = (E, F)$. *Setting* $k = 1 = s$ *in Proposition 7.15* (page 224), *we see that it suffices to prove the rest of the lemma for* $h = \nu \circ f \circ \lambda^{-1}$ *at* $\mathbf{0}$. Indeed, differentiable equivalence preserves critical points, rank, and transverasality. Hence, we have $x \in S_1(f) \iff \mathbf{0} \in S_1(h)$; $x \in {}_{\mathrm{T}}S_1(f) \iff \mathbf{0} \in {}_{\mathrm{T}}S_1(h)$; $x \in S_{1,0}(f) \iff \mathbf{0} \in S_{1,0}(h)$.

(2) Consider the charts $\lambda : U \subseteq X \to \mathbf{R}^2$ and $\nu : V \subseteq Y \to \mathbf{R}^2$ at the end of part (1) of the proof. As noted then, it suffices to prove part (2) of the lemma for $h = \nu \circ f \circ \lambda^{-1}$ at $\mathbf{0} \in S_1(h)$. Setting $\lambda(U) = U'$, $\nu(V) = V'$, and using the diffeomorphism

$$J^1(U', \mathbf{R}^2) \approx U' \times V' \times \mathbf{R}^4,$$

let us employ Proposition 7.7 (page 218) with $X = U'$, $Y = J^1(U', \mathbf{R}^2)$, $Q = S_1(U', \mathbf{R}^2) \approx U' \times V' \times S_1(2, 2)$, and $y = j^1h(\mathbf{0})$. A point of $J^1(U', \mathbf{R}^2)$ is of the form $\sigma = (s, t(s), w_{11}, w_{12}, w_{21}, w_{22})$, where $s = (s_1, s_2)$ and $w_{ij} = \frac{\partial t_i}{\partial s_j}$ with $t = (t_1, t_2)$ a representative of σ. Note that by part (1) of the proof we have $y = j^1h(\mathbf{0}) = (\mathbf{0}, \mathbf{0}, 1, 0, C_{\mathrm{u}}(\mathbf{0}), 0)$ (since $C_{\mathrm{v}}(\mathbf{0}) = 0$).

Choose an open set $\mathcal{V} \subseteq J^1(U', \mathbf{R}^2)$ about y such that all $\sigma \in \mathcal{V}$ have $w_{11} \neq 0$. [This is always possible for $\mathcal{V}$ sufficiently small since the w_{11}-component of y is nonzero.] Define a function

$$\mathcal{H} : \mathcal{V} \subseteq J^1(U', \mathbf{R}^2) \to \mathbf{R}$$

by

$$\mathcal{H}(\sigma) = w_{22} - w_{21} w_{11}^{-1} w_{12}.$$

Note that $\mathcal{H}(j^1 h(\mathbf{0})) = 0$. Then

$$d_y \mathcal{H} = [\mathbf{0},\ \mathbf{0},\ w_{21} w_{11}^{-2} w_{12},\ -w_{21} w_{11}^{-1},\ -w_{11}^{-1} w_{12},\ 1].$$

Hence, $\operatorname{rank}[d_y \mathcal{H}] = 1$. Now, all $\sigma \in g^{-1}(0)$ are such that $w_{11} \neq 0$ and

$$\det[d_{\mathbf{0}} t] = w_{11} w_{22} - w_{12} w_{21} = w_{11}(w_{21} w_{11}^{-1} w_{12}) - w_{12} w_{21} = 0,$$

i.e., $\operatorname{rank}[d_{\mathbf{0}} t] = 1$. Thus, $\mathcal{H}^{-1}(0) = S_1(U', \mathbf{R}^2) \cap \mathcal{V}$ (which contains $y = j^1 h(\mathbf{0})$).

Consider the composition $\mathcal{H} \circ (j^1 h) : (j^1 h)^{-1}(\mathcal{V}) \to \mathbf{R}$, which is given by $(\mathcal{H} \circ j^1 h)(s) = C_{\mathrm{v}}(s)$. Since $\operatorname{codim}_{J^1(U', \mathbf{R}^2)} S_1(U', \mathbf{R}^2) = 1$ and $j^1 h(\mathbf{0}) \in S_1(U', \mathbf{R}^2) \cap \mathcal{V}$, Proposition 7.7 (page 218) yields that $j^1 h \pitchfork S_1(U', \mathbf{R}^2)$ at $\mathbf{0}$ if and only if $\operatorname{rank}[d_{\mathbf{0}}(\mathcal{H} \circ j^1 h)] = 1$. But

$$d_{\mathbf{0}}(\mathcal{H} \circ j^1 h) = [C_{\mathrm{vu}}(\mathbf{0}),\ C_{\mathrm{vv}}(\mathbf{0})].$$

Thus, $\operatorname{rank}[d_{\mathbf{0}}(\mathcal{H} \circ j^1 h)] = 1$ if and only if $[C_{\mathrm{vu}}(\mathbf{0}), C_{\mathrm{vv}}(\mathbf{0})] \neq \mathbf{0}$.

(3) Since ${}_{\mathrm{T}}S_1(h)$ is the smooth curve given by $C_{\mathrm{v}}(\mathrm{u}, \mathrm{v}) = 0$ with normal vector field $(C_{\mathrm{vu}},\ C_{\mathrm{vv}}) \neq \mathbf{0}$, the vector field $Z = (-C_{\mathrm{vv}},\ C_{\mathrm{vu}})$ is tangent to (and nonvanishing on) ${}_{\mathrm{T}}S_1(h)$. Now recall that

$$S_{1,0}(h) = \{s \in {}_{\mathrm{T}}S_1(h) :\ \operatorname{rank}[d_s(h|_{{}_{\mathrm{T}}S_1(h)})] \ = \ 1\}.$$

Since $\dim[{}_{\mathrm{T}}S_1(h)] = 1$ and Z is tangent to ${}_{\mathrm{T}}S_1(h)$, it suffices to consider $d_s(h|_{{}_{\mathrm{T}}S_1(h)})$ at the vector Z_s. Moreover, $d_s(h|_{{}_{\mathrm{T}}S_1(h)})$ coincides with $d_s h$ on vectors tangent to ${}_{\mathrm{T}}S_1(h)$ at s. At points s of ${}_{\mathrm{T}}S_1(h)$, we have

$$\begin{aligned} d_s h(Z_s) &= [-C_{\mathrm{vv}}(s),\ -C_{\mathrm{u}}(s)\, C_{\mathrm{vv}}(s) + C_{\mathrm{v}}(s)\, C_{\mathrm{vu}}]^T \\ &= [-C_{\mathrm{vv}}(s),\ -C_{\mathrm{u}}(s)\, C_{\mathrm{vv}}(s)]^T. \end{aligned}$$

Hence, $\operatorname{rank}[d_s(h|_{{}_{\mathrm{T}}S_1(h)})] = 1$ if and only if $C_{\mathrm{vv}}(s) \neq 0$.

(4,5) Proposition 7.7 (page 218) will be applied again. Consider

$$J^2(U', \mathbf{R}^2) \approx U' \times V' \times \mathbf{R}^{10}$$

and

$${}_{\mathrm{T}}S_{1,1}(h) = \{s \in S_{1,1}(h) :\ j^2 h \pitchfork S_{1,1}(U', \mathbf{R}^2) \text{ at } s\},$$

where

$$S_{1,1}(h) = \{s \in {}_{\mathrm{T}}S_1(h) :\ \operatorname{rank}[d_s(h|_{{}_{\mathrm{T}}S_1(h)})] \ = \ 0\}.$$

Write the elements τ of $J^2(U', \mathbf{R}^2)$ as

$$\tau = (s,\ t(s),\ w_1^1,\ w_2^1,\ w_1^2,\ w_2^2,\ z_{11}^1,\ z_{12}^1,\ z_{22}^1,\ z_{11}^2,\ z_{12}^2,\ z_{22}^2),$$

where $s = (s_1, s_2)$, $t = (t_1, t_2)$ is a representative of τ, $w^i_j = \frac{\partial t_i}{\partial x_j}$, and $z^i_{jk} = \frac{\partial^2 t_i}{\partial x_j \partial x_k}$. Redundancies due to symmetry are taken into account in the above expression for τ. The map $j^2 h : U' \to J^2(U', \mathbf{R}^2)$ is given by

$$j^2 h(s) = (s, h(s),\ \ 1, 0, C_{\mathrm{u}}(s), C_{\mathrm{v}}(s),\ \ 0, 0, 0, C_{\mathrm{uu}}(s), C_{\mathrm{uv}}(s), C_{\mathrm{vv}}(s)).$$

Define a map $\mathcal{H} : J^2(U', \mathbf{R}^2) \to \mathbf{R}^2$ by $\mathcal{H}(\tau) = (w^2_2, z^2_{22})$. Set $y = j^2 h(\mathbf{0})$. Then

$$d_y \mathcal{H} = [\mathbf{O}_{2\times 5}\ \hat{\mathbf{i}}\ \mathbf{O}_{2\times 7}\ \hat{\mathbf{j}}],$$

where $\mathbf{O}_{n\times m}$ is the n by m zero matrix, $\hat{\mathbf{i}} = [1, 0]^T$, and $\hat{\mathbf{j}} = [0, 1]^T$. Hence, $\operatorname{rank}[d_y h] = 2$.

By end of part (3), we have $\operatorname{rank}[d_s(h|_{{}_T S_1(h)})] = 0$ if and only if $C_{\mathrm{vv}}(s) = 0$. In other words, elements of $S_{1,1}(h)$ are determined by the conditions $(C_{\mathrm{v}}(s), C_{\mathrm{vv}}(s)) = \mathbf{0}$ and $C_{\mathrm{vu}}(s) \neq 0$. Consequently, $\mathbf{0} \in S_{1,1}(h) \iff \mathcal{H}(j^2 h(\mathbf{0})) = \mathbf{0}$ and $C_{\mathrm{vu}}(\mathbf{0}) \neq 0$. In addition, $\mathcal{H}^{-1}(\mathbf{0}) = S_{1,1}(U', \mathbf{R}^2) \cap \mathcal{V}$ for some (sufficiently small) neighborhood $\mathcal{V} \subseteq J^2(U', \mathbf{R}^2)$ about $j^2 h(\mathbf{0})$. Now, the composition $\mathcal{H} \circ (j^2 h) : U' \to \mathbf{R}^2$ is given by $\mathcal{H} \circ (j^2 h)(s) = (C_{\mathrm{v}}(s), C_{\mathrm{vv}}(s))$. Since $\operatorname{codim}_{J^2(U',\mathbf{R}^2)} S_{1,1}(U', \mathbf{R}^2) = 2$ (by Proposition 7.41) and $\mathbf{0} \in S_{1,1}(h) \iff j^2 h(\mathbf{0}) \in \mathcal{H}^{-1}(\mathbf{0})$ and $C_{\mathrm{vu}}(\mathbf{0}) \neq 0$, Proposition 7.7 implies that for $\mathbf{0} \in S_{1,1}(h)$, we have $j^2 h \pitchfork S_{1,1}(U', \mathbf{R}^2)$ at $\mathbf{0}$ if and only if $\operatorname{rank}[d_{\mathbf{0}}(\mathcal{H} \circ j^2 h)] = 2$. But

$$d_{\mathbf{0}}(\mathcal{H} \circ j^2 h) = \begin{bmatrix} C_{\mathrm{vu}}(\mathbf{0}) & C_{\mathrm{vv}}(\mathbf{0}) \\ C_{\mathrm{vvu}}(\mathbf{0}) & C_{\mathrm{vvv}}(\mathbf{0}) \end{bmatrix}.$$

Since $\mathbf{0} \in {}_T S_1(h) \iff C_{\mathrm{v}}(\mathbf{0}) = 0$ and $[C_{\mathrm{vu}}(\mathbf{0}), C_{\mathrm{vv}}(\mathbf{0})] \neq \mathbf{0}$, we obtain $\mathbf{0} \in S_{1,1}(h) \iff C_{\mathrm{v}}(\mathbf{0}) = 0 = C_{\mathrm{vv}}(\mathbf{0})$ and $C_{\mathrm{vu}}(\mathbf{0}) \neq 0$. Hence, $\mathbf{0} \in {}_T S_{1,1}(h) \iff \mathbf{0} \in S_{1,1}(h)$ and $\operatorname{rank}[d_{\mathbf{0}}(\mathcal{H} \circ j^2 h)] = 2 \iff C_{\mathrm{v}}(\mathbf{0}) = 0 = C_{\mathrm{vv}}(\mathbf{0})$, $C_{\mathrm{vu}}(\mathbf{0}) \neq 0$, and $C_{\mathrm{vvv}}(\mathbf{0}) \neq 0$. □

Theorem 7.49 *Let f be in $C^\infty(X, Y)$, where* $\dim X = \dim Y = 2$. *At each point x in X there are charts $\lambda : U \subseteq X \to \mathbf{R}^2$ and $\nu : V \subseteq Y \to \mathbf{R}^2$ centered at x and $f(x)$, respectively, such that if $\lambda(p) = (\mathrm{u}, \mathrm{v})$ for $p \in U$, then:*

(1) (Regular Point) *The point x is a regular point of f if and only if*

$$(\nu \circ f \circ \lambda^{-1})(\mathrm{u}, \mathrm{v}) = (\mathrm{u}, \mathrm{v})$$

in a neighborhood of the origin in $\mathbf{R}^2$.

(2) (Whitney Fold) *The point x is a fold critical point of f if and only if*

$$(\nu \circ f \circ \lambda^{-1})(\mathrm{u}, \mathrm{v}) = (\mathrm{u}, d\mathrm{v}^2)$$

in a neighborhood of the origin in $\mathbf{R}^2$ with $|d| = 1$.

(3) (Whitney Pleat/Cusp) *The point x is a cusp critical point of f if and only if*

$$(\nu \circ f \circ \lambda^{-1})(\mathrm{u}, \mathrm{v}) = (\mathrm{u}, e\mathrm{v}^3 + d\,\mathrm{uv})$$

in a neighborhood of the origin in $\mathbf{R}^2$ *with* $|d| = |e| = 1$.

Applications of Theorem 7.49 to gravitational lensing appear in Section 9.1 (page 328). The latter also pictorially represent the action of Whitney folds and pleats.

Proof of Theorem 7.49: (1) Let $\lambda' : U \subseteq X \to \mathbf{R}^2$ and $\nu : V \subseteq Y \to \mathbf{R}^2$ be charts centered at x and $f(x)$, resp. Set $\lambda'(U) = U'$, $\nu(V) = V'$, and $h = \nu \circ f \circ \lambda'^{-1}$. Note that $h : U' \to V'$ satisfies $h(\mathbf{0}) = \mathbf{0}$ and $d_{\mathbf{0}}h$ is invertible. (since x is a regular point). By the Inverse Function Theorem, the mapping h diffeomorphically maps an open set about $\mathbf{0} \in U'$ onto an open set about $\mathbf{0} \in V'$. Then $(h \circ (h|_{U''}))(\mathrm{u}, \mathrm{v}) = (\mathrm{u}, \mathrm{v})$. Hence, $\nu \circ f \circ \lambda^{-1}(\mathrm{u}, \mathrm{v}) = (\mathrm{u}, \mathrm{v})$, where $\lambda = h|_{U''} \circ \lambda'$ is a chart centered at x.

(2) Choose coordinates as in Lemma 7.48. Then $h = \nu \circ f \circ \lambda^{-1}$ is defined about $\mathbf{0} \in {}_{\mathrm{T}}S_1(h)$ by

$$\begin{cases} E = \mathrm{u} \\ F = C(\mathrm{u}, \mathrm{v}), \end{cases}$$

where $C(\mathrm{u}, 0) = 0$ and $C_{\mathrm{v}}(\mathbf{0}) = 0$. Moreover, the set ${}_{\mathrm{T}}S_1(h)$ is given by $C_{\mathrm{v}} = 0$ and $(C_{\mathrm{vu}}, C_{\mathrm{vv}}) \neq \mathbf{0}$. By Lemma 7.48(3), a point x is a fold critical point of f if and only if $C_{\mathrm{v}}(\mathbf{0}) = 0$ and $C_{\mathrm{vv}}(\mathbf{0})) \neq 0$. Since $C_{\mathrm{vv}}(\mathbf{0})) \neq 0$, the pair $(\mathrm{u}, \hat{\mathrm{v}}) = (\mathrm{u}, C_{\mathrm{v}}(\mathrm{u}, \mathrm{v}))$ defines new coordinates in a neighborhood of $\mathbf{0}$ in the domain of h. In terms of these new coordinates, we have

$$\begin{cases} E = \mathrm{u} \\ F = B(\mathrm{u}, \hat{\mathrm{v}}), \end{cases}$$

where $B(\mathrm{u}, 0) = 0$ and $B_{\hat{\mathrm{v}}}(\mathrm{u}, \hat{\mathrm{v}}) = 0$ on ${}_{\mathrm{T}}S_1(h)$. We have that the coordinate, $\hat{\mathrm{v}} = C_{\mathrm{v}}(\mathrm{u}, \mathrm{v})$, vanishes on ${}_{\mathrm{T}}S_1(h)$. Since $B_{\hat{\mathrm{v}}}(\mathrm{u}, 0) = 0$, we see that $\hat{\mathrm{v}}$ is a factor, i.e., $B_{\hat{\mathrm{v}}}(\mathrm{u}, \hat{\mathrm{v}}) = \hat{\mathrm{v}}A(\mathrm{u}, \hat{\mathrm{v}})$. In addition:

$$\begin{aligned} B(\mathrm{u}, \hat{\mathrm{v}}) &= B_{\hat{\mathrm{v}}}(\mathrm{u}, \hat{\mathrm{v}}) - B_{\hat{\mathrm{v}}}(\mathrm{u}, 0) \\ &= \int_0^1 \frac{d}{dt} B(\mathrm{u}, t\hat{\mathrm{v}})dt \\ &= \hat{\mathrm{v}} \int_0^1 B_{\hat{\mathrm{v}}}(\mathrm{u}, t\hat{\mathrm{v}})dt \\ &= \hat{\mathrm{v}}^2 \int_0^1 A(\mathrm{u}, t\hat{\mathrm{v}})dt. \end{aligned}$$

Set $B(\mathrm{u}, \hat{\mathrm{v}}) = r\hat{v}^2 G(\mathrm{u}, t\hat{\mathrm{v}})$, where $G(\mathrm{u}, t\hat{\mathrm{v}}) = \int_0^1 A(\mathrm{u}, t\hat{\mathrm{v}})dt$. Notice that

$$G(\mathrm{u}, 0) = \int_0^1 A(\mathrm{u}, 0)dt = A(\mathrm{u}, 0) = B_{\hat{\mathrm{v}}\hat{\mathrm{v}}}(\mathrm{u}, 0) \neq 0.$$

The latter holds since $S_1(h)$ is the curve (u, 0) in the (u, $\hat{v}$)–plane and $C_{vv}(u, v) \neq 0$ on a neighborhood of $\mathbf{0} \in {}_TS_1(h)$ (by Lemma 7.48(3)). Since $G(u, 0) \neq 0$, continuity of G implies that G is either positive or negative in a sufficiently small neighborhood of the origin.

Define new coordinates in the domain of h by $(u, \tilde{v}) = (u, \hat{v}\sqrt{|G(u, \hat{v})|})$. Then

$$\begin{cases} E = u \\ F = \frac{G(u,\hat{v})}{|G(u,\hat{v})|}\tilde{v} = \pm\tilde{v}. \end{cases}$$

Dropping the tilde notationally yields the desired result.

(3) By Lemma 7.48, the map f at x is differentiably equivalent to the following map at $\mathbf{0} \in \mathbf{R}^2$:

$$\begin{cases} E = u \\ F = C(u, v), \end{cases}$$

where $C(0) = C_v(\mathbf{0}) = C_{vv}(\mathbf{0}) = 0$, $C_{vu}(\mathbf{0}) \neq 0$, and $C_{vvv}(\mathbf{0}) \neq 0$. Set $c(v) = C(0, v)$. Since $c(0) = C_v(0) = C_{vv}(0)$, we have

$$c(v) = v^3 d(v),$$

where $d(0) \neq 0$ (because $C_{vvv} \neq 0$). Let $\phi(v) = v[d(v)]^{1/3}$. Then $\phi_v(0) \neq 0$ (since $d(0) \neq 0$) and ϕ is a local diffeomorphism at the origin. By continuity, we have $d(v) \neq 0$ in a neighborhood of the origin. Consequently,

$$c(v) = \left(\frac{\phi(v)}{[d(v)]^{1/3}}\right)^3 d(v) = [\phi(v)]^3.$$

Hence, for $v' = \phi(v)$ we get

$$c \circ \phi^{-1}(v') = (v')^3.$$

Set $\hat{f} = (E, F)$. In terms of v', we get

$$\hat{f}(u, v') = (u, \hat{C}(v', u)),$$

where $\hat{C}(v', u) = C(u, \phi^{-1}(v'))$ and $\hat{C}(v', 0) = (v')^3$.

Define an *r-parameter unfolding* of a germ $f : (\mathbf{R}, 0) \to (\mathbf{R}, f(0))$ to be a germ $F : (\mathbf{R} \times \mathbf{R}^r, (0, 0)) \to (\mathbf{R}, F(0, 0))$, where $(x, y) \mapsto F(x, y)$ and $F(x, 0) = f(x)$ for all x in a neighborhood of $0 \in \mathbf{R}$. For simplicity, we shall also denote the previous germs by $[f]$ and $[F]$, respectively. It follows that we can treat $\hat{C}(v', ru)$ as a 1-parameter unfolding of v'^3.

An s-parameter unfolding $G : (\mathbf{R} \times \mathbf{R}^s, (0,0)) \to (\mathbf{R}, F(0,0))$ is *induced by* $[F]$ if there are germs

$$\begin{aligned} \varphi &: (\mathbf{R} \times \mathbf{R}^s, 0) \to (\mathbf{R}, 0), \\ h &: (\mathbf{R}^s, 0) \to (\mathbf{R}^r, 0), \\ g &: (\mathbf{R}^s, 0) \to (\mathbf{R}, 0), \\ \lambda &: (\mathbf{R} \times \mathbf{R}^s, (0,0)) \to (\mathbf{R} \times \mathbf{R}^s, (0,0)) \end{aligned}$$

such that

$$G(x,t) = F(\lambda(x, h(t))) + g(t),$$

where λ is a local diffeomorphism at $(0,0) \in \mathbf{R} \times \mathbf{R}$ with

$$\lambda(x,t) = (\varphi(x,t), t), \qquad \varphi(x,0) = x.$$

Let $\pi : \mathbf{R} \times \mathbf{R}^s \to \mathbf{R}^s$ be the projection $\pi(x,t) = t$. Then $\pi = \pi \circ \lambda$ and we can express G as follows:

$$G = (h^* F + g \circ \pi) \circ \lambda,$$

where the *pull-back* $h^* F$ is defined by $(h^* F)(x,t) = F(x, h(t))$. The unfolding $[F]$ of $[f]$ is said to be *versal* if all other unfoldings of $[f]$ are induced by $[F]$. If r is the smallest integer for which a versal unfolding of $[f]$ exists, then $[F]$ is a *universal unfolding* of $[f]$.

It can be shown that

$$\mathrm{U}(v', \mathrm{u}) = (v')^3 + \mathrm{u}v'$$

is a universal unfolding of $(v')^3$ (e.g., [Mart, p. 46]). This result also follows from the general Universal Unfolding Theorem of Thom to be presented later (see Theorem 7.71 and Table 7.1 with $r = 1$, page 273). Hence, the unfolding $\hat{C}$ is induced by U, that is,

$$\hat{C} = (h^*\mathrm{U} + g \circ \pi) \circ \lambda.$$

Here,

$$\lambda(v', \mathrm{u}) = (\varphi(v', \mathrm{u}), \mathrm{u}), \qquad \varphi(v', 0) = v', \qquad \pi(v', \mathrm{u}) = \mathrm{u}.$$

Set

$$\hat{f}_{cusp} = (\pi, h^*\mathrm{U} + g \circ \pi).$$

Explicitly,

$$\hat{f}_{cusp}(\mathrm{u}, v') = (\mathrm{u}, (v')^3 + h(\mathrm{u})v' + g(\mathrm{u})).$$

Write $\hat{f}_{cusp} = (E, \tilde{F})$ and set

$$\tilde{f}(v', \mathrm{u}) = \hat{f} \circ \hat{\lambda}(v', \mathrm{u}),$$

where $\hat{\lambda}(v', \mathrm{u}) = (\mathrm{u}, v')$ (which is a coordinate change). Since $\pi = \pi \circ \lambda$, we have

$$\tilde{f} = (\pi, \hat{C}) = (\pi, (h^*U + g \circ \pi)) \circ \lambda = \hat{f}_{cusp} \circ \lambda.$$

Hence f at x is differentiably equivalent to $\hat{f}_{cusp}$ at $\mathbf{0} \in \mathbf{R}^2$.

Define a coordinate change by $\Psi(E, \tilde{F}) = (E, \tilde{F} - g(\mathrm{u}))$, so

$$\Psi \circ \hat{f}_{cusp}(v', \mathrm{u}) = (\mathrm{u}, (v')^3 + h(\mathrm{u})v')).$$

Set $\hat{f}_{cusp}(v', \mathrm{u}) = (\mathrm{u}, (v')^3 + h(\mathrm{u})v'))$. The condition $c_{V\mathrm{u}}(0) \neq 0$ implies that $\tilde{F}_{v'\mathrm{u}}(0) \neq 0$, which yields $h_{\mathrm{u}}(0) \neq 0$. Consequently, the function h is a local diffeomorphism at $0 \in \mathbf{R}$. Let $h(\mathrm{u}) = u'$ and define new coordinates by

$$\tilde{\varphi}(v', \mathrm{u}) = (h(\mathrm{u}), v') = (u', v')$$

and

$$\tilde{\Psi}(E, \tilde{F}) = (h(E), \tilde{F}).$$

Then

$$\tilde{\Psi} \circ \tilde{f}_{cusp} \circ \tilde{\varphi}^{-1}(u', v') = (u', v'^3 + u'v') = f_{cusp}(u', v').$$

Therefore, the map f at x is differentiably equivalent to f_{cusp} at $\mathbf{0} \in \mathbf{R}^2$. □

Other Characterizations of Folds and Cusps

We shall determine folds and cusps of $f : X \to Y$ in terms of Kernel$(d_p f)$, $T_p[{}_{\mathrm{T}}S_1(f)]$, and a local function on ${}_{\mathrm{T}}S_1(f)$. For the case of maps between planes, a characterization of folds, cusps, and caustic metamorphoses using directional derivatives appears in Sections 9.4 (page 367) and 9.5.2 (page 378).

Proposition 7.50 *Let $f \in C^\infty(X, Y)$, where* $\dim X = \dim Y = 2$, *and $p_0 \in {}_{\mathrm{T}}S_1(f)$. Then there is an open set $U \subseteq X$ about p_0 on which there are a smooth nonvanishing vector field ξ on U satisfying $\xi_p \in$* Kernel$(d_p f)$ *for all p on the curve $c = U \cap {}_{\mathrm{T}}S_1(f)$, and smooth function $J : U \subseteq X \to \mathbf{R}$ such that $J = 0$ on c and $d_{p_0} J \neq 0$.*

Proof: Before determining ξ and J, we need some preliminary results. Consider the tangent bundle $\pi_X : TX \to X$. Let $\tilde{c}$ be a sub-arc of ${}_{\mathrm{T}}S_1(f)$ that is diffeomorphic to $\mathbf{R}$ and passing through p_0. Set $\pi_X^{-1}(\tilde{c}) = TX_{\tilde{c}}$ and let $\pi_{\tilde{c}} : TX_{\tilde{c}} \to \tilde{c}$ be the vector bundle restriction of π_X to $\tilde{c}$. Let $df_{\tilde{c}} : TX_{\tilde{c}} \to TY$ be the restriction of $df : TX \to TY$ to $TX_{\tilde{c}}$. Set Kernel$(df_{\tilde{c}}) = \dot{\bigcup}_{p \in \tilde{c}}$Kernel$(d_p f)$ (disjoint union). Since $\tilde{c}$ is simply connected and rank$(d_p f) = 1$ for all p on $\tilde{c}$, we obtain an orientable vector bundle $\pi_c|_{\mathrm{Kernel}(df_c)}$: Kernel$(df_c) \to c$ (e.g., [Hir, p. 104]). In addition, since $\pi_{\tilde{c}}|_{\mathrm{Kernel}(df_{\tilde{c}})}$ is a 1-plane vector bundle and c is diffeomorphic to **R**, there is also a smooth nonvanishing section $\xi_{\tilde{c}} : \tilde{c} \to \text{Kernel}(df_{\tilde{c}})$ of $\text{Kernel}(df_{\tilde{c}}) \subseteq TX_{\tilde{c}}$ (cf. [Hir, Ex. 6(c), p. 139]). The section $\xi_{\tilde{c}}$ is a smooth nonvanishing vector field along the immersed curve $\tilde{c}$ (recall that $\tilde{c}$ is an open 1-submanifold in X). Furthermore, the vector field $\xi_{\tilde{c}}$ has a smooth extension to a vector field ξ' in a neighborhood U' in X about p_0 (since a smooth vector field along an immersion always has a local smooth extension – e.g., [War, p. 39]). Because ξ' is nonvanishing along $\tilde{c}$, it is also nonvanishing in a neighborhood $U'' \subseteq U'$ of p_0.

We can now determine ξ and J. Let $\tilde{U} \subseteq U''$ be an open set about p_0 such that $\tilde{U} \cap [{}_{\mathsf{T}}S_1(f)] \equiv \tilde{c}'$ is a sub-arc of $\tilde{c}$, (so, $\tilde{c}'$ is the only part of ${}_{\mathsf{T}}S_1(f)$ that meets $\tilde{U}$). Since every submanifold is locally the common zeros of independent functions (e.g., [Gui-P, p. 23–24]), there is an open set $\tilde{U}' \subseteq X$ about p_0 and a smooth function $J' : \tilde{U}' \to \mathbf{R}$ such that $\tilde{U}' \cap \tilde{c}'$ is the level curve $J' = 0$ and rank $(d_{p_0}J') = 1$ (i.e., $d_{p_0}J' \neq 0$). Let $U = \tilde{U} \cap \tilde{U}$ and set $c = U \cap \tilde{c}'$. Then the desired vector field is $\xi \equiv \xi'|_{\tilde{U}}$, while the function is $J : U \subseteq X \to \mathbf{R}$ is given by $J = J'|_U$. Note that J is such that c corresponds to $J = 0$ and $d_{p_0}J \neq 0$ on c. □

Consider ξ, J, and c in Proposition 7.50. Then the triple (ξ, J, p_0), where $p_0 \in {}_{\mathsf{T}}S_1(f)$, induces a function

$$F_{(\xi,J,p_0)} : c \to \mathbf{R}$$

defined by

$$F_{(\xi,J,p_0)}(p) = \xi_p(J).$$

Definition 7.51 The function $F_{(\xi,J,p_0)}$ is called the *fold-cusp function* induced by the triple (ξ, J, p_0).

Remark: We have $F_{(\xi,J,p_0)}(p) = (d_pJ)(\xi_p)$ with $\text{Kernel}[d_{p_0}J] = T_{p_0 \mathsf{T}}S_1(f)$. To see the latter result, note that $d_{p_0}J \neq 0$ if and only if $\text{rank}[d_{p_0}J] = 1$ if and only if $\dim[\text{Kernel}(d_{p_0}J)] = 1$. Hence, if we represent c by a curve $c(t)$ with $c(0) = p_0$, then $J(c(t)) = 0$ and $(d_{p_0}J)(\frac{dc}{dt}(0)) = 0$, which gives $T_{p_0 \mathsf{T}}S_1(f) = \text{Kernel}[d_{p_0}J]$.

The dependency of $F_{(\xi,J,p_0)}$ on ξ and J is addressed below.

Proposition 7.52 *Each of the following conditions on* $F_{(\xi,J,p_0)} : c \to \mathbf{R}$ *is independent of the choice of* ξ *and* J*:*

(1) $F_{(\xi,J,p_0)}(p_0) \neq 0$.

(2) $F_{(\xi,J,p_0)}(p_0) = 0$ *and* $d_{p_0}F_{(\xi,J,p_0)} \neq 0$.

Proof: Let ξ' be another smooth nonvanishing vector field on $U' \subseteq X$ satisfying $\xi'_p \in \text{Kernel}(d_pf)$ for all p in c, and J' a smooth function $J' : U' \to \mathbf{R}$ such that $J' = 0$ on c and $d_{p_0}J' \neq 0$. It suffices to show that there are nonzero constants r

and r' such that $F_{(\xi,J,p_0)}(p_0) = rF_{(\xi',J',p_0)}(p_0)$ and $d_{p_0}F_{(\xi,J,p_0)} = r'd_{p_0}F_{(\xi',J',p_0)}$. Since $\dim \text{Kernel}(d_{p_0}f) = 1$, there is a nonzero real number r_1 such that $\xi'_{p_0} = r_1\xi_{p_0}$. Because $\text{Kernel}(d_{p_0}J)$ and $\text{Kernel}(d_{p_0}J')$ are 1-dimensional and each tangent vector to c at p_0 belongs to both kernels, we have $\text{Kernel}(d_{p_0}J) = \text{Kernel}(d_{p_0}J') = T_{p_0}c = T_{p_0}[{}_{\top}S_1(f)]$. In addition, the maps $d_{p_0}J$ and $d_{p_0}J'$ belong to the vector subspace

$$K_0 = \{\sigma \in \text{Hom}(T_{p_0}X, T_0\mathbf{R}) : \sigma(v) = 0 \text{ for all } v \in T_{p_0}[{}_{\top}S_1(f)]\}.$$

Let $v_1 \in T_{p_0}[{}_{\top}S_1(f)]$ and $v_2 \in \left(T_{p_0}X - T_{p_0}[{}_{\top}S_1(f)]\right)$, and let w_1 and w_2 be distinct nonzero vectors in $T_0\mathbf{R}$. The linear transformation that maps the basis $\{v_1, v_2\}$ of $T_{p_0}X$ onto $\{w_1, w_2\}$ is a nonzero element of $\text{Hom}(T_{p_0}X, T_0\mathbf{R}) - K_0$. It follows that K_0 is a proper subspace and, hence, $\dim K_0 = 1$. The latter implies that there is a nonzero constant r_2 such that $d_{p_0}J = r_2 d_{p_0}J'$, or, $d_{p_0}(J - r_2J') = 0$. Consequently, the function $J - r_2J'$ is constant on an open neighborhood U'' of p_0. The vanishing of J and J' on c, yields $J = r_2J'$ on U''. This implies that

$$F_{(\xi',J',p_0)}(p_0) = \xi'_{p_0}(J') = r_1\,\xi_{p_0}(J') = r_1r_2\xi_{p_0}(J) = r_1r_2F_{(\xi,J,p_0)}(p_0).$$

[For the third equality, recall that if $v_p \in T_pX$ and $f_1, f_2 \in C^\infty(X, \mathbf{R})$ are equal on a neighborhood of p, then $v_p(f_1) = v_p(f_2)$ (e.g., [ONe, p.7]).] Finally, since $\dim[\text{Hom}(T_{p_0}{}_{\top}S_1(f), T_0\mathbf{R})] = 1$, there is nonzero real number r_3 such that $d_{p_0}F_{(\xi',J',p_0)} = r_3 d_{p_0}F_{(\xi,J,p_0)}$. The proposition now follows. □

Theorem 7.53 *Suppose* $f \in C^\infty(X, Y)$, *where* $\dim X = \dim Y = 2$, *and* $p_0 \in {}_{\top}S_1(f)$.

(1) **(Fold)** *The following are equivalent:*

- (i) *The point* p_0 *is a fold critical point of* f.
- (ii) $F_{(\xi,J,p_0)}(p_0) \neq 0$.
- (iii) *The kernel* $\text{Kernel}(d_{p_0}f)$ *is not tangent to* ${}_{\top}S_1(f)$, *that is,* $T_{p_0}[{}_{\top}S_1(f)] \oplus \text{Kernel}(d_{p_0}f) = T_{p_0}X$.

(2) **(Cusp)** *The following are equivalent:*

- (i) *The point* p_0 *is a cusp critical point of* f.
- (ii) *The point* p_0 *is a simple zero of the fold-cusp function* $F_{(\xi,J,p_0)}$: $F_{(\xi,J,p_0)}(p_0) = 0$ *and* $d_{p_0}F_{(\xi,J,p_0)} \neq 0$.
- (iii) *The kernel* $\text{Kernel}(d_{p_0}f)$ *is tangent to* ${}_{\top}S_1(f)$ (i.e., $T_{p_0}[{}_{\top}S_1(f)] = \text{Kernel}(d_{p_0}f)$) *and* $d_{p_0}F_{(\xi,J,p_0)} \neq 0$.

In Theorem 7.53, conditions (1,ii) and (2,ii, iii) are independent of the choice of ξ and J (Proposition 7.52). Applications of Theorem 7.53 to gravitational lensing will be given in Section 9.2 (see the proof of Theorem 9.1, page 345).

Proof of Theorem 7.53: We first show that $(1, i) \iff (1, ii)$ and $(2, i) \iff (2, ii)$. Emnploying the notation of Lemma 7.48 (page 247), set $\tilde{f} = \nu \circ f \circ \lambda^{-1}$, where $\lambda(p_0) = \mathbf{0}$ and $\nu(f(p_0)) = \mathbf{0}$. Then $\tilde{f}$ maps an open subset of $\mathbf{R}^2$ into $\mathbf{R}^2$ with $\tilde{f}(\mathbf{0}) = \mathbf{0}$, and has the form $\tilde{f}(\mathrm{u}, \mathrm{v}) = (\mathrm{u}, C(\mathrm{u}, \mathrm{v}))$. Let $\tilde{J}(\mathrm{u}, \mathrm{v}) = \det[\mathrm{Jac}\, \tilde{f}](\mathrm{u}, \mathrm{v})$, so the curve ${}_{\mathrm{T}}S_1(\tilde{f})$ is given by

$$\tilde{J}(\mathrm{u}, \mathrm{v}) = C_{\mathrm{v}}(\mathrm{u}, \mathrm{v}) = 0$$

with nonvanishing normal $\nabla \tilde{J} = (C_{\mathrm{vu}}, C_{\mathrm{vv}})$ and tangent $Z = (-C_{\mathrm{vv}}, C_{\mathrm{vu}})$. The vector field $\tilde{\xi} = \partial/\partial\mathrm{v}$ is nonvanishing on a neighborhood $\tilde{U}$ of $\mathbf{0}$. Throughout the rest of the proof, we shall restrict $\tilde{J}$ to $\tilde{U}$.

At points $\mathbf{x}$ on the critical curve $\tilde{c} = \tilde{U} \cap {}_{\mathrm{T}}S_1(\tilde{f})$, we have $\tilde{\xi}_{\mathbf{x}} = (0, 1) \in \mathrm{Kernel} d_{\mathbf{x}} \tilde{f}$. Note that $\tilde{J} = 0$ on $\tilde{c}$ and $d_{\mathbf{0}}\tilde{J} = \nabla \tilde{J}(\mathbf{0}) \neq \mathbf{0}$. Moreover, if we express $\tilde{c}$ as $t \mapsto \tilde{c}(t)$ with $\tilde{c}(0) = \mathbf{0}$, then the fold-cusp function induced by $(\tilde{\xi}, \tilde{J}, \mathbf{0})$ is given by

$$\tilde{F}_{(\tilde{\xi}, \tilde{J}, \mathbf{0})}(\tilde{c}(t)) = \tilde{\xi}_{\tilde{c}(t)}(\tilde{J}) = C_{\mathrm{vv}}(\tilde{c}(t)).$$

Note that $\tilde{F}_{(\tilde{\xi}, \tilde{J}, \mathbf{0})}(\mathbf{0}) = 0 \iff C_{\mathrm{vu}}(\mathbf{0}) \neq 0$ (since $\nabla \tilde{J}(\mathbf{0}) \neq \mathbf{0}$). In addition:

$$d_{\mathbf{0}}\tilde{F}_{(\tilde{\xi}, \tilde{J}, \mathbf{0})} = \nabla C_{\mathrm{vv}}(\mathbf{0}) \cdot \frac{d\tilde{c}}{dt}(0) = r C_{\mathrm{vu}}(\mathbf{0})\, C_{\mathrm{vvv}}(\mathbf{0}),$$

where $(d\tilde{c}/dt)(0) = rZ(\mathbf{0})$ for some nonzero constant r. Hence, the conditions $\mathbf{0} \in {}_{\mathrm{T}}S_1(\tilde{f})$ and $\tilde{F}_{(\tilde{\xi}, \tilde{J}, \mathbf{0})}(\mathbf{0}) \neq 0$ are equivalent to $C_{\mathrm{v}}(\mathbf{0}) = 0$ and $C_{\mathrm{vv}}(\mathbf{0}) \neq 0$; similarly, we have that $\mathbf{0} \in {}_{\mathrm{T}}S_1(\tilde{f})$, $\tilde{F}_{(\tilde{\xi}, \tilde{J}, \mathbf{0})}(\mathbf{0}) = 0$, and $d_{\mathbf{0}}\tilde{F}_{(\tilde{\xi}, \tilde{J}, \mathbf{0})} \neq 0$ are equivalent to $C_{\mathrm{v}}(\mathbf{0}) = 0$, $C_{\mathrm{vv}}(\mathbf{0}) \neq 0$, $C_{\mathrm{vu}}(\mathbf{0}) \neq 0$, and $C_{\mathrm{vvv}}(\mathbf{0}) \neq 0$. By Lemma 7.48, it follows that p_0 *is a fold critical point of* $f \iff \mathbf{0} \in {}_{\mathrm{T}}S_1(\tilde{f})$ *and* $\tilde{F}_{(\tilde{\xi}, \tilde{J}, \mathbf{0})}(\mathbf{0}) \neq 0$*; and* p_0 *is a cusp critical point of* $f \iff \mathbf{0} \in {}_{\mathrm{T}}S_1(\tilde{f})$*,* $\tilde{F}_{(\tilde{\xi}, \tilde{J}, \mathbf{0})}(\mathbf{0}) = 0$*, and* $d_{\mathbf{0}}\tilde{F}_{(\tilde{\xi}, \tilde{J}, \mathbf{0})} \neq 0$.

Now, since differentiable equivalence preserves critical points, rank, and transversal, the curve $c = \lambda^{-1} \circ \tilde{c}$ is in ${}_{\mathrm{T}}S_1(f)$ and passes through p_0 at $t = 0$. Set $U = \lambda^{-1}(\tilde{U})$ and define $J : U \subseteq X \to \mathbf{R}$ by $J = \tilde{J} \circ \lambda$. Then $J = 0$ on the curve c and $d_{p_0}J = d_{\mathbf{0}}\tilde{J} \circ d_{p_0}\lambda \neq 0$ (because $d_{p_0}\lambda$ is a linear isomorphism and $d_{\mathbf{0}}\tilde{J} \neq \mathbf{0}$). Define a vector field ξ on U by $\xi = (d\lambda^{-1}) \circ \tilde{\xi} \circ \lambda$. Then ξ is nonvanishing on U. Indeed, since $\tilde{\xi}_{\lambda(p)} \neq 0$ and $d_{\lambda(p)}\lambda^{-1}$ is a linear isomorphism, where $p \in U$, we have $\xi_p = (d_{\lambda(p)}\lambda^{-1})(\tilde{\xi}_{\lambda(p)}) \neq 0$. Also, the vector ξ_p lies in $\ker(d_p f)$ for all p on the curve c. To see this, note that since for p on c the vector $\tilde{\xi}_{\lambda(p)}$ lies in $\mathrm{Kernel}(d_{\lambda(p)}\tilde{f})$ and $d_{\tilde{f}(\lambda(p))}\nu^{-1}$ is a linear isomorphism, it follows that

$$\begin{aligned}
d_p f(\xi_p) &= d_p f\left[(d_{\lambda(p)}\,\lambda^{-1}))(\tilde{\xi}_{\lambda(p)})\right] \\
&= \left(d_{\tilde{f}(\lambda(p))}\,\nu^{-1} \circ d_{\lambda(p)}\tilde{f} \circ d_p\lambda \circ d_{\lambda(p)}\,\lambda^{-1}\right)(\tilde{\xi}_{\lambda(p)}) \\
&= \left(d_{\tilde{f}(\lambda(p))}\,\nu^{-1}\right)\left(d_{\lambda(p)}\,\tilde{f}(\tilde{\xi}_{\lambda(p)})\right) \\
&= \left(d_{\tilde{f}(\lambda(p))}\,\nu^{-1}\right)(\mathbf{0}) = 0.
\end{aligned}$$

because $\tilde{\xi}_{\lambda(p)} \in \text{Kernel}(d_{\lambda(p)}\tilde{f})$ and $d_{\tilde{f}(\lambda(p))}\nu^{-1}$ is a linear isomorphism.]

Consider the fold-cusp function $F_{(\xi,J,p_0)} : c \to \mathbf{R}$ induced by (ξ, J, p_0). Then

$$\begin{aligned} F_{(\xi,J,p_0)}(p) &= \xi_p(J) = d_p J(\xi_p) = \left(d_{\lambda(p)}]\,\tilde{J} \circ d_p\lambda\right)\left(d_{\lambda(p)}\,\lambda^{-1}(\tilde{\xi}_{\lambda(p)})\right) \\ &= d_{\lambda(p)}\,\tilde{J}(\tilde{\xi}_{\lambda(p)}) = \tilde{\xi}_{\lambda(p)}\,(\tilde{J}) = \tilde{F}_{(\tilde{\xi},\tilde{J},\mathbf{0})}\,(\lambda(p)), \end{aligned}$$

where $\lambda(p_0) = \mathbf{0}$. Moreover, $d_{p_0}F_{(\xi,J,p_0)} = d_{\mathbf{0}}\tilde{F} \circ d_{p_0}\lambda$. It follows $F_{(\xi,J,p_0)}(p_0) = 0 \iff \tilde{F}_{(\tilde{\xi},\tilde{J},\mathbf{0})}(\mathbf{0}) = 0$; and, since $d_{p_0}\lambda$ is a linear isomorphism, we get $d_{p_0}F_{(\xi,J,p_0)} \neq 0 \iff d_{p_0}\tilde{F}_{(\tilde{\xi},\tilde{J},\mathbf{0})} \neq 0$. Thus, the critical point p_0 is a fold of $f \iff F_{(\xi,J,p_0)}(p_0) \neq 0$; and p_0 is a cusp critical point of $f \iff F_{(\xi,J,p_0)}(p_0) = 0$ and $d_{p_0}F_{(\xi,J,p_0)} \neq 0$. Note that by Proposition 7.52 these conditions on $F_{(\xi,J,p_0)}$ are independent of the choice of ξ and J.

We now prove that $(1, i) \iff (1, iii)$. The kernel $\text{Kernel}[d_p(f|_{{}_\top S_1(f)})]$, where $p \in {}_\top S_1(f)$, is a subspace of $\text{Kernel}(d_p f)$ and $T_p[{}_\top S_1(f)]$. Since $T_p({}_\top S_1(f))$ is one dimensional, we have:

$$[\text{Kernel}(d_p f)] \cap [T_{p\top}S_1(f)] = \text{Kernel}[d_p(f|_{{}_\top S_1(f)})].$$

But $\dim[\text{Kernel}[d_p(f|_{{}_\top S_1(f)})]] = \dim[T_p({}_\top S_1(f))] - \text{rank}[d_p(f|_{{}_\top S_1(f)})]$. Hence: $p_0 \in S_{1,0}(f) \iff p_0 \in {}_\top S_1(f)$ and $[\text{Kernel}(d_{p_0}f)] \cap [T_{p_0}({}_\top S_1(f))] = \{0\}$ (since we have the result $\dim[\text{Kernel}[d_{p_0}(f|_{{}_\top S_1(f)})]] = 0) \iff p_0 \in {}_\top S_1(f)$ and $T_{p_0}[{}_\top S_1(f)] \oplus \text{Kernel}(d_{p_0}f) = T_{p_0}X$.

Let us show that $(2, i) \iff (2, iii)$. We first prove that $T_{p_0\top}S_1(f) = \text{Kernel}(d_{p_0}f) \iff F_{(\xi,J,p_0)}(p_0) = 0$. Since $\xi_{p_0}(J) = (dJ)(\xi_{p_0})$, it follows that $F_{(\xi,J,p_0)}(p_0) \neq 0$ if and only if $d_{p_0}J(\xi_{p_0}) \neq 0$, or, equivalently, $\xi_{p_0} \notin \text{Kernel}(d_{p_0}J) = T_{p_0\top}S_1(f)$. Since $\text{Kernel}(d_{p_0}f) = \text{Span}(\xi_{p_0})$, it follows that for $p_0 \in {}_\top S_1(f)$ we have ξ_{p_0} is not in $T_{p_0}[{}_\top S_1(f)]$ if and only if $T_{p_0}[{}_\top S_1(f)] \oplus \text{Kernel}(d_{p_0}f) = T_{p_0}X$. Since $(2, i) \iff (2, ii)$ we obtain $(2, i) \iff (2, iii)$. □

Remarks:

(1) We have $F_{(\xi,J,p_0)}(p_0) = 0 \iff T_{p_0\top}S_1(f) = \text{Kernel}(d_{p_0}f) \iff p_0 \in S_{1,1}(f)$. The first equivalence was established in the proof of Theorem 7.53(1) (see the proof of $(1, i) \iff (1, iii)$). To obtain the second equivalence, note that $p_0 \in S_{1,1}(f) \iff p_0 \in {}_\top S_1(f)$ and $\text{Kernel}(d_{p_0}f) = T_{p_0}[{}_\top S_1(f)]$ since

$$\dim[\text{Kernel}[d_{p_0}(f|_{{}_\top S_1(f)})]] = \dim[T_p({}_\top S_1(f))] = \dim[\text{Kernel}(d_p f)] = 1.$$

(2) Theorem 7.53(2) yields that the transversality condition $j^2 f \pitchfork S_{1,1}$ at $p_0 \in S_{1,1}(f)$ is equivalent to $d_{p_0}F_{(\xi,J,p_0)} \neq 0$ (since by definition, the point being p_0 a cusp means that $p_0 \in S_{1,1}(f)$ and $j^2 f \pitchfork S_{1,1}$ at p_0).

7.5.3 Stability of Maps between 2-Manifolds

This section characterizes the stability of maps between two dimensional manifolds.

Let $f \in C^\infty(X, Y)$, where $\dim X = \dim Y = 2$. Then f is *1-generic* if $j^1 f \pitchfork S_r(X, Y)$ for $r = 0, 1, 2$ (Definition 7.37, page 235). If f is 1-generic and $j^2 f \pitchfork S_{r,s}(X, Y)$ for $r = 0, 1, 2$ and $0 \le s \le r$, then f is called *2-generic* (Definition 7.40, page 236).

Lemma 7.54 *Let $f \in C^\infty(X, Y)$, where $\dim X = \dim Y = 2$. Then:*

(1) *The map f is 1-generic if and only if each point of X is either a regular point, fold critical point, cusp critical point, or a degenerate critical point in ${}_{\mathrm{T}}S^c_{1,1}(f)$.*

(2) *The map f is 2-generic if and only if each point of X is either a regular point, fold critical point, or cusp critical point.*

Proof: (1) *Preliminaries:* Recall that $x \in S_r(f)$ if and only if $j^1 f(x) \in S_r(X, Y)$ for $r = 0, 1, 2$. It follows that $j^1 f \pitchfork S_1(X, Y)$ if and only if $S_1(f) = {}_{\mathrm{T}}S_1(f)$ (i.e., ${}_{\mathrm{T}}S^c_1(f) = \emptyset$). In addition, $j^1 f \pitchfork S_2(X, Y)$ if and only if $S_2(f) = \emptyset$. [In fact, if $S_2(f) = \emptyset$, then (because $(j^1 f)^{-1}[S_2(X, Y)] = S_2(f)$) $j^1 f(X) \cap S_2(X, Y) = \emptyset$, i.e., $j^1 f \pitchfork S_2(X, Y)$. Conversely, suppose $j^1 f \pitchfork S_2(X, Y)$. If $S_2(f)$ is nonempty, then (by Theorem 7.34): $\operatorname{codim}_X S_2(f) = \operatorname{codim}_{J^1(X,Y)} S_2(X, Y) = 4 > \dim X = 2$, which is impossible. Consequently, $S_2(f) = \emptyset$.]

We now prove the first part of the lemma. If f is 1-generic, then $X = S_0(f) \dot\cup \operatorname{Folds}(f) \dot\cup \operatorname{Cusps}(f) \dot\cup {}_{\mathrm{T}}S^c_{1,1}(f)$. Conversely, suppose that the latter holds. Then $\operatorname{Crit}(f) = {}_{\mathrm{T}}S_1(f)$. To show that $j^1 f$ is 1-generic, we must consider transversality of f to $S_0(X, Y)$, $S_1(X, Y)$, and $S_2(X, Y)$. The condition $j^1 f \pitchfork S_0(X, Y)$ is immediate (because $S_0(X, Y)$ is open in $J^1(X, Y)$). Since $\operatorname{Crit}(f) = {}_{\mathrm{T}}S_1(f)$, we have $j^1 f \pitchfork S_1(X, Y)$, and since $S_2(f) = \emptyset$ it follows that $j^1 f \pitchfork S_2(X, Y)$. Thus, the map f is 1-generic.

(2) Suppose f is 2-generic. Then f is 1-generic, which is equivalent to $X = S_0(f) \dot\cup \operatorname{Folds}(f) \dot\cup \operatorname{Cusps}(f) \dot\cup {}_{\mathrm{T}}S^c_{1,1}(f)$ (by Part (1)), and $j^2 f \pitchfork S_{1,1}(X, Y)$. The latter forces ${}_{\mathrm{T}}S^c_{1,1}(f) = \emptyset$. Hence, $X = S_0(f) \dot\cup \operatorname{Folds}(f) \dot\cup \operatorname{Cusps}(f)$. Conversely, assume $X = S_0(f) \dot\cup \operatorname{Folds}(f) \dot\cup \operatorname{Cusps}(f)$. To prove the 2-genericity of f, we must show that f is 1-generic and $j^2 f$ is transverse to $S_{0,0}(X, Y)$, $S_{1,0}(X, Y)$, $S_{1,1}(X, Y)$, and $S_{2,s}(X, Y)$ for $s = 0, 1, 2$. By the first part of the lemma, we have that f is 1-generic. Since $S_{0,0}(X, Y)$ is open in $J^2(X, Y)$, we obtain $j^2 f \pitchfork S_{0,0}(X, Y)$. Because $\operatorname{Crit}(f) = S_{1,0}(X, Y) \dot\cup {}_{\mathrm{T}}S_{1,1}(X, Y)$, it follow that $j^2 f \pitchfork S_{1,1}(X, Y)$. For $s = 0, 1, 2$ note that we have $\emptyset = S_{2,s}(f) = (j^2 f)^{-1}[S_{2,s}(X, Y)]$. In other words, we get $j^2 f(X) \cap S_{2,s}(X, Y) = \emptyset$, which yields $j^2 f \pitchfork S_{2,s}(X, Y)$ $(s = 0, 1, 2)$.

It remains to show that $j^2 f \pitchfork S_{1,0}(X, Y)$. As in the proof of Lemma 7.48 (page 247), it suffices to consider f as a local map $U \subseteq \mathbf{R}^2 \to \mathbf{R}^2$ mapping the origin to itself. Consider the natural projection $\pi_{2,1} : J^2(U, \mathbf{R}^2) \to J^1(U, \mathbf{R}^2)$ and the decomposition

$$J^2(U, \mathbf{R}^2) \approx J^1(U, \mathbf{R}^2) \times D^2(U, \mathbf{R}^2),$$

where $D^2(U, \mathbf{R}^2)$ is the 6-dimensional space of all second partial derivatives of the two component functions of a map between planes. We have

$$\pi_{2,1}^{-1}[S_1(U, \mathbf{R}^2)] = S_1'(U, \mathbf{R}^2) = S_1(U, \mathbf{R}^2) \times D^2(U, \mathbf{R}^2)].$$

The set $S_{1,0}(U, \mathbf{R}^2)$ is an open subset of $S_1(U, \mathbf{R}^2) \times D^2(U, \mathbf{R}^2)$ (see page 237). Hence

$$\begin{aligned} \operatorname{codim}_{J^2(U,\mathbf{R}^2)}[S_1(U, \mathbf{R}^2) \times D^2(U, \mathbf{R}^2)] &= \operatorname{codim}_{J^2(U,\mathbf{R}^2)}[S_{1,0}(U, \mathbf{R}^2)] \\ &= \nu_{1,0}(2, 2) \\ &= 1. \end{aligned}$$

Recall that $\nu_{1,0}(2, 2)$ is given by (7.5) on page 235. Now, near a point of $S_1(U, \mathbf{R}^2) \times D^2(U, \mathbf{R}^2)$, its defining equation in $J^2(U, \mathbf{R}^2)$ is the same as the defining equation for $S_1(U, \mathbf{R}^2)$ in $J^1(U, \mathbf{R}^2)$. Suppose that the function $\mathcal{H} : J^1(U, \mathbf{R}^2) \to \mathbf{R}$ determines the equation for $S_1(U, \mathbf{R}^2)$, namely, $S_1(U, \mathbf{R}^2)$ coincides locally with $\mathcal{H}^{-1}(0)$ — see page 250. Then locally $S_{1,0}(U, \mathbf{R}^2)$ is just $(\mathcal{H} \circ \pi_{2,1})^{-1}(0)$. In other words, the local defining equation for $S_{1,0}(U, \mathbf{R}^2)$ is determined by $\mathcal{H} \circ \pi_{2,1} : J^2(U, \mathbf{R}^2) \to \mathbf{R}$. Since $j^1 f$ is transverse at $\mathbf{0}$, we have that $\operatorname{rank}[d_{\mathbf{0}}(\mathcal{H} \circ j^1 f)] = 1$ (see page 250). But $\mathcal{H} \circ j^1 f = \mathcal{H} \circ \pi_{2,1} \circ j^2 f$, hence $\operatorname{rank}[d_{\mathbf{0}}(\mathcal{H} \circ \pi_{2,1} \circ j^2 f) = 1$. Proposition 7.7 now implies that $j^2 f \pitchfork S_{1,0}(X, Y)$. $\square$

Theorem 7.55 *Let f be in $C^\infty(X, Y)$, where* $\dim X = \dim Y = 2$. *Then the following are equivalent:*

(1) *The map f is locally stable.*

(2) *The map f is 2-generic.*

(3) *Each point p in X is either a regular point, fold critical point, or cusp critical point.*

Proof: Assume f is locally stable, i.e., for each $p \in U$, there is an open set $V \subseteq U$ such that for each open subset $U' \subseteq V$, there is a neighborhood $\mathcal{F}$ of f in $C^\infty(U, \mathbf{R}^2)$ such that for every $f' \in \mathcal{F}$, there is a point $p' \in U'$ with f' at p' differentiably equivalent to f at p. Lemma 7.15 and Corollary 7.18 imply that the set of 2-generic functions in $C^\infty(U, \mathbf{R}^2)$ is dense. Consequently, there is a 2-generic function h in $\mathcal{F}$ and h at p' is differentiably equivalent to f at p. Hence, Lemma 7.54(2) yields that the point p is either a regular point, fold critical point, or cusp critical point.

Conversely, suppose f is 2-generic. If $p \in U$ is regular, then f is stable at p (by Proposition 7.22). Assume p is either a fold or cusp critical point. Consider

$$V^k_{\epsilon, B_\epsilon(p)}(f) = \{h \in C^\infty(U, \mathbf{R}^2) : |j^k f(p') - j^k h(p')| < \epsilon \text{ for all } p' \in B_\epsilon(p)\}.$$

Let $f' \in V^k_{\epsilon, B_\epsilon(p)}(f)$. If k is sufficiently large and ϵ sufficiently small, then the values and partials up to order k if f' at points of $B_\epsilon(p)$ closely approximate those of $f|_{B_\epsilon(p)}$.

Consequently, f' has either a fold or regular cusp critical point p' near p. By Theorem 7.49, there is an open neighborhood U' of p' such that $f'|_{U'}$ is differentiably equivalent to either Whitney fold or pleat. But for ϵ sufficiently small, $f|_{B_\epsilon(x)}$ is also differentiably equivalent to the previous Whitney fold or pleat. Hence, f' at p' is differentiably equivalent to f at p. Thus, f is stable at p. □

Theorem 7.55 yields that for a locally stable map f between 2-manifolds, the set of critical points is given by

$$\text{Crit}(f) = {}_T S_1(f) = \text{Folds}(f) \dot{\bigcup} \text{Cusps}(f).$$

In other words, the critical points form curves, which we call *critical curves*. The corresponding critical values are called *caustic curves*.

Theorem 7.56 *Let $f \in C^\infty(U, \mathbf{R}^2)$, where $U \subseteq \mathbf{R}^2$ is open. Then:*

(1) *The map f is transverse stable if and only if the following five conditions hold:*

  (i) *The map f is locally stable.*

  (ii) *The restriction of f to its fold critical curves* (i.e., $f|_{S_{1,0}(f)}$) *is an immersion with transverse crossings. In other words, all self-intersections of the fold caustic caustic curves occur at nonzero angles.*

  (iii) *No fold and cusp critical points are mapped by f to the same point.*

  (iv) *No two cusp critical points are mapped by f to the same point.*

  (v) *No more than two fold critical points are mapped by f to the same point.*

(2) *Suppose that f is proper. Then f is globally stable if and only if f is transverse stable.*

Proof:

(1) Transverse stability of f means the following hold:

  (a) $j^2 f \pitchfork \mathcal{D}_\sigma$ for all $\sigma \in J^2(U, \mathbf{R}^2)$.

  (b) $j_2^2 f \pitchfork \mathcal{D}_\sigma^2$ for all diagonal elements $\sigma \in J_2^2(U, \mathbf{R}^2)$.

  (c) $j_3^2 f \pitchfork \mathcal{D}_\sigma^3$ for all diagonal elements $\sigma \in J_3^2(U, \mathbf{R}^2)$.

In conditions (a)–(c), it suffices to consider $j_s^2 f \pitchfork \mathcal{D}_\sigma^s$ at $z \in U^{(s)}$ for all diagonal elements σ in the orbit $\mathcal{D}_{j_s^2 f(z)}$; otherwise, $j_s^2 f(z) \notin \mathcal{D}_\sigma^s$, which means $j_s^2 f \pitchfork \mathcal{D}_\sigma^s$ at z. It is then no restriction to assume that $\sigma = j_s^2 f(z)$.

We first show that condition (a) is equivalent to f being 2-generic. Assume that (a) holds. Let $\pi_{2,1} : J^2(U, \mathbf{R}^2) \to J^1(U, \mathbf{R}^2)$ be the natural projection. Then $S_r'(U, \mathbf{R}^2) = \pi_{2,1}^{-1}[S_r(U, \mathbf{R}^2)]$ is a singularity manifold; in fact, it is a union of disjoint orbits (a single orbit when U is connected). Condition (a) yields that $j^2 f \pitchfork S'_r(U, \mathbf{R}^2)$ for $r = 0, 1, 2$. Since $j^1 f = \pi_{2,1} \circ j^2 f$ and $\pi_{2,1}$ is a submersion, Proposition 7.11 implies that for $r = 0, 1, 2$, $j^2 f \pitchfork S'_r(U, \mathbf{R}^2)$ for $r = 0, 1, 2$, if and only if $j^1 f \pitchfork S_r(U, \mathbf{R}^2)$, which means that f is 1-generic. But condition (a) also yields $j^2 f \pitchfork S_{r,s}(U, \mathbf{R}^2)$ if f is 1-generic. Hence, the map f is 2-generic.

Conversely, assume that f is 2-generic. Then the domain of f can be expressed as

$$U = S_{0,0}(f) \,\dot{\cup}\, \mathrm{Crit}(f) = S_0(f) \,\dot{\cup}\, S_{1,0}(f) \,\dot{\cup}\,_T S_{1,1}(f), \tag{7.11}$$

where $S_{0,0}(f) = S_0(f)$. Since f is 1-generic, Theorem 7.38 yields that $x \in U$ if and only if $\sigma = j^2 f(x)$ lies in $S_{r,s}(U, \mathbf{R}^2)$ for some $0 \leq s \leq r \leq 2$. By (7.11), it then suffices to consider the more restricted case $0 \leq s \leq r \leq 1$. In fact, the sets $S_{2,s}(f)$, where $0 \leq s \leq 2$, are empty (see proof of Lemma 7.54). Note that Proposition 7.39 yields

$$J^2(U, \mathbf{R}^2) = \left[\dot{\bigcup}_{r,s} S_{r,s}(U, \mathbf{R}^2)\right] \dot{\bigcup} D(U, \mathbf{R}^2),$$

where $0 \leq s \leq r \leq 2$. We shall not have to consider all these components of $J^2(U, \mathbf{R}^2)$. It suffices to work with the orbits $\mathcal{D}_\sigma$ lying in some 2-singularity $S_{r,s}(U, \mathbf{R}^2)$, where $0 \leq s \leq r \leq 1$. Such an orbit is a component of $S_{r,s}(U, \mathbf{R}^2)$ (because every 2-singularity in $J^2(U, \mathbf{R}^2)$ is a union of disjoint orbits $\mathcal{D}_\sigma$ — see page 232). By definition of 2-genericity, we have $j^2 f \pitchfork S_{r,s}(U, \mathbf{R}^2))$ — hence, $j^2 f \pitchfork \mathcal{D}_\sigma$.

Condition (a) is then equivalent to f being 2-generic, which is equivalent to the local stability of f (Theorem 7.55).

Before considering conditions (b) and (c), we must introduce additional notation. Assume that f is 2-generic. We have

$$J^2(U, \mathbf{R}^2) \approx U \times \mathbf{R}^2 \times J^2(2, 2)$$

and

$$S_{r,s}(U, \mathbf{R}^2) \approx U \times \mathbf{R}^2 \times S_{r,s}(2, 2).$$

Moreover,

$$\mathrm{codim}_{J^2(U,\mathbf{R}^2)}\, S_{r,s}(2, 2) = \nu_{r,s}(2, 2) = r^2 + \frac{s}{2}\,[2r^2 - rs + 2s - r], \tag{7.12}$$

which is from (7.5) on page 235. The orbits $\mathcal{D}^2_\sigma$, where $\sigma = (j^2 f(x_1), j^2 f(x_2))$ with $f(x_1) = f(x_2)$, are components of

$$S^{[2]}_{(r_1,s_1;r_2,s_2)} = U^{(2)} \times \triangleright(\mathbf{R}^2) \times S_{r_1,s_1}(2, 2) \times S_{r_2,s_2}(2, 2).$$

As before, the domain (7.11) shows that it suffices to restrict the indices r_i and s_i to $0 \leq s_i \leq r_i \leq 1$, where $i = 1, 2$.

Notation:

(i) $S^{[2]}_{(1,0;1,0)} = S^{[2]}_{F,F}$

(ii) $S^{[2]}_{(r_1,s_1;r_2,s_2)} = S^{[2]}_{F,C}$, where $(r_1, s_1, r_2, s_2) = (1, 1, 1, 0)$ or $(1, 0, 1, 1)$.

(iii) $S^{[2]}_{(1,1;1,1)} = S^{[2]}_{C,C}$

(iv) $S^{[3]}_{F,F,F} = U^{(3)} \times \rhd^3(\mathbf{R}^2) \times S_{1,0}(2,2) \times S_{1,0}(2,2) \times S_{1,0}(2,2)$ where $\rhd^3(\mathbf{R}^2) = \{(y, y, y) \in (\mathbf{R}^2)^3 : y \in \mathbf{R}^2\}$.

Note that the preimages of $S^{[2]}_{F,F}$, $S^{[2]}_{F,C}$, and $S^{[2]}_{C,C}$ under $j^2_2 f$ are, respectively, pairs of distinct folds, a fold and cusp, and distinct cusp critical points (since f is 2-generic). Elements of $(j^2_3 f)^{-1}$ $(S^{[3]})$ are distinct triples consisting of fold and/or cusp critical points.

We now investigate conditions (b) and (c), where (a) is assumed.

Case I: The f is 2-generic and $j^2_2 f \pitchfork S^{[2]}_{(r_1,s_1;r_2,s_2)}$, where either $(r_1, s_1) = (0, 0)$ or $(r_2, s_2) = (0, 0)$.

This transversality condition is equivalent to x_1 and x_2 being either regular points, or, a fold/cusp critical point and regular point. This is statement is then redundant since f is 2-generic.

Case II: The map f is 2-generic and $j^2_2 f \pitchfork S^{[2]}_{F,F}$.

We shall show that Case II is equivalent to f being 2-generic with the restriction, $\hat{f} = f|_{S_{1,0}(f)}$, an immersion such that all crossings in $\hat{f}(S_{1,0}(f))$ occur at nonzero angles. In fact, let

$$Q_1 = [S_{1,0}(U, \mathbf{R}^2) \times S_{1,0}(U, \mathbf{R}^2)] \cap J^2_2(U, \mathbf{R}^2).$$

Then Q_1 is a submanifold of $J^2_2(U, \mathbf{R}^2)$. Moreover, we have $j^2_2 f \pitchfork Q_1$ if and only if $j^2 f \pitchfork S_{1,0}(U, \mathbf{R}^2)$ (by Proposition 7.16, page 225). Set $X_1 = (j^2_2 f)^{-1}(Q_1)$ and $Q_2 = S^{[2]}_{F,F}$, which is a submanifold of Q_1. Note that

$$X_1 = [S_{1,0}(f) \cap S_{1,0}(f)] \cap U^{(2)} = S_{1,0}(f) \times S_{1,0}(f) - \rhd S_{1,0}(f).$$

Set $\widehat{j^2_2 f} = (j^2_2 f)|_{X_1}$. By Proposition 7.10 (page 220), $j^2_2 f \pitchfork Q_2$ if and only if $\widehat{j^2_2 f} : X_1 \to Q_1$ is transverse to Q_2. Let $\hat{\pi}$ be the restriction of the projection $J^2_2(U, \mathbf{R}^2) \to \mathbf{R}^2 \times \mathbf{R}^2$ to Q_1. Let $\hat{f}^{(2)} : X_1 \to \mathbf{R}^2 \times \mathbf{R}^2$ be the restriction $(\hat{f} \times \hat{f})|_{X_1}$. Since $\hat{f}^{(2)} = \hat{\pi} \circ \widehat{j^2_2 f}$ and $Q_2 = \hat{\pi}^{-1}[\rhd(\mathbf{R}^2)]$, where $\hat{\pi}$ is a submersion, Proposition 7.11 (page 220) yields that $\hat{f}^{(2)} \pitchfork \rhd(\mathbf{R}^2)$ if and only if $\widehat{j^2_2 f} \pitchfork Q_2$. Since $\hat{f}^{(2)}$ is an immersion, Proposition 7.12 (page 221) yields $\hat{f}^{(2)} \pitchfork \rhd(\mathbf{R}^2)$ if and only if for distinct points p_1, p_2 in $\hat{f}^{-1}(q)$, we have

$$d_{p_1}\hat{f}\,[T_{p_1}S_{1,0}(f)] + d_{p_2}\hat{f}\,[T_{p_2}S_{1,0}(f)] = T_q\mathbf{R}^2.$$

If c_i denotes the fold curve in $S_{1,0}(f)$ passing through p_i, then the above is equivalent to image curves $f \circ c_1$ and $f \circ c_2$ having linearly independent tangents at $q = f(p_1) = f(p_2)$.

Case III: The map f is 2-generic and $j_2^2 f \pitchfork S_{F,C}^{[2]}$.

In this case, the points x_1 and x_2 are a pair of fold and cusp critical points. Using (7.12), we get

$$\begin{aligned}\operatorname{codim}_{J_2^2(U,\mathbf{R}^2)} S_{F,C}^{[2]} &= 2 + \operatorname{codim}_{J_2^2(U,\mathbf{R}^2)} S_{1,0}(2,2) + \operatorname{codim}_{J_2^2(U,\mathbf{R}^2)} S_{1,1}(2,2) \\ &= 2 + 2 + 1 > 4 = \dim U^{(2)}.\end{aligned}$$

Note that since f is 2-generic, the set $S_{1,1}(f)$ coincides with the set of cusp critical points of f. Hence, Proposition 7.6 (page 218) yields

$$[(j_2^2 f)(U^{(2)})] \cap S_{F,C}^{[2]} = \emptyset.$$

Thus, we get $f(x_1) \neq f(x_2)$. Conversely, suppose that f is 2-generic with $f(x_1) \neq f(x_2)$ for $x_1 \in S_{1,1}(f)$ and $x_2 \in S_{1,0}(f)$. Then $j_2^2 f(x_2, x_2) \notin S_{F,C}^{[2]}$ (i.e., $j_2^2 f \pitchfork S_{F,C}^{[2]}$ at (x_1, x_2)). Hence, Case III is equivalent to f being 2-generic with no fold and cusp critical points having the same f-image.

The last two cases will employ the codimension argument used in Case III.

Case IV: The map f is 2-generic and $j_2^2 f \pitchfork S_{C,C}^{[2]}$.

Here, the points x_1 and x_2 are cusp critical points. By (7.12), we have

$$\operatorname{codim}_{J_2^2(U,\mathbf{R}^2)} S_{C,C}^{[2]} = 2 + 2\operatorname{codim}_{J_2^2(U,\mathbf{R}^2)} S_{1,1}(2,2) = 2 + 2(2) > 4 = \dim U^{(2)},$$

which yields

$$[(j_2^2 f)(U^{(2)})] \cap S_{C,C}^{[2]} = \emptyset.$$

Consequently, we have $f(x_1) \neq f(x_2)$. Conversely, if f is 2-generic and x_1, x_2 are cusp critical points with $f(x_1) \neq f(x_2)$, then $j_2^2 f(x_1, x_2) \notin S_{C,C}^{[2]}$. Thus, Case IV is equivalent to f being 2-generic with no two cusps getting mapped to the same point.

Case V: The map f is 2-generic and $j_3^2 f \pitchfork S_{F,F,F}^{[3]}$.

It suffices to consider distinct triples x_1, x_2, x_3 of fold critical points; otherwise, we reduce to either Case III or IV. Hence, we shall investigate transversality to $S_{F,F,F}^{[3]}$. Since

$$\begin{aligned}\operatorname{codim}_{J_2^2(U,\mathbf{R}^2)} S_{F,F,F}^{[3]} &\geq \operatorname{codim}_{\mathbf{R}^6} \triangleright^3 (\mathbf{R}^2) + 3\operatorname{codim}_{J_2^2(U,\mathbf{R}^2)} S_{1,0}(2,2) \\ &= 4 + 3(1) > 6 = \dim U^{(3)},\end{aligned}$$

we have $(j_3^2 f)(U^{(3)}) \cap S_{F,F,F}^{[3]} = \emptyset$. In other words, no three fold critical points are mapped to the same point. Conversely, if f is 2-generic and does not map three fold critical points to the same point, then $j_3^2 f(x_1, x_2, x_3) \notin S_{F,F,F}^{[3]}$. Case V is then equivalent to f being 2-generic and mapping at most two fold critical points to the same point.

(2) This is a consequence of Theorem 7.31. □

7.6 Thom Catastrophe Theory

Thom's catastrophe theory deals with the critical point theory of families of functions. A full mathematical treatment of this fascinating theory would occupy a separate book. We recommend the detailed mathematical treatments by Castrigiano and Hayes [Cas-H], Majthay [Maj], and Martinet [Mart]. Also, Berry and Upstill [Berr-U], Gilmore [Gil], and Poston and Stewart [Pos-S] give excellent introductions with physical applications for non-mathematicians. In addition, see the popular review of catastrophe theory by Arnold [Arn92].

The current section will summarize some of the basic mathematical concepts and results of catastrophe theory, which culminates with *Thom's Catastrophe Theorem*. Consult Section 8.4 (page 310) for applications to gravitational lensing.

7.6.1 Germs of Functions and Universal Unfoldings

Determinacy and Codimension

In this subsection, we shall present one of the key theorems of catastrophe theory, namely, the classification of germs at critical points (Theorem 7.61).

Let $\mathcal{E}_n$ and $\mathfrak{M}_n$ be, respectively, the set of germs $f : (\mathbf{R}^n, 0) \to (\mathbf{R}, f(0))$ and $h : (\mathbf{R}^n, 0) \to (\mathbf{R}, 0)$ (so each representative of such germs is a smooth function on a neighborhood of the origin in $\mathbf{R}^n$). Then $\mathcal{E}_n$ is a *local ring* (i.e., a commutative ring with multiplicative identity and unique maximal ideal) with $\mathfrak{M}_n$ its maximal ideal (e.g., [Zee-T, Lems. 2.1 and 2.2, pp. 270–71]). For simplicity, denote each germ in $\mathcal{E}_n$ by $[f]$, so f is a smooth function on a neighborhood of the origin in $\mathbf{R}^n$. Germs $[f]$ and $[h]$ in $\mathcal{E}_n$ are *k-equivalent* if the kth-order Taylor polynomials of f and h coincide at $0 \in \mathbf{R}^n$. The *k-jet* of $[f]$, denoted $j^k[f]$, is the equivalence class of germs in $\mathcal{E}_n$ that are k-equivalent to $[f]$.

Definition 7.57 A germ $[f] \in \mathcal{E}_n$ is *k-determined* if every germ in $\mathcal{E}_n$ that is k-equivalent to $[f]$ is differentiably equivalent to $[f]$. The *determinacy* of a germ $[f]$, denoted $\operatorname{detc}[f]$, is the smallest k such that $[f]$ is k-determined. If $\operatorname{detc}[f] < \infty$, then $[f]$ is *finitely determined.*

If $[f] \in \mathcal{E}_n$ is k-determined and $[h] \in \mathcal{E}_n$ is differentiably equivalent to $[f]$, then $[h]$ is k-determined (e.g., [Cas-H, Cor 12, p. 84]).

Let

$$\mathfrak{M}_n^k = \{[f] \in \mathcal{E}_n : j^{k-1}[f] = 0\} = \{[f] \in \mathcal{E}_n : \frac{\partial^{|\tau|} f}{\partial x^\tau}(0) = 0 \text{ for all } 0 \le |\tau| < k\}.$$

For example,

$$\mathfrak{M}_n^2 = \{[f] \in \mathcal{E}_n : f(0) = 0, \operatorname{grad} f(0) = 0.\}.$$

Consequently, *if $[f]$ is in $\mathfrak{M}_n^2$, then the origin is a critical point of f.* In this case, we shall also call the origin a *critical point* of $[f]$.

The set $\mathfrak{M}_n^k$ is the ideal generated by germs of monomials in $x_1, \dots, x_n$ of degree k (e.g., [Zee-T, Cor. 2.4, p. 272]). The *Jacobi ideal* $\mathcal{I}[f]$ of $[f] \in \mathcal{E}_n$ is the ideal of $\mathcal{E}_n$ generated by germs of the partials

$$\frac{\partial f}{\partial x_1}, \dots, \frac{\partial f}{\partial x_n}.$$

Definition 7.58 The *codimension* of $[f] \in \mathfrak{M}_n^2$ is defined by

$$\text{cod}[f] = \dim[\mathfrak{M}_n/\mathcal{I}[f]].$$

If $[f] \in \mathfrak{M}_n^2$ is differentiably equivalent to $[h] \in \mathcal{E}_n$, then $\text{cod}[f] = \text{cod}[h]$ (e.g., [Cas-H, Thm. 5, p. 128]). In addition, *for* $[f] \in \mathfrak{M}_n^2$ *the origin is a degenerate critical point if and only* $\text{cod}[f] \geq 1$ [Cas-H, Thm. 6, p. 129].

Theorem 7.59 *If* $[f]$ *lies in the ideal* $\mathfrak{M}_n^2$*, then* $\text{cod}[f] < \infty$ *if and only if* $\text{detc}[f] < \infty$.

A proof of Theorem 7.59 is in [Cas-H, Cor.8, p.130].

Define the *corank* of $[f] \in \mathcal{E}_n$ by: $\text{corank}[f] = n - r$, where $r = \text{rank}(\text{Hess}\, f)(0)$. For $0 \leq s \leq r \leq n$, set

$$q_{sr}(x_1, \dots, x_r) = -x_1^2 - \dots - x_s^2 + x_{s+1}^2 + \dots + x_r^2,$$

where $s + r = n$. The corank of q_{sr} is s, while the rank is r. The next result generalizes Morse's Lemma (page 239).

Theorem 7.60 (Splitting Lemma) *Every* $[f] \in \mathcal{E}_n$ *of corank* s *is differentiably equivalent to the germ at* $0 \in \mathbf{R}^n$ *of*

$$g = q_{sr}(x_1, \dots, x_r) + h(x_{r+1}, \dots, x_n),$$

where $r + s = n$ *and* $h(0) = 0$.

See [Cas-H, p. 64] and [Mart, p. 45] for proofs of the Splitting Lemma.

If the germs $[f]$ and $[g] = [q_{sr}] + [h]$ in the Splitting Lemma are such that $s > 0$, then $\text{cod}[f] = \text{cod}[g]$ (e.g., [Cas-H, Lem. 18, p. 134]). Hence, a classification of the germs $[f]$ according to codimension, reduces to a classification of $[h]$.

Notation: Let q_i be a quadratic of the form

$$-x_i^2 - \dots - x_s^2 + x_{s+1}^2 + \dots + x_n^2,$$

where $i - 1 \leq s \leq n$.

Theorem 7.61 (Classification of Germs at Critical Points) *A germ* $[f] \in \mathfrak{M}_n^2$ *has* $1 \leq \operatorname{cod}[f] \leq 5$ *(equivalently, the origin is a degenerate critical point of* $[f]$ *with* $\operatorname{cod}[f] \leq 5$*) if and only if* $[f]$ *is differentiably equivalent to one of the following germs at the origin:*

Corank 1 Germs	**cod**
$[x_1^3 + q_2]$	*1*
$[x_1^4 + q_2]$, $[-x_1^4 + q_2]$	2
$[x_1^5 + q_2]$	*3*
$[x_1^6 + q_2]$, $[-x_1^6 + q_2]$	*4*

Corank 2 Germs	**cod**
$[x_1^3 - x_1x_2^2 + q_3]$	*3*
$[x_1^3 + x_2^3 + q_3]$	*3*
$[x_1^2x_2 + x_2^4 + q_3]$, $[-x_1^2x_2 - x_2^4 + q_3]$	*4*

Codimension 5 Germs
$[x_1^7 + q_2]$
$[x_1^3 + x_2^4 + q_3]$, $[-x_1^3 - x_2^4 + q_3]$
$[x_1^2x_2 + x_2^5 + q_3]$
$[x_1^2x_2 - x_2^5 + q_3]$

A detailed proof of Theorem 7.61 is given in [Cas-H, Chap. 6, p. 147].

Unfoldings

Having presented the classification of germs, we now turn to the notion of "unfoldings." The "unfoldings" of most interest to us are those that are "versal" and "universal." The existence and classification of the latter will be presented at the end of the section (Theorems 7.70(3) and 7.71).

Definition 7.62 An *r-parameter unfolding* (or *deformation*) of $[f] \in \mathcal{E}_n$ is a germ $F : (\mathbf{R}^n \times \mathbf{R}^r, (0,0)) \to (\mathbf{R}, F(0,0))$, where $(x, y) \mapsto F(x, y)$ and $F(x, 0) = f(x)$ for all x in a neighborhood of $0 \in \mathbf{R}^n$. Hence, $x = (x_1, \ldots, x_n)$ and $y = (y_1, \ldots, y_r)$ are called, respectively, the *state variables* and *control parameters* of F.

The unfolding $[F] \in \mathcal{E}_{n+r}$ satisfying $F(x, y) = f(x)$ is called the *constant r-parameter unfolding* of f.

Consider the germ $h : (\mathbf{R}^s, 0) \to (\mathbf{R}^r, 0) : t \mapsto h(t)$ and an r-parameter unfolding $[F] \in \mathcal{E}_{n+r}$ of $[f] \in \mathcal{E}_n$. The *pull-back* of F by $[h]$ is the s-parameter unfolding h^*F of f defined by

$$(h^*F)(x, t) = F(x, h(t)).$$

Definition 7.63 Let $[F] \in \mathcal{E}_{n+r}$ be an r-parameter unfolding of $[f] \in \mathcal{E}_n$. An unfolding $[G] \in \mathcal{E}_{n+s}$ is *induced* by F if there are three germs $\varphi : (\mathbf{R}^n \times \mathbf{R}^s, 0) \to (\mathbf{R}^n, 0)$ (*variable transformation*), $h : (\mathbf{R}^s, 0) \to (\mathbf{R}^r, 0)$ (*parameter transformation*), and $g : (\mathbf{R}^s, 0) \to (\mathbf{R}, 0)$ (*shearing function*) such that

$$G(x, t) = F(\varphi(x, t), h(t)) + g(t),$$

where $\varphi(x, 0) = x$.

Let $\pi : \mathbf{R}^n \times \mathbf{R}^s \to \mathbf{R}^s$ be the projection $\pi(x, t) = t$. Define a local map λ of $\mathbf{R}^n \times \mathbf{R}^s$ at 0 into $\mathbf{R}^n \times \mathbf{R}^s$ by: $\lambda(x, t) = (\varphi(x, t), t)$. Note that $\lambda(0, 0) = (0, 0)$ and $\pi = \pi \circ \lambda$. The condition on $[G]$ in Definition 7.63 becomes: $G = (h^* F + g \circ \pi) \circ \lambda$.

Proposition 7.64 *The mapping λ is a local diffeomorphism at the origin in $\mathbf{R}^n \times \mathbf{R}^s$.*

Proof: We have $\det[d_{(x,t)}\lambda] = \det[d_x\varphi^t]$, where $\varphi^t(x) = \varphi(x, t)$. Since $\varphi^0(x) = \varphi(x, 0) = x$, it follows that $\det[d_{(0,0)}\lambda] = \det[d_0\varphi^0] = 1$. By the Inverse Function Theorem, λ is a local diffeomorphism at $(0, 0) \in \mathbf{R}^n \times \mathbf{R}^s$. □

Two unfoldings $[F] \in \mathcal{E}_{n+r}$ and $[G] \in \mathcal{E}_{n+s}$ of $[f] \in \mathcal{E}_n$ are *equivalent* if $r = s$ and $[G]$ is induced by $[F]$ such that the parameter transformation germ $[h] \in \mathcal{E}_0(r, r)$ is a local diffeomorphism at $0 \in \mathbf{R}^r$.

Definition 7.65 An r-parameter unfolding $[F] \in \mathcal{E}_{n+r}$ of $[f] \in \mathcal{E}_n$ is *versal* if all other unfoldings of $[f]$ are induced by $[F]$. If r is the smallest integer for which a versal unfolding of $[f]$ exists, then $[F]$ is a *universal unfolding* of $[f]$.

Proposition 7.66 *If $[F] \in \mathcal{E}_{n+r}$ is a universal unfolding of $[f] \in \mathcal{E}_n$, then the constant unfolding*

$$\hat{F} : (\mathbf{R}^n \times \mathbf{R}^r \times \mathbf{R}^p, 0) \longrightarrow (\mathbf{R}, \hat{F}(0))$$

given by

$$\hat{F}(x, u, v) = F(x, u),$$

where $u \in \mathbf{R}^r$ and $v \in \mathbf{R}^p$, is a versal unfolding of $[f]$.

Proof: Let $G \in \mathcal{E}_{n+s}$ be an unfolding of $[f] \in \mathcal{E}_n$. Then G is induced by F, say,

$$G(x, t) = F(\varphi(x, t), h(t)) + g(t),$$

where $\varphi(x, 0) = x$. Define $\hat{h} : \mathbf{R}^s \to \mathbf{R}^r \times \mathbf{R}^p$ by $\hat{h}(t) = (h(t), l(t))$ for some smooth function $l : \mathbf{R}^s \to \mathbf{R}^p$ with $l(0) = 0$. Then:

$$\hat{F}(\varphi(x, t), \hat{h}(t)) + g(t) = F(\varphi(x, t), h(t)) + g(t) = G(x, t). \square$$

k-Transversal Unfoldings and Universal Unfoldings

This subsection presents several relationships between transversality and versality. In fact, transversality will provide the primary tool used to obtain the existence of versal and universal unfoldings [Cas-H, Chap. 8, 10].

Let $[F] \in \mathcal{E}_{n+r}$ be an r-parameter unfolding of $[f] \in \mathfrak{M}_n^2$. Let $\mathcal{V}[F]$ be the vector subspace of $\mathfrak{M}_n$ spanned by the germs at be the vector subspace of $\mathfrak{M}_n$ spanned by the germs at $(0,0) \in \mathbf{R}^n \times \mathbf{R}^r$ of the local functions

$$\frac{\partial F}{\partial y_i}(x,0) - \frac{\partial F}{\partial y_i}(0,0),$$

where $i = 1, \dots, r$.

Definition 7.67 The germ $[F] \in \mathcal{E}_{n+r}$ is a *k-transversal unfolding* of $[f] \in \mathcal{E}_n$ for an integer $k > 0$ if

$$\mathfrak{M}_n = \mathcal{I}[f] + \mathcal{V}[F] + \mathfrak{M}_n^{k+1}.$$

First, we shall present a result (see Theorem 7.68) that shows k-transversality is equivalent to a transversality condition.

Let $\mathcal{G}_n$ be the group of germs

$$\lambda : (\mathbf{R}^n, 0) \to (\mathbf{R}^n, 0)$$

of local diffeomorphisms at $0 \in \mathbf{R}^n$. Then $\mathcal{G}_n$ acts on $\mathcal{E}_n$ by

$$([f], [\lambda]) \mapsto [f][\lambda] = [f \circ \lambda],$$

where $[f] \in \mathcal{E}_n$ and $[\lambda] \in \mathcal{G}_n$. Denote by $\mathcal{G}_n^k$ the group of all k-jets of germs in $\mathcal{G}_n$. This group acts on $J^k(n)$ (page 228) as follows:

$$(j^k[f], j^k[\lambda]) \mapsto j^k[f]j^k[\lambda] = j^k([f \circ \lambda]),$$

where $[f] \in \mathcal{E}_n$ and $[\lambda] \in \mathcal{G}_n$. The orbit through $j^k[f] \in J^k(n)$ will be denoted by $(\mathcal{G}_n^k)_{j^k[f]}$; see [Cas-H, pp. 98–107] for more on $(\mathcal{G}_n^k)_{j^k[f]}$.

Let $[F] \in \mathcal{E}_{n+r}$ be an r-parameter unfolding of a germ $[f] \in \mathfrak{M}_n$. Denote by $F^{x,y}$ a local function on $\mathbf{R}^n$ at 0 defined by

$$F^{x,y}(u) = F(x+u, y) - F(x, y).$$

Note that $F^{0,0}(u) = f(u)$. Define a local map F_{pro}^k of $\mathbf{R}^n \times \mathbf{R}^r$ at 0 into $J^k(n)$ by

$$F_{pro}^k(x, y) = j^k[F^{x,y}].$$

The germ

$$F_{pro}^k : (\mathbf{R}^n \times \mathbf{R}^r, 0) \to (J^k(n), j^k[f])$$

is called the *k-jet prolongation* (or *k-extension*) of the unfolding $[F]$.

Theorem 7.68 *An r-parameter unfolding $[F] \in \mathcal{E}_{n+r}$ of $[f] \in \mathcal{E}_n$ is k-transversal if and only if $F^k_{pro} \pitchfork (\mathcal{G}^k_n)_{j^k[f]}$.*

See [Zee-T, p. 305] and [Cas-H, pp. 193–196] for proofs of Theorem 7.68.

The next two theorems illustrate the usefulness of k-transversality as a tool for studying versal unfoldings.

Theorem 7.69 *Suppose that the germ $[f]$ lies in $\mathfrak{M}^2_n$. Then:*

(1) *Every versal unfolding of $[f]$ is k-transversal for every positive integer k. If $[f]$ is k-determined, then every k-transversal unfolding of $[f]$ is versal.*

(2) *Let $[F]$ be an r-parameter unfolding of $[f]$. If $[F]$ is k-transversal for an integer $k \geq r+1$, then $[f]$ is $(r+2)$-determined with* $\mathrm{cod}[f] \leq r$.

A proof of Theorem 7.69 is in [Cas-H, Lem. 3, p. 192; Lems. 2,3, p. 229]. The proof uses a deep result — the *Malgrange–Mather Preparation Theorem* (see [Cas-H, Chap. 9, p. 199]).

Employing Theorem 7.69 yields the following important result about versal unfoldings — see [Cas-H, p. 230] for proof:

Theorem 7.70 *Suppose that the germ $[f]$ lies in $\mathfrak{M}^2_n$. Then:*

(1) *An r-parameter unfolding $[F]$ of $[f]$ is versal precisely when $[F]$ is k-transversal for all integers $k \geq 1$.*

(2) *An unfolding $[F]$ of the germ $[f]$ is versal if and only if $\mathfrak{M}_n = \mathcal{I}[f] + \mathcal{V}[F]$.*

(3) **(Existence)** *The germ $[f]$ has a versal unfolding if and only if $[f]$ is finitely determined. Moreover, any two r-parameter versal unfoldings of $[f]$ are equivalent.*

Application of Theorems 7.69 and 7.70 now give another key result of catastrophe theory, namely, the classification of universal unfoldings:

Theorem 7.71 (The Universal Unfolding Theorem) *Let $[f]$ be a germ in $\mathfrak{M}^2_n$.*

(1) **(Existence)** *The germ $[f]$ has a universal unfolding if and only if* $\mathrm{cod}[f]$ *is finite.*

(2) *If* $\mathrm{cod}[f] = r < \infty$, *then the germ in $\mathfrak{M}^2_{n+r}$ of*

$$F(x,t) = f(x) + t_1 e_1(x) + \cdots + t_r e_r(x)$$

is a universal unfolding of $[f]$ for each basis

$$\{[e_i] + \mathcal{I}[f]\}_{i=1}^r$$

of $\mathfrak{M}_n/\mathcal{I}[f]$.

(3) *Every versal unfolding of* $[f]$ *is equivalent to an unfolding of the form* $[F] + [C]$, *where* $[F]$ *is given above in (2) and* $[C]$ *is a constant unfolding of* $[f]$.

We refer the reader to [Cas-H, p. 231] for a proof of Theorem 7.71.

7.6.2 Elementary Catastrophes and Thom's Theorem

The Elementary Catastrophes

Applications of Theorems 7.61 and 7.71 yield the classification of r-parameter universal unfoldings of germs $[f] \in \mathfrak{M}_n^2$ with $1 \leq \operatorname{cod}[f] = r \leq 5$ (see [Cas-H, pp. 173, 184] for details).

Theorem 7.72 *Let* $[f]$ *be a germ in* $\mathfrak{M}_n^2$ *with* $1 \leq \operatorname{cod}[f] = r \leq 5$ *(equivalently, the germ* $[f]$ *has a degenerate critical point at* $0 \in \mathbf{R}^n$ *and* $\operatorname{cod}[f] = r \leq 5$*). Every* r*-parameter universal unfolding of* $[f]$*, say*

$$F : (\mathbf{R}^n \times \mathbf{R}^r, (0, 0)) \to (\mathbf{R}, 0) : (x, t) \mapsto F(x, t),$$

where $x = (x_1, \ldots, x_n)$ *and* $t = (t_1, \ldots, t_r)$*, is equivalent to one of the universal unfoldings in the rightmost column of Table 7.1.*

Each universal unfolding F in Table 7.1 is called an *elementary catastrophe unfolding*. In addition, the terms q_2 and q_3 are explicitly as follows:

$$q_2 = -x_2^2 - \cdots - x_i^2 + x_{i+1}^2 + \cdots + x_n^2, \qquad (1 \leq i \leq n),$$

and

$$q_3 = -x_3^2 - \cdots - x_i^2 + x_{i+1}^2 + \cdots + x_n^2, \qquad (2 \leq i \leq n),$$

respectively.

Catastrophe Maps

Let $\mathcal{F} = C^\infty(\mathbf{R}^n \times \mathbf{R}^r, \mathbf{R})$, which is assume to have the Whitney C^∞-topology. Given a family of functions

$$F : \mathbf{R}^n \times \mathbf{R}^r \to \mathbf{R},$$

we call $\mathbf{R}^n$ the *state space* and the parameter space, $\mathbf{R}^r$, the *control space*. We take as coordinates (x, t) in $\mathbf{R}^n \times \mathbf{R}^r$, where $x = (x_1, x_2, \ldots, x_n)$ and $t = (t_1, t_2, \ldots, t_r)$. For any $F \in \mathcal{F}$, we set

$$M_F = \{(x, t) \in \mathbf{R}^n \times \mathbf{R}^r : \frac{\partial F}{\partial x_i}(x, t) = 0 \text{ for } i = 1, 2, \ldots, n\},$$

Name	Germ $[f]$	Universal Unfolding $[F]$
fold	$[x_1^3 + q_2]$	$[x_1^3 + t_1x_1 + q_2]$ $r = 1$
cusp	$[\pm x_1^4 + q_2]$	$[\pm(x_1^4 - t_1x_1^2 + t_2x_1) + q_2]$ $r = 2$
swallowtail	$[x_1^5 + q_2]$	$[x_1^5 + t_1x_1^3 + t_2x_1^2 + t_3x_1 + q_2]$ $r = 3$
butterfly	$[\pm x_1^6 + q_2]$	$[\pm(x_1^6 + t_1x_1^4 + t_2x_1^3 + t_3x_1^2 + t_4x_1) + q_2]$ $r = 4$
elliptic umbilic	$[x_1^3 - x_1x_2^2 + q_3]$	$[x_1^3 - x_1x_2^2 + t_1(x_1^2 + x_2^2) - t_2x_1 - t_3x_2 + q_3]$ $r = 3$
hyperbolic umbilic	$[x_1^3 + x_2^3 + q_3]$	$[x_1^3+x_2^3+t_1x_1x_2-t_2x_1-t_3x_2+q_3]$ $r = 3$
parabolic umbilic	$[\pm(x_1^2x_2 + x_2^4) + q_3]$	$[\pm(x_1^2x_2+x_2^4+t_1x_1^2+t_2x_2^2-t_3x_1-t_4x_2)+q_3]$ $r = 4$
wigwam	$[x_1^7 + q_2]$	$[x_1^7 + t_1x_1^5 + t_2x_1^4 + t_3x_1^3 + t_4x_1^2 + t_5x_1 + q_2]$ $r = 5$
symbolic umbilic	$[\pm(x_1^3 + x_2^4) + q_3]$	$[\pm(x_1^3 + x_2^4 + t_1x_1x_2^2 + t_2x_1x_2 + t_3x_2^2 + t_4x_2 + t_5x_1) + q_3]$ $r = 5$
2nd hyperbolic umbilic	$[x_1^2x_2 + x_2^5 + q_3]$	$[x_1^2x_2 + x_2^5 + t_1x_2^4 + t_2x_2^3 + t_3x_2^2 + t_4x_2 + t_5x_1 + q_3]$ $r = 5$
2nd elliptic umbilic	$[x_1^2x_2 - x_2^5 + q_3]$	$[x_1^2x_2 - x_2^5 + t_1x_2^4 + t_2x_2^3 + t_3x_2^2 + t_4x_2 + t_5x_1 + q_3]$ $r = 5$

Table 7.1. Classification of universal unfoldings $[F]$ of germs $[f]$, where $1 \le \text{cod}[f] = r \le 5$ and F has variables $x = (x_1, \ldots, x_n)$ and parameters $t = (t_1, \ldots, t_r)$. Each q_i is a generalized saddles in the x variables.

which is called the *catastrophe manifold* of F. *It is not a manifold in general. However, the corresponding catastrophe manifold for each elementary catastrophe unfolding (see tables above) is a manifold* (e.g., [Maj, Chaps. 7, 8]).

Let χ_F, *the catastrophe map of* F, be the restriction to M_F of the projection of $\mathbf{R}^n \times \mathbf{R}^r$ onto $\mathbf{R}^r$. The sets $\text{Crit}(\chi_F)$ and $\chi_F\ [\text{Crit}(\chi_F)]$ are called, respectively, the *catastrophe set* and *bifurcation set* of F.

If $[E]$ is an elementary catastrophe unfolding, then $[\chi_E]$ is called an *elementary catastrophe map germ* and $\text{Crit}(\chi_E)$ is termed an *elementary catastrophe*.

Example (Fold Catastrophe)**:** Consider the constant unfolding

$$F_{fold}(x_1, \ldots, x_n, t_1, t_2) = x_1^3 + t_1x_1 - x_2^2 - \cdots - x_s^2 + x_{s+1}^2 + \cdots + x_n^2$$

of the fold universal unfolding. By Proposition 7.66, the unfolding $[F]$ is versal. The corresponding catastrophe manifold M_{fold} is defined by

$$\begin{aligned}\frac{\partial}{\partial x_1}F_{fold} &= 3x_1^2 + t_1 = 0\\ \frac{\partial}{\partial x_i}F_{fold} &= \pm 2x_i = 0,\ i = 2, \ldots, n,\end{aligned}$$

that is,

$$M_{fold} = \{(x_1, 0, \ldots, 0, -3x_1^2, t_2)\} \subset \mathbf{R}^n \times \mathbf{R}^2.$$

The catastrophe map $\chi_{fold} : M_{fold} \to \mathbf{R}^2$ is given by

$$\chi_{fold}(x_1, 0, \ldots, 0, -3x_1^2, t_2) = (3x_1^2, t_2).$$

Example (Cusp Catastrophe)**:** The constant unfolding

$$F_{cusp}(x_1, \ldots, x_n, t_1, t_2) = \pm(x_1^4 - t_1x_1^2 + t_2x_1) - x_2^2 - \cdots - x_s^2 + x_{s+1}^2 + \cdots + x_n^2$$

of the cusp universal unfolding is versal (Proposition 7.66). The equations

$$\begin{aligned}\frac{\partial}{\partial x_1}F_{cusp} &= 4x_1^3 - 2t_1x_1 + t_2 = 0\\ \frac{\partial}{\partial x_i}F_{cusp} &= \pm 2x_i = 0,\ i = 2, \ldots, n,\end{aligned}$$

yield the catastrophe manifold

$$M_{cusp} = \{(x_1, 0, \ldots, 0, t_1, \mp(4x_1^3 - 2t_1x_1))\} \subset \mathbf{R}^n \times \mathbf{R}^2.$$

The catastrophe map $\chi_{cusp} : M_{cusp} \to \mathbf{R}^2$ is given by

$$\chi_{cusp}(x_1, 0, \ldots, 0, t_1, \mp(4x_1^3 - 2t_1x_1)) = (t_1, \mp(4x_1^3 - 2t_1x_1)).$$

Define mappings $\lambda_{fold} : \mathbf{R}^2 \to M_{fold}$ and $\lambda_{cusp} : \mathbf{R}^2 \to M_{cusp}$ by

$$\lambda_{fold}(x_1, t_2) = (x_1, 0, \ldots, 0, -3x_1^2, t_2)$$

and

$$\lambda_{cusp}(x_1, t_1) = (x_1, 0, \ldots, 0, t_1, \mp(4x_1^3 - 2t_1x_1)),$$

respectively. Let $h_{fold} : \mathbf{R}^2 \to \mathbf{R}^2$ and $h_{cusp} : \mathbf{R}^2 \to \mathbf{R}^2$ be defined by

$$h_{fold}(x_1, t_2) = (-3x_1^2, t_2), \qquad h_{cusp}(x_1, t_1) = (t_1, \mp(4x_1^3 - 2t_1x_1)).$$

The mappings h_{fold} and h_{cusp} are called a *Whitney fold* and *pleat* respectively. Since λ_{fold} and λ_{cusp} are diffeomorphisms, we obtain:

Proposition 7.73 *Using the previous mappings, we have:*

(1) *The catastrophe fold χ_{fold} is differentiably equivalent to the Whitney fold h_{fold}, that is, $h_{fold} = \chi_{fold} \circ \lambda_{fold}$.*

(2) *The catastrophe cusp χ_{cusp} is differentiably equivalent to the Whitney pleat h_{cusp}, that is, $h_{cusp} = \chi_{cusp} \circ \lambda_{cusp}$.*

Catastrophe Maps and Lensing Maps

In gravitational lensing, the time delay families $T : L_{(k)} \times S \to \mathbf{R}$, where $L_{(k)} = L_1 \times \cdots \times L_k$, are the unfoldings of interest. The catastrophe map $\chi_T : M_T \to S$ associated with T is (globally) differentiably equivalent to the lensing map $\boldsymbol{\eta} : P \to S$ induced by T. In fact, consider the map $\boldsymbol{\Upsilon}_S : P \to P \times L_2 \times \cdots \times L_k \times S$ given by

$$\boldsymbol{\Upsilon}_S(\mathbf{x}_1) = (\mathbf{x}_1, \boldsymbol{\eta}_2(\mathbf{x}_1), \ldots, \boldsymbol{\eta}_k(\mathbf{x}_1), \boldsymbol{\eta}_{k+1}(\mathbf{x}_1),$$

where $\boldsymbol{\eta}_{k+1} = \boldsymbol{\eta}$. Then $\boldsymbol{\Upsilon}_S$ is a diffeomorphism onto its image (see Section 6.5, page 203). Now, elements of M_T are all points $(\mathbf{x}_1, \ldots, \mathbf{x}_k, \mathbf{y})$ satisfying

$$\text{grad}_{\mathbf{x}_i} T(\mathbf{x}_1, \ldots, \mathbf{x}_k, \mathbf{y}) = \mathbf{0}$$

for $i = 1, \ldots, k$. By Theorem 6.13 (pages 199, 201), these gradient conditions are equivalent to

$$\mathbf{x}_2 = \boldsymbol{\eta}_2(\mathbf{x}_1), \ \ldots, \mathbf{x}_k = \boldsymbol{\eta}_k(\mathbf{x}_1), \mathbf{y} = \boldsymbol{\eta}(\mathbf{x}_1).$$

Consequently, points of M_T are given by $\boldsymbol{\Upsilon}_S(\mathbf{x}_1)$, that is, $M_T = \boldsymbol{\Upsilon}_S(P)$. Hence: $\boldsymbol{\eta} = \chi_T \circ \boldsymbol{\Upsilon}_S$. Thus, we showed the following:

Proposition 7.74 *Let $T : L_{(k)} \times S \to \mathbf{R}$ be a k-plane time delay family with $\chi_T : M_T \to S$ and $\boldsymbol{\eta} : P \to S$ the corresponding catastrophe and lensing maps, respectively. Then $\boldsymbol{\eta} = \chi_T \circ \boldsymbol{\Upsilon}_S$, that is, the map χ_T is differentiably equivalent to $\boldsymbol{\eta}$.*

Corollary 7.75 *Let T be a k-plane time delay family with lensing map $\boldsymbol{\eta} : P \to S$. Then the map $\boldsymbol{\Upsilon}_S$ induces the following:*

(1) *The catastrophe manifold M_T of a k-plane time delay family T is a 2-manifold diffeomorphic to the planar region P.*

(2) *There is a bijection between the catastrophe set of χ_T and set of critical points of $\boldsymbol{\eta}$.*

(3) *The bifurcation set of χ_T coincides with the set of caustics of $\boldsymbol{\eta}$.*

Thom's Catastrophe Theorem

Given a family of functions $F : \mathbf{R}^n \times \mathbf{R}^r \to \mathbf{R}$ whose catastrophe manifold is a manifold, the catastrophe map χ_F is called *F-locally stable* at $(\mathbf{x}_0, \mathbf{y}_0) \in M_F$ if given a neighborhood N of $(\mathbf{x}_0, \mathbf{y}_0) \in \mathbf{R}^n \times \mathbf{R}^r$ there is a neighborhood $\mathcal{G}$ of F in $\mathcal{F}$ such that given $g \in \mathcal{G}$ there is $(\mathbf{x}_1, \mathbf{y}_1) \in N \cap M_g$ such that the germ of χ_F at $(\mathbf{x}_0, \mathbf{y}_0)$ is equivalent to the germ of χ_g at $(\mathbf{x}_1, \mathbf{y}_1)$. We now present the central theorem of catastrophe theory:

Theorem 7.76 (Thom's Catastrophe Theorem) *If $r \leq 5$, then there is an open dense subset, $\mathcal{F}_* \subset \mathcal{F}$ such that for each $F \in \mathcal{F}_*$ we have:*

(1) *The catastrophe manifold M_F is an r-dimensional submanifold of* $\mathbf{R}^n$.

(2) *For each critical point p of χ_F, the germ $[\chi_F]_p$ is differentiably equivalent to an elementary catastrophe map germ $[\chi_E]$.*

(3) *The catastrophe map χ_F is F-locally stable.*

A proof of Theorem 7.76 is in [Zee-T]. This theorem can be extended to codimension $r \leq 6$, as well as $r = 7$ if the corank is 2. In the latter case, there are 17 elementary catastrophes (not counting duals); see [Cas-H, pp. 147,155–156, 231] for details. Theorem 7.76 also yields an open-density result for maps satisfying a certain stability condition and then locally classifies those maps about critical points using coordinate forms (namely, the elementary catastrophes). Results of this sort are a major theme in singularity theory. This can be seen from Mather's nice dimensions (Theorem 7.32, page 231), Morse theory (Theorem 7.42, page 239; Theorem 7.46(1), page 242), and Whitney theory (Theorem 7.49, page 251; Theorem 7.55, page 261).

In addition, Proposition 7.73 and Theorem 7.76 imply that if $r = 2$, then each family $F : \mathbf{R}^n \times \mathbf{R}^2 \to \mathbf{R}$ in $\mathcal{F}_*$ has a catastrophe map $\chi_F : M_F \to \mathbf{R}^2$ such that *for any critical point $p \in M_F$, the catastrophe map χ_F at p is differentiably equivalent to either a Whitney fold or a pleat at* $0 \in \mathbf{R}^2$. In other words, each critical point of χ_F for $F \in \mathcal{F}^*$ is either a fold or cusp. Hence, a study of Whitney folds and pleats is important in this case. In addition, Theorem 7.55 (page 261) shows that *since the catastrophe maps $\chi_F : M_F \to \mathbf{R}^2$ of Theorem 7.76 with $r = 2$ have only folds and cusps as possible critical points, such catastrophe maps are also locally stable.*

Remark: Let $T : L_{(k)} \times S \to \mathbf{R}$ be a k-plane time delay family that is a universal unfolding, so its catastrophe map $\chi_T : M_T \to S$ is differentiably equivalent to an elementary catastrophe map at each critical point of χ_T. The previous discussion plus differentiable equivalence of catastrophe and lensing maps (Proposition 7.74) yield that the caustics of T in the light-source plane S are either folds or cusps. However, we *cannot* conclude from Thom's catastrophe theorem that there is an open dense set $\mathfrak{D}^*$ in the space $\mathfrak{D}$ of k-plane time delay families such that for each $T \in \mathfrak{D}^*$ the caustics are folds or cusps. Indeed, Thom's theorem applies to a space $\mathcal{F}$ containing $\mathfrak{D}$ as a proper subset. There is no guarantee that the induced open subset $\mathcal{F}^* \cap \mathfrak{D}$ of $\mathfrak{D}$ is dense in $\mathfrak{D}$.[11] This issue will be taken up in Section 8.4, where the existence of the desired open dense set $\mathfrak{D}^*$ will be established.

7.7 Arnold Singularity Theory

The theory of singularities due to Arnold and his students is beautiful and powerful in its far-reaching applications. Arnold made "spectacular advances in the techniques of classification, and in understanding the structure of the objects and the [classification] lists he

[11] For example, let $\mathcal{F} = \mathbf{R}$, $\mathcal{F}^* = \mathbf{R} - I$, and $\mathfrak{D} = I$, where I is the set of integers. Then $\mathcal{F}^*$ is open and dense in $\mathcal{F}$, but $\mathcal{F}^* \cap \mathfrak{D} = \emptyset$ cannot be dense in $\mathcal{F}$.

found. He also introduced his notions of Lagrangian and Legrendrian singularities, which systematize many of the mathematical ideas behind Catastrophe Theory" — Bruce and Mond [Bru-M, p. xi]. The Arnold theory yields a classification of the critical points of stable simple Lagrangian maps (and 1-parameter evolutions) that includes and goes beyond Thom's list of elementary catastrophes. In particular, the theory leads to a classification (with moduli) for dimensions $n \leq 10$, while Thom's theory was originally for $n \leq 4$, which has since been generalized to $n \leq 6$ (as well as to families with $n = 7$ and corank 2; e.g., [Cas-H, pp. 147, 231]). Arnold's theory can also be employed to classify generic 1-parameter evolutions of Lagrangian maps. Applications to lensing are given in Section 9.5.1 (page 375).

References: Introductions to Arnold's singularity theory are given by Arnold et al. [Arn73], [Arn91], [Arn-GVI]. We also recommend the recent, excellent review for physicists by Ehlers and Newman [Ehl-N]. We add that Arnold's theory can also be applied to optical systems in a general relativistic setting; consult Friedrich and Stewart [Fri-S] and Stewart [Stew].

7.7.1 Lagrangian Maps and Generating Families

Symplectic Manifolds and Lagrangian Submanifolds

Let M be an even-dimensional manifold. A *symplectic structure* on M is a closed nondegenerate 2-form ω. The pair (M, ω) is called a *symplectic manifold.* A familiar example of a symplectic manifold is $\mathbf{R}^{2n}$. In fact, cover $\mathbf{R}^{2n}$ by coordinates $(p_1, \ldots, p_n; q_1, \ldots, q_n)$. Then $\omega = \sum_{i=1}^{n} dp_i \wedge dq_i$ is a symplectic structure on $\mathbf{R}^{2n}$. If $n = 1$, then $\omega = dp \wedge dq$, which is the oriented area element of $\mathbf{R}^2 = \{(p, q)\}$. Consequently, by expressing $\mathbf{R}^{2n}$ as $\mathbf{R}^2 \times \cdots \times \mathbf{R}^2 = \{(p_1, q_1)\} \times \cdots \times \{(p_n, q_n)\}$, the symplectic structure $\omega = \sum_{i=1}^{n} dp_i \wedge dq_i$ becomes a sum of the oriented area elements $dp_i \wedge dq_i$ of the planes $\mathbf{R}^2 = \{(p_i, q_i)\}$. The pair $(\mathbf{R}^{2n}, \sum_{i=1}^{n} dp_i \wedge dp_i)$ is called a *standard symplectic manifold.* By identifying the cotangent bundle $T^*\mathbf{R}^{2n}$ with $\mathbf{R}^{2n}$, we see that $T^*\mathbf{R}^{2n}$ becomes a symplectic manifold with the standard symplectic structure.

Diffeomorphisms $f : (M_1^{2n}, \omega_1) \longrightarrow (M_2^{2n}, \omega_2)$ between symplectic manifolds that preserve the symplectic structures (i.e., the pull-back f^* of f satisfies $f^*(\omega_2) = \omega_1$) are called *symplectomorphisms*. Using this natural notion of equivalence for symplectic manifolds, Darboux proved that all symplectic manifolds of a fixed dimension $2n$ are locally symplectomorphic to the standard $2n$-symplectic manifold (e.g., [Arn-GVI, p. 287]). Thus, locally all symplectic structures measure sum of oriented area elements.

A *Lagrangian submanifold* of a symplectic manifold (M^{2n}, ω) is an n-dimensional submanifold $L \subseteq M^{2n}$ such that $\iota^*(\omega) = 0$, where $\iota : L \longrightarrow M^{2n}$ is the inclusion map. For example, the planes $p =$ constant are Lagrangian submanifolds of the standard symplectic manifold $\mathbf{R}^{2n}$. As in the case for symplectic manifolds, there is a Darboux type theorem which states that all Lagrangian submanifolds of a fixed dimension are locally equivalent (e.g., [Arn91, pp. 7, 14], [Arn-GVI, p. 292]). This means that for any two Lagrangian submanifolds L_1 and L_2 of symplectic manifolds (M_1^{2n}, ω_1) and (M_2^{2n}, ω_2), respectively, there is a local symplectomorphism of (M_1^{2n}, ω_1) and (M_2^{2n}, ω_2) which induces a local diffeomorphism $\psi : \mathcal{U}_1 \subseteq L_1 \longrightarrow \psi(\mathcal{U}_1) \subseteq L_2$ such that for each $p \in \mathcal{U}_1$ we have $(\psi^{-1})^*(\omega_1\big|_p) = \omega_2\big|_{\psi^{-1}(p)}$. Hence, all Lagrangian submanifolds are locally of the form $p =$ constant.

Generating Families of Lagrangian Submanifolds

An n-parameter family of functions

$$F : \mathbf{R}^k \times \mathbf{R}^n \to \mathbf{R} : (x, q) \mapsto F(x, q)$$

satisfying

$$\operatorname{rank}\left[\frac{\partial^2 F}{\partial x_i \partial x_j} \quad \frac{\partial^2 F}{\partial x_i \partial q_m}\right] = k \qquad \text{at } (x, q) \text{ such that } \frac{\partial F}{\partial x}(x, q) = 0$$

is called a *Morse family*. For simplicity, let $F_x(x, q)$ and $F_q(x, q)$ be, respectively, the gradients of F relative to x and q. Also, let

$$F_{xx} = \left[\frac{\partial^2 F}{\partial x_i \partial x_j}\right], \qquad F_{xq} = \left[\frac{\partial^2 F}{\partial x_i \partial q_m}\right].$$

Now, define two submanifolds to be *germ-equivalent* at some point p belonging to both, if they coincide in a neighborhood of p. A *submanifold germ* is any of the associated equivalence classes. The next theorem yields that germs of Lagrangian submanifolds are generated by germs of Morse families. In addition, we shall see that "Lagrangian equivalence" classes of Lagrangian submanifold germs can be identified with "$\mathbf{R}^+$-stable equivalence" classes of germs of Morse families of real-valued functions (see Theorem 7.79).

Theorem 7.77

(1) *Let* $\mathbf{F}$ *be a map-germ of a Morse family* $F : \mathbf{R}^k \times \mathbf{R}^n \to \mathbf{R} : (x, q) \mapsto F(x, q)$. *Then*

$$\mathbf{L_F} = \left\{(p, q) : \text{ there is } x \text{ such that } F_x(x, q) = 0,\ p = F_q(x, q),\ \ F \in \mathbf{F}\right\}$$

is a Lagrangian submanifold germ of $(\mathbf{R}^{2n}, \sum_{i=1}^{n} dp_i \wedge dq_i)$, *and* $\mathbf{F}$ *is called a* **generating family** *of* $\mathbf{L_F}$.

(2) *Conversely, if* $\mathbf{L}$ *is a Lagrangian submanifold germ of* $(\mathbf{R}^{2n}, \sum_{i=1}^{n} dp_i \wedge dp_i)$*, then for some natural number k, there is a map-germ* $\mathbf{F}$ *of a Morse family of functions* $F : \mathbf{R}^k \times \mathbf{R}^n \to \mathbf{R} : (x, q) \mapsto F(x, q)$ *satisfying rank* $[F_{xx}\ F_{xq}] = k$ *and* $\mathbf{L_F} = \mathbf{L}$*.*

A proof of Theorem 7.77 appears in [Arn-GVI, p. 300].

Corollary 7.78 *Every k-plane time delay family* $T : L_{(k)} \times S \to \mathbf{R}$*, where S is the light source plane, is a Morse family and, hence, generate a Lagrangian submanifold germ of the symplectic manifold* T^*S*.*

Proof: Denote the values of T by $T(x_1, \ldots, x_{2k}, y_1, y_2)$. We consider the following $2k \times 2(k+1)$ matrix:

$$\left[T_{x_i x_j}\ \ T_{x_i y_m}\right] = \begin{bmatrix} \mathbf{T}_{11} & \mathbf{I}_1 & \mathbf{0} & \mathbf{0} & \mathbf{0} & \cdots & \mathbf{0} & \mathbf{0} & \mathbf{0} \\ \mathbf{I}_1 & \mathbf{T}_{22} & \mathbf{I}_2 & \mathbf{0} & \mathbf{0} & \cdots & \mathbf{0} & \mathbf{0} & \mathbf{0} \\ \mathbf{0} & \mathbf{I}_2 & \mathbf{T}_{33} & \mathbf{I}_3 & \mathbf{0} & \cdots & \mathbf{0} & \mathbf{0} & \mathbf{0} \\ \mathbf{0} & \mathbf{0} & \mathbf{I}_3 & \mathbf{T}_{44} & \mathbf{I}_4 & \cdots & \mathbf{0} & \mathbf{0} & \mathbf{0} \\ \vdots & \vdots & \vdots & \vdots & \vdots & \cdots & \vdots & \vdots & \vdots \\ \mathbf{0} & \mathbf{0} & \mathbf{0} & \mathbf{0} & \mathbf{0} & \cdots & \mathbf{T}_{k-1,k-1} & \mathbf{I}_{k-1} & \mathbf{0} \\ \mathbf{0} & \mathbf{0} & \mathbf{0} & \mathbf{0} & \mathbf{0} & \cdots & \mathbf{I}_{k-1} & \mathbf{T}_{k,k} & \mathbf{I}_k \end{bmatrix}.$$

For $i = 1, \ldots, k$,

$$\mathbf{T}_{i,i} = (\vartheta_{i-1} + \vartheta_i)\mathbf{I} - \beta_i \vartheta_i (\text{Hess}\ \psi_i), \qquad \mathbf{I}_i = -\vartheta_i \mathbf{I},$$

where $\vartheta_0 = 0$ and $\mathbf{I}$ is the 2×2 identity matrix. Since $\vartheta_i \neq 0$, the matrix has rank $2k$. Hence T is a Morse family. Theorem 7.77 implies that T generates a Lagrangian submanifold germ of $T^*S \approx \mathbf{R}^4$. □

Lagrangian Map-Germs and Germs of Generating Families

First, a *Lagrangian fibration* of a symplectic manifold E^{2n} is a fibre bundle $\pi : E^{2n} \to B^n$ whose total space has a symplectic structure and fibres are Lagrangian submanifolds. The *standard Lagrangian fibration* is

$$(\mathbf{R}^{2n}, \sum_{i=1}^{n} dp_i \wedge dq_i) \to \mathbf{R}^n : (p, q) \mapsto q.$$

Two Lagrangian fibrations are *symplectomorphic* if there are local diffeomorphism between their base spaces, as well as between their total spaces, where fibres are mapped onto fibres and symplectic structures are mapped into each other. There is also a Darboux type theorem for Lagrangian fibrations: All Lagrangian fibrations of fixed dimension are locally symplectomorphic to the standard Lagrangian fibration for that dimension [Arn-GVI, p. 292].

A "Lagrangian map" is an inclusion followed by a Lagrangian fibration. More precisely, let $\iota : L \to E$ be the inclusion of a Lagrangian submanifold L into the total space E of a Lagrangian fibration $\pi : E \to B$. The composition $\pi \circ \iota : L \to B$ is called a *Lagrangian map*. For example, let π be the fibration $\mathbf{R}^{2n} \to \mathbf{R}^n : (p, q) \mapsto p$ and consider a Lagrangian submanifold $L_F = \{(p, q)\}$ of $\mathbf{R}^{2n}$ generated by a Morse family F. Then $L_F \to \mathbf{R}^n : (p, q) \mapsto q$ is a Lagrangian map. Gradient, normal, and Gauss mappings are also all Lagrangian maps — see [Arn92, p. 79–84] for details. The critical values of a Lagrangian map are called *Lagrangian caustics*. Theorem 7.81 will show that the notion of Lagrangian caustics coincide locally with caustics in gravitational lensing.

A *Lagrangian equivalence* of Lagrangian map-germs $(L_i, x_i) \to (E_i, x_i) \to (\mathrm{B}_i, \pi_i(x_i))$, where $i = 1, 2$, and the first arrow is an inclusion with the second a Lagrangian fibration, is defined by the following commutative diagram

$$\begin{array}{ccccc} (L_1, x_1) & \to & (E_i, x_1) & \to & (\mathrm{B}_1, \pi_1(x_1)) \\ \downarrow & & \downarrow & & \downarrow \\ (L_2, x_2) & \to & (E_2, x_2) & \to & (\mathrm{B}_2, \pi_2(x_2)) \end{array} .$$

The middle vertical arrow is a local symplectomorphism and the remaining vertical arrows are local diffeomorphism. Roughly speaking, a Lagrangian equivalence of two Lagrangian map-germs is a symplectomorphism of the total spaces that sends one Lagrangian fibration to the other and one Lagrangian inclusion to the other. We add that every Lagrangian map-germ is Lagrangian equivalent to a gradient map-germ, as well as a normal map-germ, and also a Gauss map-germ (e.g., [Arn-GVI, p. 296]). For more on Gauss mappings, readers are referred to Banchoff, Gaffney and McCrory [Ban-GM], who geometrically characterize cusps of Gauss mappings.

Now, two families of functions, say, $F_1(x, q)$ and $F_2(x, q)$, where $x \in \mathbf{R}^k$ and $q \in \mathbf{R}^n$, are called *R^+-equivalent* if there is a diffeomorphism $\phi : \mathbf{R}^n \to \mathbf{R}^n : q \mapsto \phi(q)$ and a smooth map $h : \mathbf{R}^k \times \mathbf{R}^n \to \mathbf{R}^k : (x, q) \mapsto h(x, q)$ such that for each $q \in \mathbf{R}^n$ the map $h(\cdot, q)$ is a diffeomorphism with

$$F_1(x, q) = F_2(h(x, q), \phi(q)) + \Phi(q),$$

where $\Phi : \mathbf{R}^n \to \mathbf{R}$ is smooth. In other words, R^+-equivalence of families of functions roughly means that there is a smooth change of the independent variables x and parameters q of one family which will transform that family into the other if a smooth real-valued function of q is added. Germs of n-parameter families of functions of not necessarily the same number of variables are *stably R^+-equivalent* if adding nondegenerate quadratic forms in new variables to their respective representatives makes the representatives R^+-equivalent. For example, the two-parameter families $F(x; q_1, q_2) = x^4 + q_1x^2 + q_2x$ and $G(x, y; q_1, q_2) = x^4 + q_1x^2 + q_2x + y^2$ of one and two variables, respectively, are stably R^+-equivalent.

The next result identifies Lagrangian map-germs with germs of Morse families of functions using the above equivalences:

Theorem 7.79 *Two Lagrangian map-germs are Lagrangian equivalent if and only if the germs of the Morse families generating them are stably R^+-equivalent.*

Consult [Arn-GVI, p. 305] for a proof of Theorem 7.79. Qualitatively speaking, this theorem translates the local classification of Lagrangian maps and, hence, Lagrangian caustics, into that of Morse families of functions. The result is given in the next subsection (see Theorem 7.80).

Classification of Stable Simple Lagrangian Singularities

Several preliminary concepts are needed before we can state Arnold's classification theorem:

(1) Define a *Lagrangian singularity* to be a Lagranginan map-germ, considered up to Lagrangian equivalence, at a critical point.

(2) A Lagrangian map-germ $(\mathbf{f}, x_0)$ at x_0 is *Lagrange stable* if for any representative Lagrangian map f of $(\mathbf{f}, x_0)$, there is a neighborhood $\mathcal{F}$ in the space of Lagrangian maps (with the Whitney C^∞-topology) where for all $g \in \mathcal{F}$ there is an open set U about x_0 and a point $y_0 \in U$ such that the germ of g at y_0 is Lagrange equivalent to the germ $(\mathbf{f}, x_0)$. In other words, a Lagrangian map-germ $(\mathbf{f}, x_0)$ is Lagrangian stable if for a sufficiently small perturbation of any of its representatives, there is a point y_0 near x_0 such that the germ at y_0 of the perturbation of the representative is Lagrange equivalent to $(\mathbf{f}, x_0)$. Hence, small perturbations of representatives of a Lagrange stable map-germ $(\mathbf{f}, x_0)$ do not "change" $(\mathbf{f}, x_0)$.

(3) A Lagrangian map-germ $(\mathbf{f}, x_0)$ is *simple* if there is a finite set $\mathfrak{G}$ of Lagrangian map-germs such that any finite-parameter family f_λ of Lagrangian maps containing a representative of $(\mathbf{f}, x_0)$ for $\lambda = 0$ has, for all λ sufficiently small and all points in a sufficiently small neighborhood of x_0, only germs which are Lagrangian equivalent to germs of $\mathfrak{G}$. Loosely speaking, the germ $(\mathbf{f}, x_0)$ is simple if the set of all Lagrangian map-germs at points near x_0, of small perturbations of representatives of $(\mathbf{f}, x_0)$, can be "covered" by finitely many Lagrange equivalence classes of Lagrangian map-germs (no "moduli").

(4) Let $f : (\mathbf{R}^k, 0) \to (\mathbf{R}, 0)$ denote a smooth map-germ. A *deformation* F of f with base $\mathbf{R}^n$ is a map-germ $F : (\mathbf{R}^k \times \mathbf{R}^n, 0) \to (\mathbf{R}, 0)$ satisfying $F(x, 0) = f(x)$. Deformations of f are called *versal* if every other deformation of f is *equivalent* to F, that is, any other deformation $G : (\mathbf{R}^k \times \mathbf{R}^{\bar{n}}, 0) \to \mathbf{R}$ of f is expressible in the form $G(x, \bar{q}) = F(h(x, \bar{q}), \phi(\bar{q}))$, where $\phi : (\mathbf{R}^{\bar{n}}, 0) \to (\mathbf{R}^n, 0)$ and $h : (\mathbf{R}^k \times \mathbf{R}^{\bar{n}}, 0) \to (\mathbf{R}^k, 0)$ are smooth maps-germs with $h(x, 0) = x$ and $\det[d_x h(x, \bar{q})](0, 0) \neq 0$.

Arnold's classification of the stable simple Lagrangian singularities is given by the following:

Theorem 7.80 (Stable Simple Lagrangian Singularities)

(1) *There is an open dense set in the space of Lagrangian maps $L \to \mathbf{R}^n$ for $n \leq 5$, such that the corresponding germs are stable simple Lagrangian map-germs.*

(2) *The Lagrangian singularities or caustics of stable simple Lagrangian map-germs of n-dimensional Lagrangian submanifolds are exhausted by the Lagrangian singularities determined from the following list of generating families, up to stably R^+-equivalence:*

(i) A_ℓ $(1 \le \ell \le n+1)$:
$$F(x,q) = \pm x_1^{\ell+1} + q_1 x_1^{\ell-1} + \cdots + q_{\ell-1}x_1 + Q.$$

(ii) D_ℓ $(4 \le \ell \le n+1)$:
$$F(x,q) = x_1^2 x_2 \pm x_2^{\ell-1} + q_1 x_2^{\ell-2} + \cdots + q_{\ell-2}x_2 + q_{\ell-1}x_1 + Q.$$

(iii) E_6 $(6 \le n)$:
$$F(x,q) = x_1^3 \pm x_2^4 + q_1 x_1 x_2^2 + q_2 x_1 x_2 + q_3 x_2^2 + q_4 x_1 + q_6 x_2 + Q.$$

(iv) E_7 $(6 \le n)$:
$$F(x,q) = x_1^3 + x_1 x_2^3 + q_1 x_1^2 x_2 + q_2 x_1^2 + q_3 x_1 x_2 + q_4 x_2^2 + q_6 x_1 + q_6 x_2 + Q.$$

(v) E_8 $(7 \le n)$:
$$F(x,q) = x_1^3 + x_2^6 + q_1 x_1 x_2^3 + q_2 x_1 x_2^2 + q_3 x_2^3 + q_4 x_1 x_2 + q_6 x_2^2 + q_6 x_1 + q_7 x_2 + Q,$$

where $x = (x_1, \ldots, x_k) \in \mathbf{R}^k$, $q = (q_1, \ldots, q_n)$, *and* $Q = -x_2^2 - \cdots - x_s^2 + \cdots + x_k^2$. *Furthermore, the families* $F(x,q)$ *are versal deformations of the associated germs* $f(x) = F(x,0)$.

Readers are referred to [Arn73, pp. 269–271] and [Arn-GVI, p. 330] for a proof of Theorem 7.80. The A, D, E notation was introduced by Arnold because of the connections he found between his list of Lagrangian singularities and the Coxeter–Dynkin diagrams of the simple Lie algebras A_ℓ, D_ℓ, E_ℓ [Arn73].

A generalization of Theorem 7.80 to $n \le 10$ was carried out by Zakalyukin [Zaka76, Thm. 7, p. 29] (also, see [Arn-GVI, Chap. 21]). Moduli appear in this higher dimensional context.

Though the classification list in Theorem 7.80 applies to the case of *general Lagrangian singularities*, which includes those arising from non-optical systems, Guckenheimer showed that the theorem also enumerates the list for Lagrangian singularities arising from *general optical systems* [Guc73, p. 54], [Guc74]. We remark, however, that since gravitational lens systems form a special class of optical systems, we must be careful when applying Theorem 7.80. This issue will be discussed in the next subsection.

Lagrangian Caustics and Caustics in Gravitational Lensing

The next result yields that the concept of Lagrangian caustics coincides locally with our notion of caustics in gravitational lensing (since every k-plane time delay family is a Morse family — Corollary 7.78).

Theorem 7.81 *Let* $\mathbf{F}$ *be a germ of a Morse family* $F(x,q)$, *where* $x \in \mathbf{R}^k$ *and* $q \in \mathbf{R}^n$. *Then the set of Lagrangian caustics of the Lagrangian map-germ* $\mathbf{L_F} \to \mathbf{R}^n : (p,q) \mapsto q$ *is given by*

$$C_{\mathbf{F}} = \left\{ q \in \mathbf{R}^n : \text{ there is } x_0 \text{ where } F_x(x_0,q) = 0, \det F_{xx}(x_0,q) = 0,\ F \in \mathbf{F} \right\}.$$

Proof: The submanifold germ $\mathbf{L_F}$ is determined by projections onto (p,q) of solutions $(p,x,q) \in \mathbf{R}^n \times \mathbf{R}^k \times \mathbf{R}^n$ of the equations

$$F_x(x,q) = 0, \qquad p - F_q(x,q) = 0,$$

where F is a representative of $\mathbf{F}$. Set

$$\Lambda = \{(y,p,x,q) : y = F_x,\ p = F_q\}, \qquad Q = \{(0,p,x,q)\}.$$

Then

$$\Lambda \cap Q = \{(p,x,q) : F_x(x,q) = 0,\ p = F_q(x,q)\}$$

is a manifold (see [Arn-GVI, pp. 299-302] and [Arn90, p. 93]). As (p,x,q) varies over a sufficiently small neighborhood $\mathcal{W}$ of $\Lambda \cap Q$, the projection π of $\mathcal{W}$ onto the (p,q) variables is

$$L_F = \left\{ (p,q) : \text{ there is } x_0 \text{ where } F_x(x_0,q) = 0,\ p = F_q(x_0,q) \right\}.$$

The space L_F is an n-dimensional local Lagrangian submanifold of $\mathbf{R}^{2n}$ (Theorem 7.77). The submersion $\pi : \mathcal{W} \to L_F$ is also an immersion [Arn-GVI, pp. 300–301]; so π is a diffeomorphism for $\mathcal{W}$ sufficiently small. The restrictions to $\Lambda \cap Q$ of the one-forms dx, dp, dq defined on $\mathbf{R}^{2n+k} = \{(p,x,q)\}$ satisfy

$$F_{xx}dx + F_{xq}dq = 0, \qquad dp - F_{qx}dx - F_{qq}dq = 0 \tag{7.13}$$

on tangent vectors to $\mathcal{W}$. The critical points of the projection $\phi : L_F \to \mathbf{R}^n : (p,q) \mapsto q$ are those $(p,q) \in L_F$ at which $d_{(p,q)}\phi$ is not onto, that is, where the one-forms $dq_1, \ldots, dq_n$ are linearly dependent. At such a point, say (p_0, q_0), there is a $v \neq 0$ in $T_{(p_0,q_0)}L_F \approx T_{(p_0,x_0,q_0)}\mathcal{W}$ for some x_0, where $dq_i(v) = 0$ for $i = 1, \ldots, n$. Viewing v as an element of $T_{(p_0,x_0,q_0)}\mathbf{R}^{2n+k}$, write

$$v = v_1\frac{\partial}{\partial p} + v_2\frac{\partial}{\partial x} + v_3\frac{\partial}{\partial q}.$$

The conditions $dq_i(v) = 0$ imply $v_3 = 0$, and the right equation in (7.13) yields $v_1 = F_{qx}v_2$. Hence

$$v = v_2\left(F_{qx}\frac{\partial}{\partial p} + \frac{\partial}{\partial x}\right). \tag{7.14}$$

Since $v \neq 0$, we have $v_2 \neq 0$. By the left equation in (7.13), it follows that $F_{xx}v_2 = 0$. Because $v_2 \neq 0$, the condition $\det F_{xx} = 0$ is forced. Hence, the critical values of ϕ consists of all local q such that $F_x(x, q) = 0$ and $[\det F_{xx}](x, q) = 0$. Conversely, let $q_0 \in \mathbf{R}^n$ be such that there is an $\mathbf{x}_0 \in \mathbf{R}^k$ with $F_x(\mathbf{x}_0, q_0) = 0$ and $[\det F_{xx}](x_0, q_0) = 0$. Then there is some $v_2 \neq 0$ such that $F_{xx}v_2 = 0$. Define a vector v at (p_0, x_0, q_0), where $p_0 = F_x(x_0, q_0)$, by equation (7.14). Then v is tangent to L_F at a critical point of ϕ. Thus, the point (p_0, q_0) lies in the set of Lagrangian caustics of the Lagrangian map ϕ. □

Remarks: Our comments will parallel those made at the end of Section 7.6.2 (page 276) concerning catastrophe theory.

(1) Theorem 7.80 with $n = 2$ locally characterizes the caustics of k-plane time delay families T whose Lagrangian map-germs are Lagrange stable and simple. Such a family T is stably R^+-equivalent in a neighborhood of $(\mathbf{x}_1, \ldots, \mathbf{x}_k, \mathbf{y})$, where $(\mathbf{x}_1, \ldots, \mathbf{x}_k)$ is a degenerate light ray of a light source at $\mathbf{y}$,[12] to one of the following families F with variables $x = (x_1, \ldots, x_{2k})$ and parameters $q = (q_1, q_2)$:

(i) A_2, **Fold:**
$F(x, q) = \pm x_1^3 + q_1 x_1 + Q(x_2, \ldots, x_{2k})$.

(ii) $A_3^{\pm}$, **Cusp:**
$F(x, q) = \pm x_1^4 + q_1 x_1^2 + q_2 x_1 + Q(x_2, \ldots, x_{2k})$,

where Q is of the form $-x_2^2 - \cdots - x_s^2 + \cdots + x_{2k}^2$. These are the fold and cusp elementary catastrophe unfoldings in Table 7.1 (page 273). Hence, the discussion after Thom's catastrophe theorem (page 276) shows that the point $\mathbf{y}$ is either a fold or cusp caustic point of T

(2) Though Theorem 7.81 shows that locally the concept of caustics in gravitational lensing coincides with that of Lagrangian caustics, the Lagrangian caustics in lensing come from a specific class of optical systems. In addition, the classification list in Theorem 7.80 applies to the case of *general Lagrangian singularities*, which includes those arising from non-optical systems. Guckenheimer showed that the results of Theorem 7.80 also enumerates the list for Lagrangian singularities arising from *general optical systems* [Guc73, p. 54], [Guc74].[13] However, since gravitational lens systems form a special class of optical systems, we are not assured from Theorem 7.80 and Guckenheimer's result that there is an open dense set in the space of Lagrangian maps due to gravitational lensing, such that the corresponding germs are stable simple Lagrangian map-germs (i.e., the caustics are folds or cusps). We shall prove in Section 8.4 that there is an open dense set $\mathfrak{M}$ in the space of k-plane

[12] That is, the point $(\mathbf{x}_1, \ldots, \mathbf{x}_k)$ is a degenerate critical point of the time delay function $T_{\mathbf{y}}(\cdot) = T(\cdot, \mathbf{y})$.

[13] There are special restrictions on the global topology of Lagrangian caustics due to optical systems (e.g., [Arn90, p. 35]).

time delay such that for each $T \in \mathfrak{M}$, if $(\mathbf{x}_1, \ldots, \mathbf{x}_k)$ is a degenerate light ray of a light source at $\mathbf{y}$, then the point $\mathbf{y}$ is either a fold or cusp caustic point of T.

7.7.2 1-Parameter Families of Lagrangian Caustics

This section presents an application of Arnold's characterization of stable simple Lagrangian maps (Theorem 7.80) to the classification of 1-parameter families of Lagrangian caustics. Section 9.5.1 (page 375) will discuss these results in the context of caustic metamorphoses in gravitational lensing.

Let L_t be a smooth 1-parameter family of Lagrangian submanifolds of $T^*\mathbf{R}^n$. By Theorem 7.78, the manifolds L_t are given locally by a generating family of functions $F_t(x, q)$ depending smoothly on $t \in \mathbf{R}$, $q \in \mathbf{R}^n$, and $x \in \mathbf{R}^k$. The function $F(x, q, t) = F_t(x, q)$ viewed as a family of functions with variables x and parameters (t, q) is a generating family of some Lagrangian submanifold L_F in $T^*\mathbf{R}^{n+1}$. The Lagrangian caustic C of the associated Lagrangian map $L_F \to \mathbf{R}^{n+1} : (x, q, t) \mapsto (q, t)$ is given by the Lagrangian caustics C_t of the Lagrangian maps $L_t \to \mathbf{R}^n$ with each C_t lying in the level hypersurface, $t =$ constant, of $\mathbf{R}^{n+1} = \{(q, t)\}$. The Lagrangian caustic C in $\mathbf{R}^{n+1}$ is called a *big caustic*. It depicts the "instantaneous" caustics C_t in the "space-time" $\{(q, t)\}$, where $\{t\}$ and $\{q\}$ play the roles of "time" and "space," respectively.

Define two families of Lagrangian caustics C_t^1 and C_t^2 of the Lagrangian maps $L_t^1 \to \mathbf{R}^n$ and $L_t^2 \to \mathbf{R}^n$, respectively, to have *equivalent evolutions* or *metamorphoses* (or *perestroikas*) if there is a diffeomorphism $h : \mathbf{R} \to \mathbf{R} : t \mapsto h(t)$ and a family of diffeomorphisms $\phi_t : \mathbf{R}^n \to \mathbf{R}^n$ depending smoothly on t, such that $\phi_t(C_t^1) = C_{h(t)}^2$. In other words, two caustics evolutions C_t^1 and C_t^2 are equivalent if there is a diffeomorphism of $\mathbf{R}^{n+1}$ that maps a big caustic of L_t^1 into a big caustic of L_t^2, and preserves the partitioning of $\mathbf{R}^{n+1}$ into level hypersurfaces of t. This is precisely the notion of equivalence needed for applications to 1-parameter families $T_t(\mathbf{x}_1, \ldots, \mathbf{x}_k, \mathbf{y})$ of time delay families in gravitational lensing. The reason is because the parameter t (e.g., external shear) is physically different from the light source position $\mathbf{y} = (y_1, y_2)$. Thus, the previous definition of equivalence of metamorphoses distinguishes the "time coordinate" t from the "spatial" ones y_1 and y_2.

By Theorem 7.80, there is an open dense set in the space of Lagrangian maps $L \to \mathbf{R}^{n+1}$, where $n \leq 4$, such that the corresponding germs are stable simple Lagrangian map-germs, which have big caustics in $\mathbf{R}^{n+1}$ of type A_k $(1 \leq k \leq 6)$, D_k $(4 \leq k \leq 6)$, or E_6. Using a nondegenerate time function to slice the latter big caustics, Zakalyukin [Zaka84, p. 2727] reduced the time function to simplest form using diffeomorphisms of $\mathbf{R}^{n+1}$ $(n \leq 4)$ that preserve the latter big caustics. He then to obtain the following local classification of the corresponding 1-parameter families of caustics:

Theorem 7.82 (1-Parameter Families of Lagrangian Caustics) *There is an open dense set $\mathcal{D}$ in the space of 1-parameter families of Lagrangian maps $L_t \to \mathbf{R}^2$ depending smoothly on t, such that the associated 1-parameter families of Lagrangian caustics due to germs of* *maps in $\mathcal{D}$ are exhausted, up to equivalence of evolutions, by the 1-parameter families of Lagrangian caustics arising from the following generating families in a neighborhood of the origin in $\mathbf{R}^3 = \{(q_1, q_2, t)\}$:*

(1) A_2 *(**Fold**):*
$F(x, q_1, q_2, t) = x^3 + q_1 x.$

(2) A_3 *(**Cusp**):*
$F(x, q_1, q_2, t) = x^4 + q_2 x^2 + q_1 x.$

(3) $A_3(+)$ *(**Lips**):*
$F(x, q_1, q_2, t) = x^4 + (t + q_2^2)x^2 + q_1 x.$

(4) $A_3(-)$ *(**Beak-to-Beak**):*
$F(x, q_1, q_2, t) = x^4 + (t - q_2^2)x^2 + q_1 x.$

(5) A_4 *(**Swallowtail**):*
$F(x, q_1, q_2, t) = x^6 + tx^3 + q_2 x^2 + q_1 x.$

(6) D_4^- *(**Elliptic Umbilic or Pyramid**):*
$F(x_1, x_2, q_1, q_2, t) = -x_2^2 x_1 + x_1^3 + (t - q_1 + aq_2)x_1^2 + q_1 x_1 + q_2 x_2.$

(7) D_4^+ *(**Hyperbolic Umbilic or purse**):*
$F(x_1, x_2, q_1, q_2, t) = x_2^2 x_1 + x_1^3 + (t + q_1 + aq_2)x_1^2 + q_1 x_1 + q_2 x_2.$

Readers should consult Zakalyukin [Zaka84] for a proof of Theorem 7.82, as well as a classification for dimensions $n = 3, 4$.

Chapter 8

Classification and Genericity of Stable Lens Systems

Though changed I shall arise the same.

JACQUES BERNOULLI

She has her ambassadors in every quarter of the world — her couriers upon every road. Her officers march along with armies, and her envoys walk into statesmen's cabinets. They are ubiquitous.

WILLIAM THACKERAY

In gravitational lensing, the functional relationships (e.g., time delay and lensing maps) between physical quantities are approximations, and should ideally be expressed using maps resistant to small perturbations. This begs investigating the classification and genericity of stable lens systems. The problem then carries over to the time delay families, including the induced time delay functions and lensing maps, used to model gravitational lens systems. We shall show that such an analysis yields nontrivial progress on problem **M3**, which calls for a characterization of the local and global properties of critical curves and caustics due to generic gravitational lens systems.

The first part of the chapter deals with characterizing time delay functions and lensing maps under local, transverse, and global stabilities. In the latter part of the chapter, we investigate the genericity of time delay families with transverse stable, as well as globally stable, lensing maps.

Outline: Section 8.1 presents several cautionary remarks concerning the use of singularity theory in gravitational lensing. This is followed by applications of the results in

Chapter 7 to time delay and lensing maps. In particular, Section 8.2 characterizes time delay functions and lensing maps under various stabilities (local, transverse, and global). In preparation for studying the genericity of local, transverse, and global stability of lensing maps, we present a generalization of Thom's catastrophe theorem for two parameters in Section 8.3. Proceeding in parallel to Section 8.3, we show in Section 8.4 that the time delay families with transverse stable (hence, locally stable) lensing maps are dense among all time delay families. In addition, among the time delay families with proper lensing maps, those whose lensing maps are globally stable form an open dense set.

8.1 Some Cautionary Remarks

This section collects several remarks that should be kept in mind when applying singularity theory to gravitational lensing. The first two deal with the concept of stability and genericity for time delay functions and lensing maps. The next two remarks touch on the issue of coordinate changes as it relates to the notion of equivalence of maps and families of functions. The final remark considers the quantitative aspects of gravitational lensing.

Remark 1: Stability

Let $\mathcal{T}(L_{(k)}; \mathbf{y})$ be the space of k-plane time delay functions $T_{\mathbf{y}} : L_{(k)} \to \mathbf{R}$ at $\mathbf{y}$, where $L_{(k)} = L_1 \times \cdots \times L_k$ (product of lens planes) and $\mathbf{y}$ is a fixed point of the light source plane S. The entire space of k-plane time delay functions is then $\mathcal{T}(L_{(k)}) = \bigcup_{\mathbf{y}\in S} \mathcal{T}(L_{(k)}; \mathbf{y})$. Denote the space of k-plane lensing maps $\boldsymbol{\eta} : P \to S$ by $Ł(P, S)$. Then $\mathcal{T}(L_{(k)})$ and $Ł(P, S)$ are, respectively, proper subsets of $C^\infty(L_{(k)}, \mathbf{R})$ and $C^\infty(P, S)$. The reader should note that *local, transverse, and global stability of time delay functions $T_{\mathbf{y}}$ and lensing maps $\boldsymbol{\eta}$ are defined relative to neighborhoods in the ambient spaces $C^\infty(L_{(k)}, \mathbf{R})$ and $C^\infty(P, S)$, respectively. It is not in terms of $\mathcal{T}(L_{(k)})$ and $Ł(P, S)$.*

For example, if a lensing map $\boldsymbol{\eta} \in Ł(P, S)$ is locally stable, then the map-germ $[\boldsymbol{\eta}]_x$ is stable at each x in P. In other words, for every neighborhood U of x, there is a neighborhood $\mathfrak{F}$ of $\boldsymbol{\eta}$ in $C^\infty(P, S)$ such that for each map h in $\mathfrak{F}$ there is a point x' in U with $[h]_{x'}$ differentiably equivalent to $[\boldsymbol{\eta}]_x$. There is no guarantee that $\mathfrak{F}$ consists entirely of lensing maps. Nonetheless, we shall see that in the space of time delay families, there is an open dense set, say $\mathfrak{M}^*$, such that for each time delay family in $\mathfrak{M}^*$ the associated lensing map is not only locally stable, but also transverse stable.

Remark 2: Genericity

The word "generic" is often used informally in singularity theory to designate a property that is typical among mappings in a given space of maps. In particular, most authors would call a property "generic" if it is shared by members of an open dense subset of a space. For example, Theorem 7.32 and Figure 7.2 (page 232) yield that global stability is a "generic" property among proper mappings between spaces with nice dimensions. Since all proper k-plane lensing maps $\boldsymbol{\eta} : P \to S$ lie in the nice dimensions $(2, 2)$, Theorem 7.32 shows that in the space of smooth proper maps $P \to S$ there is an arbitrarily small perturbation of each proper lensing map $\boldsymbol{\eta}$ yielding a stable proper map $\tilde{\boldsymbol{\eta}}$. As noted earlier, there is no guarantee that $\tilde{\boldsymbol{\eta}}$ is a lensing map. In other words, *the fact that proper lensing maps fall in nice dimensions does not imply that stable proper lensing maps are "generic" in the space of proper lensing maps $Ł(P, S)$. Similar remarks hold for k-plane time delay functions.* The desired "genericity" of stable proper lensing maps requires additional analysis within the context of gravitational lensing.

One way to define a general notion of "genericity" is as follows: Let $\mathfrak{F}$ be a topological subspace of $C^\infty(X, Y)$ relative to the Whitney C^∞-topology. A property is *generic* in $\mathfrak{F}$ if the set of elements of $\mathfrak{F}$ possessing the property contains a residual subset[1] of $\mathfrak{F}$. This would carry an even stronger sense of "typicality" if $\mathfrak{F}$ is a Baire space, that is, all residual subsets are dense (e.g., if $\mathfrak{F} = C^\infty(X, Y)$). We caution that not all authors abide by this formulation of genericity. Consequently, we shall not use the term "generic" in the statement of theorems. In fact, we shall explicitly state whether a set of elements with a given property is dense, or, open and dense, leaving it to the reader to decide whether to interpret such a property as generic.

Remark 3: Differentiable Equivalence and Mixing of Coordinates

The notions of differentiable equivalence employed for local and global stability are natural when applied to k-plane lensing maps $\boldsymbol{\eta} : P \longrightarrow S$. It keeps the coordinate changes on P and S separated, that is, there is no mixing of the coordinates on P with those on S, though curvilinear coordinates are not ruled out. Mixing is not permitted because P and S are physically distinct. We can then classify the critical points and caustics of (locally, transverse, or globally) stable lensing maps using Whitney's characterization of stable maps between planes.

In the case of time delay functions $T_{\mathbf{y}} : L_{(k)} \to \mathbf{R}$, if there is only one lens plane (i.e., $k = 1$), then differentiable equivalence applies naturally. When there are at least two lens planes, differentiable equivalence does not preserve the partitioning of $L_{(k)} = L_1 \times \cdots \times L_k$ into distinct lens planes. It allows mixing of coordinates from different lens planes. This violates the assumption that all lens planes are physically independent (no gravitational interaction between them). Strictly speaking, we should only allow coordinate changes on $L_1 \times \cdots \times L_k$ of the form $\lambda = \lambda_1 \times \cdots \times \lambda_k$, where λ_i is a coordinate change on L_i. However, there is no need to restrict to such coordinate changes when classifying light rays of time delay maps. This is because the index and nullity of a light rays of $T_{\mathbf{y}}$, as well as the total number of light rays, is independent of coordinates. Thus, from the point of view

[1] A subset of a topological space F is *residual* if it is the intersection of a countable number of open dense sets in F.

of characterizing light rays of (locally, transverse, or globally) stable time delay functions, our notion of differentiable equivalence is appropriate — Morse theory can be applied.

Remark 4: Equivalence for Families

An approach to classifying caustics locally is to apply Thom's catastrophe theory to time delay families $T : L_1 \times \cdots \times L_k \times S \longrightarrow \mathbf{R}$. Employing catastrophe theory's notion of equivalence for unfoldings will not mix coordinates of the state space $L_1 \times \cdots \times L_k$ into those of the control space S, but will mix those of the control space into those of the state space.[2] In addition, mixing of coordinates from different lens planes is allowed. Consequently, the partitioning of $L_1 \times \cdots \times L_k$ into distinct lens planes is not preserved. This in fact contributes to Thom's catastrophes having "normal forms" with a reduced number of variables (Table 7.1, page 273). Nonetheless, since the bifurcation set of the catastrophe map χ_T due to T coincides with the caustics of T, Thom's theory still yields that the caustics of a time delay family that is a universal unfolding will be folds or cusps (see the Remark on page 276).[3] Similar remarks apply when using Arnold's singularity theory to classify caustics locally. In this case, the caustics of T locally coincide with the critical values of the Lagrangian map-germ generated by T. Furthermore, those time delay families generating stable simple Lagrangian map-germs will have folds or cusps as caustics (see the Remark on page 284).

Despite coordinate mixing, the theories of Thom and Arnold still yield that certain time delay families will have only fold or cusp caustics. The issue of the prevalence (in the context of gravitational lensing) of time delay families with only fold or cusp caustics cannot be answered directly from the classification theorems of Thom and Arnold. We shall see in Section 8.4 that there is an open dense set of time delay families T in the space of time delay families, such that the caustics of T are either folds or cusps.

Remark 5: Quantitative Aspects of Lensing

The quantitative aspects of gravitational lensing is important in physical applications. For example, the shapes and sizes of lensed images, the areas of regions of given magnification, and the lengths, curvature, etc., of caustics become important. Also, in physical applications one is interested in how matter on different lens planes factor quantitatively into the lensing effect. Thus, when studying the quantitative features of gravitational lensing, the use of curvilinear coordinates and mixing of coordinates is not allowed (since such transformations distort regions, etc.). Instead, we have to restrict to orthogonal transformations of rectangular coordinates — see Section 9.2 (page 341). Bear in mind, however, that the

[2] Wasserman [Was] further distinguishes coordinates of the control space (cf. [Zaka84, p. 2731]).

[3] It may interest the reader to know that for r-parameter families of maps with $3 \leq r \leq 5$, Thom's list of elementary catastrophes and Whitney's singularities do not coincide in general. For example, if $r = 3$, then Whitney's list consists of folds, cusps, and swallowtails, while Thom's list includes Whitney's plus elliptic and hyperbolic umbilics (see Zeeman [Zee-T, p. 265]. The two lists (accidentally) coincide for $r = 2$, yielding folds and cusps.

quantitative aspects of lensing are very difficult (often analytically intractable). For this reason, astrophysicists often resort to sophisticated numerical schemes when addressing these issues.

8.2 Stability of Time Delay Functions and Lensing Maps

By Chapters 3 and 6, every k-plane gravitational lens system can be described by a time delay family $T : L_{(k)} \times S \to \mathbf{R}$, where $L_{(k)} = L_1 \times \cdots \times L_k$ with L_i the ith lens plane and S the light source plane. The family T induces a time delay function $T_{\mathbf{y}} : L_{(k)} \to \mathbf{R}$ for each $\mathbf{y} \in S$, and a lensing map $\boldsymbol{\eta} : P \subseteq \mathbf{R}^2 \to S$. The current section investigates $T_{\mathbf{y}}$ and $\boldsymbol{\eta}$ under local, transverse, and global stability.

Requisite Mathematics: Sections 7.2, 7.4, and 7.5.

Local, Transverse, and Global Stability of Time Delay Functions

Theorem 8.1 *Let $T_{\mathbf{y}} : L_{(k)} \to \mathbf{R}$ be a k-plane time delay function at* $\mathbf{y}$. *Then:*

(1) *The function $T_{\mathbf{y}}$ is locally stable if and only if all light rays of $T_{\mathbf{y}}$ are nondegenerate (i.e., either a minimum, maximum, or generalized saddle).*

(2) *The function $T_{\mathbf{y}}$ is transverse stable if and only if $T_{\mathbf{y}}$ is locally stable and distinct light rays of $T_{\mathbf{y}}$ have distinct time delays.*

(3) *Suppose that $T_{\mathbf{y}}$ is proper. Then $T_{\mathbf{y}}$ is globally stable if and only if $T_{\mathbf{y}}$ is transverse stable.*

Note that if $T_{\mathbf{y}}$ is proper, then the regular values of $T_{\mathbf{y}}$ form an open set in $\mathbf{R}$.

Proof: This is a consequence of Theorem 7.46 (page 242) and Mather's stability theorem (Theorem 7.31, page 230). □

Example: Consider a singular isothermal sphere time delay function (Section 6.3.2):

$$T_{\mathbf{y}}^{sis}(\mathbf{x}) = \frac{|\mathbf{x} - \mathbf{y}|^2}{2} - C|\mathbf{x}|.$$

If $0 < |\mathbf{y}| < C$, then $\mathrm{Crit}(T_{\mathbf{y}}^{sis}) = \{\mathbf{x}_+, \mathbf{x}_-\}$, where $\mathbf{x}_+$ and $\mathbf{x}_-$ are minimum and saddle light rays, respectively. In addition, the light rays $\mathbf{x}_+$ and $\mathbf{x}_-$ have distinct time delays (since $|\mathbf{x}_+| > |\mathbf{x}_-|$). Consequently, the time delay function $T_{\mathbf{y}}^{sis}$ is transverse stable for $0 < |\mathbf{y}| < C$. When $|\mathbf{y}| > C$, the set $\mathrm{Crit}(T_{\mathbf{y}}^{sis})$ consists of one minimum light ray and, hence, is also transverse stable.

Let us consider the subset of points $\mathbf{y}$ in S for which $T_{\mathbf{y}}$ is locally stable.

Theorem 8.2 *Let* $T : L_{(k)} \times S \to \mathbf{R}$ *be a k-plane time delay family. Then the set*

$$\{\mathbf{y} \in S : T_{\mathbf{y}} \text{ is locally stable}\}$$

is dense in S. It is the complement of the measure zero set Caustic(T).

It is worth noting that *for any measure zero subset Z of a manifold X, the complement of Z is dense in X* (e.g., [Gib, p. 220]).

Proof: Let

$$N = \{\mathbf{y} \in S : T_{\mathbf{y}} \text{ is nondegenerate}\}.$$

Then Caustic$(T) = S - N$. Theorem 6.17(4) (page 204) yields Caustic$(T) =$ Caustic$(\boldsymbol{\eta})$, where $\boldsymbol{\eta}$ is the k-plane lensing map induced by T. Equivalently, the set N consists of all regular values of $\boldsymbol{\eta}$. By Sard's theorem (Theorem 6.3, page 176), the set Caustic$(\boldsymbol{\eta})$ has measure zero. Hence N is dense in S (cf. Brown's theorem, Corollary 6.4, page 176). Theorem 8.1(1) now shows that $N = \{\mathbf{y} \in S : T_{\mathbf{y}} \text{ is locally stable}\}$. □

We now consider all $\mathbf{y} \in S$ such that $T_{\mathbf{y}}$ is transverse stable. In this case, the light rays of $T_{\mathbf{y}}$ have distinct time delays, so the difference of time delays between distinct light rays is nonzero. This difference is inversely related to Hubble's constant (Section 4.4, page 134).

Before stating the next theorem, we introduce the notion of a "Maxwell set" and construct a mapping Φ_T using the time delay family T.

Define the *Maxwell set* of a k-plane time delay family $T : L_{(k)} \times S \to \mathbf{R}$ to be the set of all $\mathbf{y}$ in S such that $T_{\mathbf{y}}$ has at least two light rays with the same time delay.[4] Denote the Maxwell set of T by Maxwell(T). Then the set $S -$ Maxwell(T) consists of every $\mathbf{y} \in S$ such that all the light rays of $T_{\mathbf{y}}$ have distinct time delays. Note that Maxwell(T) may contain caustic points.

Now, consider the 2-fold 1-jet bundle (Section 7.4.2, page 240)

$$J_2^1(L_{(k)}, \mathbf{R}) \approx L_{(k)}^{(2)} \times (\mathbf{R} \times \mathbf{R}) \times (\mathbf{R}^{2k} \times \mathbf{R}^{2k}),$$

where $L_{(k)}^{(2)} = L_{(k)} \times L_{(k)} - (\triangleright L_{(k)})$, and the following submanifold of $J_2^1(L_{(k)}, \mathbf{R})$:

$$S_1^{[2]}(L_{(k)}, \mathbf{R}) \approx L_{(k)}^{(2)} \times (\triangleright \mathbf{R}) \times \{0\} \times \{0\}.$$

Define a map $\Phi_T : L_{(k)}^{(2)} \times S \to J_2^1(L_{(k)}, \mathbf{R})$ by $\Phi_T(\mathbf{x}, \mathbf{y}) = (j_2^1 T_{\mathbf{y}})(\mathbf{x})$. We have the following:

Theorem 8.3 *Let* $T : L_{(k)} \times S \to \mathbf{R}$ *be a time delay family such that* $\Phi_T \pitchfork S_1^{[2]}(L_{(k)}, \mathbf{R})$. *Then:*

[4] cf. Arnold [Arn93b, p. 98] and Gilmore [Gil, p. 89].

(1) *The set* Maxwell(T) *has measure zero and (hence) its complement,* $S -$ Maxwell(T), *is dense in* S.

(2) *The set*

$$\{\mathbf{y} \in S : T_{\mathbf{y}} \text{ is transverse stable}\}$$

is dense in S. *It is the complement of a measure zero set in* S.

Proof: (1) Define

$$j : S \to C^{\infty}(L_{(k)}^{(2)}, J_2^1(L_{(k)}, \mathbf{R}))$$

by $j(\mathbf{y}) = j_2^1 T_{\mathbf{y}}$. Then $\Phi(\mathbf{x}, \mathbf{y}) = j(\mathbf{y})(\mathbf{x})$. Using the maps j and Φ_T, apply Lemma 7.13 (page 222) with $\phi = \Phi_T$, $Q = S_1^{[2]}(L_{(k)}, \mathbf{R})$, and $D = S$. Then

$$M = \{\mathbf{y} \in S : j_2^1 T_{\mathbf{y}} \pitchfork S_1^{[2]}(L_{(k)}, \mathbf{R})\}$$

is dense in S and the complement of a measure zero set. Lemma 7.45(1) (page 241) implies that $j_2^1 T_{\mathbf{y}} \pitchfork S_1^{[2]}(L_{(k)}, \mathbf{R})$ if and only if all critical points of $T_{\mathbf{y}}$ have distinct critical values. The latter is equivalent to $\mathbf{y} \notin$ Maxwell(T). Hence: $M = S -$ Maxwell(T). Since M is the complement of a measure zero set, we see that Maxwell(T) has measure zero.

(2) Theorem 8.1(2) shows:

$$S - [\text{Caustic}(T) \cup \text{Maxwell}(T)] = \{\mathbf{y} \in S : T_{\mathbf{y}} \text{ is transverse stable}\}.$$

Since Caustic(T) and Maxwell(T) have measure zero, the union Caustic$(T) \cup$ Maxwell(T) has measure zero (e.g., [Gol-G, p. 30, Lem. 1.2]). Moreover, we noted earlier that the complement of a measure zero subset of a manifold is dense (e.g., [Gib, p. 220]). Hence $S - [\text{Caustic}(T) \cup \text{Maxwell}(T)]$ is dense in S. □

Local, Transverse, and Global Stability of Lensing Maps

For any k-plane lensing map, we have the following:

Theorem 8.4 *Let* $\eta : P \to S$ *be a k-plane lensing map. Then:*

(1) *The set* Crit(η) *is closed and* Caustic(η) *has measure zero. If* Crit(η) *is bounded, then* Crit(η) *and* Caustic(η) *are compact.*

(2) *The set of regular values of* η *is dense in* S. *If* η *is onto or proper, then the set of regular values of* η *is also open.*

Proof: We saw in the proof of Theorem 8.2 that Caustic(η) has measure zero and (hence) the set of regular values of η, i.e., $S -$ Caustic(η), is dense in S. If $\mathbf{x}$ is a regular point of η, then η is a local diffeomorphism at $\mathbf{x}$. It follows that there is an open set of regular points of η about about $\mathbf{x}$, so the set of regular points is open, i.e., the set Crit(η)

of critical points is closed. If Crit(η) is also bounded, then $\eta[\text{Crit}(\eta)] = \text{Caustic}(\eta)$ is compact. Now, the local diffeomorphism ξ yields an open set of regular values about $\eta(\mathbf{x})$. However, this does not guarantee that the set R_η of regular values of η is open in S (since the regular values include points outside of the range $\eta(P)$). If $\eta(P) = S$, then R_η is open; also, Proposition 6.5 on page 176 implies that R_η is open if η is proper. □

Let us consider lensing maps under local, transverse, and global stability. The next theorem applies to the critical points of η under local stability. It is a consequence of Theorem 7.55 on page 261 and Theorem 7.49, page 251 (also, see page 245):

Theorem 8.5 *Let $\eta : P \to S$ be a locally stable k-plane lensing map with* Crit(η) *nonempty. Then:*

(1) *The map η is locally stable if and only if each point p in P is either a regular point, fold critical point, or cusp critical point.*

(2) *The set of critical points*

$$\text{Crit}(\eta) = \text{Folds}(\eta) \bigcup \text{Cusps}(\eta)$$

forms a disjoint union of simple closed curves and/or open arcs without end points in P.

(3) *The set of fold critical points,* Folds(η)*, is a disjoint union of simple closed curves and/or open arcs (possibly with end points in P). The cusp critical points in* Cusps(η) *are isolated with each abutted by two arcs of fold critical points.*

The following addresses transverse and global stability, and is an immediate corollary of Theorem 7.56 (page 262):

Theorem 8.6 *Let $\eta : P \to S$ be a k-plane lensing map. Then:*

(1) *The map η is transverse stable if and only if η is locally stable and the following hold:*

  (i) *The η-image of fold critical curves have normal crossings.*

  (ii) *No more than two fold critical points are mapped to the same point.*

  (iii) *No fold and cusp critical points are mapped by η to the same point.*

  (iv) *No two cusp critical points are mapped by η to the same point.*

(2) *Suppose that η is proper. Then η is globally stable if and only if η is transverse stable.*

The genericity of lensing maps under local, transverse, and global stability will be taken up in Section 8.4.1.

8.3 Generalization of Thom's Theorem for Two Parameters

Before addressing the genericity of time delay functions whose lensing maps are local, transverse, or globally stable, we generalize Thom's Catastrophe Theorem (Theorem 7.76, page 275) for the case of two control parameters. The techniques of this section will be used in Section 8.4 to investigate the previous genericity issue for time delay families. The proof of our result will extend the treatment of Thom's Catastrophe Theorem found in the published notes of lectures of E.C. Zeeman as written and revised by D.J. A. Trotman [Zee-T, pp. 263–327].

Requisite Mathematics: Sections 7.1, 7.2, and 7.6.

8.3.1 Statement and Application of Generalization

Statement of the Result

We begin by reviewing some basic terminologies from catastrophe theory (see *Elementary Catastrophes* on page 272).

Consider an n-dimensional manifold, X, $(n > 0)$ and r-dimensional manifold, Y. Let $\mathcal{F} = C^\infty(X \times Y, \mathbf{R})$ with the Whitney C^∞-topology. For each family of functions, $f : X \times Y \to \mathbf{R}$, in $\mathcal{F}$ we call X and Y, respectively the *state space* and *control space* of f.

Consider the cotangent bundle T^*X of X and let $\mathbb{O}$ be the image of the zero section in the vector bundle T^*X. For $f \in C^\infty(X \times Y, \mathbf{R})$, let $d_x f : X \times Y \to T^*X : (x, y) \to df_y(x)$, where $f_y = f|_{(X\times\{y\})}$. The subset $M_f = (d_x f)^{-1}(\mathbb{O}) \subset X \times Y$ is the *catastrophe manifold* of f. The *catastrophe map* of f is $\chi_f = \pi|_{M_f}$, where π is the projection of $X \times Y$ onto Y. For the Euclidean case $f \in \mathcal{F} = C^\infty(\mathbf{R}^n \times \mathbf{R}^r, \mathbf{R})$, we have

$$M_f = \left\{(x, y) \in \mathbf{R}^n \times \mathbf{R}^r : \frac{\partial f}{\partial x_i}(x, y) = 0,\ i = 1, \ldots, n\right\}.$$

A catastrophe map χ_f is called *f-locally stable* at $(\mathbf{x}_0, \mathbf{y}_0) \in M_f$ if for each neighborhood N of $(\mathbf{x}_0, \mathbf{y}_0) \in X \times Y$, there exists a neighborhood $\mathcal{G} \subset \mathcal{F}$ of f such that for every $g \in \mathcal{G}$ there is a point $(\mathbf{x}_1, \mathbf{y}_1) \in N \cap M_g$ where the germ of χ_g at $(\mathbf{x}_1, \mathbf{y}_1)$ is differentiably equivalent to that the germ of χ_f at $(\mathbf{x}_0, \mathbf{y}_0)$. Thom's theorem (Theorem 7.76, page 275) then states:

Thom's Catastrophe Theorem: *If $r \leq 5$, then there is an open dense subset $\mathcal{F}_* \subset \mathcal{F}$ such that for every $f \in \mathcal{F}_*$ the catastrophe manifold M_f is an r-dimensional submanifold of $\mathbf{R}^n \times \mathbf{R}^r$, the critical points of χ_f are only elementary catastrophes, and χ_f is f-locally stable.*

Note. The f-local stability of the catastrophe maps in this theorem is appropriate for our lensing application. The catastrophe map of a time delay family is its lensing map, up to a diffeomorphism. The perturbations of the lensing maps that interest us are those that arise from perturbations of the time delay families. The justification for this is that a time delay family completely determines its lensing map.

We are interested in the case where $r = 2$ and shall content ourselves with the description of the catastrophes that arise in that case. The catastrophe manifold M_f is then a surface and χ_f maps the surface to $\mathbf{R}^2$, and the only critical points are folds and cusps. In this case, the map χ_f is also locally stable (see Theorem 7.55, page 261; cf. page 276). For such a catastrophe map χ_f to be globally stable we would require that only transversal double points of the fold caustic curves would arise, the image of no cusp critical point would lie on the image of a fold critical curve, no two cusp critical points and no more than two fold critical points are mapped to the same point, and that χ_f be proper (Theorem 7.56, page 262). Thus, as a step towards identifying a subset of $\mathcal{F}$ having globally stable catastrophe maps, we must extend Thom's theorem to account for multiple points of χ_f restricted to its set of critical points. We want a theorem as above whose final phrase is: "and χ_f is transverse stable." If we then restrict to those f for which χ_f is proper, then shall see that there is an open dense subset of those maps have globally stable χ_f.

Note. There is a notion of global stability of catastrophe maps which defines χ_f to be a "globally stable catastrophe map" if there is a neighborhood V of f in $\mathcal{F}$ such that for each $h \in V$, the spaces M_f and M_h are "naturally" diffeomorphic and using that diffeomorphism, the map χ_f is differentiably equivalent to χ_h. Our result for $r = 2$ refers to the *stronger* notion of global stability introduced above, whereby χ_f is equivalent to every nearby map from M_f into $\mathbf{R}^r$. The distinction is that these nearby maps do not necessarily arise as catastrophe maps.

Our goal is to prove the following generalization of Thom's Catastrophe Theorem for two-parameter families of functions:

Theorem 8.7 (Generalization of Thom's Theorem) *Let X be an n-dimensional manifold and Y a 2-dimensional manifold. Suppose that $\mathcal{F} = C^\infty(X \times Y, \mathbf{R})$ with the Whitney C^∞-topology. Then:*

(1) *There is an open and dense subset $\mathcal{F}^\star$ of $\mathcal{F}$ consisting of families multi-transverse to all singularities in $J_s^k(X) \times Y^s$. In particular, the set $\mathcal{F}^\star$ consists of all $f \in \mathcal{F}$ whose catastrophe manifold M_f is a smooth 2-dimensional surface and the corresponding catastrophe map χ_f is transverse stable.*

(2) *Let $\mathcal{F}_P$ be the subset of elements in $\mathcal{F}$ whose catastrophe maps are proper and set $\mathcal{F}_P^\star = \mathcal{F}^\star \cap \mathcal{F}_P$. Then $\mathcal{F}_P$ is a nonempty open subset of $\mathcal{F}$ and $\mathcal{F}_P^\star$ is open and dense in $\mathcal{F}_P$. If $f \in \mathcal{F}_P^\star$, then the catastrophe map $\chi_f : M_f \to Y$ is globally stable as an element of $C^\infty(M_f, Y)$.*

A proof of Theorem 8.7 will be given in Section 8.3.2 (see page 308).

Application to Generalized Time Delay Families

We now apply Theorem 8.7 to a generalization of time delay families.

The time delay family $T : L_{(k)} \times S \to \mathbf{R}$, where $L_{(k)} = L_1 \times \cdots \times L_k$ is the product of k lens planes L_i, is given explicitly as follows (see Section 6.4.1, page 195):

$$T(\mathbf{x}_1, \ldots, \mathbf{x}_k, \mathbf{y}) = \sum_{i=1}^{k} \vartheta_i \left[\frac{|\mathbf{x}_i - \mathbf{x}_{i+1}|^2}{2} - \beta_i \psi_i(\mathbf{x}_i) \right].$$

We shall assume that $\Delta_{\mathbf{x}_i} \psi_i > 0$ for $i = 1, \ldots, k$. This is a reasonable physical assumption since we are not considering weak lensing by the large scale structure (see Section 3.1.4, page 46). It follows that

$$\Delta_{\mathbf{x}_i} T(\mathbf{x}_1, \ldots, \mathbf{x}_k, \mathbf{y}) = 2(\theta_{i-1} + \theta_i) - \theta_i \beta_i \Delta_{\mathbf{x}_i} \psi_i(\mathbf{x}_i) \; < \; 2(\theta_{i-1} + \theta_i).$$

Here, the Laplacian $\Delta_{\mathbf{x}_i} T$ of T is with respect to the coordinates $\mathbf{x}_i$ on the ith lens plane L_i, each lens plane being a punctured plane, $L_i = \mathbf{R}^2 - A_i$ with A_i a finite set of points. In addition, note that either $T(\mathbf{x}_1, \ldots, \mathbf{x}_k, \mathbf{y}) \to \infty$ or $\Delta_{\mathbf{x}_i} T(\mathbf{x}_1, \ldots, \mathbf{x}_k, \mathbf{y}) \to -\infty$ as any $\mathbf{x}_i \to A_i$ (cf. Definition 6.1, page 172).

We now abstract the previous properties. Define a *generalized k-plane time delay family* T to be an element of $C^\infty(L_{(k)} \times S, \mathbf{R})$ such that for $i = 1, \ldots, k$ we have

$$\Delta_{\mathbf{x}_i} \mathsf{T} < \Theta_i,$$

where each Θ_i is a positive constant, and if $(\mathbf{x}_1, \ldots, \mathbf{x}_k) \in L_{(k)}$, then as $\mathbf{x}_i \to A_i$ either $\mathsf{T}(\mathbf{x}_1, \ldots, \mathbf{x}_k, \mathbf{y}) \to \infty$ or $\Delta_{\mathbf{x}_i} \mathsf{T}(\mathbf{x}_1, \ldots, \mathbf{x}_k, \mathbf{y}) \to -\infty$. Denote the set of generalized k-plane time delay families by $\mathcal{T}$. The set $\mathcal{T}$ contains all k-plane time delay families (with $\Delta_{\mathbf{x}_i} \psi_i > 0$ for $i = 1, \ldots, k$).

The set $\mathcal{T}$ is open in $\mathcal{F}$ relative to the Whitney C^∞-topology. In fact, suppose that T' lies in the following open "ball" in $\mathcal{F}$ centered at T with radius $\delta > 0$ (see page 215):

$$V_\delta^2(\mathsf{T}) = \{ f \in \mathcal{F} : |j^2\mathsf{T}' - j^2\mathsf{T}| < \delta \}.$$

For δ sufficiently small, we have $\Delta_{\mathbf{x}_i} \mathsf{T}' < \Theta_i$ if and only if $\Delta_{\mathbf{x}_i} \mathsf{T} < \Theta_i$. Consequently, $\mathsf{T}' \in \mathcal{T}$ and (hence) $\mathcal{T}$ is open in $\mathcal{F}$.

Let $\mathcal{F} = C^\infty(L_{(k)}, \mathbf{R})$. Then there is an open dense subset $\mathcal{F}^\star$ of $\mathcal{F}$ consisting of all $f \in \mathcal{F}$ whose catastrophe map χ_f is smooth and transverse stable (Theorem 8.7(1)). Let $\mathcal{T}^\star = \mathcal{F}^\star \cap \mathcal{T}$, which is the set of all generalized k-plane time delay families whose catastrophe maps are transverse (hence, locally) stable. Also, let $\mathcal{T}_P = \mathcal{F}_P \cap \mathcal{T}$, where $\mathcal{F}_P$ is the set of all $f \in \mathcal{F}$ whose catastrophe map χ_f is proper. By Theorem 8.7 the set $\mathcal{F}_P$ is open in $\mathcal{F}$ and since $\mathcal{T}$ is open in $\mathcal{F}$, the set $\mathcal{T}_P$ is open in $\mathcal{F}_P$.

Using the fact that $\mathcal{F}^\star$ is open and dense in $\mathcal{F}$, and $\mathcal{T}$ is a nonempty open subset of $\mathcal{F}$, we see that the set, $\mathcal{T}^\star = \mathcal{F}^\star \cap \mathcal{T}$, is open and dense in $\mathcal{T}$. Similarly, the open-density of $\mathcal{F}_P^\star$ in $\mathcal{F}_P$ and the openness of the nonempty set $\mathcal{T}_P$ in $\mathcal{F}_P$ yields that $\mathcal{T}_P^\star$ is open and dense in $\mathcal{T}_P$. We state these results below:

Theorem 8.8 *The sets $\mathcal{T}^\star$ and $\mathcal{T}_P^\star$ are open dense subsets of $\mathcal{T}$ and $\mathcal{T}_P$, respectively.*

Theorem 8.8 yields that the set of all families in $\mathcal{T}$ whose catastrophe maps are locally stable, is dense in $\mathcal{T}$.

Results similar to that in Theorem 8.8 will be obtained for time delay families — see Theorem 8.17 (page 311).

8.3.2 Proof of Generalization

We first present some open-density results for families of functions under multijet transversality conditions. These conditions are then interpreted in the case of 2-parameter families. At the end of the section, we use these results to prove Theorem 8.7. The methods of this section will be employed in Section 8.4.2 to prove an analog of Theorem 8.8 for time delay families.

Open-Density for Multijet Transversality of Families

The results of this subsection will yield a proof of the open-density part of Theorem 8.7.

Let X be an n-dimensional manifold and Y an r-dimensional manifold. As usual, we denote by $J^k(X, Y)$ the bundle over $X \times Y$ of k-jets of map-germs from X to Y. The fibre over a point $(x, y) \in X \times Y$ is the Euclidean space of k-jets of germs of maps with source x and target y; that is, a space diffeomorphic to $J^k(n, r)$ — the space of k-jets of germs mapping $(\mathbf{R}^n, 0)$ to $(\mathbf{R}^p, 0)$. We define the projection $(\alpha, \beta) : J^k(X, Y) \to X \times Y$ as follows: For $j^k f(x)$ a k-jet of a C^∞-germ, $f : (X, x) \to (Y, f(x))$, we have $\alpha(j^k f(x)) = x$ and $\beta(j^k f(x)) = f(x)$. For any open subset U of X and a C^∞-mapping $f : U \to Y$, we let $j^k f : U \to J^k(X, Y) : x \to j^k f(x)$, where $j^k f(x)$ is the k-jet of the germ at x defined by f.

Notation: In what follows, we shall freely use the following interchangeably: $C^\infty(X)$ and $C^\infty(X, \mathbf{R})$; $J^k(X)$ and $J^k(X, \mathbf{R})$; and $J_s^k(X)$ and $J_s^k(X, \mathbf{R})$.

In $J^k(X \times Y, \mathbf{R})$ we want to consider only those jets that arise from partials with respect to the coordinates in X. That is, the subset of $J^k(X \times Y, \mathbf{R})$ consisting of $J^k(X, \mathbf{R}) \times Y$. This makes sense as follows: Locally, the jet bundle $J^k(X \times Y, \mathbf{R})$ is $(X \times Y) \times \mathbf{R} \times J^k(n+r) \supset X \times \mathbf{R} \times J^k(n) \times Y$, which is what $J^k(X, \mathbf{R}) \times Y$ is locally. Here, as usual, we denote $J^k(m, 1)$ by $J^k(m)$.

For any $k \in \mathbb{N}$ and any function $f \in \mathcal{F}$ we let $f^k : X \times Y \to J^k(X, \mathbf{R}) \times Y$ be defined by $f^k(x, y) = (j^k f_y(x), y)$ where $f_y : X \to \mathbf{R} : x \to f(x, y)$. To simplify the notation a bit, we shall write $f^k(x, y) = j^k f_y(x)$ as a point of $J^k(X, \mathbf{R}) \times Y$ that sits over $((x, f_y(x)), y)$. Since we are interested in the situation where several critical points have the same image as well as simultaneous transversality to more than one singularity, we shall work with and prove our results in the context of multijets.

If $X^s = X \times \cdots \times X$ (s-fold product), then let $X^{(s)} = \{(x_1, x_2, \ldots, x_s) \in X^s | x_i \neq x_j, i \neq j\}$. If $\alpha : J^k(X, \mathbf{R}) \to X$ is the source map and $\alpha^s : (J^k(X, \mathbf{R}))^s \to X^s$, we let $J_s^k(X, \mathbf{R}) = (\alpha^s)^{-1}(X^{(s)})$, the usual s-fold k-jet bundle. The spaces we shall work with are $J_s^k(X, \mathbf{R}) \times Y^s$. Analogous to the map f^k, we define $f_s^k : X^{(s)} \times Y^s \to J_s^k(X, \mathbf{R}) \times Y^s$ by

$$f_s^k((x_1, x_2, \ldots, x_s), (y_1, y_2, \ldots, y_s)) = (j^k f_{y_1}(x_1), j^k f_{y_2}(x_2), \ldots, j^k f_{y_s}(x_s)).$$

Note. Our notation does not respect the order of the factors in the products involved in the spaces considered. We hope that this abuse of notation will not cause any problems for the reader.

Definition 8.9 Let Z be a manifold. A *stratification* of a closed subset S of Z is a finite, disjoint collection of submanifolds $\{W_i\}$ of Z such that:

(1) $S = \bigcup\limits_i W_i$.

(2) $\partial W_i = \overline{W}_i - W_j$ *is a union of* W_j *of lower dimension.*

(3) *If* $\sigma \in W_j \subset \partial W_i$ *and a submanifold* M *is transverse to* W_j *at* σ, *then* M *is transverse to* W_i *in neighborhood for* σ.

We say that S is a *stratified subset* of Z and the W_i are called *strata* of S. If S is a closed submanifold of Z, it is a stratification of S with S as its only stratum.

Transversality to S means simultaneous transversality to all strata of W_i of S.

We now consider a stratified subset S of $J_s^k(X, \mathbf{R}) \times Y^s$. The stratified set S will be fixed, so until further notice our notation will contain no reference to it. Given any subset $A \subset X^{(s)} \times Y^s$, we let

$$\mathcal{F}^A = \{f \in \mathcal{F} : f_s^k \text{ is transverse to } S \text{ at each point of } A\}.$$

Our goal is to prove that the set

$$\mathcal{F}^{(X^{(s)} \times Y^s)} = \{f \in \mathcal{F} : f_s^k \text{ is transverse to } S\}$$

is open and dense in $\mathcal{F}$.

To deal with the above multitransversality situation, it is convenient to consider the following subsets of $C^\infty(X \times Y, \mathbf{R})$: Let $\mu : (X \times Y)^s \to \mathbf{R}_+$ be continuous; we define the set $V_\mu^k(f_s)$, to be the set of all $g \in C^\infty(X \times Y, \mathbf{R})$ such that

$$|(j^k g(x_1, y_1), \ldots, j^k g(x_s, y_s)) - (j^k f(x_1, y_1), \ldots, j^k f(x_s, y_s))| < \mu((x_1, y_1), \ldots, (x_s, y_s))$$

for all $((x_1, y_1), \ldots, (x_s, y_s)) \in (X \times Y)^s$. Similarly, for a subset $A \subset (X \times Y)^s$, we define the set $V^k_{\mu,A}(f_s)$ to be the set of all $g \in C^\infty(X \times Y, \mathbf{R})$ such that

$$|(j^k g(x_1, y_1), \ldots, j^k g(x_s, y_s)) - (j^k f(x_1, y_1), \ldots, j^k f(x_s, y_s))| < \mu((x_1, y_1), \ldots, (x_s, y_s))$$

for all $((x_1, y_1), \ldots, (x_s, y_s)) \in A$.

Lemma 8.10 *The sets $V^k_\mu(f_s)$ and $V^k_{\mu,A}(f_s)$ are C^k-open neighborhoods of f in $\mathcal{F}$.*

Proof: The mapping $\tau_s : C^\infty(X \times Y, \mathbf{R})^s \to C^\infty((X \times Y)^s, \mathbf{R}^s)$: $(f_1, \ldots, f_s) \mapsto (f_1 \times \cdots \times f_s)$ is continuous in all the Whitney C^k-topologies (i.e., the map is continuous in the Whitney C^∞-topology) [Gol-G, Prop. 3.10, p. 49]. Furthermore, the diagonal mapping $\triangleright_s : C^\infty(X \times Y, \mathbf{R}) \to C^\infty(X \times Y, \mathbf{R})^s : f \mapsto (f, \ldots, f)$ is also continuous. This last statement follows from the definition of the product topology. We consider $(\tau_s \circ \triangleright_s)(f) = (f \times \cdots \times f)$. In $C^\infty((X \times Y)^s, \mathbf{R}^s)$, a C^k-open neighborhood of $(f \times \cdots \times f)$ is $V^k_\mu(f \times \cdots \times f)$, where $\mu : (X \times Y)^s \to \mathbf{R}_+$ is a continuous map. As defined above, this set consists of all $(g_1 \times \cdots \times g_s) \in C^\infty((X \times Y)^s, \mathbf{R}^s)$ such that

$$|(j^k g_1(\zeta) \times \cdots \times j^k g_s(\zeta)) - (j^k f(x_1, y_1) \times \cdots \times j^k f(x_s, y_s))| < \mu(\zeta)$$

for all $\zeta \in (X \times Y)^s$, where $\zeta = ((x_1, y_1), \ldots, (x_s, y_s))$. The open set $(\tau_s)^{-1}$ $(V^k_\mu(f \times \cdots \times f))$ is the set of all $(h_1, \ldots, h_s) \in C^\infty((X \times Y), \mathbf{R})^s$ such that

$$|(j^k h_1(x_1, y_1), \ldots, j^k h_s(x_s, y_s)) - (j^k f(x_1, y_1), \ldots, j^k f(x_s, y_s))| < \mu(\zeta)$$

for all $\zeta \in (X \times Y)^s$ with $\zeta = ((x_1, y_1), \cdots, (x_s, y_s))$. Finally, the open set $(\tau_s \circ \triangleright_s)^{-1}(V^k_\mu(f \times \cdots \times f))$ is the set of all $e \in C^\infty((X \times Y), \mathbf{R})$ such that

$$|(j^k e(x_1, y_1), \ldots, j^k e(x_s, y_s)) - (j^k f(x_1, y_1), \ldots, j^k f(x_s, y_s))| < \mu(\zeta)$$

for all $\zeta \in (X \times Y)^s$, where $\zeta = ((x_1, y_1), \cdots, (x_s, y_s))$. Thus, the set $V^k_\mu(f_s)$ coincides with the set, $(\tau_s \circ \triangleright_s)^{-1}$ $(V^k_\mu(f \times \cdots \times f))$, and so is open. The set $V^k_{\mu,A}(f_s)$ is also open, being the union $\bigcup_\nu V^k_\nu(f_s)$ over all continuous functions $\nu : (X \times Y)^s \to \mathbf{R}_+$ such that $\nu|_A \leq \mu|_A$. $\square$

Lemma 8.11 *Let $K \subset X^{(s)} \times Y^s$ be a compact set. Then $\mathcal{F}^K$ is C^{k+1}-open in $\mathcal{F}$.*

Proof: Let $f \in \mathcal{F}^K$ and let $p \in K$. Let U_p be a compact neighborhood of p in $X^{(s)} \times Y^s$ on which f^k_s is transverse to S. For any $g \in C^\infty(X \times Y, \mathbf{R})$ whose multijet map g^k_s and the differential map dg^k_s are sufficiently close on U_p to f^k_s and df^k_s, respectively, we know that g^k_s is also transverse to S on U_p. In particular, there is an $\epsilon > 0$ such that $V^{k+1}_{\epsilon,U_p}(f_s) \subset \mathcal{F}^{U_p}$. Cover the compact set K with a finite number of such compact neighborhoods $U_{p_i}, i = 1, \ldots, m$, with ϵ_i the corresponding constants. If $\epsilon = \min\{\epsilon_i\}$, then $V^{k+1}_{\epsilon,K}(f_s) \subset \bigcap_{i=1}^{m} V^{k+1}_{\epsilon_1,U_{p_i}}(f_s) \subset \bigcap_{i=1}^{m} \mathcal{F}^{U_{p_i}} \subset \mathcal{F}^K$. □

Note. In the preceding, although f_s^k involves partials of f with respect to the coordinates of X only, the differential map df_s^k includes partials involving one derivative with respect to the coordinates of Y as well. Also, although the conditions for membership of g in $V^{k+1}_{\epsilon,K}(f_s)$ depend on all $(k+1)^{st}$ partials of g, its membership in $\mathcal{F}^K$ depends only on the behavior of g_s^k and its first derivatives.

Lemma 8.12 *Let $M = \bigcup_{i=1}^{\infty} K_i$ be a countable union of disjoint, compact sets K_i in $X^{(s)} \times Y^s$ with disjoint open neighborhoods $\mathcal{L}_i$. Then $\mathcal{F}^M$ is C^{k+1}-open in $\mathcal{F}$.*

Proof: For each i, choose a smooth bump function $b_i : ((X \times Y)^s) \to [0, 1]$ that is 1 on K_i and 0 outside of $\mathcal{L}_i$. In addition, let $b_0 = 1 - \sum_{i=1}^{\infty} b_i$. Let $f \in \mathcal{F}^M$; then $f \in \mathcal{F}^{K_i}$ for all i. Since K_i are compact, there are constants $\epsilon_i > 0$ such that $V^{k+1}_{\epsilon_i,K_i}(f) \subset \mathcal{F}^{K_i}$ by Lemma 8.11. Letting $\mu = b_0 + \sum_{i=1}^{\infty} \epsilon_i b_i$, we have $V^{k+1}_{\mu}(f) \subset \bigcap_{i=1}^{\infty} V^{k+1}_{\epsilon_i,K_i}(f) \subset \bigcap_{i=1}^{\infty} \mathcal{F}^{K_i} = \mathcal{F}^M$. □

Lemma 8.13 *For all $p \in X^{(s)} \times Y^s$ and each $f \in \mathcal{F} = C^\infty(X \times Y, \mathbf{R})$, there is a compact neighborhood N of p in $X^{(s)} \times Y^s$ and a neighborhood $\mathcal{V}$ of $f \in \mathcal{F}$ such that $\mathcal{F}^N$ is C^∞-dense in $\mathcal{V}$.*

Proof: Recall that S is a stratified subset of $J_s^k(X, \mathbf{R}) \times Y^s$. For simplicity of notation, we consider W as a (not necessarily closed) stratum of S. Let $f_s^k(p) = z$. If $z \notin \overline{W}$, let N be such that $f_s^k(N)|_W$ and choose $\mathcal{V}$ such that if $g \in \mathcal{V}$, then $g_s^k(N) \parallel W$. Thus $\mathcal{V} \subset \mathcal{F}^N$, so $\mathcal{F}^N$ is C^∞-dense in $\mathcal{V}$.

We turn to the case in which $z \in W$. Let $p = (x_1, \ldots, x_s; y_1, \ldots, y_s)$ and let $(\overline{U}_i, \ldots, \overline{U}_s,; \overline{V}_1, \ldots, \overline{V}_s)$, where $(x_i; y_i) \in U_i \times V_i$, and $\{\overline{U}_i\}$ are pairwise disjoint, and the $\overline{U}_i$ and $\overline{V}_i$ are compact, coordinate patches with origins at x_i and y_i, respectively. Without loss of generality, we may assume that $J_s^k(X, \mathbf{R})|_{(\overline{U}_1 \times \cdots \times \overline{U}_s)}$ is trivial, which we write as $\prod_{i=1}^{s} (\bar{U}_i \times \mathbf{R} \times J^k(n))$. In terms of this trivialization, we have

$$f_s^k(x, y) = ((x_1, y_1, f_{y_1}^k(x_1)), \ldots, (x_s, y_s, f_{y_s}^k(x_s))).$$

Choose coordinate neighborhoods B_i of $f_{y_i}^k(x_i) \in J^k(n)$ such that

$$f_s^k(\prod_{i=1}^{s} \overline{U}_i \times \prod_{i=1}^{s} \overline{V}_i) \subset \prod_{i=1}^{s} (\overline{U}_i \times \overline{V}_i \times \mathbf{R} \times B_i).$$

Note that

$$f_s^k(x, y) = ((x_1, y_1, f(x_1, y_1), j^k f_{y_1}(x_1)), \dots, (x_s, y_s, f(x_s, y_s), j^k f_{y_s}(x_s))).$$

Since $p \in \left(\prod_{i=1}^{s} \overline{U}_i \times \prod_{i=1}^{s} \overline{V}_i\right)$ and we shall choose the neighborhood N of p to be $\left(\prod_{i=1}^{s} \overline{U}_i \times \prod_{i=1}^{s} \overline{V}_i\right)$, we may assume that $W \subset \prod_{i=1}^{s}(\overline{U}_i \times \overline{V}_i \times \mathbf{R} \times B_i)$. Let

$$C = \left(J_s^k(X, \mathbf{R}) \times Y^s - \pi^{-1}\left(\prod_{i=1}^{s} \overline{U}_i \times \prod_{i=1}^{s} \overline{V}_i\right)\right) \bigcup \prod_{i=1}^{s}(\overline{U}_i \times \overline{V}_i \times \mathbf{R} \times B_i),$$

where π is the projection of $J_s^k(X, \mathbf{R}) \times Y^s$ onto $X^{(s)} \times Y^s$. (We hope that the abuses of notation are neither confusing nor offensive.) The set C is open since the set in the first parentheses is open and if E_i and F_i are any open sets containing $\bar{U}_i$ and $\bar{V}_i$, respectively, then

$$C = \left(J_s^k(X, \mathbf{R}) \times Y^s - \pi^{-1}\left(\prod_{i=1}^{s} \bar{U}_i \times \prod_{i=1}^{s} \bar{V}_i\right)\right) \bigcup \prod_{i=1}^{s}(E_i \times F_i \times \mathbf{R} \times B_i),$$

which is open.

Let $M(C)$ be our neighborhood $\mathcal{V}$ of f. By the choice of B_i and the definition of C, we know that $f \in M(C)$. Recall that we have chosen our neighborhood of p to be $N = \prod_{i=1}^{S}(\bar{U}_i \times \bar{V}_i)$. For each $i = 1, \dots, s$, let us suppose that our coordinate patches are contained in slightly larger, open, precompact (or totally bounded) coordinate patches, U_i' and V_i' also centered at x_i and y_i, respectively, such that

$$\overline{U}_i' \supset U_i' \supset \overline{U}_i \quad \text{and} \quad \bar{V}_i' \supset V_i' \supset \overline{V}_i \supset V_i.$$

Furthermore, we assume that the $\overline{U}_i'$ are pairwise disjoint.

Consider polynomials p_i and q_i defined on $\bar{U}_i'$ and $\bar{V}_i'$, respectively. We take bump functions, say ρ_i and σ_i, which are 1 on $\overline{U}_i$ and $\overline{V}_i$, respectively, and vanishing off of U_i' and V_i', respectively. By means of these bump functions, we extend the polynomials p_i and q_i to be smooth functions defined on all of X and Y, respectively, vanishing off of U_i' and V_i' and being unchanged on $\overline{U}_i$ and $\overline{V}_i$, respectively. Whenever convenient, we shall make no distinction between polynomials defined on $\bar{U}_i'$ and $\bar{V}_i'$ and their extensions by zero using the functions ρ_i and σ_i. Now take any $g \in M(C)$, we shall show that arbitrarily close to it there is an element in $M(C)$ transverse to W on N.

Now, let P be the set of all polynomials in n-variables of degree less than or equal to k. Consider the mapping:

$$G : \prod_{i=1}^{s}(\overline{U}_i \times \bar{V}_i \times P) \to \prod_{i=1}^{s}(\overline{U}_i \times \overline{V}_i \times \mathbf{R} \times J^k(n)) :$$

$$(x, y, p) \mapsto (x_1, y_1, (g + p_1)(x_1, y_1), j^k(g + p_1)_{y_1}(x_1)), \ldots ,$$

$$(x_s y_s, (g + p_s)(x_s, y_s), j^k(g + p_s)_{y_s}(x_s)),$$

where $(x, y, p) = (x_1, y_1, p_1, \ldots , x_s, y_s, p_s) \in \prod_{i=1}^{s}(\overline{U}_i \times \overline{V}_i \times P)$. Since $g \in M(C)$, there is a neighborhood $(P_1, P_2, \ldots , P_s)$ of 0 in P^s such that $G\left(\prod_{i=1}^{s}(\bar{U}_i \times \bar{V}_i \times P_i)\right) \subset C$. Note that $G(x, y, p)$ is a translation of $g_s^k(x, y)$ by $(p_1(x_1), j^k(p_1)(x_1)), \ldots , (p_s(x_s), j^k(p_s)(x_s))$ in the direction of $\prod_{i=1}^{s}(\mathbf{R} \times J^k(n))$. Since for any $x \in \mathbf{R}^n$, the mapping $P \to (\mathbf{R} \times J^k(n))$: $p \mapsto (p(x), j^k(p)(x))$ is a diffeomorphism, we see that G is also a diffeomorphism. Thus G is transverse to any submanifold of $\prod_{i=1}^{s}(\overline{U}_i \times \overline{V}_i \times \mathbf{R} \times J^k(n))$. In particular, the map G is transverse to W and the restriction $G|\prod_{i=1}^{s}(\overline{U}_i \times \overline{V}_i \times P_i)$ is transverse to W. To this situation, we apply Lemma 7.13 (page 222), which we restate for the reader's convenience:

> Let X, B, Y be smooth manifolds and let W be a submanifold of Y. Let $j : B \to C^\infty(X, Y)$ be such that the map $\Phi : X \times B \to Y : (x, b) \mapsto j(b)(x)$ is smooth. If Φ is transverse to W, then $\{b \in B | j(b)$ is transverse to $W\}$ is dense in B. In fact, it is the complement of a set of measure zero.

By this lemma, the set of s-tuples of polynomials in $(p_1, p_2, \ldots , p_s) \in (P_1, P_2, \ldots , P_s)$ such that $(j^k(g + p_1), \ldots , j^k(g + p_s)) : \prod_{i=1}^{s}(\overline{U}_i \times \overline{V}_i) \to \prod_{i=1}^{s}(\overline{U}_i \times \overline{V}_i \times \mathbf{R} \times J^k(n))$ is transverse to W, is dense in $(P_1, P_2, \ldots , P_s)$. Now since the $\overline{U}_i$ (introduced earlier) are pairwise disjoint, using our convention about the bump functions, the s-tuple of polynomials $(p_1, p_2, \ldots , p_s)$ can be amalgamated to an element of $C^\infty(X) : p = \sum_{i=1}^{s} p_i$, where $p|\overline{U}_i = p_i|\overline{U}_i$. We can pick an arbitrarily small such p such that $(g + p)$ is in $M(C)$, arbitrarily close to g and this $(g + p)$ is transverse to W on $N = \prod_{i=1}^{s}(\overline{U}_i \times \overline{V}_i)$. This completes the case in which $z \in W$.

Finally, suppose that $z \in \partial W$. Since S is a closed union of strata, we have $z \in W' \subset \partial W$. By the preceding proof, given any $g \in \mathcal{V}$ there is an $h \in \mathcal{V}$ arbitrarily close to g transverse to W' on N. Again, by the preceding proof and Lemma 8.11, we can find an $h' \in \mathcal{V}$ arbitrarily close to h (hence, also close to g) such that h' is still transverse to W' on N, but is now transverse to W on N as well. This completes the proof for the current case since we can repeat this argument for the finite number of strata of S in ∂W. *Note.* In this last case, no use was made of property (2) in the definition of stratifications (Definition 8.9). □

Lemma 8.14 *If $K \subset X^{(s)} \times Y^s$ is compact, then $\mathcal{F}^K$ is C^∞-dense and open in $\mathcal{F}$.*

Proof: For each $f \in \mathcal{F}$, using the preceding Lemmas 8.11 and 8.13, we can find a finite number of compact neighborhoods N_i that cover K and neighborhoods $\mathcal{V}_i$ of f such that $\mathcal{F}^{N_i}$ is open and also C^∞-dense in $\mathcal{V}_i$. Let $\mathcal{V} = \bigcap \mathcal{V}_i$. Then $\mathcal{F}^{N_i}$ is C^∞-dense in $\mathcal{V}$ as well. Thus $\mathcal{F}^{\cup N_i} = \cap \mathcal{F}^{N_i}$ is also dense in $\mathcal{V}$ and open. Since $K \subset \cup N_i$, we have $\mathcal{F}^K \supset \mathcal{F}^{\cup N_i}$ and is therefore dense in $\mathcal{V}$. Thus $\mathcal{F}^K$ is dense in $\mathcal{F}$. Openness of $\mathcal{F}^K$ is guaranteed by Lemma 8.11. □

Lemma 8.15 *Let $M = \bigcup_{i=1}^\infty K_i$ be a countable union of disjoint, compact sets K_i in $X^{(s)} \times Y^s$ with disjoint open, precompact neighborhoods $\mathcal{L}_i$. Then for any k, the set $\mathcal{F}^M$ is C^{k+1}-open and C^∞-dense in $\mathcal{F}$.*

Proof: Openness is given by Lemma 8.12. Take any $f \in \mathcal{F}$ and any basic C^∞-neighborhood $V_\mu^k(f)$. We show that $V_\mu^k(f) \bigcup \mathcal{F}^M \neq \emptyset$. As in Lemma 8.12, choose C^∞-bump functions b_i with support in L_i and which are 1 on K_i. Since $\overline{L_i}$ is compact, for each i we choose an e_i so small that if $h \in V_{e_i, L_i}^k(f)$, then $b_i h \in V_\mu^k$. By Lemma 8.14, for each i we can find an $f_i \in V_{e_i, L_i} \cap \mathcal{F}^{K_i}$. Now let

$$g = b_0 f + \sum_{i=1}^\infty b_i f_i,$$

where $b_0 = 1 - \sum_{i=1}^\infty b_i$. Outside of $\overset{\infty}{\underset{i=1}{\cup}} L_i$, we have $g = f$. On each L_i, we see that $g = (1 - b_i) f + b_i f_i = f + b_i (f_i - f)$. Since we chose $(f_i - f) \in V_{e_i, L_i}$, we know that $b_i (f_i - f) \in V_\mu^k(f)$. Thus $g \in V_\mu^k(f)$. Finally, since on K_i we have $g = f_i$ and $f_i \in \mathcal{F}^{K_i}$ for each i, we see that $g \in \mathcal{F}^{K_i}$ for every i. Hence $g \in \overset{\infty}{\underset{i=1}{\cap}} \mathcal{F}^{K_i} = \mathcal{F}^{(\overset{\infty}{\underset{i=1}{\cup}} K_i)} = \mathcal{F}^M$. □

Lemma 8.16 *The set $\mathcal{F}^{X^{(s)} \times Y^s}$ is C^{k+1}-open (hence C^∞-open) and C^∞-dense in $\mathcal{F}$.*

Proof: Lemma 8.15 covers the case when $X^{(s)} \times Y^s$ is a countable union of disjoint compact components. Otherwise, for this proof we use the existence of a smooth proper function on $X^{(s)} \times Y^s$ [Gol-G, Prop. 5.11, p. 25]. Let $\lambda : X^{(s)} \times Y^s \to \mathbf{R}$ be such a proper function. If λ is proper, then so is $\lambda^2 + 1$. Hence, it is no restriction to assume that $\lambda(X^{(s)} \times Y^s) \subset [1, \infty)$. Since at least one component of $X^{(s)} \times Y^s$ is noncompact, the λ image of each such component is a connected, noncompact subset of $[1, \infty)$ so we may assume that their images all equal $[1, \infty)$. Since the λ images of the compact components can be moved into $[1, \infty)$, we may conclude that $\lambda(X^{(s)} \times Y^s) = [1, \infty)$.

Let $K_i = \lambda^{-1}([i, i+1])$ for $i \in \mathbb{N}$, and let $\mathcal{L}_i = \lambda^{-1}\left((i - \frac{1}{2}, i + \frac{3}{2})\right)$. Since λ is proper, each K_i is compact and every $\mathcal{L}_i$ is a precompact open neighborhood of K_i. Furthermore, $\bigcup_{i=1}^{\infty} K_i = X^{(s)} \times Y^s$. Thus $\bigcup_{i=1}^{\infty} K_{2i} = Z_e$ is a countable union of disjoint compact sets K_{2i} with disjoint precompact neighborhoods $\mathcal{L}_{2i}$. Similarly, $\bigcup_{i=1}^{\infty} K_{2i-1} = Z_0$ is a countable union of disjoint compact sets, K_{2i-1}, with disjoint precompact neighborhoods $\mathcal{L}_{2i-1}$. By Lemma 8.15, each $\mathcal{F}^{Z_e}$ and $\mathcal{F}^{Z_0}$ is C^{k+1}-open and C^∞-dense on $\mathcal{F}$. Hence, since $X^{(s)} \times Y^s = Z_e \cup Z_0$ we have $\mathcal{F}^{X^{(s)} \times Y^s} = \mathcal{F}^{Z_e} \cap \mathcal{F}^{Z_0}$ is C^{k+1}-open and C^∞-dense on $\mathcal{F}$. □

Lemma 8.16 proves the open-density claim in part (1) of Theorem 8.7.

Interpreting Multijet Transversality for Two Parameter Families

This section establishes that the multijet transversality conditions in Theorem 8.7 on each family of functions f in $\mathcal{F}^\star$ guarantee that the catastrophe map of f is transverse stable. This will prove the latter claim in part (1) of Theorem 8.7.

We now restrict Y to be 2-dimensional (i.e., there are two control parameters). In this case, the only critical points that catastrophe maps can display transversally are cusps and folds.[5] By Lemma 8.16 applied to the case of $s = 1$ (or appealing to Thom's theorem), we know that there is an open dense subset of $\mathcal{F}$ whose catastrophe maps display only those critical points. That is, for a function f in this dense set, the 1-jet $j^1 f$ is transverse to the singularity defined by the vanishing of the gradient of f with respect to the x-variables. This, of course, guarantees that the catastrophe manifold M_f is a submanifold of $X \times Y$ of codimension n (i.e., M_f is two dimensional). If the matrix of second partials $\{\frac{\partial^2 f}{\partial x_i \partial x_j}, i, j = 1, \ldots, n\}$ has rank n, then the catastrophe manifold M_f is defined locally by $x_i = 0$, where $i = 1, \ldots, n$. Hence, the catastrophe map is locally just the projection of $(\{0\} \times Y)$ onto Y, which is clearly a nonsingular map. If we consider the orbit $J^2(X)$ of those 2-jets, then the matrix of second partials, $\{\frac{\partial^2 f}{\partial x_i \partial x_j}, i, j = 1, \ldots, n\}$ has rank $n - \lambda$. Using the Splitting Lemma (Theorem 7.60, page 267), we can introduce coordinates

$$x_1, \ldots, x_\lambda, z_1, \ldots, z_{n-\lambda}$$

in X so that locally, $f(x, z, y) = Q(z) + g(x, y)$, where Q is a nonsingular quadratic form in the z's and the matrix $\{\frac{\partial^2 g}{\partial x_i \partial x_j}, i, j = 1, \ldots, \lambda\}$ vanishes. In these coordinates, this

[5] See Thom's Catastrophe Theorem — Theorem 7.76, page 275 — and the discussion following the theorem.

singularity is defined by

$$\left\{\frac{\partial g}{\partial x_1} = 0, \dots, \frac{\partial g}{\partial x_\lambda} = 0, \frac{\partial g}{\partial z_1} = 0, \dots, \frac{\partial f}{\partial z_{n-\lambda}} = 0, \right.$$
$$\left. \text{and } \frac{\partial^2 g}{\partial x_i \partial x_j} = 0, \; i, j = 1, \dots, \lambda \right\}.$$

Since the codimension of this singularity is $n + \frac{1}{2}\lambda(\lambda+1)$ in $J^2(X) \times Y$, we see that $j^2 f$ transversally misses all such singularities except that of $\lambda = 1$. Thus, the only singularities we have to consider are those defined by the vanishing of

$$\left(z_1, \dots, z_{n-1}, \frac{\partial g}{\partial x}, \frac{\partial^2 g}{\partial x^2}\right).$$

The transversality condition that our map must satisfy is that the following matrix must have rank 2:

$$\begin{pmatrix} \frac{\partial^2 g}{\partial x^2} & \frac{\partial^3 g}{\partial x^3} \\ \frac{\partial^2 g}{\partial x \partial y_1} & \frac{\partial^3 g}{\partial x^2 \partial y_1} \\ \frac{\partial^2 g}{\partial x \partial y_2} & \frac{\partial^3 g}{\partial x^2 \partial y_2} \end{pmatrix}.$$

This singularity has two parts, a relatively open subset where $\frac{\partial^3 g}{\partial x^3} \neq 0$ and its complement, a closed subset where $\frac{\partial^3 g}{\partial x^3} = 0$. In both cases, since we are working where $\frac{\partial^2 g}{\partial x^2} = 0$, we know that at least one of $\frac{\partial^2 g}{\partial x \partial y_1}$ and $\frac{\partial^2 g}{\partial x \partial y_2}$ does not vanish. By a change of coordinates we may assume that $\frac{\partial^2 g}{\partial x \partial y_1} = -1$ and $\frac{\partial^2 g}{\partial x \partial y_2} = 0$.

In case $\frac{\partial^3 g}{\partial x^3} \neq 0$, we may assume that $\frac{\partial^3 g}{\partial x^3} = \frac{1}{3}$ and $g(x, y_1, y_2) = \frac{x^3}{3} - xy_1 + H(x, y_1, y_2)$, where H is at least cubic in the variables. If we consider $g(x, 0, 0) = \frac{x^3}{3} + H(x, 0, 0)$, we know that $H(x, 0, 0) = x^4 A(x)$. Thus $g(x, 0, 0) = x^3(\frac{1}{3} + xA(x))$. We can make the change of the x-coordinate so that $g(x, 0, 0) = x^3$ in the new x-coordinate. Since f is in our open dense set of k-transversal mappings, the germ of f at 0 is a k-transversal unfolding of $Q(z) + g(x, 0, 0)$ (Theorem 7.68, page 271). Furthermore, the germ of $Q(z) + g(x, 0, 0)$ is finitely determined and the codimension of $g(x, 0, 0)$, $\operatorname{cod} g(x, 0, 0) = 1$. Thus, by Theorem 7.69 (page 271) the unfolding f at 0 is versal. Thus we know that $Q(z) + g(x, y_1, y_2)$ can be transformed by a change of coordinates into the polynomial form: $g(x, y_1, y_2) = \frac{x^3}{3} - xy_1$ (Theorem 7.71, page 271). *The orbit in* $J^3(X) \times Y$ *of this singularity produces the smooth fold curves of the catastrophe map.*

Similarly, if $\frac{\partial^3 g}{\partial x^3} = 0$, then our singularity is given by the vanishing of

$$\left(z_1, \dots, z_{n-1}, \frac{\partial g}{\partial x}, \frac{\partial^2 g}{\partial x^2}, \frac{\partial^3 g}{\partial x^3}\right)$$

and, hence, our transversality condition is nonsingularity of the matrix:

$$\begin{pmatrix} \frac{\partial^2 g}{\partial x^2} & \frac{\partial^3 g}{\partial x^3} & \frac{\partial^4 g}{\partial x^4} \\ \frac{\partial^2 g}{\partial x \partial y_1} & \frac{\partial^3 g}{\partial x^2 \partial y_1} & \frac{\partial^4 g}{\partial x^3 \partial y_1} \\ \frac{\partial^2 g}{\partial x \partial y_2} & \frac{\partial^3 g}{\partial x^2 \partial y_2} & \frac{\partial^4 g}{\partial x^3 \partial y_2} \end{pmatrix}.$$

This forces $\frac{\partial^4 g}{\partial x^4} \neq 0$ and the nonsingularity of the matrix:

$$\begin{pmatrix} \frac{\partial^2 g}{\partial x \partial y_1} & \frac{\partial^3 g}{\partial x^2 \partial y_1} \\ \frac{\partial^2 g}{\partial x \partial y_2} & \frac{\partial^3 g}{\partial x^2 \partial y_2} \end{pmatrix}.$$

Again, using an argument which follows precisely as that given in the preceding case, it is no restriction to assume that $g(x, y_1, y_2) = \frac{x^4}{4} + \frac{x^2 y_2}{2} - x y_1$. The only significant difference is that the codimension of $Q(z) + \frac{x^4}{4}$ is 2. Hence, its universal unfolding involves both y_1 and y_2. *The orbit in $J^4(X) \times Y$ of this singularity produces the cusp points.*

In these cases the catastrophe manifolds in the zxy-space are given by $z = 0$ and $y_1 = x_1^2$ in the fold case and $y_1 = x^3 + x y_2$ in the cusp case. Thus, the catastrophe maps are the projections onto the last two factors of the subspaces $(0, x, x^2, y_2)$ and of $(0, x, x y_2 + x^3, y_2)$, or more simply, $(x, y_2) \mapsto (x^2, y_2)$ and $(x, y_2) \mapsto (x y_2 + x^3, y_2)$. Thus, as already announced, the two singularities produce folds and cusps as critical points of the catastrophe maps, respectively.

We have described these singularities as orbits in $J^3(n)$ and $J^4(n)$ instead of $J^2(n)$ and $J^3(n)$, respectively, because we incorporated the transversality in the given normal forms requiring in the fold case, $\frac{\partial^3 f}{\partial x^3} \neq 0$ and in the cusp case $\frac{\partial^4 f}{\partial x^4} \neq 0$. For convenience, by lifting them we can consider both of these singularities as orbits in $J^4(n+2, 1)$. Denote these two singularities as F and C of codimension $(n+1)$ and $(n+2)$, respectively. For simplicity of exposition we write the jet bundle as a trivial bundle,

$$J^4(X^{(2)} \times Y^2) = X^{(2)} \times Y^2 \times \mathbf{R}^2 \times J^4(n+2, 1) \times J^4(n+2, 1),$$

where $X^{(2)} = X \times X - \triangleright X$. To account for the double points on the fold curves, we let

$$D = X^{(2)} \times \triangleright Y \times \mathbf{R}^2 \times F \times F$$

and to account for the (non)-coincidence of cusps on folds, we let

$$E = X^{(2)} \times \triangleright Y \times \mathbf{R}^2 \times F \times C.$$

In $J^4(X^{(2)} \times Y^2)$, the codimension of D is $2 + 2(n+1) = 2n+4$, the dimension of $X^{(2)} \times Y^2$. Thus, if $j_2^4 f$ is transverse to D, then $(j_2^4 f)^{-1}(D)$ consists of isolated points. Furthermore, the codimension of E in $J^4(X^{(2)} \times Y^2)$ is $2 + (n+1) + (n+2) = 2n+5$, which is greater than the dimension of $X^{(2)} \times Y^2$. Hence, if $j_2^4 f$ is transverse to E, then $(j_2^4 f)^{-1}(E)$ is empty. Thus, for such multi-transverse maps, the restriction of the map to the fold curves can have only transverse double points and no cusp can be a double point of this restriction. That is, under such a map, if two distinct fold critical points have the same image, then the images of the two fold critical curves cross transversally at that point. Furthermore, no cusp point can have the same image as a fold point. Of course, as we have remarked above, for dimensional reasons no two cusp points can have the same image. Similarly, such a map restricted to its set of critical points — fold critical curves and cusp critical points lying on them — have no triple points. Theorem 7.56 (page 262) now yields that the catastrophe map of f is transverse stable. This proves the transverse stability claim in part (1) of Theorem 8.7.

Completing the Proof of Theorem 8.7:

For convenience, we restate Theorem 8.7 below:

Theorem:

(1) *Let X be an n-dimensional manifold, Y a 2-dimensional manifold, and $\mathcal{F} = C^\infty(X \times Y, \mathbf{R})$ in the Whitney C^∞-topology. Then there is an open and dense subset $\mathcal{F}^\star$ consisting of families multi-transverse to all singularities in $J_s^k(X) \times Y^s$. In particular, the set $\mathcal{F}^\star$ consists of all f in $\mathcal{F}$ whose catastrophe manifold M_f is a smooth 2-dimensional surface and the corresponding catastrophe map χ_f is transverse stable.*

(2) *Let $\mathcal{F}_P$ be the subset of elements in $\mathcal{F}$ whose catastrophe maps are proper and set $\mathcal{F}_P^\star = \mathcal{F}^\star \cap \mathcal{F}_P$. Then $\mathcal{F}_P$ is a nonempty open subset of $\mathcal{F}$ and $\mathcal{F}_P^*$ is open and dense in $\mathcal{F}_P$. If $f \in \mathcal{F}_P^*$, then the catastrophe map $\chi_f : M_f \to Y$ is globally stable as an element of $C^\infty(M_f, Y)$.*

Proof: All of part (1) is already proven as well as practically all of part (2) (see the previous two subsections). The global stability statement follows from Mather stability theorem (Theorem 7.31, page 230), according to which a map that is both transverse stable and proper is globally stable.

The only point to check is the openness of $\mathcal{F}_P^\star$ in $\mathcal{F}$. It suffices to prove that the subset $\mathcal{F}_P$ of $C^\infty(X \times Y, \mathbf{R})$ of those maps whose catastrophe maps are proper is a nonempty open set. In fact, since $\mathcal{F}^\star$ is an open dense subset of $\mathcal{F}$ by part (1) of the theorem, if we have that $\mathcal{F}_P$ is a nonempty open subset of $\mathcal{F}$, then the set $\mathcal{F}_P^* = \mathcal{F}^\star \cap \mathcal{F}_P$ is open and dense in $\mathcal{F}_P$.

The mapping $d_x : C^\infty(X \times Y, \mathbf{R}) \to C^\infty(X \times Y, T^*X) : f \to d_x f$ is continuous. [Note that for $f \in \mathcal{F}_P$ we do not require that the catastrophe sets $M_f = (d_x f)^{-1}(\mathbb{O})$ be a manifold.] Hence, it suffices to prove that the set, $\{\sigma \in C^\infty(X \times Y, T^*X) : \pi|_{\sigma^{-1}(\mathbb{O})}$ is proper $\}$, is open.

Introduce a vector bundle metric $\| \ . \ \|$ in T^*X. Now consider the mapping, $C^\infty(X \times Y, T^*X) \to C^\infty(X \times Y, \mathbf{R} \times Y) : \sigma \mapsto (\| \sigma \|^2, \pi)$. This is also continuous.

Claim: It suffices to prove that the proper maps in $C^\infty(X \times Y, \mathbf{R} \times Y)$ form an open set. In fact, suppose that $(\| \sigma \|^2, \pi) \in C^\infty(X \times Y, \mathbf{R} \times Y)$ is proper. Then for any compact set K in Y, the set $(\| \sigma \|^2, \pi)^{-1}(0, K)$ is compact in $X \times Y$. But $(\| \sigma \|^2, \pi)^{-1}(0, K) = \sigma^{-1}(\mathbb{O}) \bigcap \pi^{-1}K = (\pi|\sigma^{-1}(\mathbb{O}))^{-1}(K)$. Thus $\pi|\sigma^{-1}(\mathbb{O})$ is proper. This proof is complete with the following:

Lemma 8.1 *For X and Y manifolds, the set of proper maps in $C^\infty(X, Y)$ is open.*

Proof: If X is compact, then this is immediate since all maps in $C^\infty(X, Y)$ are proper. If Y is compact and X is not, there are no proper maps in $C^\infty(X, Y)$. Thus, we treat the only nontrivial case, namely the case in which both X and Y are noncompact. For the rest of the proof, we assume that both X and Y are noncompact.

Lemma 8.2 *On any noncompact manifold there is a metric with the property that a set not contained in any compact has infinite diameter.*

Proof: Let Z be a noncompact manifold. We know by [Gol-G, p. 25, Prop. 5.11] that on Z there is a proper, real valued function, ρ. If we are given a metric on Z, we denote the distance between z and z' by $\| z - z' \|$. A new metric, given by dist$(z, z'(=\| z - z' \| + |\rho(z) - \rho(z')|$ has the desired property. For suppose $S \subset Z$ is contained in no compact set and assume that the diam$S = k < \infty$, then $|\rho(z) - \rho(z')| < k$ for all z, z' in S. Thus, for $z_0 \in S$, $\rho(z) \in [\rho(z_0) - k, \rho(z_0 + k]$, for all $z \in S$. That is $S \subset \rho^{-1}[\rho(z_0) - k, \rho(z_0) + k]$. Thus, since ρ is proper this implies that S is contained in a compact set. Contradiction.

□ **(Lemma 8.2)**

Continuing with the proof of Lemma 8.1, let $f \in C^\infty(X, Y)$ be proper and let $\delta > 0$

$$V_\delta(f) = \{g \in C^\infty(X, Y) : |f(x) - g(x)| < \delta\}.$$

Here, again, we denote by $|y - y'|$ the distance y and y' with respect to some complete, metric in Y which has the property given by the preceding Lemma 8.2.

Claim: All maps in $V_\delta(f)$ are proper. Let $g \in V_\delta(f)$ and suppose that g is not proper. Then for some compact K in Y the set $g^{-1}(K)$ is not compact. Moreover, the set $g^{-1}(K)$ is nonempty, otherwise it would be compact. So there is a sequence of points $x_1, x_2, \ldots$ in $g^{-1}(K)$ with no limit point in X; denote their g images by $y_i = g(x_i)$. Since f is proper, the sequence $f(x_i) = z_i$ does not lie in a compact set in Y. Hence the set of distances, $\{|z_i - y_i|\} = \{|f(x_i)|\}$ is unbounded which contradicts our choice of neighborhood $V_\delta(f)$.

□ **(Lemma 8.1)**

This completes the proof of Theorem 8.7. □

8.4 Most Time Delay Families have Stable Lensing Maps

We shall investigate the genericity of multiplane time delay families with local, transverse, and globally stable lensing maps. The methods of Section 8.3.2 will be applied.

Requisite Mathematics: Sections 7.1–7.3, 7.6, and 8.3.

8.4.1 Statement of Main Theorem

We first recall the definitions of a gravitational lens potential and time delay family.

Definition (Definition 6.1, page 172). Let A be a finite subset of $\mathbf{R}^2$ and set $L = \mathbf{R}^2 - A$ (lens plane) . Then a smooth function $\psi : L \to \mathbf{R}$ is called a *gravitational lens potential on L with singularities A* if:

(1) $\Delta\psi \geq 0$, where Δ is the Laplace operator.

(2) As $\mathbf{x} \to A$, either $\psi(\mathbf{x}) \to -\infty$ or $\Delta\psi(\mathbf{x}) \to \infty$.

Suppose that $\Delta\psi_i > 0$ for $i = 1, \ldots, k$ (which is physically reasonable in strong lensing). Let $L_{(k)} = L_1 \times \cdots \times L_k$ be the product of k lens planes, each lens plane being of the form $L_i = \mathbf{R}^2 - A_i$, where A_i is a finite set of points. The rectangular coordinates on $L_{(k)}$ are $(\mathbf{x}_1, \ldots, \mathbf{x}_k)$, where $\mathbf{x}_i = (\mathrm{u}_i, \mathrm{v}_i)$ for $i = 1, \ldots, k$. Furthermore, the set S (light source plane) will be a copy of $\mathbf{R}^2$ with coordinates $\mathbf{y}$. To simply the notation, we shall also let $\mathbf{y} = \mathbf{x}_{k+1} = (\mathrm{u}_{k+1}, \mathrm{v}_{k+1})$.

Definition (Definition 6.12, page 196). A *k-plane time delay family* is a smooth function $T : L_{(k)} \times S \to \mathbf{R}$ such that[6]

$$T(\mathbf{x}_1, \ldots, \mathbf{x}_k, \mathbf{y}) = \sum_{i=1}^{k} \vartheta_i \left(\frac{|\mathbf{x}_i - \mathbf{x}_{i+1}|^2}{2} - \beta_i \psi_i(\mathbf{x}_i) \right),$$

where ϑ_i and β_i are positive constants, and ψ_i are gravitational lens potentials on L_i with singularities A_i.

For each time delay family T, the associated lensing map $\boldsymbol{\eta}$ and catastrophe map χ_T are differentiably equivalent (Proposition 7.74, page 275). Hence, stability of χ_T is equivalent to stability of $\boldsymbol{\eta}$.

We denote the set of all k-plane time delay families $\mathbf{T} : \mathbf{L}_{(\mathbf{k})} \times \mathbf{S} \to \mathbf{R}$ by $\mathcal{M}$.

Let $\mathcal{M}^\star$ be the subset of $\mathcal{M}$ of all k-plane time delay families whose lensing (or catastrophe) maps are transverse (hence, locally) stable. Denote by $\mathcal{M}_P$ the subset of families in $\mathcal{M}$ whose lensing maps are proper and set $\mathcal{M}^\star_P = \mathcal{M}^\star \cap \mathcal{M}_P$.

Our main theorem is as follows:

[6] Note that $\beta_k = 1$.

Theorem 8.17 *The sets $\mathcal{M}^\star$ and $\mathcal{M}^\star_P$ are open and dense in $\mathcal{M}$ and $\mathcal{M}_P$, respectively.*

Note that Theorem 8.17 yields: *Among the k-plane time delay families $T : L_{(k)} \times S \to \mathbf{R}$ whose lensing (or catastrophe) maps are proper, those whose lensing maps are also globally stable (equivalently, proper plus transverse stable) form an open dense set in $\mathcal{M}_P$.*

8.4.2 Proof of Main Theorem

We present a proof of Theorem 8.17.

Let $\mathcal{F} = C^\infty(L_{(k)}, \mathbf{R})$. Then there is an open dense subset $\mathcal{F}^\star$ of $\mathcal{F}$ consisting of all $f \in \mathcal{F}$ whose catastrophe map χ_f is smooth and transverse stable (Theorem 8.7(1)).

We know that $\mathcal{M}^\star = \mathcal{T}^\star \cap \mathcal{M}$, where $\mathcal{M} \subset \mathcal{T}$. This is immediate since they are both subsets of $C^\infty(L_{(k)}, \mathbf{R})$ and hence for elements of either to be transverse stable their lensing (or catastrophe) maps must be multi-transverse to the same multijet singularities. Now, Theorem 8.8 (page 298) yields that $\mathcal{T}^\star$ is open and dense in $\mathcal{T}$. Hence, the set $\mathcal{M}^\star$ is surely open in $\mathcal{M}$. However, we must still check the density of $\mathcal{M}^\star$ in $\mathcal{M}$. Since $\mathcal{M}$ is a rather small subset of $\mathcal{T}$, the density result does not follow trivially from Theorem 8.8.

The set $\mathcal{M}_P$ is open in $\mathcal{M}$ (since $\mathcal{M} \subset \mathcal{T}$, $\mathcal{T}_P$ is open in $\mathcal{T}$, and $\mathcal{M}_P = \mathcal{T}_P \cap \mathcal{M}$). Consequently, the set, $\mathcal{M}^\star_P = \mathcal{M}^\star \cap \mathcal{M}_P$, is open in $\mathcal{M}$ and (hence) $\mathcal{M}_P$. If we assume that $\mathcal{M}^\star$ is open and dense in $\mathcal{M}$, then since the nonempty set $\mathcal{M}_P$ is also open in $\mathcal{M}$, we see that $\mathcal{M}^\star_P$ is open and dense in $\mathcal{M}_P$. Hence, it suffices to establish:

The set $\mathcal{M}^\star$ is open and dense in $\mathcal{M}$.

The remainder of this sections deals with proving the above statement.

Openness/Density Criterion for Subsets of $\mathcal{M}$ via the Map G

The following reasonable physical condition will be assumed (since we are not investigating weak lensing by large scale structure density perturbations; see Section 3.1.4, page 46):

Each gravitational lens potential $\psi_i : L_i \to \mathbf{R}$ obeys $\Delta_{x_i}\psi_i > 0$, i $= 1, \ldots, k$.

Recall the Notational Conventions:

$$J^p(\mathbf{R}^n, \mathbf{R}) = J^p(\mathbf{R}^n) = J^p(n), \quad C^\infty(X, \mathbf{R}) = C^\infty(X), \quad J^p(X, \mathbf{R}) = J^p(X).$$

Then the set of all gravitational lens potentials ψ_i, each of which has a strictly positive Laplacian by assumption, is open in $C^\infty(L_i)$.

We call the functions ψ_i *normalized* if $\psi_i(0) = 0$ $(i = 1, \ldots, k)$. This can always be achieved for any gravitational lens potential by adding an appropriate constant. Adding such a constant does not alter the lensing map. Without loss of generality, we shall consider in the remainder of the proof only gravitational lens potentials that are normalized.

A time delay family $T \in \mathcal{M}$ is uniquely determined by the k-tuple of normalized functions $(\psi_1, \ldots, \psi_k) \in \prod_{i=1}^{k} C^\infty(L_i)$. We then have a canonical "inclusion" map

$$E : \prod_{i=1}^{k} C^\infty(L_i) \longrightarrow C^\infty(L_{(k)} \times S)$$

given by

$$E(\psi_1, \ldots, \psi_k) = \sum_{i=1}^{k} \vartheta_i \left(\frac{|\mathbf{x}_i - \mathbf{x}_{i+1}|^2}{2} - \beta_i \psi_i(\mathbf{x}_i) \right) = T.$$

We also have a map G in the other direction such that $G \circ E$ is the identity map on normalized k-tuples in $\prod_{i=1}^{k} C^\infty(L_i)$. Explicitly, it is the map

$$G : C^\infty(L_{(k)} \times S) \longrightarrow \prod_{i=1}^{k} C^\infty(L_i)$$

defined by (with $\beta_k = 1$)

$$\begin{aligned} G(f(\mathbf{x}_1, \ldots, \mathbf{x}_{k+1})) &= \left(\frac{g(\mathbf{x}_1, \mathbf{0}, \ldots, \mathbf{0})}{\vartheta_1 \beta_1}, \frac{g(\mathbf{0}, \mathbf{x}_2, \mathbf{0}, \ldots, \mathbf{0})}{\vartheta_2 \beta_2}, \right. \\ &\qquad \left. \ldots, \frac{g(\mathbf{0}, \ldots, \mathbf{0}, \mathbf{x}_{k-1}, \mathbf{0}, \mathbf{0})}{\vartheta_{k-1} \beta_{k-1}}, \frac{g(\mathbf{0}, \ldots, \mathbf{0}, \mathbf{x}_k, \mathbf{0})}{\vartheta_k} \right), \end{aligned}$$

and

$$g(\mathbf{x}_1, \ldots, \mathbf{x}_k, \mathbf{x}_{k+1}) = \sum_{i=1}^{k} \vartheta_i \left(\frac{|\mathbf{x}_i - \mathbf{x}_{i+1}|^2}{2} \right) - f(\mathbf{x}_1, \ldots, \mathbf{x}_{k+1}).$$

Lemma 8.18 *The mappings $E : \prod_{i=1}^{k} C^\infty(L_i) \longrightarrow C^\infty(L_{(k)} \times S)$ and $G : C^\infty(L_{(k)} \times S) \longrightarrow \prod_{i=1}^{k} C^\infty(L_i)$ are both continuous in the Whitney C^∞-topology.*

Proof: To show that E is continuous it suffices to show the continuity of the map

$$H : \prod_{i=1}^{k} C^\infty(L_i) \longrightarrow C^\infty(L_{(k)} \times S)$$

given by

$$H(\psi_1, \ldots, \psi_k) = \sum_{i=1}^{k} (\psi_i \circ \pi_i),$$

where $\pi_i : L_{(k)} \times S \to L_i$ is the projection map. Since the map

$$\prod_{i=1}^{k} C^\infty(L_{(k)} \times S) \longrightarrow C^\infty(L_{(k)} \times S) : (f_1, \ldots, f_k) \mapsto \sum_{i=1}^{k} f_i$$

is continuous [Gol-G, p. 48, Cor. 3.8], it suffices to prove that the map

$$\prod_{i=1}^{k} C^\infty(L_i) \longrightarrow C^\infty(L_{(k)} \times S) : f \mapsto f \circ \pi_i$$

is continuous. **N.B.** In general, composition on the right is not continuous. Nevertheless:

Lemma 8.19 *Let X and Y be manifolds and let $\pi : X \times Y \to X$ be the projection. Then the map $\pi^* : C^\infty(X) \to C^\infty(X \times Y) : f \mapsto f \circ \pi$ is continuous in the Whitney C^∞-topology.*

Proof: Let $\dim X = n$ and $\dim Y = p$. For this proof, without loss of generality we may assume that X and Y are Euclidean space $\mathbf{R}^n$ and $\mathbf{R}^p$, respectively. Thus $J^k(X) = X \times \mathbf{R} \times J^k(n)$ and $J^k(X \times Y) = X \times Y \times \mathbf{R} \times J^k(n+p)$. Furthermore, we may write $J^k(n+p)$ as $J^k(n) \times M^k(n+p)$, where $M^k(n+p)$ are those components of $J^k(n+p)$ which involve a partial in any of the Y-directions. Since $J^k(n)$, $J^k(n+p)$, and $M^k(n+p)$ are all Euclidean spaces, they all have origins (0 in all components). Now, let $\mathbf{U} = U \times V \times W \times P \times Q \subseteq X \times Y \times \mathbf{R} \times J^k(n) \times M^k(n+p) = J^k(X \times Y)$ be an open set such that $j^k(f \circ \pi)(X \times Y) \subseteq \mathbf{U}$, where $f \in C^\infty(X)$. Note that from each $(x, y) \in X \times Y$, we have $j^k(f \circ \pi)(x, y) = (x, y, f(x), j^k f(x), 0) \in \mathbf{U}$. Now let

$$\mathbf{V} = U \times W \times P \subseteq X \times \mathbf{R} \times J^k(n) = J^k(X).$$

We claim that $\pi^*(M(\mathbf{V})) \subseteq M(\mathbf{U})$. In fact, let $g \in M(\mathbf{V})$ and let $(x, y) \in X \times Y$. Then $j^k(g \circ \pi)(x, y) = (x, y, g(x), j^k g(x), 0) \in \mathbf{U}$. □ **(Lemma 8.19)**

We now prove that $G : C^\infty(L_{(k)} \times S) \to \prod_{i=1}^{k} C^\infty(L_i)$ is also continuous. It suffices to prove the continuity of

$$K : C^\infty(L_{(k)} \times S) \longrightarrow \prod_{i=1}^{k} C^\infty(L_i) : f \mapsto (f \circ \sigma_1, \ldots, f \circ \sigma_k),$$

where

$$\sigma_i : L_i \longrightarrow L_{(k)} \times S : \mathbf{x}_i \mapsto (\mathbf{0}, \ldots, \mathbf{0}, \mathbf{x}_i, \mathbf{0}, \ldots, \mathbf{0}).$$

In fact, it suffices to prove the continuity of $C^\infty(L_{(k)} \times S) \to C^\infty(L_1) : f \mapsto f \circ \sigma_1$.

Lemma 8.20 *Let* $X = \mathbf{R}^n$ *and* $Y = \mathbf{R}^p$ *and let* $\sigma : X \to X \times Y : x \mapsto (x, 0)$. *Then the map* $\sigma^* : C^\infty(X \times Y) \to C^\infty(X) : f \mapsto f \circ \sigma$ *is continuous in the Whitney* C^∞*-topology.*

Proof: We shall use some of the notations and conventions in the proof of Lemma 8.19. Let $f \in C^\infty(X \times Y)$ and consider $\mathbf{U}$ and $\mathbf{V}$, where $j^k(f \circ \sigma)(X) \subseteq \mathbf{V}$. The latter means that for each $x \in X$, we have $j^k(f \circ \sigma)(x) = (x, f(x, 0)) \in \mathbf{V}$. We claim that $\sigma^*(M(\mathbf{U})) \subseteq M(\mathbf{V})$. In fact, if $g \in M(\mathbf{U})$, then for all $(x, y) \in X \times Y$ we have that $j^k g(x, y) = (x, y, g(x, y), j^k g(x, y)) \in \mathbf{U}$. In particular, we see that $g(x, 0) \in W$ and $j^k(g|X \times 0)(x, 0) \in P$. The fact that $\sigma^*(g)$ is in $M(\mathbf{V})$ follows from $j^k(\sigma^*(g))(x) = j^k(g(x, 0)) = (x, g(x, 0), j^k(g|X \times 0)(x, 0)) \in \mathbf{V} \subseteq X \times \mathbf{R} \times J^k(n)$. □ **(Lemma 8.20)**

With Lemma 8.20 established, we have completed the proof of Lemma 8.18. □

Let

$$\mathcal{M}_\times = \{(\psi_1, \ldots, \psi_k) \in \prod_{i=1}^{k} C^\infty(L_i) : \psi_i \text{ is normalized gravitational lens potential on } L_i \text{ with singularities } A_i,\ i = 1, \ldots, k\}.$$

By virtue of the definition of multiplane time delay families and our normalization, we have $\mathcal{M} = E(\mathcal{M}_\times)$. Thus, we have not only $G \circ E$ equaling the identity map on $\mathcal{M}_\times$, but also $E \circ G|_\mathcal{M}$ being the identity on $\mathcal{M}$. Thus:

Corollary 8.21 *The map* $E : \mathcal{M}_\times \to \mathcal{M}$ *and* $G : \mathcal{M} \to \mathcal{M}_\times$ *are homeomorphisms and* $G|_\mathcal{M} = (E|_{\mathcal{M}_\times})^{-1}$.

The following is a consequence of Corollary 8.21:

> **To prove that a subset $\mathcal{S} \subseteq \mathcal{M}$ is dense or open in $\mathcal{M}$ is equivalent to proving openness or density of $G(\mathcal{S}) \subseteq \mathcal{M}_\times$.**

Transversality and Stratified Sets for the Setting $\mathcal{M}$

The first singularity that an element T in $\mathcal{M}$ may encounter is $S_1(L_{(k)}, \mathbf{R}) \times S$ in $J^1(L_{(k)}, \mathbf{R}) \times S$, where S is the light source plane ($S = \mathbf{R}^2$).

Lemma 8.22 *For any* $T \in \mathcal{M}$, *if* $j^1 T$ *meets an element of* $S_1(L_{(k)}, \mathbf{R}) \times S$ *at point of* $L_{(k)} \times S$, *then* $j^1 T$ *is transverse to* $S_1(L_{(k)}, \mathbf{R}) \times S$ *at that point.*

Proof: The 1-jet map $j^1 T$ meets $S_1(L_{(k)}, \mathbf{R}) \times S$ wherever all k gradients of T vanish. Such a point is given by solutions of

$$(\vartheta_{i-1} + \vartheta_i)\mathbf{x}_i - (\vartheta_{i-1}\mathbf{x}_{i-1} + \vartheta_i \mathbf{x}_{i+1}) - \vartheta_i \beta_i \operatorname{grad}_i \psi_i(\mathbf{x}_i) = \mathbf{0},$$

where $\vartheta_0 = 0$ and $i = 1, \ldots, k$. These $2k$ functions of $2(k+1)$ variables define a map from $\mathbf{R}^{2(k+1)}$ into $\mathbf{R}^{2k}$. To prove the transversality claimed in the lemma it suffices to see that the Jacobian of this map has rank $2k$ wherever the above equation holds. Computing the Jacobian of this map we obtain the following $2k \times 2(k+1)$ matrix:

$$\begin{bmatrix} T_{\mathbf{x}_i\mathbf{x}_j} & T_{\mathbf{x}_i\mathbf{y}} \end{bmatrix} = \begin{bmatrix} \mathbf{T}_{11} & \mathbf{I}_1 & \mathbf{0} & \mathbf{0} & \mathbf{0} & \ldots & \mathbf{0} & \mathbf{0} & \mathbf{0} \\ \mathbf{I}_1 & \mathbf{T}_{22} & \mathbf{I}_2 & \mathbf{0} & \mathbf{0} & \ldots & \mathbf{0} & \mathbf{0} & \mathbf{0} \\ \mathbf{0} & \mathbf{I}_2 & \mathbf{T}_{33} & \mathbf{I}_3 & \mathbf{0} & \ldots & \mathbf{0} & \mathbf{0} & \mathbf{0} \\ \mathbf{0} & \mathbf{0} & \mathbf{I}_3 & \mathbf{T}_{44} & \mathbf{I}_4 & \ldots & \mathbf{0} & \mathbf{0} & \mathbf{0} \\ \vdots & \vdots & \vdots & \vdots & \vdots & \ldots & \vdots & \vdots & \vdots \\ \mathbf{0} & \mathbf{0} & \mathbf{0} & \mathbf{0} & \mathbf{0} & \ldots & \mathbf{T}_{k-1,k-1} & \mathbf{I}_{k-1} & \mathbf{0} \\ \mathbf{0} & \mathbf{0} & \mathbf{0} & \mathbf{0} & \mathbf{0} & \ldots & \mathbf{I}_{k-1} & \mathbf{T}_{k,k} & \mathbf{I}_k \end{bmatrix}.$$

The entries of the matrix are

$$\mathbf{T}_{i,i} = (\vartheta_{i-1} + \vartheta_i)\mathbf{I} - \beta_i\vartheta_i(\text{Hess}\,\psi_i), \qquad \mathbf{I}_i = -\vartheta_i\mathbf{I},$$

where $i = 1, \ldots, k$ and $\mathbf{I}$ is the 2×2 identity matrix. All the ϑ_i are nonzero, so the rank of the matrix is $2k$. □

Lemma 8.22 guarantees that *for every element T of $\mathcal{M}$, the catastrophe manifold of T is a smooth surface* (cf. Corollary 7.75, page 275).

Notational Convention. **The symbol for any mapping h : $\mathbf{J}^{\mathrm{p}}_{\mathrm{s}}(\mathbf{X}, \mathbf{Y}) \to \mathbf{Z}$ will also be used for the composition $\mathbf{h} \circ \pi : \mathbf{J}^{\mathrm{k}}_{\mathrm{s}}(\mathbf{X}, \mathbf{Y}) \to \mathbf{J}^{\mathrm{p}}_{\mathrm{s}}(\mathbf{X}, \mathbf{Y}) \to \mathbf{Z}$, where $\mathbf{k} \geq \mathbf{p}$ and π is the projection. In other words, we shall suppress the π. Also, if $\mathbf{W} \subseteq \mathbf{J}^{\mathrm{p}}_{\mathrm{s}}(\mathbf{X}, \mathbf{Y})$, then we shall also denote π^{-1} by W (i.e., we shall suppress the π^{-1}).**

In the subsection, *Interpreting Multijet Transversality for Two Parameter Families* (page 305), we described a stratified set in $J^4(n+1, 1)$ and derived stratified set in $J^4_2(X, \mathbf{R}) \times Y^2$. We take precisely the same stratified sets in our setting, namely, in the bundle $J^4(2k + 2, 1)$, which gives a stratified set in $J^4(L_{(k)}, \mathbf{R}) \times S$ and one in $J^4_2(L_{(k)}, \mathbf{R}) \times S^2$. The singularities in the stratified set in $J^4(L_{(k)}, \mathbf{R}) \times S$ determine the fold and cusp critical points of the catastrophe (or lensing) map. The singularities in the stratified set in $J^4_2(L_{(k)}, \mathbf{R}) \times S^2$ are the pairs of fold critical points of the catastrophe map with the same image in S and the pairs of points — one fold and one cusp critical point — of the catastrophe map with the same image in S. And in $J^4_3(L_{(k)}, \mathbf{R}) \times S^3$, we have three fold critical points with the same image in S. We shall prove that the jets, respectively the multijets, of the elements of an open dense subset of $\mathcal{M}$ are transverse to these stratified sets.

For later proofs, it is convenient to treat the above stratified sets in some detail. If $T = E(\psi_1, \ldots, \psi_k) \in \mathcal{M}$, then the catastrophe manifold $\mathcal{M}_T$ of T is the subset of $L_{(k)} \times S$

defined by the pairs of equations ($i = 1, \ldots, k$):

$$\begin{aligned} F_i\left(\mathbf{x}_{i-1}, \mathbf{x}_i, \mathbf{x}_{i+1}, \frac{\partial \psi_i}{\partial \mathbf{x}_i}(\mathbf{x}_i)\right) &= (\vartheta_{i-1} + \vartheta_i)\mathbf{x}_i - (\vartheta_{i-1}\mathbf{x}_{i-1} + \vartheta_i \mathbf{x}_{i+1}) \\ &\qquad -\vartheta_i \beta_i \frac{\partial \psi_i}{\partial \mathbf{x}_i}(\mathbf{x}_i) \\ &= \mathbf{0}. \end{aligned} \tag{8.1}$$

Notice that we can interpret the previous equations as defining a subset of

$$\left(\prod_{i=1}^{k} J^k(L_i)\right) \times S = L_{(k)} \times S \times \mathbf{R}^k \times \left(\prod_{i=1}^{k} J^1(2_i)\right),$$

where by $J^p(2_i)$ we mean the space of p-jets of real valued functions of the two space variables $\mathbf{x_i} = (\mathrm{u_i}, \mathrm{v_i})$ of L_i. The defining equations for this subset being: $F_i(\mathbf{x}_{i-1}, \mathbf{x}_i, \mathbf{x}_{i+1}, \frac{\partial \psi_i}{\partial \mathbf{x}_i}(\mathbf{x}_i)) = \mathbf{0}$, where we interpret the partials as coordinate functions in $J_i^1(2)$; *we obtain a linear subspace of $L_{(k)} \times S \times \mathbf{R}^k \times \left(\prod_{i=1}^{k} J_i^1(2)\right)$ of codimension $2k$.*

We denote this subspace of $\left(\prod_{i=1}^{k} J^1(L_i)\right) \times S$ by K.

Letting $F : \left(\prod_{i=1}^{k} J^1(L_i)\right) \times S \to \mathbf{R}^{2k}$, we obtain the set defined by (8.1) by taking the preimage of $\mathbf{0}$ of the composition

$$F \circ \left[\left(\prod_{i=1}^{k} j^1\psi_i\right) \times 1_S\right] : \left(\prod_{i=1}^{k} L_i\right) \times S \to \left(\prod_{i=1}^{k} J^1(L_i)\right) \times S \to \mathbf{R}^{2k}.$$

That is,

$$\left[\left(\prod_{\mathbf{i=1}}^{\mathbf{k}} j^1\psi_\mathbf{i}\right) \times \mathbf{1_S}\right]^{-1} (\mathbf{K})$$

is the catastrophe set $\mathcal{M}_T$ for $T = E(\psi_1, \ldots, \psi_k) \in \mathcal{M}$. By Lemma 8.22, the set so defined is always a 2-dimensional submanifold of $L_{(k)} \times S \times \mathbf{R}^k \times \left(\prod_{i=1}^{k} J^1(2_i)\right)$.

The first transversality condition, which is automatic for $T \in \mathcal{M}$, is that the $2k \times 2(k+1)$ matrix in the proof of Lemma 8.22 has rank $2k$. Inside the preimage of K via the projection $\left(\prod_{i=1}^{k} J^2(L_i)\right) \times S \longrightarrow \left(\prod_{i=1}^{k} J^1(L_i)\right) \times S$, this condition on the rank is always true. Thus, the open set $K_0 \subseteq \left(\prod_{i=1}^{k} J^2(L_i)\right) \times S$ on which the condition holds is just the whole lift of K.

Important Remark 1. The part of the mapping

$$\left[\left(\prod_{i=1}^{k} j^2\psi_i\right) \times 1_S\right] : L_{(k)} \times S \to \left(\prod_{i=1}^{k} J^2(L_i)\right) \times S = L_{(k)} \times S \times \mathbf{R}^k \times \prod_{i=1}^{k} J^2(2_i)$$

that guarantees the transversality is in the factor $\prod_{i=1}^{k} J^2(2_i)$. In particular, the transversality is guaranteed by the nonvanishing of any one of the $(2k+1)(k+1)$ $2k$-minors for which polynomials in the second partials of the $\psi's$ can be interpreted as being defined on $\prod_{i=1}^{k} J^2(2_i)$. But one of these minors is upper diagonal with 1's down the diagonal — hence always nonsingular. Thus, this transversality condition is always satisfied.

Lemma 8.22 yields:

For any $T = E(\psi_1, \ldots, \psi_k) \in \mathcal{M}$, we have
$$\left[\left(\prod_{i=1}^{k} j^2\psi_i\right) \times 1_S\right](L_{(k)} \times S) \subseteq K_0.$$

The equation for the critical points on the catastrophe manifold of T is the vanishing of the determinant of the $2k \times 2k$ Hessian matrix:

$$\begin{bmatrix} T_{\mathbf{x}_i\mathbf{x}_j} & T_{\mathbf{x}_i\mathbf{y}} \end{bmatrix} = \begin{bmatrix} \mathbf{T}_{11} & \mathbf{I}_1 & \mathbf{0} & \mathbf{0} & \mathbf{0} & \ldots & \mathbf{0} & \mathbf{0} \\ \mathbf{I}_1 & \mathbf{T}_{22} & \mathbf{I}_2 & \mathbf{0} & \mathbf{0} & \ldots & \mathbf{0} & \mathbf{0} \\ \mathbf{0} & \mathbf{I}_2 & \mathbf{T}_{33} & \mathbf{I}_3 & \mathbf{0} & \ldots & \mathbf{0} & \mathbf{0} \\ \mathbf{0} & \mathbf{0} & \mathbf{I}_3 & \mathbf{T}_{44} & \mathbf{I}_4 & \ldots & \mathbf{0} & \mathbf{0} \\ \vdots & \vdots & \vdots & \vdots & \vdots & \ldots & \vdots & \vdots \\ \mathbf{0} & \mathbf{0} & \mathbf{0} & \mathbf{0} & \mathbf{0} & \ldots & \mathbf{T}_{k-1,k-1} & \mathbf{I}_{k-1} \\ \mathbf{0} & \mathbf{0} & \mathbf{0} & \mathbf{0} & \mathbf{0} & \ldots & \mathbf{I}_{k-1} & \mathbf{T}_{k,k} \end{bmatrix}.$$

If we call this polynomial D, then it is a polynomial in the second partials of the potentials ψ_i $(i = 1, \ldots, k)$. Thus, interpreting this function as one of the 2-jets, we have a mapping

$$F \times D : \left(\prod_{i=1}^{k} J^2(L_i)\right) \times S \to \mathbf{R}^{2k+1}.$$

We denote by $K_{0,1}$ the preimage of $\mathbf{0}$ *via* $(F \times D)$.

The preimage of $\mathbf{0}$ via composition with the jet mapping, namely, the composition

$$(F \times D) \circ \left[\left(\prod_{i=1}^{k} j^2\psi_i \right) \times 1_S \right] : \left(\prod_{i=1}^{k} L_i \right) \times S \to \left(\prod_{i=1}^{k} J^2(L_i) \right) \times S \to \mathbf{R}^{2k+1},$$

gives the subset of the catastrophe manifold that is the *Critical-Point Set* for $T = E(\psi_1, \ldots, \psi_k) \in \mathcal{M}$. That is,

The Critical-Point Set in $\mathcal{M}$ for the catastrophe map χ_T of $T \in \mathcal{M}$ is

$$\left[\left(\left(\prod_{i=1}^{k} j^2\psi_i \right) \right) \times 1_S \right]^{-1} (K_{0,1}).$$

Important Remark 2. The function $D : \left(\prod_{i=1}^{k} J^2(L_i) \right) \times S \to \mathbf{R}$ depends only on the fibres $\prod_{i=1}^{k} J^2(2_i)$ of $\prod_{i=1}^{k} J^2(L_i)$. That is, there is a function $\mathcal{D} : \prod_{i=1}^{k} J^2(2_i) \to \mathbf{R}$ such that D is the composition $D = \mathcal{D} \circ \hat{\pi}$, where $\hat{\pi}$ is the projection of $\left(\prod_{i=1}^{k} J^2(L_i) \right) \times S = \left(\prod_{i=1}^{k} (L_i \times J^2(2_i)) \right) \times S$ onto $\prod_{i=1}^{k} J^2(2_i)$. Let $\mathfrak{R} = \mathcal{D}^{-1}(0) \subseteq \prod_{i=1}^{k} J^2(2_i)$. We have $\mathcal{D}^{-1}(0) = \left(\prod_{i=1}^{k} L_i \right) \times \mathfrak{R} \times S$ and $K_{0,1} = (F \times D)^{-1}(\mathbf{0}) = K_0 \cap \left[\left(\prod_{i=1}^{k} L_i \right) \times \mathfrak{R} \times S \right]$.

The equations $(F \times D) \circ \left[\left(\prod_{i=1}^{k} j^2\psi_i \right) \times 1_S \right] = \mathbf{0}$ define smooth curves of critical points on M_T if the rank of this composition is $2k + 1$. Such a requirement is our second transversality condition. To check the rank condition, consider the following $(2k + 1) \times (2k + 2)$ matrix:

$$\begin{bmatrix}
\mathbf{T}_{11} & \mathbf{I}_1 & \mathbf{0} & \mathbf{0} & \mathbf{0} & \cdots & \mathbf{0} & \mathbf{0} & \mathbf{0} \\
\mathbf{I}_1 & \mathbf{T}_{22} & \mathbf{I}_2 & \mathbf{0} & \mathbf{0} & \cdots & \mathbf{0} & \mathbf{0} & \mathbf{0} \\
\mathbf{0} & \mathbf{I}_2 & \mathbf{T}_{33} & \mathbf{I}_3 & \mathbf{0} & \cdots & \mathbf{0} & \mathbf{0} & \mathbf{0} \\
\mathbf{0} & \mathbf{0} & \mathbf{I}_3 & \mathbf{T}_{44} & \mathbf{I}_4 & \cdots & \mathbf{0} & \mathbf{0} & \mathbf{0} \\
\vdots & \vdots & \vdots & \vdots & \vdots & \cdots & \vdots & \vdots & \vdots \\
\mathbf{0} & \mathbf{0} & \mathbf{0} & \mathbf{0} & \mathbf{0} & \cdots & \mathbf{T}_{k-1,k-1} & \mathbf{I}_{k-1} & \mathbf{0} \\
\mathbf{0} & \mathbf{0} & \mathbf{0} & \mathbf{0} & \mathbf{0} & \cdots & \mathbf{I}_{k-1} & \mathbf{T}_{k,k} & \mathbf{I}_k \\
\mathbf{D}_1 & \mathbf{D}_2 & \mathbf{D}_3 & \mathbf{D}_4 & \mathbf{D}_5 & \cdots & \mathbf{D}_{k-1} & \mathbf{D}_k & \mathbf{0}
\end{bmatrix},$$

where $\mathbf{D}_i = [\partial D/\partial \mathbf{x}_i]^T$ (i.e., 1×2 matrix). If $K_{0,1}$ is lifted to $\left(\prod_{i=1}^{k} J^3(L_i)\right) \times S$, this transversality condition defines a set $K_{0,1,\tau}$. If we have

$$\left[\left(\prod_{i=1}^{k} j^3\psi_i\right) \times 1_S\right]\left[\left(\prod_{i=1}^{k} J^3(L_i)\right) \times S\right] \subseteq K_{0,1,\tau},$$

then the Critical-Point Set is a collection of smooth curves in M_T.

Note that the open set $K_{0,1,\tau}$ sitting in the 3-jets over $K_{0,1}$ is defined by an open condition in $\prod_{i=1}^{k} J^3(2_i)$.

The set $K_{0,1,\tau}$ breaks up into two pieces, an open part $K_{0,1,0}$ and closed part $K_{0,1,1}$, where $K_{0,1,0}$ (respectively, $K_{0,1,1}$) is the subset of $K_{0,1,\tau}$ with the first $2k$ rows of the preceding matrix having rank $2k$ (resp., $2k-1$):

$$\begin{bmatrix} \mathbf{T}_{11} & \mathbf{I}_1 & \mathbf{0} & \mathbf{0} & \mathbf{0} & \ldots & \mathbf{0} & \mathbf{0} \\ \mathbf{I}_1 & \mathbf{T}_{22} & \mathbf{I}_2 & \mathbf{0} & \mathbf{0} & \ldots & \mathbf{0} & \mathbf{0} \\ \mathbf{0} & \mathbf{I}_2 & \mathbf{T}_{33} & \mathbf{I}_3 & \mathbf{0} & \ldots & \mathbf{0} & \mathbf{0} \\ \mathbf{0} & \mathbf{0} & \mathbf{I}_3 & \mathbf{T}_{44} & \mathbf{I}_4 & \ldots & \mathbf{0} & \mathbf{0} \\ \vdots & \vdots & \vdots & \vdots & \vdots & \ldots & \vdots & \vdots \\ \mathbf{0} & \mathbf{0} & \mathbf{0} & \mathbf{0} & \mathbf{0} & \ldots & \mathbf{T}_{k-1,k-1} & \mathbf{I}_{k-1} \\ \mathbf{0} & \mathbf{0} & \mathbf{0} & \mathbf{0} & \mathbf{0} & \ldots & \mathbf{I}_{k-1} & \mathbf{T}_{k,k} \\ \mathbf{D}_1 & \mathbf{D}_2 & \mathbf{D}_3 & \mathbf{D}_4 & \mathbf{D}_5 & \ldots & \mathbf{D}_{k-1} & \mathbf{D}_k \end{bmatrix}.$$

If $\left[\left(\prod_{i=1}^{k} j^3\psi_i\right) \times 1_S\right]\left[\left(\prod_{i=1}^{k} J^3(L_i)\right) \times S\right] \subseteq K_{0,1,\tau}$, *then* $\left[\left(\prod_{i=1}^{k} j^3\psi_i\right) \times 1_S\right]^{-1}$ *($K_{0,1,0}$) is the set of fold critical points of the catastrophe map χ_T and* $\left[\left(\prod_{i=1}^{k} j^3\psi_i\right) \times 1_S\right]^{-1}$ *($K_{0,1,1}$) contains the set of cusp critical points of χ_T.*

The condition in $K_{0,1,\tau} \subseteq \left(\prod_{i=1}^{k} J^3(L_i)\right) \times S$ defining $K_{0,1,1}$ is the vanishing of all $2k+1$, $2k \times 2k$ minors $(E_1, \ldots, E_{2k+1})$ of the $(2k+1) \times 2k$ matrix immediately above. Let $E = (E_1, \ldots, E_{2k+1}) : \left(\prod_{i=1}^{k} J^3(L_i)\right) \times S \to \mathbf{R}^{2k+1}$ be the list of values of these $2k+1$ minors (in some arbitrary, but fixed order).

Important Remark 3. The function $E : \left(\prod_{i=1}^{k} J^3(L_i)\right) \times S \to \mathbf{R}^{2k+1}$ depends only on the fibres $\prod_{i=1}^{k} J^3(2_i)$ of $\prod_{i=1}^{k} J^3(L_i)$. That is, there is a function $\mathcal{C} : \prod_{i=1}^{k} J^3(2_i) \to \mathbf{R}^{2k+1}$ such that E is the composition $E = \mathcal{C} \circ \tilde{\pi}$, where $\tilde{\pi}$ is the projection of $\left(\prod_{i=1}^{k} J^3(L_i)\right) \times S = \left(\prod_{i=1}^{k}(L_i \times J^3(2_i))\right) \times S$ onto $\prod_{i=1}^{k} J^3(2_i)$. Let $\wp = (\mathcal{C} \times \mathcal{D})^{-1}(0) \subseteq \prod_{i=1}^{k} J^3(2_i)$. We have $(E \times \mathcal{D})^{-1}(0) = \left(\prod_{i=1}^{k} L_i\right) \times \wp \times S$ and $K_{0,1,1} = (F \times D \times E)^{-1}(\mathbf{0}) = K_0 \cap \left[\left(\prod_{i=1}^{k} L_i\right) \times \wp \times S\right]$.

The additional transversality condition for cusps is that the mapping

$$(F \times D \times E) \circ \left[\left(\prod_{i=1}^{k} j^3\psi_i\right) \times 1_S\right] : \left(\prod_{i=1}^{k} L_i\right) \times S \to \mathbf{R}^{4k+2}$$

has rank $2k+2$. The condition on the 4-jets is then given by the maximal rank of the matrix:

$$\begin{bmatrix}
\mathbf{T}_{11} & \mathbf{I}_1 & \mathbf{0} & \mathbf{0} & \mathbf{0} & \cdots & \mathbf{0} & \mathbf{0} & \mathbf{0} \\
\mathbf{I}_1 & \mathbf{T}_{22} & \mathbf{I}_2 & \mathbf{0} & \mathbf{0} & \cdots & \mathbf{0} & \mathbf{0} & \mathbf{0} \\
\mathbf{0} & \mathbf{I}_2 & \mathbf{T}_{33} & \mathbf{I}_3 & \mathbf{0} & \cdots & \mathbf{0} & \mathbf{0} & \mathbf{0} \\
\mathbf{0} & \mathbf{0} & \mathbf{I}_3 & \mathbf{T}_{44} & \mathbf{I}_4 & \cdots & \mathbf{0} & \mathbf{0} & \mathbf{0} \\
\vdots & \vdots & \vdots & \vdots & \vdots & \cdots & \vdots & \vdots & \vdots \\
\mathbf{0} & \mathbf{0} & \mathbf{0} & \mathbf{0} & \mathbf{0} & \cdots & \mathbf{T}_{k-1,k-1} & \mathbf{I}_{k-1} & \mathbf{0} \\
\mathbf{0} & \mathbf{0} & \mathbf{0} & \mathbf{0} & \mathbf{0} & \cdots & \mathbf{I}_{k-1} & \mathbf{T}_{k,k} & \mathbf{I}_k \\
\mathbf{D}_1 & \mathbf{D}_2 & \mathbf{D}_3 & \mathbf{D}_4 & \mathbf{D}_5 & \cdots & \mathbf{D}_{k-1} & \mathbf{D}_k & \mathbf{0} \\
\mathbf{E}_{1,1} & \mathbf{E}_{1,2} & \mathbf{E}_{1,3} & \mathbf{E}_{1,4} & \mathbf{E}_{1,5} & \cdots & \mathbf{E}_{1,k-1} & \mathbf{E}_{1,k} & \mathbf{0} \\
\vdots & \vdots & \vdots & \vdots & \vdots & \cdots & \vdots & \vdots & \vdots \\
\mathbf{E}_{2k+1,1} & \mathbf{E}_{2k+1,2} & \mathbf{E}_{2k+1,3} & \mathbf{E}_{2k+1,4} & \mathbf{E}_{2k+1,5} & \cdots & \mathbf{E}_{2k+1,k-1} & \mathbf{E}_{2k+1,k} & \mathbf{0}
\end{bmatrix},$$

where $\mathbf{E}_{i,j} = [\partial E_i/\partial \mathbf{x}_j]^T$ (i.e., a 1×2 matrix). This defines an open set in $\left(\prod_{i=1}^{k} J^4(L_i)\right) \times S$, which we can label $K_{0,1,1,0}$, the subset in the jet space which gives rise to cusp critical points.

Let $S^i = S \times \cdots \times S$ (i-fold product of the light source plane S with itself) and $\rhd_{S^i} = \{(s, \ldots, s) \in S^i\}$ (i.e., the diagonal in S^i). If we lift all the singularities to $\left(\prod_{i=1}^{k} J^4(L_i)\right) \times S$, we have

$$K \supseteq K_{0,1} \supseteq K_{0,1,1}.$$

In

$$\left(\prod_{i=1}^{k} J_2^2(L_i)\right) \times S^2 = L_{(k)}^{(2)} \times \mathbf{R}^{2k} \times \left(\prod_{i=1}^{k} [J^2(2_i)]^2\right) \times S^2,$$

we obtain

$$(K_0 \times K_0) \cap (L_{(k)}^{(2)} \times \mathbf{R}^{2k} \times \mathfrak{R} \times \mathfrak{R} \times \triangleright_{S^2})$$

for the fold double points, and in

$$\left(\prod_{i=1}^{k} J_2^3(L_i)\right) \times S^2 = L_{(k)}^{(2)} \times \mathbf{R}^{2k} \times \left(\prod_{i=1}^{k} [J^3(2_i)]^2\right) \times S^2,$$

we get

$$(K_0 \times K_0) \cap (L_{(k)}^{(2)} \times \mathbf{R}^{2k} \times \mathfrak{R} \times \wp \times \triangleright_{S^2})$$

for the coincident image of a fold and cusp critical point.

Finally, in

$$\left(\prod_{i=1}^{k} J_3^2(L_i)\right) \times S^3 = L_{(k)}^{(3)} \times \mathbf{R}^{3k} \times \left(\prod_{i=1}^{k} [J^2(2_i)]^3\right) \times S^3,$$

we have

$$(K_0 \times K_0 \times K_0) \cap (L_{(k)}^{(3)} \times \mathbf{R}^{3k} \times \mathfrak{R} \times \mathfrak{R} \times \mathfrak{R} \times \triangleright_{S^3})$$

for fold triple points.

As a consequence of the three "Important Remarks" these singularities are defined by the vanishing of polynomials in $\left(\prod_{i=1}^{k} J^2(2_i)\right)^s$ *or* $\left(\prod_{i=1}^{k} J^3(2_i)\right)^s$, *where* $s = 1, 2,$ *or* 3. *Furthermore, the conditions for transversality to any of these singularities is given in terms of the nonvanishing of polynomial expressions in* $\left(\prod_{i=1}^{k} J^3(2_i)\right)^s$ *or* $\left(\prod_{i=1}^{k} J^4(2_i)\right)^s$, *here* $s = 1, 2,$ *or* **3**. *With respect to these singularities, any expression of transversality for* $T = E(\psi_1, \ldots, \psi_k) \in \mathcal{M}$ *is equivalent to an expression of transversality statement for* $\psi = (\psi_1, \ldots, \psi_k)$. *This will be useful in the proof of Lemma 8.25 below.*

The Key Lemmas for Openness and Density

The lemmas in the current subsection are analogous to those in *Open-Density for Multijet Transversality of Families* (page 298).

In this setting, the product of lens planes, $L_{(k)} = L_1 \times \cdots \times L_k$, replaces X and the light source plane S replaces Y. The stratified sets just described above replace the arbitrary ones in the subsection *Open-Density for Multijet Transversality of Families* on page 298.

Lemma 8.23 (cf. Lemma 8.11, page 300) *Let* $K \subseteq L_{(k)}^{(s)} \times S^s$ *be a compact set. Then* $\mathcal{M}^K$ *is* C^{k+1}*-open in* $\mathcal{M}$.

The proof of Lemma 8.23 is immediate since $\mathcal{M}^K = \mathcal{M} \cap \mathcal{F}^K$.

Lemma 8.24 (cf. Lemma 8.12, page 301) *Let* $L_{(k)} = \bigcup_{i=1}^{\infty} K_i$, *be a countable union of disjoint, compact sets* K_i *in* $L_{(k)}^{(s)} \times S^s$ *with disjoint open neighborhoods* $\mathcal{J}_i$. *Then for any* n, *the set* $\mathcal{M}^{L_{(k)}}$ *is* C^{p+1}*-open for any integer* p.

This lemma is a consequence of Lemma 8.12 (replacing $\mathcal{F}$ by $\mathcal{M}$). Note that we have changed the notation for the disjoint open neighborhoods containing the compact sets K_i from $\mathcal{L}_i$ to $\mathcal{J}_i$.

Lemma 8.25 (cf. Lemma 8.13, page 301) *For all* $p \in L_{(k)}^{(s)} \times S^s$ *and* $T = E(\psi) \in \mathcal{M}$, *there is a compact neighborhood* N *of* p *in* $L_{(k)}^{(s)} \times S^s$ *and neighborhood* $\mathcal{V}$ *of* T *in* $\mathcal{M}$ *such that* $\mathcal{M}^N$ *is* C^∞*-dense in* $\mathcal{V}$.[7]

Proof: For this lemma, we only have to consider the cases $s = 1, 2$ and 3 because it suffices to check: (1) If the image of a pair of folds meet, then they do so transversally, (2) no cusp critical point and fold critical point can have the same image, and (3) no three fold critical points can have the same image.

By remarks at the end of the previous subsection, we know that the singularities to which the elements of $\mathcal{M}$ must be transverse are defined in $\prod_{i=1}^{k} J^3(2_i) \subseteq J^3(2k)$. We can establish this local density of the elements of $\mathcal{M}$ multi-transverse to the required singularities by showing that the elements of $\mathcal{M}_\times$ which are multi-transverse to these singularities, are locally dense in $\mathcal{M}$. However, this follows since the local perturbations that we make of an element $(\psi_1, \ldots, \psi_k)$ are made by adding an arbitrary polynomial locally as in the proof of Lemma 8.13. The set of such local polynomials is rich enough so that the jet map indexed by them is multi-transverse to any submanifold of $\prod_{i=1}^{k} J_s^3(2_i)$. The end of the proof uses Lemma 8.13, which states that for any submanifold or stratified set, there is a dense set of the local polynomial indexing set such that perturbing by each one produces an element of $\mathcal{M}_\times$ multi-transverse to the given stratified set. $\square$

Lemma 8.26 (cf. Lemma 8.14, page 304) *If* $K \subseteq L_{(k)}^{(s)} \times S^s$ *is compact, then* $\mathcal{M}^K$ *is* C^∞*-dense and open in* $\mathcal{M}$.

This is immediate from Lemma 8.14.

The version of Lemma 8.15 that we shall prove will require instead that compact sets in the countable collection are products.

[7]cf. Theorem 7.17, page 225.

Lemma 8.27 (cf. Lemma 8.15, page 304) *Let $M = \bigcup_{i=1}^{\infty} K_i$, be a countable union of disjoint, compact sets K_i in $X^{(s)} \times Y^s$ with disjoint open, precompact (or totally bounded) neighborhoods $\mathcal{J}_i$. We assume further that the compact sets K_i and their neighborhoods $\mathcal{J}_i$ are products in $L_{(k)}^{(s)}$, say $K_j^{i,r} \subseteq \mathcal{J}_j^{i,r} \subseteq \mathcal{L}_i$, where $j = 1, \ldots, \infty$, $i = 1, \ldots, k$, and $r = 1, \ldots, s$. Then for any integer p, the set $\mathcal{M}^M$ is C^{p+1}-open and C^{∞}-dense in $\mathcal{M}$.*

Proof: Openness is again given by Lemma 8.12. Here we shall use the fact that $\mathcal{M}$ and $\mathcal{M}_\times$ are homeomorphic via the homeomorphisms $G : \mathcal{M} \to \mathcal{M}_\times$ and $E : \mathcal{M}_\times \to \mathcal{M}$. For any subset $A \subseteq L_{(k)}^{(s)} \times S^s$, we shall denote by $\mathcal{M}_\times^A$ the image under the homeomorphism G of $\mathcal{M}^A$. To prove the density part, we take any $f \in \mathcal{M}$. Let $G(f) = \psi = (\psi_1, \ldots, \psi_k)$ and let $V_{\mu_i}^p((\psi_i)_s)$ be an open C^∞-neighborhood about ψ_i in $C^\infty(L_i)$ for $i = 1, \ldots, k$ (see page 299 for notation).

We show that $\left(\prod_{i=1}^{k} V_{\mu_i}^p((\psi_i)_s)\right) \bigcap \mathcal{M}_\times^M \neq \emptyset$.

Note. For fixed i and j, since $\prod_{r=1}^{s} K_j^{i,r} \subseteq \prod_{r=1}^{s} \mathcal{J}_j^{i,r} \subseteq L_i^{(s)}$, we know that $\mathcal{J}^{i,r} \cap \mathcal{J}_j^{i,r'} = \emptyset$ for $r \neq r'$.

Let $K_j^i = \bigcup_{r=1}^{s} K_j^{i,r}$ and $\mathcal{J}_j^i = \bigcup_{r=1}^{s} \mathcal{J}_j^{i,r}$. We work in

$$\prod_{i=1}^{k} \prod_{r=1}^{s} K_j^{i,r} = K_j \subseteq \mathcal{J}_j = \prod_{i=1}^{k} \prod_{r=1}^{s} \mathcal{J}_j^{i,r}.$$

Choose smooth bump functions $\beta_j^{i,r}$ with support in $\mathcal{J}_j^{i,r}$ and equal to 1 on $K_j^{i,r}$. Let $\beta_j^i = \sum_{r=1}^{s} \beta_j^{i,r}$, that is, β_j^i is a smooth bump function which is 1 on K_j^i and vanishes outside of $\mathcal{J}_i$. For each j, i, choose e_j^i sufficiently small such that $h \in V_{e_j^i, \mathcal{J}_j^i}^p((\psi_i)_s)$; then $\beta_j^i h \in V_{\mu_i}^p((\psi_i)_s)$. By Lemma 8.14, we can find $\psi_{i,j} \in V_{e_j^i, \mathcal{J}_j^i}^p((\psi_i)_s)$ such that $\psi_{(j)} \equiv (\psi_{1,j}, \ldots, \psi_{k,j}) = G(f_j) \in \mathcal{M}_\times^{K_j}$, that is, $f_j \in \mathcal{M}^{K_j}$. Now let $\phi_i = \beta_0^i \psi_i + \sum_{j=1}^{\infty} \beta_j^i \psi_{i,j}$, where $\beta_0^i = 1 - \sum_{j=1}^{\infty} \beta_j^i$. On $\mathcal{J}_j^i$, we have $\phi_i = (1 - \beta_j^i)\psi_i + \beta_j^i \psi_{i,j} = \psi_i + \beta_j^i(\psi_{i,j} - \psi_i)$. Since we chose $(\psi_{i,j} - \psi_i) \in V_{e_j^i, \mathcal{J}_j^i}^p((\psi_i)_s)$, we know that $\beta_j^i(\psi_{i,j} - \psi_i) \in V_{\mu_i}^p((\psi_i)_s)$. Thus $\phi_i \in V_{\mu_i}^p((\psi_i)_s)$. Finally, since on K_j^i we have for each j that $\phi_i = \psi_{i,j}$ for $i = 1, \ldots, k$ and $\psi_{(j)} = (\psi_{1,j}, \ldots, \psi_{k,j}) \in \mathcal{M}_\times^{K_j}$, it follows that $\Phi = (\phi_1, \ldots, \phi_k) \in \mathcal{M}_\times^{K_j}$ for all $j = 1, \ldots, \infty$. Thus

$$\Phi \in \bigcap_{j=1}^{\infty} \mathcal{M}_\times^{K_j} = \mathcal{M}_\times^{\bigcup_{j=1}^{\infty} K_j} = \mathcal{M}_\times^M. \ \Box$$

We now prove an analog of Lemma 8.16 using Lemma 8.27. To this end we need the following elementary set theoretic lemma:

Lemma 8.28 *Let $X_1, \dots, X_n$ be sets each of which can be written as $X_i = \bigcup_{j=1}^{\infty} U_{i,j}$ with $U_{i,2j} \cap U_{i,2j'} = \emptyset$ and $U_{i,2j+1} \cap U_{i,2j'+1} = \emptyset$ for $1 \leq j \neq j' < \infty$. Then $\prod_{i=1}^{n} X_i$ can be written as a finite union of sets each of which is a countable union of disjoint sets of the form $\prod_{i=1}^{n} U_{i,j_k}$.*

Proof: We shall prove the lemma by induction. Suppose that $n = 1$. By hypothesis, $X = \bigcup_{j=1}^{\infty} U_j$, where $U_2 j \cap U_2 j' = \emptyset$ and $U_{2j+1} \cap U_{2j'+1} = \emptyset$ for $1 \leq j \neq j' < \infty$. Then

$$X = \left(\bigcup_{j=1}^{\infty} U_{2j} \right) \bigcup \left(\bigcup_{k=1}^{\infty} U_{2k+1} \right).$$

Next, suppose that the lemma is proven for $1 \leq k < n$. Then we can write $\prod_{i=2}^{n} X_i = \bigcup_{r=1}^{p} M_r$, where M_r is countable union of disjoint sets, say, $M_r = \bigcup_{k=1}^{\infty} V_k$ with each V_k a product as required by the lemma. Now $X_1 = \bigcup_{j=1}^{\infty} U_{1,j}$. Each of the sets $M_{r,0} = \bigcup_{j,k=1}^{\infty} (U_{1,2j} \times V_k)$ and $M_{r,1} = \bigcup_{j,k=1}^{\infty} (U_{1,2j+1} \times V_k)$ is a countable union of disjoint sets, each of which is a product as required. Thus

$$\begin{aligned} \prod_{i=1}^{n} X_i &= X_1 \times \left(\bigcup_{r=1}^{p} M_r \right) = \left(\bigcup_{r=1}^{p} M_{r,0} \right) \bigcup \left(\bigcup_{r=1}^{p} M_{r,1} \right) \\ &= \left[\bigcup_{r=1}^{p} \left(\bigcup_{j=1}^{\infty} (U_{1,2j} \times M_r) \right) \right] \bigcup \left[\bigcup_{r=1}^{p} \left(\bigcup_{j=1}^{\infty} (U_{1,2j+1} \times M_r) \right) \right]. \quad \square \end{aligned}$$

Lemma 8.29 (cf. Lemma 8.16, page 304) *For any p, the set $\mathcal{M}^{L^{(s)}_{(k)} \times S^s}$ is C^{p+1}-open (hence, C^∞-open) and C^∞-dense in $\mathcal{M}$.*

Proof: The statement of the lemma is equivalent to the set $\mathcal{M}^{L^{(s)}_{(k)}}$ being C^{p+1}-open and C^∞-dense in $\mathcal{M}_\times$. We have $L^{(s)}_{(k)} = \prod_{i=1}^{k} L^{(s)}_i$. By Lemma 8.28 and Lemma 8.27, it suffices to show that: (1) We can write $L^{(s)}_i$ as a union of compact sets, say, $\bigcup_{j=1}^{\infty} K_{i,j}$, where $K_{i,2j} \cap$

$K_{i,2j'+1} = \emptyset$ for $1 \leq j \neq j' < \infty$, and (2) we can find open, precompact neighborhoods $\mathcal{J}_{i,j}$ of $K_{i,j}$ such that $\mathcal{J}_{i,2j} \cap \mathcal{J}_{i,2j'} = \emptyset$ and $\mathcal{J}_{i,2j+1} \cap \mathcal{J}_{i,2j'+1} = \emptyset$ for $1 \leq j \neq j' < \infty$. But this is shown exactly as in Lemma 8.16 (page 304) by using a proper function $\lambda_i : L_i^{(s)} \to [1, \infty)$, defining $K_{i,j} = \lambda_i^{-1}[j, j+1]$ and $\mathcal{J}_{i,j} = \lambda_i^{-1}(j - \frac{1}{2}, j + \frac{3}{2})$. □

Lemma 8.29 then yields that $\mathcal{M}^\star$ is open and dense in $\mathcal{M}$. □ **(Theorem 8.17)**

Chapter 9

Local Lensing Geometry

One of the most important deductions of singularity theory is the universality *of certain simple forms like folds and cusps which one can expect to encounter everywhere and which it is useful to learn to recognise.*

VLADIMIR ARNOLD

We showed in Chapter 8 that the critical points of a generic k-plane lensing map are either folds or cusps. The current chapter discusses in detail the qualitative features of k-plane lensing maps near fold and cusp critical points, including quantitative results for the single and double-plane cases. This is followed by a study of the local geometry of caustics under 1-parameter evolutions. These results bear on Problem **M3**, which includes a characterization of the local properties of critical curves and caustics in gravitational lensing.

Outline: Section 9.1 *qualitatively* treats multiplane lensing in a neighborhood of folds (Section 9.1.1) and cusps (Section 9.1.2), as well as signed magnification, sources on caustics, and orientation of caustics (Section 9.1.3). By analyzing the Taylor expansions of a locally stable lensing map at fold and cusp critical points, we are seeking a "best approximation" to the lensing map's local *quantitative* behavior. Though such an approach seems unmanageably complicated for general multiplane lens systems, we have some results in the single and double-plane cases (Sections 9.2 and 9.3, resp.). Using orthogonal coordinate changes to simplify the Taylor expansions, we discuss the local convexity of fold caustics in Section 9.3 (general definition — Section 9.3.1). This deals with whether fold caustic curves appear convex locally when viewed from the side with more lensed images. In single-plane lensing, local convexity holds if the surface mass density is locally constant (Section 9.3.2). In the process, we treat locally stable double-plane lensing maps near its fold critical points using the Taylor coefficients of the potentials and orthogonal coordinate changes (Section 9.3.3). The violation of local convexity in the double-plane case is also proven (Section 9.3.4). Explicit equations characterizing folds and cusps are given in Section 9.4 using directional derivatives. Section 9.5 presents the classification of local 1-parameter evolutions of planar caustics (Section 9.5.1), the equations determining caustic metamorphoses (Section 9.5.2), and illustrations of caustic metamorphoses in single and double-plane lensing (Sections 9.5.3 and 9.5.4). The latter also highlights differences between caustics in single- and double-plane lensing.

9.1 Qualitative Features of Multiplane Lensing Near Caustics

In a neighborhood of a critical point, every locally stable k-plane lensing map $\boldsymbol{\eta}$ is differentiably equivalent to one of the following (see Theorem 7.49 (2,3), page 251):

(1) **Whitney fold**: $\boldsymbol{\eta}_F : \mathbf{R}^2 \longrightarrow \mathbf{R}^2$ given by $\boldsymbol{\eta}_F(\mathrm{u}, \mathrm{v}) = (\mathrm{u}, a\mathrm{v}^2)$, where $a = \pm 1$.

(2) **Whitney pleat/cusp**: $\boldsymbol{\eta}_C : \mathbf{R}^2 \longrightarrow \mathbf{R}^2$ given by $\boldsymbol{\eta}_C(\mathrm{u}, \mathrm{v}) = (\mathrm{u}, a\mathrm{uv} + b\mathrm{v}^3)$, where $|a| = |b| = 1$.

In other words, the qualitative features of $\boldsymbol{\eta}$ in a sufficiently small neighborhood of a critical point and its $\boldsymbol{\eta}$-image (caustic) is captured by either $\boldsymbol{\eta}_F$ or $\boldsymbol{\eta}_C$. The target space $\mathbf{R}^2$ of $\boldsymbol{\eta}_F$ and $\boldsymbol{\eta}_C$ will be covered by rectangular coordinates (s_1, s_2). We shall also transfer some of the basic terminologies about $\boldsymbol{\eta}$ to $\boldsymbol{\eta}_F$ and $\boldsymbol{\eta}_C$. For example, the domain of either $\boldsymbol{\eta}_F$ or $\boldsymbol{\eta}_C$ is viewed as an open subset of the *first lens plane* (counting in the direction from observer to light source plane), while the target is the *light source plane*. Furthermore, elements of $\boldsymbol{\eta}_F^{-1}(s_1, s_2)$ and $\boldsymbol{\eta}_C^{-1}(s_1, s_2)$ are called *lensed images*, and concepts like *magnification*, *critical curves*, *caustics*, etc. also carry over. The notions of "lensed-image parity" and "$\mathbf{C}^2$-lensed images" will now be introduced.

Lensed-Image Parity: The *parity* of a lensed image $\mathbf{x}$ is the sign of $\det[\mathrm{Jac}\,\boldsymbol{\eta}](\mathbf{x})$. The notation $r_\oplus$ and $r_\ominus$ will denote, respectively, lensed images of positive and negative parities. By Lemma 6.16 (page 204), even index lensed images (includes minima and maxima) have positive parity, while those of odd index have negative parity. Degenerate lensed images will be denoted by $r_\bigcirc$ (i.e., $\det[\mathrm{Jac}\,\boldsymbol{\eta}](\mathbf{x}) = 0$ at such points).

$\mathbf{C}^2$-Lensed Images: The k-plane lens equation

$$\mathbf{y} = \mathbf{x}_1 - \sum_{\ell=1}^{k} (\boldsymbol{\alpha}_\ell \circ \boldsymbol{\eta}_\ell)(\mathbf{x}_1),$$

is a pair of equations of the form: $y_1 = u - h_1(u, x)$, $y_2 = x - h_2(u, x)$. Fix the light source position $\mathbf{y} = (y_1, y_2)$ and treat $\mathbf{x}_1 = (u, x)$ as the unknown variables. A $\mathbf{C}^2$-*lensed image* is a solution (u, x) in $\mathbf{C}^2$ of the lens equation such that the imaginary parts of u and x are nonzero. Note that such solutions are not observable. If there is a $\mathbf{C}^2$-lensed image

(w_1, w_2), where $w_\ell = c_\ell + ie_\ell$ for $\ell = 1, 2$, then we can associate two complex numbers z_1 and z_2 given by

$$\begin{aligned} z_1 &= w_1 + i\,w_2 = (c_1 - e_2) + i\,(c_2 + e_1) \\ z_2 &= \bar{w}_1 + i\,\bar{w}_2 = (c_1 + e_2) + i\,(c_2 - e_1). \end{aligned}$$

Of course, this association is not unique. We shall call $\{z_1, z_2\}$ a *virtual-lensed-image pair* induced by the $\mathbf{C}^2$-lensed image by (w_1, w_2). Note that $z_1 \neq z_2$ since $e_\ell \neq 0$ for $\ell = 1, 2$.

9.1.1 Lensing Near Folds

Let p_0 be a fold critical point of a locally stable k-plane lensing map $\boldsymbol{\eta}$. Then $\boldsymbol{\eta}$ at p_0 is differentiably equivalent to the Whitney fold $\boldsymbol{\eta}_F$ at 0,

$$\boldsymbol{\eta}_F : \mathbf{R}^2 \longrightarrow \mathbf{R}^2 : \quad \boldsymbol{\eta}_F(\mathrm{u}, \mathrm{v}) = (\mathrm{u}, a\mathrm{v}^2), \quad \text{with } |a| = 1.$$

We shall qualitatively investigate $\boldsymbol{\eta}$ near p_0 using $\boldsymbol{\eta}_F$.

Before studying the lensing properties of $\boldsymbol{\eta}_F$, we take note of a basic fact to be used when treating lensed-image magnification locally. Suppose that $\boldsymbol{\eta} = \lambda_2^{-1} \circ \boldsymbol{\eta}_F \circ \lambda_1$ in a neighborhood $\mathcal{U}$ of p_0. Then *for all* $p \in \mathcal{U}$, *we have* $\det[d_p\boldsymbol{\eta}] = 0$ *if and only if* $\det[d_{\lambda_1(p)}\,\boldsymbol{\eta}_F] = 0$.

Without loss of generality, we shall assume that for all $p \in \mathcal{U}$, $\det[d_p\boldsymbol{\eta}] > 0$ *if and only if* $\det[d_{\lambda_1(p)}\,\boldsymbol{\eta}_F] > 0$.

In fact, $\det[d_p\boldsymbol{\eta}] = \det[d_{\boldsymbol{\eta}_F(\lambda_1(p))}\,(\lambda_2^{-1})]\det[d_p\lambda_1]\det[d_{\lambda_1(p)}\,\boldsymbol{\eta}_F]$ for all $p \in \mathcal{U}$. Since λ_1 and λ_2 are charts, the factor $\det[d_{\boldsymbol{\eta}_F(\lambda_1(p))}\,(\lambda_2^{-1})]\det[d_p\lambda_1]$ is either everywhere positive or everywhere negative in $\mathcal{U}$. Suppose that the latter occurs. Write $\lambda_1(p) = (\mathrm{u}, \mathrm{v})$ and set $\boldsymbol{\eta}_F^* = \boldsymbol{\eta}_F \circ \lambda_1^*$, where $\lambda_1^* : \lambda_1(\mathcal{U}) \to \mathbf{R}^2$ is given by $\lambda_1^*(\mathrm{u}, \mathrm{v}) = (\mathrm{u}, -\mathrm{v})$. Then $\boldsymbol{\eta}_F^*(\mathrm{u}, \mathrm{v}) = (\mathrm{u}, a\mathrm{v}) = \boldsymbol{\eta}_F(\mathrm{u}, \mathrm{v})$; hence, $\boldsymbol{\eta}_F^*$ is a Whitney fold on $\lambda_1(\mathcal{U})$. On an open set $\mathcal{U}' \subseteq \mathcal{U}$, we have $\boldsymbol{\eta} = \lambda_2^{-1} \circ \boldsymbol{\eta}_F^* \circ (\lambda_1^* \circ \lambda_1)$ (since $\boldsymbol{\eta}_F^* \circ (\lambda_1^* \circ \lambda_1) = \boldsymbol{\eta}_F \circ \lambda_1$) and $\det[d_p\boldsymbol{\eta}] = -\det[d_{\boldsymbol{\eta}_F(\lambda_1(p))}\,(\lambda_2^{-1})]\det[d_p\lambda]\det[d_{(\lambda_1^*\circ\lambda_1)(p)}\,\boldsymbol{\eta}_F^*]$. Hence, $\det[d_p\boldsymbol{\eta}]$ and $\det[d_{(\lambda_1^*\circ\lambda_1)(p)}\,\boldsymbol{\eta}_F^*]$ have the same sign for all $p \in \mathcal{U}'$.

Let us consider the critical-point properties of $\boldsymbol{\eta}_F$ for $a = 1$. This will yield the local qualitative behavior of a locally stable k-plane lensing map $\boldsymbol{\eta}$ in a neighborhood of a fold critical point. Each critical point of $\boldsymbol{\eta}_F$ is a fold and its set of critical points coincides with the u-axis. The set of caustics of $\boldsymbol{\eta}_F$ consists of the points $\boldsymbol{\eta}_F(\mathrm{u}, 0) = (\mathrm{u}, 0)$, where $-\infty < \mathrm{u} < +\infty$; that is, Caustic($\boldsymbol{\eta}_F$) is the s_1–axis. To depict this pictorially, let $\tilde{f}_F(\mathrm{u}, \mathrm{v}) = (\mathrm{u}, \mathrm{v}, a\mathrm{v}^2)$ and $\pi(\mathrm{u}, \mathrm{v}, \mathrm{w}) = (\mathrm{u}, \mathrm{w})$. Then $\boldsymbol{\eta}_F = \pi \circ \tilde{f}_F$. The map $\tilde{f}_F$ folds the plane, while π projects the folded plane onto the upper half of the light source plane (i.e., target of $\boldsymbol{\eta}_F$). This "folding" action of $\boldsymbol{\eta}_F$ is illustrated in Figure 9.1.

Suppose that a light source is located at (s_1, s_2). We shall study the lensed images as the light source moves parallel to the s_2-axis (so s_1 = constant). The lensed images are the real solutions (u, v) of the lens equation, $(\mathrm{u}, \mathrm{v}^2) = (s_1, s_2)$. If $s_2 < 0$, then there are no

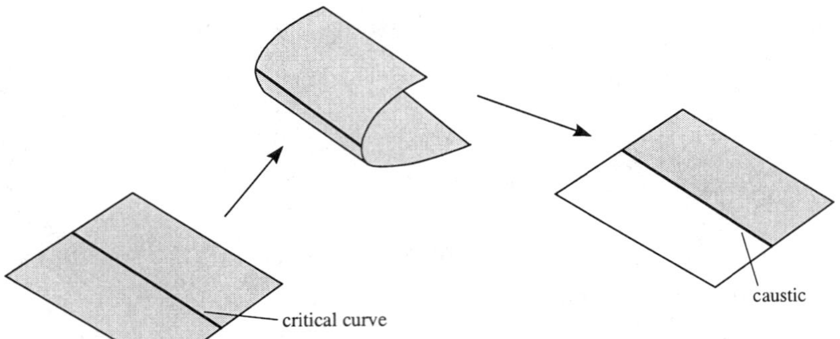

Figure 9.1. Action of a Whitney fold. A local region of the first lens plane (left) is folded (middle) and projected into a local region of the light source plane (right).

real solutions. Consequently, there are no lensed images, locally, of a light source at (s_1, s_2) with $s_2 < 0$. On the other hand, there are two $\mathbf{C}^2$-lensed images: $w_\pm = (s_1, \pm i\sqrt{|s_2|}) \in \mathbf{C}^2$. As $s_2 \to 0^-$ the $\mathbf{C}^2$-lensed images w_+ and w_- merge, creating a real infinitely magnified lensed image $r_\bigcirc = (s_1, 0)$ when the light source reaches the fold caustic (i.e., the s_1-axis).

As s_2 passes through the origin and attains positive values, the lensed image $r_\bigcirc$ splits into two opposite parity lensed images, $r_\oplus = (s_1, \sqrt{s_2})$ and $r_\ominus = (s_1, -\sqrt{s_2})$; see

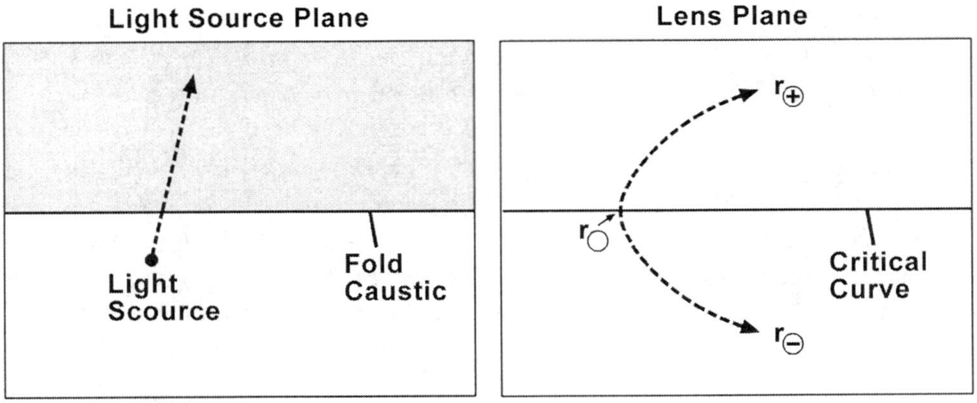

Figure 9.2. Local qualitative motion of lensed images as light source traverses a fold caustic curve. When the light source reaches the caustic, an infinitely magnified lensed image $r_\bigcirc$ suddenly appears. This lensed image splits into two opposite parity lensed images — $r_\oplus$ and $r_\ominus$ — both with finite magnification. The shaded region is doubly covered, locally, by the corresponding multiplane lensing map. The right panel is a local region of the first lens plane (counting in the direction from observer to light source plane).

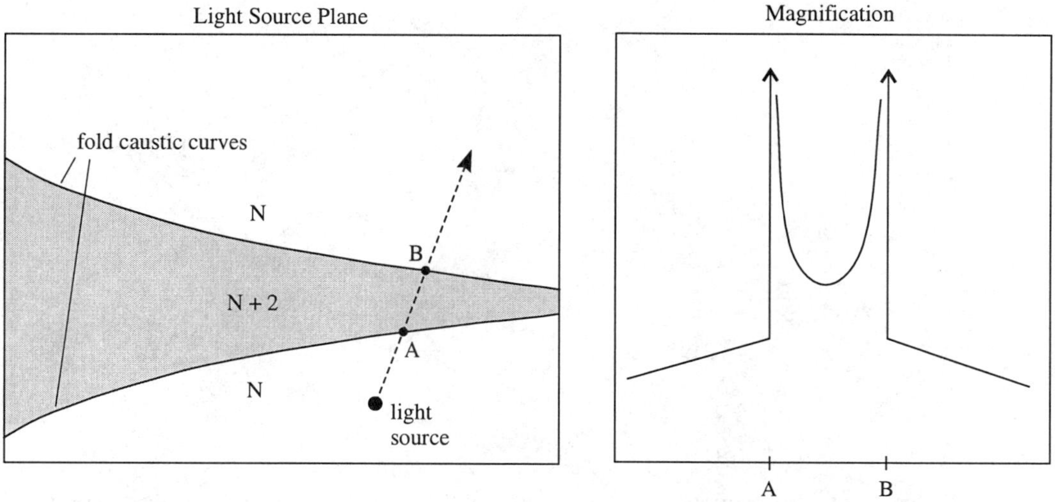

Figure 9.3. As a light source traverses two fold caustics (left panel), its light curve (i.e., plot of a moving source's total magnification as a function of position or time) qualitatively displays a "U" shape (right panel). An extra pair of lensed images is produced when the light source is in the shaded region. The total number of lensed images in a given region is indicated. Light sources on the $N+2$ side that are close to one of a caustic curve is more magnified than those on the other side near the caustic.

Figure 9.2. Since

$$\frac{1}{\det[\text{Jac}\,\boldsymbol{\eta}_F](s_1,\, \pm\sqrt{s_2})} = \frac{1}{2}\frac{1}{\sqrt{s_2}}, \tag{9.1}$$

the lensed images $r_{\oplus}$ and $r_{\ominus}$ are highly magnified for s_2 sufficiently small and positive. When s_2 increases along positive values, the magnifications of $r_{\oplus}$ and $r_{\ominus}$ decrease. The reversed picture is obtained when s_2 is varied from positive to negative values. Hence, *for a locally stable k-plane lensing map, if in some neighborhood exactly two lensed images are merging (leading to a reduction in the number of lensed images), then the light source must approach a fold caustic point. Moreover, the lensed images must have opposite parity.* In addition, notice that one side near a fold caustic curve is such that light sources there exhibit more total magnification than on the other side close to the caustic — see Figure 9.3.

Remark: If we had made the choice $\boldsymbol{\eta}_F(\mathrm{u}, \mathrm{v}) = (\mathrm{u}, -\mathrm{v}^2)$, then the results of this section would essentially be the same. A main exception is that the lower-half plane (i.e., $s_2 < 0$) would now be doubly covered by $\boldsymbol{\eta}_F$.

Observational evidence supporting the crossing of fold caustics by a light source is depicted in Figure 5.7 (page 161).

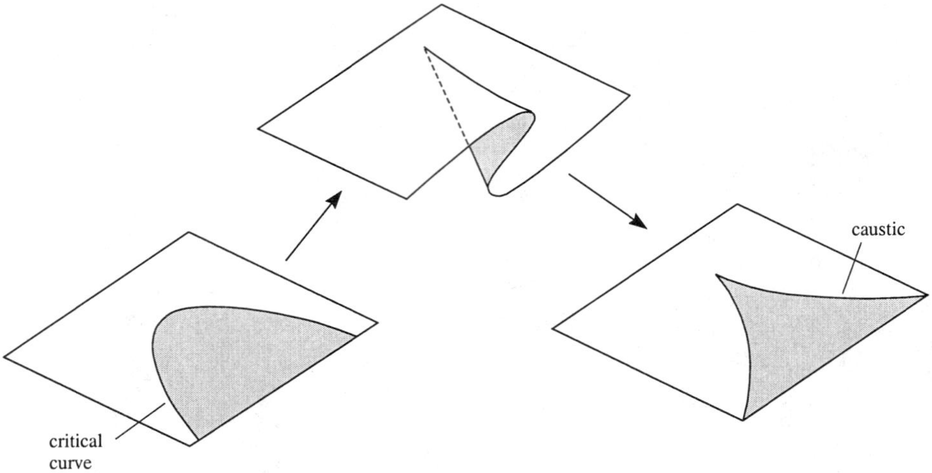

Figure 9.4. Action of a Whitney Pleat. A local region of the first lens plane (left) is pleated (middle) and projected into the light source plane (right).

9.1.2 Lensing Near Positive and Negative Cusps

Let p_0 be a cusp critical point of locally stable k-plane lensing map $\boldsymbol{\eta}$. Then the multiplane lensing map $\boldsymbol{\eta}$ at p_0 is differentiably equivalent to a Whitney pleat $\boldsymbol{\eta}_C$ at 0, say, $\boldsymbol{\eta} = \xi_2^{-1} \circ \boldsymbol{\eta}_C \circ \xi_1$ in a neighborhood $\mathcal{V}$ of p_0, where $\boldsymbol{\eta}_C : \mathbf{R}^2 \to \mathbf{R}^2 : \boldsymbol{\eta}_C(\mathrm{u}, \mathrm{v}) = (\mathrm{u}, a\mathrm{uv} + b\mathrm{v}^3)$, with $|a| = |b| = 1$.

Without loss of generality, we shall assume that for all p in $\mathcal{V}$, $\det[d_p\boldsymbol{\eta}] > 0$ *if and only if* $\det[d_{\xi_1(p)}\, \boldsymbol{\eta}_C] > 0$.

To see this, it suffices to consider the case when $\det[d_{\boldsymbol{\eta}_F(\xi_1(p))}\,(\xi_2^{-1})]\det[d_p\xi_1]$ is everywhere negative in $\mathcal{V}$. Now, write $\xi_1(p) = (\mathrm{u}, \mathrm{v})$ and replace $\boldsymbol{\eta}_C$ by $\boldsymbol{\eta}_C^* = \boldsymbol{\eta}_C \circ \xi_1^*$, where $\xi_1^* : \xi_1(\mathcal{V}) \to \mathbf{R}^2$ is given by $\xi_1^*(\mathrm{u}, \mathrm{v}) = (\mathrm{u}, -\mathrm{v})$. Then $\boldsymbol{\eta}_C^*(\mathrm{u}, \mathrm{v}) = (\mathrm{u}, -a\mathrm{uv} - b\mathrm{v}^3)$, which is also a Whitney pleat on $\xi_1(\mathcal{V})$. Note that $\boldsymbol{\eta}_C^* \circ \xi_1^* \circ \xi_1 = \boldsymbol{\eta}_C \circ \xi_1$ on an open subset $\mathcal{V}'$ of $\mathcal{V}$. Consequently, $\boldsymbol{\eta} = \xi_2^{-1} \circ \boldsymbol{\eta}_C^* \circ (\xi_1^* \circ \xi_1)$ and

$$\det[d_p\boldsymbol{\eta}] = -\det[d_{\boldsymbol{\eta}_C(\xi_1(p))}\,(\xi_2^{-1})]\det[d_p\xi]\det[d_{(\xi_1^*\circ\xi_1)(p)}\,\boldsymbol{\eta}_C^*]$$

on $\mathcal{V}'$. Hence, $\det[d_p\boldsymbol{\eta}]$ and $\det[d_{(\xi_1^*\circ\xi_1)(p)}\,\boldsymbol{\eta}_C^*]$ have the same sign for all $p \in \mathcal{V}'$.

Let us first consider the behavior of the Whitney pleat $\boldsymbol{\eta}_C$, that is, the local qualitative behavior of a locally stable k-plane lensing map $\boldsymbol{\eta}$ in a neighborhood of a cusp critical point. Since $\det[\mathrm{Jac}\,\boldsymbol{\eta}_C](\mathrm{u}, \mathrm{v}) = a\mathrm{u} + 3b\mathrm{v}^2$, the set of critical points, $\mathrm{Crit}(\boldsymbol{\eta}_C)$, coincides with the parabola $\mathrm{u} = -(3b/a)\mathrm{v}^2$. The vertex of the parabola, that is, the origin, is a cusp critical point. All other points on the parabola are fold critical points. The corresponding caustic curve is given parametrically by $s_1 = -(3b/a)\mathrm{v}^2$ and $s_2 = -2b\mathrm{v}^3$. Eliminating v, we obtain the cusped curve $4s_1^3 + 27(b/a)\, s_2^2 = 0$. We can graphically display the map $\boldsymbol{\eta}_C$ by noting that $\boldsymbol{\eta}_C = \pi \circ \tilde{\boldsymbol{\eta}}_C$, where $\tilde{\boldsymbol{\eta}}_C(\mathrm{u}, \mathrm{v}) = (\mathrm{u}, \mathrm{v}, a\,\mathrm{uv} + b\,\mathrm{v}^3)$ and $\pi(\mathrm{u}, \mathrm{v}, \mathrm{w}) = (\mathrm{u}, \mathrm{w})$.

Then $\tilde{\eta}_C$ pleats the (u, v)-plane and π projects the pleated plane onto the light source plane (i.e., target of f_p). Figure 9.4 captures the pleating action of $\boldsymbol{\eta}_C$. The region interior to the cusped curve is triply covered by $\boldsymbol{\eta}_C$, while the remainder of the light source plane in the figure is singly covered.

We shall now consider lensing of a light source at position (s_1, s_2) as the light source moves along the s_1-axis.

Case I: Positive Cusps (a = −1 and b = 1)

The lensed images of a light source at $(s_1, 0)$ are solutions u, v of the equations $\mathrm{u} = s_1$ and $-\mathrm{uv} + \mathrm{v}^3 = 0$. If $s_1 < 0$, there is one lensed image $r_\oplus = (s_1, 0)$ and two (non-observable) $\mathbf{C}^2$-lensed images, $w_\pm = (s_1, \pm i\sqrt{|s_1|})$. Since

$$\frac{1}{\det[\mathrm{Jac}\,\boldsymbol{\eta}_C](s_1, 0)} = \frac{1}{-s_1},$$

as the light source approaches the cusp caustic from outside (i.e., as $s_1 \to 0^-$), the magnification of the positive parity lensed image $r_\oplus$ increases without bound. In fact, there is a region of high magnification outside the cusp caustic (Figure 9.5). To show this, let us determine the lensed images of a light source at (s_1, s_2), that is, find points the (u, v) obeying the lens equation

$$s_1 = \mathrm{u}, \quad s_2 = -\mathrm{uv} + \mathrm{v}^3.$$

Set

$$C(s_1, s_2) = -(s_1^3/27) + (s_2^2/4).$$

The cusped caustic curve due to $\boldsymbol{\eta}_C$ is the level curve $C(s_1, s_2) = 0$. If (s_1, s_2) is not a caustic point, then the lensed images (u, v) of a light source at (s_1, s_2) are determined by the cubic

$$\mathrm{v}^3 - s_1\mathrm{v} - s_2 = 0$$

(since $\mathrm{u} = s_1$). This equation has one real root (and a pair of complex conjugate roots) if $C(s_1, s_2) > 0$, and three real roots if $C(s_1, s_2) < 0$. The regions of the light source plane that are singly and triply covered by $\boldsymbol{\eta}_C$ are then determined by $C(s_1, s_2) > 0$ and $C(s_1, s_2) < 0$, respectively — see Figure 9.5. A light source in the singly covered region (i.e., "outside" the cusp) has a lensed image at (u, v), where $\mathrm{u} = s_1$ and v is the real root given by

$$\mathrm{v}(s_1, s_2) = \left(\frac{s_2}{2} + \sqrt{C(s_1, s_2)}\right)^{1/3} + \left(\frac{s_2}{2} - \sqrt{C(s_1, s_2)}\right)^{1/3}. \tag{9.2}$$

Since $s_1 < 0$, we have

$$\frac{1}{\det[\mathrm{Jac}\,\boldsymbol{\eta}_C]\,(s_1, \mathrm{v}(s_1, s_2))} = \frac{1}{-s_1 + 3\,[\mathrm{v}(s_1, s_2)]^2} > 0 \tag{9.3}$$

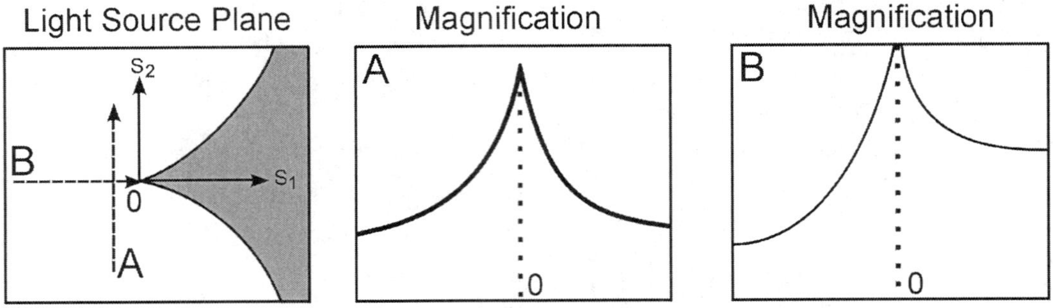

Figure 9.5. Leftmost panel: cusped caustic curve $C(s_1, s_2) = 0$. The shaded region ($C(s_1, s_2) < 0$) is triply covered locally by the lensing map η_C, while the unshaded part ($C(s_1, s_2) > 0$) is singly covered. Middle and rightmost panels: light curves. As a light source moves along track A, the source's total magnification increases to a maximum value as it passes near the cusp and then decreases (middle panel). Along track B (s_1-axis), the magnification increases without bound as the cusp is approached and then decreases inside the shaded region to values higher than those for points sufficiently far outside the cusp (right panel). Consequently, there is a high magnification "lobe" outside cusp caustics.

and, if we take the limit $(s_1, s_2) \to (0, 0)$ within the singly covered region $C(s_1, s_2) > 0$, then the magnification of the positive parity lensed image $r_\oplus = (s_1, \mathrm{v}(s_1, s_2))$ increases without bound — see Figure 9.5. Hence, *there is a "lobe" of high magnification outside the cusp* (e.g., Figure 9.6).

Observational evidence supporting a passing through a cusp caustic by a light source is given in Figure 5.9 (page 163).

Let us now consider the lensed images of a light source as the source moves along the s_1-axis from the singly to triply covered region. As $s_1 \to 0^-$, the lensed image $r_\oplus$ merges with the $\mathbf{C}^2$-lensed images w_+ and w_-, forming an infinitely magnified lensed image, $r_\bigcirc = (0, 0)$, when the light source is on the cusp. Increasing s_1 from zero to positive values, the lensed image $r_\bigcirc$ splits into one negative parity and two positive parity lensed images: $r_\ominus = (s_1, 0)$ and $r_\oplus^\pm = (s_1, \pm\sqrt{s_1})$; see Figure 9.7. As s_1 attains positive values, the magnifications of the three lensed images decrease.

Reversing the direction of the light source in Figure 9.7, we see that two positive parity and one negative parity lensed image merge, leaving behind a positive parity lensed image. When such merges occur, the cusp caustic is called a *positive cusp*.

Remark: *The case with $a = 1$ and $b = 1$ is similar to Case I.* They differ primarily in that the parabolic critical curve now opens towards the negative u–axis (i.e, $\mathrm{u} = -3\mathrm{v}^2$) and the cusped caustic curve opens towards the negative s_1-axis (i.e., $s_1 = -3\mathrm{v}^2$, $s_2 = -2\mathrm{v}^3$). Consequently, there is a positive parity lensed image for $s_1 > 0$, while one negative parity and two positive parity lensed images occur for $s_1 < 0$. The latter three lensed images merge as $s_1 \to 0^-$, leaving behind a positive parity lensed image for $s_1 > 0$. The origin is then also a positive cusp.

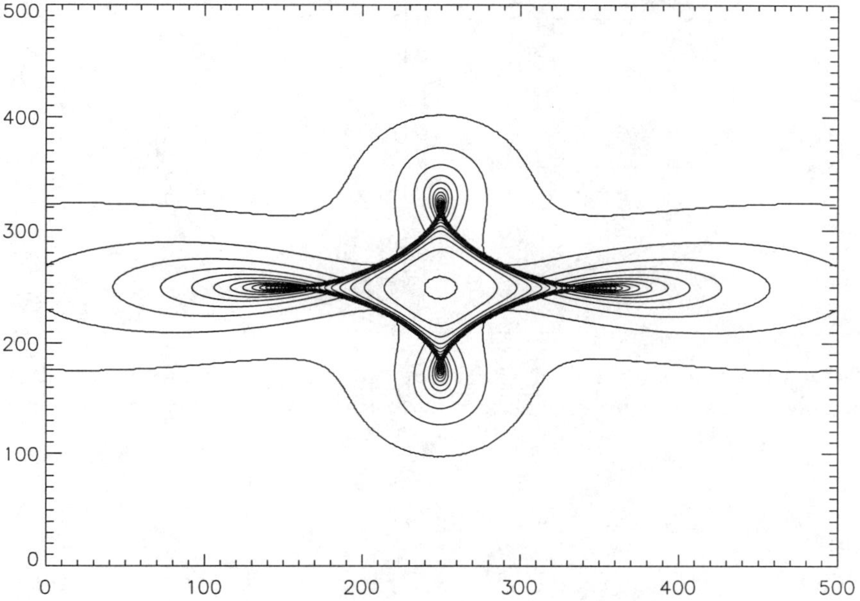

Figure 9.6. Caustic (astroid) and contours of constant total magnification due to a point mass lens with shear γ from infinity. Lobes of high magnification occur outside the cusps. Notice the higher magnification (darkening) on the side of the fold caustic interior to the astroid.

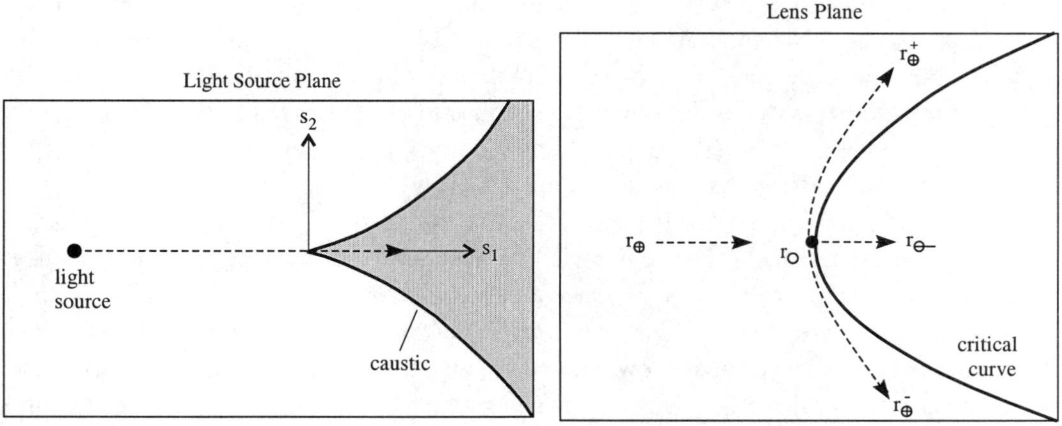

Figure 9.7. Light source passing through a positive cusp. Outside the cusp there is, locally, one lensed image $r_\oplus$. As the light source approaches the cusp, the positive parity lensed image $r_\oplus$ transforms into the infinitely magnified lensed image $r_\bigcirc$. When the light source moves to positive s_1 values, the lensed image $r_\bigcirc$ splits into three lensed images with finite magnification, $r_\oplus^\pm$ (positive parity) and $r_\ominus$ (negative parity). Their magnifications decrease as the light source moves away from the cusp along the positive s_1-axis.

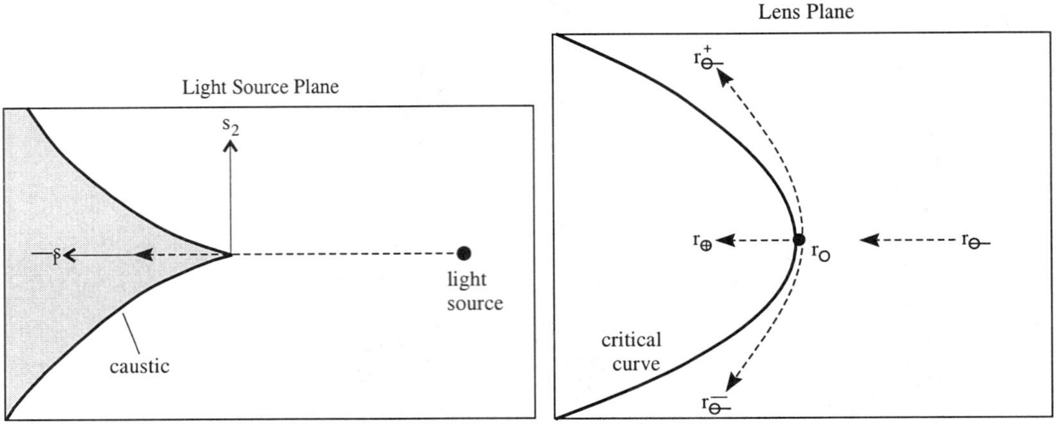

Figure 9.8. Light source passing through a negative cusp. Reversing the direction of the light source, we see that two negative parity ($r_{\ominus}^{\pm}$) and one positive parity ($r_{\oplus}$) lensed image merge, leaving behind a negative parity lensed image ($r_{\ominus}$).

Case II: Negative Cusps (a = −1 and b = −1)

This situation is analogous to Case I: There is a parabolic critical curve and cusped caustic curve. In the current case, however, the critical curve opens towards the negative u-axis (i.e., $u = -3v^2$) and the caustic curve opens to the negative s_1-axis (i.e., $s_1 = -3v^2$, $s_2 = 2v^3$)). Let us move the light source along the s_1-axis from positive to negative s_1 values — see Figure 9.8. For $s_1 > 0$, there is one lensed image $r_{\ominus} = (s_1, 0)$, along with two $\mathbf{C}^2$-lensed images $w_{\pm} = (s_1, \pm i\sqrt{s_1})$. The magnification of the negative parity lensed image $r_{\ominus}$ increases without bound as $s_1 \to 0^+$. Also, note that w_+ and w_- merge with $r_{\ominus}$ when $s_1 \to 0^+$. At $s_1 = 0$, the lensed image $r_{\ominus}$ becomes a degenerate lensed image, $r_{\bigcirc} = (0, 0)$. As s_1 assumes negative values, we see that $r_{\bigcirc}$ splits into two negative parity and one positive parity lensed image: $r_{\ominus}^{\pm} = (s_1, \pm\sqrt{-s_1})$ and $r_{\oplus} = (s_1, 0)$. The magnifications of these lensed images decrease as the light source travels away from the origin along the negative s_1-axis. Moving the light source from negative to positive s_1 values, we have two negative parity and one positive parity lensed image merging to leave behind a negative parity lensed image. In this case, we call the cusp caustic a *negative cusp*.

Remark: *The case $a = 1$, $b = -1$ is similar to Case II.* The main differences are that the parabolic critical curve opens towards the positive u-axis and the cusped caustic curve opens towards the positive s_1-axis. There is a negative parity lensed image for $s_1 < 0$, and, one positive parity and two negative parity lensed images appear for $s_1 > 0$. The latter three lensed images merge as $s_1 \to 0^+$. Hence, the origin is a negative cusp.

Example (Positive/Negative Cusps)**:** Positive and negative cusps due to an elliptical potential are shown in Figure 9.9.

Example (Light Curves, Lensed Images, Virtual Lensed Images)**:** A sample of the light curves and lensed-image/virtual-lensed-image paths due to a point mass lens with

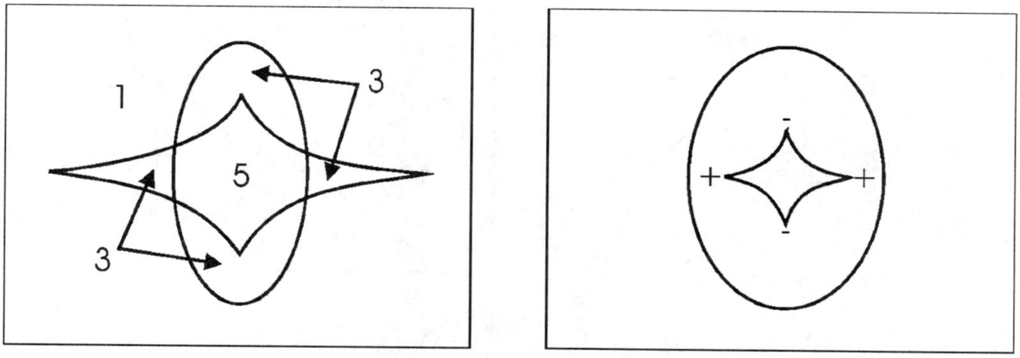

Figure 9.9. Caustics of an elliptical potential. The right panel shows a caustic with positive cusps (+) and negative cusps (−) (cf. [Bla-N86, p. 573]). The numbers in the left panel indicate how many lensed images are seen of a light source in the given region. Note that there are four regions with the same lensed image multiplicity (namely, 3).

shear appears in Figure 9.10. Along track B in Figure 9.10, the light source traverses the high magnification "lobe" outside the cusp. The lower panels in Figure 9.10 display the lensed-image trajectories — including those of the virtual-lensed-image pairs induced by the $\mathbf{C}^2$-lensed images — as the light source moves along track A. When the light source travels from b to c along track A, a virtual-lensed-image pair merges to create a degenerate lensed image at the critical curve (dashed curve). This degenerate lensed image then splits into two nondegenerate lensed images that move away from each other. As the light source moves from c to d, two nondegenerate lensed images merge into a degenerate lensed image at the critical curve. The degenerate lensed image then vanishes as it splits into a virtual-lensed-image pair that moves apart.

Remarks: First, suppose that a light source is moved along a curve segment whose initial and end points are noncaustic points. Assume that the total number of lensed images when the light source is at the initial point is different from that at the endpoint. Then the curve segment must intersect a caustic. However, the number of lensed images at the initial and final points can coincide for curve segments passing through a caustic. In other words, the complement of the caustics can have different components with the same lensed-image multiplicity (e.g., Figure 9.9). Second, if $\eta : P \to S$ is of a locally stable k-plane lensing map, and if in some neighborhood of P three lensed images are merging (leading to a reduction in the number of lensed images), then the light source must approach a cusp caustic from the interior (i.e., the triply covered part). Either one of the merging lensed images has negative parity and the other two have positive parity (positive cusp), or, one has positive parity with the other two having negative parity (negative cusp). Hence, if two odd index lensed images merge, then this must occur together with an even index lensed image. Third, the complement in P of the critical curves of η consists of disjoint open sets (Theorem 8.4, page 293) and det[Jac η] has opposite signs on any two such adjacent open sets.

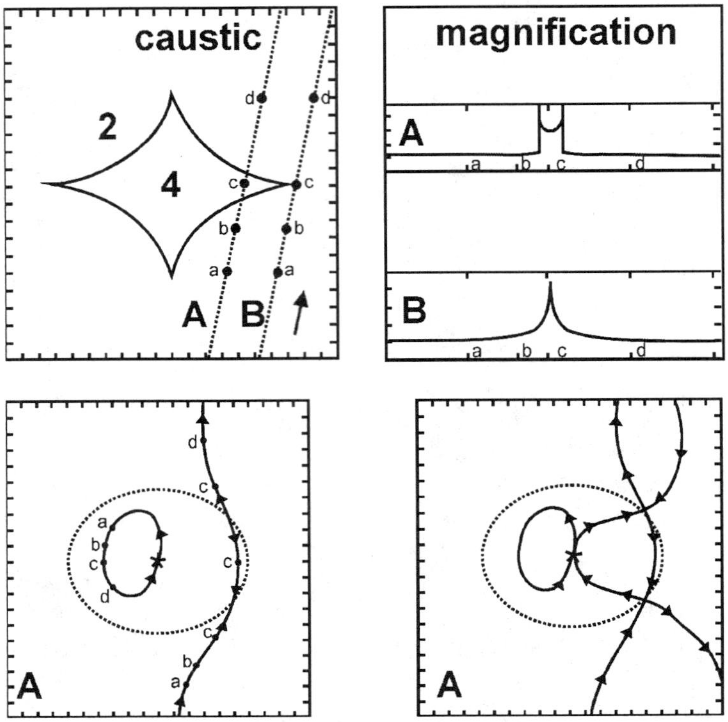

Figure 9.10. *Upper left panel*: An astroid caustic due to a point mass lens with shear from infinity. The numbers show how many lensed images are seen of a light source in the given region. The total magnifications of light sources traversing tracks A and B are shown in the upper-right panel. *Bottom left panel*: trajectories of the lensed images as the light source moves along track A. The dashed curve is the critical curve and the asterisk is the point mass position. The paths of the virtual-lensed-image pairs induced by the $\mathbf{C}^2$-lensed images are included in the bottom right panel. (After Witt [Wit93].)

9.1.3 Magnification, Lensed Caustics, and Orientation

Local Signed Magnification of Light Sources Interior to a Cusp

Consider the Whitney cusp $\boldsymbol{\eta}_C(\mathrm{u}, \mathrm{v}) = (\mathrm{u}, a\mathrm{uv} + b\mathrm{v}^3)$, where $|a| = |b| = 1$. This mapping qualitatively decribes the action of a locally stable k-plane lensing map in a neighorhood of a cusp critical point. For the map $\boldsymbol{\eta}_C$, the signed magnification of a lensed image $\mathbf{x}$ of a light source at $\mathbf{y}$ is defined by (cf. page 182):

$$\mu_{sgn}(\mathbf{x}; \mathbf{y}) = \frac{1}{\det[\mathrm{Jac}\, \boldsymbol{\eta}_C](\mathbf{x})} = (-1)^i \mathrm{Mag}(\mathbf{x}; \mathbf{y}),$$

where $\boldsymbol{\eta}_C(\mathbf{x}) = \mathbf{y}$ and i is the index of the lensed image. Our goal is to establish a general relationship obeyed by the total signed magnification of light sources in the triply covered region interior to the cusp of $\boldsymbol{\eta}_C$. Readers interested in the behavior of the "magnification cross section" near caustics should see Section 13.3.2 (page 479).

Without loss of generality, we shall assume that $\boldsymbol{\eta}_C$ yields a positive cusp critical point at the origin (i.e., $a = -1$, $b = 1$). For simplicity, set $\mu_{sgn}(\mathbf{x}) = \mu_{sgn}(\mathbf{x}; \mathbf{y})$, where $\mathbf{x} = (\mathrm{u}, \mathrm{v})$ and $\mathbf{y} = (s_1, s_2)$. Suppose that the light source is at $(s_1, 0)$ inside the triply covered region for the cusped curve

$$C(s_1, s_2) = -(s_1^3/27) + (s_2^2/4) = 0.$$

We saw that the lensed images are (see Figure 9.7, page 335):

$$r_\ominus = (s_1, 0), \quad r_\oplus^+ = (s_1, \sqrt{s_1}), \quad r_\oplus^- = (s_1, -\sqrt{s_1}).$$

Since $\mu_{sgn}(\mathrm{u}, \mathrm{v}) = 1/(-\mathrm{u} + 3\mathrm{v}^2)$, we obtain

$$\mu_{sgn}(r_\ominus) + \mu_{sgn}(r_\oplus^+) + \mu_{sgn}(r_\oplus^-) = 0. \tag{9.4}$$

We now prove that (9.4) holds for any light source position (s_1, s_2) inside the triply covered region (i.e., where $C(s_1, s_2) < 0$). Our presentation will parallel that of Zakharov [Zakh99, p. 1500], who considered this issue in the context of single plane lensing. Denote the three lensed images of a light source at (s_1, s_2) by $\mathbf{r}_i = (\mathrm{u}_i, \mathrm{v}_i)$, where $i = 1, 2, 3$. Here $\mathrm{u}_i = s_1$ for $i = 1, 2, 3$ and v_i are solutions of

$$\mathrm{v}^3 - s_1 \mathrm{v} - s_2 = 0.$$

Assume that $\mathbf{r}_1$ has negative parity, while $\mathbf{r}_2$ and $\mathbf{r}_3$ have positive parity. We have

$$0 = \mathrm{v}^3 - s_1\mathrm{v} - s_2 = (\mathrm{v} - \mathrm{v}_1)(\mathrm{v} - \mathrm{v}_2)(\mathrm{v} - \mathrm{v}_3) = \mathrm{v}^3 - \mathcal{S}_1\mathrm{v}^2 + \mathcal{S}_2\mathrm{v} - \mathcal{S}_3,$$

where

$$\begin{aligned}
\mathcal{S}_1 &= \mathrm{v}_1 + \mathrm{v}_2 + \mathrm{v}_3 = 0 \\
\mathcal{S}_2 &= \mathrm{v}_1\mathrm{v}_2 + \mathrm{v}_1\mathrm{v}_3 + \mathrm{v}_2\mathrm{v}_3 = -s_1 \\
\mathcal{S}_3 &= \mathrm{v}_1\mathrm{v}_2\mathrm{v}_3 = s_2.
\end{aligned}$$

The result $\mathcal{S}_1^2 = 0$ yields

$$\mathrm{v}_1^2 + \mathrm{v}_2^2 + \mathrm{v}_3^2 = 2s_1, \tag{9.5}$$

while $0 = \mathcal{S}_1\mathcal{S}_3 = \mathrm{v}^2\mathrm{v}_2\mathrm{v}_3 + \mathrm{v}_1\mathrm{v}_2^2\mathrm{v}_3 + \mathrm{v}_1\mathrm{v}_2\mathrm{v}_3^2$ and $\mathcal{S}_2^2 = s_2^2$ yield

$$(\mathrm{v}_1\mathrm{v}_2)^2 + (\mathrm{v}_1\mathrm{v}_3)^2 + (\mathrm{v}_2\mathrm{v}_3)^2 = s_1^2. \tag{9.6}$$

By (9.5) and (9.6), we obtain the desired result:

$$\mu_{sgn}(\mathbf{r}_1) + \mu_{sgn}(\mathbf{r}_2) + \mu_{sgn}(\mathbf{r}_3) = \frac{1}{-s_1 + 3\mathrm{v}_1^2} + \frac{1}{-s_1 + 3\mathrm{v}_2^2} + \frac{1}{-s_1 + 3\mathrm{v}_3^2} = 0.$$

This relationship naturally carries over to the case of negative cusps.

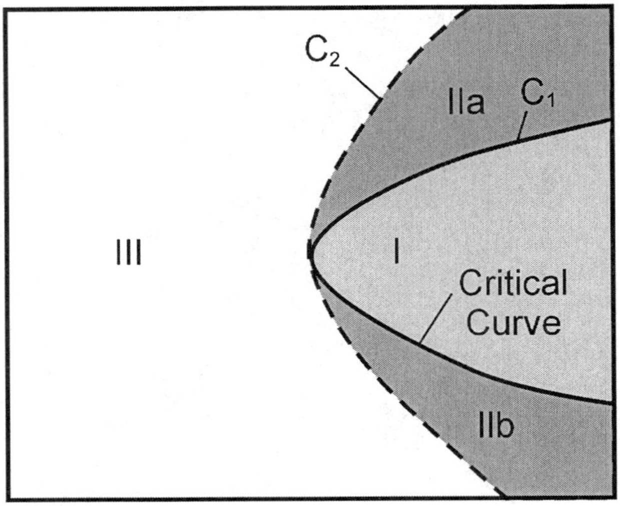

Figure 9.11. For a locally stable multiplane lensing map $\boldsymbol{\eta}$, the curves C_1 and C_2 depict the local qualitative form into which a cusped caustic curve c is lensed (i.e., $\boldsymbol{\eta}^{-1}(c) = C_1 \cup C_2$). A light source interior to the caustic curve has three lensed images — one in region I, one in region IIa, and one in IIb. Source outside the cusp have one lensed image in region III (cf. [Schn-W92, p. 4]).

Local Lensed Images of Light Sources on a Cusped Caustic Curve

Without loss of generality, we consider the local situation of a positive cusp caustic curve, namely,

$$C(s_1, s_2) = -(s_1^3/27) + (s_2^2/4) = 0.$$

The map producing such a caustic curve is the Whitney pleat $\boldsymbol{\eta}_C(\mathrm{u}, \mathrm{v}) = (s_1, s_2)$, where $s_1 = \mathrm{u}$ and $s_2 = -\mathrm{uv} + \mathrm{v}^3$. We wish to find the points (u, v) such that $\boldsymbol{\eta}_C(\mathrm{u}, \mathrm{v})$ lies on the caustic curve $C(s_1, s_2) = 0$. Since $\mathrm{u} = s_1$, we must find the solutions v to the cubic $\mathrm{v}^3 - s_1\mathrm{v} - s_2 = 0$, where (s_1, s_2) is subject to $C(s_1, s_2) = 0$. There are two real solutions:

$$\mathrm{v}_1 = 2^{2/3} s_2^{1/3}, \qquad \mathrm{v}_2 = -\left(\frac{s_2}{2}\right)^{1/3},$$

where v_2 is a double root. The $\boldsymbol{\eta}_C$-preimage of the caustic curve, $(s_2/2)^2 = (s_1/3)^3$, consists of the two curves

$$C_1(s_1, s_2) = (s_1, -(s_2/2)^{1/3}), \quad C_2(s_1, s_2) = (s_1, 2^{2/3} s_2^{1/3}).$$

Using the equation for the caustic, we see that C_1 is the parabola $\mathrm{u} = 3\mathrm{v}^2$, which coincides with the critical curve of $\boldsymbol{\eta}_C$. The curve C_2 is the wider parabola $u = (3/4)\mathrm{v}^2$ outside C_1. Figure 9.11 depicts these curves. See Figure 11.1 (page 427) for an example of these curves in the case of a lens with an elliptical potential.

A light source on the caustic curve, $C : (s_2/2)^2 = (s_1/3)^3$, has two lensed images — one on the solid curve (critical curve) and the other one on the dashed curve; see Figure 9.11. Light sources inside the caustic C have their three lensed images distributed with one in region I, one in IIa, and the other in IIb. A source outside C has one lensed image in region III. Moreover, the map $\boldsymbol{\eta}_C$ transforms regions I and III injectively onto the interior and exterior, respectively, of the caustic C (cf. Figure 9.5).

Orientation of Caustics

The local action of Whitney folds and pleats also allow us to orient caustics. Let c be a critical curve of a locally stable k-plane lensing map $\boldsymbol{\eta}$. Figures 9.1 and 9.4 show that about each point of c there is a neighborhood V such that $\boldsymbol{\eta}$ maps V to one side of the caustic curve $\boldsymbol{\eta} \circ c$. *Henceforth, we shall assume that $\boldsymbol{\eta} \circ c$ is oriented so that $\boldsymbol{\eta}[V]$ lies to the left of $\boldsymbol{\eta} \circ c$* (e.g., [Lev-PW, p. 4796–4799]). Loosely speaking, caustic curves with this orientation are such that light sources close to the left side have two more lensed images than those close to the right. A more precise and detailed treatment of the given orientation will be given in Section 14.3 (page 495). This will be needed when we investigate the global geometry of caustics (Section 15.4.2, page 550).

The previous orientation is depicted in Figure 9.12. It provides a convenient way of assessing, locally, from which side of a caustic curve a light source has more lensed images. *Generically, there is a change of ± 2 in the number of lensed images as a light source crosses a fold caustic curve.*[1] This is true of caustics due to transverse stable k-plane lensing map $\boldsymbol{\eta}$, that is, the map $\boldsymbol{\eta}$ is (i) locally stable, (ii) fold caustic curves cross at a nonzero angle, (iii) no fold and cusp, or, no two cusp critical points are mapped by $\boldsymbol{\eta}$ to the same point, and (iv) no more than two fold critical points are mapped to the same point (Theorem 8.6, page 294).

Figure 9.12 also includes two transversely intersecting oriented fold caustic curves. If a light source crosses such an intersection of caustics, then the number of lensed images changes by at least ± 2 and at most ± 4.

9.2 Folds and Cusps via Taylor Coefficients of the Potential

The goal of this section is to relate a gravitational lens to the folds and cusps it produces. Explicitly, we shall link the stable critical points of a single-plane lensing map to the Taylor coefficients of the associated gravitational lens potential. This is useful when studying the quantitative aspects of lensing. We shall employ these results in Section 9.3 to investigate the local shape of fold caustic curves due to a lens with locally constant surface mass density.

[1] Of course, in a typical crossing we are excluding passes through an intersection point of fold caustics (Figure 9.12).

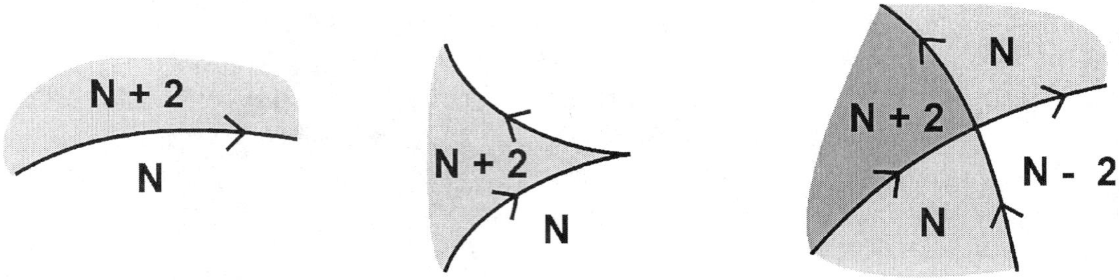

Figure 9.12. Transverse stable caustics are oriented such that a light source close to the left side has two more lensed images than one close to the right side. When fold caustics intersect at an angle, the number of lensed images can change by ± 4 for crossings between the $N-2$ and $N+2$ regions that pass through the intersection point.

Requisite Mathematics: Sections 7.2 and 7.5.

First, it is important to note that orthogonal coordinate changes must be used when studying quantitative issues like the shapes and sizes of lensed images, magnification cross sections (i.e., areas of regions in the light source plane with magnification greater than a given amount), multiple imaging cross sections, etc. Since Whitney folds and cusps result from curvilinear coordinate transformations, they can distort regions. For this reason, we shall resort to orthogonal coordinate transformations and Taylor expansions for local quantitative properties.

The single-plane time delay family due to a gravitational lens potential ψ is

$$T(\mathbf{x}, \mathbf{y}) = T_{\mathbf{y}}(\mathbf{x}) = \frac{1}{2}|\mathbf{x}-\mathbf{y}|^2 - \psi(\mathbf{x}).$$

For the single plane case, we admit orthogonal changes of variables that are the same in both the light source plane and the lens plane. This way, the form of the time delay function is preserved!

In fact, let O be a real 2×2 orthogonal matrix. Suppose that $\mathbf{x} = O\tilde{\mathbf{x}}$ and $\mathbf{y} = O\tilde{\mathbf{y}}$. Then

$$\begin{aligned} T(\mathbf{x}, \mathbf{y}) &= \frac{1}{2}|\mathbf{x}-\mathbf{y}|^2 - \psi(\mathbf{x}) = T(O\tilde{\mathbf{x}}, O\tilde{\mathbf{y}}) \\ &= \frac{1}{2}|O(\tilde{\mathbf{x}}-\tilde{\mathbf{y}})|^2 - \psi(O\tilde{\mathbf{x}}) = \frac{1}{2}|\tilde{\mathbf{x}}-\tilde{\mathbf{y}}|^2 - \tilde{\psi}(\tilde{\mathbf{x}}) \\ &\equiv \tilde{T}(\tilde{\mathbf{x}}, \tilde{\mathbf{y}}), \end{aligned}$$

where $\tilde{\psi}(\tilde{\mathbf{x}}) = \psi(O\tilde{\mathbf{x}})$. Write $\mathbf{x} = (\mathrm{u}, \mathrm{v})$, $\tilde{\mathbf{x}} = (\tilde{\mathrm{u}}, \tilde{\mathrm{v}})$, and set

$$\operatorname{grad}\psi(\mathbf{x}) = (\psi_{\mathrm{u}}(\mathbf{x}), \psi_{\mathrm{v}}(\mathbf{x})), \qquad \operatorname{grad}\tilde{\psi}(\tilde{\mathbf{x}}) = (\tilde{\psi}_{\tilde{\mathrm{u}}}(\tilde{\mathbf{x}}), \tilde{\psi}_{\tilde{\mathrm{v}}}(\tilde{\mathbf{x}})).$$

If we let

$$O = \begin{bmatrix} \alpha & \beta \\ -\beta & \alpha \end{bmatrix},$$

where $\alpha^2 + \beta^2 = 1$, then $\text{grad}\,\tilde{\psi}(\tilde{\mathbf{x}}) = (\alpha\psi_u(\mathbf{x}) - \beta\psi_v(\mathbf{x}), \beta\psi_u(\mathbf{x}) - \alpha\psi_v(\mathbf{x}))$. Hence, $O[\text{grad}\,\tilde{\psi}(\tilde{\mathbf{x}})] = \text{grad}\,\psi(\mathbf{x})$. Setting the gradient of $T_{\mathbf{y}}$ to zero then yields

$$\mathbf{0} = \mathbf{x} - \mathbf{y} - \text{grad}\,\psi(\mathbf{x}) = O(\tilde{\mathbf{x}} - \tilde{\mathbf{y}}) - O(\text{grad}\,\tilde{\psi}(\tilde{\mathbf{x}})).$$

Therefore, the lens equation

$$\mathbf{y} = \mathbf{x} - \text{grad}\,\psi(\mathbf{x})$$

is preserved under orthogonal coordinate changes with our definition of $\tilde{\psi}$:

$$\tilde{\mathbf{y}} = \tilde{\mathbf{x}} - \text{grad}\,\tilde{\psi}(\tilde{\mathbf{x}}).$$

The corresponding lensing map, $\boldsymbol{\eta}(\mathbf{x}) = \mathbf{y}$, is given in component form by

$$y_1 = \mathrm{u} - \frac{\partial\psi}{\partial \mathrm{u}}(\mathbf{x}), \quad y_2 = \mathrm{v} - \frac{\partial\psi}{\partial \mathrm{v}}(\mathbf{x}),$$

where $\mathbf{y} = (y_1, y_2)$. To avoid subscripts as much as possible, we relabel our coordinates so that the lensing map in *rectangular coordinates* becomes

$$\begin{aligned} U(u,x) &= u - \frac{\partial\psi}{\partial u}(u,x) \\ X(u,x) &= x - \frac{\partial\psi}{\partial x}(u,x). \end{aligned}$$

In other words, $\boldsymbol{\eta}(u,x) = (U(u,x), X(u,x))$. Note that the Jacobian of the map, $\boldsymbol{\eta} : (u,x) \to (U(u,x), X(u,x))$, is *symmetric in these coordinates*.

Conventions: We shall assume here and in Section 9.3 that $\boldsymbol{\eta}$ (even for multiple lens planes) maps the origin to itself (which can always be accomplished by a translation in the light source plane). In addition, no distinction will be made, notationally, between a map-germ and its representatives.

Using the fact that stable map-germs $(\mathbf{R}^2, \mathbf{0}) \mapsto (\mathbf{R}^2, \mathbf{0})$ *are determined by their* 3*-jets (Theorem 7.25, page 229), we shall consider the Taylor expansion of the time delay function up to order four. In other words, the Taylor expansion of the corresponding lensing map will be treated up to order three.*

Write the Taylor expansion of the potential as follows:

$$\psi(u,x) = A(u,x) + B(u,x) + C(u,x) + D(u,x) + E(u,x), \tag{9.7}$$

where

$$
\begin{aligned}
A(u,x) &= su + tx \\
B(u,x) &= au^2 + bux + cx^2 \\
C(u,x) &= eu^3 + fu^2x + gux^2 + hx^3 \\
D(u,x) &= ku^4 + mu^3x + nu^2x^2 + pux^3 + rx^4
\end{aligned}
\tag{9.8}
$$

with E of higher order in u and x. For example,

$$a = \frac{1}{2}\psi_{uu}(\mathbf{0}), \qquad b = \psi_{ux}(\mathbf{0}), \qquad c = \frac{1}{2}\psi_{xx}(\mathbf{0}). \tag{9.9}$$

The Taylor expansion of $\boldsymbol{\eta}$ up to order 3 is

$$
\begin{aligned}
U &= u - (A + B + C + D)_u\,(u,x) \\
X &= x - (A + B + C + D)_x\,(u,x)\,.
\end{aligned}
$$

Since $\boldsymbol{\eta}$ maps the origin to itself, we may assume that ψ has no linear part (i.e., $A = 0$). We shall give conditions on B, C, and D that guarantee that $\boldsymbol{\eta}$ has a fold or cusp critical point at the origin.

We now choose O to be the orthogonal matrix that diagonalizes the Jacobian matrix of $\boldsymbol{\eta}$ at the origin:

$$[\mathrm{Jac}\,\boldsymbol{\eta}](\mathbf{0}) = \begin{bmatrix} 1-2a & -b \\ -b & 1-2c \end{bmatrix}. \tag{9.10}$$

If $(1-2a) \neq 0$, then

$$O = \frac{1}{\sqrt{(1-2a)^2 + b^2}} \begin{bmatrix} 1-2a & -b \\ b & 1-2a \end{bmatrix}, \tag{9.11}$$

while if $(1-2c) \neq 0$, then

$$O = \frac{1}{\sqrt{(1-2c)^2 + b^2}} \begin{bmatrix} 1-2c & -b \\ b & 1-2c \end{bmatrix}. \tag{9.12}$$

Let

$$K_a = \sqrt{(1-2a)^2 + b^2}, \qquad K_c = \sqrt{(1-2c)^2 + b^2}.$$

Set

$$R(u,x) = (C + D + E)(u,x)$$

and

$$\tilde{R}(\tilde{u}, \tilde{x}) = R(u(\tilde{u}, \tilde{x}), x(\tilde{u}, \tilde{x})),$$

where

$$\begin{bmatrix} u \\ x \end{bmatrix} = O^T \begin{bmatrix} \tilde{u} \\ \tilde{x} \end{bmatrix}.$$

Explicitly, $\tilde{R}$ is given to order 4 by

$$\tilde{R}(\tilde{u}, \tilde{x}) \quad = \quad \tilde{e}\tilde{u}^3 + \tilde{f}\tilde{u}^2\tilde{x} + \tilde{g}\tilde{u}\tilde{x}^2 + \tilde{h}\tilde{x}^3 + \tilde{k}\tilde{u}^4 + \tilde{m}\tilde{u}^3\tilde{x} + \tilde{n}\tilde{u}^2\tilde{x}^2 + \tilde{p}\tilde{u}\tilde{x}^3 + \tilde{r}\tilde{x}^4. \quad (9.13)$$

We now write out $\tilde{g}$, $\tilde{h}$, and $\tilde{r}$ for the case $(1-2a) \neq 0$ and $\tilde{e}$, $\tilde{f}$, and $\tilde{k}$ for the case $(1-2c) \neq 0$, in terms of the original Taylor coefficients (see (9.7) and (9.8)):

$$\begin{aligned}
K_a^3\,\tilde{g} &= -fb^3 + (3e-2g)b^2(1-2a) + (2f-3h)b(1-2a)^2 + g(1-2a)^3. \\
K_a^3\,\tilde{h} &= eb^3 + fb^2(1-2a) + gb(1-2a)^2 + h(1-2a)^3. \\
K_a^4\,\tilde{r} &= kb^4 + mb^3(1-2a) + nb^2(1-2a)^2 + pb(1-2a)^3 + r(1-2a)^4. \\
K_c^3\,\tilde{e} &= e(1-2c)^3 - f(1-2c)^2 b + g(1-2c)b^2 - hb^3. \\
K_c^3\,\tilde{f} &= f(1-2c)^3 + (2g-3e)(1-2c)^2 b + (3h-2f)(1-2c)b^2 - gb^3. \\
K_c^4\,\tilde{k} &= k(1-2a)^4 - m(1-2a)^3 b + n(1-2c)^2 b^2 - p(1-2c)b^3 + rb^4.
\end{aligned} \quad (9.14)$$

The following summarizes the results of this section:

Theorem 9.1 *Let* $\eta : (\mathbf{R}^2, \mathbf{0}) \longrightarrow (\mathbf{R}^2, \mathbf{0})$ *be a single-plane lensing map-germ induced by a gravitational lens potential* ψ.

(1) *The origin is a rank 1 critical point of* η *if and only if:* $(1-2a)(1-2c) = b^2$ *and at least one of* $(1-2a)$, $(1-2c)$, *or* b^2 *is nonzero, where* a, b, *and* c *are given by* (9.9).

(2) *If the origin is a rank 1 critical point, then there is an orthogonal coordinate change that is the* **same** *in the lens and light source planes, namely,*

$$\begin{bmatrix} \tilde{U} \\ \tilde{X} \end{bmatrix} = O \begin{bmatrix} U \\ X \end{bmatrix} \quad \text{and} \quad \begin{bmatrix} \tilde{u} \\ \tilde{x} \end{bmatrix} = O \begin{bmatrix} u \\ x \end{bmatrix},$$

such that, in these coordinates, the Taylor expansion of η at the origin has the form

$$\begin{bmatrix} \tilde{U} \\ \tilde{X} \end{bmatrix} = K \begin{bmatrix} \tilde{u} \\ 0 \end{bmatrix} - \begin{bmatrix} \tilde{R}_{\tilde{u}} \\ \tilde{R}_{\tilde{x}} \end{bmatrix} (\tilde{u}, \tilde{x}), \tag{9.15}$$

where

$$K = 2(1 - (a + c)) \neq 0.$$

Note that K involves only the coefficients of the quadratic part of the Taylor expansion of ψ.

(3) *If the germ η is stable, then there are only two types of critical points that η can display at the origin: a fold or cusp.*

   (i) **Fold:** *η has a fold critical point at the origin if and only if: $(1-2a)(1-2c) = b^2$ and for $(1-2a) \neq 0$ the condition $\tilde{h} \neq 0$ holds (or,* mutatis mutandis, *for $(1-2c) \neq 0$, we have $\tilde{e} \neq 0$).*

   (ii) **Cusp:** *η has a cusp critical point at the origin if and only if: $(1-2a)(1-2c) = b^2$ and for $(1-2a) \neq 0$ the conditions $\tilde{g} \neq 0$, $\tilde{h} = 0$, and $\tilde{r} \neq 0$ hold (or,* mutatis mutandis, *for $(1-2c) \neq 0$, we have $\tilde{f} \neq 0$, $\tilde{e} = 0$, and $\tilde{k} \neq 0$).*

A generalization of Theorem 9.1(3) to folds of a general stable map-germ between planes appears in Theorem 9.2 (page 354).

Before proving Theorem 9.1, we consider how our Taylor expansion in (9.15) relates to the usual approximations employed by astrophysicists for the lensing map in a neighborhood of a fold or cusp critical point.

Using (9.13), let us explicitly write out (9.15) up to order 3:

$$\begin{cases} \tilde{U}(\tilde{u}, \tilde{x}) = K\tilde{u} - (3\tilde{e}\tilde{u}^2 + 2\tilde{f}\tilde{u}\tilde{x} + \tilde{g}\tilde{x}^2 + 4\tilde{k}\tilde{u}^3 + 3\tilde{m}\tilde{u}^2\tilde{x} + 2\tilde{n}\tilde{u}\tilde{x}^2 + \tilde{p}\tilde{x}^3) \\ \tilde{X}(\tilde{u}, \tilde{x}) = -(\tilde{f}\tilde{u}^2 + 2\tilde{g}\tilde{u}\tilde{x} + 3\tilde{h}\tilde{x}^2 + \tilde{m}\tilde{u}^3 + 2\tilde{n}\tilde{u}^2\tilde{x} + 3\tilde{p}\tilde{u}\tilde{x}^2 + 4\tilde{r}\tilde{x}^3). \end{cases} \tag{9.16}$$

Since the origin is a rank 1 critical point, the coefficients in (9.16) are such that $(1-2a)(1-2c) = b^2$ and at least one of $(1-2a)$, $(1-2c)$, or b^2 is nonzero.

Note. We do not assume that the Jacobian matrix $[\mathrm{Jac}\,\eta](\mathbf{0})$ is already in diagonal form (as is done in [Schn-EF, p. 185]). By (9.10), such an assumption requires $b = 0$.

For the case with $(1-2a) \neq 0$ and $1-2c = 0 = b$. Equation (9.11) becomes $O = s_a\mathbf{I}$, where s_a is the sign of $(1-2a)$ and $\mathbf{I}$ is the 2×2 identity matrix. Consequently, the coefficients in (9.16) are then given by

$$\begin{aligned}
K &= 1-2a = 1-\psi_{uu}(\mathbf{0}),\ \tilde{e} = s_a e = s_a\frac{\psi_{uuu}(\mathbf{0})}{6},\ \tilde{f} = s_a f = s_a\frac{\psi_{uux}(\mathbf{0})}{2},\\
\tilde{g} &= s_a g = s_a\frac{\psi_{uxx}(\mathbf{0})}{2},\quad \tilde{h} = s_a h = s_a\frac{\psi_{xxx}(\mathbf{0})}{6},\\
\tilde{k} &= k = \frac{\psi_{uuuu}(\mathbf{0})}{24},\quad \tilde{m} = m = \frac{\psi_{uuux}(\mathbf{0})}{6},\quad \tilde{n} = n = \frac{\psi_{uuxx}(\mathbf{0})}{4},\\
\tilde{p} &= p = \frac{\psi_{uxxx}(\mathbf{0})}{6},\quad \tilde{r} = r = \frac{\psi_{xxxx}(\mathbf{0})}{24}.
\end{aligned} \tag{9.17}$$

Similarly, for the case with $(1-2c) \neq 0$ and $1-2a = 0 = b$, equation (9.12) shows that the same results hold, except that s_a is now replaced by the sign of $(1-2c)$.

Now sufficiently close to the origin various degrees of approximation to (9.15) can be employed. Since we shall not delve into the thorny issue of what is meant by a "good approximation," it will be shown that by ignoring the contributions of certain terms, the Taylor polynomial (9.16) can be reduced to the forms commonly used by astrophysicists, namely, (9.18) and (9.19) below. We then leave it to the reader to decide whether this leads to an acceptable approximation. Greater accuracy can always be obtained by including more terms from (9.16) (or (9.15)).[2]

Fold: Assume that $1-2c = 0 = b$, $(1-2a) \neq 0$, and $\tilde{h} \neq 0$. Then Theorem 9.1(3i) yields that the origin is a fold critical point. If in (9.16) all terms of degree 3 or higher are dropped and if, in the expression for $\tilde{U}$, we assume that near the origin $\tilde{u}$ dominates $\tilde{u}^2$, then (9.16) transforms into the following form:

$$\begin{cases} \tilde{U}(\hat{u},\tilde{x}) = K\tilde{u} - 2\tilde{f}\tilde{u}\tilde{x} - \tilde{g}\tilde{x}^2 \\ \tilde{X}(\tilde{u},\tilde{x}) = -\tilde{f}\tilde{u}^2 - 2\tilde{g}\tilde{u}\tilde{x} - 3\tilde{h}\tilde{x}^2. \end{cases} \tag{9.18}$$

where K, $\tilde{f}$, $\tilde{g}$, and $\tilde{h}$ are given in (9.17). Equation (9.18) is identical to (6.13) on page 187 of [Schn-EF].

Cusp: Assume that $1-2c = 0 = b$, $(1-2a) \neq 0$, $\tilde{g} \neq 0$, $\tilde{h} = 0$, and $\tilde{r} \neq 0$. Theorem 9.1(3ii) implies that the origin is a cusp critical point. For the expression of $\tilde{U}$ in (9.16), suppose that near the origin $\tilde{u}$ dominates $\tilde{u}^2$, the term $\tilde{u}\tilde{x}$ dominates $\tilde{u}^2\tilde{x}$ and

[2] We are implicitly assuming that the potential ψ can be approximated by its Taylor polynomials. Though this is the case when ψ is analytic, we remind the reader that there are smooth functions for which such approximations fail. In fact, their Taylor expansions can converge to the wrong function or even be divergent! Consult Majthay [Maj, pp. 137–141] for further discussion. In practice, it is difficult to verify analyticity for most functions.

$\tilde{u}\tilde{x}^2$, while $\tilde{x}^2$ dominates $\tilde{x}^3$. Also, in the expression for $\tilde{X}$, assume that near the origin $\tilde{u}^2$ dominates $\tilde{u}^3$, while $\tilde{u}\tilde{x}$ dominates $\tilde{u}^2\tilde{x}$ and $\tilde{u}\tilde{x}^2$. Under these assumptions, equation (9.16) becomes

$$\begin{cases} \tilde{U}(\hat{u}, \tilde{x}) = K\tilde{u} - 2\tilde{f}\tilde{u}\tilde{x} - \tilde{g}\tilde{x}^2 \\ \tilde{X}(\tilde{u}, \tilde{x}) = -\tilde{f}\tilde{u}^2 - 2\tilde{g}\tilde{u}\tilde{x} - 4\tilde{r}\tilde{x}^3, \end{cases} \tag{9.19}$$

where K, $\tilde{f}$, $\tilde{g}$, and $\tilde{r}$ are given in (9.17). Equation (9.19) coincides with (6.27) on page 193 of [Schn-EF].

Proof of Theorem 9.1: First, we recall the conditions such that the germ at the origin of the lensing map $\boldsymbol{\eta}$ is stable[3] with the origin a critical point. In other words, the origin is either a fold or cusp critical point — Theorem 7.55 (page 261). Below we list the conditions determining a fold or cusp critical point — see Theorem 7.53 (page 257).

(1) The map-germ $\boldsymbol{\eta}$ has a critical point at the origin, i.e., the rank of the differential map, $d\boldsymbol{\eta}$, at the origin is 0 or 1. In other words, *the determinant of the Jacobian of $\boldsymbol{\eta}$ vanishes at the origin.*

(2) The rank of the Jacobian of $\boldsymbol{\eta}$ is 1 at the origin and the 1-jet map $j^1\boldsymbol{\eta}$ is transverse to $S_1(2, 2)$ (i.e., $\mathbf{0} \in {}_{\mathrm{T}}S_1(\boldsymbol{\eta})$). Note that ${}_{\mathrm{T}}S_1(\boldsymbol{\eta})$ is a smooth curve. Setting

$$J = \det[\mathrm{Jac}\,\boldsymbol{\eta}],$$

we see that condition (2) is equivalent to $J(\mathbf{0}) = 0$, $(\mathrm{Jac}\,\boldsymbol{\eta})(\mathbf{0}) \neq \mathbf{O}$, where $\mathbf{O}$ is the 2×2 zero matrix, and $\mathrm{grad}\, J(\mathbf{0}) \neq \mathbf{0}$.

(3) This condition divides into two cases:

Let ξ be a smooth nonvanishing vector field in a neighborhood of the origin such that at each point where J vanishes, ξ is in the kernel of $\mathrm{Jac}\,\boldsymbol{\eta}$.

(3i) The kernel of $\mathrm{Jac}\,\boldsymbol{\eta}$ at the origin is not tangent to ${}_{\mathrm{T}}S_1(\boldsymbol{\eta})$ at the origin. Theorem 7.53(1) implies that (3i) is equivalent to the fold-cusp function

$$F_{(\xi, J, \mathbf{0})} = (\xi J)\,|_{\mathrm{T}}S_1(\boldsymbol{\eta})$$

not vanishing at the origin.

Conditions (1), (2) *and* (3i) *state that $\boldsymbol{\eta}$ has a fold critical point at the origin,* that is, the origin is an element of $S_{1,0}(\boldsymbol{\eta})$.

[3] See *Stable Map-Germs*, page 226.

(3ii) The kernel of Jac $\boldsymbol{\eta}$ at the origin is tangent to ${}_{\mathrm{T}}S_1(\boldsymbol{\eta})$ at the origin. This states that the origin is an element of $S_{1,1}(\boldsymbol{\eta})$. For this critical point to be a cusp, the 2-jet $j^2\boldsymbol{\eta}$ must be transverse to $S_{1,1}(2,2)$ at the origin.

Conditions (1), (2) *and* (3ii) *with the transversality condition state that* $\boldsymbol{\eta}$ *displays a cusp at the origin* (i.e., $\mathbf{0} \in {}_{\mathrm{T}}S_{1,1}(\boldsymbol{\eta})$).

Theorem 7.53(2) implies that *the origin lies in* $S_{1,1}(\boldsymbol{\eta})$ *if and only if*

$$\text{Kernel}[\text{Jac}\,\boldsymbol{\eta}](\mathbf{0}) = T_{\mathbf{0}}\left({}_{\mathrm{T}}S_1(\boldsymbol{\eta})\right),$$

which is equivalent to $F_{(\xi,J,0)}$ *vanishing at the origin.* The same theorem yields that condition (3ii) with *the transversality requirement is equivalent to the origin being a simple zero of* $F_{(\xi,J,0)}$. Namely, the function $(\xi J)\,|_{\mathrm{T}}S_1(\boldsymbol{\eta})$ vanishes at the origin and its differential does not vanish at the origin.

We now express conditions (1–3) in terms of the Taylor coefficients of the potential. The lensing map is given by

$$\begin{bmatrix} U(u,x) \\ X(u,x) \end{bmatrix} = \begin{bmatrix} 1-2a & -b \\ -b & 1-2c \end{bmatrix} \begin{bmatrix} u \\ x \end{bmatrix} - \begin{bmatrix} (C_u + D_u + E_u)(u,x) \\ (C_x + D_x + E_x)(u,x) \end{bmatrix}.$$

The Jacobian of this map has rank 1 at the origin if and only if

$$(1-2a)(1-2c) - b^2 = 0, \textit{ and at least one of } (1-2a),\ (1-2c),\ b \textit{ does not vanish.}$$

Both $(1-2a)$ *and* $(1-2c)$ *cannot vanish.* It is then no essential restriction to assume that $(1-2a) \neq 0$. The argument assuming that $(1-2c) \neq 0$ is analogous.

Since the symmetric matrix

$$H = \begin{bmatrix} 1-2a & -b \\ -b & 1-2c \end{bmatrix}$$

is assumed to have rank 1, we know that it is orthogonally equivalent to

$$\begin{bmatrix} d & 0 \\ 0 & 0 \end{bmatrix},$$

where $d = 2(1-(a+c))$, the sum of the diagonal elements of H. The orthogonal matrix that diagonalizes H is

$$O = \frac{1}{\sqrt{(1-2a)^2 + b^2}} \begin{bmatrix} 1-2a & -b \\ b & 1-2a \end{bmatrix}.$$

We define new coordinates $(\tilde{U}, \tilde{X})$ and $(\tilde{u}, \tilde{x})$ by

$$\begin{bmatrix} \tilde{U} \\ \tilde{X} \end{bmatrix} = O \begin{bmatrix} U \\ X \end{bmatrix} \quad \text{and} \quad \begin{bmatrix} \tilde{u} \\ \tilde{x} \end{bmatrix} = O \begin{bmatrix} u \\ x \end{bmatrix}.$$

The lensing map becomes

$$\begin{bmatrix} \tilde{U} \\ \tilde{X} \end{bmatrix} = OHO^T \begin{bmatrix} \tilde{u} \\ \tilde{x} \end{bmatrix} - O \begin{bmatrix} R_u \\ R_x \end{bmatrix},$$

or,

$$\begin{bmatrix} \tilde{U} \\ \tilde{X} \end{bmatrix} = 2(1 - (a+c)) \begin{bmatrix} \tilde{u} \\ 0 \end{bmatrix} - O \begin{bmatrix} R_u \\ R_x \end{bmatrix} (u\,(\tilde{u}, \tilde{x}),\ x\,(\tilde{u}, \tilde{x})),$$

where $R = C + D + E$. If we let $\tilde{R}(\tilde{u}, \tilde{x}) = R\,(u(\tilde{u}, \tilde{x}),\ x\,(\tilde{u}, \tilde{x}))$, then

$$\begin{bmatrix} R_u \\ R_x \end{bmatrix} (u\,(\tilde{u}, \tilde{x}),\ x(\tilde{u}, \tilde{x})) = O^T \begin{bmatrix} \tilde{R}_{\tilde{u}} \\ \tilde{R}_{\tilde{x}} \end{bmatrix} (\tilde{u}, \tilde{x}).$$

Our final form is

$$\begin{bmatrix} \tilde{U} \\ \tilde{X} \end{bmatrix} = 2\,(1 - (a + c)) \begin{bmatrix} \tilde{u} \\ 0 \end{bmatrix} - \begin{bmatrix} \tilde{R}_{\tilde{u}} \\ \tilde{R}_{\tilde{x}} \end{bmatrix} (\tilde{u}, \tilde{x}).$$

Using this notation and dropping all tildes, our lensing map transforms to

$$\begin{bmatrix} U \\ X \end{bmatrix} = K \begin{bmatrix} u \\ 0 \end{bmatrix} - \begin{bmatrix} R_u \\ R_x \end{bmatrix} (u, x), \tag{9.20}$$

where

$$K = 2\,(1 - (a+c)) \neq 0.$$

Notation: *For simplicity, the coordinate expression of the lensing map given in (9.20) will also be denoted by $\boldsymbol{\eta}$ in the remainder of the proof.*

The Jacobian of (9.20) is

$$\operatorname{Jac} \boldsymbol{\eta} = \begin{bmatrix} K - R_{uu} & -R_{xu} \\ -R_{ux} & -R_{xx} \end{bmatrix}.$$

At the origin, the Jacobian reduces to

$$[\operatorname{Jac} \boldsymbol{\eta}](\mathbf{0}) = \begin{bmatrix} K & 0 \\ 0 & 0 \end{bmatrix}.$$

Note that the rank 1 condition on $(\text{Jac}\,\boldsymbol{\eta})(\mathbf{0})$ requires K to be nonzero.

The equation for the critical points of $\boldsymbol{\eta}$ in these coordinates is

$$J = -[(K - R_{uu})\, R_{xx} + (R_{ux})^2] = 0. \tag{9.21}$$

The condition $\text{grad}\, J(\mathbf{0}) \neq \mathbf{0}$ yields that ${}_{\top}S_1(\boldsymbol{\eta})$ is a smooth curve. If $J_x(\mathbf{0}) \neq 0$, then by the Implicit Function Theorem $x = M(u)$ with $M(0) = 0$ and $(dM/du)(0) = -J_u(\mathbf{0})/J_x(\mathbf{0})$. The tangent line to ${}_{\top}S_1(\boldsymbol{\eta})$ at $\mathbf{0}$ is then given by

$$J_u(\mathbf{0})u + J_x(\mathbf{0})x = 0. \tag{9.22}$$

Similarly, if $J_u(\mathbf{0}) \neq 0$, then $u = N(x)$ with $N(0) = 0$, $(dN/dx)(0) = -J_x(\mathbf{0})/J_u(\mathbf{0})$; hence, equation (9.22) also holds for the tangent line. Taking the Taylor expansion of J about the origin, equation (9.21) becomes

$$0 = J(u, x) = J_u(\mathbf{0})u + J_x(\mathbf{0})x + (\text{terms at least quadratic in } u, x).$$

In other words, *the tangent line to* ${}_{\top}S_1(\boldsymbol{\eta})$ *at the origin is obtained by setting the linear part of (9.21) equal to zero. This is the same as setting the linear part of* R_{xx} *equal to zero.*

A vector field ξ that is in Kernel[Jac $\boldsymbol{\eta}$] along ${}_{\top}S_1(\boldsymbol{\eta})$ is

$$\xi\,(u, x) = \xi = R_{ux}\,\frac{\partial}{\partial u} + (K - R_{uu})\,\frac{\partial}{\partial x}.$$

The vector field ξ is nonvanishing on ${}_{\top}S_1(\boldsymbol{\eta})$. Evaluating the (ξJ) gives

$$\begin{aligned} -(\xi J) &= 3(K - R_{uu})R_{ux}R_{uxx} + R_{ux}(2R_{ux}R_{uux} - R_{uuu}R_{xx}) \\ &\qquad +(K - R_{uu})\,[(K - R_{uu})R_{xxx} - R_{uux}R_{xx}] \\ &= (K - R_{uu})\,[3R_{ux}R_{uxx} + (K - R_{uu})R_{xxx} - R_{uux}R_{xx}] \\ &\qquad +R_{ux}[2R_{ux}R_{uux} - R_{uuu}R_{xx}]. \end{aligned} \tag{9.23}$$

Depending on which component of $\text{grad}\, J(\mathbf{0})$ is nonzero, restricting (ξJ) to ${}_{\top}S_1(\boldsymbol{\eta})$, then means plugging either $x = M(u)$ or $u = N(x)$ into (9.23). Consequently, the value of the fold-cusp function at the origin is

$$F_{(\xi, J, \mathbf{0})}(\mathbf{0}) = [(\xi J)|_{\top}S_1(\boldsymbol{\eta})](\mathbf{0}) = -K^2 R_{xxx}(\mathbf{0}).$$

Hence, *the kernel of* Jac $\boldsymbol{\eta}$ *at the origin is not tangent to* ${}_{\top}S_1(\boldsymbol{\eta})$ *at the origin (condition (3i)) if and only if*

$$R_{xxx}(\mathbf{0}) \neq 0.$$

In order to link these results to the Taylor coefficients of the potential, we write

$$R(u,x) = \tilde{e}u^3 + \tilde{f}u^2x + \tilde{g}ux^2 + \tilde{h}x^3 + S(u,x).$$

where $S(u,x)$ consists of terms of at least 4th degree.

N.B. *This is the Taylor expansion of* $\tilde{R}(\tilde{u},\tilde{x})$ *where we dropped all of the tildes on the* u*'s and* x*'s.*

Putting conditions (1), (2), and (3i) together, it follows that since

$$R_{xxx}(\mathbf{0}) = 6\tilde{h},$$

we have shown:

For $\boldsymbol{\eta}$ *to display a fold at the origin we require that in terms of the coefficients of the Taylor expansion of* ψ: $(1-2a)(1-2c) = b^2$, *and* $(1-2a) \neq 0$ *and* $\tilde{h} \neq 0$ *(or,* mutatis mutandis, *if we had chosen* $(1-2c) \neq 0$, *then we would require* $\tilde{e} \neq 0$*).*

We now return to condition (3ii) with the transversality requirement. The condition that the kernel of Jac $\boldsymbol{\eta}$ is tangent to ${}_{T}S_1(\boldsymbol{\eta})$ at the origin is equivalent to $R_{xxx}(\mathbf{0}) = 0$ (i.e., $\tilde{h} = 0$). To see what is required for the transversality part, consider equation (9.23). Since the terms in R are at least cubic in coordinates centered at the origin, to find out if

$$d_{\mathbf{0}}F_{(\xi,J,\mathbf{0})} = d_{\mathbf{0}}\left[(\xi J)\,|\,{}_{T}S_1(\boldsymbol{\eta})\right] \neq 0,$$

we need only keep track of the terms in (ξJ) of degree 1. This truncated expression for $-(\xi J)$ is given as follows, after factoring out a common nonzero factor of K:

$$3R_{ux}R_{uxx} + KR_{xxx} - 2R_{uu}R_{xxx} - R_{xx}R_{uux} \equiv (*).$$

To compute this, we use the terms of R through 4th degree, namely,

$$R(u,x) = \tilde{e}u^3 + \tilde{f}u^2x + \tilde{g}ux^2 + \tilde{k}u^4 + \tilde{m}u^3x + \tilde{n}u^2x^2 + \tilde{p}ux^3 + \tilde{r}x^4.$$

Note that we have omitted the $\tilde{h}x^3$ term as required by condition 3ii.

Computing expression (*) yields

$$\begin{aligned} &3(2\tilde{f}u + 2\tilde{g}x + 3\tilde{m}u^2 + 4\tilde{n}ux + 3\tilde{p}x^2)(2\tilde{g} + 4\tilde{n}u + 6\tilde{p}x) \\ &\quad + K(6\tilde{p}u + 24\tilde{r}x) - (2\tilde{g}u + 2\tilde{n}u^2 + 6\tilde{p}ux + 12\tilde{r}x^2)(2\tilde{f} + 6\tilde{m}u + 4\tilde{n}x) \\ &\qquad - 2(6\tilde{e}u + 2\tilde{f}x + 12\tilde{k}u^2 + 6\tilde{m}ux + 2\tilde{n}x^2)(6\tilde{p}u + 24\tilde{r}x) \\ &= K(6\tilde{p}u + 24\tilde{r}x) + 8\tilde{f}\tilde{g}u + \text{(terms at least quadratic in } u, x) \\ &\equiv (**). \end{aligned}$$

In order to restrict this to ${}_{T}S_1(\boldsymbol{\eta})$, we write out (9.21):

$$\begin{aligned} 0 &= -J = (K - R_{uu})R_{xx} - (R_{ux})^2 \\ &= [K - (6\tilde{e}u + 2\tilde{f}x + 12\tilde{k}u^2 + 6\tilde{m}ux + 2\tilde{n}x^2)][2\tilde{g}u + 2\tilde{n}u^2 + 6\tilde{p}ux \\ &\qquad + 12\tilde{r}x^2] + (2\tilde{f}u + 2\tilde{g}x + 3\tilde{m}u^2 + 4\tilde{n}ux + 3\tilde{p}x^2)^2. \end{aligned}$$

Simplifying, in order to isolate the lowest order terms:

$$0 = -J(u, x) = 2K\tilde{g}u + (\text{quadratic or higher terms in } u \text{ and } x).$$

Note that $J_x(\mathbf{0}) = 0$. Since grad $J(\mathbf{0}) \neq \mathbf{0}$, we get

$$J_u(\mathbf{0}) = -2K\tilde{g} \neq 0 \qquad\qquad (\text{i.e., } \tilde{g} \neq 0).$$

By the Implicit Function Theorem, we can write the equation for ${}_{\mathrm{T}}S_1(\boldsymbol{\eta})$ near the origin as $u = N(x)$. We know that $N(0) = 0$ and $(dN/dx)(0) = 0$ (since $(dN/dx)(0) = -J_x(\mathbf{0})/J_u(\mathbf{0})$). Thus, if we substitute $N(x)$ for u in (**), we see that

$$\begin{aligned} F_{(\xi,J,\mathbf{0})}(x) &= [(\xi J) \mid {}_{\mathrm{T}}S_1(\boldsymbol{\eta})](x) \\ &= K[6\tilde{p}N(x) + 24\tilde{r}x] + 8\tilde{f}\tilde{g}N(x) \\ &\qquad + (\text{terms at least quadratic in } N(x), x). \end{aligned}$$

Hence, the transversality condition becomes

$$0 \neq \frac{dF_{(\xi,J,\mathbf{0})}}{dx}(0) = 24K\tilde{r}, \qquad\qquad (\text{i.e., } \tilde{r} \neq 0).$$

Therefore, *for $\boldsymbol{\eta}$ to display a cusp at the origin, we require that in terms of the coefficients of the Taylor expansion of ψ: $(1-2a)(1-2c) = b^2$ and if $(1-2a) \neq 0$, then $\tilde{g} \neq 0$, $\tilde{h} = 0$, and $\tilde{r} \neq 0$ (or, mutatis mutandis, if we had chosen $(1-2c) \neq 0$, then we would require $\tilde{f} \neq 0$, $\tilde{e} = 0$, and $\tilde{k} \neq 0$).* □

9.3 Local Convexity of Fold Caustics

Loosely speaking, a fold caustic curve is "locally convex" if, locally, the curve looks convex when viewed from near the side of the caustic having two more lensed images (e.g., from the shaded region in Figure 9.3, page 331). In gravitational lensing, this property was suspected on numerical grounds by Grieger et al. [Gri-KRS, p. 198] and only noted in Schneider, Ehlers and Falco [Schn-EF, p. 192]. However, "local convexity" was observed much earlier by Berry in his study of caustics due to water droplets [Berr76, p. 14]. Our goal is to precisely define "local convexity" in general and prove that it holds in the single-plane case for potentials with locally constant Laplacians (e.g., point mass lens systems), but fails in double-plane lensing for certain potentials whose Laplacians are locally constant.

Requisite Mathematics: Sections 7.2 and 7.5.

9.3.1 General Definition of Local Convexity

Before defining local convexity, we need to consider general stable map-germs $f : (\mathbf{R}^2, \mathbf{0}) \longrightarrow (\mathbf{R}^2, \mathbf{0})$ with the origin as a fold critical point. All results about f will apply to any stable k-plane lensing map-germ $\boldsymbol{\eta} : (\mathbf{R}^2, \mathbf{0}) \longrightarrow (\mathbf{R}^2, \mathbf{0})$.

Write the Taylor expansion of f as follows:

$$\begin{cases} U = \mathtt{A}u - \mathtt{B}x + \mathcal{S}(u,x) \\ X = -\mathtt{B}'u + \mathtt{C}x + \mathcal{T}(u,x), \end{cases} \tag{9.24}$$

where $\mathtt{A}, \mathtt{B}, \mathtt{B}', \mathtt{C}$ are constants and $\mathcal{S}(u,x)$, $\mathcal{T}(u,x)$ are quadratic or higher in u and x.

Theorem 9.2 *Let* $f : (\mathbf{R}^2, \mathbf{0}) \to (\mathbf{R}^2, \mathbf{0})$ *be a map-germ with the origin a fold critical point of* f. *Without loss of generality, we may assume that* $\mathtt{A} \neq 0$ *and there are orthogonal coordinate changes in the domain and target such that the Taylor expansion of* f *has the form*

$$\begin{cases} \tilde{U}(\tilde{u},\tilde{x}) = k\tilde{u} + \mathsf{A}(\tilde{u},\tilde{x}) + \mathsf{R}(\tilde{u},\tilde{x}) \\ \tilde{X}(\tilde{u},\tilde{x}) = a\tilde{x}^2 - 2b\tilde{u}\tilde{x} + c\tilde{u}^2 - \mathsf{S}(\tilde{u},\tilde{x}) \end{cases}, \tag{9.25}$$

where (see (9.24))

$$\mathtt{AC} - \mathtt{BB}' = 0, \quad k = \sqrt{(1+\Lambda'^2)\,(\mathtt{A}^2+\mathtt{B}^2)} \neq 0$$

with $\Lambda' = -\mathtt{B}'/\mathtt{A}$; $a \neq 0$, b, *and* c *are constants;* A *is quadratic, and* R, S *are cubic or higher in* u *and* x.

Proof: We proceed in a manner similar to the proof of Theorem 9.1.

The condition that the lensing map in (9.24) has a rank 1 critical point at the origin is that

$$\mathtt{AC} - \mathtt{BB}' = 0, \textit{ and at least one of } \mathtt{A}, \mathtt{C}, \mathtt{B}, \mathtt{B}' \textit{ does not vanish.}$$

Here, since the Jacobian is not symmetric, the preceding condition would be satisfied with $\mathtt{B}$ or $\mathtt{B}' \neq 0$ and the other three numbers vanishing. The case in which one of $\mathtt{B}$ or $\mathtt{B}' \neq 0$ and both $\mathtt{A}$ and $\mathtt{C}$ vanish is simpler than the case in which we assume $\mathtt{A}$ or $\mathtt{C} \neq 0$. Since assuming either $\mathtt{A}$ or $\mathtt{C} \neq 0$ leads to exactly analogous computations and all cases lead to the *same result*, we carry out the more involved one, assuming that $\mathtt{A} \neq 0$.

Letting

$$\Lambda' = -\frac{\mathtt{B}'}{\mathtt{A}},$$

rewrite the Taylor expansion (9.24) of f as follows:

$$\begin{bmatrix} U \\ X \end{bmatrix} = \begin{bmatrix} 1 & 0 \\ 0 & \Lambda' \end{bmatrix} \begin{bmatrix} \mathtt{A} & -\mathtt{B} \\ \mathtt{A} & -\mathtt{B} \end{bmatrix} \begin{bmatrix} u \\ x \end{bmatrix} + \begin{bmatrix} \mathcal{S} \\ \mathcal{T} \end{bmatrix}. \tag{9.26}$$

We now define new coordinates $\tilde{U}$ and $\tilde{X}$ in the target by an orthogonal coordinate change:

$$\begin{bmatrix} \tilde{U} \\ \tilde{X} \end{bmatrix} = \frac{1}{\sqrt{1+\Lambda'^2}} \begin{bmatrix} 1 & \Lambda' \\ -\Lambda' & 1 \end{bmatrix} \begin{bmatrix} U \\ X \end{bmatrix}.$$

Explicitly,

$$\sqrt{1+\Lambda'^2} \begin{bmatrix} \tilde{U} \\ \tilde{X} \end{bmatrix} = \begin{bmatrix} 1 & \Lambda'^2 \\ -\Lambda' & \Lambda' \end{bmatrix} \begin{bmatrix} \mathtt{A} & -\mathtt{B} \\ \mathtt{A} & -\mathtt{B} \end{bmatrix} \begin{bmatrix} u \\ x \end{bmatrix} + \begin{bmatrix} 1 & \Lambda' \\ -\Lambda' & 1 \end{bmatrix} \begin{bmatrix} \mathcal{S} \\ \mathcal{T} \end{bmatrix}.$$

We also define new coordinates $\tilde{u}$ and $\tilde{x}$ in the domain by an orthogonal coordinate change

$$\frac{1}{\sqrt{\mathtt{A}^2+\mathtt{B}^2}} \begin{bmatrix} \mathtt{A} & \mathtt{B} \\ -\mathtt{B} & \mathtt{A} \end{bmatrix} \begin{bmatrix} \tilde{u} \\ \tilde{x} \end{bmatrix} = \begin{bmatrix} u \\ x \end{bmatrix}. \tag{9.27}$$

Substituting, we obtain

$$\begin{bmatrix} \tilde{U} \\ \tilde{X} \end{bmatrix} = \frac{1}{\sqrt{(1+\Lambda'^2)(\mathtt{A}^2+\mathtt{B}^2)}} \begin{bmatrix} 1 & \Lambda'^2 \\ -\Lambda' & \Lambda' \end{bmatrix} \begin{bmatrix} \mathtt{A} & -\mathtt{B} \\ \mathtt{A} & -\mathtt{B} \end{bmatrix} \begin{bmatrix} \mathtt{A} & \mathtt{B} \\ -\mathtt{B} & \mathtt{A} \end{bmatrix} \begin{bmatrix} \tilde{u} \\ \tilde{x} \end{bmatrix}$$
$$+ \frac{1}{\sqrt{1+\Lambda'^2}} \begin{bmatrix} 1 & \Lambda' \\ -\Lambda' & 1 \end{bmatrix} \begin{bmatrix} \tilde{\mathcal{S}}(\tilde{u},\tilde{x}) \\ \tilde{\mathcal{T}}(\tilde{u},\tilde{x}) \end{bmatrix},$$

where

$$\begin{bmatrix} \tilde{\mathcal{S}}(\tilde{u},\tilde{x}) \\ \tilde{\mathcal{T}}(\tilde{u},\tilde{x}) \end{bmatrix} = \begin{bmatrix} \mathcal{S}(u(\tilde{u},\tilde{x}),\ x(\tilde{u},\tilde{x})) \\ \mathcal{T}(u(\tilde{u},\tilde{x}),\ x(\tilde{u},\tilde{x})) \end{bmatrix}.$$

Thus, simplifying

$$\begin{bmatrix} \tilde{U} \\ \tilde{X} \end{bmatrix} = \sqrt{(1+\Lambda'^2)\,(\mathtt{A}^2+\mathtt{B}^2)} \begin{bmatrix} \tilde{u} \\ 0 \end{bmatrix} + \frac{1}{\sqrt{1+\Lambda'^2}} \begin{bmatrix} \tilde{\mathcal{S}}+\Lambda'\tilde{\mathcal{T}} \\ -\Lambda'\tilde{\mathcal{S}}+\tilde{\mathcal{T}} \end{bmatrix}. \tag{9.28}$$

Set

$$k = \sqrt{(1+\Lambda'^2)\,(\mathtt{A}^2+\mathtt{B}^2)}, \qquad \tilde{M} = \frac{\tilde{\mathcal{S}}+\Lambda'\tilde{\mathcal{T}}}{\sqrt{1+\Lambda'^2}}, \qquad \tilde{N} = \frac{-\Lambda'\tilde{\mathcal{S}}+\tilde{\mathcal{T}}}{\sqrt{1+\Lambda'^2}}.$$

Then (9.28) becomes

$$\begin{bmatrix} \tilde{U} \\ \tilde{X} \end{bmatrix} = \begin{bmatrix} k\tilde{u}+\tilde{M} \\ \tilde{N} \end{bmatrix}. \tag{9.29}$$

Note that $\tilde{M}$ and $\tilde{N}$ are at least quadratic in $\tilde{u}$ and $\tilde{x}$.

Notation: *Throughout the remainder of the proof, the orthogonally transformed coordinate form of the map f in (9.29) will also be denoted by f.*

The Jacobian matrix of (9.29) is

$$\text{Jac}\, f = \begin{bmatrix} k + \tilde{M}_{\tilde{u}} & \tilde{M}_{\tilde{x}} \\ \\ \tilde{N}_{\tilde{u}} & \tilde{N}_{\tilde{x}} \end{bmatrix},$$

which has rank 1 at the origin if and only of $k \neq 0$. The critical points of f are determined by

$$J = (k + \tilde{M}_{\tilde{u}})\tilde{N}_{\tilde{x}} - \tilde{M}_{\tilde{x}}\tilde{N}_{\tilde{u}} = 0. \tag{9.30}$$

Enforcing the condition grad $J(\mathbf{0}) \neq \mathbf{0}$, we get that ${}_{\mathrm{T}}S_1(f)$ is a smooth curve. As in the proof of Theorem 9.1 (page 351), we have that the tangent line to ${}_{\mathrm{T}}S_1(f)$ at the origin is obtained by setting the linear part of (9.30) equal to zero. This is equivalent to setting the linear part of $\tilde{N}_{\tilde{x}}$ equal to zero.

The vector field

$$\xi(\tilde{u}, \tilde{x}) = \xi = -\tilde{M}_{\tilde{x}}\frac{\partial}{\partial \tilde{u}} + (k + \tilde{M}_{\tilde{u}})\frac{\partial}{\partial \tilde{x}}$$

is nonvanishing on ${}_{\mathrm{T}}S_1(\boldsymbol{\eta})$ and lies in Kernel[Jac f] along ${}_{\mathrm{T}}S_1(f)$. It follows that

$$\begin{aligned}(\xi J) \;=\; & -\tilde{M}_{\tilde{x}}\left\{-\tilde{M}_{\tilde{u}\tilde{u}}\tilde{N}_{\tilde{x}} + (k + \tilde{M}_{\tilde{u}})\tilde{N}_{\tilde{x}\tilde{u}} + \tilde{M}_{\tilde{x}\tilde{u}}\tilde{N}_{\tilde{u}} - \tilde{M}_{\tilde{x}}\tilde{N}_{\tilde{u}\tilde{u}}\right\} \\ & +(k + \tilde{M}_{\tilde{u}})\left\{\tilde{M}_{\tilde{u}\tilde{x}}\tilde{N}_{\tilde{x}} + (k + \tilde{M}_{\tilde{u}})\tilde{N}_{\tilde{x}\tilde{x}} - \tilde{M}_{\tilde{x}\tilde{x}}\tilde{N}_{\tilde{u}} - \tilde{M}_{\tilde{x}}\tilde{N}_{\tilde{u}\tilde{x}}\right\}.\end{aligned} \tag{9.31}$$

The Implicit Function Theorem yields that depending on which component of grad $J(\mathbf{0})$ is nonzero, the restriction $(\xi J)|_{\mathrm{T}}S_1(f)$ requires plugging a functional form $x = g_1(u)$ (or $u = g_2(x)$) into (9.31), where $g_1(0) = 0$ (or $g_2(0) = 0$). Hence, evaluating the fold-cusp function at the origin yields

$$F_{(\xi, J, \mathbf{0})}(\mathbf{0}) = [(\xi J)|_{\mathrm{T}}S_1(f)](\mathbf{0}) = k^2\tilde{N}_{\tilde{x}\tilde{x}}(\mathbf{0}).$$

The kernel of Jac f at the origin is not tangent to ${}_{\mathrm{T}}S_1(f)$ at the origin if and only if

$$\tilde{N}_{\tilde{x}\tilde{x}}(\mathbf{0}) \neq 0.$$

Express $\tilde{N}$ as follows:

$$\tilde{N}(\tilde{u}, \tilde{x}) = a\tilde{x}^2 - 2b\tilde{u}\tilde{x} + c\tilde{u}^2 - \mathsf{S}(\tilde{u}, \tilde{x}),$$

where a, b, and c are constants and S is cubic or higher in $\tilde{u}$ and $\tilde{x}$. Thus, if the origin is a fold critical point of f, then there are orthogonal coordinate changes in the domain and target such that the form (9.29) of f becomes

$$\begin{bmatrix} k\tilde{u} + \tilde{M} \\ \\ \tilde{N} \end{bmatrix} = \begin{bmatrix} k\tilde{u} + \mathsf{A}(\tilde{u}, \tilde{x}) + \mathsf{R}(\tilde{u}, \tilde{x}) \\ \\ a\tilde{x}^2 - 2b\tilde{u}\tilde{x} + c\tilde{u}^2 - \mathsf{S}(\tilde{u}, \tilde{x}) \end{bmatrix}$$

with the $k \neq 0$ and $a \neq 0$. □

Let $f : (\mathbf{R}^2, \mathbf{0}) \to (\mathbf{R}^2, \mathbf{0})$ be a map-germ with the origin a fold critical point. We first consider the f-image of ${}_{\mathsf{T}}S_1(f)$. By Theorem 9.2, we can express f in the form (9.25), where all the tildes are dropped:

$$\begin{cases} U(u, x) = ku + \mathsf{A}(u, x) + \mathsf{R}(u, x) \\ \\ X(u, x) = ax^2 - 2bux + cu^2 - \mathsf{S}(u, x) \end{cases} \tag{9.32}$$

with $k \neq 0$ and $a \neq 0$. The line tangent to ${}_{\mathsf{T}}S_1(f)$ at the origin is given by (page 351):

$$[Linear\ part\ of\ X_x(u, x)] = 2(ax - bu) = 0.$$

To find the f-image of this curve, we substitute $ax = bu$ into (9.32) and obtain

$$\begin{aligned} U &= ku + \mathsf{C}(u) \\ aX &= (-b^2 + ac)\, u^2 + \mathsf{D}(u), \end{aligned}$$

where C is at least quadratic and D at least cubic. Solving for u in terms of U, gives

$$u = \frac{U}{k} + \mathsf{E}(U),$$

where E is at least quadratic in U. Substituting into the equation for X gives the local equation for the f-image of the tangent line to critical curve $f[{}_{\mathsf{T}}S_1(f)]$:

$$aX = (ac - b^2)\left(\frac{U}{k}\right)^2 + \mathsf{F}(U), \tag{9.33}$$

where F is at least cubic in U. Since $[d_{\mathbf{0}}f](t(u)) = (ku, 0)$, where $t(u) = (u, bu/a)$ is the tangent line to ${}_{\mathsf{T}}S_1(f)$, we see that the tangent line to $f[{}_{\mathsf{T}}S_1(f)]$ is the U-axis. Hence, the curve (9.33) is tangent to $f[{}_{\mathsf{T}}S_1(f)]$ at the origin in the (U, X)-plane. In addition, the parabola

$$aX = (ac - b^2)\left(\frac{U}{k}\right)^2 \tag{9.34}$$

is tangent to $f[{}_{\mathrm{T}}S_1(f)]$ at the origin in the (U, X)-plane and approximates the f-image of the line tangent to ${}_{\mathrm{T}}S_1(f)$ up to third order in U. Thus, this curve has the same curvature at the origin as the f-image of the tangent line to ${}_{\mathrm{T}}S_1(f)$ at the origin.

Let us now determine the f-image of a neighborhood of the origin in the (u, x)-plane near the origin in the (U, X)-plane. In what follows we work near the origin in both the (u, x) and the (U, X)-planes. Write the first equation of (9.32) as $F(U, x, u) = 0$. Since $F_u(0, 0, 0) = k \neq 0$, by the Implicit Function Theorem we express u as a function of U and x, say

$$u = \frac{U}{k} + \mathsf{V}(U, x),$$

where V is at least quadratic in U and x. This tells us that for x near 0, if U is the first coordinate of the f-image of (u, x), then U, x, u must satisfy an equation of the preceding form. We now substitute this expression for u into the second equation of (9.32). This yields an expression for X in terms of U and x that describes which pairs (U, X) can be in the f-image of (u, x) for some u:

$$X = ax^2 - 2b\frac{U}{k}x + c\left(\frac{U}{k}\right)^2 + \mathsf{W}(U, x),$$

where W is at least cubic. Rewriting the preceding:

$$\begin{aligned} aX - (ac - b^2)\left(\frac{U}{k}\right)^2 &= a^2\left[x^2 - 2\frac{b}{a}x\frac{U}{k} + \left(\frac{b}{a}\right)^2\left(\frac{U}{k}\right)^2\right] + a\mathsf{W}(U, x) \\ &= a^2\left(x - \frac{b}{a}\frac{U}{k}\right)^2 + a\mathsf{W}(U, x). \end{aligned}$$

This equation has solutions for (U, x) near $(0, 0)$ if and only if

$$aX \geq (ac - b^2)\left(\frac{U}{k}\right)^2.$$

Thus, locally f double covers the plane where

$$aX > (ac - b^2)\left(\frac{U}{k}\right)^2 + \mathsf{F}(U).$$

Moreover, the curve that is tangent to $f[{}_{\mathrm{T}}S_1(f)]$ at the origin and which approximates $f[{}_{\mathrm{T}}S_1(f)]$ to third order is parabola (9.34). Thus: *Locally, when viewed from the region in the (U, X)-plane that is double covered by f, the curve $f[S_1(f)]$ is convex. The signed curvature function along such fold caustics does not change sign. That is, locally convex fold caustics curves have no inflection points.*

Definition 9.3 We shall say that the f-image of the fold critical curve is *convex* if $(ac - b^2) < 0$.

N.B. Since we allow orthogonal coordinate changes to bring our map displaying a fold critical curve into the given form (9.32), this definition has coordinate free sense. Namely, when viewed from the double covered portion of the target plane, the f-image of the fold critical curve is curved outwards.

9.3.2 Local Convexity in the Single-Plane Case

Consider a single-plane lensing map-germ $\boldsymbol{\eta} : (\mathbf{R}^2, \mathbf{0}) \to (\mathbf{R}^2, \mathbf{0})$. We shall show that if the Laplacian of the potential inducing $\boldsymbol{\eta}$ is locally constant, then local convexity holds.

Choose the orthogonal coordinates $(\tilde{u}, \tilde{x})$ and $(\tilde{U}, \tilde{X})$ in the domain and target of $\boldsymbol{\eta}$, respectively, as in Theorem 9.1(2) or Theorem 9.2. Since our coordinate change in u and x is orthogonal, given any smooth function $F(u, x)$, if we let $\tilde{F}(\tilde{u}, \tilde{x}) = F(u(\tilde{u}, \tilde{x}), x(\tilde{u}, \tilde{x}))$, then

$$(\tilde{F}_{\tilde{u}\tilde{u}} + \tilde{F}_{\tilde{x}\tilde{x}})(\tilde{u}, \tilde{x}) = (F_{uu} + F_{xx})(u(\tilde{u}, \tilde{x}), x(\tilde{u}, \tilde{x})).$$

Consequently, if the Laplacian ΔF of F is locally constant, then so is the Laplacian $\Delta \tilde{F}$ of $\tilde{F}$. In our situation, if

$$\Delta \psi = \Delta (A + B) + \Delta (\psi - (A + B))$$

is locally constant, then so is $\Delta (\psi - (A + B))$ (since $(A + B)$ is quadratic — see page 344 for notation). Hence, when $\Delta \psi$ is locally constant, any partial of $\Delta (\psi - (A + B))$ is locally zero. Thus, the coefficients of the nonconstant terms in the Taylor expansion of $\Delta (\psi - (A + B))$ vanish. In particular, $\frac{\partial}{\partial u}(\Delta R) = 0$ and $\frac{\partial}{\partial x}(\Delta R) = 0$, where $R = C + D$. By our preceding remark, $\Delta \tilde{R}$ is also locally constant, so $\frac{\partial}{\partial \tilde{u}}(\Delta \tilde{R}) = 0$ and $\frac{\partial}{\partial \tilde{x}}(\Delta \tilde{R}) = 0$.

As before, we shall drop the tildes from u, x, U, X, and R. For example, $\tilde{R}(\tilde{u}, \tilde{x})$ now becomes

$$R(u, x) = \tilde{e}u^3 + \tilde{f}u^2x + \tilde{g}ux^2 + \tilde{h}x^3 + S(u, x), \tag{9.35}$$

where $S(u, x)$ consists of terms of at least 4th, degree. Also, if we assume that the Laplacian $\Delta \psi$ of the potential ψ is constant near the origin, then

$$R_{xxx}(\mathbf{0}) + R_{uux}(\mathbf{0}) = 0. \tag{9.36}$$

Suppose that the origin is a fold critical point of $\boldsymbol{\eta}$. Write out equation (9.15) in the form given in (9.32), dropping tildes as above:

$$\begin{cases} U = Ku - [3\tilde{e}u^2 + 2\tilde{f}ux + \tilde{g}x^2 + S_u(u, x)] \\ X = -[\tilde{f}u^2 + 2\tilde{g}ux + 3\tilde{h}x^2 + S_x(u, x)]. \end{cases}$$

Since $\mathbf{0}$ is a fold critical point, Theorem 9.1(3i) (or Theorem 9.2) yields that $K \neq 0$ and for the assumed choice $\mathrm{A} = (1-2a) \neq 0$, we have $\tilde{h} \neq 0$. Comparing with (9.32), we see that

$$a = -3\tilde{h}, \quad b = \tilde{g}, \quad c = -\tilde{f}.$$

Evaluating (9.36), we get $2\tilde{f} = -3\tilde{h}$, so $\tilde{f} \neq 0$ (since $\tilde{h} \neq 0$). It follows that

$$ac - b^2 = 3\tilde{h}\tilde{f} - \tilde{g}^2 = -(2\tilde{f} + \tilde{g}^2) < 0.$$

In light of Definition 9.3, we then obtain:

Theorem 9.4 *Let $\boldsymbol{\eta}$ be a locally stable single-plane lensing map induced by a gravitational lens potential ψ with locally constant surface mass density $\kappa = \Delta\psi/2$. Then the fold caustic curves of $\boldsymbol{\eta}$ are locally convex.*

The potential due to g point masses on a lens plane L with continuous matter κ_c and shear γ from infinity is given by (see page 175):

$$\psi_{ptcs}(\mathbf{x}) = \frac{\kappa_c}{2}|\mathbf{x}|^2 - \frac{\gamma}{2}(\mathrm{u}^2 - \mathrm{v}^2) + \sum_{\ell=1}^{g} m_\ell \ln|\mathbf{x} - \boldsymbol{\xi}_\ell|,$$

where $\kappa_c \geq 0$, $\gamma \geq 0$, $m_\ell \geq 0$, $\mathbf{x} = (\mathrm{u}, \mathrm{v})$, and $L = \mathbf{R}^2 - \{\boldsymbol{\xi}_1, \ldots, \boldsymbol{\xi}_g\}$. The corresponding surface mass density on L is constant: $\kappa_{ptcs}(\mathbf{x}) = \kappa_c$.

Corollary 9.5 *Let $\boldsymbol{\eta}_{ptcs}$ be the single-plane lensing map induced by the gravitational lens potential ψ_{ptcs}. If $\boldsymbol{\eta}$ is locally stable, then the fold caustic curves of $\boldsymbol{\eta}$ are locally convex.*

An example of the caustics due to two point masses on a single lens plane with shear is shown in Figure 9.13. Locally, the caustics appear *convex* when viewed from near the side with more lensed images.

Local convexity is especially useful when trying to determine the number of lensed images of a light source in different regions of a caustic network. Roughly speaking, if, locally, a fold caustic curve looks convex when viewed from one side of the caustic, then crossing to the concave side decreases the number of lensed images by two.

Note: If the Laplacian of a potential is not locally constant, then local convexity need not hold. For example, it is violated in Figure 9.9 (page 337), where an elliptical potential displays a caustic component that is a simple closed fold curve.

9.3.3 Double-Plane Lensing Maps Displaying Folds

Before we can show that local convexity fails in double-plane lensing (for certain potentials whose Laplacians are locally constant), we must first express the condition for a fold critical point in terms of the potentials on the two lens planes. This is the aim of the current section.

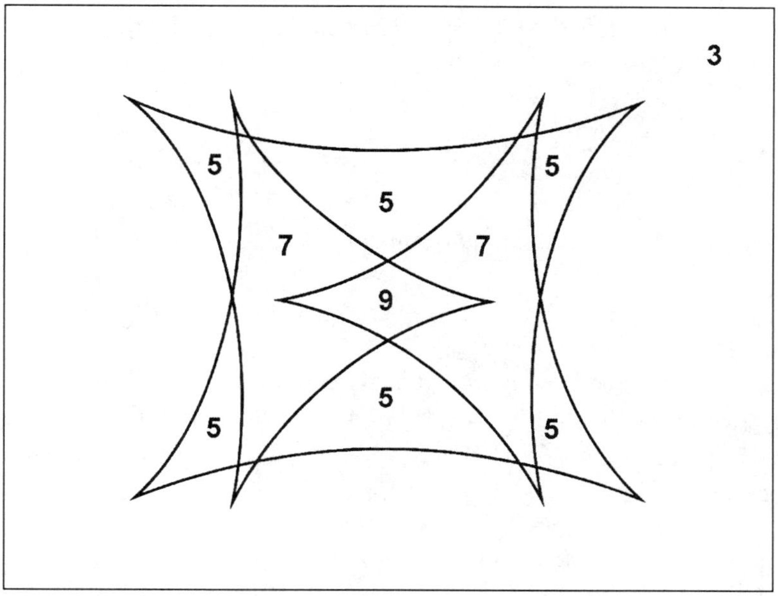

Figure 9.13. Caustics due to two point masses on the same lens plane with shear [Wit-P, p. 4109]. The numbers indicate how many lensed images are seen of a light source in the given region. Locally, the fold arcs appear convex when viewed from a region with more lensed images.

Consider the time delay function for a two lens plane system:

$$T_y(\mathbf{x}_1, \mathbf{x}_2) = \theta_{1,2}\left[\frac{|\mathbf{x}_2 - \mathbf{x}_1|^2}{2} - \beta_{1,2}\,\psi_1(\mathbf{x}_1)\right] + \theta_{2,3}\left[\frac{|\mathbf{y} - \mathbf{x}_2|^2}{2} - \beta_{2,3}\,\psi_2(\mathbf{x}_2)\right].$$

Recall that $\beta_{2,3} = 1$. For simplicity, *we absorb* $\beta_{1,2}$ *into* ψ_1. To obtain the lensing map for this system, we set the $\text{grad}_{\mathbf{x}_1} T_{\mathbf{y}}$ and $\text{grad}_{\mathbf{x}_2} T_{\mathbf{y}}$ equal to zero:

$$\text{grad}_{\mathbf{x}_1} T_{\mathbf{y}} = \theta_{1,2}\left[\mathbf{x}_1 - \mathbf{x}_2 - \text{grad}_{\mathbf{x}_1}\psi_1(\mathbf{x}_1)\right] = \mathbf{0}$$

and

$$\text{grad}_{\mathbf{x}_2} T_{\mathbf{y}} = \theta_{1,2}\,[\mathbf{x}_2 - \mathbf{x}_1] + \theta_{2,3}\left[\mathbf{x}_2 - \mathbf{y} - \text{grad}_{\mathbf{x}_2}\psi_2(\mathbf{x}_2)\right] = \mathbf{0}.$$

The first equation gives

$$\mathbf{x}_2 = \mathbf{x}_1 - \text{grad}_{\mathbf{x}_1}\psi_1(\mathbf{x}_1).$$

Remark: The preceding is just the equation of a lensing map for a single lens plane if we interpret $\mathbf{x}_2$ as coordinates on the light source plane. We shall use this later to obtain single-plane local convexity from a double lensing map (see page 365).

Substituting into the second equation gives us the lensing map from the first lens plane into the light source plane:

$$\mathbf{y} = \mathbf{x}_1 - \left(1 + \frac{\theta_{1,2}}{\theta_{2,3}}\right) \operatorname{grad}_{\mathbf{x}_1} \psi_1(\mathbf{x}_1) - \operatorname{grad}_{\mathbf{x}_2} \psi_2(\mathbf{x}_1 - \operatorname{grad}_{\mathbf{x}_1} \psi_1(\mathbf{x}_1)).$$

Once again, *for simplicity, we rename the constant* $(1 + \frac{\theta_{1,2}}{\theta_{2,3}})$ *merely* Θ. Thus

$$\mathbf{y} = \mathbf{x}_1 - \Theta \operatorname{grad}_{\mathbf{x}_1} \psi_1(\mathbf{x}_1) - \operatorname{grad}_{\mathbf{x}_2} \psi_2(\mathbf{x}_1 - \operatorname{grad}_{\mathbf{x}_1} \psi_1(\mathbf{x}_1)) \tag{9.37}$$

The associated double-plane lensing map $\boldsymbol{\eta}$ is given by $\boldsymbol{\eta}(\mathbf{x}_1) = \mathbf{y}$. Suppose that the double lens plane map-germ $\boldsymbol{\eta} : (\mathbf{R}^2, \mathbf{0}) \to (\mathbf{R}^2, \mathbf{0})$ has a fold critical point at the origin.

So to avoid subscripts, we let $\mathbf{y} = \boldsymbol{\eta}(\mathbf{x}_1) = (U, X)$, $\mathbf{x}_1 = (u, x)$ *and* $\mathbf{x}_2 = (\mathrm{v}, z)$. *Also, we let* $\psi_1 = \psi$ *and* $\psi_2 = \phi$. *The two equations of (9.37) become*

$$\begin{cases} U = u - \Theta\psi_u(u, x) - \phi_v(u - \psi_u(u, x), x - \psi_x(u, x)) \\ \\ X = x - \Theta\psi_x(u, x) - \phi_z(u - \psi_u(u, x), x - \psi_x(u, x)). \end{cases} \tag{9.38}$$

Assume that the potential on the first lens plane has the form

$$\psi(u, x) = au^2 + bux + cx^2 + R(u, x), \tag{9.39}$$

where $R(u, x)$ is the remainder in the Taylor expansion of ψ.

N.B. The function R coincides with the usage on page 344, but not in (9.35) (which is actually $\tilde{R}$).

Suppose that the potential on the second lens plane is expressed as

$$\phi(\mathrm{v}, z) = \rho \mathrm{v}^2 + \sigma \mathrm{v} z + \tau z^2 + \Pi(\mathrm{v}, z), \tag{9.40}$$

where the Taylor expansion of Π begins with cubic terms. Rewriting (9.38) in these terms, we obtain

$$\begin{cases} U = (1 - 2\Theta a)\, u - \Theta bx - (2\rho \mathrm{v} + \sigma z) - \Theta R_u - \Pi_{\mathrm{v}}(\mathrm{v}, z) \\ \\ X = (1 - 2\Theta c)\, x - \Theta bu - (\sigma \mathrm{v} + 2\tau z) - \Theta R_x - \Pi_z(\mathrm{v}, z). \end{cases}$$

Here, as in what follows, the arguments (v, z) are given by

$$\begin{bmatrix} \mathrm{v}(u, x) \\ z(u, x) \end{bmatrix} = \begin{bmatrix} u - \psi_u(u, x) \\ x - \psi_x(u, x) \end{bmatrix}. \tag{9.41}$$

The lensing map becomes

$$\begin{aligned} U &= \{(1-2\Theta a) - 2\rho(1-2a) + \sigma b\}\,u - (\Theta b - 2\rho b + \sigma\,(1-2c))x \\ &\quad +(2\rho - \Theta)R_u + \sigma R_x - \Pi_{\mathrm{v}}(\mathrm{v}, z) \\ X &= \{(1-2\Theta c) + 2\tau(1-2c) + \sigma b\}\,x - (\Theta b + \sigma(1-2a) - 2\tau b)u \\ &\quad +\sigma R_u + (2\tau - \Theta)R_x - \Pi_z(\mathrm{v}, z)\,. \end{aligned} \tag{9.42}$$

Set

$$\begin{aligned} \mathrm{A} &= (1-2\Theta a)\; - 2\rho(1-2a) + \sigma b. \\ \mathrm{B} &= \Theta b - 2\rho b + \sigma\,(1-2c). \\ \mathrm{B}' &= \Theta b + \sigma(1-2a) - 2\tau b. \\ \mathrm{C} &= (1-2\Theta c) + 2\tau(1-2c) + \sigma b. \end{aligned}$$

and

$$\begin{bmatrix} \mathcal{S}(u,x) \\ \mathcal{T}(u,x) \end{bmatrix} = \begin{bmatrix} (2\rho-\Theta) & \sigma \\ \sigma & (2\tau-\Theta) \end{bmatrix} \begin{bmatrix} R_u(u,x) \\ R_x(u,x) \end{bmatrix}$$

$$- \begin{bmatrix} \Pi_{\mathrm{v}}(\mathrm{v}(u,x), z(u,x)) \\ \Pi_z(\mathrm{v}(u,x), z(u,x)) \end{bmatrix},$$

where $(\mathrm{v}(u,x), z(u,x))$ is defined by (9.41). This simplifies the double-plane lensing map to the form (9.24) on page 354:

$$\begin{cases} U \;=\; \mathrm{A}u \;-\; \mathrm{B}x \;+\; \mathcal{S}(u,x) \\ X \;=\; -\,\mathrm{B}'u \;+\; \mathrm{C}x \;+\; \mathcal{T}(u,x). \end{cases} \tag{9.43}$$

Since the map (9.43) has a fold critical point at the origin, Theorem 9.2 (page 354) yields, without loss of generality, that

$$\mathrm{AC} - \mathrm{BB}' = 0, \qquad \mathrm{A} \neq 0, \tag{9.44}$$

and (via (9.28) on page 355) there are orthogonal coordinate changes transforming (9.43) to

$$\begin{bmatrix} \tilde{U} \\ \tilde{X} \end{bmatrix} = \begin{bmatrix} k\tilde{u} + \frac{\tilde{\mathcal{S}} + \Lambda'\tilde{\mathcal{T}}}{\sqrt{1+\Lambda'^2}} \\ \frac{-\Lambda'\tilde{\mathcal{S}} + \tilde{\mathcal{T}}}{\sqrt{1+\Lambda'^2}} \end{bmatrix}.$$

Here

$$k = \sqrt{(1 + \Lambda'^2)\,(\mathrm{A}^2 + \mathrm{B}^2)} \neq 0, \qquad (\Lambda = -\mathrm{B}'/\mathrm{A}), \tag{9.45}$$

$$\begin{bmatrix} \tilde{\mathcal{S}}(\tilde{u}, \tilde{x}) \\ \tilde{\mathcal{T}}(\tilde{u}, \tilde{x}) \end{bmatrix} = \begin{bmatrix} \mathcal{S}(u(\tilde{u}, \tilde{x}),\ x(\tilde{u}, \tilde{x})) \\ \mathcal{T}(u(\tilde{u}, \tilde{x}),\ x(\tilde{u}, \tilde{x})) \end{bmatrix}, \tag{9.46}$$

and

$$\tilde{X}_{\tilde{x}\tilde{x}}(\mathbf{0}) \neq 0. \tag{9.47}$$

This result will be used in the next section to establish the violation of local convexity in double-plane lensing.

9.3.4 Violation of Convexity in Double-Plane Case

Though local convexity holds in single-plane lensing with potentials having locally constant Laplacians, we shall now show that local convexity can fail in double-plane lensing with such potentials. Numerical evidence for the latter was given by [Erd-S, p. 463] for a double-plane two point mass lens, and [Pet-Wic95] for a double-plane two point mass lens with continuous matter and shear (see Figure 9.14).

Consider a double-plane lensing map-germ with the origin as a fold critical point. Employing the notation of Section 9.3.3, consider the case where both potential functions, ψ and ϕ, have constant Laplacians, or what is the same: $R_{uu} + R_{xx}$ and $\Pi_{vv} + \Pi_{zz}$ are constants — see (9.39) and (9.40) for R and Π. By Definition 9.3 (page 359), it suffices to consider the quadratic part of $\tilde{X}$. We introduce notation for the cubic parts of R and Π:

$$\begin{aligned} R(u, x) &= eu^3 + fu^2x + gux^2 + hx^3 + \text{(higher order)}, \\ \Pi(\mathrm{v}, z) &= \alpha \mathrm{v}^3 + \beta v^2 z + \gamma \mathrm{v} z^2 + \delta x^3 + \text{(higher order)}, \end{aligned}$$

$$R_{uu} \;+\; R_{xx} = 2\{(3e \,+\, g)\,u + (f + 3h)x\} + \text{(higher order)} = \text{constant},$$

$$\Pi_{\mathrm{vv}} \;+\; \Pi_{zz} \;=\; 2\{(3\alpha + \gamma)\,\mathrm{v} + (\beta + 3\delta)\,z\} + \text{(higher order)} = \text{constant}.$$

For the coefficients of the cubic terms in R and Π, we then have

$$g \;=\; -\,3e,\; f \;=\; -\,3h,\; \gamma \;=\; -\,3\alpha,\; \beta \;=\; -\,3\delta.$$

Thus, writing only the quadratic parts:

$$R_u(u,x) = 3eu^2 + 2fux + gx^2 = g(x^2 - u^2) + 2fux,$$

$$R_x(u,x) = fu^2 + 2gux + 3hx^2 = f(u^2 - x^2) + 2gux,$$

$$\Pi_{\mathrm{v}}(\mathrm{v},z) = 3\alpha\mathrm{v}^2 + 2\beta\mathrm{v}z + \gamma z^2 = \gamma(z^2 - \mathrm{v}^2) + 2\beta\mathrm{v}z,$$

$$\Pi_z(\mathrm{v},z) = \beta\mathrm{v}^2 + 2\gamma\mathrm{v}z + 3\delta z^2 = \beta(\mathrm{v}^2 - z^2) + 2\gamma\mathrm{v}z.$$

In the expression,

$$\begin{bmatrix} \mathcal{S} \\ \mathcal{T} \end{bmatrix} = \begin{bmatrix} (2\rho - \Theta) & \sigma \\ \sigma & (2\tau - \Theta) \end{bmatrix} \begin{bmatrix} R_u \\ R_x \end{bmatrix} - \begin{bmatrix} \Pi_{\mathrm{v}} \\ \Pi_z \end{bmatrix}$$

note that the quadratic parts of

$$\begin{bmatrix} (2\rho - \Theta) & \sigma \\ \sigma & (2\tau - \Theta) \end{bmatrix} \begin{bmatrix} R_u \\ R_x \end{bmatrix}$$

are linear functions of $(x^2 - u^2)$ and ux. Further, when we replace (u,x) by $(u(\tilde{u},\tilde{x}), x(\tilde{u},\tilde{x}))$ [see (9.27) on page 355], the quadratic parts of the resulting functions $\tilde{\mathcal{S}}$, $\tilde{\mathcal{T}}$ [see (9.46)] will be linear functions of $(\tilde{x}^2 - \tilde{u}^2)$ and $\tilde{u}\tilde{x}$. Thus, since the product of the coefficients of the $\tilde{x}^2$ and $\tilde{u}^2$ terms are always nonpositive, Definition 9.3 (page 359) yields the following earlier result:

In single-plane lensing due to a potential ψ with locally constant Laplacian, the fold caustic curves are always locally convex.

We shall now show that:

In double-plane lensing, if the lenses have potentials with locally constant Laplacians, then the fold caustic curves can violate local convexity.

Rather than trying to determine the exact conditions on the coefficients which produce nonconvex fold curves, we shall content ourselves with an example.

To simplify matters as much as possible let us assume that

$$2\Theta c = \beta = 1, \qquad b = g = \gamma = \rho = \sigma = \tau = \mathtt{B} = \mathtt{B}' = \Lambda' = 0.$$

Since we are assuming that the origin is a fold critical point, the constraints (9.44), (9.45), and (9.47) must hold. Note that because $\mathtt{C} = 0$, the condition $\mathtt{AC} = \mathtt{BB}' = 0$ is automatic.

Also, the condition $\mathtt{A} \neq 0$ yields $k = |\mathtt{A}| \neq 0$. Hence, our main constraints are $\mathtt{A} \neq 0$ and (9.47). The latter is equivalent to

$$\Theta f + (1 - 2c)^2 \neq 0.$$

From Definition 9.3 (page 359), we see that for local convexity it suffices to consider the quadratic part of $\tilde{X}$, or, equivalently, of

$$-\Lambda' \mathcal{S} + \mathcal{T} = -\Theta R_x - \Pi_z.$$

Writing only the quadratic parts:

$$\begin{aligned} -\Theta R_x &= \Theta f \mathtt{A}^2(\tilde{x}^2 - \tilde{u}^2), \\ \Pi_z &= \mathrm{v}^2 - z^2 = [u(1-2a)]^2 - [x(1-2c)]^2 = [(1-2a)\mathtt{A})\tilde{u}]^2 - [\mathtt{A}\tilde{x}(1-2c)]^2. \end{aligned}$$

Thus,

$$-\Theta R_x - \Pi_z = \tilde{x}^2[\Theta f + (1 - 2c)^2]\mathtt{A}^2 + \tilde{u}^2[-\Theta f - (1 - 2a)^2]\mathtt{A}^2.$$

The convexity condition is then that the product of these coefficients be negative. *Can this product, namely,*

$$-[\Theta f + (1 - 2c)^2][\Theta f + (1 - 2a)^2],$$

be positive?

Recalling that $2c\Theta = 1$, we want to know whether the condition

$$-[\Theta f + (1 - \frac{1}{\Theta})^2][\Theta f + (1 - 2a)^2] > 0$$

can hold. This inequality can readily be satisfied by choosing $(-f)$ large enough so that $\Theta f + \left(1 - \frac{1}{\Theta}\right)^2 < 0$, that is, $f < -\frac{1}{\Theta}\left(1 - \frac{1}{\Theta}\right)^2$, and by choosing a large enough so that

$$\Theta f + (1 - 2a)^2 > 0 \quad \text{or} \quad 2a > 1 + \sqrt{|\Theta f|}.$$

Therefore, for such cases *local convexity is violated.*

We conclude this section with an illustration of a violation of local convexity by point masses on distinct lens planes with continuous matter and shear. Such a lens system can produce a teardrop caustic — see Figure 9.14 (A.P. and Wicklin [Pet-Wic95, p. 1403]). This caustic violates local convexity even though the Laplacians of the respective gravitational lens potentials are locally constant.

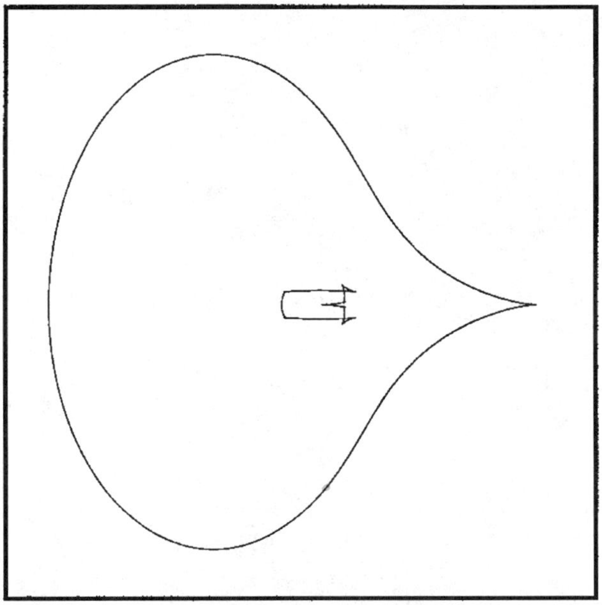

Figure 9.14. Caustic due to a double-plane two point mass gravitational lens with continuous matter and shear [Pet-Wic95, p. 1403]. The teardrop caustic violates local convexity — see Figure 9.26 (page 392) for another example.

9.4 Folds and Cusps via Directional Derivatives

We shall use directional derivatives to write down explicitly the equations for fold, cusp, and degenerate critical points. The results will be formulated to apply not just to k-plane lensing maps, but any smooth map between planes. In Section 9.5.2, we further decompose the set of degenerate critical points in order to characterize caustic metamorphoses.

Requisite Mathematics: Section 7.5.

When $f \in C^\infty(U, \mathbf{R}^2)$, where $U \subseteq \mathbf{R}^2$ is open, we shall write $f = (a, b)$ and assume rectangular coordinates (u, v) on U, unless stated to the contrary. In this case, the set of critical points of f is

$$\mathrm{Crit}(f) = \{p \in U : J(p) = 0\}, \qquad J = \det(\mathrm{Jac}\, f) = a_{\mathrm{u}}\, b_{\mathrm{v}} - a_{\mathrm{v}}\, b_{\mathrm{u}}.$$

Each element of Crit(f) is either a fold, cusp, or degenerate critical point.

The fold and cusp critical points of f are defined explicitly as follows (Section 7.5.1, page 244):

$$\mathrm{Folds}(f) = \{p \in {}_{\mathrm{T}}S_1(f) : \mathrm{rank}[d_p(f|_{{}_{\mathrm{T}}S_1(f)})] = 1\}$$

$$\mathrm{Cusps}(f) = \{x \in {}_{\mathrm{T}}S_1(f) : \mathrm{rank}[d_p(f|_{{}_{\mathrm{T}}S_1(f)})] = 0\},$$

where

$$_{\mathrm{T}}S_1(f) = \{p \in S_1(f) : j^1 f \pitchfork S_1(X, Y) \text{ at } p\}.$$

with

$$S_1(f) = \{p \in U : \mathrm{rank}(d_p f) = 1\}.$$

We must first determine the equations characterizing elements of $S_1(f)$ and $_{\mathrm{T}}S_1(f)$. The rank 1 critical points of f are determined as follows:

$$S_1(f) = \{p \in U : J(p) = 0, (\mathrm{Jac}\, f)(p) \neq \mathbf{O}\},$$

where $\mathbf{O}$ is the zero matrix. Note that $\mathrm{Jac}\, f = \begin{bmatrix} a_{\mathrm{u}} & a_{\mathrm{v}} \\ b_{\mathrm{u}} & b_{\mathrm{v}} \end{bmatrix}$ and the components of grad J are

$$\begin{aligned} J_{\mathrm{u}} &= b_{\mathrm{v}}\, a_{\mathrm{uu}} + a_{\mathrm{u}}\, b_{\mathrm{vu}} - b_{\mathrm{u}}\, a_{\mathrm{vu}} - a_{\mathrm{v}}\, b_{\mathrm{uu}} \\ J_{\mathrm{v}} &= b_{\mathrm{v}}\, a_{\mathrm{uv}} + a_{\mathrm{u}}\, b_{\mathrm{vv}} - a_{\mathrm{vv}}\, b_{\mathrm{u}} - a_{\mathrm{v}}\, b_{\mathrm{uv}}. \end{aligned} \tag{9.48}$$

It follows that if $(\mathrm{Jac}\, f)(p) = \mathbf{O}$, then $J(p) = 0$ and grad $J(p) = \mathbf{0}$. The set of transverse rank 1 critical points of f is

$$_{\mathrm{T}}S_1(f) = \{p \in S_1(f) : j^1 f \pitchfork S_1(U, \mathbf{R}^2) \text{ at } p\}.$$

The following proposition characterizes elements of $_{\mathrm{T}}S_1(f)$:

Proposition 9.6 *Assume $f \in C^\infty(U, \mathbf{R}^2)$, where $U \subseteq \mathbf{R}^2$ is open. Then*

$$_{\mathrm{T}}S_1(f) = \{p \in U : J(p) = 0, \ (\mathrm{Jac}\, f)(p) \neq \mathbf{O}, \ \mathrm{grad}\, J(p) \neq \mathbf{0}\}.$$

Proof: Let p be an element of $_{\mathrm{T}}S_1(f)$. Consider the transversality condition $j^1 f \pitchfork S_1(U, \mathbf{R}^2)$ at p. We proceed as in the proof of Lemma 7.48(2) (page 249). Using the charts λ and ν of the lemma, where $\lambda(p) = \mathbf{0}$ and $\nu(f(p) = \mathbf{0}$, it suffices to prove the proposition for

$$h = \nu \circ f \circ \lambda^{-1}$$

at $\mathbf{0} \in S_1(h)$. Note that $h(\mathbf{0}) = \mathbf{0}$. Let $\lambda(U) = U'$, $\nu(V) = V'$, and consider

$$J^1(U', \mathbf{R}^2) \approx U' \times V' \times \mathbf{R}^4.$$

The basic idea is to apply Proposition 7.7 (page 218) with $y = j^1 h(\mathbf{0})$, $X = U'$, $Y = J^1(U, \mathbf{R}^2)$, and

$$Q = S_1(U, \mathbf{R}^2) \approx U' \times V' \times S_1(2, 2).$$

Elements of $J^1(U, \mathbf{R}^2)$ are of the form

$$\sigma = (s, t(s), \frac{\partial t_1}{\partial s_1}, \frac{\partial t_1}{\partial s_2}, \frac{\partial t_2}{\partial s_1}, \frac{\partial t_2}{\partial s_2}),$$

where $s = (s_1, s_2)$ and $t = (t_1, t_2)$ is a representative of σ. Without loss of generality, assume that $\frac{\partial t_1}{\partial s_1} \neq 0$. Let $\mathcal{V} \subseteq J^1(U, \mathbf{R}^2)$ be an open set about y such that all $\sigma \in \mathcal{V}$ have $\frac{\partial t_1}{\partial s_1} \neq 0$. Define

$$\mathcal{H} : \mathcal{V} \subseteq J^1(U', \mathbf{R}^2) \to \mathbf{R}$$

by

$$\mathcal{H}(\sigma) = \frac{\partial t_1}{\partial s_1}\frac{\partial t_2}{\partial s_2} - \frac{\partial t_1}{\partial s_2}\frac{\partial t_2}{\partial s_1}.$$

Note that $\mathcal{H}(j^1 h(\mathbf{0})) = 0$ since $\mathbf{0} \in S_1(h)$. In addition, $\mathcal{H}^{-1}(0)$ equals $S_1(U', \mathbf{R}^2) \cap \mathcal{V}$, which contains $y = j^1 h(\mathbf{0})$. The composition $\mathcal{H} \circ (j^1 h) : (j^1 h)^{-1}(\mathcal{V}) \to \mathbf{R}$ is given by

$$(\mathcal{H} \circ j^1 h)(s) = \det(\mathrm{Jac}\, h).$$

Now, $\mathrm{codim}_{J^1(U', \mathbf{R}^2)} S_1(U', \mathbf{R}^2) = 1$ and $j^1 h(\mathbf{0}) \in S_1(U', \mathbf{R}^2) \cap \mathcal{V}$, so Proposition 7.7 (page 218) yields that $j^1 h \pitchfork S_1(U', \mathbf{R}^2)$ at $\mathbf{0}$ if and only if $\mathrm{rank}[d_{\mathbf{0}}(\mathcal{H} \circ j^1 h)] = 1$. But

$$d_{\mathbf{0}}(\mathcal{H} \circ j^1 h) = \mathrm{grad}[\det(\mathrm{Jac}\, h)].$$

Hence, if $\mathbf{0} \in S_1(h)$, then $\mathrm{rank}[d_{\mathbf{0}}(\mathcal{H} \circ j^1 h)] = 1$ if and only if $\widetilde{\nabla}[\det(\mathrm{Jac}\, h)] \neq \mathbf{0}$, where $\widetilde{\nabla}$ is the gradient operator relative to the coordinate system λ.

Finally, we have $(\mathrm{Jac}\, h)(\mathbf{0}) \neq \mathbf{O} \iff (\mathrm{Jac}\, f)(p) \neq \mathbf{O}$, and $\det(\mathrm{Jac}\, h)(\mathbf{0}) = 0 \iff \det(\mathrm{Jac}\, f)(p) = 0$. Consequently, $\mathbf{0} \in S_1(h) \iff p \in S_1(f)$. If $p \in S_1(f)$ (i.e., $\det(\mathrm{Jac}\, f)(p) = 0$ and $(\mathrm{Jac}\, f)(p) \neq \mathbf{O}$), then $\widetilde{\nabla}[\det(\mathrm{Jac}\, h)](\mathbf{0}) \neq \mathbf{0} \iff \nabla[\det(\mathrm{Jac}\, f)](p) \neq \mathbf{0}$, where ∇ is the gradient operator relative to the usual rectangular coordinates on U. Thus, if $p \in S_1(f)$, then $j^1 f \pitchfork S_1(U, \mathbf{R}^2)$ at p if and only if $\nabla[\det(\mathrm{Jac}\, f)](p) \neq \mathbf{0}$. □

With Proposition 9.6 available, we now turn to the equations for fold, cusp, and degenerate critical points . Suppose that ${}_{\mathrm{T}}S_1(f) \neq \emptyset$. Then ${}_{\mathrm{T}}S_1(f)$ is (up to diffeomorphism) a union of disjoint circles and/or real lines (see page 245). The vector field

$$Z = -J_{\mathrm{v}} \frac{\partial}{\partial \mathrm{u}} + J_{\mathrm{u}} \frac{\partial}{\partial \mathrm{v}}$$

is perpendicular to $\mathrm{grad}\, J$ at all points of U. Consequently, the field Z is a nonvanishing tangent vector field on ${}_{\mathrm{T}}S_1(f)$. Now, for each p on a component curve c of $S_1^*(f)$, there is a locally unique solution[4] φ of

$$\frac{d\varphi}{dt}(t) = Z(\varphi(t) \neq \mathbf{0} \qquad (9.49)$$

[4]See Abraham and Marsden [Ab-M, p. 62].

with $\varphi(0) = p$. Combining each such solutions at all $p \in c$ gives a parametrization ϕ of c satisfying (9.49) with $\phi(0) = p$. The chain rule yields the velocity and acceleration of $f \circ \phi$ in terms of directional derivatives relative to Z:

$$\frac{d(f \circ \phi)}{dt}(0) = \nabla_Z f(p), \qquad \frac{d^2(f \circ \phi)}{dt^2}(0) = \nabla_Z(\nabla_Z f(p).$$

Recall that the directional derivative is defined as follows:

$$\nabla_Z f(p) = ((\nabla_Z a)(p), (\nabla_Z b)(p))$$

and

$$\nabla_Z^{(2)} f(p) \equiv \nabla_Z(\nabla_Z f)(p) = ((\nabla_Z(\nabla_Z a))(p), (\nabla_Z(\nabla_Z b))(p)),$$

where $(\nabla_Z a)(p) = Z_p \cdot [(\operatorname{grad} a)(p)]$. Explicitly,

$$\nabla_Z f(p) = \begin{bmatrix} -J_{\mathrm{V}}(p)\, a_{\mathrm{U}}(p) + J_{\mathrm{U}}(p)\, a_{\mathrm{V}}(p), \\ -J_{\mathrm{V}}(p)\, b_{\mathrm{U}}(p) + J_{\mathrm{U}}\, b_{\mathrm{V}}(p)) \end{bmatrix} \tag{9.50}$$

and

$$\nabla_Z^{(2)} f(p) = \begin{bmatrix} -J_{\mathrm{V}}(p)(-J_{\mathrm{V}} a_{\mathrm{U}} + J_{\mathrm{U}} a_{\mathrm{V}})_{\mathrm{U}}\,(p) + J_{\mathrm{U}}(p)(-J_{\mathrm{V}} a_{\mathrm{U}} + J_{\mathrm{U}} a_{\mathrm{V}})_{\mathrm{V}}\,(p) \\ -J_{\mathrm{V}}(p)\,(-J_{\mathrm{V}} b_{\mathrm{U}} + J_{\mathrm{U}} b_{\mathrm{V}})_{\mathrm{U}}\,(p) + J_{\mathrm{U}}(p)\,(-J_{\mathrm{V}} b_{\mathrm{U}} + J_{\mathrm{U}} b_{\mathrm{V}})_{\mathrm{V}}\,(p) \end{bmatrix}.$$

Theorem 9.7 *Suppose $f \in C^\infty(U, \mathbf{R}^2)$, where $U \subseteq \mathbf{R}^2$ is open. If p is a point of U, then*

(1) *p is a fold critical point of f if and only if $p \in {}_{\mathrm{T}}S_1(f)$ and $\nabla_Z f(p) \neq \mathbf{0}$.*

(3) *p is a cusp critical point of f if and only if $p \in {}_{\mathrm{T}}S_1(f)$, $\nabla_Z f(p) = \mathbf{0}$, and $\nabla_Z^{(2)} f(p) \neq \mathbf{0}$.*

(3) *p is a degenerate critical point of f if and only if $p \in {}_{\mathrm{T}}S_1(f)$, $\nabla_Z f(p) = \nabla_Z^{(2)} f(p) = \mathbf{0}$.*

The directional derivative conditions in Theorem 9.7 express how Whitney originally defined folds and cusps [Whi55, pp. 378, 379]. A detailed treatment of folds and cusps using directional derivatives and following the development in Whitney's paper is given by Lu [Lu, pp. 33–51].

Theorem 9.7 shows that if p is a fold critical point of a locally stable map f, then the velocity vector $\mathbf{v} = d(f \circ \phi)/dt$ of the caustic curve c traced out by $f \circ \phi$ is nonzero at $f(p)$, where $p = \phi(0)$. However, if p is a cusp critical point, then the velocity vanishes at $f(p)$, but the acceleration $d^2(f \circ \phi)/dt^2$ is nonzero. In particular, the speed $|\mathbf{v}|$ becomes positive at a nonzero rate as one moves away from the cusp along the caustic.

By Theorem 9.7, we see that equations (9.48), (9.50), and (9.51) yield: *Fold critical points of $f = (a, b)$ are determined by partials derivatives of a and b up to order two,* *while cusps require partials up to order three – no higher!* This dependence of f on its Taylor polynomial of order 3 is reflective of Theorem 7.25 (page 229).

Proof of Theorem 9.7: Without loss of generality, suppose that we are given a smooth map $F : U \to \mathbf{R}^2$, where $U \subseteq \mathbf{R}^2$ is open, defined by $F(\mathrm{u}, \mathrm{v}) = (A(\mathrm{u}, \mathrm{v}), B(\mathrm{u}, \mathrm{v}))$ and such that $F(\mathbf{0}) = \mathbf{0}$ and $\mathrm{rank}(\mathrm{Jac}\, F)(\mathbf{0}) \neq 0$. At least one of the partials of A and B does not vanish at 0. Assume that $A_{\mathrm{u}}(\mathbf{0}) \neq 0$. Then in a neighborhood of $\mathbf{0}$, the map $\phi(\mathrm{u}, \mathrm{v}) = (A(\mathrm{u}, \mathrm{v}), \mathrm{v})$ is a diffeomorphism. In addition, $B(\mathrm{u}, \mathrm{v}) = (B \circ \phi^{-1} \circ \phi)(\mathrm{u}, \mathrm{v}) = (B \circ \phi^{-1})(A(\mathrm{u}, \mathrm{v}), \mathrm{v})$. If we let $C = B \circ \phi^{-1}$, then F has the form $F(\mathrm{u}, \mathrm{v}) = (A(\mathrm{u}, \mathrm{v}), C(A(\mathrm{u}, \mathrm{v}), \mathrm{v}))$. Our choice that $A_{\mathrm{u}}(\mathbf{0}) \neq 0$ is not important; any other nonzero first partial of A or B would work. If we introduce new coordinates

$$\begin{aligned} \mathrm{u}' &= A(\mathrm{u}, \mathrm{v}) \\ \mathrm{v}' &= \mathrm{v}, \end{aligned}$$

then the map F becomes $F'(\mathrm{u}', \mathrm{v}') = (\mathrm{u}', C(\mathrm{u}', \mathrm{v}'))$.

Applying Lemma 7.48 (page 247) to F (with $h = F'$), we obtain the following conditions for folds and cusps critical points of F:

(1) $\mathbf{0} \in S_1(F) \iff C_{\mathrm{v}'}(\mathbf{0}) = 0.$

(2) $\mathbf{0} \in {}_{\top}S_1(F) \iff C_{\mathrm{v}'}(\mathbf{0}) = 0,\ [C_{\mathrm{v}'\mathrm{u}'}(\mathbf{0}), C_{\mathrm{v}'\mathrm{v}'}(\mathbf{0})] \neq \mathbf{0}.$

(3) **Fold:** $\mathbf{0} \in S_{1,0}(F) \iff C_{\mathrm{v}'}(\mathbf{0}) = 0$ and $C_{\mathrm{v}'\mathrm{v}'}(\mathbf{0}) \neq 0.$

(4) $\mathbf{0} \in S_{1,1}(F) \iff C_{\mathrm{v}'}(\mathbf{0}) = C_{\mathrm{v}'\mathrm{v}'}(\mathbf{0}) = 0$ and $C_{\mathrm{v}'\mathrm{u}'}(\mathbf{0}) \neq 0.$

(5) **Cusp:** $\mathbf{0} \in {}_{\top}S_{1,1}(F) \iff C_{\mathrm{v}'}(\mathbf{0}) = 0 = C_{\mathrm{v}'\mathrm{v}'}(\mathbf{0})),\ C_{\mathrm{v}'\mathrm{u}'}(\mathbf{0}) \neq 0$, and $C_{\mathrm{v}'\mathrm{v}'\mathrm{v}'}(\mathbf{0}) \neq 0.$

Recall that

$$\begin{aligned} S_{1,0}(F) &= \{p \in {}_{\top}S_1(F) : \mathrm{rank}\left[d_p\left(F|_{{}_{\top}S_1(F)}\right)\right] = 1\} \\ S_{1,1}(F) &= \{p \in {}_{\top}S_1(F) : \mathrm{rank}\left[d_p\left(F|_{{}_{\top}S_1(F)}\right)\right] = 0\} \\ {}_{\top}S_{1,1}(F) &= \{p \in S_{1,1}(F) : j^2 f \pitchfork S_{1,1}(U, \mathbf{R}^2) \text{ at } p\}. \end{aligned}$$

The rest of the proof is now concerned with showing that the above conditions convert into constraints on the directional derivatives of F.

Letting $H(\mathrm{u}', \mathrm{v}') = (\mathrm{u}', C(\mathrm{u}', \mathrm{v}'))$, we see that $F = H \circ \phi$ and the Jacobian matrix of F is

$$\begin{bmatrix} 1 & 0 \\ C_{\mathrm{u}'} & C_{\mathrm{v}'} \end{bmatrix} \begin{bmatrix} A_{\mathrm{u}} & A_{\mathrm{v}} \\ 0 & 1 \end{bmatrix} = \begin{bmatrix} A_{\mathrm{u}} & A_{\mathrm{v}} \\ C_{\mathrm{u}'} A_{\mathrm{u}} & C_{\mathrm{u}'} A_{\mathrm{v}} + C_{\mathrm{v}'} \end{bmatrix}$$

with determinant

$$J(\mathrm{u}, \mathrm{v}) = C_{\mathrm{v}'}(A(\mathrm{u}, \mathrm{v}), \mathrm{v})\, A_{\mathrm{u}}(\mathrm{u}, \mathrm{v}).$$

Thus, the equation of $S_1(F)$ is $J = 0$, which yields $C_{v'}(A(u, v), v) = 0$ (since by assumption $A_u \neq 0$ near $\mathbf{0}$). In particular, $\mathbf{0} \in S_1(F)$.

Hereafter we shall suppress the arguments for C, *writing* C *rather than* $C(A(u, v), v)$ *and* $C_{v'}$ *rather than* $C_{v'}(A(u, v), v)$, *etc.*

The gradient of J is the vector field

$$\begin{aligned} \text{grad } J &= J_u \frac{\partial}{\partial u} + J_v \frac{\partial}{\partial v} \\ &= \left(C_{v'u'} A_u^2 + C_{v'} A_{uu}\right) \frac{\partial}{\partial u} \\ &\qquad + \left(C_{v'u'} A_v A_u + C_{v'v'} A_u + C_{v'} A_{uv}\right) \frac{\partial}{\partial v}. \end{aligned}$$

The following vector field is orthogonal to grad J:

$$\begin{aligned} Z &= -J_v \frac{\partial}{\partial u} + J_u \frac{\partial}{\partial v} \\ &= -\left(C_{v'u'} A_v A_u + C_{v'v'} A_u + C_{v'} A_{uv}\right) \frac{\partial}{\partial u} \\ &\qquad + \left(C_{v'u'} A_u^2 + C_{v'} A_{uu}\right) \frac{\partial}{\partial v}. \end{aligned} \tag{9.51}$$

On $S_1(f)$, we have

$$\begin{aligned} \text{grad } J &= \left(C_{v'u'} A_u^2\right) \frac{\partial}{\partial u} + \left(C_{v'u'} A_v A_u + C_{v'v'} A_u\right) \frac{\partial}{\partial v} \\ &= A_u \left(\left(C_{v'u'} A_u\right) \frac{\partial}{\partial u} + \left(C_{v'u'} A_v + C_{v'v'}\right) \frac{\partial}{\partial v} \right). \end{aligned}$$

Since $A_u \neq 0$, the vector field grad J is nonzero along $S_1(F)$ if and only if at least one of $C_{v'u'}$ or $C_{v'v'}$ is nonzero on $S_1(F)$ near $\mathbf{0}$. The restriction of Z to $S_1(f)$ is

$$Z|_{S_1(F)} = A_u \left(-\left(C_{v'u'} A_v + C_{v'v'}\right) \frac{\partial}{\partial u} + \left(C_{v'u'} A_u\right) \frac{\partial}{\partial v} \right).$$

Hence, the condition $Z|_{S_1(F)} \neq \mathbf{0}$ is equivalent to at least one of $C_{v'u'}$ or $C_{v'v'}$ being nonzero. In others words, the nonvanishing of $Z|_{S_1(F)}$ characterizes elements of ${}_{\top}S_1(F)$. Thus, $\mathbf{0} \in {}_{\top}S_1(F)$ if and only if $J(\mathbf{0}) = 0$ and grad $J(\mathbf{0}) \neq \mathbf{0}$.

We now compute the directional derivatives $\nabla_Z F = dF(Z)$ and $\nabla_Z^{(2)} F = d(dF(Z))(Z)$ on ${}_{\top}S_1(F)$. First,

$$\begin{aligned} dF(Z) &= -\left(C_{v'u'} A_v A_u + C_{v'v'} A_u + C_{v'} A_{uv}\right) dF\left(\frac{\partial}{\partial u}\right) \\ &\qquad + \left(C_{v'u'} A_u^2 + C_{v'} A_{uu}\right) dF\left(\frac{\partial}{\partial v}\right). \end{aligned}$$

Choosing rectangular coordinates (U, V) on the target of F, we get

$$dF\left(\frac{\partial}{\partial \mathrm{u}}\right) = A_{\mathrm{u}}\left(\frac{\partial}{\partial \mathrm{U}} + C_{\mathrm{u}'}\frac{\partial}{\partial \mathrm{V}}\right).$$

and

$$dF\left(\frac{\partial}{\partial \mathrm{v}}\right) = A_{\mathrm{V}}\frac{\partial}{\partial \mathrm{U}} + (C_{\mathrm{u}'}A_{\mathrm{V}} + C_{\mathrm{V}'})\frac{\partial}{\partial \mathrm{V}}.$$

Hence

$$\begin{aligned} dF(Z) &= \left[-(C_{\mathrm{V}'\mathrm{V}'}A_{\mathrm{u}}^2 + C_{\mathrm{V}'}A_{\mathrm{uv}}A_{\mathrm{u}}) + C_{\mathrm{V}'}A_{\mathrm{uu}}A_{\mathrm{V}})\right]\frac{\partial}{\partial \mathrm{U}} \\ &\quad + [-(C_{\mathrm{V}'\mathrm{V}'}A_{\mathrm{u}} + C_{\mathrm{V}'}A_{\mathrm{uv}})(C_{\mathrm{u}'}A_{\mathrm{u}}) + (C_{\mathrm{V}'}A_{\mathrm{uu}})(C_{\mathrm{u}'}A_{\mathrm{V}} + C_{\mathrm{V}'}) \\ &\qquad + C_{\mathrm{V}'\mathrm{u}'}C_{\mathrm{V}'}A_{\mathrm{u}}^2]\frac{\partial}{\partial \mathrm{V}}. \end{aligned} \tag{9.52}$$

Restricting to ${}_{\mathrm{T}}S_1(F)$, we obtain

$$[dF(Z)]_{S_1(F)} = -A_{\mathrm{u}}^2 C_{\mathrm{V}'\mathrm{V}'}\left(\frac{\partial}{\partial \mathrm{U}} + C_{\mathrm{u}'}\frac{\partial}{\partial \mathrm{V}}\right).$$

The point $\mathbf{0}$ lies in $S_{1,0}(F)$ if and only if $C_{\mathrm{V}'}(\mathbf{0}) = 0$ and $C_{\mathrm{V}'\mathrm{V}'}(\mathbf{0}) \neq 0$, which is equivalent to $[dF(Z)]|_{S_1(F)}$ not vanising at $\mathbf{0}$ (i.e., $[dF(Z)](\mathbf{0}) \neq \mathbf{0}$).

The condition $\mathbf{0} \in S_{1,1}(F)$ is equivalent to $C_{\mathrm{V}'}(\mathbf{0}) = C_{\mathrm{V}'\mathrm{V}'}(\mathbf{0}) = 0$, which is equivalent to $dF(Z)|_{S_1(F)}$ vanishing at $\mathbf{0}$. We now proceed to the transversality condition to guarantee that $\mathbf{0} \in {}_{\mathrm{T}}S_{1,1}$. Let $G = dF(Z) : U \to \mathbf{R}^2$. Then $dG = d(dF(Z)) : TU \to T(T\mathbf{R}^2))$. Since $Z : U \to \mathbf{R}^2$, we have $d(dF(Z))(Z) = dG(Z) : U \to T(T\mathbf{R}^2)$. By (9.51), we obtain

$$\begin{aligned} dG(Z) &= -(C_{\mathrm{V}'\mathrm{u}'}A_{\mathrm{V}}A_{\mathrm{u}} + C_{\mathrm{V}'\mathrm{V}'}A_{\mathrm{u}} + C_{\mathrm{V}'}A_{\mathrm{uv}})\; dG\left(\frac{\partial}{\partial \mathrm{u}}\right) \\ &\qquad + \left(C_{\mathrm{V}'\mathrm{u}'}A_{\mathrm{u}}^2 + C_{\mathrm{V}'}A_{\mathrm{uu}}\right) dG\left(\frac{\partial}{\partial \mathrm{v}}\right). \end{aligned}$$

Since we are working at $\mathbf{0} \in S_{1,1}(F)$, we know that $C_{\mathrm{V}'}$ and $C_{\mathrm{V}'\mathrm{V}'}$ vanish. Hence, at $\mathbf{0} \in S_{1,1}(F)$ we have

$$dG(Z) = A_{\mathrm{u}}C_{\mathrm{V}'\mathrm{u}'}\left[-A_{\mathrm{V}}\, dG\left(\frac{\partial}{\partial \mathrm{u}}\right) + A_{\mathrm{u}}\, dG\left(\frac{\partial}{\partial \mathrm{v}}\right)\right]. \tag{9.53}$$

We now compute $dG\left(\frac{\partial}{\partial \mathrm{u}}\right)$ and $dG\left(\frac{\partial}{\partial \mathrm{v}}\right)$. The only possible nonzero part of $dG(Z)$ is tangent to the space spanned by $\frac{\partial}{\partial \mathrm{U}}$ and $\frac{\partial}{\partial \mathrm{V}}$. We denote the coordinates of this subspace of

$T\mathbf{R}^2$ by U and V, and the natural basis of the tangent space to this subspace by $\frac{\partial}{\partial \mathsf{U}}$ and $\frac{\partial}{\partial \mathsf{V}}$. Write

$$G = \mathsf{R}\frac{\partial}{\partial \mathsf{U}} + \mathsf{S}\frac{\partial}{\partial \mathsf{V}}.$$

By (9.52), we have

$$\begin{aligned} \mathsf{R} &= -(C_{\mathrm{v'v'}} A_{\mathrm{u}}^2 + C_{\mathrm{v'}} A_{\mathrm{uv}} A_{\mathrm{u}}) + C_{\mathrm{v'}} A_{\mathrm{uu}} A_{\mathrm{v}}) \\ \mathsf{S} &= -(C_{\mathrm{v'v'}} A_{\mathrm{u}} + C_{\mathrm{v'}} A_{\mathrm{uv}})(C_{\mathrm{u'}} A_{\mathrm{u}}) + (C_{\mathrm{v'}} A_{\mathrm{uu}})(C_{\mathrm{u'}} A_{\mathrm{v}} + C_{\mathrm{v'}}) \\ &\qquad + C_{\mathrm{v'u'}} C_{\mathrm{v'}} A_{\mathrm{u}}^2. \end{aligned}$$

Now,

$$dG\left(\frac{\partial}{\partial u}\right) = \mathsf{R}_{\mathrm{u}}\frac{\partial}{\partial \mathsf{U}} + \mathsf{S}_{\mathrm{u}}\frac{\partial}{\partial \mathsf{V}}, \qquad dG\left(\frac{\partial}{\partial v}\right) = \mathsf{R}_{\mathrm{v}}\frac{\partial}{\partial \mathsf{U}} + \mathsf{S}_{\mathrm{v}}\frac{\partial}{\partial \mathsf{V}}.$$

Also, $\mathbf{0} \in S_{1,1}(F)$ if and only if $C_{\mathrm{v'}}(\mathbf{0}) = C_{\mathrm{v'v'}}(\mathbf{0}) = 0$ and $C_{\mathrm{v'u'}}(\mathbf{0}) \neq 0$. Hence, at the origin we obtain

$$\begin{aligned} \mathsf{R}_{\mathrm{u}} &= -(C_{\mathrm{v'v'u'}} A_{\mathrm{u}}^3 + C_{\mathrm{v'u'}} A_{\mathrm{uv}} A_{\mathrm{u}}^2) + C_{\mathrm{v'u'}} A_{\mathrm{u}} A_{\mathrm{uu}} A_{\mathrm{v}}), \\ \mathsf{S}_{\mathrm{u}} &= -(C_{\mathrm{v'v'u'}} A_{\mathrm{u}}^2 + C_{\mathrm{v'u'}} A_{\mathrm{u}} A_{\mathrm{uv}})(C_{\mathrm{u'}} A_{\mathrm{u}}) + (C_{\mathrm{u'v'}} A_{\mathrm{u}} A_{\mathrm{uu}})(C_{\mathrm{u'}} A_{\mathrm{v}}) \\ &\qquad + C_{\mathrm{v'u'}} C_{\mathrm{v'u'}} A_{\mathrm{u}}^3, \end{aligned}$$

and

$$\begin{aligned} \mathsf{R}_{\mathrm{v}} &= -((C_{\mathrm{v'v'u'}} A_{\mathrm{v}} + C_{\mathrm{v'v'v'}}) A_{\mathrm{u}}^2 + C_{\mathrm{v'u'}} A_{\mathrm{v}} A_{\mathrm{uv}} A_{\mathrm{u}}) + C_{\mathrm{v'u'}} A_{\mathrm{v'}} A_{\mathrm{uu}} A_{\mathrm{v}}, \\ \mathsf{S}_{\mathrm{v}} &= -(C_{\mathrm{v'v'u'}} A_{\mathrm{v}} A_{\mathrm{u}} + C_{\mathrm{v'v'v'}} A_{\mathrm{u}} + C_{\mathrm{v'u'}} A_{\mathrm{v}} A_{\mathrm{uv}})(C_{\mathrm{u'}} A_{\mathrm{u}}) \\ &\quad + (C_{\mathrm{v'u'}} A_{\mathrm{v}} A_{\mathrm{uu}})(C_{\mathrm{u'}} A_{\mathrm{v}}) + C_{\mathrm{v'u'}}^2 A_{\mathrm{v}} A_{\mathrm{u}}^2. \end{aligned}$$

By (9.53), the following holds at the origin:

$$\begin{aligned} \frac{dG(Z)}{A_{\mathrm{u}} C_{\mathrm{v'u'}}} &= (-A_{\mathrm{v}} R_{\mathrm{u}} + A_{\mathrm{u}} R_{\mathrm{v}})\frac{\partial}{\partial \mathsf{U}} + (-A_{\mathrm{v}} S_{\mathrm{u}} + A_{\mathrm{u}} S_{\mathrm{v}})\frac{\partial}{\partial \mathsf{V}} \\ &= -A_{\mathrm{u}}^3\, C_{\mathrm{v'v'v'}}\left(\frac{\partial}{\partial \mathsf{U}} + C_{\mathrm{u'}}\frac{\partial}{\partial \mathsf{V}}\right). \end{aligned}$$

It follows that $d(dF(Z))(Z) = dG(Z) \neq \mathbf{0}$ if and only if $C_{\mathrm{v'v'v'}}(\mathbf{0}) \neq 0$. Therefore, we have $\mathbf{0} \in {}_{\mathrm{T}}S_{1,1}(F)$ if and only if $J(\mathbf{0}) = 0$, grad $J(\mathbf{0})$, $(\nabla_Z F)(\mathbf{0}) = [dF(Z)](\mathbf{0}) = \mathbf{0}$, and $(\nabla_Z^{(2)} F)(\mathbf{0}) = [d(dF(Z))(Z)](\mathbf{0}) \neq \mathbf{0}$.

This completes the proof of Theorem 9.7. □

9.5 Caustic Metamorphoses in Lensing

9.5.1 Local Planar Caustic Metamorphoses

This section applies Arnold's singularity theory (Section 7.7) to the study of local caustic metamorphoses in gravitational lensing. Our presentation will follow the application given by A.P. in [Pet93, p. 3576–3580].

List of Generic Local Planar Caustic Metamorphoses

Time delay families $T_t(\mathbf{x}_1, \ldots, \mathbf{x}_k, \mathbf{y})$ depending on an extra physical parameter λ are often considered in gravitational lensing. For example, the parameter t may be external shear, core radius, redshift, etc. Varying a parameter t of a lensing map causes caustic curves c_t in the light source plane S to evolve with t. Each evolving caustic traces out a surface, called a *big caustic*, in the space $S \times \mathbf{R}$, where $\mathbf{R} = \{t\}$. Caustic evolutions are then obtained by slicing a big caustic with level planes, t = constant. The corresponding paths traced out by the cusps as t varies are called *bicaustics*. Since the parameter λ of $T_t(\mathbf{x}_1, \ldots, \mathbf{x}_k, \mathbf{y})$ is physically different from the two parameters corresponding to the light source position $\mathbf{y}$, we require that two caustic evolutions in $S \times \mathbf{R}$ are equivalent if there is a diffeomorphism of $S \times \mathbf{R}$ that maps the big caustic due to one evolution into the big caustic due to the other, and preserves the partitioning of $S \times \mathbf{R}$ into level surfaces t = constant. This is exactly the notion of equivalence used to classify 1-parameter families of Lagrangian caustics in Arnold's singularity theory (Section 7.7.2, page 285). Moreover, since the concept of Lagrangian caustics coincides with that of caustics in gravitational lensing (Theorem 7.81, page 282), Arnold's theory applies.

Now, Theorem 7.82 (page 285) of Arnold's singularity theory essentially states that *there are generically five types of local* 1*-parameter metamorphoses of planar caustics due to Lagrangian maps* $L_t \to \mathbf{R}^2$. Figure 9.15 presents Arnold's depiction of these five metamorphoses, including the big caustics and bicaustics, for the parameter t. They correspond to *lips* $A_3(+)$, *beak-to-beaks* $A_3(-)$, *swallowtails* A_4, *elliptic umbilics* D_4^-, and *hyperbolic umbilics* D_4^+. It follows that if T_t generates a stable simple Lagrangian map $L_{T_t} \to \mathbf{R}^3$, where $\mathbf{R}^3 = S \times \mathbf{R}$, then slicing the big caustic C of this Lagrangian map by the level planes, t = constant, of a nondegenerate time function yields a 1-parameter family of caustics of T that is locally equivalent to one of the types in Theorem 7.82.

Remark: It would take us too far afield to attempt an analysis along the lines of Section 8.4 (page 310) to prove that there is an open dense set $\mathfrak{T}$ in the space of 1-parameter families of k-plane time delay families such that the local metamorphoses due to each $T_t \in \mathfrak{T}$ is equivalent to one of the five basic types of planar metamorphoses cited above. We shall content ourselves with numerically illustrating that these five metamorphoses occur in gravitational lensing — see Section 9.5.3.

In Figure 9.15, at $t < 0$ there is no caustic for the lips metamorphosis $A_3(+)$. A point caustic appears at $t = 0$. The point grows into lips of length $t^{1/2}$ and width $t^{3/2}$ as t

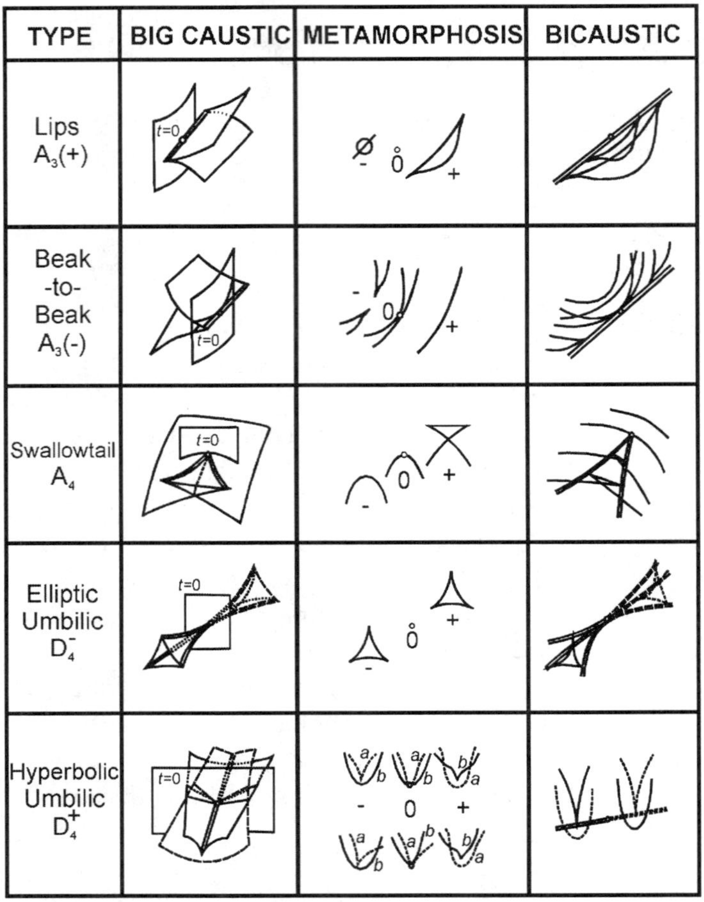

Figure 9.15. Big caustics, caustics metamorphoses and bicaustics due to Arnold [Arn86, p. 236]. The middle column shows the five types of generic local 1-parameter metamorphoses of planar caustics. The evolutions are from negative to positive t and vice-versa. Note that the point "o" should *not* be omitted; it is part of the caustics in each panel. The left and right columns show the big caustics (i.e., caustics in "spacetime" $\{t\} \times S$) and bicaustics (i.e., paths traced by cusps during an evolution). The A, D (and E) notation was introduced by Arnold because of links to the Coxeter–Dynkin diagrams of simple Lie algebras (see [Arn73]).

becomes positive [Arn86, p. 235]. The rate of increase lengthwise along the lips' bicaustic is $v = t^{-1/2}$, which is unbounded at $t = 0$. In other words, the lips evolves at infinite speed in the transition from a point into finite size.

For the beak-to-beak metamorphosis $A_3(-)$, the two cusps are at positions $\pm(-t)^{1/2}$ on the bicaustic when $t < 0$ [Arn86, p. 235]. The two cusps move towards each other with speed $v = (-t)^{-1/2}$, which becomes unbounded as $t \to 0^-$. As t becomes positive, the two cusps disappear, leaving behind two fold arcs. Also, note that the hyperbolic umbilic D_4^+ has two evolutions, each of which exhibits an exchange of a cusp between the solid and dashed caustic curves.

The lips and beak-to-beak metamorphoses arise from slices through cusp-ridge surfaces (cf. [Broc-L, p. 168]). In addition, the big caustics for the swallowtail, elliptic umbilic, and hyperbolic umbilic are the bifurcation sets of the corresponding elementary catastrophes (see Section 7.6.2).

The two D_4^+ evolutions result because for a big caustic D_4^+ along the t axis, the slices by the light source plane producing the metamorphoses are not required to occur exactly perpendicular to the t-axis. Such perturbations can happen, for example, if the t axis is the line of sight to a deflector, earth's revolution around the sun creates small perturbations about such a direction. Further treatments of the previous bifurcation sets and metamorphoses appear, respectively, in [Maj, pp. 89–102], [Gil, Chap. 5] and [Arn91, pp. 31–37], [Schn-EF, pp. 198–211].

Caustic evolutions due to big caustic crossings: Transversal intersections of big caustics can lead to additional evolutions of caustics, resulting from the "global" aspects of a caustic network. In this case, crossings by "independent" parts of caustics can occur. For example, Figure 9.16 shows three transversal crossings of big caustics. Slicing the surfaces by the planes, t = constant, produces the evolutions in the right column. Notice the occurrences of planar Reidemeister moves of types II (fold-penetrating-a-fold) and III (triple-fold-crossing) — cf. Adams [Ada, p. 12].

Examples: Butterfly and Parabolic Umbilic Evolutions

The importance of the evolutions in Figures 9.15 and 9.16 is that typical 1-parameter evolutions of planar caustics exhibit these basic metamorphoses. We shall illustrate this by taking slices of the bifurcation sets of the butterfly and parabolic umbilic unfoldings.

By the table of universal unfoldings on pages 273, the *butterfly unfolding* can be expressed as

$$F_b(x, u, v, w, t) = x^6 + tx^4 + ux^3 + vx^2 + wx,$$

where x is the variable and u, v, w, t are the parameters. Holding (t, u) fixed and slicing the bifurcation set of F_b with the (v, w)-plane yields a curve. Figure 9.17 shows the resulting evolution for parameter values on the unit circle in the (t, u)-plane, where the circle is traversed counterclockwise starting at $(1, 0)$. Observe the occurrences of swallowtail (3–5, 11–13) and cusp-penetrating-a-fold (5–7, 9–11) evolutions. See Bröcker and Lander [Broc-L, p. 163] and Majthay [Maj, pp. 103–111] for more on the butterfly catastrophe.

The *parabolic unfolding* is (see the table on page 273):

$$F_{pu}(x, u, v, w, t) = x^6 + tx^4 + ux^3 + vx^2 + wx.$$

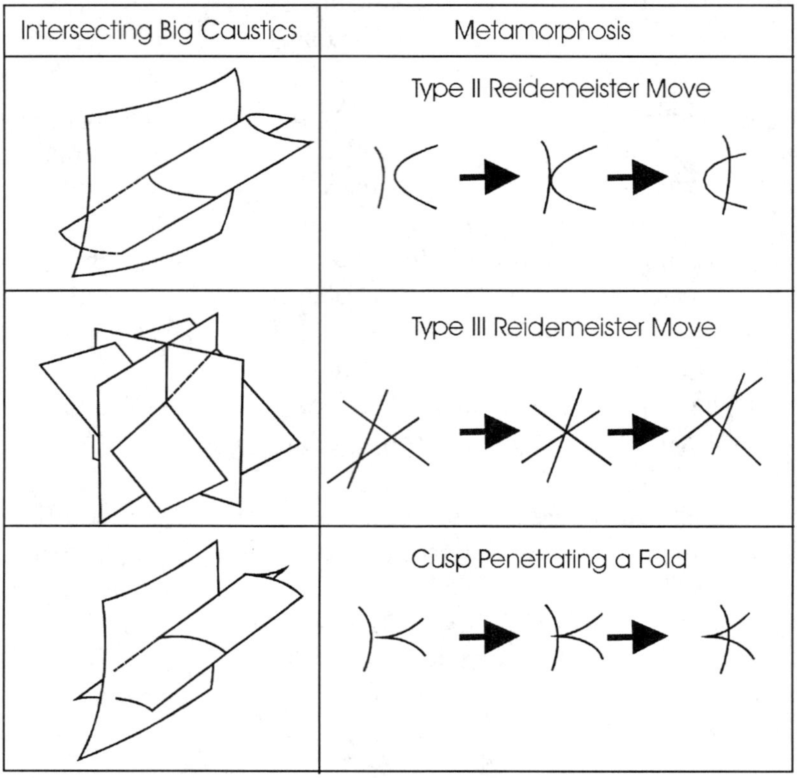

Figure 9.16. Evolutions of planar caustics induced by slices of transversal crossings of big caustics; see Figure 9.26, page 392. Arrows are reversible.

Figure 9.18 shows the corresponding evolution for parameter values on the unit circle in the (t, w)-plane with slices by the (u, v)-plane. The circle is followed counterclockwise starting from (1, 0). Notice the occurrence of lips (18-20), beak-to-beaks (3-5), swallowtails (2-4), elliptic umbilic (7-9), hyperbolic umbilic (12-14), and cusp-penetrating-a-fold (10-11, 14-17) evolutions. Readers are referred to Bröcker and Lander [Broc-L, pp. 162–167] and Majthay [Maj, pp. 111–125] for more.

9.5.2 Equations for Caustics and Their Metamorphoses

Requisite Mathematics: Section 9.4.

Let $\boldsymbol{\eta} : P \to S$ be a k-plane lensing map, which we express as $\boldsymbol{\eta}(\mathrm{u}, \mathrm{v}) = (a(\mathrm{u}, \mathrm{v}), b(\mathrm{u}, \mathrm{v}))$ relative to rectangular coordinates (u, v) on P. The set of critical points of $\boldsymbol{\eta}$ is

$$\mathrm{Crit}(\boldsymbol{\eta}) = \{\mathbf{x} \in P : J(\mathbf{x}) = 0\},$$

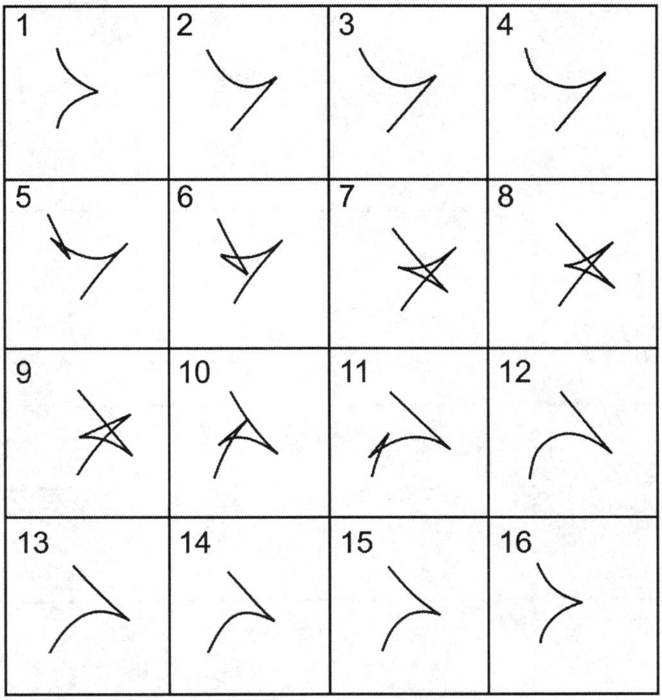

Figure 9.17. A caustic evolution due to slices through the butterfly bifurcation set.

where

$$J = \det(\mathrm{Jac}\,\boldsymbol{\eta}).$$

Each element of Crit($\boldsymbol{\eta}$) is a fold, cusp, or degenerate critical point (Section 7.5.1, page 244):

$$\mathrm{Crit}(f) = \mathrm{Folds}(f) \dot{\bigcup} \mathrm{Cusps} \dot{\bigcup} \mathrm{Degen}(f),$$

where Degen(f) is the set of degenerate critical points of f. Theorem 9.7 (page 370) and Proposition 9.6 (page 368) yield

$$\begin{aligned}
\mathrm{Folds}(\boldsymbol{\eta}) &= \{\mathbf{x} \in P : J(\mathbf{x}) = 0,\ (\mathrm{Jac}\,\boldsymbol{\eta})(\mathbf{x}) \neq \mathbf{O},\ (\mathrm{grad}\,J)(\mathbf{x}) \neq \mathbf{0}, \\
&\qquad (\nabla_Z \boldsymbol{\eta})(\mathbf{x}) \neq \mathbf{0}\} \\
\mathrm{Cusps}(\boldsymbol{\eta}) &= \{\mathbf{x} \in P : J(\mathbf{x}) = 0,\ (\mathrm{Jac}\,\boldsymbol{\eta})(\mathbf{x}) \neq \mathbf{O},\ (\mathrm{grad}\,J)(\mathbf{x}) \neq \mathbf{0}, \\
&\qquad (\nabla_Z \boldsymbol{\eta})(\mathbf{x}) = \mathbf{0},\ (\nabla_Z^{(2)} \boldsymbol{\eta})(\mathbf{x}) \neq \mathbf{0}\Big\} \\
\mathrm{Degen}(\boldsymbol{\eta}) &= \{\mathbf{x} \in P : J(\mathbf{x}) = 0,\ (\mathrm{Jac}\,\boldsymbol{\eta})(\mathbf{x}) \neq \mathbf{O},\ (\mathrm{grad}\,J)(\mathbf{x}) \neq \mathbf{0}, \\
&\qquad (\nabla_Z \boldsymbol{\eta})(\mathbf{x}) = (\nabla_Z^{(2)} \boldsymbol{\eta})(\mathbf{x}) = \mathbf{0}\Big\}.
\end{aligned}$$

Here,

$$Z = -J_{\mathrm{v}} \frac{\partial}{\partial \mathrm{u}} + J_{\mathrm{u}} \frac{\partial}{\partial \mathrm{v}}$$

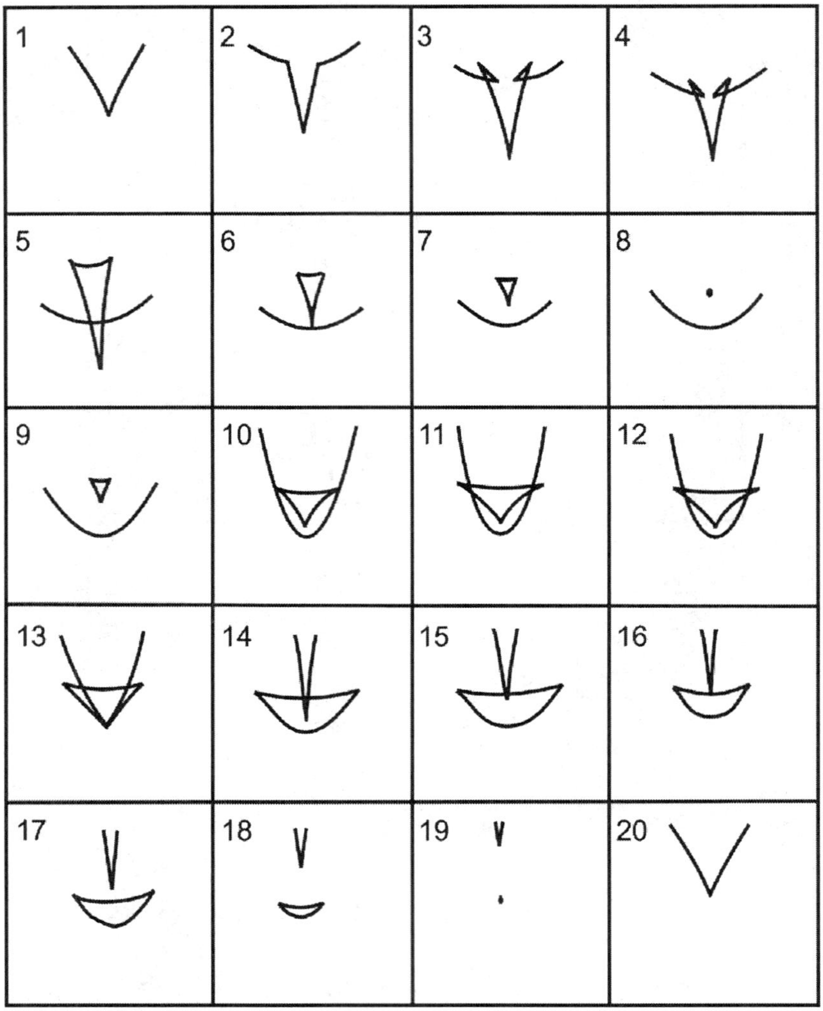

Figure 9.18. A caustic evolution due to slices through the parabolic umbilic bifurcation set.

and

$$\nabla_Z^{(n)}\boldsymbol{\eta} = \nabla_Z(\nabla_Z^{(n-1)}\boldsymbol{\eta}) = (Z \cdot \text{grad}(\nabla_Z^{(n-1)}a),\, Z \cdot \text{grad}(\nabla_Z^{(n-1)}b))$$

for all positive integers n. Caustic metamorphoses are determined by elements of Degen$(\boldsymbol{\eta})$. Table 9.1 summarizes the equations characterizing generic planar caustics and their metamorphoses (cf. [Schn-EF, p. 214]); also, see Figure 9.19. Note: Equations (9.50) and (9.51) yield that the condition $(\text{grad}\, J)(\mathbf{x}) = \mathbf{0}$ in Table 9.1 implies that $(\nabla_Z\boldsymbol{\eta})(\mathbf{x}) = (\nabla_Z^{(2)}\boldsymbol{\eta})(\mathbf{x}) = \mathbf{0}$. Also, the condition $(\text{Jac}\,\boldsymbol{\eta})(\mathbf{x}) = \mathbf{O}$ in the table implies that $J(\mathbf{x}) = 0$ and (by (9.48)) $(\text{grad}\, J)(\mathbf{x}) = \mathbf{0}$; so $\mathbf{x}$ is a degenerate critical point. Hence, excluding folds and cusps, the remaining critical points in the Table 9.1 are all degenerate.

Critical Point	**Equations**
Fold	$J(\mathbf{x}) = 0, \quad (\mathrm{Jac}\,\boldsymbol{\eta})(\mathbf{x}) \neq \mathbf{O}, \quad (\mathrm{grad}\,J)(\mathbf{x}) \neq \mathbf{0},$ $(\nabla_Z \boldsymbol{\eta})(\mathbf{x}) \neq \mathbf{0}$
Cusp	$J(\mathbf{x}) = 0, \quad (\mathrm{Jac}\,\boldsymbol{\eta})(\mathbf{x}) \neq \mathbf{O}, \quad (\mathrm{grad}\,J)(\mathbf{x}) \neq \mathbf{0},$ $(\nabla_Z \boldsymbol{\eta})(\mathbf{x}) = \mathbf{0}, \quad (\nabla_Z^{(2)} \boldsymbol{\eta})(\mathbf{x}) \neq \mathbf{0}$
Lips	$J(\mathbf{x}) = 0, \quad (\mathrm{Jac}\,\boldsymbol{\eta})(\mathbf{x}) \neq \mathbf{O}, \quad (\mathrm{grad}\,J)(\mathbf{x}) = \mathbf{0},$ $\det[\mathrm{Hess}\,J](\mathbf{x}) > 0$
Beak-to-Beak	$J(\mathbf{x}) = 0, \quad (\mathrm{Jac}\,\boldsymbol{\eta})(\mathbf{x}) \neq \mathbf{O}, \quad (\mathrm{grad}\,J)(\mathbf{x}) = \mathbf{0},$ $\det[\mathrm{Hess}\,J](\mathbf{x}) < 0$
Swallowtail	$J(\mathbf{x}) = 0, \quad (\mathrm{Jac}\,\boldsymbol{\eta})(\mathbf{x}) \neq \mathbf{O}, \quad (\mathrm{grad}\,J)(\mathbf{x}) \neq \mathbf{0},$ $(\nabla_Z \boldsymbol{\eta})(\mathbf{x}) = (\nabla_Z^{(2)} \boldsymbol{\eta})(\mathbf{x}) = \mathbf{0}, \quad (\nabla_Z^{(3)} \boldsymbol{\eta})(\mathbf{x}) \neq \mathbf{0}$
Elliptic Umbilic	$(\mathrm{Jac}\,\boldsymbol{\eta})(\mathbf{x}) = \mathbf{O}, \quad \det[\mathrm{Hess}\,J](\mathbf{x}) > 0$
Hyperbolic Umbilic	$(\mathrm{Jac}\,\boldsymbol{\eta})(\mathbf{x}) = \mathbf{O}, \quad \det[\mathrm{Hess}\,J](\mathbf{x}) < 0$

Table 9.1. Equations characterizing certain critical points $\mathbf{x}$ of a k-plane lensing map $\boldsymbol{\eta}$ (cf. [Schn-EF, p. 214]). The critical points $\mathbf{x}$ designated by lips, beak-to-beak, swallowtail, elliptic umbilic, and hyperbolic umbilic are mapped by $\boldsymbol{\eta}$ to caustic points $\boldsymbol{\eta}(\mathbf{x})$ that mark the transition points of the corresponding caustic metamorphoses — see Figure 9.15. The caustics $\boldsymbol{\eta}(\mathbf{x})$ correspond to the points "○" in the Metamorphoses column of Figure 9.15 or the points "•" in Figure 9.19.

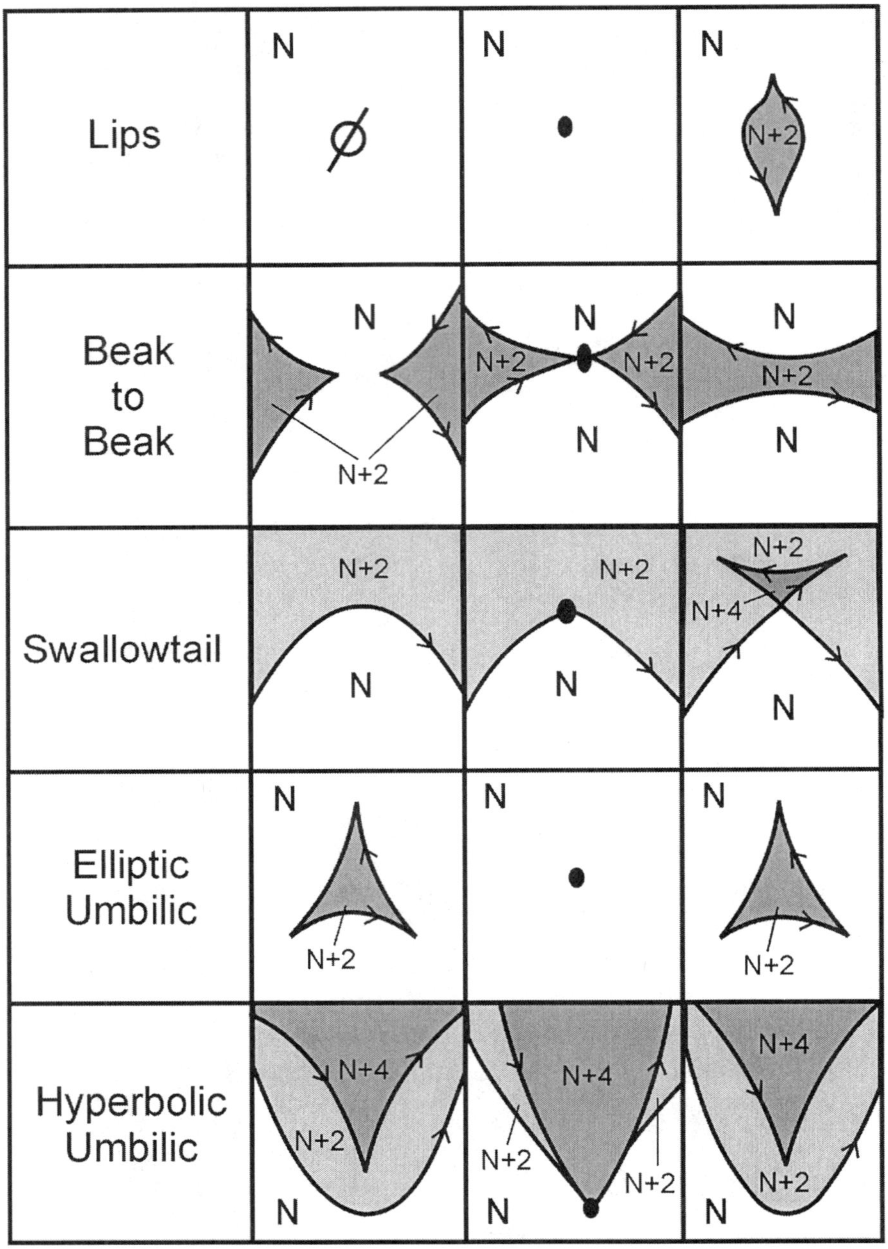

Figure 9.19. Caustic metamorphoses and their lensed image multiplicities in the light source plane. Each evolution runs either left to right, or, vice versa along a row. The equations in Table 9.1 determine the points • in the middle column. These points mark the occurrence of the given metamorphosis. The numbers N, N+2, and N+4 indicate how many lensed images are seen of a light source in the given region. The caustic curves are oriented so such that, locally, light sources to the left generically have two more lensed images than those to the right (cf. Figure 9.12).

9.5.3 Illustrations of Single-Plane Caustic Metamorphoses

In this section, we illustrate that all five of the caustic metamorphoses in Figure 9.15 or Figure 9.19 occur in gravitational lensing.

Consider a single-plane lens systems with point masses, continuous matter, and shear from infinity:

$$\psi_{ptcs}(\mathbf{x}) = \frac{\kappa_c}{2}|\mathbf{x}|^2 - \frac{\gamma}{2}(\mathrm{u}^2 - \mathrm{v}^2) + \sum_{\ell=1}^{g} m_\ell \ln |\mathbf{x} - \boldsymbol{\xi}_\ell|,$$

where $\kappa_c \geq 0$, $\gamma \geq 0$, $m_\ell \geq 0$, and $\mathbf{x} = (\mathrm{u}, \mathrm{v})$. *The caustics due to this potential can produce beak-to-beak, swallowtail, and elliptic umbilic metamorphoses.* Figures 9.20 and 9.21 show some examples. Notice from these figures that the before-and-after topologies of critical curves are changed by a beak-to-beak metamorphosis and left unchanged by swallowtails and elliptic umbilics.

Lips and hyperbolic umbilics cannot be generated by single-plane lens systems having point masses with continuous matter and shear; otherwise, local convexity is violated (Corollary 9.5, page 360). However, elliptical potentials

$$\psi_{ep}(\mathbf{x}) = C\sqrt{x_{cr}^2 + (1-\epsilon)\mathrm{u}^2 + (1+\epsilon)\mathrm{v}^2},$$

where $x_{cr} \geq 0$ is the (dimensionless) core radius and $0 \leq \epsilon \lesssim 0.2$ is the ellipticity, can produce lips and hyperbolic umbilic metamorphoses — Figures 9.20 and 9.22. As seen in these figures, the before-and-after topologies of the critical curves are changed by a lips metamorphosis, but unchanged by hyperbolic umbilics.

Butterfly metamorphoses can be generated by two point masses on a lens plane with continuous matter and shear — Figure 9.23. In the figure, these metamorphoses do not change the before-and-after topologies of the critical curves. Some features of a parabolic umbilic metamorphosis are also produced by this lens system — Figure 9.23. Note, however, that the lips features of a parabolic umbilic metamorphosis (see Figure 9.18) cannot be captured (otherwise, local convexity is violated). The topology of the critical curves changes in this metamorphosis.

The section concludes with examples of the bicaustics due to a point mass with varying shear — Figure 9.24. The associated bicaustic equations were determined by A.P. and Witt [Pet-Wit, p. 2933].

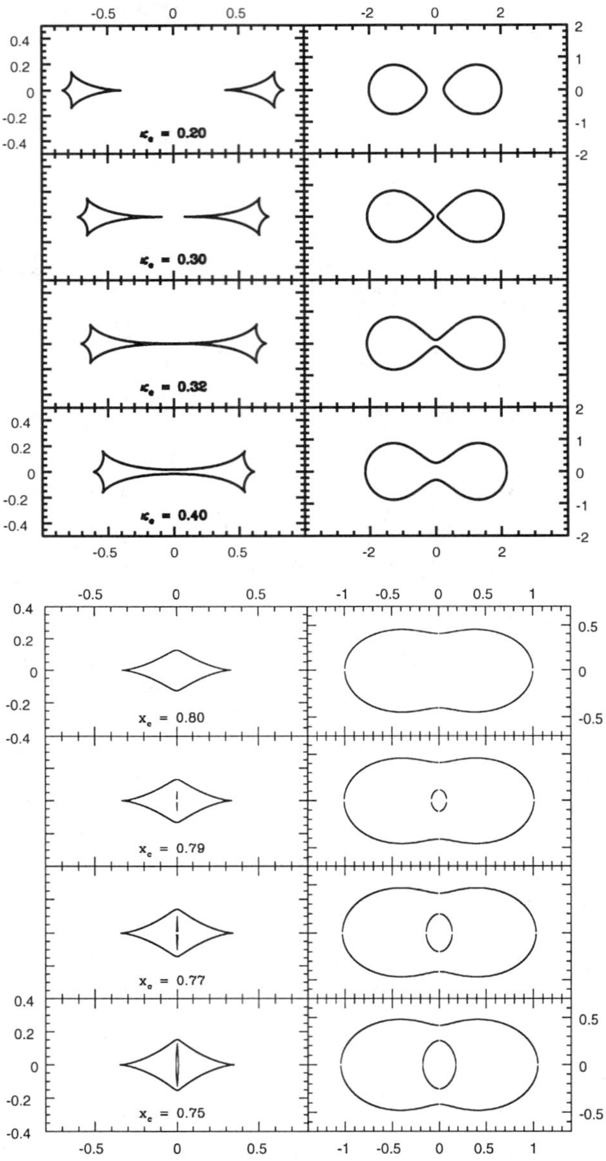

Figure 9.20. *Top figure:* Lips metamorphosis due to an elliptical potential with fixed ellipticity $\epsilon = 0.2$ and varying core radius x_c. The left panels depict the caustics, while the critical curves are shown on the right. Notice the appearance/disappearance of the lips caustic and critical curve during the metamorphosis. *Bottom figure:* Beak-to-beak metamorphoses due to two point masses on a single lens plane with fixed separation $d = 2.4$ and varying continuous matter density κ_c. A change occurs in the topology of the critical curves. Consult Kassiola, Kovner and Blandford [Kas-KB] for applications of the lips and beak-to-beak metamorphoses to the "straight arc" lensing phenomenon in the cluster of galaxies Abell 2390.

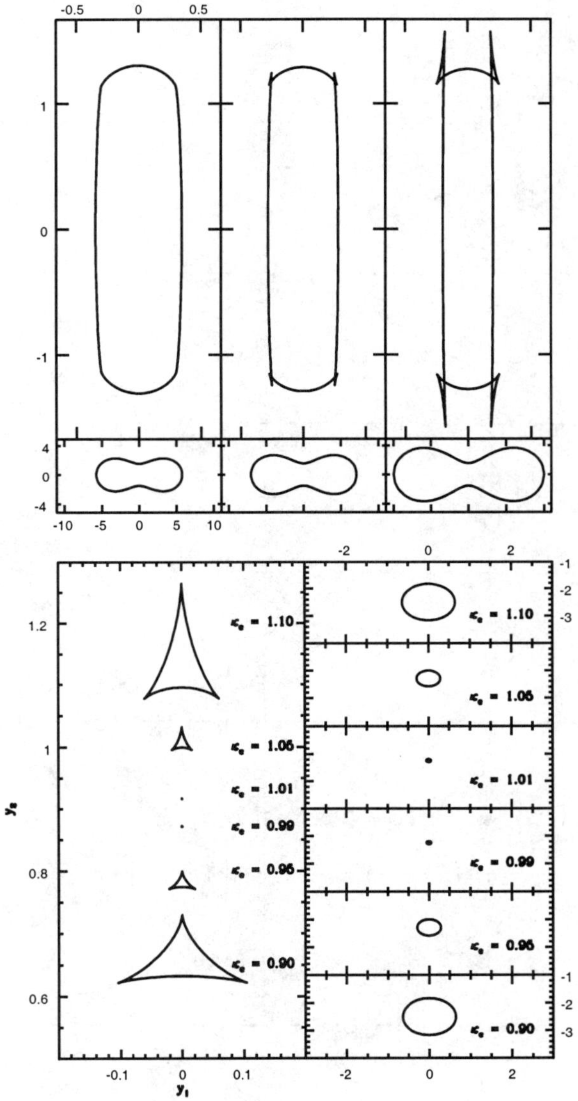

Figure 9.21. *Top figure:* Swallowtail metamorphosis due to a point mass lens with fixed shear from infinity, $\gamma = 0.2$, and continuous matter with varying density κ_c. The caustic can evolve from having eight cusps associated with four swallowtail structures, to no cusp. The bottom panels show that no change occurs in the topology of the critical curve. *Bottom figure:* Elliptic umbilic metamorphosis due to a point mass lens with fixed shear from infinity, $\gamma = 0.2$, and continuous matter with varying density κ_c. Both the caustic (left panel) and critical curve (right panels) shrink to a point as κ_c passes through the critical value $\kappa_c = 1$, but the before-and-after topology of the curves is the same.

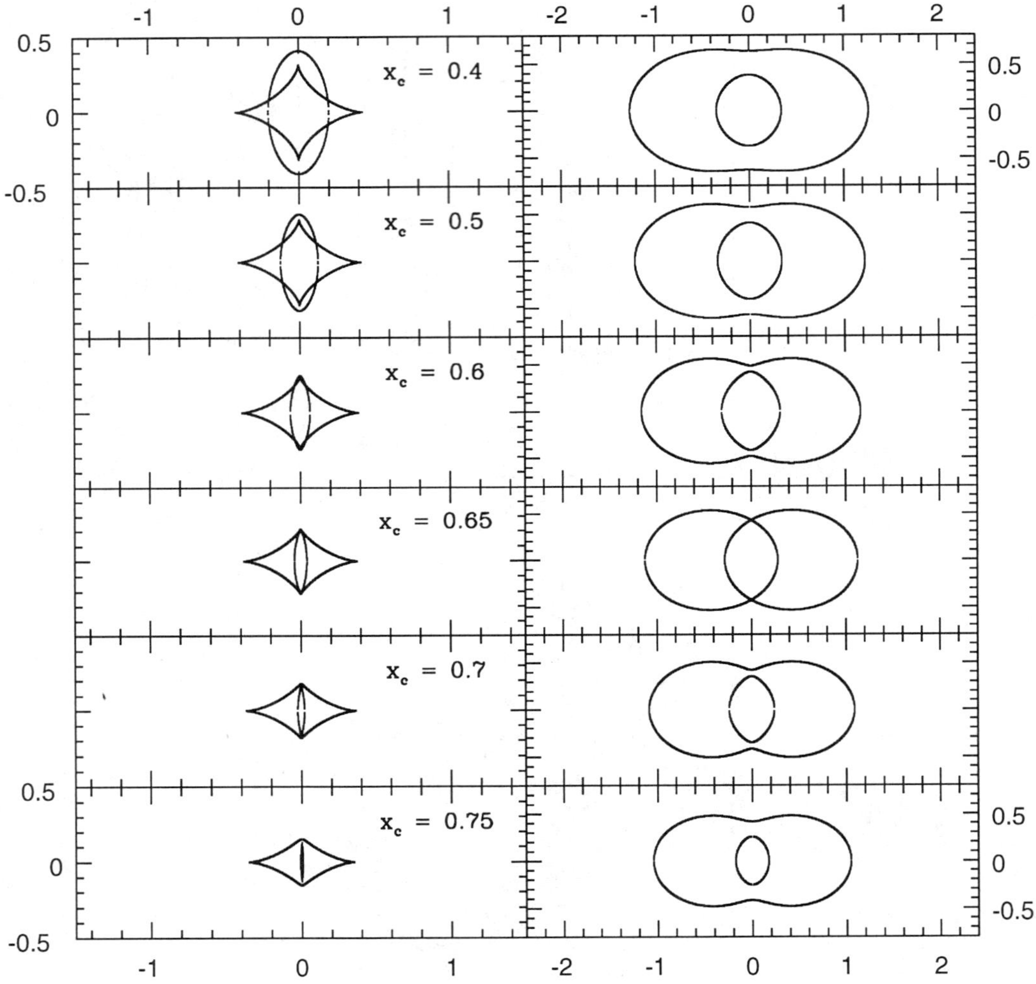

Figure 9.22. Hyperbolic umbilic metamorphosis due to an elliptical potential with fixed ellipticity $\epsilon = 0.2$ and varying core radius x_c. The caustics are shown in the left panels, while the critical curves are on the right. Though the critical curves merge at the instant of the metamorphosis, their before-and-after topologies are the same.

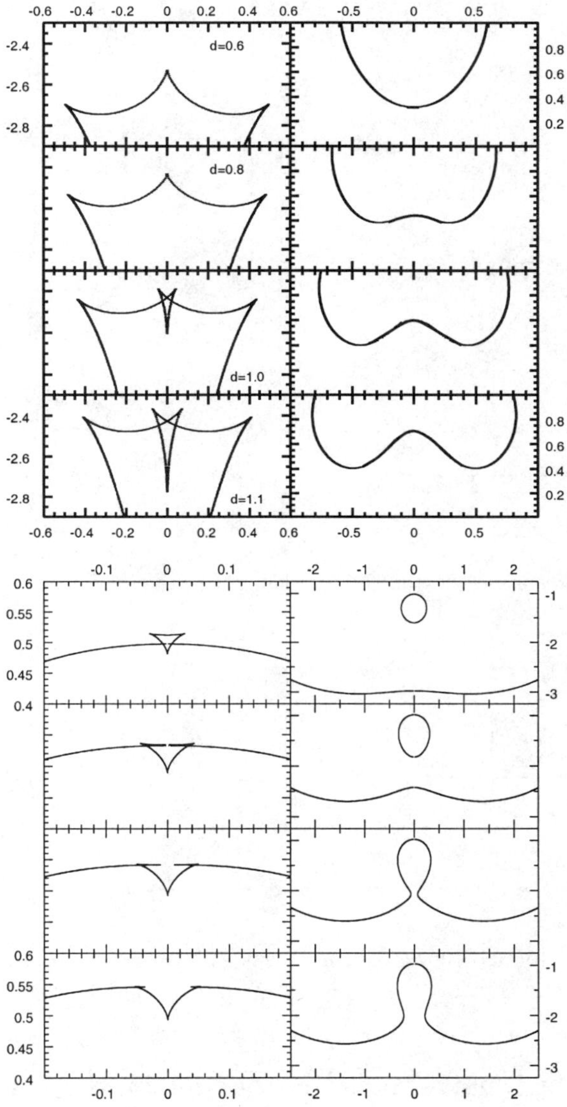

Figure 9.23. *Top figure:* Butterfly metamorphosis due to two point masses of equal mass, $m_1 = m_2 = 0.5$, on a single lens plane with fixed separation continuous matter density $\kappa_c = 2.2$ and shear $\gamma = 1.8$, and varying separation d between the masses. No change occurs in the topology of the critical curve. *Bottom figure:* Portion of a parabolic umbilic metamorphosis due to two point masses of equal mass, $m_1 = m_2 = 0.5$, and fixed separation, $d = 2.4$, on a single lens plane with varying continuous matter density $\kappa_c = 1.083, 1.085, 1.087$ (from top to bottom). A change occurs in the topology of the critical curves. Not all features of the parabolic umbilic metamorphosis can be produced by this lens. For example, the lips component is impossible since it violates local convexity (cf. Figure 9.18).

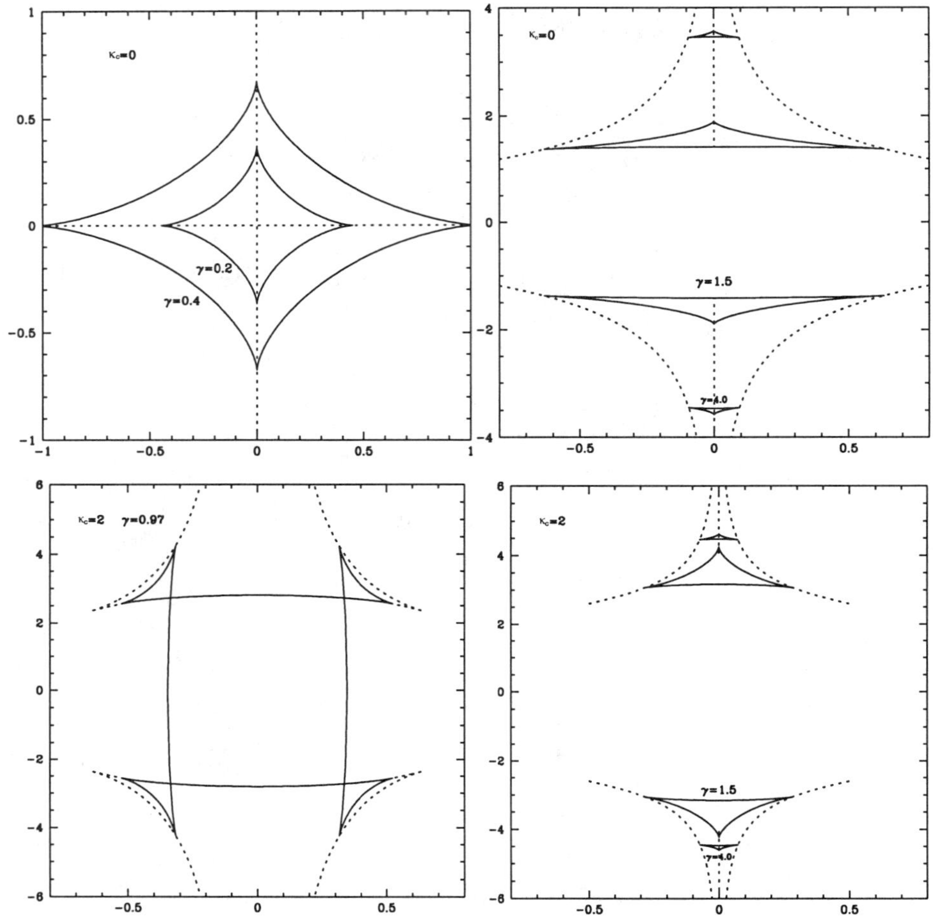

Figure 9.24. Bicaustics (dotted curves) due to caustics produced by a point mass lens with varying shear γ and fixed continuous matter κ_c (After [Pet-Wit, p. 2931].)

9.5.4 Single versus Double-Plane Caustic Metamorphoses

In the current section, we highlight some of the distinctions between caustics due to single and double-plane lensing. We shall see that caustics due to two point masses on distinct planes with continuous matter and shear from infinity display a rich and striking variety of geometric forms. This includes caustic features that cannot be generated by single-plane lensing with point masses (including continuous matter and shear from infinity) and double-plane two point mass lenses. Some of these features involve caustic metamorphoses. This section is based on work by A.P. and Wicklin [Pet-Wic95].

Preliminaries: Two point masses on distinct lens planes with continuous matter and shear. Consider a light source at $\mathbf{y}$ undergoing gravitational lensing by two point masses m_1 and m_2 on different lens planes L_1 and L_2, respectively. Suppose the on L_i there is a constant shear γ_i from infinity and continuous matter with constant density κ_i, where $i = 1, 2$. Let $(\mathrm{u}_i, \mathrm{v}_i)$ be rectangular coordinates on L_i, where $i = 1, 2$. Assume that m_1 is positioned at $(0, 0)$ in L_1 and m_2 at $\mathbf{d}$ on L_2. Relative to ray tracing from the observer to the light source plane, suppose that the shear γ_1 is along the u_1-axis and γ_2 is along the v_2-axis. Then the lens equation for this lens system is (cf. pages 101 and 175):

$$\begin{aligned}\mathbf{y} \;=\; & \mathbf{x}_1 - \left(m_1 \frac{\mathbf{x}_1}{|\mathbf{x}_1|^2} + \kappa_1 \mathbf{x}_1 + \gamma_1(-\mathrm{u}_1, \mathrm{v}_1) \right) \\ & - \left(m_2 \frac{\mathbf{x}_2 - \mathbf{d}}{|\mathbf{x}_2(\mathbf{x}_1) - \mathbf{d}|^2} + \kappa_2 \mathbf{x}_2 + \gamma_2(\mathrm{u}_2, -\mathrm{v}_2) \right), \end{aligned} \tag{9.54}$$

where

$$\mathbf{x}_2 = \mathbf{x}_1 - \beta \left(m_1 \frac{\mathbf{x}_1}{|\mathbf{x}_1|^2} + \kappa_1 \mathbf{x}_1 + \gamma_1(-\mathrm{u}_1, \mathrm{v}_1) \right)$$

with $\mathbf{x}_i = (\mathrm{u}_i, \mathrm{v}_i)$ and $\beta = d_{1,2} d_S / d_2 d_{1,S}$. The system has nine parameters:

$$\mathbf{d} = (\delta_1, \delta_2), \quad \beta, \quad m_1, m_2, \quad \kappa_1, \kappa_2, \quad \gamma_1, \gamma_2.$$

All figures in this section will be described relative to the above nine parameters. Note that if $\beta = 0$, $\kappa_2 = 0$, and $\gamma_2 = 0$, then we obtain a single-plane two point mass lens configuration with continuous matter κ_1 and shear γ_1 from infinity along the u_1-axis, relative tracing light rays from the observer to light source plane (cf. equation (3.123), page 104):

$$\mathbf{y} = \begin{bmatrix} 1 - \kappa_1 + \gamma_1 & 0 \\ 0 & 1 - \kappa_1 - \gamma_1 \end{bmatrix} \mathbf{x}_1 - m_1 \frac{\mathbf{x}_1}{|\mathbf{x}_1|^2} - m_2 \frac{\mathbf{x}_1 - \mathbf{d}}{|\mathbf{x}_1 - \mathbf{d}|^2}. \tag{9.55}$$

Notation: Let $\boldsymbol{\eta}^{(3D)}_{ptcs}(\mathbf{x}_1) = \mathbf{y}$ and $\boldsymbol{\eta}^{(2D)}_{ptcs}(\mathbf{x}_1) = \mathbf{y}$ be the lensing maps defined by (9.54) and (9.55), respectively.

Single versus double-plane binary point mass lensing. We now list some of the differences between the caustic metamorphoses due to the lensing maps $\boldsymbol{\eta}^{(2D)}_{ptcs}$ and $\boldsymbol{\eta}^{(3D)}_{ptcs}$ (also see Figure 15.12 on page 527 for more):

(1) We saw in Section 9.5.3 that *a single-plane binary point mass system with continuous matter and shear* (i.e., the system $\boldsymbol{\eta}^{(2D)}_{ptcs}$) *can produce beak-to-beak, swallowtail, and elliptic umbilic metamorphoses, but lips and hyperbolic umbilics are impossible* (otherwise, local convexity is violated — Section 9.3, page 353). *However, lips and hyperbolic umbilics, as well as beak-to-beaks, swallowtails, and elliptic umbilics, can be generated by the double-plane binary system* $\boldsymbol{\eta}^{(3D)}_{ptcs}$. Examples of lips and hyperbolic umbilics due to $\boldsymbol{\eta}^{(3D)}_{ptcs}$ are depicted in Figure 9.25.

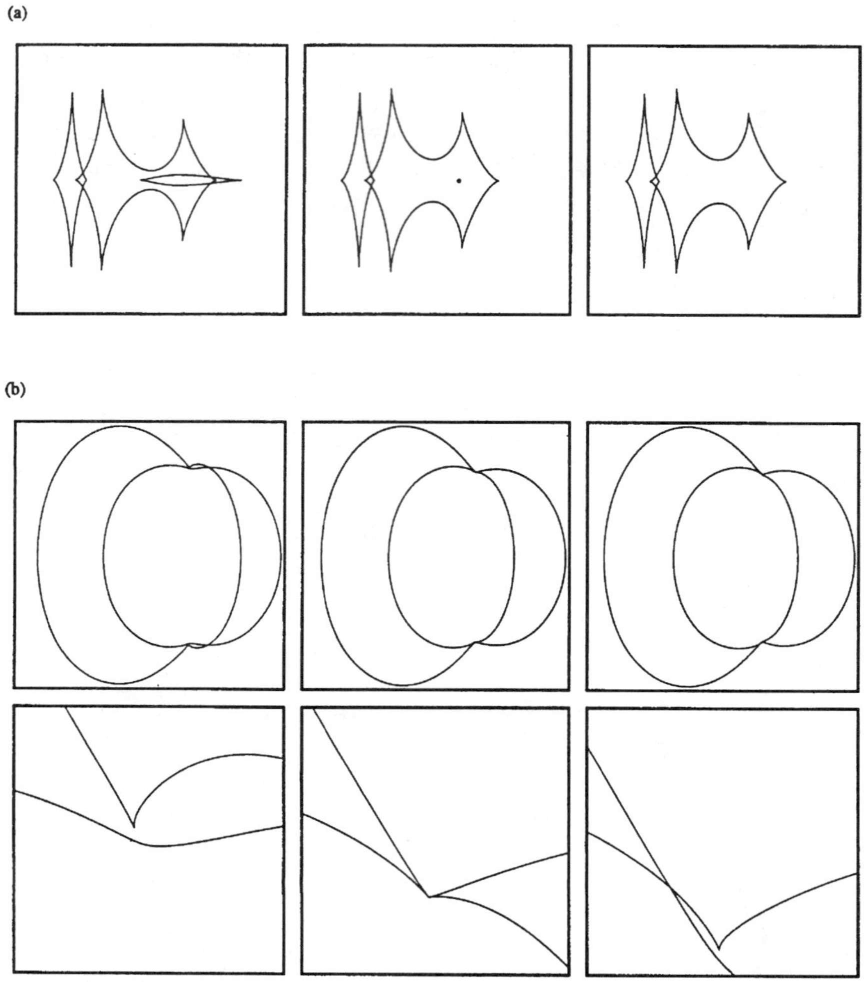

Figure 9.25. Lips and hyperbolic umbilic caustic metamorphoses produced by the double-plane point mass lens system $\boldsymbol{\eta}^{(3D)}_{ptcs}$. (a) Lips metamorphosis for $\mathbf{d} = (\delta_1, 0)$ with $\delta_1 = 0.46, 0.493$, and 0.495; $\beta = 0.6$, $m_1 = 1$, $m_2 = 1$, $\kappa_1 = 0.6$, $\kappa_2 = 0.4$, $\gamma_1 = 0.2$, and $\gamma_2 = 0.3$. The lip caustic shrinks to a point and then vanishes. (b) Hyperbolic umbilic metamorphosis for $\mathbf{d} = (\delta_1, 0)$ with $\delta_1 = 1.8, 1.869$, and 1.92; $\beta = 0.6$, $m_1 = 1$, $m_2 = 1$, $\kappa_1 = 3.2$, $\kappa_2 = 3$, $\gamma_1 = 0.2$, and $\gamma_2 = 0.3$. A close-up of the caustics is shown in the bottom row. (After [Pet-Wic95, p. 1401].)

(2) *Two point masses on a single lens plane* (i.e., the system $\boldsymbol{\eta}^{(2D)}_{ptcs}$ with $\kappa_1 = \gamma_1 = 0$) *can produce only beak-to-beak metamorphoses* (e.g., [Erd-S, p. 466]). Let us show why this is true. We already know that lips and hyperbolic umbilics are not possible. Swallowtails cannot be produced since transverse crossings of caustics would occur. In fact, such crossings can cause a change of ± 4 in the number of lensed images of a light source (see Figure 9.19, page 382). However, this contradicts the fact that

two point masses on a single lens plane produce either three or five lensed images of a light source (see Theorem 11.11, page 439). Finally, elliptic umbilics cannot be produced by $\boldsymbol{\eta}^{(2D)}_{ptcs}$. More generally, let $\boldsymbol{\eta}$ be a lensing map due to g point masses on a single lens plane ($g \geq 1$). Then $\kappa(\mathbf{x}) = 0$ in the domain of $\boldsymbol{\eta}$ since each point mass contributes a Dirac delta surface mass density at the mass's position and such positions are excised from the domain of $\boldsymbol{\eta}$ (see page 175). Now, if $\mathbf{x}$ is an elliptic umbilic, then Table 9.1 (page 381) shows that $(\text{Jac}\,\boldsymbol{\eta})(\mathbf{x}) = \mathbf{O}$. This implies that $1 - \psi_{\text{uu}} = 0 = 1 - \psi_{\text{vv}}$ and hence $\kappa(\mathbf{x}) = 1$, which is a contraction. Thus, *elliptic umbilics cannot be produced by g point masses on a single lens plane.*

Two point masses on distinct lens planes (i.e., the system $\boldsymbol{\eta}^{(2D)}_{ptcs}$ *with* $\kappa_i = \gamma_i = 0$, $i = 1, 2$*) can produce only beak-to-beak and elliptic umbilic metamorphoses* (one more than in the single plane case). To see this, we first use the fact that such systems produce either four or six lensed images (see Theorem 12.10, page 460). This rules out transverse intersections of caustics — e.g., swallowtails — and hyperbolic umbilics since such metamorphoses lead to changes of ± 4 in the number of lensed images — Figure 9.19 (page 382). Lips cannot occur because no critical curve of the lens system can vanish — see [Erd-S, pp. 457–465] for details. The interested reader is referred to [Erd-S, pp. 463] for an illustration of the occurrences of beak-to-beak and elliptic umbilic metamorphoses due to double-plane binary point mass lens systems. Our discussion also shows that two point masses on distinct lens planes with continuous matter $\kappa_i > 0$ and shears $\gamma_i > 0$ can generate metamorphoses that are impossible for a pure binary point mass system on distinct lens planes (namely, swallowtails, lips, and hyperbolic umbilics).

(3) The system $\boldsymbol{\eta}^{(3D)}_{ptcs}$ can also generate fold-penetrating-a-fold and cusp-penetrating-a-fold caustic evolutions. Two of these evolutions are shown in Figure 9.26. Notice that the evolution involves a *kidney bean* caustic, that is, the simple closed without cusps. This caustic violates local convexity. Consequently, *the caustic network evolution of Figure 9.26 cannot be generated by single-plane point mass lenses with continuous matter and shear from infinity.* In addition, *the fold-penetrating-a-fold and cusp-penetrating-a-fold evolutions are impossible for a double-plane two point mass lens* (since the caustics of such lenses cannot have self-intersections).

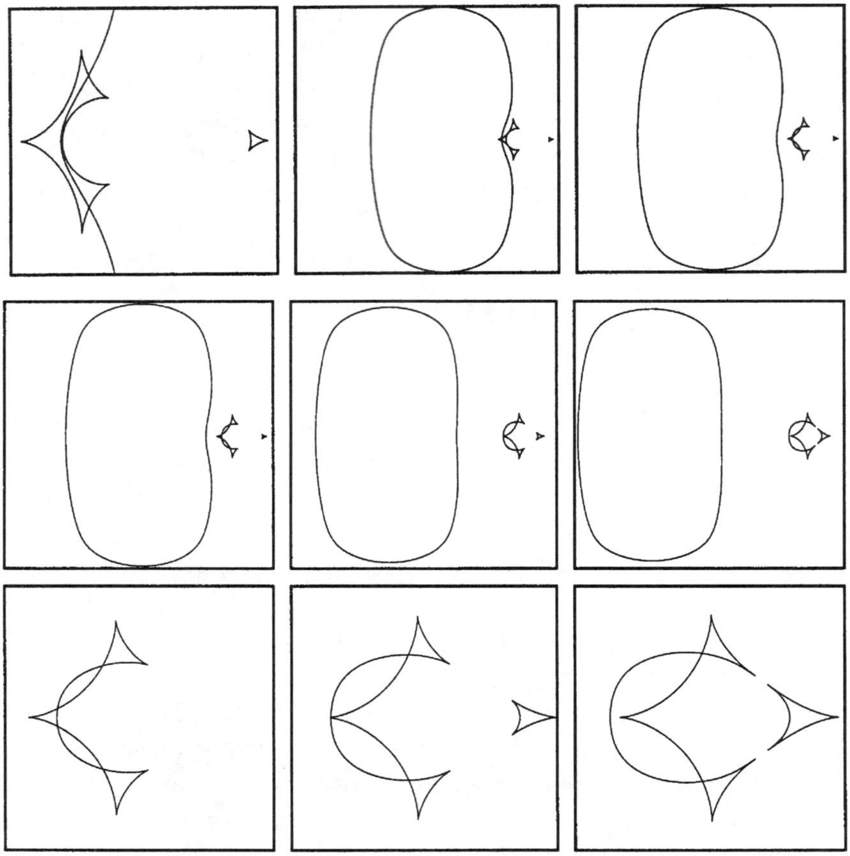

Figure 9.26. Caustics due to two point masses on distinct lens planes with continuous matter and shear. These metamorphoses are for varying $\mathbf{d} = (\delta_1, 0)$ with $\beta = 0.6$, $m_1 = 1$, $m_2 = 1$, $\kappa_1 = 1.4$, $\kappa_2 = 1.2$, $\gamma_1 = 0.2$, and $\gamma_2 = 0.3$. The middle row of panels show a cusp-penetrating-a-fold on the same caustic. A close-up of the evolution is shown in the bottom row of panels. The parameter values are $\delta_1 = 1.5, 2.2$, and 2.6. — see [Pet-Wic95]. Notice the violation of local convexity by the kidney bean caustic. (After [Pet-Wic95, p. 1402].)

Chapter 10

Morse Inequalities

In retrospect all great ideas take on a certain simplicity and inevitability, partly because they shape the whole subsequent development of the subject. And so to us, today, Morse theory seems natural and inevitable.

RAOUL BOTT

In the early 1920s Marston Morse [Mor] began the development of what has become known as *Morse theory*.[1] This highly acclaimed work introduced a set of celebrated inequalities called the *Morse inequalities*. We shall see in subsequent chapters that the Morse inequalities impact powerfully on several central issues in gravitational lensing: counting lensed images (Chapters 11 and 12), finding lower bounds on total magnification (Chapter 13), and linking the total curvature of caustics to the lens (Chapters 14 and 15). The Morse inequalities also apply to counting fixed points[2] of gravitational lens systems (A.P. and Wicklin [Pet-Wic98]). Our goal is to present the Morse inequalities in a context relevant to gravitational lensing. Readers already familiar with Morse theory should browse through Section 10.2 for our notational conventions.

Outline: Section 10.1 introduces Betti numbers both heuristically and precisely, and gives some theorems for calculating them. The Morse inequalities are presented in Section 10.2. They are divided according to boundary conditions A and B, with Morse inequalities B generalizing A. In Section 10.3, we present an informal tour of the basic ideas behind the proof of the Morse inequalities. Section 10.3.1 discusses topological changes in certain subspaces of the domain of the time delay function of an elliptical lens, as critical values are crossed. This implicitly shows a connection between topology and critical point theory (cf. Figure 10.7). In order to express this link explicitly, relative Betti numbers are introduced

[1] Historical accounts appear in [Bot], [Gores-M, p. 19], and [Die, p. 227].

[2] These are points from which a light source has a lensed image that — despite gravitational lensing — corresponds to the original position of the source.

in Section 10.3.2. These numbers are shown to directly connect with the number of critical points of a given index (see equation (10.4)). The results of Sections 10.3.1 and 10.3.2 are then employed in Section 10.3.3 to prove Morse inequalities A and B.

10.1 Betti Numbers

Betti numbers are named after the nineteenth century mathematical physicist Enrico Betti. They play a fundamental role in the Morse inequalities. We shall introduce these numbers and present some useful facts for calculating them in situations of relevance to gravitational lensing.

10.1.1 Heuristic Treatment of Betti Numbers

This section presents a heuristic introduction to homology groups and Betti numbers (i.e., "dimensions" of homology groups) that we hope will give the reader an intuitive feel for the subject. A precise treatment of Betti numbers will be presented in Section 10.1.2. For additional intuitive treatments of the ideas underlying Betti numbers, see the excellent book by Geroch [Ger, pp. 198–210] (who also intuitively discusses many other topics in mathematics).

Before defining the Betti numbers of a topological space X, we must associate "homology groups" to X. The construction of these groups requires several preliminary concepts: simplices, chains, boundary operators, and cycles. We start by defining the *standard k-simplex* in $\mathbf{R}^{k+1}$ to be the topological subspace

$$\Delta_k = \left\{ \mathbf{x} \in \mathbf{R}^{k+1} : \mathbf{x} = \sum_{i=0}^{k} a_i \mathbf{v}_i, \ a_i \geq 0, \ \sum_{i=0}^{k} a_i = 1 \right\},$$

where $\mathbf{v}_i$ is the point $(0, \ldots, 0, 1, 0, \ldots, 0)$ in $\mathbf{R}^{k+1}$ having 1 in the $(i+1)$st slot and zeros in the remaining slots. The points $\mathbf{v}_0, \mathbf{v}_1, \ldots, \mathbf{v}_k$ are the *vertices* of Δ_k. The standard simplices Δ_0, Δ_1, Δ_2, and Δ_3 are a point, line segment, equilateral triangular region, and solid tetrahedron, respectively. They are depicted in Figure 10.1, which also includes an orientation of the simplices. To see how the latter are introduced, define two orderings of the vertices of Δ_k to be *equivalent* if the permutation transforming one into the other is even. This divides the $(k+1)!$ orderings of the vertices of Δ_k into two equivalence classes, each of which is an *orientation* of Δ_k. For example, in the case of Δ_2 the orderings $\mathbf{v}_0, \mathbf{v}_1, \mathbf{v}_2$ and $\mathbf{v}_2, \mathbf{v}_0, \mathbf{v}_1$ determine the same orientation, while $\mathbf{v}_0, \mathbf{v}_2, \mathbf{v}_1$ and $\mathbf{v}_1, \mathbf{v}_0, \mathbf{v}_2$ give the other orientation. Choosing an orientation of Δ_k then means selecting an ordering of the vertices of Δ_k, up to an even permutation. *We shall assume that Δ_k has the natural orientation determined by $\mathbf{v}_0, \mathbf{v}_1, \ldots, \mathbf{v}_k$.* In particular, the standard simplex Δ_1 is the line segment in the direction from $\mathbf{v}_0$ to $\mathbf{v}_1$. Δ_2 has the orientation induced by circulating counterclockwise along its boundary, and Δ_3 has a right-handed orientation — see Figure 10.1.

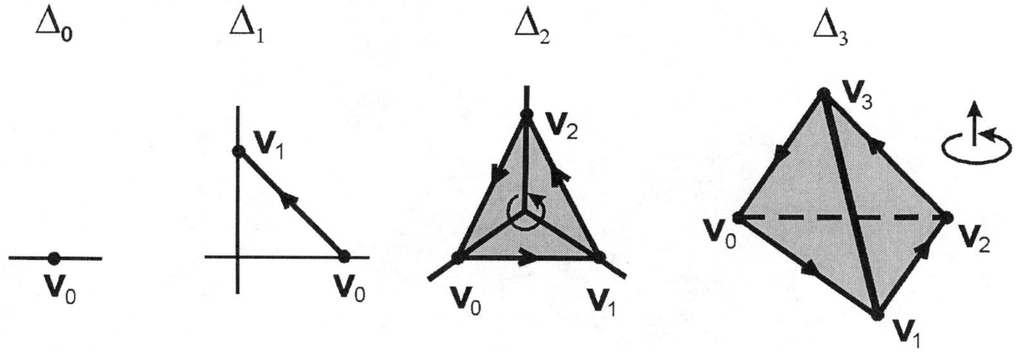

Figure 10.1. Some standard simplices with their natural orientation. The simplex Δ_2 is bounded by an equilateral triangle, while Δ_3 is a solid tetrahedron.

A *singular k-simplex* in a topological space X is a continuous map $e : \Delta_k \to X$. The term "singular" is used since a singular k-simplex may be a constant map (i.e., the range $e[\Delta_k]$ is a point). The range of a singular 0-simplex is a point. Examples of the ranges of singular 1- and 2-simplices are shown in Figure 10.2. It is useful to keep these examples in mind. Note that, in general, the ranges of singular simplices are connected (since the spaces Δ_k are connected).

By "stringing" together finitely many singular simplices, we can construct more complicated objects in X, namely, "chains." A *singular k-chain* c in X is a finite collection of singular k-chains, each with a "multiplicity." If c is an empty collection, then write $c = o$, which is called the *zero* k-chain. When c is nonempty, we express c formally as follows: $c = r_1 \cdot e_1 + \cdots + r_n \cdot e_n$, where $r_i \in \mathbf{R}$ and e_i is a singular k-simplex. In particular, all singular k-simplices are k-chains. For r a positive integer and any singular k-chain e, we think of $r \cdot e$ as "r copies of e superimposed on e" and call r the "multiplicity" of $r \cdot e$. By convention, set $0 \cdot c = o$. We also think of negative multiplicities as allowing for the removal of chains: $r \cdot e - r \cdot e = (r - r) \cdot e = o$. If we imagine that $e(\Delta_k)$ is oriented (e.g., Figure 10.2), then we interpret the effect of multiplying by -1 as reversing the orientation of $e(\Delta_k)$. Also, we see intuitively that the following relation holds: $r \cdot e + r' \cdot e = (r + r') \cdot e$, where r, r' are integers. In general, the algebra of chains using coefficients in $\mathbf{R}$ will just formally mimic the intuitively clear situation where the coefficients are integers.

Let $S_k(X; \mathbf{R})$ be the set of all singular k-chains in X (i.e., the "span" of all singular k-simplices in X). We shall convert the set $S_k(X; \mathbf{R})$ into a real vector space. Let $c = r_1 \cdot e_1 + \cdots + r_n \cdot e_n$ and $c' = r'_1 \cdot e'_1 + \cdots + r'_m \cdot e'_m$ be any two singular k-chains in X. Define "vector addition" and "scalar multiplication" for singular k-chains by: $c + c' = r_1 \cdot e_1 + \cdots + r_n \cdot e_n + r'_1 \cdot e'_1 + \cdots + r'_m \cdot e'_m$ and $\alpha \cdot c = (\alpha r_1) \cdot e_1 + \cdots + (\alpha r_n) \cdot e_n$ where $\alpha \in \mathbf{R}$. Our "vector addition" is commutative: $c + c' = c' + c$. In fact, the latter holds intuitively since singular simplices obey $r \cdot e + r' \cdot e' = r' \cdot e' + r \cdot e$. For example, in the space D_1

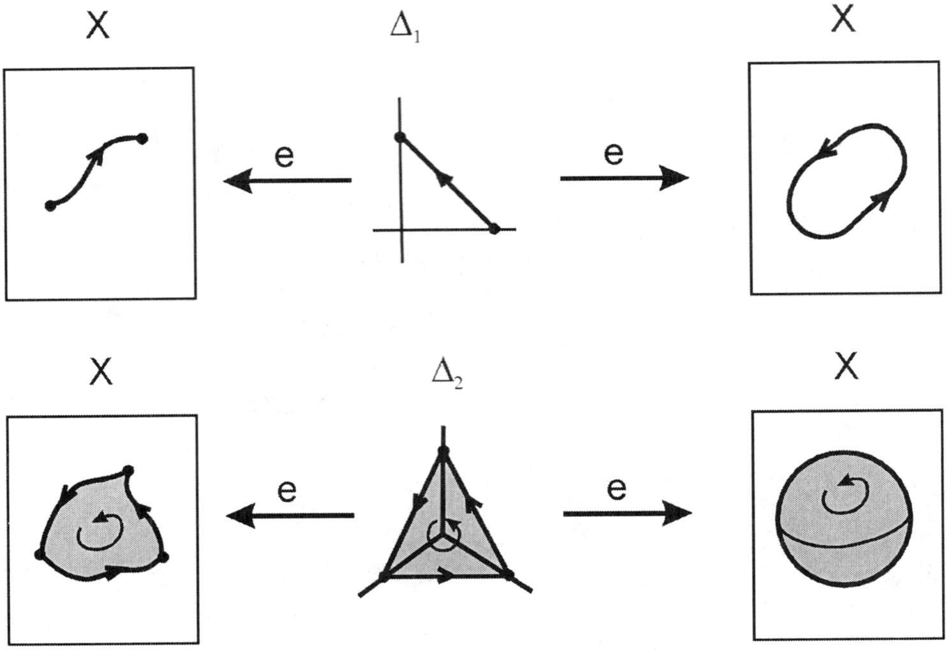

Figure 10.2. Top row: Two singular 1-simplices $e : \Delta_1 \to X$ mapping the line segment Δ_1 into a simple arc (top left panel) and simple closed curve (top right panel). Bottom row: Two singular 2-simplices $e : \Delta_2 \to X$ mapping the region Δ_2 bounded by an equilateral triangle into a curvilinear triangle (bottom left panel) and a surface homeomorphic to a sphere (bottom right panel). Note that it is possible for a singular k-simplex $e : \Delta_k \to X$ to map the standard k-simplex Δ_k into a point.

of Figure 10.3, the 1-chain $+a+b$ is the union of the two curves going around the hole; consequently, $+a+b = +b+a$. (It is assumed that each curve in the figure goes around exactly once.) We shall suppose that $S_k(X; \mathbf{R})$ is endowed with the vector space structure induced by the previous operations.

Let us now consider the "boundary" of singular k-chains. We begin with the boundary of a standard k-simplex. For $k \geq 1$ and $i = 0, 1, \dots, k$, define $k+1$ continuous maps $f_k^i : \Delta_{k-1} \to \Delta_k$ by $f_k^i(a_0, \dots, a_i, \dots, a_{k-1}) = (a_0, \dots, a_{i-1}, 0, a_i, a_{i+1}, \dots, a_{k-1})$, that is, add an extra zero at the ith slot. Observe that

$$f_k^i \circ f_{k-1}^{j-1} = f_k^j \circ f_{k-1}^i, \qquad 0 \leq i < j \leq k. \tag{10.1}$$

In the case $\Delta_0 \to \Delta_1$, the ranges of $f_1^0 : a_0 \mapsto (0, a_0)$ and $f_1^1 : a_0 \mapsto (a_0, 0)$ give the end points of the line segment Δ_1. For $\Delta_1 \to \Delta_2$, the ranges of $f_2^0 : (a_0, a_1) \mapsto (0, a_0, a_1)$, $f_2^1 : (a_0, a_1) \mapsto (a_0, 0, a_1)$, and $f_2^2 : (a_0, a_1) \mapsto (a_0, a_1, 0)$ yield the three edges of the triangle Δ_2. Similarly, the ranges of the maps $f_k^i : \Delta_{k-1} \to \Delta_k$, where $i = 0, 1, \dots, k$ determine the $k+1$ faces of the polyhedron Δ_k. Note that f_k^i leaves the vertices $\mathbf{v}_0, \dots, \mathbf{v}_{i-1}$

unchanged, but maps $\mathbf{v}_i, \ldots, \mathbf{v}_{k-1}$ into $\mathbf{v}_{i+1}, \ldots, \mathbf{v}_k$, respectively. In other words, the map f_k^i sends Δ_{k-1} into the $(k-1)$-face of Δ_k that is opposite to the vertex $\mathbf{v}_i$. The *boundary* of a singular k-simplex e is then defined by

$$\partial_k e = e \circ f_k^0 - e \circ f_k^1 + \cdots + (-1)^k e \circ f_k^k.$$

Since each map $e \circ f_k^i : \Delta_{k-1} \to X$ is continuous, it is a singular $(k-1)$-simplex. Consequently, the boundary $\partial_k e$ is a singular $(k-1)$-chain in X. In the case of a singular 0-simplex e, we set $\partial_0 e = o$. If e is a singular 1-simplex that is a simple arc from a point $p_0 = e(\mathbf{v}_0)$ to $p_1 = e(\mathbf{v}_1)$, then $\partial_1 e = p_1 - p_0$. When e is a simple closed curve, we get $\partial_1 e = o$.

The *boundary* of a singular k-chain $c = r_1 \cdot e_1 + \cdots + r_k \cdot e_k$ is defined to be the singular $(k-1)$-chain

$$\partial_k c = r_1 \cdot \partial_k e_1 + \cdots + r_k \cdot \partial_k e_k.$$

For any two singular k-chains c and c', we have $\partial_k(c+c') = \partial_k c + \partial_k c'$ and $\partial_k(r \cdot c) = r \cdot \partial_k c$, where $r \in \mathbf{R}$. In other words, the boundary operator ∂_k is a linear map from $S_k(X; \mathbf{R})$ into $S_{k-1}(X; \mathbf{R})$. Also, equation (10.1) yields that the boundary of the boundary is zero, that is,

$$\partial_{k-1}(\partial_k c) = o. \tag{10.2}$$

A *k-cycle* is a singular k-chain c without boundary (i.e., $\partial_k c = o$). For example, singular 0-chains are 0-cycles. A simple closed curve in X (viewed as a singular 1-chain) is a 1-cycle; similarly, every subset of X that is homeomorphic to a sphere is a 2-cycle. Our convention is to assign positive multiplicities to counterclockwise 1-cycles, 2-cycles with a right-handed orientation at each point, and higher dimensional k-cycles with the usual orientation of $\mathbf{R}^k$ at each point (cf. Figure 10.2).

Denote the set of all k-cycles in X by $Z_k(X; \mathbf{R})$. If c and c' are k-cycles, then so are $c + c'$ and $r \cdot c$, where $r \in \mathbf{R}$. Consequently, $Z_k(X; \mathbf{R})$ is a vector subspace of $S_k(X; \mathbf{R})$. A cycle c in $Z_k(X; \mathbf{R})$ is *bounding* if $c = \partial_{k+1} d$, where d is a singular $(k+1)$-chain in X; otherwise, c is *nonbounding*. For example, the zero k-cycle o is bounding since it is the boundary of a singular $(k+1)$-chain without boundary. The 1-cycle $+c$ in D_1 of Figure 10.3 is bounding (because it bounds the region it encloses). Note that equation (10.2) yields that the set $Bd_k(X; \mathbf{R})$ of bounding cycles in X forms a vector subspace of $Z_k(X; \mathbf{R})$ (since $Bd_k(X; \mathbf{R})$ is the kernel of ∂_k).

Two k-cycles c and c' are *homologous*, written $c \sim c'$, if $c - c'$ is bounding. It follows that a bounding k-cycle cannot be homologous to a nonbounding k-cycle. Also, $c \sim o$ when c is bounding. For example, $+c \sim o$ in D_1 of Figure 10.3. Furthermore, $+a \sim +b$ (since $+a - b$ is the boundary of the annular region between the two curves). In D_2, note that $+c \sim +a + b$. The relation $\sim$ is an equivalence relation on $Z_k(X; \mathbf{R})$. We then divide $Z_k(X; \mathbf{R})$ into equivalence classes of mutually homologous cycles and call each class a *k-homology class*. Denote the k-homology class of a k-cycle c by $[c]$. Observe that the class $[o]$, called the *zero* k-homology class, consists of all bounding k-cycles in X.

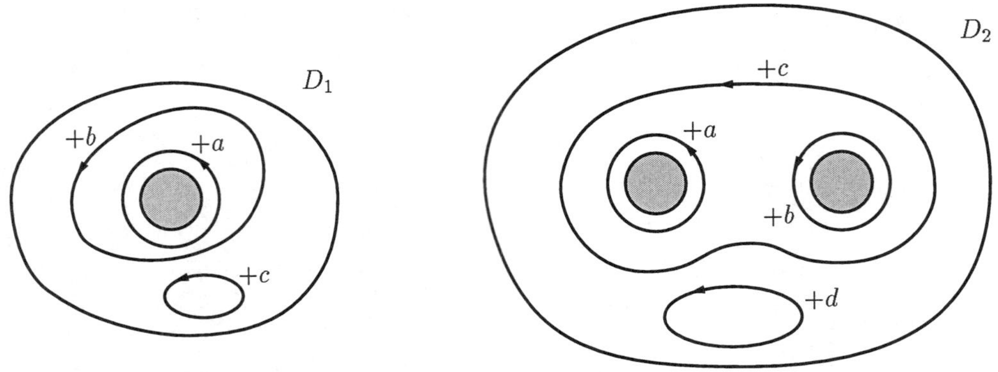

Figure 10.3. The above spaces show some 1-cycles.

Let $H_k(X; \mathbf{R})$ be the set of all k-homology classes in X. Define "vector addition" and "scalar multiplication" on $H_k(X; \mathbf{R})$ as follows: $[c]+[d] = [c+d]$ and $\alpha[c] = [\alpha \cdot c]$, where $\alpha \in \mathbf{R}$. These operations are well defined since if $c \sim c'$ and $d \sim d'$, then $c + d \sim c' + d'$ and $\alpha \cdot c \sim \alpha \cdot c'$ (i.e., $[c] + [d] = [c'] + [d']$ and $\alpha[c] = \alpha[c']$). Hence, the set $H_k(X; \mathbf{R})$ becomes a vector space, called the kth *homology vector space*, or, *homology group* of X.[3] Notice that every homology class of $H_k(X; \mathbf{R})$, except the zero class $[o]$, consists of nonbounding k-cycles in X. In other words, the vector space $H_k(X; \mathbf{R})$ is the quotient vector space $Z_k(X; \mathbf{R})/Bd_k(X; \mathbf{R})$, where $Bd_k(X; \mathbf{R})$ is the subspace of all bounding cycles in $Z_k(X; \mathbf{R})$. For simplicity, we shall often write $H_k(X)$ instead of $H_k(X; \mathbf{R})$.

The dimension of $H_k(X)$ is called the kth *Betti number* of X and denoted $B_k(X)$. A fundamental property of Betti numbers is that these numbers remain unchanged when X is deformed via a homeomorphism (see Theorem 10.1). In fact, Betti numbers are among the earliest topological invariants discovered. Another basic property of Betti numbers is that if X is a compact n-manifold, then all Betti numbers of X are finite and $B_k(X) = 0$ for all $k > n$ (see Theorem 10.3). The latter follows intuitively since X cannot contain a k-cycle when $k > n$. Thus, the manifold X has $n + 1$ Betti numbers: $B_0(X), B_1(X), \ldots, B_n(X)$.

We now prove that *if X is path connected, then $B_0(X) = 1$*. Fix a point p_0 (i.e., 0-cycle) in X. Since X is a connected manifold, it is path connected. Consequently, every singular 0-cycle p in X is homologous to the singular 0-cycle p_0 (since a simple arc s from p_0 to p is a 1-chain with $\partial s = p - p_0$). Indeed, let $p = r_1 \cdot p_1 + \cdots + r_n \cdot p_n$ be any 0-chain in X. Then $c \sim r_1 \cdot p_0 + \cdots + r_n \cdot p_0 = (\sum_{i=1}^{n} r_i) p_0$ (since $p_i \sim p_0$). In other words, every 0-cycle in X is homologous to a 0-cycle of the form $r \cdot p_0$, where $r \in \mathbf{R}$. The bounding property of 0-cycles is then determined by that of $r \cdot p_0$ (since two homologous cycles are either both bounding or both nonbounding). We claim that $r \cdot p_0$ is bounding if and only if $r = 0$. If $r = 0$, then $r \cdot p_0 = o$, which is bounding. Conversely, suppose that $r \cdot p_0 = \partial c$ for some singular 1-chain

[3]This notion developed out of the seminal works of Poincaré and his predecessors Riemann and Betti (see Chapter 1 of Dieudonné [Die]).

c, say, $c = \alpha_1 \cdot e_1 + \cdots + \alpha_m \cdot e_m$, where e_i is a simple arc from some point q_i to q_i'. Note that if the singular 1-chain e_i is a loop, then $q_i' = q_i$. Then $\partial c = \alpha_1 \cdot (q_1' - q_1) + \cdot + \alpha_m \cdot (q_m' - q_m)$. Define a map $\epsilon : C_0(X) \to \mathbf{R}$ by $\epsilon(p) = r_1 + \cdots + r_n$. Then ϵ is linear. Consequently, $r = \epsilon(rp_0) = \epsilon(\partial c) = \alpha_1 \cdot (\epsilon(q_1') - \epsilon(q_1)) + \cdot + \alpha_m \cdot (\epsilon(q_m') - \epsilon(q_m))$. But $\epsilon(q_i') = 1 = \epsilon(q_i))$, which yields $\epsilon(\partial c) = 0$. Hence, $r = 0$. We can now conclude that all nonbounding 0-cycles in X belong to a k-homology class of the form $[r \cdot p_0]$, where $r \neq 0$. Thus, $H_k(X) = \{r[p_0] : r \in \mathbf{R}\}$ and $B_k(X) = 1$ (since $\{[p_0]\}$ is a basis). Notice that the map ϵ induces a vector space isomorphism $H_k(X) \approx \mathbf{R}$.

Examples:

(1) **Disk:** Let D_0 be a disc in the plane. Since D_0 is path-connected, $B_0(D_0) = 1$. We claim that $B_1(D_0) = 0$. In fact, every 1-cycle c in D_0 is homologous to o (since c bounds the region it encloses). Consequently, all 1-cycles in D_0 are also homologous to o (since they are linear combinations of 1-cycles). Hence, $H_1(D_0) = \{[o]\}$, which has no basis.[4] Finally, since no 2-cycle is a subset of a planar region, we have $B_2(D_0) = 0$.

(2) **Annulus:** In Figure 10.3, the annulus D_1 has $B_0(D_1) = 1$ and $B_2(D_1) = 0$ (since 2-cycles cannot exist in D_1). Let us show that $B_1(D_1) = 1$. Each 1-cycle e in D_1 either goes around the hole (e.g., $+a$) or does not (e.g., $+c$). In the latter case, we have $e \sim o$ (i.e., e is bounding), which yields that all 1-cycles that do not go around the hole belong to the class $[o]$. If e goes around the hole, then e is nonbounding since the presence of the hole makes it impossible for e to form the entire boundary of a singular 2-chain in D_1. But any two counterclockwise 1-cycles around the hole are homologous (e.g., $+a \sim +b$ since they bound the annular region between them). A similar result holds for clockwise 1-cycles. Consequently, if e does go around the hole, then $e \sim \pm a$. Hence, every 1-cycle about the hole is of the form $r \cdot a$, where $r \in \mathbf{R}$. Thus, $H_1(D_0) = \{r[a] : r \in \mathbf{R}\}$, which is one dimensional.

(3) **Disc with g holes:** First, consider the two-holed disc D_2 in Figure 10.3. We have $B_0(D_2) = 1$ and $B_2(D_2) = 0$. The reader may guess that $B_1(D_2) = 2$. In fact, any 1-cycle e in D_2 either goes around the left hole (e.g., $+a$), right hole (e.g., $+b$), both holes (e.g., $+c$), or none of the holes (e.g., $+d$). For the latter, we have $e \sim o$; hence, each 1-cycle that does not go around any hole lies in $[o]$. If e goes around the left hole, then $e \sim \pm a$ and e is nonbounding (because the hole prevents e from being the whole boundary of a 2-chain in D_2). For a similar reason, if e goes around the right hole, then e is nonbounding and $e \sim \pm b$. In addition, any 1-cycle going around the left hole is not homologous to one about the right one (since their difference is not the boundary of a 2-chain in D_2). If e goes around both holes, then e is nonbounding and homologous to $\pm(a \pm b)$ (note a figure "8" looping about both holes). Hence, every 1-cycle in D_2 is homologous to a 1-cycle of the form $r \cdot a + r' \cdot b$, where $r, r' \in \mathbf{R}$. $H_1(D_0) = \{r[a] + r'[b] : r, r' \in \mathbf{R}\}$, which is two dimensional with $\{[a], [b]\}$ as a basis. Using analogous arguments, we see that *if D_g is a disc with g holes, then $B_0(D_g) = 1$, $B_1(D_g) = g$, and $B_2(D_g) = 0$.*

[4] In a vector space, any finite set containing the zero vector is linearly dependent.

The above examples suggest an intuitive way to think about Betti numbers. For example, the zeroth Betti number $B_0(X)$ is the number of path connected components of X. For $k \geq 1$, the kth Betti number $B_k(X)$ is, loosely speaking, the number of "$(k+1)$-dimensional holes" in X, that is, "holes" bounded by k-cycles. In other words, the Betti numbers $B_k(X)$ with $k \geq 1$, count the number of "holes" of certain types in X. For example, the disc D_0 has no holes, so $B_1(D_0) = 0$. If we remove g discs from the interior of D_0, what remains is a disc D_g with g holes and $B_1(D_g) = g$. Similarly, if $\mathcal{B}$ is a 3-dimensional ball, then $B_2(\mathcal{B}) = 0$. Removing balls, solid tori, etc., produces 3-dimensional holes that increase $B_2(\mathcal{B})$. Note that introducing such holes may also increase $B_1(\mathcal{B})$.

An important integer that is constructed out of Betti numbers is

$$\chi(X) \equiv B_0(X) - B_1(X) + \cdots (-1)^n B_n(X),$$

which is called the *Euler characteristic* of X. For example, if X is path connected and has no "holes," then $\chi(X) = 1$. The Euler characteristic appears in the Morse inequalities and plays an important role in lensed-image counting, magnification, and the global properties of caustic networks.

10.1.2 Precise Treatment of Betti Numbers

Fix a topological space X. First, we reintroduce standard k-simplices Δ_k without the restriction that they must lie in $\mathbf{R}^{k+1}$. Let $\mathbf{R}^\infty$ be a product of countably many copies of $\mathbf{R}$ and set $\epsilon_i = (0, \ldots, 0, 1, 0, \ldots)$ with a 1 in the $(i+1)$ slot and zeros in all the remaining slots ($i = 0, 1, 2, \ldots$). The *standard k-simplex* in $\mathbf{R}^\infty$ with vertices in the natural order $\epsilon_0, \epsilon_1, \ldots, \epsilon_k$, is the following topological subspace of $\mathbf{R}^\infty$:

$$\Delta_k = \left\{ \mathbf{x} \in \mathbf{R}^\infty : \mathbf{x} = \sum_{i=0}^{k} a_i \epsilon_i, \; a_i \geq 0, \; \sum_{i=0}^{k} a_i = 1 \right\}.$$

Define a *singular k-simplex* in X to be a continuous map $\sigma : \Delta_k \to X$. Denote the set of all singular k-simplices in X by $\mathcal{S}_k(X)$. Let $S_k(X; \mathbf{R})$ be the set of all maps $c : \mathcal{S}_k(X) \to \mathbf{R}$ such that $c(s) \neq 0$ for only finitely many s in $\mathcal{S}_k(X)$. Elements of $S_k(X; \mathbf{R})$ are called *singular k-chains*. If $c, c' \in S_k(X; \mathbf{R})$ and $r \in \mathbf{R}$, then defining addition and scalar multiplication by $(c + c')(s) = c(s) + c'(s)$ and $(r \cdot c)(s) = rc(s)$, respectively, transform $S_k(X; \mathbf{R})$ into a real vector space (i.e., the free $\mathbf{R}$-module generated by $\mathcal{S}_k(X)$). Given $\sigma \in \mathcal{S}_k(X)$, define a characteristic function c_σ at σ by $c_\sigma : \mathcal{S}_k(X) \to \mathbf{R}$ with $c_\sigma(s) = 0$ if $s \neq \sigma$ and $c_\sigma(s) = 1$ if $s = \sigma$. Then the collection $\{c_\sigma : \sigma \in \mathcal{S}_k(X)\}$ is a basis for $S_k(X; \mathbf{R})$. Each singular k-chain $c \in S_k(X; \mathbf{R})$ can be written uniquely as $c = \sum r_\sigma c_\sigma$, where $r_\sigma = c(\sigma)$ and the sum runs over all $\sigma \in \mathcal{S}_k(X)$ for which $c(\sigma) \neq 0$. Notationally, we shall identify c_σ with σ and write simply $c = \sum r_\sigma \sigma$.

For $k \geq 1$ and $i = 0, 1, \ldots, k$, define continuous maps $F_k^i : \Delta_{k-1} \to \Delta_k$ by $F_k^i(a_0, a_1, \ldots, a_{k-1}, 0, \ldots) = (a_0, a_1, \ldots, a_{i-1}, 0, a_i, \ldots, a_{k-1}, 0, \ldots)$. The *boundary* of a singular

k-simplex s is the singular $(k-1)$-chain in X given by

$$\partial_k s = \sum_{i=0}^{k} (-1)^i s \circ F_k^i.$$

If $c = \sum r_\sigma \sigma$ is a singular k-chain in X, then the *boundary* of c is defined by

$$\partial_k c = \sum r_\sigma \partial \sigma,$$

so ∂_k is a linear map $S_k(X; \mathbf{R}) \to S_{k-1}(X; \mathbf{R})$. Note that since $F_k^i \circ F_{k-1}^{j-1} = F_k^j \circ F_{k-1}^i$ for $0 \le i < j \le k$ (cf. equation (10.1)), we obtain $\partial_{k-1}(\partial_k c) = o$, where o is the zero vector in $S_k(X; \mathbf{R})$.

A singular k-chain c is called a *k-cycle* if $\partial c = o$. *Bounding* k-cycles c satisfy $c = \partial_{k+1} d$ for some singular $(k+1)$-chain d; otherwise, c is *nonbounding*. Let $Z_k(X; \mathbf{R})$ be the vector subspace of $S_k(X; \mathbf{R})$ consisting of all singular k-cycles in X. Two singular k-cycles c and c' in $Z_k(X; \mathbf{R})$ are *homologous*, written $c \sim c'$, if there is a singular $(k+1)$-chain d in $C_{k+1}(X; \mathbf{R})$ such that $c - c' = \partial d$. The set $Bd_k(X; \mathbf{R})$ of bounding k-cycles on X forms a vector subspace of $Z_k(X; \mathbf{R})$ (since $\partial_{k-1}(\partial_k c) = o$ yields that $Bd_k(X; \mathbf{R})$ is contained in Kernel(∂_k)). The quotient vector space $H_k(X; \mathbf{R}) = Z_k(X; \mathbf{R})/Bd_k(X; \mathbf{R})$ is called the kth *(absolute) singular homology vector space* (or *singular homology group*) of the topological space X. The space $H_k(X; \mathbf{R})$ consists of equivalence classes of mutually homologous k-cycles. The dimension of $H_k(X; \mathbf{R})$ is called the kth *(absolute) Betti number* of X and denoted by $B_k(X; \mathbf{R})$ or $B_k(X)$ for simplicity.

The theorems below are useful to keep in mind when computing Betti numbers. However, their proofs assume a knowledge of algebraic topology — see Munkres [Mun84] for an excellent introduction. We encourage the reader to apply these theorems to the spaces in Figure 10.4.

Theorem 10.1 *If X and X' are homeomorphic topological spaces, then $B_k(X) = B_k(X')$ for all non-negative integers k.*

Proof: See [Gre-H, p. 49]. □

Theorem 10.2 *If X is a topological space with m path-connected components, then $B_0(X) = m$.*

Proof: See [Gre-H, p. 48, Prop. 9.6]. □

Theorem 10.3 *If X is a compact n-manifold with or without boundary, then all Betti numbers of X are finite and $B_k(X) = 0$ for all integers $k \ge n+1$.*

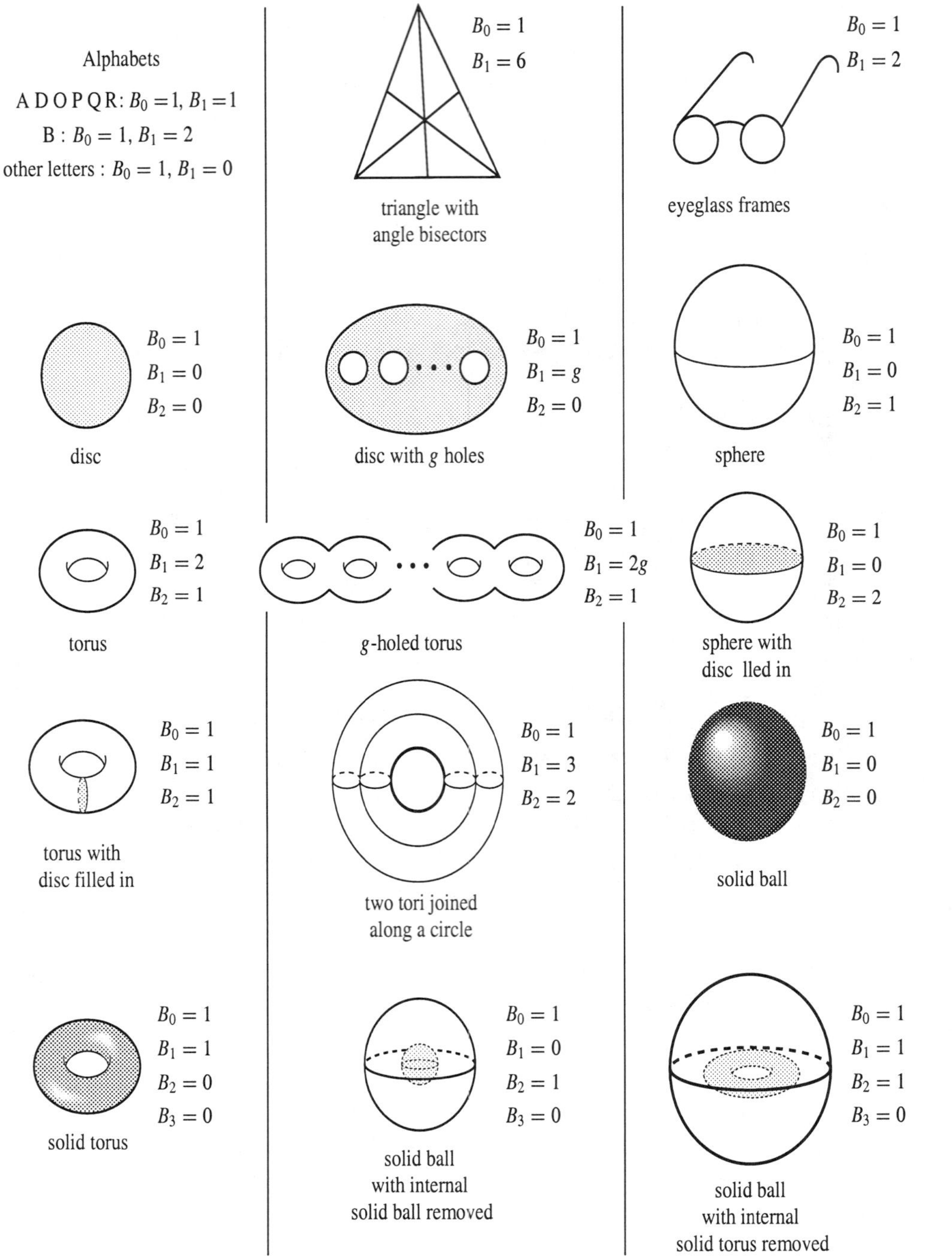

Figure 10.4. A gallery of topological spaces and their Betti numbers. (Examples courtesy of Robert MacPherson.)

Proof of Theorem 10.3: For the case $\partial X = \emptyset$, see [Mor-C, p. 267, Thm 30.1 (i)]. When $\partial X \neq \emptyset$, consult [Mas, p. 372] for the finiteness of the Betti numbers.[5] □

Theorem 10.4 (Alexander Duality) *If X is a compact subset of $\mathbf{R}^{k+1}$, then*

$$B_k(X) = B_0(\mathbf{R}^{k+1} - X) - 1$$

for all nonnegative integers k.

Proof: See [Spa, p. 296, Thm. 16] and [Gre-H, p. 48, (9.7); p. 194, (23.40)]. □

Theorem 10.5 (Künneth's Formula) *For topological spaces X_1 and X_2, we have*

$$B_k(X_1 \times X_2) = \sum_{\ell+m=k} B_\ell(X_1) B_m(X_2)$$

for all nonnegative integers k.

Proof: See [Gre-H, p. 256, Cor. 29.11.1]. □

10.2 Morse Inequalities A and B

The current section presents Morse inequalities A and B, while Section 10.3 deals with their proofs.

Motivation: Let f be a function on an open subset of $\mathbf{R}^n$. Suppose that the critical points of f which are of interest to us lie in a compact set C whose boundary ∂C is an $(n-1)$-manifold. The Morse inequalities relate the number of such critical points of a given type to the Betti numbers of C, taking into account the behavior of grad f at points of ∂C. This gives rise to the so-called *Morse inequalities under boundary conditions A and B.* Since critical points of time delay functions $T_{\mathbf{y}}$ are identified with light rays, these inequalities are relevant to gravitational lensing. The light rays lie in a compact region D that experiences gravitational pulling by surrounding matter. Such external tugs affect the behavior of grad $T_{\mathbf{y}}$ on the boundary of D. If the ambient matter's gravitational tug is negligible, then grad $T_{\mathbf{y}}$ is outward pointing on ∂D, that is, Morse boundary conditions A hold. For nontrivial tugging, the gradient grad $T_{\mathbf{y}}$ points inward on parts of ∂D, a situation invoking Morse boundary conditions B.

Let D be a compact subset of $\mathbf{R}^2$ with topological boundary ∂D a smooth 1-manifold. Suppose that $T_{\mathbf{y}}$ is a single-plane time delay function defined on D such that $\mathbf{y}$ is not a caustic point and no light ray of $T_{\mathbf{y}}$ lies on ∂D. Assume that $T_{\mathbf{y}}$ satisfies *Morse boundary*

[5] To show that $B_k(X) = 0$ for $k > n$ when $\partial X \neq \emptyset$, use the fact that the relative homology group $H_k(X, \partial X)$ vanishes for $k > n$ [Eil-S, p. 311] and employ the exact homology sequence of the pair $(X, \partial X)$. See Section 10.3.2 (page 411) for more on relative homology groups.

conditions A relative to D, that is, at each point of ∂D the gradient of T_y is directed into the exterior of D. For example, the single-plane time delay function due to a point mass potential ψ_{pt} satisfies Morse boundary conditions A relative to some closed punctured disc. Denote the respective numbers of local minima, saddles, and local maxima of T_y in D by M_0, M_1, and M_2. Note that M_i is the number of critical points in D of index i. *Morse inequalities A* apply and state that M_i and the Betti numbers $B_i(D)$ of D obey

$$\begin{aligned} M_0 &\geq B_0(D) \\ M_1 - M_0 &\geq B_1(D) - B_0(D) \\ M_2 - M_1 + M_0 &= B_2(D) - B_1(D) + B_0(D) = \chi(D). \end{aligned}$$

Consequently, $M_i \geq B_i(D)$ for $i = 0, 1, 2$. Morse inequalities A also yield counting formulas and lower bounds for the number of lensed images of different types due to T_y as time delay function. In fact, the total number N of lensed images in D satisfies

$$\begin{aligned} N &= M_0 + M_1 + M_2 = 2(M_0 + M_2) - \chi(D) = 2M_1 + \chi(D) \\ N &\geq B_0(D) + B_1(D) + B_2(D). \end{aligned}$$

If T_y describes a gravitational lens system consisting of point masses on a single-plane with a strong external shear or sufficiently dense continuous matter, then at certain boundary points of D the gradient of T_y points into the interior of D. Situations like this (where Morse boundary conditions A fail) are handled by Morse boundary conditions B. In fact, let D_- be the union of all open subsets of ∂D where grad T_y points into the interior of D. Then T_y is said to obey Morse boundary conditions B if the restriction of T_y to ∂D is a map with only nondegenerate critical points. Let M_0' and M_1' be, resp., the number of local minima and maxima of $T_y|_{D_-}$. Then *Morse inequalities B* apply and yield the following generalization of Morse inequalities A:

$$\begin{aligned} M_0 &\geq (B_0(D) - M_0') \\ M_1 - M_0 &\geq (B_1(D) - M_1') - (B_0(D) - M_0') \\ M_2 - M_1 + M_0 &= B_2(D) - (B_1(D) - M_1') + (B_0(D) - M_0'). \end{aligned}$$

Consequently, we obtain $M_0 \geq B_0(D) - M_0'$, $M_1 \geq B_1(D) - M_1'$, and $M_2 \geq B_2(D)$. The total number of lensed images satisfies

$$\begin{aligned} N &= 2(M_0 + M_2) - \chi(D) + (M_0' - M_1') = 2M_1 + \chi(D) - (M_0' - M_1') \\ N &\geq B_0(D) + B_1(D) + B_2(D) - (M_0' + M_1'). \end{aligned}$$

For gravitational lensing by k lens planes, the domain of T_y has dimension $2k$, which requires the full higher dimensional Morse inequalities. We now precisely present these inequalities. Define a *regular (smooth) domain* in $\mathbf{R}^n$ to be a compact subset $C \subset \mathbf{R}^n$

whose topological boundary is a smooth $(n-1)$-dimensional manifold, denoted by ∂C.[6] Second, let $f : U \subseteq \mathbf{R}^n \to \mathbf{R}$, where U is open, be a twice continuously differentiable function. Assume that U contains a regular domain C in $\mathbf{R}^n$. If f has only nondegenerate critical points in C and no critical point on ∂C, then f is called *admissible relative* to C. The function f obeys *Morse boundary conditions B* relative to a regular domain C in $\mathbf{R}^n$ if f is admissible relative to C and the restricted function $f|_{\partial C}$ is nondegenerate.

Suppose that $f : U \subseteq \mathbf{R}^n \to \mathbf{R}$ obeys Morse boundary conditions B relative to a regular domain C in $\mathbf{R}^n$. By Corollaries 7.43 (page 240) and 7.44 (page 240), *the critical points of f in C, as well as those of the restricted function $f|_{\partial C}$, are isolated and finite in number.*

Notation:

(1) For $0 \leq i \leq n$, let M_i be the number of index i critical points of f lying in C. For example, M_0 and M_n are, respectively, the total number of local minima and maxima of f in C. Set:

$$N = M_0 + M_1 + \cdots + M_n = \text{total \# of critical points in } C.$$

$$M_+ = \sum_{i(\text{even})} M_i = \text{total \# of critical points in } C \text{ of even index.}$$

$$M_- = \sum_{i(\text{odd})} M_i = \text{total \# of critical points in } C \text{ of odd index.}$$

(2) Let C_- be the union of all open submanifolds of the boundary ∂C of C on which the gradient of f is directed into the interior of $C \subset \mathbf{R}^n$.

(3) Let M_i' be the number of critical points of index i of the restriction $f|_{C_-}$, where $i = 0, \ldots, n-1$.

Morse's inequalities under boundary conditions B are given precisely as:

Theorem 10.6 (Morse Inequalities B) *Let $f : U \subseteq \mathbf{R}^n \to \mathbf{R}$, where U is open, be a twice continuously differentiable function. Suppose that f satisfies Morse boundary conditions* B *relative to a regular smooth domain C in U. If $B_0, B_1, \ldots, B_n$ with $B_n = 0$ are the Betti numbers of C, then*

[6] Note that a regular domain need not be connected.

$$\begin{aligned}
M_0 &\geq (B_0 - M_0') \\
M_1 - M_0 &\geq (B_1 - M_1') - (B_0 - M_0') \\
M_2 - M_1 + M_0 &\geq (B_2 - M_2') - (B_1 - M_1') + (B_0 - M_0') \\
&\vdots \\
M_{n-1} - \cdots (-1)^{n-1} M_0 &\geq (B_{n-1} - M_{n-1}') - \cdots (-1)^{n-1}(B_0 - M_0') \\
M_n - M_{n-1} + \cdots (-1)^n M_0 &= B_n - (B_{n-1} - M_{n-1}') + \cdots (-1)^n (B_0 - M_0').
\end{aligned}$$

The above inequalities are sometimes called the *strong Morse inequalities.*

Remark: Though the full Morse inequalities appeared in 1925 [Mor], they were preceded by an equality of Poincaré and an inequality of Birkhoff. Poincaré showed in 1885 that for a vector field on a compact orientable two-dimensional surface S of genus g, we have $\mathcal{N} - \mathcal{C} + \mathcal{F} = 2 - 2g$, where $\mathcal{N}$, $\mathcal{C}$, and $\mathcal{F}$ are, resp., the number of nodes, saddles, and foci of the vector field [Poi, p. 121]. Observe that since S is a g-holed torus, we have $2 - 2g = \chi(S)$. However, it was Birkhoff's discovery in 1917 [Bir, p. 42] of the *Minimax Principle* that gave the first indication of a link between the critical points of a function and the Betti numbers of its domain. This principle states that the number of minimax critical points (i.e., saddles of index 1) is bounded below by $M_0 + B_1 - 1$. In other words, Birkhoff found the inequality, $M_1 - M_2 \geq B_1 - B_0$, with $B_0 = 1$. It was the Minimax Principle that inspired Morse's work (see [Mor, p. 346]).

Corollary 10.7 (Weak Morse Inequalities) *Under the hypotheses of Theorem 10.6, we have* $M_i \geq B_i - M_i'$ *for* $i = 0, 1, \ldots, n-1$, *and* $M_n \geq 0$.

The latter lower bound in Corollary 10.7 is not a trivial statement; there are gravitational lens systems with $M_n = 0$ (e.g., Figure 11.3, page 436).

If $M_i' = 0$ for $i = 1, \ldots, n-1$, then Morse inequalities B are called *Morse inequalities* A. For example, this is the case when $C_- = \emptyset$. We also have:

Corollary 10.8 *Every nondegenerate function on a compact n-manifold without boundary obeys Morse inequalities* A.

Now consider a k-plane time delay function $T_{\mathbf{y}} : L_{(k)} \to \mathbf{R}$, where $L_{(k)} = L_1 \times \cdots \times L_k \subseteq \mathbf{R}^{2k}$ (product of lens planes), that has finitely many light rays. Suppose that $T_{\mathbf{y}}$ satisfies the hypotheses of Theorem 10.6 relative to a regular smooth domain D in $L_{(k)}$ containing all its light rays. Then the total number N of lensed images of a light source at $\mathbf{y}$ satisfies:[7]

[7] By Theorem 6.17(1) (page 204), the total number of lensed images of a light source at $\mathbf{y}$ coincides with the number of light rays of $T_{\mathbf{y}}$.

$$
\begin{aligned}
N &= 2\sum_{i\,(\text{even})}^{2k} M_i - \chi(D) + \sum_{i=0}^{2k-1} (-1)^i M_i' \\
&= 2\sum_{i\,(\text{odd})}^{2k-1} M_i + \chi(D) - \sum_{i=0}^{2k-1} (-1)^i M_i' \\
N &\geq \sum_{i=0}^{2k} B_i - \sum_{i=0}^{2k-1} M_i'.
\end{aligned}
$$

10.3 Proof of the Morse Inequalities

Before we can sketch the proof of Morse inequalities A and B, several preliminary concepts and facts are needed. They are collected in Sections 10.3.1 and 10.3.2. These sections yield initial links between the critical points of a function and the topology of the function's domain. The actual proofs of Morse inequalities A and B appear in Section 10.3.3. Our main reference for these inequalities is Morse and Cairns [Mor-C, p. 64, Chap. 9; p. 69, Chap. 10].

10.3.1 Critical Values and Change of Topology

Consider the single-plane time delay function due to an elliptical potential:

$$
T_{\mathbf{y}}^{ep}(\mathbf{x}) = \frac{|\mathbf{x}-\mathbf{y}|^2}{2} - C\sqrt{x_{cr}^2 + (1-\epsilon)\mathrm{u}^2 + (1+\epsilon)\mathrm{v}^2}, \qquad (10.3)
$$

where $\mathbf{x} = (\mathrm{u}, \mathrm{v})$, $C > 0$, $x_c > 0$ (core radius), $0 \leq \epsilon \lesssim 0.2$ (ellipticity) — see Section 3.3.2 (page 105). For appropriate values of the latter parameters, the time delay surface of $T_{\mathbf{y}}^{ep}$ (i.e., the graph of $T_{\mathbf{y}}^{ep}$) can have the overall "sombrero" shape in Figure 10.5 with three critical points — a minimum, saddle, and maximum (e.g., [Bla-N86, p. 571]).[8]

The time delay surface of $T_{\mathbf{y}}$ is depicted in Figure 10.5 over a closed disc D. The function $T_{\mathbf{y}}^{ep}$ always obeys Morse boundary conditions A relative to a sufficiently large D (since the time delay surface of $T_{\mathbf{y}}^{ep}$ increases without bound as $|\mathbf{x}| \to \infty$). Also, there is a regular value r such that $D = (T_{\mathbf{y}}^{ep})^{-1}(-\infty, r]$. If the ellipticity ϵ is sufficiently large, then the time delay surface over points near the boundary ∂D can "saddle," that is, the gradient grad $T_{\mathbf{y}}^{ep}$ is inward pointing on a union D_- of opposite arcs of the circle ∂D. In this case Morse inequalities A do not apply. However, after an arbitrarily small linear perturbation, the restriction of $T_{\mathbf{y}}^{ep}$ to ∂D has only nondegenerate critical points — Morse boundary conditions B hold.

[8] See page 426 for a detailed discussion of multiple imaging due to elliptical potentials.

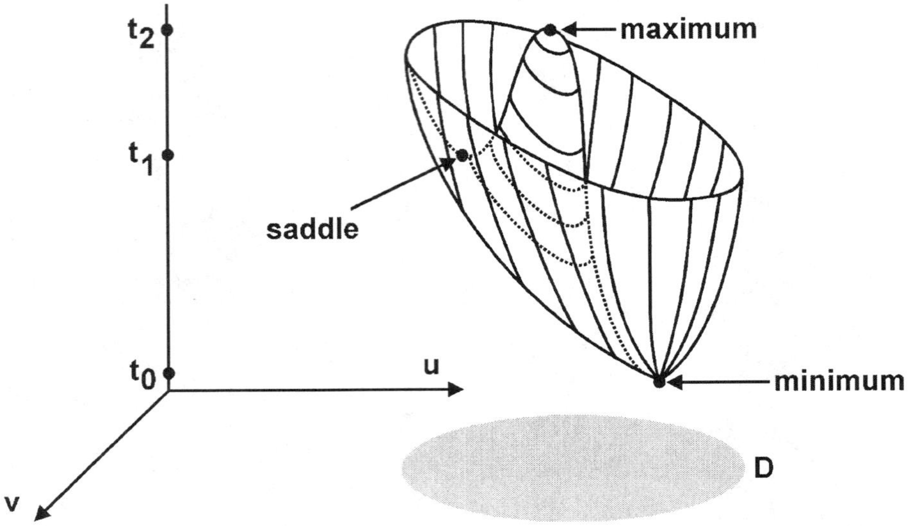

Figure 10.5. "Sombrero" time delay surface due to an elliptical lens. The graph is depicted above a closed disc D.

We now consider the link between critical values and change of topology without using the detailed form of $T_{\mathbf{y}}^{ep}$. In fact, let $f : U \subseteq \mathbf{R}^2 \to \mathbf{R}$, where U is open, be a general twice continuously differentiable function satisfying Morse boundary conditions A relative to a closed disc $D \subset U$ and whose graph over D has the "sombrero" form of Figure 10.5. Then there is a regular value t_r such that

$$D = f^{-1}(-\infty, t_r] .$$

Define a one parameter family of compact subsets of D by

$$D_t = f^{-1}(-\infty, t],$$

where $t \leq t_r$. Note that for any closed interval $[t_\alpha, t_\beta]$ with $t_\beta \leq t_r$, the set $f^{-1}[t_\alpha, t_\beta]$ is compact. We shall study the topology of the "half-spaces" D_t as the "time" t varies.

The following three facts implicitly show a link between the critical values of f and the topology of D_t (consult Milnor [Mil, pp. 12–24] and Hirsch [Hir, pp. 156–160] for details and the general case for manifolds):

Fact 1: *If f has no critical values in the range $[t_\alpha, t_\beta]$, then D_{t_α} is diffeomorphic to D_{t_β}* — see Figure 10.6 and [Mil, p. 12].

This diffeomorphism is essentially constructed by pushing D_{t_β} down onto D_{t_α} along the flow of the gradient of f, that is, along the trajectories orthogonal to level lines $f(\mathbf{x}) = t_c$, where t_c is a constant. Such a deformation is possible because the gradient flow has no singularities for t_c in the interval (t_α, t_β). It follows that D_{t_α} and D_{t_β} have the same Betti numbers.

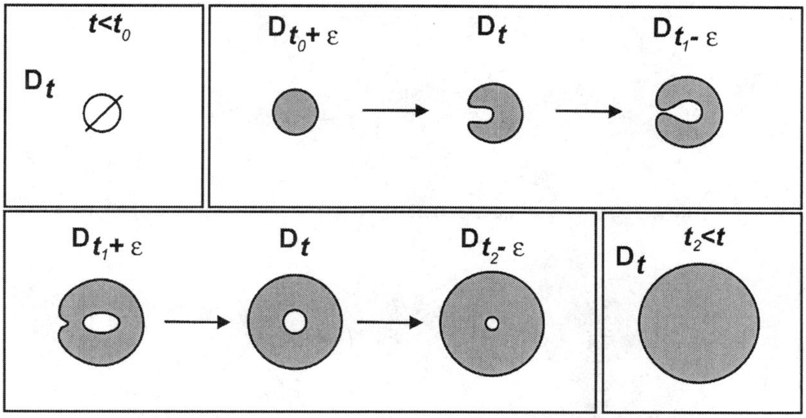

Figure 10.6. Evolution of D_t as t varies through noncritical values for the "sombrero" surface. Observe that no change occurs in the topology of D_t. Arrows can be reversed.

Fact 2: Figure 10.7 yields that for $i = 0, 1, 2$, *as t passes through the critical value* t_i *— which corresponds to a critical point of index i — the topology of* D_t *changes, that is, at least one Betti number varies after each critical value is traversed.*

Define a *k-cell* e_k to be a closed k-dimensional ball (e.g., e_0, e_1, and e_2 are a point, line segment, and closed disc, resp.). A subspace A of a topological space X is a "deformation retract" of X if X can continuously be deformed onto A in such a way that each point of A remains fixed during the collapse – see Figure 10.8. Precisely, the subspace A is a *deformation retract* of X if there is a continuous map $F : X \times I \to X$, where $I = [0, 1]$, such that $F(x, 0) = x$ and $F(x, 1) \in A$ for all $x \in X$, and $F(a, t) = a$ for all $a \in A$ and $t \in I$. Note that in Figure 10.8, the spaces X and A have the same Betti numbers. In general, *if a subspace A of a topological space X is a deformation retract of X, then the Betti numbers of A and X coincide.* [A deformation retract of X onto A yields a homotopy equivalence of X and A. Homotopically equivalent spaces have the same Betti numbers [Mun84, p. 108].] Note that if the interval (t_α, t_β) has no critical value of f, then D_{t_α} is a deformation retract of D_{t_β}.

Fact 3: In Figure 10.9, *the spaces denoted by* $D_{t_i-\epsilon} \bigcup_g e_i$, *where for* $i = 0, 1, 2$, *are deformation retracts of* $D_{t_i+\epsilon}$, *so they have the same Betti numbers.*

The Betti numbers of $D_{t_i+\epsilon}$ can be obtained from $D_{t_i-\epsilon}$ by "gluing" an i-cell e_i to $D_{t_i-\epsilon}$ to form the space $D_{t_i-\epsilon} \bigcup_g e_i$ (the "g" is for "gluing") and then using the fact that

$$B_k(D_{t_i+\epsilon}) = B_k(D_{t_i-\epsilon} \textstyle\bigcup_g e_i).$$

Choose values $\tau_0 < \tau_1 < \tau_2 < \tau_3$ such that D_{τ_ℓ} contains exactly ℓ-critical points and $D_{\tau_3} = D$. For example, set $\tau_0 = t_0 - \epsilon$, $\tau_1 = t_0 + \epsilon$, $\tau_2 = t_1 + \epsilon$, and $\tau_3 = t_2 + \epsilon_0$ for some large enough $\epsilon_0 > 0$ (see Figure 10.7). Then

$$B_k(D_{\tau_\ell}) = B_k(D_{\tau_{\ell-1}} \textstyle\bigcup_g e_{i(\ell)}),$$

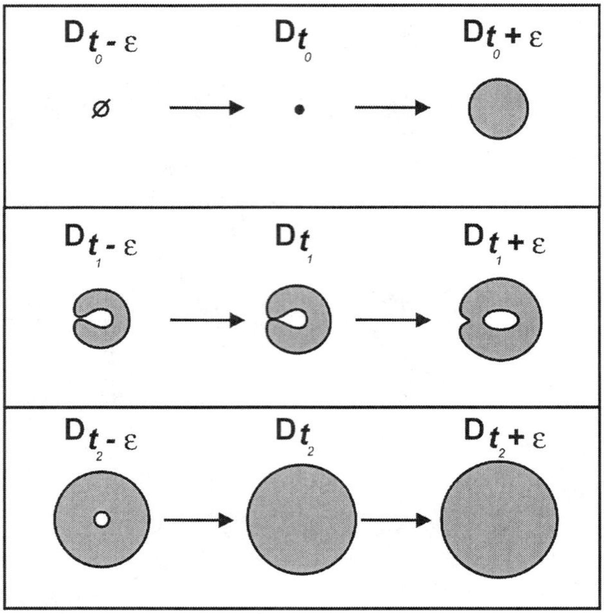

Figure 10.7. Change in topology of D_t as t varies through critical values for the "sombrero" surface. The arrows can be reversed.

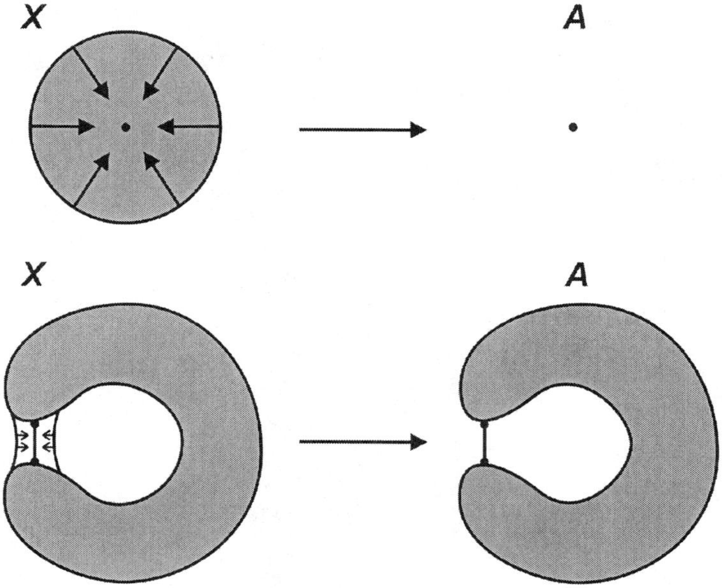

Figure 10.8. The space A is a deformation retract of X.

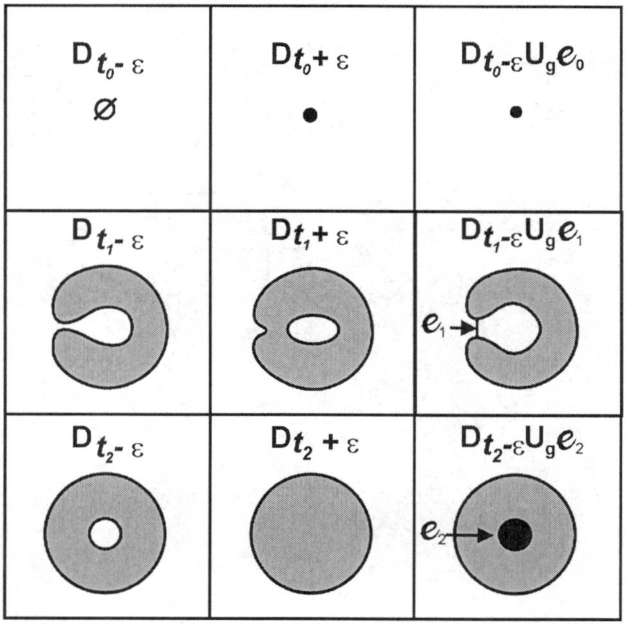

Figure 10.9. The space $D_{t_i-\epsilon} \bigcup_g e_i$ is a deformation retract of $D_{t_i+\epsilon}$ and can be obtained from $D_{t_i-\epsilon}$ by "gluing" on an i-cell e_i.

where $i(\ell)$ is the index of the critical point in $D_{\tau_\ell} - D_{\tau_{\ell-1}}$. It is no accident that an i-cell is "attached" when a transition is made through a critical point of index i — see [Mil, pp. 14-24].

10.3.2 Relative Betti Numbers

In Section 10.3.1, we indirectly showed that for a "sombrero" function $f : U \subseteq \mathbf{R}^2 \to \mathbf{R}$, the topology of the spaces D_t in the closed disc $D \subset U$ is related to the indices of the critical points of f in D. In this section, a direct relationship will be expressed between the critical points (with a given index) of f in D and the "relative Betti numbers" of the spaces D_t — see equation (10.4) below. We shall also present some fundamental properties of these numbers. Readers should compare our discussion with Section 10.1.2 (page 400).

Fix a topological space X and subspace A in X. Consider the real vector space $S_k(X; \mathbf{R})$ of all singular k-chains in X. Let $S_k(A; \mathbf{R})$ be the vector subspace of $S_k(X; \mathbf{R})$ consisting of all singular k-chains $\Delta_k \to X$ whose range is a subset of A. A *relative k-cycle in X mod A* is a k-chain c with $\partial c \in S_k(A; \mathbf{R})$ (i.e., either $\partial c = o$ or the range of ∂c lies in A). For example, simple arcs with endpoints in A and simple closed curves are the ranges of relative 1-cycles in X mod A. In Figure 10.10, the space X is an annulus and the curves are relative 1-cycles in the annulus mod the boundary A of the annulus. Generally, the set of all relative k-cycles in X mod A form a vector subspace $Z_k(X, A; \mathbf{R})$ of $S_k(X; \mathbf{R})$.

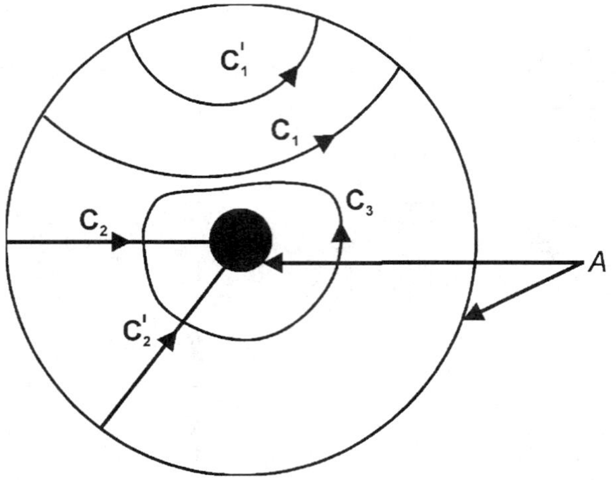

Figure 10.10. Relative 1-cycles in an annulus. The set A consists of the inner and outer boundary circles of the annulus.

If c is a relative k-cycle in X mod A, then c is *bounding* if there is a chain $d \in S_{k+1}(X; \mathbf{R})$ such that $(c - \partial d) \in S_k(A; \mathbf{R})$, that is, c along with some k-chain in A form the boundary of a $(k + 1)$-chain in X. If c is not bounding, then it is called *nonbounding*. The relative 1-cycles c_2 and c_3 in Figure 10.10 are nonbounding and bounding, respectively. Two relative k-cycles in X mod A, say c and c', are *homologous*, written $c \sim c'$, if $c - c'$ is bounding. The cycles c_1 and c_1' (resp., c_2 and c_2') in Figure 10.10 are homologous. Note that all bounding relative k-cycles in X mod A are homologous to the zero relative k-cycle in X mod A. Using the $\sim$ equivalence relation, partition $Z_k(X, A; \mathbf{R})$ into equivalence classes, called *relative k-homology classes in X mod A*. The set of all such classes forms a vector space $H_k(X, A; \mathbf{R})$ called the kth *relative homology vector space* (or, *homology group*) *of X mod A*. Every nonzero relative homology class of $H_k(X, A; \mathbf{R})$ consists of nonbounding relative k-cycles in X mod A. The kth *relative Betti number $B_k(X, A)$ of X mod A* is the dimension of $B_k(X, A)$. Note that for $A = \emptyset$ we have $B_k(X, \emptyset) = B_k(X)$, which are the (absolute) Betti numbers of Section 10.1.2 (page 400). *If there is a homeomorphism mapping X onto another topological space X' mapping A onto a subspace $A' \subseteq X'$, then $H_k(X, A; \mathbf{R})$ is isomorphic to $H_k(X', A'; \mathbf{R})$* (e.g., Wallace [Wall, p. 16, Thm. 1.6]).

Example: Choose X to be the annulus in Figure 10.10 and let A be its boundary. The relative 0-cycles of X are the points of X. They are all homologous since there is a path connecting any two of them. In addition, there is a path c from some point a in A to any relative 0-cycle, say x_0, of X. The boundary of c consists of x_0 and a, so the difference between the boundary of c and x_0 is a 0-dimensional region in A, namely, a. In other words, every relative 0-cycle is bounding. Hence, $B_0(X, A) = 0$. In general, *if X is a path connected topological space and A a nonempty subspace of X, then $B_0(X, A) = 0$* (e.g., [Gre-H, p. 73, Prop. (13.11)]). The relative 1-cycle c_1 is bounding since there is a

2-dimensional region in X (i.e., the region between c_1 and the outer boundary circle) whose boundary differs from c_1 by a 1-dimensional subspace of A. However, the relative cycle c_2 is nonbounding and its relative homology class forms a basis for $H_1(X, A; \mathbf{R})$ (cf. [Mun84, p. 49]), so $B_1(X, A) = 1$. Finally, the interior of the annulus is a relative 2-cycle. It is nonbounding because no 3-dimensional space can exist in D. Thus, $B_2(D, A) = 1$.

It should be apparent from the above example that calculating a relative Betti number directly from its definition can be taxing on one's intuition. We now list some theorems that are helpful in determining the relative Betti numbers appearing in the proof of the Morse inequalities. For example, Theorem 10.11 below gives an easy way to compute the Betti numbers of an annulus relative to its boundary.

Theorem 10.9 (Deformation Retracts) *Suppose A and A' are, respectively, subspaces of topological spaces X and X'. Let A' and X' be deformation retracts of A and X, respectively. Then $B_k(X, A) = B_k(X', A')$ for all nonnegative integers k.*

Proofs of Theorem 10.9 are in [Mun84, pp. 108, 174] and [Mor-C, p. 251].

Example: $B_k(D_{\tau_\ell}, D_{\tau_{\ell-1}}) = B_k(D_{\tau_{\ell-1}} \bigcup_g e_{i(\ell)}, D_{\tau_{\ell-1}})$ by Theorem 10.9 and Figure 10.9. The critical point in $D_{\tau_\ell} - D_{\tau_{\ell-1}}$ has index $i(\ell)$.

Theorem 10.10 (Excision Theorem) *Let A be a subspace of a topological space X. If U is an open set in X having closure contained in the interior of A, then for all nonnegative integers k we obtain*

$$B_k(X - U, A - U) = B_k(X, A).$$

The Excision Theorem states that the open set U can be excised from the subset A of X without affecting the relative Betti numbers. See [Wall, p. 36] and [Mun84, p. 180] for proofs.

Example: By the Excision Theorem and Figure 10.9, if U is the interior of $D_{\tau_{\ell-1}} - (D_{\tau_{\ell-1}} \bigcap e_{i(\ell)})$, where $D_{\tau_{\ell-1}} \bigcap e_{i(\ell)} = \partial e_{i(\ell)}$, then

$$B_k(D_{\tau_{\ell-1}} \bigcup_g e_{i(\ell)}, D_{\tau_{\ell-1}}) = B_k((D_{\tau_{\ell-1}} \bigcup_g e_{i(\ell)}) - U, D_{\tau_{\ell-1}} - U) = B_k(e_{i(\ell)}, \partial e_{i(\ell)}).$$

Theorem 10.11 (Lefschetz Duality) *Suppose X is a compact n-manifold with oriented boundary ∂X.[9] For all non-negative integers k we get*

$$B_k(X, \partial X) = B_{n-k}(X - \partial X) = B_{n-k}(X).$$

[9] This theorem actually occurs in the more general context of *topological n-manifolds with boundary*, that is, a paracompact Hausdorff space X such that every point of X has a neighborhood homeomorphic to an open subset of Euclidean half-space $\bar{\mathbf{R}}^n_+ = \{(x_1, x_2, \ldots, x_n) \in \mathbf{R}^n : x_n \geq 0\}$.

The Lefschetz duality theorem expresses relative Betti numbers in terms of the usual (absolute) Betti numbers — see [Gre-H, p. 194; p. 242], [Vic, p. 171], and [Spa, p. 298] for proofs. Note how quickly this theorem yields the Betti numbers of an annulus relative to its boundary.

Example: Continuing with Figure 10.9 and the function f (whose spaces D_t are depicted), the Lefschetz duality theorem and the fact that $e_i - \partial e_i$ has no "holes" imply

$$B_k(e_{i(\ell)}, \partial e_{i(\ell)}) = B_{i(\ell)-k}(e_{i(\ell)} - \partial e_{i(\ell)}) = B_{i(\ell)-k}(e_{i(\ell)}) = \delta_{k,i(\ell)},$$

where $\delta_{k,i(\ell)}$ is the Kronecker delta. We can immediately conclude from this and the examples above that $B_k(D_{\tau_\ell}, D_{\tau_{\ell-1}}) = \delta_{k,i(\ell)}$, where $i(\ell)$ is the index of the critical point in $D_{\tau_\ell} - D_{\tau_{\ell-1}}$. In other words, the number $B_k(D_{\tau_\ell}, D_{\tau_{\ell-1}})$ equals 1 whenever an index k critical point occurs in $D_{\tau_\ell} - D_{\tau_{\ell-1}}$, and equals 0 otherwise. Consequently, the number M_k of critical points of f of index k in D is given by

$$M_k = \sum_{\ell=1}^{3} B_k(D_{\tau_\ell}, D_{\tau_{\ell-1}}). \tag{10.4}$$

This expresses for f a direct link between the number of critical points of a given type and relative Betti numbers.

10.3.3 Derivation of Morse Inequalitites A and B

Morse inequalities A are given precisely as follows: Let $f : \mathcal{U} \longrightarrow \mathbf{R}$, where $\mathcal{U}$ is open in $\mathbf{R}^n$, be a twice continuously differentiable function. Suppose that f is admissible relative to a regular domain C in $\mathcal{U}$ and satisfies Morse boundary condition A. If $B_0, B_1, \ldots, B_n$ with $B_n = 0$ are the Betti numbers of C, then

$$\begin{aligned}
M_0 &\geq B_0 \\
M_1 - M_0 &\geq B_1 - B_0 \\
M_2 - M_1 + M_0 &= B_2 - B_1 + B_0 \\
&\vdots \\
M_{n-1} - \cdots + (-1)^{n-1} M_0 &\geq B_{n-1} - \cdots + (-1)^{n-1} B_0 \\
M_n - M_{n-1} + \cdots + (-1)^n M_0 &= B_n - B_{n-1} + \cdots + (-1)^n B_0.
\end{aligned}$$

This will be established by using a generalization of equation (10.4) and Theorem 10.12 below.

For Y a subspace of a topological space X, set

$$E_k(X, Y) = B_k(X, Y) - B_{k-1}(X, Y) + \cdots + (-1)^k B_0(X, Y).$$

The *Euler characteristic*, $\chi(X, Y)$, of the pair (X, Y) defined by an alternating sum:

$$\chi(X, Y) = \sum_{i=0}^{\infty}(-1)^i B_i(X, Y),$$

where it is assumed that all, except finitely many, $B_k(X, Y)$ are zero.

Theorem 10.12 (Sub-Additivity) *Consider topological spaces $X_0 \subset X_1 \subset \cdots \subset X_{m-1} \subset X_m$. Then*

$$E_k(X_m, X_0) \leq \sum_{\ell=1}^{m} E_k(X_\ell, X_{\ell-1})$$

and

$$\chi(X_m, X_0) = \sum_{\ell=1}^{m} \chi_i(X_\ell, X_{\ell-1}).$$

A proof of Theorem 10.12 is in [Mil, pp. 28–30].

Let $C_\tau = \{x \in C : f(x) \leq \tau\}$ and choose values $\tau_0 < \tau_1 < \cdots < \tau_{n+1}$ such that C_{τ_ℓ} contains exactly ℓ critical points and $C_{\tau_{n+1}} = C$ (compare with the spaces D_{τ_ℓ} of Sections 10.3.1 and 10.3.2). By construction of C_{τ_ℓ} we have

$$\emptyset = C_{\tau_0} \subset C_{\tau_1} \subset \cdots \subset C_{\tau_n} \subset C_{\tau_{n+1}} = C.$$

Moreover, using the Lefschetz duality theorem, equation (10.4) extends as follows:

$$M_k = \sum_{\ell=1}^{n+1} B_k(C_{\tau_\ell}, C_{\tau_{\ell-1}}). \tag{10.5}$$

Theorem 10.12 now implies

$$E_k(C) \leq \sum_{\ell=1}^{n+1} E_k(C_{\tau_\ell}, C_{\tau_{\ell-1}}) \tag{10.6}$$

and

$$\chi(C) = \sum_{\ell=1}^{n+1} \chi(C_{\tau_\ell}, C_{\tau_{\ell-1}}) \tag{10.7}$$

where $E_k(C) \equiv E_k(C, \emptyset)$ and $\chi(C) \equiv \chi(C, \emptyset)$.

Morse inequalities A are contained in (10.6) and (10.7). In fact, equation (10.6) yields

$$\begin{aligned} E_k(C) \quad &\leq \quad \sum_{\ell=1}^{n+1} B_k(C_{\tau_\ell}, C_{\tau_{\ell-1}}) - \sum_{\ell=1}^{n+1} B_{k-1}(C_{\tau_\ell}, C_{\tau_{\ell-1}}) && (10.8)\\ &\quad + \cdots + (-1)^k \sum_{\ell=1}^{n+1} B_0(C_{\tau_\ell}, C_{\tau_{\ell-1}}) \\ &= \quad M_k - M_{k-1} + \cdots + (-1)^k M_0. && (10.9) \end{aligned}$$

The last equality above is a consequence of equation (10.5). Set $B_k(C) = B_k(C, \emptyset)$. Then $E_k(C) = B_k(C) - B_{k-1}(C) + \cdots + (-1)^k B_0(C)$. Hence

$$M_k - M_{k-1} + \cdots + (-1)^k M_0 \;\geq\; B_k(C) - B_{k-1}(C) + \cdots + (-1)^k B_0(C), \tag{10.10}$$

where k is any nonnegative integer. For $k = n$ and $k = n+1$, we have

$$M_n - M_{n-1} + \cdots (-1)^n M_0 \geq B_n(C) - B_{n-1}(C) + \cdots + (-1)^n B_0(C)$$

$$M_{n+1} - [M_n - M_{n-1} + \cdots (-1)^n M_0] \geq B_{n+1}(C) - [B_n(C) - B_{n-1}(C) + \cdots + (-1)^n B_0(C)]. \tag{10.11}$$

Since $M_{n+1} = 0 = B_{n+1}(C)$, the inequalities in (10.11) yield

$$M_n - M_{n-1} + \cdots + (-1)^n M_0 \;=\; B_n(C) - B_{n-1}(C) + \cdots + (-1)^n B_0(C). \tag{10.12}$$

Hence, equations (10.10) and (10.12) yield Morse inequalities A.

We now turn to Morse inequalities B. If f obeys Morse boundary conditions B, then we claim that there is a function $\hat{f}$ defined on an open set containing C such that:

(1) All critical points of $\hat{f}$ in C are nondegenerate and do not occur on the boundary of C.

(2) $\hat{f}$ satisfies Morse boundary condition A relative to C.

(3) $\hat{f} = f$ on $C - V$, where V is an open neighborhood of ∂C containing no critical points of f.

(4) If the total number of critical points of $\hat{f}$ in C of index k is $\hat{M}_k$, then $\hat{M}_k = M_k + M_k'$, where M_k' is the total number of index k critical points of $f|_{C_-}$.

Application of Morse inequalities A to the critical points of $\hat{f}$ yields Morse inequalities B, establishing Theorem 10.6:

$$\hat{M}_k - \hat{M}_{k-1} + \cdots + (-1)^k \hat{M}_0 \geq B_k(C) - B_{k-1}(C) + \cdots + (-1)^k B_0(C),$$

where $k = 0, \ldots, n-1$, and

$$\hat{M}_n - \hat{M}_{n-1} + \cdots + (-1)^n \hat{M}_0 = B_n(C) - B_{n-1}(C) + \cdots (-1)^n B_0(C).$$

We conclude this section with a construction of $\hat{f}$ in the 2-dimensional case — see [Mor-C, p. 74] for proof of the n dimensional analog. Let f be a twice continuously differentiable function f defined everywhere in a regular smooth domain C in the plane. Recall that we denote by C_- the union of all open subsets of points on the boundary ∂C where the vector field $X = \text{grad } f$ points into the region. We assume that f has no singularities in a neighborhood of ∂C and that the restriction $f|_{C_-}$ has only nondegenerate critical points — simple maxima and minima. Consequently, wherever $f' = 0$, $f'' \neq 0$. Here, the prime is the derivative along C_-.

Proposition 10.13 *We can modify f in an arbitrarily small neighborhood of ∂C so that the gradient of the altered f, denoted $\hat{f}$, is outward pointing at each point of ∂C and $\hat{f}$ introduces additional nondegenerate critical points interior to C, one minimum for each minimum of $f|_{C_-}$ and one saddle for each maximum of $f|_{C_-}$.*

Proof: We work on one component of ∂C at a time. For convenience we orient ∂C and the plane, respectively, so that the interior of C lies to its left and the outward direction is positive.

If $f|_{C_-}$ has no critical points, then at each point of C, grad f is either pointing outward or its tangential component is nonzero. This case of the proposition is proven by adding to f a function g, where grad g is outward pointing with support in a (small) collar neighborhood about ∂C such that grad $g|_{\partial C}$ is larger than the inward pointing component of any vector of grad $f|_{\partial C}$.

Now, suppose that $f|_{C_-}$ has at least one critical point. To reduce to the previous case, we remove the critical points of $f|_{C_-}$ as follows. We shall modify f in a neighborhood of each critical point of $f|_{C_-}$ so as to push the critical points of $f|_{C_-}$ into the interior of C, replacing each minimum of $f|_{C_-}$ with a nondegenerate minimum of the modified f and replacing each maximum of $f|_{C_-}$ with a nondegenerate saddle of the modified f. These are the only critical points to be introduced by our local modifications. The modified f will agree with the original f except in the neighborhoods of the critical points $f|_{C_-}$. We shall deal with two cases at the same time, i.e., where $f|_{C_-}$ has a minimum and where $f|_{C_-}$ has a maximum.

Model a neighborhood of an isolated critical point of $f|_{C_-}$ as a half-open rectangular neighborhood of the origin in $\mathbf{R}^2$, with the open interval about the origin on the x-axis playing the role of C_-, and C lying below the x-axis. The origin is a critical point of grad $f|_{\partial C}$, where $f|_{\partial C}$ has a maximum or minimum.

Consult Figure 10.11 for a scheme of the proof to follow. In a neighborhood of the origin, consider the set at which $\frac{\partial f}{\partial x} = 0$. This set contains the origin. Since $f|_{\partial C}$ has only nondegenerate critical points, we know that, locally, this set is a smooth curve, transverse to the x-axis near the origin. Choose coordinates in a neighborhood of the origin so that the set at which $\frac{\partial f}{\partial x} = 0$ is just the little half-open interval of the y-axis below the origin. Without loss of generality, we may assume that at every point of the little interval, $\frac{\partial^2 f}{\partial^2 x} \neq$

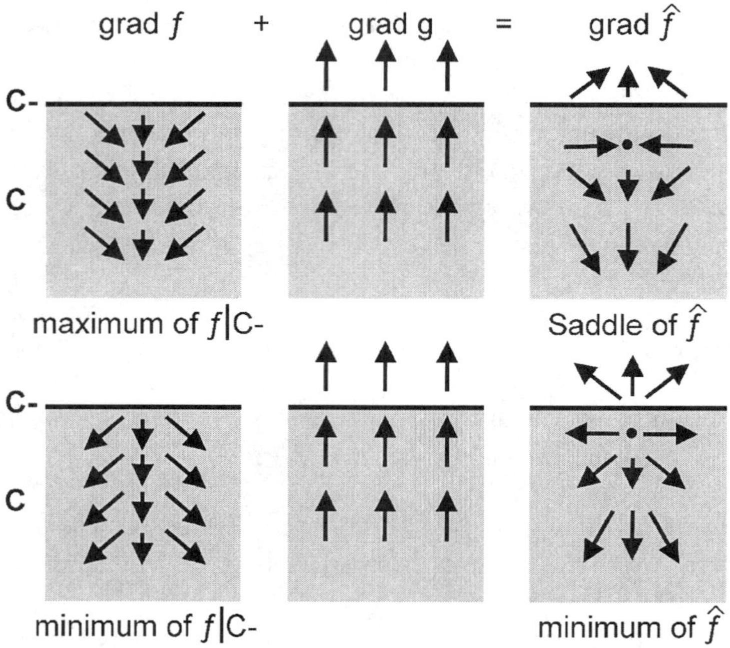

Figure 10.11. Behavior of the gradient of the function $\hat{f} = f + g$ in a small rectangular region (shaded) about a maximum and minimum of the restriction of f to C_-.

0, and that $\frac{\partial f}{\partial y} < 0$ there. For notational convenience, assume that this interval is $I = (-1, 0]$ on the y-axis. Let h be a smooth function on I that agrees with $f(0, y)$ for $y \in (-1, -\frac{2}{3})$ and such that $\frac{\partial h}{\partial y}(y) > 0$ for $y \in (-\frac{2}{3}, 0]$. For example, we may take $\frac{\partial h}{\partial y}(y) = -\frac{\partial f}{\partial y}(0, -(1+y))$, where $y \in (-\frac{1}{3}, 0]$. Note that $\frac{\partial h}{\partial y}(y)$ is negative on $(-1, -\frac{2}{3})$ and positive on $(-\frac{1}{3}, 0]$. We complete the description of $\frac{\partial h}{\partial y}$ by smoothly pasting in an arbitrary smooth, strictly increasing function on $(-\frac{2}{3}, -\frac{1}{3})$ with the following conditions. For some point $y_0 \in (-\frac{2}{3}, -\frac{1}{3})$, we have $\frac{\partial h}{\partial y}(y_0) = 0$ and the slope of the graph of $\frac{\partial h}{\partial y}$ at y_0, i.e., $\frac{\partial^2 h}{\partial y^2}(y_0)$, is strictly larger than $\left[\frac{\partial^2 f}{\partial x \partial y}(0, y_0)\right]^2 / \left|\frac{\partial^2 f}{\partial x^2}(0, y_0)\right|$. The desired function h is now obtained by integration, choosing the constant of integration so that $h(y)$ agrees with $f(0, y)$ for $y \in (-1, -\frac{2}{3})$.

Next, define $g(y) = h(y) - f(0, y)$. In the little rectangle, let $\hat{f} = f + g$. Notice that along I, we have $\hat{f}(0, y) = h(y)$, so $\hat{f}(0, y)$ has only one critical point in I, namely, a minimum at y_0. Since g depends only on y, we get $\frac{\partial \hat{f}}{\partial x} = \frac{\partial f}{\partial x}$. Hence, at every point of I, we obtain $\frac{\partial \hat{f}}{\partial x} = 0$ and $\frac{\partial^2 \hat{f}}{\partial x^2} \neq 0$. In particular, the only critical point of $\hat{f}$ in the rectangle occurs at $(0, y_0)$ and the sign of det[Hess $\hat{f}$]$(0, y)$ is the same as that of $\frac{\partial^2 f}{\partial x^2}$. Thus, $\hat{f}$ has a minimum if $f|_{C_-}$ has a minimum at $(0, 0)$; $\hat{f}$ has a saddle if $f|_{C_-}$ has a maximum at $(0, 0)$.

□

Chapter 11

Counting Lensed Images: Single-Plane Case

"The light coming from a star A *traverses the gravitational field of another star* B*, ... the observer will see* A *as two point-like light-sources, which are deviated from the true geometrical position of* A *. . . . there is no great chance of observing this phenomenon . . . ."*

ALBERT EINSTEIN, 1936

"**0957+561 A, B: twin quasistellar objects or gravitational lens?** *. . . . Their spectra leave little doubt that they are associated. Difficulties arise in describing them as two distinct objects and the possibility that they are two images of the same object formed by a gravitational lens is discussed."*

WALSH, CARSWELL, AND WEYMANN, 1979

"We believe that we have observed the gravitational lens that is responsible for producing the double quasar. . . . We conclude that the double quasar is almost certainly the multiple image of a single object produced by a gravitational lens."

YOUNG, GUNN, KRISTIAN, OKE, AND WESTPHAL, 1980

A striking feature of gravitational lensing is the formation of multiple images of a background light source. Within our mathematical framework, *lensed images* of a light source at position $\mathbf{y}$ are solutions $\mathbf{x}$ of the lens equation (see (6.30) on page 200 and Section 3.2.3, page 77):

$$\mathbf{y} = \mathbf{x} - \boldsymbol{\alpha}_1 \circ \boldsymbol{\eta}_1(\mathbf{x}) - \cdots - \boldsymbol{\alpha}_k \circ \boldsymbol{\eta}_k(\mathbf{x}), \tag{11.1}$$

where $\boldsymbol{\alpha}_i$ is the bending angle vector field on the ith lens plane and $\boldsymbol{\eta}_i$ the ith partial lensing map. For a gravitational lens consisting of a single point mass (e.g., star or black hole), there are exactly *two* lensed images produced of a light source not on the line of sight (Section 6.3.3, page 187). This result was known to Einstein as early as 1912 (before he completed the general theory of relativity); see page 6. For two point masses on the same lens plane, equation (11.1) can be solved directly (after a lot of tedious algebra) to show that there are either *three* or *five* lensed images for sources not on caustics (see Schneider and Weiss [Schn-W86]). If there are three or more point masses on one or several lens planes, it is practically not feasible to determine analytically the solutions of (11.1). In such cases, numerical methods are employed with simple gravitational lens potentials — a task that is also nontrivial and has the difficulty of knowing whether all lensed images have been found. We are then naturally led to the **Lensed-Image Counting Problem**: For a generic gravitational lens system, determine the number of lensed images of a light source not on a caustic.

This chapter addresses the single-plane Lensed-Image Counting Problem following an approach by A.P. [Pet91], [Pet92], [Pet95a]. The basic idea is to convert the counting of solutions of (11.1) into the counting of light rays of the associated time delay function, and then apply Morse inequalities B to obtain counting formulas and lower bounds on the number of lensed images of different types. These results are independent of the lens plane geometry and the detailed form of the potential used to model the lens. We shall also present upper bounds on the number of lensed images and discuss lensed-image locations due to point masses on a lens plane with continuous matter and shear. In other words, the chapter centers on Problem **M1** (counting/locating lensed images) for single-plane lensing.

Outline: In Section 11.1 we determine counting formulas and lower bounds on the total number of lensed images of various types due to a general single-plane gravitational lens. The section also presents certain criteria for having finitely many lensed images and treats the locations of lensed images relative to a general lens. Sections 11.2, 11.3, and 11.4 discuss single-plane image-counting for, respectively, general isolated gravitational lenses, nonsingular isolated gravitational lenses, and point masses lenses with continuous matter and shear. In addition, Section 11.3 includes the Odd Number Image Theorem, while Section 11.4 subdivides as follows. Section 11.4.1 presents results for subcritical point mass lensing, while Sections 11.4.2 and 11.4.3 focus, respectively, on the strong external shear and supercritical cases. Section 11.5 employs resultants (Section 11.5.1) to obtain upper bounds on the number of images due to point masses with continuous matter and shear (Section 11.5.2). Finally, Section 11.6 investigates the location of lensed images generated by the latter lenses.

11.1 General Single-Plane Gravitational Lens

Our counting results will apply to time delay functions obeying Morse boundary conditions B relative to a regular domain.[1] This will enable us to apply Morse inequalities B to count lensed images in such domains.

Requisite Mathematics: Sections 7.4.1 and 10.2.

Let $T_{\mathbf{y}} : L \to \mathbf{R}$ be a single-plane time delay function induced by a gravitational lens potential ψ:

$$T_{\mathbf{y}}(\mathbf{x}) = \frac{|\mathbf{x} - \mathbf{y}|^2}{2} - \psi(\mathbf{x}),$$

where $L = \mathbf{R}^2 - A$ with A the (finite) set of singularities of ψ — see Sections 6.1 (page 172) and 6.2.2 (page 177). Define a regular domain D in $\mathbf{R}^2$ to be *punctured by* a subset A' of A if: (1) D is diffeomorphic to a closed g-holed disc and contains no element of A', (2) inside each inner boundary curve of D there is exactly one element of A', no element of $A - A'$, and each element of A' lies inside an inner boundary curve of D.

The next result shows that Morse boundary conditions B is a "typical" property of time delay functions:

Theorem 11.1 *Let $T_{\mathbf{y}} : L \to \mathbf{R}$ be a single-plane time delay function. Suppose that $T_{\mathbf{y}}$ is admissible relative to a regular domain D in L that is punctured by A'. Then for almost all $\mathbf{a} \in \mathbf{R}^2$ (except in a measure-zero set) with $|\mathbf{a}|$ sufficiently small, the linear perturbation*

$$T_{\mathbf{y}}^{\mathbf{a}}(\mathbf{x}) = T_{\mathbf{y}}(\mathbf{x}) - \mathbf{a} \cdot \mathbf{x}, \quad \mathbf{x} \in L,$$

of $T_{\mathbf{y}}$ is a time delay function satisfying:

(1) *$T_{\mathbf{y}}^{\mathbf{a}}$ obeys Morse boundary conditions* B *relative to D.*

(2) *As $|\mathbf{a}| \to 0$, the light rays of $T_{\mathbf{y}}^{\mathbf{a}}$ in D tend uniformly and bijectively with preservation of critical point type, to those of $T_{\mathbf{y}}$ in D. In particular, $T_{\mathbf{y}}^{\mathbf{a}}$ and $T_{\mathbf{y}}$ have the same number of minimum, maximum, and saddle light rays in D.*

Theorem 11.1 will be generalized to k-lens planes and proven in Section 12.1.2 (see Theorem 12.4, page 452).

Suppose that $T_{\mathbf{y}}$ satisfies Morse boundary conditions B with respect to a regular domain D punctured by a subset A' of A, where A' has g points ($g \geq 0$). Let D_- be the union of all open submanifolds of ∂D where grad $T_{\mathbf{y}}$ points into the interior of D. The set D_- is either

[1]Recall from Section 10.2 (page 403) that a *regular domain* in $\mathbf{R}^n$ is a compact set D in $\mathbf{R}^n$ whose topological boundary is a smooth $(n-1)$-dimensional manifold. Moreover, a smooth map $f : U \subseteq \mathbf{R}^n \to \mathbf{R}$, where U is open, satisfies *Morse boundary conditions* B *relative to* D if f has only nondegenerate critical points in D with none on ∂D, and the restricted map $f|_{\partial D}$ is nondegenerate.

empty, a disjoint union of simple closed curves, or a disjoint union of open arcs and/or simple closed curves. Moreover, the function $T_{\mathbf{y}}$ has finitely many light rays in D and $T_{\mathbf{y}}|_{\partial D}$ is nondegenerate with finitely many critical points (see Corollary 7.44, page 240). Since the lensed images of a light source at $\mathbf{y}$ coincide with the light rays of $T_{\mathbf{y}}$, we can count lensed images by counting critical points. The notation below is then useful.

Notation:

(1) N = total # of lensed images of $T_{\mathbf{y}}$ in D.
M_{min} = total # of minimum lensed images of $T_{\mathbf{y}}$ in D.
M_{sad} = total # of saddle lensed images of $T_{\mathbf{y}}$ in D.
M_{max} = total # of maximum lensed images of $T_{\mathbf{y}}$ in D.
$M_+ = M_{min} + M_{max}$.

(2) N' = total # of critical points of $T_{\mathbf{y}}|_{D_-}$.
M'_{min} = total # of minimum critical points of $T_{\mathbf{y}}|_{D_-}$.
M'_{max} = total # of maximum critical points of $T_{\mathbf{y}}|_{D_-}$.

The following result applies to a wide range of lens models and yields counting information for lensed images within sub-regions of a single-plane mass distribution (A.P. [Pet91], [Pet92]):

Theorem 11.2 (General Lens) *Let $T_{\mathbf{y}} : L \to \mathbf{R}$ be a single-plane time delay function induced by a gravitational lens potential ψ. Suppose that $T_{\mathbf{y}}$ obeys Morse boundary conditions* B *relative to a regular domain D in L, where D is punctured by g singularities of ψ.*[2] *Then the number of lensed images in D of a light source at $\mathbf{y}$ is such that:*

(1) $M_{min} \geq 1 - M'_{min}, \quad M_{sad} \geq M_{max} + g - M'_{max},$

$M_{sad} \geq M_{min} + g - 1 + (M'_{min} - M'_{max}),$

$M_+ = M_{sad} - g + 1 - (M'_{min} - M'_{max}).$

(2) $N = 2M_+ + g - 1 + (M'_{min} - M'_{max}) = 2M_{sad} - g + 1 - (M'_{min} - M'_{max}),$

$N \geq g + 1 - N'.$

Proof: Since $B_0(D) = 1$, $B_1(D) = g$, and $B_2(D) = 0$, the theorem is an immediate consequence of the strong Morse inequalities (see Theorem 10.6, page 405). The result, $N \geq g + 1 - N'$, follows from the weak Morse inequalities (Corollary 10.7, page 406). Also, since $M_{min} \geq 1 - M'_{min}$ and $M_{sad} = M_{max} + g + M_{min} - 1 + (M'_{min} - M'_{max})$, we obtain $M_{sad} \geq M_{max} + g - M'_{max}$. □

Note that Theorem 11.2 gives no information on whether the total number of lensed images (which includes those outside of D) is finite. The next result addresses this issue.

[2] The total number of singularities of ψ may exceed g (i.e., some may be outside D).

Proposition 11.3 *Let $T_{\mathbf{y}} : L \to \mathbf{R}$ be a single-plane time delay function and lensing map induced by ψ. Suppose that $\mathbf{y}$ is not a caustic point and each singularity of ψ (if any) is an infinite singularity.*[3] *If $|\operatorname{grad} T_{\mathbf{y}}(\mathbf{x})| \to \infty$ as $|\mathbf{x}| \to \infty$, then a light source at $\mathbf{y}$ has finitely many lensed images.*

Proof: Since $T_{\mathbf{y}}$ is nondegenerate, and because for every infinite singularity $\mathbf{p} \in A$ we have $|\operatorname{grad} T_{\mathbf{y}}(\mathbf{x})| \to \infty$ as $\mathbf{x} \to \mathbf{p}$ (since if $\mathbf{p}$ is an infinite singularity, then $T_{\mathbf{y}}(\mathbf{x}) \to \infty$ as $\mathbf{x} \to \mathbf{p}$), there is an open set $D(\mathbf{p})$ in $\mathbf{R}^2$ such that

$$D(\mathbf{p}) \cap A = \{\mathbf{p}\}$$

and

$$D(\mathbf{p}) \cap \operatorname{Crit}(T_{\mathbf{y}}) = \emptyset.$$

Because

$$|\operatorname{grad} T_{\mathbf{y}}(\mathbf{x})| \to \infty \text{ as } \mathbf{x} \to \mathbf{p},$$

the set $\operatorname{Crit}(T_{\mathbf{y}})$ lies in a compact subset of $\mathbf{R}^2$. Consequently, there is a regular domain (diffeomorphic to some closed g-hole disc) in L such that all critical points of $T_{\mathbf{y}}$ lie in the interior of D. In other words, $T_{\mathbf{y}}$ is admissible relative to D. Apply Corollary 7.44 (page 240). □

We are also interested in the locations of lensed images relative to a general lens (see Section 11.6, page 441, for more). In fact, notice that

$$0 < \Delta T_{\mathbf{y}}(\mathbf{x}_0) = 2[1 - \kappa(\mathbf{x}_0)]$$

if $\mathbf{x}_0$ is a minimum, while

$$0 > \Delta T_{\mathbf{y}}(\mathbf{x}_0) = 2[1 - \kappa(\mathbf{x}_0)]$$

for $\mathbf{x}_0$ a maximum. Hence:[4]

(1) *Minimum lensed images $\mathbf{x}_{min}$ are positioned where the surface mass density is subcritical: $0 \le \kappa(\mathbf{x}_{min}) < 1$.*

(2) *Maximum lensed images $\mathbf{x}_{max}$ are located where the surface mass density is supercritical: $\kappa(\mathbf{x}_{max}) > 1$.*

There is no general restriction on the location of saddle lensed images relative to a lens. For example, a saddle lensed image $\mathbf{x}_{sad}$ can get arbitrarily close to a singularity of the gravitational lens potential as in the case of a point mass when light source heads to "infinity" (i.e., $|\mathbf{y}| \to \infty$); see Theorem 11.7 (2, ii), page 431. On the other hand, we have $|\mathbf{x}_{sad}| \to \infty$ as $|\mathbf{y}| \to \infty$ for point masses with sufficiently strong shear — Theorem 11.8(3, ii), page 432.

---

[3]A singularity $\mathbf{p}$ of ψ is an *infinite singularity* if $\psi(\mathbf{x}) \to -\infty$ as $\mathbf{x} \to \mathbf{p}$ (page 172).

[4]See page 470 for a discussion of the magnification/location of lensed images.

11.2 Isolated Gravitational Lenses

Intuitively, a gravitational lens is "isolated" if the effects due to the lens are negligible at "infinity." We shall give a general and precise definition of an "isolated" gravitational lens.

Let $T : L \times S \to \mathbf{R}$ be the single-plane time delay family induced by a gravitational lens potential ψ. The eigenvalues of [Hess $T_{\mathbf{y}}](\mathbf{x})$ are

$$\lambda_1(\mathbf{x}; \mathbf{y}) = 1 - \kappa(\mathbf{x}) + \Gamma(\mathbf{x}), \qquad \lambda_2(\mathbf{x}; \mathbf{y}) = 1 - \kappa(\mathbf{x}) - \Gamma(\mathbf{x}),$$

where Γ is the magnitude of shear. Call T *subcritical at infinity* if as $|\mathbf{x}| \to \infty$ we have $\lambda_1(\mathbf{x}; \mathbf{y}) > 0$ and $\lambda_2(\mathbf{x}; \mathbf{y}) > 0$ for all $\mathbf{y} \in S - \text{Caustic}(T)$. In this case, the Gaussian curvature at each point of the time delay surface of $T_{\mathbf{y}}$ (i.e., the graph of $T_{\mathbf{y}}$) is positive for $|\mathbf{x}|$ sufficiently large (i.e., at "far away points" on the time delay surface, the sum of the angles of a triangle with geodesic sides is greater than π). Moreover, κ and Γ are also subcritical at infinity (i.e., $\kappa(\mathbf{x}) < 1$ and $\Gamma(\mathbf{x}) < 1$ as $|\mathbf{x}| \to \infty$). If T is subcritical at infinity and, for $\mathbf{y} \in S - \text{Caustic}(T)$, we have $T_{\mathbf{y}}(\mathbf{x}) \to \infty$ as $|\mathbf{x}| \to \infty$, then T is called *isolated.* Intuitively, matter "at infinity" is not strong enough to "pull down" the graph of $T_{\mathbf{y}}$ or distort its shape away from positive Gaussian curvature. We shall call a gravitational lens potential ψ *isolated* if ψ induces an isolated time delay family. The surface mass density of such a ψ will also be termed *isolated.*

Example: Time delay families T whose gravitational lens potentials are dominated by the quadratic $|\mathbf{x} - \mathbf{y}|^2/2$ as $|\mathbf{x}| \to \infty$ for $\mathbf{y} \in S - \text{Caustic}(T)$ are isolated. This includes lenses with the asymptotic behavior mentioned at the start of this section. Elliptical and soft-power-law potentials, as well as finitely many point masses, yield isolated time delay families. Note that the time delay family due to a uniform sheet with point masses,

$$T^{ptc}(\mathbf{x}, \mathbf{y}) = \frac{|\mathbf{x} - \mathbf{y}|^2}{2} - \frac{\kappa_{cts}}{2}|\mathbf{x}|^2 - \sum_{\ell=1}^{g} m_\ell \ln |\mathbf{x} - \boldsymbol{\xi}_\ell|,$$

is not isolated for $\kappa_{cts} > 1$ (since $T_{\mathbf{y}}^{ptc}(\mathbf{x}) \to -\infty$ as $|\mathbf{x}| \to \infty$). In this model, supercritical matter extends to "infinity."

If a gravitational lens is isolated and produces caustics lying in a bounded subset of the light source plane, then we can obtain some counting information on the set of all lensed images, including their locations relative to the singularities of the lens as a light source moves to "infinity."

Theorem 11.4 (General Isolated Lenses) *Let T and η be, respectively, a single-plane time delay family and lensing map induced by an isolated gravitational lens potential ψ. Suppose that ψ has g singularities, all of which are infinite singularities. Then the total number of lensed images of a light source at a noncaustic point $\mathbf{y}$ is finite and:*

(1) $M_{min} \geq 1, \quad M_{sad} \geq M_{max} + g, \quad M_{sad} \geq M_{min} + g - 1,$
$M_+ = M_{sad} - g + 1.$

(2) $N = 2M_+ + g - 1 = 2M_{sad} - g + 1, \quad N \geq g + 1.$

(3) *Suppose that $\boldsymbol{\eta}$ is locally stable with* Crit($\boldsymbol{\eta}$) *bounded.*

(i) *For $|\mathbf{y}|$ sufficiently large, we have $N = g + 1$ with $M_{min} = 1$, $M_{sad} = g$, and $M_{max} = 0$.*

(ii) *If $|\mathbf{y}| \to \infty$, then the collection of all saddle lensed images lies inside a compact set. In fact, each saddle lensed image approaches a singularity of ψ, with distinct saddle lensed images approaching distinct singularities. Also, the remaining minimum lensed image $\mathbf{x}_{min}$ satisfies $|\mathbf{x}_{min}| \to \infty$ as $|y| \to \infty$.*

Proof: Since all singularities of ψ are infinite singularities and T is isolated, Proposition 11.3(1) implies that $T_\mathbf{y}$ has finitely many critical points for $\mathbf{y} \in S$ − Caustic(T).

(1–2) By the proof of Proposition 11.3, there is a regular domain D in L such that D is punctured by A, all critical points of $T_\mathbf{y}$ lie in the interior of D, and grad $T_\mathbf{y}$ is outward pointing on ∂D. In other words, $T_\mathbf{y}$ obeys Morse boundary conditions A relative to D. Theorem 11.2 now applies with all primed quantities equal to zero.

(3) Note that by Theorem 8.4(1) (page 293), the sets Crit($\boldsymbol{\eta}$) and Caustic($\boldsymbol{\eta}$) are compact. Hence, det[Hess $T_\mathbf{y}$]($\mathbf{x}$) $\neq 0$ for all $\mathbf{x}$ outside a sufficiently large disc. Second, $|\mathbf{x}_{sad}| < \infty$ *as* $|\mathbf{y}| \to \infty$. Indeed, if $|\mathbf{x}_{sad}| \to \infty$ as $|\mathbf{y}| \to \infty$, then det[Hess $T_\mathbf{y}$]($\mathbf{x}_{sad}$) < 0 for $|\mathbf{x}_{sad}|$ sufficiently large, contradicting T being subcritical at infinity. Third, *as $|\mathbf{y}| \to \infty$, each saddle lensed image of $T_\mathbf{y}$ approaches a singularity of ψ, and every minimum lensed image obeys $|\mathbf{x}_{min}| \to \infty$.* To see this, suppose that $\mathbf{x}_0 \in$ Crit($T_\mathbf{y}$) and $|\mathbf{x}_0| < \infty$ as $|\mathbf{y}| \to \infty$. This holds in particular for saddle lensed images $\mathbf{x}_0$. Furthermore, using $|\mathbf{y}| \leq |\mathbf{x}_0| + |\boldsymbol{\alpha}(\mathbf{x}_0)|$ (which is a consequence of the lens equation), we have $|\boldsymbol{\alpha}(\mathbf{x}_0)| \to \infty$ as $|\mathbf{y}| \to \infty$. But the components $\psi_u(\mathbf{x}_0)$ and $\psi_v(\mathbf{x}_0)$ of $\boldsymbol{\alpha}(\mathbf{x}_0)$ are finite (since ψ is C^∞ in L). Hence, there is an infinite singularity $\mathbf{p}$ of ψ such that $\mathbf{x}_0 \to \mathbf{p}$ as $|\mathbf{y}| \to \infty$. Now, suppose that $\mathbf{x}_0$ is a minimum lensed image. Then $|$ grad $T_\mathbf{y}(\mathbf{x}_0)| >> 0$ for $|\mathbf{y}|$ sufficiently large (since as $|\mathbf{y}| \to \infty$, we have $\mathbf{x}_{min} \to \mathbf{p}$ and, hence, $T_\mathbf{y}(\mathbf{x}_0) \to \infty$). This contradicts the fact that grad $T_\mathbf{y}(\mathbf{x}_0) = \mathbf{0}$. Thus, if $\mathbf{x}_0$ is a minimum, then $|\mathbf{x}_0| \to \infty$ as $|\mathbf{y}| \to \infty$. Fourth, *if we use the local stability of $\boldsymbol{\eta}$, then distinct saddle lensed images of $T_\mathbf{y}$ approach distinct singularities of ψ as $|\mathbf{y}| \to \infty$.* In fact, if two saddle lensed images merge, then the merger must occur with a third lensed image that is either a minimum or maximum (see page 337). Furthermore, if three lensed images approach each other to merge, then the pointlike light source approaches a cusp caustic (from the interior of the cusp). Hence, if two saddle lensed images approach the same singularity as $|\mathbf{y}| \to \infty$, then det[Hess $T_\mathbf{y}$] ($\mathbf{x}_{sad}$) $\to 0$. This contradicts the boundedness of Crit($\boldsymbol{\eta}$). Note that the isolatedness of T was not used in the previous argument. Finally, we have shown that if $|\mathbf{y}|$ is sufficiently large and $\boldsymbol{\eta}$ is locally stable with Crit($\boldsymbol{\eta}$) bounded, then a light source at $\mathbf{y}$ has exactly g saddle lensed images. Thus, by part (1) of the theorem, $M_{min} = 1$, $M_{max} = 0$, $N = g + 1$, and the remaining minimum lensed image obeys $|\mathbf{x}_{min}| \to \infty$ as $|\mathbf{y}| \to \infty$. □

11.3 Nonsingular Isolated Lenses

We consider the number of lensed images due to a general nonsingular isolated gravitational lens, lensed-image multiplicities due to elliptical potentials, and certain criteria for multiple lensed imaging. Note that the nonsingular isolated lenses include isothermal ellipsoids with softened core.

Theorem 11.5 (Nonsingular Isolated Lenses) *Let $T_{\mathbf{y}}$ be a single-plane time delay function induced by a nonsingular isolated gravitational lens potential ψ. Then the total number of lensed images of a light source at a noncaustic point $\mathbf{y}$ is finite and satisfies:*

(1) **Odd Number Image Theorem:** $N = 2M_+ - 1 = 2M_{sad} + 1$.

(2) $M_{min} \geq 1,\ M_{sad} \geq M_{max},\ M_{sad} + 1 \geq M_{min},$
$M_{min} + M_{max} = M_{sad} + 1.$

(3) *Suppose that the associated lensing map $\boldsymbol{\eta}$ is locally stable with* Crit($\boldsymbol{\eta}$) *bounded.*

  (i) *For $|\mathbf{y}|$ sufficiently large, we have $N = 1$ with $M_{min} = 1$ and $M_{sad} = M_{max} = 0$.*

  (ii) *As $|\mathbf{y}| \to \infty$, the collection of all saddle lensed images lies within a compact set and the resulting minimum lensed image $\mathbf{x}_{min}$ satisfies $|\mathbf{x}_{min}| \to \infty$.*

Theorem 11.5 is a special case of Theorem 11.4 (set $g = 0$). Also, the single-plane Odd Number Image Theorem was first shown (heuristically) by Burke [Bur] via the Poincaré–Hopf index theorem.

Example: Elliptical Potentials

By Theorem 11.5, the possible lensed-image types in a three/five lensed-image configuration due to a general nonsingular isolated gravitational lens is as follows:

(1) **Three lensed images:**

  (i) $M_{min} = 1, M_{sad} = 1, M_{max} = 1;$

  (ii) $M_{min} = 2, M_{sad} = 1, M_{max} = 0.$

(2) **Five lensed images:**

  (i) $M_{min} = 1, M_{sad} = 2, M_{max} = 2;$

  (ii) $M_{min} = 2, M_{sad} = 2, M_{max} = 1;$

  (iii) $M_{min} = 3, M_{sad} = 2, M_{max} = 0.$

For a single-plane time delay function induced by an elliptical potential (see (10.3, page 407), a point source inside the lips caustic of Figure 11.1(b) has three lensed images with $M_{min} = 2$, $M_{sad} = 1$, and $M_{max} = 0$. In Figure 11.1(d), light sources situated in the region between the ellipse and astroid have three lensed images with $M_{min} = 1$, $M_{sad} = 1$, and $M_{max} = 1$. Inside the astroid, a source has five lensed images: $M_{min} = 2$, $M_{sad} = 2$, and $M_{max} = 1$. See Blandford and Narayan [Bla-N86, p. 571–574] for more.

We now turn to the merger/creation of lensed images, including their overall locations, as a light source traverses the caustic network due to an elliptical potential. Our discussion will center on Figure 11.1, which is based on [Bla-N86, p. 573].

In Figure 11.1 (a,b), the curve c' is the critical curve associated with the lips caustic c. A light source in region 1 of Figure 11.1(b) has one lensed image — a minimum, while for region 3 there are one saddle and two minimum lensed images. As a light source moves from region 3 to 1 passing through a fold caustic point, a minimum and saddle lensed image merge at a point of c' (when the light is on c) to form an infinitely magnified lensed image. While the light source is at a fold point of c, the remaining minimum lensed image lies on the dashed curve in (a). In fact, the dashed and solid curves are the paths traced

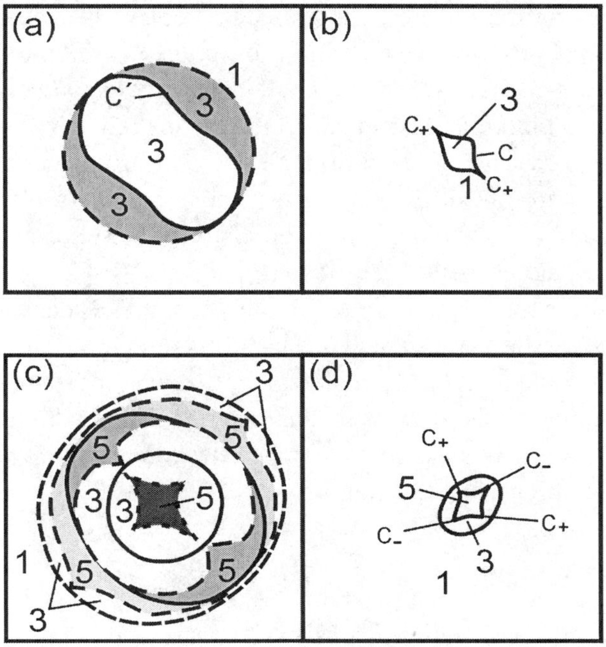

Figure 11.1. The solid curves in (a,c) and (b,d) are, respectively, critical curves and caustics due to an elliptical potential. The points $c_{\pm}$ indicate positive/negative cusps. Numbers in (b,d) indicate how many lensed images are seen of a light source in the given region, with identical numbers in (a,c) denoting possible regions where the lensed images lie. See the text for more discussion. (After Blandford and Narayan [Bla-N86, p. 573]).

out, respectively, by the minimum and infinitely magnified lensed image (formed from a merger) as the light source moves along the caustic c.

When the light source moves from region 3 into region 1 through a fold caustic point, the infinitely magnified lensed image disappears, leaving behind a minimum lensed image. For sources inside region 3, the three lensed images lie inside the dashed curve, with the saddle lensed image inside the critical curve c' and the two minima lying in one of the shaded regions outside c'. If, instead, the light source moved from region 1 to 3, then we would have sudden creations (rather than mergers) of lensed images when the source crosses a caustic point. Note that the two cusps are positive.

For Figure 11.1 (c,d), the caustics are the two curves in (d) and the corresponding critical curves are the solid curves in (c). The outer critical curve in (c) is associated with the inner caustic in (d), while the inner critical curve of (c) is for the outer caustic in (d). A light source in region 1 has one lensed image — a minimum. If the light source lies on a fold point of the outer caustic curve in (d), then an infinitely magnified lensed image suddenly appears on the inner critical curve in (c). In fact, the outer dashed curve and inner solid curve in (c) form the set of the lensed images of all points on the outer caustic curve in (d).

As the light source moves into region 3, the infinitely magnified lensed image splits into a pair of saddle and maximum lensed images, giving a total of three lensed images. The possible locations for these lensed images are the regions labeled 3 in (c). If the light source lies on a fold point of the inner astroid caustic, then an infinitely magnified lensed image suddenly appears on the outer critical curve. This gives a total of four lensed images — including the original minimum, saddle, and maximum. The latter three lensed images lie, respectively, on the dashed curves in (c) with two outward pointing cusps, two inward pointing cusps, and four outward pointing cusps (the astroid). In other words, the latter three dashed curves and the outer solid curve in (c) form the set of all lensed images of a light source moving along the astroid caustic in (d).

As the source moves into region 5, the infinitely magnified lensed image splits into a pair of minimum and saddle lensed images. This gives a total of five lensed images. The two minima are located in the lighter shaded region, the two saddles in the medium shaded region, and the maximum in the darkest region (astroid). The maximum is located where the surface mass density κ is supercritical, while the minima are where κ is subcritical (page 423; see page 470 for magnifications).

Criteria for Producing Multiple Lensed Images

If a single-plane isolated gravitational lens has at least one singularity, then by Theorem 11.4(2) there are multiple lensed images. For lenses without singularities, we have the following theorem:

Theorem 11.6 *Let η be a single-plane lensing map induced by a nonsingular isolated gravitational lens potential ψ with surface mass density $\kappa = \sigma/\sigma_{crt}$.*

(1) *Suppose that* $\boldsymbol{\eta}$ *has a regular point* $\mathbf{x}_0$ *where* κ *is supercritical* (i.e., $\kappa(\mathbf{x}_0) > 1$). *Then a light source at* $\mathbf{y}_0 = \boldsymbol{\eta}(\mathbf{x}_0)$ *has multiple lensed images:* $N \geq 3$ *with* $M_{min} \geq 1$, $M_{sad} \geq 1$, *and* $M_{max} + M_{min} \geq 2$.

(2) *Suppose that* $\boldsymbol{\eta}$ *has a regular point* $\mathbf{x}_0$ *such that a light source at* $\mathbf{y}_0 = \boldsymbol{\eta}(\mathbf{x}_0)$ *has a saddle lensed image. Then the light source has multiple lensed images:* $N \geq 3$ *with* $M_{min} \geq 1$, $M_{sad} \geq 1$, *and* $M_{max} \geq 1$.

Proof: Since ψ is isolated and $\mathbf{y}_0$ is not a caustic point (since $\mathbf{x}_0$ is regular), we have $M_{min} \geq 1$. (1) Using $\kappa(\mathbf{x}_0) > 1$, it follows that $\mathbf{x}_0$ is either a local maximum or saddle lensed image. If $\mathbf{x}_0$ is a maximum, then $M_{sad} \geq M_{max} \geq 1$; hence, $N \geq 3$. If $\mathbf{x}_0$ is a saddle, then $N \geq 3$ (since N is odd and $M_{min} \geq 1$) and $M_{max} + M_{min} \geq 2$. (2) Since $M_{min} \geq 1$, $M_{sad} \geq 1$, and N is odd, we have $N \geq 3$ and $M_{max} \geq 1$. □

Theorem 11.6(1) shows that since $\kappa(\mathbf{x}_0) = \sigma(d_L\mathbf{x})/\sigma_{crt}$ with σ the physical surface mass density of the lens and σ_{crt} the critical density (see pages 42 and 79), *the condition* $\sigma(d_L\mathbf{x}) > \sigma_{crt}$ ("strong" lens) *is sufficient for producing multiple lensed images* (cf. [Schn-EF, p. 177]). *If a surface mass density* κ *is subcritical* (i.e., $\kappa(\mathbf{x}) < 1$ for all $\mathbf{x} \in \mathbf{R}^2$), *then multiple lensed images can still be generated (despite the lens being "weak").*

Example: The perturbed Plummer model,

$$\psi_{ppl}(\mathbf{x}) = \frac{\kappa_0}{2} \ln(1 + |\mathbf{x}|^2) - \frac{\gamma}{2} (\mathrm{u}^2 - \mathrm{v}^2),$$

where $\mathbf{x} = (\mathrm{u}, \mathrm{v})$, $\kappa_0 > 0$ and $\gamma > 0$, has surface mass density

$$\kappa_{ppl}(\mathbf{x}) = \frac{\kappa_0}{(1 + |\mathbf{x}|^2)^2}.$$

This density is subcritical when $0 < \kappa_0 < 1$. Figure 11.2 illustrates that in such a regime multiple lensed images can still occur (cf. [Schn-EF, p. 255]).

11.4 Point Masses with Continuous Matter and Shear

Motivation: Consider an elliptical potential and treat the associated lensed images of a light source as "macroimages." Such a potential is often used to model the lens system Q2237+0305 (Section 5.4.1, page 157), whose "macroimages" have a maximum separation of roughly 1.8 arc seconds (e.g., [Schn-EF, p. 67]). The angular separations of the "macroimages" is roughly 2 arc seconds. If stars are added to the elliptical-potential lens,

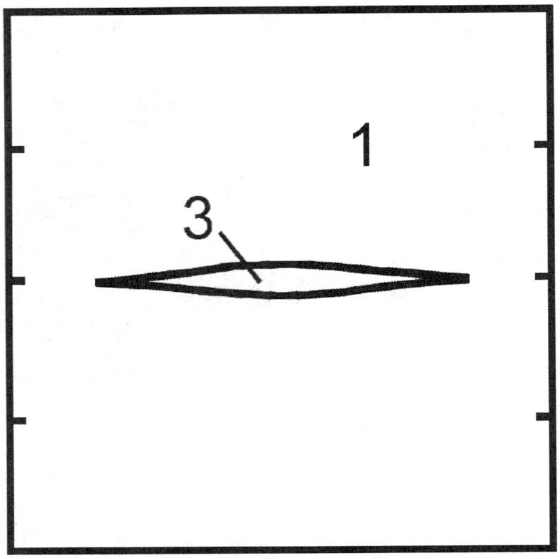

Figure 11.2. An example of lensed-image multiplicities due to a subcritical perturbed plummer gravitational lens with $\kappa_0 = 0.75$ and $\gamma = 0.5$. The indicate how many lensed images are seen of a light source in the given region. (Courtesy of Wicklin.)

then "microimages" of the stars are also produced. These "microimages" are separated by only about 10^{-6} arc seconds.[5]

Stars near the line of sight to a given "macroimage" — which may be a local minimum, maximum, or saddle — generate "microimages" that collect in a neighborhood of the "macroimage." Consequently, in the vicinity of such a "macroimage" the time delay surface is shaped overall either like a minimum, maximum, or saddle, with infinitely high "spikes" corresponding to the point mass positions.[6] Thus, if D is a region about a minimum "macroimage" (e.g., D could be a closed disc punctured by small discs about the point masses), then near the outer boundary ∂D of D the gradient of the lens system's time delay function will point outward. If the "macroimage" is a maximum, then the gradient will be inward pointing, while for a saddle, it will point inward on two opposite sides of ∂D and outward on the remaining two boundary components.

We shall approximate the matter distribution in the vicinity of a "macroimage" by point masses on a lens plane with continuous matter and shear from infinity. This naturally di-

[5] In our galaxy, a pointlike-star (point mass) produces double lensed images of a background light source near the line of sight to the lens, and these lensed images are separated roughly by the angular diameter d_E of the star's Einstein ring (Section 6.3.3, page 187). Since $d_E = 2\theta_E$, where θ_E is the angular Einstein radius, it follows that d_E is of order 10^{-6} arc seconds (see page 43).

[6] As the light source moves, the number and brightnesses of the "microimages" change. Pairs of "microimages" are created or annihilated as the source crosses caustics. This causes an overall "twinkling" of a "macroimage." The reader may recall that these effects are signatures of microlensing; see Sections 4.3 (page 131) and 5.4.1, page 157).

vides our study into three cases corresponding to whether the "macroimage" is either a minimum (subcritical case), saddle (strong shear case), or maximum (supercritical case). This section precisely investigates multiple lensed imaging in all these cases (A.P. [Pet91], [Pet92]).

Preliminaries

Consider the single-plane time delay function $T_{\mathbf{y}}^{ptcs}$ induced by g point masses on a lens plane with continuous matter κ_c and shear γ from infinity (Section 3.3.1, page 101; also, page 175):

$$T_{\mathbf{y}}^{ptcs}(\mathbf{x}) = \frac{|\mathbf{x}-\mathbf{y}|^2}{2} - \frac{\kappa_c}{2}|\mathbf{x}|^2 + \frac{\gamma}{2}(\mathrm{u}^2 - \mathrm{v}^2) - \sum_{\ell=1}^{g} m_\ell \ln|\mathbf{x} - \boldsymbol{\xi}_\ell|, \tag{11.2}$$

where $\mathbf{x} = (\mathrm{u}, \mathrm{v})$, $\kappa_c \geq 0$, and $\gamma \geq 0$. The lens equation is

$$\mathbf{y} = \boldsymbol{\eta}_{ptcs}(\mathbf{x}) = \begin{bmatrix} 1-\kappa_c+\gamma & 0 \\ 0 & 1-\kappa_c-\gamma \end{bmatrix}\mathbf{x} - \sum_{\ell=1}^{g} m_\ell \frac{\mathbf{x}-\boldsymbol{\xi}_\ell}{|\mathbf{x}-\boldsymbol{\xi}_\ell|^2}, \tag{11.3}$$

where $\boldsymbol{\eta}_{ptcs}$ is the associated lensing map. Consequently,

$$|\mathbf{y}| \leq |\mathbf{x}| + (\kappa_c+\gamma)^2\mathrm{u}^2 + (\kappa_c-\gamma)^2\mathrm{v}^2 + \sum_{\ell=1}^{g} \frac{m_\ell}{|\mathbf{x}-\boldsymbol{\xi}_\ell|}. \tag{11.4}$$

We have $\det[\mathrm{Hess}(T_{\mathbf{y}})](\mathbf{x}) = [1-\kappa_c+\Gamma(\mathbf{x})][1-\kappa_c-\Gamma(\mathbf{x})]$, where $1-\kappa_c+\Gamma(\mathbf{x})$ and $1-\kappa_c-\Gamma(\mathbf{x})$ are the eigenvalues of $\mathrm{Hess}(T_{\mathbf{y}})(\mathbf{x})$ with Γ the magnitude of shear (Section 3.2.7, page 95). Explicitly, $\Gamma^2(\mathbf{x}) = (\gamma + \phi_{\mathrm{uu}}(\mathbf{x}))^2 + \phi_{\mathrm{uu}}^2(\mathbf{x})$ with $\phi(\mathbf{x}) = \sum_{i=1}^{g} m_\ell \ln|\mathbf{x}-\boldsymbol{\xi}_\ell|$. We then have $\lim_{|\mathbf{x}|\to\infty} \det[\mathrm{Hess}\, T_{\mathbf{y}}](\mathbf{x}) = (1-\kappa_c+\gamma)(1-\kappa_c-\gamma)$.

Notation: Set $a = 1-\kappa_c+\gamma$ and $b = 1-\kappa_c-\gamma$.

Suppose that $ab \neq 0$. Then $|\,\mathrm{grad}\, T_{\mathbf{y}}^{ptcs}(\mathbf{x})| \to (a^2\mathrm{u}^2 + b^2\mathrm{v}^2)^{1/2} \to \infty$ and $\det[\mathrm{Hess}\, T_{\mathbf{y}}](\mathbf{x}) \to ab \neq 0$ as $|\mathbf{x}| \to \infty$. Hence, Proposition 11.3 (page 423) and Theorem 8.4(1) (page 293) yield that $T_{\mathbf{y}}^{ptcs}$ *with* $\mathbf{y}$ *not a caustic point and* $ab \neq 0$ *has finitely many lensed images, and* $\mathrm{Crit}(\boldsymbol{\eta}_{ptcs})$ *and* $\mathrm{Caustic}(\boldsymbol{\eta}_{ptcs})$ *are compact.*

11.4.1 Subcritical Case

The time delay function $T_{\mathbf{y}}^{ptcs}$ is called *subcritical* if $a > 0$ and $b > 0$ (i.e., $1-\kappa_c+\gamma > 0$ and $1-\kappa_c-\gamma > 0$). In this case, $0 \leq \kappa_c < 1$, $0 \leq \gamma < 1$, and the graph of $T_{\mathbf{y}}^{ptcs}$ has a "minimum" form for $|\mathbf{x}|$ sufficiently large.

Theorem 11.7 (Subcritical) *Let the time delay function* $T_{\mathbf{y}}^{ptcs}$ *be subcritical. For almost all points* $\mathbf{a}$ *in* $\mathbf{R}^2$ *(except in a measure-zero set) with* $|\mathbf{a}|$ *sufficiently small, the linear perturbation,*

$$[T_{\mathbf{y}}^{ptcs}]^{\mathbf{a}}(\mathbf{x}) = T_{\mathbf{y}}^{ptcs}(\mathbf{x}) - \mathbf{a}\cdot\mathbf{x},$$

satisfies the following, where all quantities are defined relative to $[T_{\mathbf{y}}^{ptcs}]^{\mathbf{a}}$:

(1) (i) $M_{min} \geq 1,\ M_{sad} \geq g,\ M_{max} = 0.$

(ii) $M_{min} = M_{sad} + 1 - g.$

(iii) $N = 2M_{min} + g - 1 = 2M_{sad} - g + 1.$

(iv) $N \geq g + 1.$

(2) *Suppose that the lensing map* $\boldsymbol{\eta}^{\mathbf{a}}_{ptcs}$ *for* $[T_{\mathbf{y}}^{ptcs}]^{\mathbf{a}}$ *is locally stable.*

(i) *For* $|\mathbf{y}|$ *sufficiently large, we have* $N = g + 1$ *with* $M_{min} = 1$ *and* $M_{sad} = g$.

(ii) *If* $|\mathbf{y}| \to \infty$, *then the resulting minimum lensed image* $\mathbf{x}_{min}$ *satisfies* $|\mathbf{x}_{min}| \to \infty$. *Moreover, each of the remaining g saddle lensed images approaches a point mass, with distinct saddle lensed images approaching distinct point masses.*

Proof: Using the subcriticality of $T_{\mathbf{y}}^{ptcs}$, we see that the time delay family $[T_{\mathbf{y}}^{ptcs}]^{\mathbf{a}}$ is isolated and $\mathrm{Crit}(\boldsymbol{\eta}^{\mathbf{a}}_{ptcs})$ is bounded. Also, there is no maximum since the Laplacian of $T_{\mathbf{y}}^{ptcs}]^{\mathbf{a}}$ satisfies: $\Delta[T_{\mathbf{y}}^{ptcs}]^{\mathbf{a}}(\mathbf{x}_0) = \Delta T_{\mathbf{y}}^{ptcs}(\mathbf{x}_0) = 2(1-\kappa_c) > 0$. The results now follow using Theorems 11.1 (page 421) and 11.4 (page 424). □

Observe that N and g have *opposite parity*. Also, since the minimum value $g+1$ for N is attained when $M_{min} = 1$ and $M_{sad} = g$, the relation $M_{min} = M_{sad} - g + 1$ implies that M_{min} increases by one if and only if M_{sad} does as well. In other words, each additional minimum must be paired with a saddle and vice-versa.

11.4.2 Strong Shear Case

The time delay function $T_{\mathbf{y}}^{ptcs}$ has *strong shear at infinity* if $a > 0$ and $b < 0$ (i.e., $1 - \kappa_c + \gamma > 0$ and $1 - \kappa_c - \gamma < 0$). Note that the constraint, $a < 0$ and $b > 0$, is impossible (since then $\kappa_c > 1 + \gamma$ and $1 > \kappa_c + \gamma$). In the strong shear case, the graph of $T_{\mathbf{y}}^{ptcs}(\mathbf{x})$ "saddles" for $|\mathbf{x}|$ sufficiently large.

Theorem 11.8 (Strong Shear) *Let the time delay function* $T_{\mathbf{y}}^{ptcs}$ *have strong shear at infinity. For almost all points* $\mathbf{a}$ *in* $\mathbf{R}^2$ *(except in a measure-zero set) with* $|\mathbf{a}|$ *sufficiently small, the linear perturbation,*

$$[T_{\mathbf{y}}^{ptcs}]^{\mathbf{a}}(\mathbf{x}) = T_{\mathbf{y}}^{ptcs}(\mathbf{x}) - \mathbf{a} \cdot \mathbf{x},$$

satisfies the following, where all quantities are defined relative to $[T_{\mathbf{y}}^{ptcs}]^{\mathbf{a}}$:

(1) (i) $M_{min} \geq 0,\ M_{sad} \geq M_{max} + g,\ M_{sad} \geq M_{min} + g + 1,\quad M_{max} \geq 0.$

(ii) $M_{+} = M_{sad} - (g + 1).$

(iii) $N = 2M_{sad} - (g + 1) = 2M_{+} + g + 1.$

(iv) $N \geq g+1$.

(2) (i) *If* $0 \leq \kappa_c < 1$, *then* $M_{max} = 0$.

(ii) *If* $\kappa_c > 1$, *then* $M_{min} = 0$.

(iii) *If* $\kappa_c = 1$, *then* $M_{min} = M_{max} = 0$.

(3) *Suppose that the lensing map* $\boldsymbol{\eta}^{\mathbf{a}}_{ptcs}$ *for* $[T_{\mathbf{y}}^{ptcs}]^{\mathbf{a}}$ *is locally stable.*

(i) *For* $|\mathbf{y}|$ *sufficiently large, we have* $N = g+1$ *with* $M_{min} = 0$, $M_{sad} = g+1$, *and* $M_{max} = 0$.

(ii) *If* $|\mathbf{y}| \to \infty$, *then there is exactly one saddle lensed image* $\mathbf{x}_{sad}$ *satisfying* $|\mathbf{x}_{sad}| \to \infty$. *Each of the remaining g saddle images approaches a point mass, with distinct saddle images approaching distinct point masses.*

Proof: (1) We start by applying Theorem 11.1 (page 421). Since $|[T_{\mathbf{y}}^{ptcs}]^{\mathbf{a}}(\mathbf{x})| \to \infty$ as $|\mathbf{x}| \to \infty$, all critical points of $[T_{\mathbf{y}}^{ptcs}]^{\mathbf{a}}$ lie inside a closed disc D centered at the origin. Using polar coordinates $\mathrm{u} = r\cos\theta$, $\mathrm{v} = r\sin\theta$ on D with the diameter of D sufficiently large, we see that for r near the outer boundary of D, the function $[T_{\mathbf{y}}^{ptcs}]^{\mathbf{a}}$ has the following approximate form: $[T_{\mathbf{y}}^{ptcs}]^{\mathbf{a}}(r,\theta) \simeq (r^2/2)[a\cos^2\theta + b\sin^2\theta]$. The set of points on D where the gradient of $[T_{\mathbf{y}}^{ptcs}]^{\mathbf{a}}$ points inward is

$$D_- = \{(r_0,\theta) : \pi/2 - \theta_0 < \theta < \pi/2 + \theta_0,\, 3\pi - \theta_0 < \theta < 3\pi + \theta_0\}.$$

Here, r_0 is the radius of D and $\theta_0 = \cot^{-1}[\sqrt{-b/a}]$ is the principal value. The restricted function $[T_{\mathbf{y}}^{ptcs}]^{\mathbf{a}}|_{D_-}(\theta) = [T_{\mathbf{y}}^{ptcs}]^{\mathbf{a}}(r_0,\theta)$, where $\theta \in D_-$ (and $a - b > 0$), has two minima (namely, $\theta_1 = \pi/2$, $\theta_2 = 3\pi/2$) and no maxima: $M'_{min} = 2$ and $M'_{max} = 0$. Therefore, we can now employ Theorem 11.2, which yields the desired results.

(2) Suppose that $a > -b$ and $\mathbf{x}_0$ is a local maximum of $[T_{\mathbf{y}}^{ptcs}]^{\mathbf{a}}$, then $\Delta[T_{\mathbf{y}}^{ptcs}]^{\mathbf{a}}(\mathbf{x}_0) = 2(1-\kappa_c) < 0$, which contradicts $0 \leq \kappa_c < 1$. Hence, $M_{max} = 0$ if $0 \leq \kappa_c < 1$. Similarly, if $a < -b$, then $M_{min} = 0$. If $a = -b$ and $\mathbf{x}_0$ is a local minimum or maximum of $[T_{\mathbf{y}}^{ptcs}]^{\mathbf{a}}$, then contradictions also occur.

(3) First, note that *if a lensed image* $\mathbf{x}_0$ *satisfies* $|\mathbf{x}_0| < \infty$ *as* $|\mathbf{y}| \to \infty$, *then* $\mathbf{x}_0$ *approaches some point mass.* Indeed, equation (11.4), page 431, yields $\sum_{\ell=1}^{g}(m_\ell/|\mathbf{x}_0 - \mathbf{x}_\ell|^2) \to \infty$ as $|\mathbf{y}| \to \infty$. Hence, $\mathbf{x}_0$ approaches some point mass.

Now, since $\det[\mathrm{Hess}[T_{\mathbf{y}}^{ptcs}]^{\mathbf{a}}(\mathbf{x}_0) \to -\infty$ if $\mathbf{x}_0$ approaches a point mass, $\mathbf{x}_0$ cannot be a minimum or maximum. Consequently, if there exists a minimum or maximum lensed image $\mathbf{x}_0$, then $|\mathbf{x}_0| \to \infty$ as $|\mathbf{y}| \to \infty$. In this case, the eigenvalues $\lambda_1(\mathbf{x}_0;\mathbf{y})$ and $\lambda_2(\mathbf{x}_0;\mathbf{y})$ of $[\mathrm{Hess}[T_{\mathbf{y}}^{ptcs}]^{\mathbf{a}}](\mathbf{x}_0)$ are either both positive or both negative for $|\mathbf{x}_0|$ sufficiently large. However, for all $\mathbf{x}$ outside a sufficiently large disc, we have $\lambda_1(\mathbf{x};\mathbf{y}) = 1 - \kappa_c + \Gamma(\mathbf{x}) \geq a > 0$ and $\lambda_2(\mathbf{x};\mathbf{y}) = 1 - \kappa_c - \Gamma(\mathbf{x}) \leq b < 0$. This is a contradiction. Therefore, if $|\mathbf{y}|$ is sufficiently large, then $M_{min} = 0 = M_{max}$ and $M_{sad} = g+1 = N$.

Next, suppose that for each saddle lensed image $\mathbf{x}_{sad}$, we have $|\mathbf{x}_{sad}| < \infty$ as $|\mathbf{y}| \to \infty$. Then every saddle lensed image approaches a point mass and, because $M_{sad} = g + 1$, there are at least two saddle images that will have to approach the same point mass. This is a contradiction because by part (3) of the proof of Theorem 11.4, distinct saddle lensed images must approach distinct point masses (since $\boldsymbol{\eta}^{\mathbf{a}}_{ptcs}$ is locally stable and Crit$(\boldsymbol{\eta}^{\mathbf{a}}_{ptcs})$ is bounded). Thus, one saddle lensed image must tend to infinity as $|\mathbf{y}| \to \infty$. □

Note that the subcritical and strong shear cases have the same lower bound on N. However, the strong shear case has $M_{sad} \geq g + 1$, while $M_{sad} \geq g$ for the subcritical situation.

11.4.3 Supercritical Case

The time delay function $T^{ptcs}_{\mathbf{y}}$ is *supercritical*[7] if $a < 0$ and $b < 0$ (i.e., $1 - \kappa_c + \gamma < 0$ and $1 - \kappa_c - \gamma < 0$). The supercritical case implies that $\kappa_c > 1$. The graph of $T^{ptcs}_{\mathbf{y}}(\mathbf{x})$ has a "maximum" form for $|\mathbf{x}|$ sufficiently large.

Theorem 11.9 (Supercritical) *Let the time delay function $T^{ptcs}_{\mathbf{y}}$ be supercritical. For almost all points $\mathbf{a}$ in $\mathbf{R}^2$ with $|\mathbf{a}|$ sufficiently small, the linear perturbation,*

$$[T^{ptcs}_{\mathbf{y}}]^{\mathbf{a}}(\mathbf{x}) = T^{ptcs}_{\mathbf{y}}(\mathbf{x}) - \mathbf{a} \cdot \mathbf{x},$$

satisfies the following, where all quantities are defined relative to $[T^{ptcs}_{\mathbf{y}}]^{\mathbf{a}}$:

(1) (i) $M_{min} = 0, \quad M_{sad} \geq g - 1, \quad M_{max} \geq 0.$

(ii) $M_{max} = M_{sad} - g + 1.$

(iii) $N = 2M_{sad} - g + 1 = 2M_{max} + g - 1.$

(iv) $N \geq g - 1.$

(2) *Suppose that the lensing map $\boldsymbol{\eta}^{\mathbf{a}}_{ptcs}$ for $[T^{ptcs}_{\mathbf{y}}]^{\mathbf{a}}$ is locally stable.*

(i) *For $|\mathbf{y}|$ sufficiently large, we have $N = g + 1$ with $M_{min} = 0$, $M_{sad} = g$, and $M_{max} = 1$.*

(ii) *If $|\mathbf{y}| \to \infty$, then there is exactly one maximum lensed image $\mathbf{x}_{max}$ satisfying $|\mathbf{x}_{max}| \to \infty$. Each of the remaining g saddle lensed images approaches a point mass, with distinct saddle lensed images approaching distinct point masses.*

Proof: First, employ Theorem 11.1 (page 421). In fact, since $T^{ptcs}_{\mathbf{y}}(\mathbf{x})$ is supercritical, we have $[T^{ptcs}_{\mathbf{y}}]^{\mathbf{a}}(\mathbf{x}) \to -\infty$ as $|\mathbf{x}| \to \infty$. This implies that all critical points of $[T^{ptcs}_{\mathbf{y}}]^{\mathbf{a}}$ lie inside some closed disc punctured by g disjoint open discs centered at the point mass

[7] See [Got-PL] for gravitational lensing in cosmology consisting of a Robertson–Walker universe plus a galaxy with a supercritical or "over-focused" smoothed out matter distribution.

positions. If the outer radius of D is sufficiently large and the inner radii sufficiently small, then the set D_- of points of ∂D where the gradient of $[T_{\mathbf{y}}^{ptcs}]^{\mathbf{a}}$ is inward pointing is the outer boundary circle of D.

(1) Applying Corollary 10.8 (page 406) to $[T_{\mathbf{y}}^{ptcs}]^{\mathbf{a}}|_{D_-}$, we see that $M'_{min} - M'_{max} = \chi(D_-) = 0$ and $M'_{min} \geq B_0(D_-) = 1$. Except for $M_{min} = 0$, Theorem 11.2 (page 422) guarantees part (1). But there is no minimum lensed image $\mathbf{x}_0$ because then,

$$\Delta[T_{\mathbf{y}}^{ptcs}]^{\mathbf{a}}(\mathbf{x}_0) = 2(1 - \kappa_c) > 0.$$

This contradicts the supercriticality of $T_{\mathbf{y}}^{ptcs}$. Note that $N = 2M_{max} + g - 1 \geq g - 1$ and $M_{sad} \geq M_{min} + g - 1 \geq g - 1$.

(2) Each saddle lensed image $\mathbf{x}_{sad}$ satisfies $|\mathbf{x}_{sad}| < \infty$ as $|\mathbf{y}| \to \infty$. In fact, if $|\mathbf{x}_{sad}| \to \infty$ as $|\mathbf{y}| \to \infty$, then (since $\mathbf{x}_{sad}$ is a saddle and Caustic$(\boldsymbol{\eta}^{\mathbf{a}}_{ptcs})$ is bounded) we have

$$\lim_{|\mathbf{y}| \to \infty} \det[\mathrm{Hess}[T_{\mathbf{y}}^{ptcs}]^{\mathbf{a}}](\mathbf{x}_{sad}) = (1 - \kappa_c + \gamma)(1 - \kappa_c - \gamma) < 0.$$

This contradicts the fact that $ab > 0$. It follows by equation (11.4) that every saddle lensed image approaches a point mass as $|\mathbf{y}| \to \infty$. Since $\boldsymbol{\eta}^{\mathbf{a}}_{ptcs}$ is locally stable, distinct saddles approaching distinct point masses (see part (3) of the proof of Theorem 11.4, page 424). By $M_{sad} \geq g - 1$, there are either g or $g - 1$ saddle lensed images for $|\mathbf{y}|$ sufficiently large. Because Caustic$(\boldsymbol{\eta}^{\mathbf{a}}_{ptcs})$ is bounded, if $|\mathbf{y}|$ is sufficiently large, then $M_{max} = M_{sad} - g + 1$ and (hence) $M_{max} = 1$ or 0. Now, for $|\mathbf{y}|$ sufficiently large and $\mathbf{x}$ in a sufficiently small neighborhood of $\mathbf{y}$, we have

$$[T_{\mathbf{y}}^{ptcs}]^{\mathbf{a}}(\mathbf{x}) \simeq \frac{|\mathbf{x} - \mathbf{y}|^2}{2} - \frac{\kappa_c - \gamma}{2}\mathrm{u}^2 - \frac{\kappa_c + \gamma}{2}\mathrm{v}^2,$$

which has a single maximum (note: $\kappa_c - \gamma > 0$). Thus, if $|\mathbf{y}|$ is sufficiently large, there is one maximum lensed image and $M_{sad} = M_{max} + g - 1 = g$. We claim that as $|\mathbf{y}| \to \infty$, the maximum lensed image obeys $|\mathbf{x}_{max}| \to \infty$. In fact, if $|\mathbf{x}_{max}| < \infty$ as $|\mathbf{y}| \to \infty$, then $\mathbf{x}_{max}$ approaches a point mass (cf. proof of Theorem 11.7(3)), which yields $\det[\mathrm{Hess}[T_{\mathbf{y}}^{ptcs}]^{\mathbf{a}}](\mathbf{x}_{max}) = (1 - k)^2 - \Gamma^2(\mathbf{x}_{max}) \longrightarrow -\infty$. This contradicts the fact that $\det[\mathrm{Hess}[T_{\mathbf{y}}^{ptcs}]^{\mathbf{a}}](\mathbf{x}_{max}) = 0$. □

Theorem 11.9(1,iv) implies that $N \geq 0$ for $g = 1$, which may seem like a triviality. However, this must not cause one to overlook investigating whether $N = 0$ is possible. In fact, for $g = 1$ and $\kappa_c > \sqrt{4/3}\,\gamma + 1$, the associated caustic is a simple closed fold curve; see Figure 11.3. If $\mathbf{y}$ is inside this caustic, then no images occur. For $\mathbf{y}$ outside the caustic, there are two lensed images — a saddle and maximum.

Critical Case: The function $T_{\mathbf{y}}^{ptcs}$ is *critical* if either $a = 0$ or $b = 0$, which is a nongeneric situation in the (κ_c, γ)-plane. If $T_{\mathbf{y}}^{ptcs}$ is critical, then there are no minimum and maximum lensed images (since $\det[\mathrm{Hess}(T_{\mathbf{y}}^{ptcs})](\mathbf{x}) = -\Gamma^2(\mathbf{x}) \leq 0$). In addition, if $a = b = 0$ — hence, $\kappa_c = 1$, $\gamma = 0$ — and $g = 1$, then (11.3) shows that a light source at the origin has no lensed image; formally speaking, the lensed image is at "infinity."

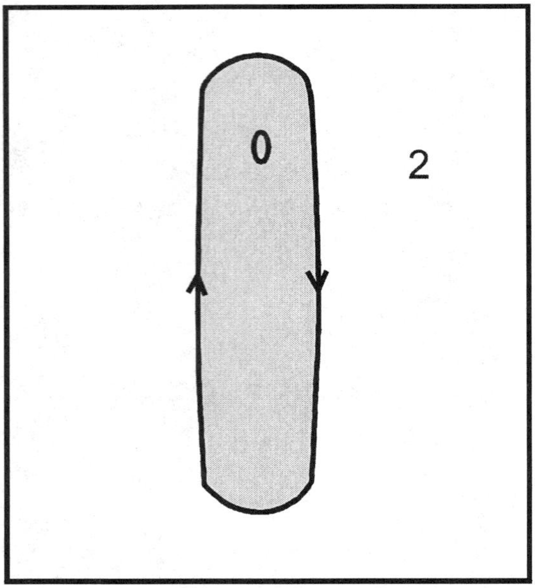

Figure 11.3. A caustic curve due to a point mass with continuous matter κ_c and shear γ, where $\kappa_c > \sqrt{4/3}\ \gamma + 1$. The numbers indicate how many lensed images are seen of a pointlike light source in the given region. The shaded area is a "blind spot" — a light source in that region is invisible.

11.5 Upper Bounds on Number of Lensed Images

There is no single upper bound on the set of all numbers of lensed images (of a light source not on a caustic) due to gravitational lens systems (though a lower bound exists and is achievable, namely, zero). For example, extra lensed images can always be created by adding an elliptical potential that is centered at a sufficiently far point.[8]

We shall show that g point masses (no shear from infinity) on the same lens plane produce at most $g^2 + 1$ lensed images of a light source not on a caustic. If there is also shear from infinity, then at most $(g + 1)^2$ lensed images occur. These upper bounds were found by Witt [Wit90] using an adroit substitution with complex quantities. Unfortunately, this trick substitution does not carry over to the multiplane case. Consequently, we shall prove the previous upper bounds following A.P. [Pet97], who employed resultants to recover

[8] In fact, let $T_{\mathbf{y}}^{iso}$ be a time delay function belonging to an isolated time delay family induced by ψ_{iso} and let $T_{\mathbf{y}}$ be the time delay function induced by $\psi = \psi_{iso} + \psi_{[ep;\mathbf{c}]}$, where

$$\psi_{[ep;\mathbf{c}]}(\mathbf{x}) = C\sqrt{x_{cr}^2 + (1-\epsilon)(\mathrm{u}-\mathrm{u}_0)^2 + (1+\epsilon)(\mathrm{v}-\mathrm{v}_0)^2}$$

with $\mathbf{x} = (\mathrm{u}, \mathrm{v})$ and $\mathbf{c} = (\mathrm{u}_0, \mathrm{v}_0)$. Loosely speaking, if $|\mathbf{c}|$ is sufficiently large and $\mathbf{y}$ a noncaustic point (relative to ψ), then the graph of $T_{\mathbf{y}}$ is similar to that of $T_{\mathbf{y}}^{iso}$, except that $T_{\mathbf{y}}^{iso}$ has additional critical points in a region near $\mathbf{c}$ (where the influence of ψ_{iso} is negligible compared to that of $\psi_{[ep;\mathbf{c}]}$).

the single-plane upper bounds and extend them to multiplane lensing (see Section 12.3.2, page 459). Section 11.5.2 presents the upper bounds, while Section 11.5.1 gathers the necessary background from the theory of resultants.

11.5.1 Resultants

Consider any two polynomials in x_1 and x_2 with coefficients in a field:

$$p(x_1, x_2) = a_0(x_2) + a_1(x_2)x_1 + \cdots + a_s(x_2)x_1^s$$

and

$$q(x_1, x_2) = b_0(x_2) + b_1(x_2)x_1 + \cdots + b_t(x_2)x_1^t.$$

The x_1-*Sylvester resultant* of p and q is defined by $\text{Res}_{(p,q)}(x_2) = \det(\mathbf{S}_{s+t})$, where

$$\mathbf{S}_{s+t} = \begin{bmatrix} a_s & a_{s-1} & a_{s-2} & \cdot & \cdot & \cdot & a_0 & 0 & 0 & \cdots & 0 \\ 0 & a_s & a_{s-1} & \cdot & \cdot & \cdot & a_1 & a_0 & 0 & \cdots & 0 \\ \vdots & \vdots & \vdots & \vdots & \vdots & \vdots & \vdots & \vdots & \vdots & \vdots & \vdots \\ 0 & \cdots & 0 & a_s & a_{s-1} & a_{s-2} & \cdot & \cdot & \cdot & a_1 & a_0 \\ b_t & b_{t-1} & \cdot & \cdot & \cdot & b_0 & 0 & \cdot & \cdot & \cdot & 0 \\ 0 & b_t & b_{t-1} & \cdot & \cdot & \cdot & b_0 & \cdot & \cdot & \cdot & 0 \\ \vdots & \vdots & \vdots & \vdots & \vdots & \vdots & \vdots & \vdots & \vdots & \vdots & \vdots \\ 0 & \cdot & \cdot & \cdot & 0 & b_t & b_{t-1} & \cdot & \cdot & \cdot & b_0 \end{bmatrix}.$$

In the matrix above, notice that the exact location of the $(s+1)$ and $(t+1)$ columns relative to each other depends on the values of s and t. The matrix $\mathbf{S}_{s+t}$ is described precisely as follows: The ℓth–row of $\mathbf{S}_{s+t}$, where $1 \leq \ell \leq t$, consists of $\ell - 1$ zeros, followed by $a_s, a_{s-1}, \ldots, a_0$, and the remainder of the row filled with zeros. For $t+1 \leq \ell \leq s+t$, the ℓth–row has $\ell - t - 1$ zeros followed by $b_t, b_{t-1}, \ldots, b_0$, and the remaining slots having zeros. Note that $\text{Res}_{(p,q)}(x_2)$ is a polynomial in x_2 (the variable x_1 was eliminated).

The degree of the resultant is bounded above as follows:

$$\deg[\text{Res}_{(p,q)}(x_2)] \leq t\left(\max\left\{\deg[a_\ell]_{\ell=0}^s\right\}\right) + s\left(\max\left\{\deg[b_\ell]_{\ell=0}^t\right\}\right). \tag{11.5}$$

To see this, first observe that since $\text{Res}_{(p,q)}(x_2)$ is the determinant of $\mathbf{S}_{s+t}$, we can view $\text{Res}_{(p,q)}(x_2)$ as a polynomial (with integer coefficients) in the variables $a_0, a_1, \ldots, a_s$, $b_0, b_1, \ldots, b_t$:

$$\text{Res}_{(p,q)}(x_2) \;=\; \sum S_{i_0\ldots i_s j_0\ldots j_t} a_0^{i_0}\ldots a_s^{i_s} b_0^{j_0}\ldots b_t^{j_t} \equiv \text{Res}(a_0, \ldots, a_s, b_0, \ldots, b_t). \tag{11.6}$$

It follows that $\text{Res}(a_0, \ldots, a_s, b_0, \ldots, b_t)$ is homogeneous of degree t in the coefficients a_ℓ and degree s in b_ℓ, that is,

$$(\tau a_0, \ldots, \tau a_s, \lambda b_0, \ldots, \lambda b_t) = \tau^t \lambda^s\, \text{Res}(a_0, \ldots, a_s, b_0, \ldots, b_t). \tag{11.7}$$

Consequently, $i_0 + \cdots + i_s = t$ and $j_0 + \cdots + j_t = s$. Hence, the degree of $\mathrm{Res}_{(p,q)}$ does not exceed $t\left(\max\left\{\deg[a_\ell]_{\ell=0}^s\right\}\right) + s\left(\max\left\{\deg[b_\ell]_{\ell=0}^t\right\}\right)$.

The following key fact about resultants will be used often:

Theorem 11.10 (Resultant Theorem) *Suppose that the coefficient field of p and q is the complex numbers (or any algebraically closed field). Then $x_2 = \beta$ satisfies $\mathrm{Res}_{(p,q)}(\beta) = 0$ if and only if, either $a_s(\beta) = 0 = b_t(\beta)$ or $p(\alpha, \beta) = 0 = q(\alpha, \beta)$ for some $x_1 = \alpha$.*

Proof: Given a polynomial p in variables x_1 and x_2, we shall set $p_{x_2}(x_1) = p(x_1, x_2)$ to emphasize that $p(x_1, x_2)$ is viewed as a polynomial in x_1 with x_2 fixed.

Assume that either $a_s(\beta) = 0 = b_t(\beta)$, or, $p(\alpha, \beta) = 0 = q(\alpha, \beta)$ for some $x_1 = \alpha$. If $a_s(\beta)$ and $b_t(\beta)$ vanish, then the first column of $\mathbf{S_{s+t}}$ vanishes, yielding $\mathrm{Res}_{(p,q)}(\beta) = 0$. In addition, if $p(\alpha, \beta) = 0 = q(\alpha, \beta)$ with either $a_s(\beta) \neq 0$ or $b_t(\beta) \neq 0$, then $p_\beta(x_1) = (x_1 - \alpha)\tilde{p}_\beta(x_1)$ and $q_\beta(x_1) = (x_1 - \alpha)\tilde{q}_\beta(x_1)$. Here, $\tilde{p}_\beta$ and $\tilde{q}_\beta$ are not both zero (since either $a_s(\beta) \neq 0$ or $b_t(\beta) \neq 0$). Moreover, $\deg[\tilde{p}_\beta] \leq s - 1$ and $\deg[\tilde{q}_\beta] \leq t - 1$. Write $\tilde{p}_\beta(x_1) = c_0 + c_1 x_1 + \cdots + c_{s-1}x_1^{s-1}$ and $\tilde{q}_\beta(x_1) = d_0 + d_1 x_1 + \cdots + d_{t-1}x_1^{t-1}$. Then $\tilde{q}_\beta p_\beta - \tilde{p}_\beta q_\beta = 0$. This is equivalent to the following system of $s + t$ linear equations in (unknowns) $c_0, \ldots, c_{s-1}, d_0, \ldots, d_{t-1}$:

$$\begin{array}{lcrcl}
a_s d_{t-1} & - & b_t c_{s-1} & = & 0 \\
a_{s-1}d_{t-1} + a_s d_{t-2} & - & b_{t-1}c_{s-1} - b_t c_{s-2} & = & 0 \\
 & \vdots & & \vdots & \\
a_0 d_0 & - & b_0 c_0 & = & 0.
\end{array}$$

Note that the top equation arises from setting the coefficient of x_1^{s+t-1} equal to zero. This system has a nonzero solution (in fact, infinitely many) if and only if the determinant of the associated coefficient matrix $\mathbf{C}$ vanishes. But $\det[\mathbf{C}] = 0$ if and only if $\mathrm{Res}_{(p,q)}(\beta) = 0$ (since $\mathbf{S}_{s+t}$ coincides with $\mathbf{C}^T$ after multiplying appropriate rows by -1). Since $\tilde{p}_\beta$ and $\tilde{q}_\beta$ are known nonzero polynomials, we have a nonzero solution of the above system. Hence, $\mathrm{Res}_{(p,q)}(\beta) = 0$.

Suppose that $\mathrm{Res}_{(p,q)}(\beta) = 0$ for some $x_2 = \beta$. Then the above system of equations has a nonzero solution. Consequently, there are polynomials $\hat{p}$ and $\hat{q}$ in x_1 such that: (1) $\hat{p}$ and $\hat{q}$ are not both zero, say $\hat{p} \neq 0$, (2) $\deg \hat{p} \leq s - 1$ and $\deg \hat{q} \leq t - 1$, and (3) $\hat{q}p - \hat{p}q = 0$. Now, assume that $p_\beta(x_1)$ and $q_\beta(x_2)$ have no common zero. This is equivalent to p_β and q_β having no common factor (since the coefficient field is algebraically closed). The greatest common divisor of p_β and q_β is then 1. Consequently, there are polynomials $\hat{p}_*$ and $\hat{q}_*$ in x_1 with $\hat{p}_* p + \hat{q}_* q = 1$. Since $\hat{q}p = \hat{p}q$, it follows that: $\hat{p} = (\hat{p}_* p + \hat{q}_* q)\hat{p} = (\hat{p}_*\hat{p} + \hat{q}_*\hat{q})p$. Hence, $\deg \hat{p} \geq \deg p = s$ (since $\hat{p} \neq 0$), which is a contradiction. Thus, p_β and q_β have a common zero. □

11.5.2 Application to Point Masses

Consider the time delay function $T_{\mathbf{y}}^{ptcs}$ and lensing map $\boldsymbol{\eta}_{ptcs}$ due to g point masses on a lens plane with continuous matter κ_c and shear γ from infinity (see (11.2) and (11.3), page 431).

Notation: Setting $a = 1 - \kappa_c + \gamma$ and $b = 1 - \kappa_c + \gamma$, let $N^*(a, b) = g + 1$ if $T_{\mathbf{y}}^{ptcs}$ is subcritical ($a > 0, b > 0$) or has strong shear at infinity ($a > 0, b < 0$), and let $N^*(a, b) = g - 1$ if $T_{\mathbf{y}}^{ptcs}$ is supercritical ($a < 0, b < 0$).

Theorem 11.11 *Suppose that $ab \neq 0$ and $\kappa_c \neq 1$. For almost all points $\mathbf{a}$ in $\mathbf{R}^2$ (except in a measure-zero set) with $|\mathbf{a}|$ sufficiently small, the linear perturbation,*

$$[T_{\mathbf{y}}^{ptcs}]^{\mathbf{a}}(\mathbf{x}) = T_{\mathbf{y}}^{ptcs}(\mathbf{x}) - \mathbf{a} \cdot \mathbf{x},$$

satisfies the following, where N is the total number of lensed images relative to $[T_{\mathbf{y}}^{ptcs}]^{\mathbf{a}}$:

$$N^*(a, b) \leq N \leq \begin{cases} (g+1)^2 & \text{if } \gamma > 0, \\ g^2 + 1 & \text{if } \gamma = 0, \end{cases}$$

where N and g have opposite parity. If there are only point masses (i.e., $\kappa_c = \gamma = 0$), *then*

$$g + 1 \leq N \leq g^2 + 1.$$

In particular, a point mass generates two lensed images (minimum and saddle), while two point masses produce either three (one minimum and two saddles) or five lensed images (two minima and three saddles) of a light source not on a caustic).[9]

Proof: Since $T_{\mathbf{y}}^{ptcs}$ has finitely many lensed images (see page 431), so does $[T_{\mathbf{y}}^{ptcs}]^{\mathbf{a}}$. The lower bound $N^*(a, b)$ and parity statement follow from Theorems 11.7 (page 431), 11.8 (page 432), and 11.9 (page 434).

In order to obtain the upper bounds on N, consider the (perturbed) lens equation: $\mathbf{y} = \boldsymbol{\eta}^{\mathbf{a}}_{ptcs}(\mathbf{x}) = \boldsymbol{\eta}_{ptcs}(\mathbf{x}) - \mathbf{a}$. Explicitly:

$$\mathbf{y} = -\mathbf{a} + \begin{bmatrix} 1 - \kappa_c + \gamma & 0 \\ 0 & 1 - \kappa_c - \gamma \end{bmatrix} \mathbf{x} - \sum_{\ell=1}^{g} m_\ell \frac{\mathbf{x} - \boldsymbol{\xi}_\ell}{|\mathbf{x} - \boldsymbol{\xi}_\ell|^2}.$$

We first convert the lens equation into complex form as in Witt [Wit90]. Then following A.P. [Pet97], we translate the problem into finding an upper bound on the total number of zeros of the resultant of certain complex polynomials.

Replace $\mathbf{y} + \mathbf{a}$ and $\mathbf{x}_\ell$ by complex variables w and z, respectively. We shall also treat $\boldsymbol{\xi}_\ell$ as a complex number. Then the lens equation becomes

$$w = (1 - \kappa_c) z + \gamma \bar{z} + \sum_{\ell=1}^{g} \frac{m_\ell}{\bar{\xi}_\ell - \bar{z}}. \tag{11.8}$$

[9] Four lensed images cannot occur since N and g have opposite parity, and g is even.

Clear fractions to obtain

$$\begin{aligned} 0 &= (1-\kappa_c)\left(\prod_{\ell=1}^{g}(\bar{\boldsymbol{\xi}}_\ell - \bar{z})\right) z + (\gamma\bar{z} - w)\prod_{\ell=1}^{g}(\bar{\boldsymbol{\xi}}_\ell - \bar{z}) + \sum_{\ell=1}^{g} m_\ell \prod_{\substack{r=1\\ r\neq\ell}}^{g}(\bar{\boldsymbol{\xi}}_r - \bar{z}) \\ &\equiv p(z,\bar{z}) \end{aligned} \qquad (11.9)$$

The corresponding conjugate equation is

$$\begin{aligned} 0 &= (\gamma z - \bar{w})\prod_{\ell=1}^{g}(\boldsymbol{\xi}_\ell - z) + (1-\kappa_c)\bar{z}\prod_{\ell=1}^{g}(\boldsymbol{\xi}_\ell - z) + \sum_{\ell=1}^{g} m_\ell \prod_{\substack{r=1\\ r\neq\ell}}^{g}(\boldsymbol{\xi}_r - z). \\ &\equiv \bar{p}(z,\bar{z}) \end{aligned} \qquad (11.10)$$

Hence, if z is a lensed image of $\mathbf{y}$ via $\boldsymbol{\eta}^{\mathbf{a}}_{ptcs}$, then $(z, \bar{z})$ is a common root of both (11.9) and (11.10).[10] Thus, the total number of lensed images obeys

$$N \leq \#(\text{common zeros } (z,\bar{z}) \text{ of } p \text{ and } \bar{p}).$$

We can now translate finding an upper bound on the total number N of lensed images into determining such a bound for the number of common zeros of $p(z,\bar{z})$ and $\bar{p}(z,\bar{z})$. The theory of resultants then applies. Let $\mathrm{Res}_{(p,\bar{p})}(\bar{z})$ be the z-resultant of p and $\bar{p}$. Write

$$p(z,\bar{z}) = a_0(\bar{z}) + a_1(\bar{z})z$$

and

$$\bar{p}(z,\bar{z}) = b_0(\bar{z}) + \cdots + b_{t(\gamma)}(\bar{z})z^{t(\gamma)},$$

where $t(\gamma) = g$ for $\gamma = 0$ and equals $g+1$ if $\gamma > 0$. Note that since $\kappa_c \neq 1$ and $\bar{z} \neq \bar{\boldsymbol{\xi}}_\ell$ for any ℓ (because $\boldsymbol{\xi}_\ell$ is not in the domain of $\boldsymbol{\eta}^{\mathbf{a}}_{ptcs}$), we get

$$a_1(\bar{z}) = (1-\kappa_c)\prod_{\ell=1}^{g}(\bar{\boldsymbol{\xi}}_\ell - \bar{z}) \neq 0.$$

The condition $a_1(\bar{z}) = 0 = b_{t(\gamma)}(\bar{z})$ is never satisfied. Hence, since the common zeros of p and $\bar{p}$ occur in conjugate pairs $(z,\bar{z})$, the Resultant Theorem (Theorem 11.10) implies that there is a bijection between the common zeros $(z,\bar{z})$ of the pair p,$\bar{p}$ and the zeros $\bar{z}$ of the resultant polynomial $\mathrm{Res}_{(p,\bar{p})}(\bar{z})$. Thus

$$N \leq \deg[\mathrm{Res}_{(p,\bar{p})}(\bar{z})].$$

[10]The converse need not hold — see $\mathbf{C}^2$-*lensed images* (page 328) and Figure 9.10 (page 338).

Equation (11.5) then yields

$$\begin{aligned} N &\leq t(\gamma)\,(\max\{\deg[a_0], \deg[a_1]\}) + (1)\left(\max\left\{\deg[b_\ell]_{\ell=0}^{g}\right\}\right) \\ &\leq [t(\gamma)]^2 + 1. \end{aligned}$$

If $\gamma = 0$, then $N \leq g^2 + 1$. For $\gamma > 0$, we have $N \leq (g+1)^2 + 1$. Since N and g have opposite parity, while g and $(g+1)^2 + 1$ have the same parity, the upper bound $(g+1)^2+1$ is never achieved by N. Hence, $N \leq (g+1)^2$. □

11.6 Location of Lensed Images due to Point Masses

We shall consider the locations of lensed images of a point source moving along a straight line and undergoing lensing by point masses with continuous matter and shear. Since there are roughly at least as many lensed images as point masses, it is of interest to investigate lensed-image positions — especially for lenses with numerous point masses. Our discussion parallels and rigorously treats that of Lewis et al. [Lew-MRW] and Witt [Wit93]. Readers should see the latter references for applications to the numerical problem of finding all lensed images.

Figure 11.4 shows the motion of the lensed images due to a point mass lens with a light source moving along a straight line. Notice that one of the lensed-image tracks passes

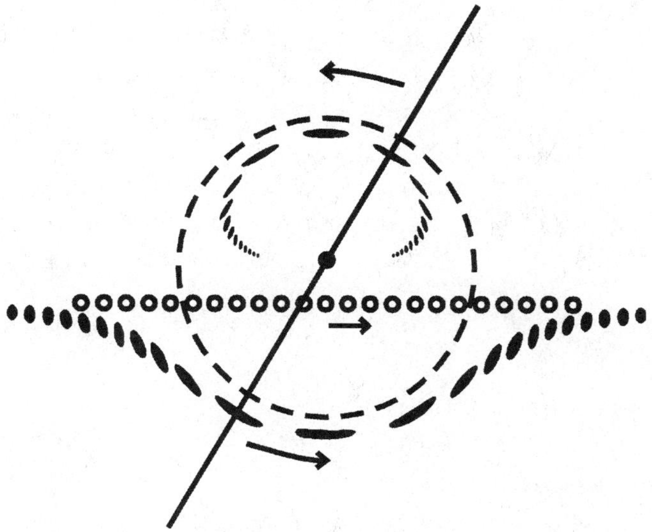

Figure 11.4. Motion of the two lensed images due to a point mass lens (• at center of circle — Einstein ring) as a light source ∘ moves to the right along the x-axis. The small circular source is lensed into an elliptical shape. (After Paczyński [Pac96, p. 425].)

through the point mass. We shall investigate this result for the case of many point masses with continuous matter and shear from infinity.

Let $\boldsymbol{\eta}_{ptcs} : L \to S$ be the single-plane lensing map induced by a gravitational lens potential ψ_{ptcs} with continuous matter κ_c, shear γ from infinity, and g point masses at positions $\boldsymbol{\xi}_1, \ldots, \boldsymbol{\xi}_g$ (see (11.3), page 431). Note that $L = \mathbf{R}^2 - \{\boldsymbol{\xi}_1, \ldots, \boldsymbol{\xi}_g\}$. Let Q be a straight line in the light source plane S. If a light source moves along Q, then the positions of the lensed images are given in the next theorem for a generic lensing map.

Suppose that $f : X \to Y$ is a smooth map between manifolds and K a submanifold of Y. Recall that f is *transverse* to K if for every $x \in X$, either $f(x) \notin K$, or, $f(x) \in K$ and $T_{f(x)}Y = df_x(T_xX) + T_{f(x)}K$. Then:

Theorem 11.12 *Suppose that $\boldsymbol{\eta}_{ptcs} : L \to S$ is locally stable and transverse to the straight line Q in the light source plane S.*[11] *Assume that $(1 - \kappa_c + \gamma)(1 - \kappa_c - \gamma) \neq 0$. Then the lensed image of Q, namely, $\boldsymbol{\eta}^{-1}(Q)$, is a boundaryless 1-manifold in L and $[\boldsymbol{\eta}^{-1}(Q)] \cup \{\boldsymbol{\xi}_1, \ldots, \boldsymbol{\xi}_g\} \subset \mathbf{R}^2$ consists of:*

(1) *an infinitely long curve passing through a finite number of point masses and extending without bound in opposite directions;*

(2) *and, a finite collection of disjoint simple closed curves, each passing through at least one point mass; there are at most g such loops.*

Proof: The condition, $(1 - \kappa_c + \gamma)(1 - \kappa_c - \gamma) \neq 0$ implies that $\mathrm{Crit}(\boldsymbol{\eta}_{ptcs})$ and $\mathrm{Caustic}(\boldsymbol{\eta}_{ptcs})$ are compact (see page 431), while the transversality of $\boldsymbol{\eta}_{ptcs}$ to Q guarantees that $\boldsymbol{\eta}^{-1}_{ptcs}(Q)$ is a 1-manifold without boundary. We claim that $\boldsymbol{\eta}^{-1}_{ptcs}(Q)$ *is a union of disjoint open arcs (possibly of infinite length), none of which has an endpoint in L.*

First, if c is a component open arc of $\boldsymbol{\eta}^{-1}(Q)$ and $\mathbf{x}_0 \in L$ an endpoint of c, then $\boldsymbol{\eta}_{ptcs}(c) \subset Q$ and $\boldsymbol{\eta}_{ptcs}(\mathbf{x}_0) \notin Q$. This contradicts the continuity of $\boldsymbol{\eta}_{ptcs}$. Second, no component of $\boldsymbol{\eta}^{-1}_{ptcs}(Q)$ is a simple closed curve in L. To see this, suppose that S has rectangular coordinates (y_1, y_2) and line Q is given by $y_2 = my_1 + b$. Writing $\boldsymbol{\eta}_{ptcs}(\mathrm{u}, \mathrm{v}) = (y_1, y_2)$, it follows that $\boldsymbol{\eta}^{-1}_{ptcs}(Q)$ coincides with the level set $p(\mathrm{u}, \mathrm{v}) = 0$, where

$$p(\mathrm{u}, \mathrm{v}) = y_2 - my_1 - b = \mathrm{v} - \psi_{\mathrm{v}}(\mathrm{u}, \mathrm{v}) - m(\mathrm{u} - \psi_{\mathrm{u}}(\mathrm{u}, \mathrm{v})) - b$$

and $\psi = \psi_{ptcs}$. Since the Laplacian of p is

$$\Delta p = -\psi_{\mathrm{uuv}} - \psi_{\mathrm{vvv}} + (\psi_{\mathrm{uuu}} + \psi_{\mathrm{vvu}})$$

and $\psi_{\mathrm{uu}} = -\psi_{\mathrm{vv}}$ on L, we obtain $\Delta p = 0$ on L. Now, suppose that $\boldsymbol{\eta}^{-1}_{ptcs}(Q)$ has a component, c, diffeomorphic to a circle. Orient c counterclockwise and let $\mathbf{n}$ be the outward

[11] The transversality condition yields that Q is not tangent to any caustic curve of $\boldsymbol{\eta}_{ptcs}$ (cf. Figure 7.1, page 217).

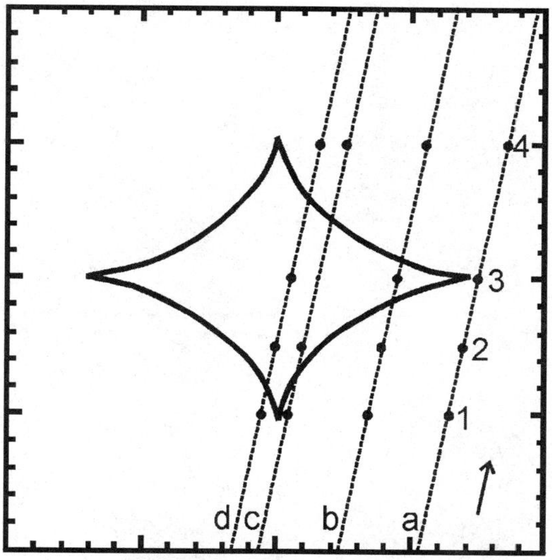

Figure 11.5. Caustic due to a point mass lens with shear at infinity. The dashed lines show the tracks of a light source in the direction indicated by the arrow. Figure 11.6 shows the paths of the lensed images for the given light source tracks. (After Witt [Wit93, pp. 531].)

unit normal to c. Let R denote the region bounded by c. Green's theorem yields

$$\oint_c (\text{grad } p) \cdot \mathbf{n}\, ds = \iint_R \text{div}(\text{grad } p)\, dA = 0.$$

However, the integrand on the left is always positive (since grad p and $\mathbf{n}$ point in the same direction). Hence, the line integral is positive, which is a contradiction. Thus, there is no component c.

Let D be a closed disc in $\mathbf{R}^2$ that properly contains $[\text{Crit}(\eta_{ptcs})] \cup \{\xi_1, \ldots, \xi_g\}$. By Theorems 11.7(2) (page 431), 11.8(3) (page 432), and 11.9(2) (page 434), which all require local stability, we have the following for arcs in $\eta_{ptcs}^{-1}(Q)$ as $|\mathbf{y}| \to \infty$ along opposite directions on Q (cf. Figures 11.5 and 11.6):

(i) Each point mass is the limiting position of exactly two arcs (formed from the motion of saddle lensed images).

(ii) There are exactly two arcs in $\mathbf{R}^2 - D$; each has infinite length and both extend without bound in opposite directions.

Now, trace the two arcs of (ii) into the interior of D. Since no component of $\eta_{ptcs}^{-1}(Q)$ has an endpoint in L, either the two arcs meet or each terminates at a point mass. If both terminate at the same mass, then we also obtain a curve of the type cited in part (1) of the theorem. This will still be the case if the two arcs terminate at different masses. For the

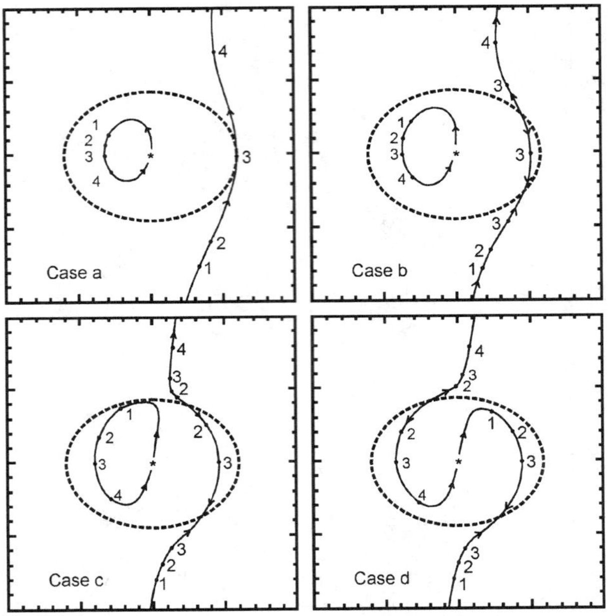

Figure 11.6. Lensed-image trajectories due to each light source track in Figure 11.5. The dashed curves are critical curves. Notice that as the light source traverses the caustic, nondegenerate lensed images are annihilated by mergers and spontaneously created out of points on the critical curve. (After Witt [Wit93, p. 532].)

latter, result (i) shows that either the two masses are formed by an arc in L, or, an arc in $\mathbf{R}^2$ that passes through at least one additional mass. Finally, given the curve in part (1) of the theorem, result (i) yields that each remaining component of $\boldsymbol{\eta}^{-1}_{ptcs}(Q)$ is a simple closed curve passing through at least one point mass. □

The results of Theorem 11.12 for the case of a point mass with shear from infinity are illustrated in Figures 11.5 and 11.6.

Chapter 12

Counting Lensed Images: Multiplane Case

"What is one and one and one and one and one and one and one and one and one and one?"
"I don't know," said Alice. "I lost count."
"She can't do addition," said the Red Queen.

LEWIS CARROLL

This chapter extends the work of Chapter 11 to a study of lensed-image counting due to gravitational lenses on any finite number of lens planes (Problem **M1**). As in Chapter 11, we follow the approach in [Pet95a], employing Morse theory to investigate the number of lensed images of different types. In particular, lensed images are counted within a compact region D not a priori required to contain the whole deflector on each lens plane. The associated time delay function T_{y} must then obey Morse boundary conditions B relative to D before we can apply Morse theory. We shall show that there is an arbitrarily small linear perturbation of T_{y} in D such that the resulting function satisfies Morse boundary conditions B relative to D and — most importantly — is also a time delay function.

All results we obtain through Morse theory depend essentially on only the topological properties induced in the lens planes by the deflectors and the behavior of the gradient of T_{y} on the boundary of D. Applications are given to nonsingular isolated multiplane lenses and multiplane point mass distributions (with possibly several point masses on each lens plane). We also use resultants to determine an upper bound on the number of lensed images due to point masses in three-dimensional arrays with a single point mass on each lens plane [Pet97]. This chapter assumes the basics of multiplane lensing (Section 6.4, page 193).

Outline: Section 12.1 addresses the Multiplane Lensed-Image Counting Problem for general deflectors. Section 12.1.1 presents counting formulas and lower bounds on the number of lensed images of various types due to multiplane lens systems obeying Morse boundary conditions B. In Section 12.1.2, we show that the latter condition is a generic property among time delay functions. Sections 12.2 and 12.3 restrict, respectively, to the cases of nonsingular isolated multiplane lenses and multiplane point mass lens systems. The former includes a multiplane Odd Number Lensed-Image Theorem. Section 12.3 divides into Section 12.3.1, which addresses lensed-image counting for multiplane point mass systems having finitely many masses on every lens plane, and Section 12.3.2, which presents an upper bound on the number of lensed images when there is one point mass on each lens plane.

12.1 General Multiplane Gravitational Lens

12.1.1 Multiplane Lensed-Image Counting Theorem

Requisite Mathematics: Sections 7.4.1 and 10.2.

We must consider the following before setting up the notation for the multiplane lensed-image counting theorem:

Smoothing Corners in a Product of Manifolds with Boundary

Let U_1 and U_2 be manifolds with boundary, where $\dim U_i = n_i$ for $i = 1, 2$. In general, $U_1 \times U_2$ is not a manifold with boundary (though it is if either U_1 or U_2 is a manifold without boundary). We now show how to transform $U_1 \times U_2$ into a manifold with boundary.[1]

Set $\bar{\mathbf{R}}^n_+ = \{(x_1, \ldots, x_n) \in \mathbf{R}^n : x_n \geq 0\}$. Let $\mathcal{A}_i$ be the maximal atlas of all charts $\nu_i : \mathcal{V}_i \longrightarrow \bar{\mathbf{R}}^{n_i}_+$ in U_i and set $\mathcal{A} = \{\nu_1 \times \nu_2 : \nu_i \in \mathcal{A}_i, i = 1, 2\}$. The map $\nu_1 \times \nu_2 : \mathcal{V}_1 \times \mathcal{V}_2 \longrightarrow \bar{\mathbf{R}}^{n_1}_+ \times \bar{\mathbf{R}}^{n_2}_+$ is not necessarily a chart in $U_1 \times U_2$. This is because the set, $\bar{\mathbf{R}}^{n_1}_+ \times \bar{\mathbf{R}}^{n_2}_+ = \mathbf{R}^{n_1+n_2-2} \times \bar{\mathbf{R}}_+ \times \bar{\mathbf{R}}_+$, which has a "corner," is not an open subset of $\bar{\mathbf{R}}^{n_1+n_2}_+$; hence, $\nu_1 \times \nu_2$ is not guaranteed to be a homeomorphism into $\bar{\mathbf{R}}^{n_1+n_2}_+$. In fact, since

$$(\bar{\mathbf{R}}^{n_1}_+ \times \bar{\mathbf{R}}^{n_2}_+)^c = \bar{\mathbf{R}}^{n_1+n_2}_+ - (\bar{\mathbf{R}}^{n_1}_+ \times \bar{\mathbf{R}}^{n_2}_+) = \mathbf{R}^{n_1+n_2-2} \times (\mathbf{R} - \bar{\mathbf{R}}_+) \times \bar{\mathbf{R}}_+$$

is open in $\bar{\mathbf{R}}^{n_1+n_2}_+$, the set $\bar{\mathbf{R}}^{n_1}_+ \times \bar{\mathbf{R}}^{n_2}_+$ is closed in $\bar{\mathbf{R}}^{n_1+n_2}_+$. If $\bar{\mathbf{R}}^{n_1}_+ \times \bar{\mathbf{R}}^{n_2}_+$ is also open, then $\bar{\mathbf{R}}^{n_1+n_2}_+$ is a disjoint union of the nonempty open sets $\bar{\mathbf{R}}^{n_1}_+ \times \bar{\mathbf{R}}^{n_2}_+$ and $(\bar{\mathbf{R}}^{n_1}_+ \times \bar{\mathbf{R}}^{n_2}_+)^c$. This contradicts the connectedness of $\bar{\mathbf{R}}^{n_1+n_2}_+$.

Define a map $f : \bar{\mathbf{R}}^2_+ \longrightarrow \bar{\mathbf{R}}_+ \times \bar{\mathbf{R}}_+$ by $f(r, \theta) = (r, \theta/2)$, where (r, θ) are polar coordinates. Then f is a homeomorphism whose inverse "straightens" the corner in the first quadrant $\bar{\mathbf{R}}_+ \times \bar{\mathbf{R}}_+$. We can employ f to remove the "corner" in $\bar{\mathbf{R}}^{n_1}_+ \times \bar{\mathbf{R}}^{n_2}_+$ and, hence,

[1] See [Broc-J, pp. 136–139] for a treatment via the collar theorem. A discussion of manifolds with corners appears in Section 1 of [MatII] and §2 of [Micho].

transform $\nu_1 \times \nu_2$ into a chart. Explicitly, let $\widetilde{\nu_1 \times \nu_2}$ be the homeomorphism given by the following composition:

$$\mathcal{V}_1 \times \mathcal{V}_2 \stackrel{\nu_1 \times \nu_2}{\longrightarrow} \mathbf{R}^{n_1+n_2-2} \times \bar{\mathbf{R}}_+ \times \bar{\mathbf{R}}_+ \stackrel{\left(1_{\mathbf{R}^{n_1+n_2-2}}\right)\times f^{-1}}{\longrightarrow} \mathbf{R}^{n_1+n_2-2} \times \bar{\mathbf{R}}_+^2 = \bar{\mathbf{R}}_+^{n_1+n_2}.$$

The map $\widetilde{\nu_1 \times \nu_2}$ is a chart in $U_1 \times U_2$ that results from a "smoothing" of $\nu_1 \times \nu_2$.

Let $\tilde{\mathcal{A}} = \{\widetilde{\nu_1 \times \nu_2} : \nu_1 \times \nu_2 \in \mathcal{A}\}$. Then $\tilde{\mathcal{A}}$ is a smooth atlas on $U_1 \times U_2$. To see smooth overlap, let $\widetilde{\nu_1 \times \nu_2}$ and $\widetilde{\xi_1 \times \xi_2}$ be elements of $\tilde{\mathcal{A}}$ with overlapping domains. Then

$$(\widetilde{\nu_1 \times \nu_2}) \circ (\widetilde{\xi_1 \times \xi_2})^{-1} = [(1_{\mathbf{R}^{n_1}}) \circ \nu_1 \circ \xi_1^{-1}] \times [((1_{\mathbf{R}^{n_2-2}}) \times f^{-1}) \circ \nu_2 \circ \xi_2^{-1}]$$

is a diffeomorphism from an open subset of $\bar{\mathbf{R}}_+^{n_1+n_2}$ onto an open subset of $\mathbf{R}^{n_1} \times (\mathbf{R}^{n_2-2} \times \bar{\mathbf{R}}_+^2) = \bar{\mathbf{R}}_+^{n_1+n_2}$. *Thus,* $(U_1 \times U_2, \tilde{\mathcal{A}})$ *is an* (n_1+n_2)*-dimensional manifold with boundary.* We shall call $(U_1 \times U_2, \tilde{\mathcal{A}})$ a *smoothing* of $U_1 \times U_2$.

Notation for Main Theorem

Let $T_{\mathbf{y}} : L_{(k)} \to \mathbf{R}$, where $L_{(k)} = L_1 \times \cdots \times L_k$, be a k-plane time delay function induced at $\mathbf{y} \in S$ by gravitational lens potentials $\psi_1, \ldots, \psi_k$. Here, $L_i = E_i - A_i$, where E_i is the Euclidean plane and A_i the (finite) set of singularities of ψ_i. Recall from equation (6.23) (page 197) that

$$\mathrm{Dom}(T_{\mathbf{y}}) = \mathbf{R}^{2k} - \bigcup_{i=1}^{k} E_1 \times \cdots \times E_{i-1} \times A_i \times E_{i+1} \times \cdots \times E_k.$$

Let A_{g_i} be a subset of A_i consisting of g_i singularities of ψ_i and let D_{g_i} be a regular domain in L_i that is punctured by A_{g_i} (see page 421). Set $D_{(k)} = D_{g_1} \times \cdots \times D_{g_k}$, $A(1) = A_{g_1}$, and

$$\mathbf{A}(k) = \bigcup_{i=1}^{k} E_1 \times \cdots \times E_{i-1} \times A_{g_i} \times E_{i+1} \times \cdots \times E_k, \qquad (k \geq 2).$$

Since D_{g_i} is a manifold with boundary, the previous subsection yields $D_{(k)}$ can be smoothed into a $2k$-dimensional manifold $\tilde{D}_{(k)}$ with boundary. Moreover, since $D_{(k)}$ is a proper subset of the $L_{(k)}$, we can always construct $\tilde{D}_{(k)}$ such that $\tilde{D}_{(k)} \subset L_{(k)}$ (by choosing the domains of charts like $\widetilde{\nu_1 \times \nu_2}$ sufficiently small). We now define a regular domain D in $L_{(k)}$ to be *punctured by* $\mathbf{A}(k)$ if D is a smoothing of a product $D_{(k)}$.

Assume that $T_{\mathbf{y}}$ satisfies Morse boundary conditions B relative to a regular domain D which is punctured by $\mathbf{A}(k)$. Let D_- be the union of all open submanifolds of ∂D on which grad $T_{\mathbf{y}}$ is directed into the interior of D. By Corollary 7.44 (page 240), the functions $T_{\mathbf{y}}$ and $T_{\mathbf{y}}|_{\partial D}$ have finitely many critical points, all of which are nondegenerate. Theorem 6.17(1) (page 204) guarantees that the lensed images of a light source at $\mathbf{y}$ are in 1-1 correspondence with the critical points of $T_{\mathbf{y}}$. Hence, *the counting of lensed images due to multiplane lensing is equivalent to a counting of critical points.*

The following notation is convenient:

(1) Let M_i, where $0 \leq i \leq 2k$, be the number of index i critical points of $T_{\mathbf{y}}$ in D (page 197). For example, M_0 and M_{2k} are, respectively, the total number of local minimum and maximum lensed images in D. Set:

$$N = M_0 + M_1 + \cdots + M_{2k} = \text{number of lensed images in } D,$$

$$M_+ = \sum_{i(\text{even})} M_i = \text{number of even index lensed images in } D,$$

$$M_- = \sum_{i(\text{odd})} M_i = \text{number of odd index lensed images in } D.$$

(2) Let M_i' be the total number of index i critical points of $T_{\mathbf{y}}|_{D_-}$, where $i = 0, \ldots, 2k-1$. Set:

$$N' = M_0' + \cdots + M_{2k-1}', \qquad M_+' = \sum_{i(\text{even})} M_i', \qquad M_-' = \sum_{i((\text{odd})} M_i',$$

$$M'' = M_0' - M_1' + \cdots - M_{2k-1}', \qquad M_j'' = M_j' - M_{j-1}' + \cdots (-1)^j M_0'.$$

(3) Set $G_0 = 1$ and, for $j = 1, \ldots, k$, let

$$G_j = \sum_{1 \leq \ell_1 < \cdots < \ell_j \leq k} g_{\ell_1} \cdots g_{\ell_j} - \sum_{1 \leq \ell_1 < \cdots < \ell_{j-1} \leq k} g_{\ell_1} \cdots g_{\ell_{j-1}} + \cdots (-1)^{j-1} \sum_{\ell_1=1}^{k} g_{\ell_1} + (-1)^j.$$

Main Theorem

With the previous notation, we have (A.P. [Pet95a, p. 4269]):

Theorem 12.1 (General Multiplane Lens) *Let $T_{\mathbf{y}} : L_{(k)} \to \mathbf{R}$, be a k-plane time delay function obeying Morse boundary conditions* B *relative to a regular domain D in $L_{(k)}$ that is punctured by $\mathbf{A}(k)$. Then the number of lensed images in D of a light source at $\mathbf{y}$ satisfies:*

(1) *Weak Inequalities:*

(i) $M_0 \geq 1 - M_0'$.

(ii) $M_j \geq \sum_{1 \leq \ell_1 < \cdots < \ell_j \leq k} g_{\ell_1} \cdots g_{\ell_j} - M_j'$, *where* $1 \leq j \leq k$.

(iii) $M_j \geq 0$, *where* $k + 1 \leq j \leq 2k$.

(2) *Strong Inequalities:*

(i) $M_j - M_{j-1} + \cdots (-1)^j M_0 \geq G_j - M''_j$, *where* $0 \leq j \leq k$.

(ii) $M_j - M_{j-1} + \cdots (-1)^j M_0 \geq (-1)^{j-k} G_k - M''_j$, *where* $k \geq 2$ *and* $k + 1 \leq j \leq 2k - 1$.

(iii) $M_+ = M_- - M'' + \prod_{i=1}^{k} (1 - g_i)$.

(3) *Total number of lensed images:*

(i) $N = 2M_+ + M'' - \prod_{i=1}^{k} (1 - g_i) = 2M_- - M'' + \prod_{i=1}^{k} (1 - g_i)$.

(ii) $N \geq \prod_{i=1}^{k} (1 + g_i) - N'$.

Theorem 12.1 testifies to the power of Morse inequalities since the stated counting results are far from apparent on physical grounds. Furthermore, given the amount of information contained in the theorem, we shall not be able to exhaust all its physical consequences. Instead, the important cases of nonsingular isolated lenses and point masses will be treated in Section 12.2 (page 455) and Section 12.3 (page 456).

Proof of Theorem 12.1: We begin by computing the Betti numbers, $B_n(D) = \dim H_n(D; \mathbf{R})$, for any non-negative integer n. First, Theorem 10.1 (page 401) implies that D and $D_{(k)}$ have the same Betti numbers for all $n \geq 0$.

Künneth's formula (Theorem 10.5, page 403) yields

$$B_n(D) = \sum_{i_1 + \cdots + i_k = n} B_{i_1}(D_{g_1}) \cdots B_{i_k}(D_{g_k}). \tag{12.1}$$

Path-connectedness of each D_{g_i} gives (Theorem 10.2, page 401):

$$B_0(D) = 1.$$

If $n > k$, then the condition $i_1 + \cdots + i_k = n$ forces $i_\ell \geq 2$ for some $\ell \in \{1, \ldots, k\}$. Consequently, each term $B_{i_1}(D_{g_i}) \cdots B_{i_k}(D_{g_k})$ will have a factor $B_{i_\ell \geq 2}(D_{g_j})$. Since $B_{i_\ell \geq 2}(D_{g_j}) = 0$ for $j = 1, \ldots, k$, equation (12.1) implies that

$$B_{n>k}(D) = 0.$$

If $1 \leq n \leq k$, then since $B_0(D_{g_i}) = 1$ and $B_1(D_{g_i}) = g_i$ for $i = 1, \ldots, k$ (e.g., Figure 10.4, page 402), equation (12.1) yields

$$B_{1 \leq n \leq k}(D) = \sum_{1 \leq \ell_1 < \cdots < \ell_n \leq k} g_{\ell_1} \cdots g_{\ell_n}.$$

Also, observe that the Euler characteristic of D is

$$\chi(D) = \prod_{i=1}^{k} \chi(D_{g_i}) = \prod_{i=1}^{k}(1 - g_i)$$

and

$$\sum_{n=0}^{2k} B_n(D) = 1 + \sum_{n=1}^{k}\left(\sum_{1\le \ell_1 < \cdots < \ell_n \le k} g_{\ell_1} \cdots g_{\ell_n} \right) = \prod_{i=1}^{k}(1 + g_i).$$

By Theorem 10.6 (page 405), the strong Morse inequalities hold:

$$M_j - M_{j-1} + \cdots (-1)^j M_0 \ \geq\ [B_j(D) - M'_j] - \cdots (-1)^j [B_0(D) - M'_0]$$

$$M_{2k} - M_{2k-1} + \cdots + M_0 \ =\ B_{2k}(D) - [B_{2k-1}(D) - M'_{2k-1}] + \cdots + [B_0(D) - M'_0],$$

where $0 \leq j \leq 2k - 1$ and $B_{2k}(D) = 0$. The weak Morse inequalities (Corollary 10.7, page 406) are immediate:

$$M_i \geq B_i(D) - M'_i,$$

where $i = 0, \ldots, 2k - 1$, and $M_{2k} \geq 0$. With the previously computed Betti numbers, the theorem follows from the strong and weak Morse inequalities. □

The next result gives useful sufficient conditions for having finitely many lensed images.

Proposition 12.2 *Let $T_{\mathbf{y}} : L_{(k)} \to \mathbf{R}$ be a k-plane time delay function induced by gravitational lens potentials ψ_i, $i = 1, \ldots, k$. Suppose that all the singularities of ψ_i (if any) are infinite singularities.*[2] *If* $\mathbf{y}$ *is not a caustic point and* $|\operatorname{grad} T_{\mathbf{y}}(\mathbf{X})| \to \infty$ *as* $|\mathbf{X}| \to \infty$, *then a light source at* $\mathbf{y}$ *has finitely many lensed images.*

Proof: We have $D_{(k)} = D_{g_1} \times \cdots \times D_{g_k}$ with $A_{g_i} = A_i$, where $i = 1, \ldots, k$. Let $D(\mathbf{a}; \epsilon)$ be an open $2k$-ball in $\mathbf{R}^{2k}$ of radius ϵ centered at $\mathbf{a}$. Since $T_{\mathbf{y}}$ is nondegenerate and for every $\mathbf{a} \in \mathbf{A}(k)$, we have $|\operatorname{grad} T_{\mathbf{y}}(\mathbf{X})| \to \infty$ as $\mathbf{X} \to \mathbf{a}$, it follows that there is a real number ϵ_0 such that no critical point of $T_{\mathbf{y}}$ lies in $D(\mathbf{a}; \epsilon)$ for all $\epsilon \in (0, \epsilon_0]$. In addition, since $|\operatorname{grad} T_{\mathbf{y}}(\mathbf{X})| \to \infty$ as $\mathbf{X} \to \infty$, the set $\operatorname{Crit}(T_{\mathbf{y}})$ lies in a compact subset of $\mathbf{R}^{2k}$. Consequently, there is a regular domain D in $L_{(k)}$ that is punctured by $\mathbf{A}(k)$ such that all critical points of $T_{\mathbf{y}}$ lie in the interior of D. Hence, $T_{\mathbf{y}}$ is admissible relative to D. Corollary 7.44 (page 240) now guarantees that $T_{\mathbf{y}}$ has finitely many critical points. □

The regions on the various lens planes where minimum and maximum light rays impact have surface mass densities obeying the following:

[2]That is, if A_i is the set of singularities of ψ_i, then $\psi_i(\mathbf{x}_i) \to -\infty$ as $\mathbf{x}_i \to \mathbf{p}_i$ for each $\mathbf{p}_i \in A_i$.

Proposition 12.3 *Suppose that $T_{\mathbf{y}}$ is a k-plane time delay function with* **y** *not a caustic point. Let* $\mathbf{X} = (\mathbf{x}_1, \ldots, \mathbf{x}_k)$ *be a light ray of $T_{\mathbf{y}}$. Then:*[3]

(1) *If* **X** *is a minimum, then* $\kappa_i(\mathbf{x}_i) < 1/\beta_i$ *for* $i = 1, \ldots, k$.

(2) *If* **X** *is a maximum, then:*

  (i) $\kappa_1(\mathbf{x}_1) > 1/\beta_1$;

  (ii) $\kappa_i(\mathbf{x}_i) > (1/\beta_i) + (\vartheta_{i-1}/\vartheta_i)$, *for* $i = 2, \ldots, k-1$;

  (iii) $\kappa_k(\mathbf{x}_k) > 1$.

Proof: (1) The proof involves some rather lengthy calculations, whose techniques will not be needed later. For this reason, we refer the reader to Seitz and Schneider [Sei-S92, p. 5], who give a detail proof.

(2) Since [Hess $T_{\mathbf{y}}$]($\mathbf{X}_0$) is negative definite for $\mathbf{X}_0$ a maximum, all diagonal entries are negative. Consequently, the following Laplacians are negative, which yields the desired result:

$$\Delta_{\mathbf{x}_1} T = 2\vartheta_1(1 - \beta_1\kappa_1) > 0 \tag{12.2}$$

$$\Delta_{\mathbf{x}_i} T = 2[\vartheta_{i-1} + \vartheta_i(1 - \beta_i\kappa_i)] > 0, \qquad (i = 2, \ldots, k-1), \tag{12.3}$$

$$\Delta_{\mathbf{x}_k} T = 2\vartheta_k(1 - \kappa_k) > 0. \tag{12.4}$$

□

Note that a minimum light ray is forced to impact on a subcritical part of the kth lens plane (since $\beta_k = 1$), while maximum light rays must impact in a supercritical region of that lens plane. Saddle light rays may impact in a subcritical or supercritical part of a lens.

12.1.2 Genericity of Morse Boundary Conditions B

Theorem 12.1 requires that a time delay function obeys Morse boundary conditions B relative to the regular domain in which light rays are counted. We shall now show that this requirement is a "generic" property of time delay functions.

Requisite Mathematics: Sections 7.1.2, 7.3.1, 7.4, and 10.2.

Let $T_{\mathbf{y}} : L_{(k)} \to \mathbf{R}$ be a k-plane time delay function, where **y** is not a caustic point. Consider any regular domain D in $L_{(k)}$ that is punctured by $\mathbf{A}(k)$. If a critical point of $T_{\mathbf{y}}$ lies on ∂D, then since $\mathrm{Crit}(T_{\mathbf{y}})$ consists of nondegenerate critical points, which are isolated, a slight smooth deformation of ∂D yields $\partial D \cap \mathrm{Crit}(T_{\mathbf{y}}) = \emptyset$. Without loss of genericity, we can then assume that $T_{\mathbf{y}}$ is admissible relative to D.

[3] Recall that the positive constants β_i and ϑ_i are defined in terms of angular diameter distances between various lens planes (see Section 6.4.1, page 195).

Theorem 12.4 *Let $T_{\mathbf{y}} : L_{(k)} \to \mathbf{R}$ be a k-plane time delay function. Suppose that $T_{\mathbf{y}}$ is admissible relative to a regular domain D in $L_{(k)}$ that is punctured by $\mathbf{A}(k)$. Then for almost all $\mathbf{a} \in \mathbf{R}^{2k}$ with $|\mathbf{a}|$ sufficiently small, the linear perturbation*

$$T_{\mathbf{y}}^{\mathbf{a}}(\mathbf{X}) = T_{\mathbf{y}}(\mathbf{X}) - \mathbf{a} \cdot \mathbf{X}, \quad \mathbf{X} \in L_{(k)},$$

of $T_{\mathbf{y}}$ is a time delay function satisfying:

(1) *$T_{\mathbf{y}}^{\mathbf{a}}$ obeys Morse boundary conditions* B *relative to D.*

(2) *As $|\mathbf{a}| \to 0$, the light rays of $T_{\mathbf{y}}^{\mathbf{a}}$ in D tend uniformly and bijectively with preservation of critical point type to those of $T_{\mathbf{y}}$ in D.*

Remark: For $i = 1, \ldots, k$, set $\psi^{\mathbf{a}_i}(\mathbf{x}_i) = \psi_i(\mathbf{a}_i) - (\mathbf{a}_i \cdot \mathbf{x}_i)/(\vartheta_i \beta_i)$. Then each $\psi_i^{\mathbf{a}}$ is a gravitational lens potential and they induce the time delay function:

$$T_{\mathbf{y}}^{\mathbf{a}}(\mathbf{X}) = \sum_{i=1}^{k} \vartheta_i \left[\frac{|\mathbf{x}_{i+1} - \mathbf{x}_i|^2}{2} - \beta_i \psi_i^{\mathbf{a}_i}(\mathbf{x}_i) \right] = T_{\mathbf{y}}^{\mathbf{a}}(\mathbf{X}) - a \cdot \mathbf{X},$$

where $\mathbf{X} = (\mathbf{x}_i, \ldots, \mathbf{x}_k)$. The surface mass densities $\kappa_i^{\mathbf{a}_i}$ and κ_i associated, respectively, with $\psi_i^{\mathbf{a}_i}$ and ψ_i are identical. By choosing $|\mathbf{a}_i|$ sufficiently small, we can make $\kappa_i^{\mathbf{a}_i}$ an arbitrarily small perturbation of the original matter distribution κ_i on L_i.

Proof of Theorem 12.4:

Lemma 12.5 *Let $H : V \times \mathbf{R}^n \to \mathbf{R}$, where $V \subseteq \mathbf{R}^m$ is open, be a smooth n-parameter family of functions. Suppose that H is a Morse family.*[4] *Then the set $\mathcal{N}$ of $a \in \mathbf{R}^n$ such that $h^a : V \subseteq \mathbf{R}^m \to \mathbf{R}$ is nondegenerate, is dense in $\mathbf{R}^n$ and the complement of $\mathcal{N}$ in $\mathbf{R}^n$ has measure zero.*

Proof: Write $H(x, a) = h^a(x)$ for all $(x, a) \in V \times \mathbf{R}^n$. Let $[H_{i,j}]$ be the $m \times (m+n)$ matrix

$$\left[\frac{\partial^2 H}{\partial x_i \partial x_j} \;\; \frac{\partial^2 H}{\partial x_i \partial q_m} \right], \qquad \text{where } x = (x_1, \ldots, x_m), \; a = (a_1, \ldots, a_n).$$

Denote by Z the set of all $(x, a) \in V \times \mathbf{R}^n$ such that

$$\frac{\partial H}{\partial x_1}(x, a) = \cdots = \frac{\partial H}{\partial x_m}(x, a) = 0.$$

[4] That is, if H is given by $(x, a) \mapsto H(x, a)$, where $x \in V \subseteq \mathbf{R}^m$ and $a \in \mathbf{R}^n$, then

$$\text{rank} \left[\frac{\partial^2 H}{\partial x_i \partial x_j} \;\; \frac{\partial^2 H}{\partial x_i \partial a_m} \right] = m \qquad \text{at } (x, a) \text{ where } \frac{\partial H}{\partial x_i}(x, a) = 0 \text{ for } i = 1, \ldots, m.$$

Let $j : \mathbf{R}^n \to C^\infty(V, J^1(V,\mathbf{R}))$ be given by $j(a) = j^1 h^a$. Define a smooth map $\Phi : V \times \mathbf{R}^n \to J^1(V,\mathbf{R})$ by $\Phi(x,a) = j(a)(x)$. Explicitly,

$$\Phi(x,a) = (x, h^a(x), \frac{\partial h^a}{\partial x_1}, \ldots, \frac{\partial h^a}{\partial x_m}).$$

Consider the singularity manifold $S_1(V,\mathbf{R})$ and set $\Phi(x,a) = y$. Then $y \in S_1(V,\mathbf{R})$ if and only if $(x,a) \in Z$.

We now show that if $(x,a) \in Z$, then $\Phi \pitchfork S_1(V,\mathbf{R})$ at (x,a) if and only if $\text{rank}[H_{i,j}](x,a) = m$. As in the proof of Lemma 7.45 (see page 241), it suffices to assume that V is a coordinate neighborhood. We then employ Proposition 7.7 (page 218) with $X = V \times \mathbf{R}^n$, $Y = J^1(V,\mathbf{R})$, $f = \Phi$, and $Q = S_1(V,\mathbf{R})$. Define $h : J^1(V,\mathbf{R}) \to \mathbf{R}^m$ by

$$h(x, f(x), \frac{\partial f}{\partial x_1}, \ldots, \frac{\partial f}{\partial x_m}) = \begin{bmatrix} \frac{\partial f}{\partial x_1} & \cdots & \frac{\partial f}{\partial x_m} \end{bmatrix}^T,$$

where $f \in C^\infty(V,\mathbf{R})$. Since $d_q h = [\text{Hess}\, f \;\; \mathbf{0}_m \;\; I_m]$, we obtain $\text{rank}[d_q h] = m$ for $q \in J^1(V,\mathbf{R})$. In addition, $h^{-1}(0) = S_1(V,\mathbf{R})$ locally. By Theorem 7.34 (page 232), we have

$$\text{codim}_{J^1(V,\mathbf{R})}\, S_1(V,\mathbf{R}) = m.$$

Proposition 7.7 then implies that $\Phi \pitchfork S_1(V,\mathbf{R})$ at $(x,a) \in Z$ if and only if $\text{rank}[d_{(x,a)}(h \circ \Phi)] = m$. But $d_{(x,a)}(h \circ \Phi) = [H_{i,j}]$.

Using the fact that H is a Morse family, we have $\text{rank}[H_{i,j}](x,a) = m$ for $(x,a) \in Z$, which is equivalent to $\Phi \pitchfork S_1(V,\mathbf{R})$ at (x,a). Applying Lemma 7.13 (page 222) with $D = \mathbf{R}^n$, $X = V$, $Y = J^1(V,\mathbf{R})$, $Q = S_1(V,\mathbf{R})$, plus j and Φ as above, we see that the set

$$\{a \in \mathbf{R}^n : j(a) \pitchfork S_1(V,\mathbf{R})\}$$

is dense in $\mathbf{R}^n$ and its complement has measure zero. In addition, since $j(a) = j^1 h^a$, Lemma 7.45(1) (page 241) implies that $j(a) \pitchfork S_1(V,\mathbf{R})$ if and only if h^a is a nondegenerate function. □ **(Lemma 12.5)**

Theorem 12.4 is now a consequence of the following lemma (Morse and Cairns [Mor-C, p. 47]):

Lemma 12.6 *Let $f : U \subset \mathbf{R}^n \to \mathbf{R}$, where U is open, be a twice continuously differentiable function and for each $a \in \mathbf{R}^n$, let*

$$f^a(x) = f(x) - a \cdot x, \quad x \in U.$$

Then:

(1) *The linear perturbation f^a of f is nondegenerate for all $a \in \mathbf{R}^n$, except on a set of measure zero.*

(2) *Suppose that f is admissible relative to a regular domain C in $\mathbf{R}^n$. For all points $a \in \mathbf{R}^n$, except on a set of measure zero, the critical points of f^a in C tend uniformly and bijectively with preservation of critical point type to those of f in C as $|a| \to 0$. Moreover, for $|a|$ sufficiently small, the restricted map $f^a|_{\partial C}$ is nondegenerate.*

Proof of Lemma 12.6: (1) Let $F = \operatorname{grad} f$. Then F is a smooth map from U into $\mathbf{R}^n$. By Sard's theorem, $F[\operatorname{Crit}(F)]$ has measure zero. Now, since $\operatorname{grad} f^a = F - a$, we have $\operatorname{Crit}(f^a) = F^{-1}(a)$. Moreover, because $\operatorname{Hess} f^a = \operatorname{Hess} f = \operatorname{Jac} F$, all points in $\operatorname{Crit}(f^a)$ are nondegenerate if and only if $a \notin F[\operatorname{Crit}(F)]$.

(2) We first show that *for $|a|$ sufficiently small, all critical points of f^a are nondegenerate and depend smoothly on a.* Let x_a be a critical point of f^a. Then $F(x_a) = a$. If a is the origin, then $F(x_0) = \mathbf{0}$ and $\operatorname{Jac} F(x_0)$ is invertible (since f is nondegenerate). By the Inverse Function Theorem, F^{-1} restricts to a smooth diffeomorphism $F_{V_0}^{-1}$ from an open set V_0 about $\mathbf{0} \in \mathbf{R}^n$ onto an open set U_0 about x_0. Hence, for $|a|$ a sufficiently small positive real number, we may assume that $a \in V_0$ and, hence, $F_{V_0}^{-1}(a) = x_a \in U_0$. In addition, x_a depends smoothly on a for $a \in V_0$, with $x_a = x_0$ for a the origin. Since $\operatorname{Hess} f^a(x_a) = \operatorname{Hess} f(x_a) \neq 0$ for a the origin, it follows that $\det[\operatorname{Hess} f^a](x_a) \neq 0$ for $|a|$ sufficiently small and positive. Thus, x_a is a nondegenerate critical point of f^a.

The above also shows that the critical points of f^a tend uniformly and bijectively to those of f as $|a| \to 0$. Furthermore, the entries of $\operatorname{Hess} f^a(x_a)$ vary continuously as $|a| \to 0$ (since x_a depends smoothly on a). But the eigenvalues of $\operatorname{Hess}(f^a)(x_a)$ also change continuously with continuous variations of the entries in $\operatorname{Hess} f^a(x_a)$ (since the eigenvalues are continuous functions of the coefficients of the characteristic polynomial).[5] Thus, the index of x_a remains constant when $|a| \to 0$.

Next, we show that $f^a|_{\partial C}$ is nondegenerate for almost all $a \in \mathbf{R}^n$ (except in a measure-zero set). To show the nondegeneracy of $f^a|_{\partial C}$, choose local coordinates $\xi : \mathcal{V} \subset \partial C \to \mathbf{R}^{n-1} : \xi(p) = r$, where $r = (r_1, \ldots, r_{n-1})$. Write $\xi^{-1}(r) = (q_1(r), \ldots, q_{n-1}(r))$. Define a map $h : \xi(\mathcal{V}) \times \mathbf{R}^n \to \mathbf{R}$ by

$$h(r, a) = (f \circ \xi^{-1})(r) - q_1(r)a_1 - \cdots - q_{n-1}(r)a_{n-1},$$

where $a = (a_1, \ldots, a_n)$. For $1 \leq i, j \leq n-1$, we have $[h_{r_i a_j}]^T = [(q_j)_{r_i}]^T = \operatorname{Jac} \xi^{-1}$. Consequently,

$$\operatorname{rank}\begin{bmatrix} h_{r_i r_j} & h_{r_i a_1} & \cdots & h_{r_i a_n} \end{bmatrix} = n - 1,$$

where $1 \leq i, j \leq n-1$. Let $h^a : \xi(\mathcal{V}) \to \mathbf{R}$ be given by $h^a(r) = h(r, a)$. Lemma 12.5 yields that for all a in $\mathbf{R}^n$, except in a measure-zero set, the function h^a is nondegenerate. Since $h^a \circ \xi = f^a|_{\mathcal{V}}$, the map $f^a|_W$ is also nondegenerate for all $a \in \mathbf{R}^n$ except in a set of measure zero. Covering the compact space ∂C by finitely many coordinate domains

[5] The zeros of a single variable polynomial depend continuously on the coefficients of the polynomial (e.g., [Cool]).

$\mathcal{V}$ (and noting that a countable union of measure-zero sets has zero measure) yields the nondegeneracy of $f^a|_{\partial C}$ for almost all a in $\mathbf{R}^n$. $\square$ **(Lemma 12.6)**

This completes the proof of Theorem 12.4. $\square$

12.2 Nonsingular Isolated Multiplane Lenses

The results here apply to gravitational lensing by any nonsingular isolated smooth matter distribution (e.g., isothermal ellipsoids with softened core and elliptical potentials) on several lens planes. The detailed nature of the gravitational-lens models are irrelevant to the counting results.

Let $T : L_{(k)} \times S \to \mathbf{R}$ be a k-plane time delay family induced by gravitational lens potentials $\psi_1, \ldots, \psi_k$. For each $\mathbf{y} \in S$, denote the eigenvalues of [Hess $T_{\mathbf{y}}$]($\mathbf{X}$) by $\lambda_i(\mathbf{X}; \mathbf{y})$, $i = 1, \ldots, 2k$. Define T to be *subcritical at infinity* if $\lambda_i(\mathbf{X}; \mathbf{y}) > 0$ for $i = 1, \ldots, 2k$ and all $\mathbf{y} \in S - \text{Caustic}(T)$. In other words, the Gauss–Kronecker curvature at each point of the time delay surface of $T_{\mathbf{y}}$ is positive for $|\mathbf{X}|$ sufficiently large (see (6.25), page 198). If T is subcritical at infinity and $T_{\mathbf{y}}(\mathbf{X}) \to \infty$ as $|\mathbf{X}| \to \infty$ for each $\mathbf{y} \in S - \text{Caustic}(T)$, then T is called *isolated.*

Our first application of Theorem 12.1 is to the case of nonsingular isolated lenses (A.P. [Pet95a, p. 4270]):

Theorem 12.7 (Nonsingular Isolated Multiplane Lenses) *Let $T_{\mathbf{y}}$ be an isolated k-plane time delay function induced by nonsingular gravitational lens potentials. Then the total number of lensed images of a light source at a noncaustic point* $\mathbf{y}$ *is finite and satisfies:*

(1) **Odd Number Image Theorem:** $N = 2M_+ - 1 = 2M_- + 1$.

(2) *Lensed images of given type:*

(i) $M_0 \geq 1$.

(ii) $M_0 + M_2 + \cdots + M_{2j} \geq M_1 + M_3 + \cdots + M_{2j-1} + 1$, *where* $j = 1, \ldots, k-1$.

(iii) $M_1 + M_3 + \cdots + M_{2k} \geq M_2 + M_4 + \cdots + M_{2k-1}$.

(iv) $M_1 + M_3 + \cdots + M_{2j-1} + 1 \geq M_0 + M_2 + \cdots + M_{2(j-1)}$, *where* $j = 1, \ldots, k$.

(v) $M_+ = M_- + 1$.

Proof: Since T is isolated, Proposition 12.2 (page 450) implies that $T_{\mathbf{y}}$ has finitely many critical points for $\mathbf{y} \in S - \text{Caustic}(T)$. Moreover, since T is nonsingular and isolated, and $T_{\mathbf{y}}$ obeys Morse boundary conditions A, Theorem 12.1 (page 448) applies with all primed quantities zero and $g_i = 0$ for $i = 1, \ldots, k$. $\square$

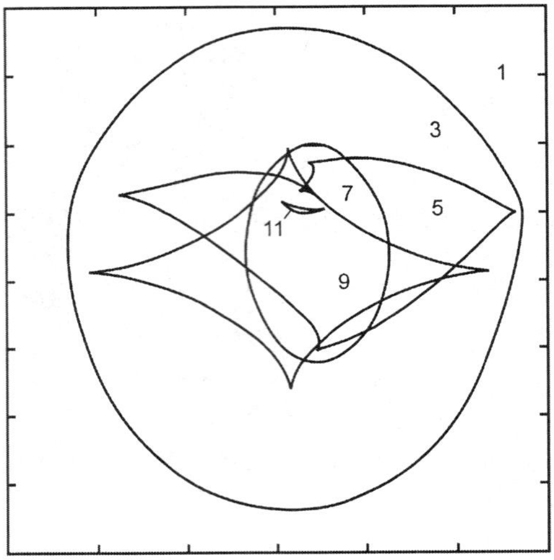

Figure 12.1. Caustics due to two elliptical potentials on distinct lens planes. The numbers indicate how many lensed images are seen of a point light source in the given region. (Courtesy of Wicklin.)

The Multiplane Odd Number Lensed-Image Theorem is guaranteed by Theorem 12.7(1). This is illustrated in Figure 12.1 for two elliptical potential on distinct lens planes. Also, notice that nonsingular isolated lenses cannot produce a five lensed-image configuration whose total number of minimum and maximum lensed images exceed three. In addition, if $M_+ = 1$ or $M_- = 0$, then no multiple lensed imaging occurs. Consequently, if there is multiple lensed imaging, then $M_- \geq 1$. In other words, it is impossible to have a multiple lensed-image configuration consisting only of minimum, maximum, and even indexed saddles, or, simply minima and maxima. There has to be at least one odd index saddle lensed image.

By Theorem 12.7(2), the sum of the first j even indexed saddles is at least one greater than the sum of the first $j-1$ odd indexed saddles, while the sum of all the even indexed saddles is exactly one greater than the sum of all the odd indexed ones ($j = 1, \dots, k$). The sum of the first j even indexed saddles never exceeds one plus the sum of the first j odd indexed saddles ($j = 1, \dots, k$). Finally, notice that part (iii) of the corollary is a consequence of parts (i) and (v). It is a generalization of the result, $M_{sad} \geq M_{max}$, in Theorem 11.5(2) (page 426).

12.3 Point Masses in Three-Dimensional Arrays

We shall first present counting relations and lower bounds on the number of lensed images due to k-plane gravitational lensing by point masses on each lens plane. The section concludes with an upper bound on the number of lensed images when there is one point mass on each lens plane.

12.3.1 Lensed-Image Counting Formulas and Lower Bounds

Let $T_{\mathbf{y}}^{pt} : L_{(k)} \to \mathbf{R}$ be the k-plane time delay function due to point masses:

$$T_{\mathbf{y}}^{pt}(\mathbf{x}_1, \ldots, \mathbf{x}_k) = \sum_{i=1}^{k} \vartheta_i \left[\frac{|\mathbf{x}_{i+1} - \mathbf{x}_i|^2}{2} - \beta_i \sum_{\ell=1}^{g_i} m_\ell^i \ln |\mathbf{x}_i - \boldsymbol{\xi}_\ell^i|, \right], \tag{12.5}$$

where $\mathbf{x}_{k+1} = \mathbf{y}$. Application of Theorem 12.1 yields (A.P. [Pet95a, p. 4272]):

Theorem 12.8 *Consider the time delay function $T_{\mathbf{y}}^{pt}$ with $\mathbf{y}$ not a caustic point. Then the total number of lensed images of a point light source at $\mathbf{y}$ is finite and satisfies:*

(1) *Weak inequalities:*

(i) $M_0 \geq 1,\ M_j \geq \sum_{1 \leq \ell_1 < \cdots < \ell_j \leq k} g_{\ell_1} \cdots g_{\ell_j},$ *where* $1 \leq j \leq k$.

(ii) $M_j \geq 0$, *where* $k + 1 \leq j \leq 2k - 1$.

(2) *Strong inequalities:*[6]

(i) $M_j - M_{j-1} + \ldots (-1)^j M_0 \geq G_j,$ *where* $0 \leq j \leq k$.

(ii) $M_j - M_{j-1} + \ldots (-1)^j M_0 \geq (-1)^{j-k} G_k,$ *where* $k \geq 2$ *and* $k + 1 \leq j \leq 2k - 1$.

(iii) $M_+ = M_- + \prod_{i=1}^{k} (1 - g_i)$.

(3) *Total number of lensed images:*

(i) $N = 2M_+ - \prod_{i=1}^{k} (1 - g_i) = 2M_- + \prod_{i=1}^{k} (1 - g_i)$.

(ii) $N \geq \prod_{i=1}^{k} (1 + g_i)$.

Proof: Because all the singularities of the potentials are infinite singularities and we have $|\operatorname{grad} T_{\mathbf{y}}^{pt}(\mathbf{X})| \to \infty$ as $|\mathbf{X}| \to \infty$, Proposition 12.2 (page 450) implies that $T_{\mathbf{y}}^{pt}$ has finitely many critical points. Also, since $|\operatorname{grad} T_{\mathbf{y}}^{pt}(\mathbf{X})| \to \infty$ as $|\mathbf{X}| \to \mathbf{p}$ for all $\mathbf{p} \in \mathbf{A}(k)$, there is a regular domain D in $L_{(k)}$ that is punctured by $\mathbf{A}(k)$ such that $\operatorname{Crit}(T_{\mathbf{y}}^{pt}) \subset D$ and $\operatorname{grad} T_{\mathbf{y}}^{pt}$ is outward pointing at all points of ∂D. In other words, $T_{\mathbf{y}}$ satisfies Morse

[6] Recall that $G_0 = 1$ and, for $j = 1, \ldots, k$, we have

$$G_j = \sum_{1 \leq \ell_1 < \cdots < \ell_j \leq k} g_{\ell_1} \cdots g_{\ell_j} - \sum_{1 \leq \ell_1 < \cdots < \ell_{j-1} \leq k} g_{\ell_1} \cdots g_{\ell_{j-1}} + \cdots (-1)^{j-1} \sum_{\ell_1 = 1}^{k} g_{\ell_1} + (-1)^j.$$

boundary conditions A relative to D. Theorem 12.1 (page 448) now holds with all primed quantities zero. To see why $M_{2k} = 0$, note that if there is a maximum critical point $\mathbf{X}_0$, then $\Delta T_{\mathbf{y}}^{pt}(\mathbf{X}_0) < 0$. However, by equations (12.2)–(12.4), page 451, we have: $\Delta T_{\mathbf{y}}^{pt}(\mathbf{X}_0) = 2\left[\vartheta_1 + \vartheta_k + \sum_{i=2}^{k-1}(\vartheta_{i-1} + \vartheta_i)\right] > 0$. This is a contradiction; no maximum exists. □

Theorem 12.8(3,i) shows that *if any lens plane has only one point mass (or an odd number), then the total number of images is* **even**. In addition, for most astrophysical situations, three-dimensional point mass distributions can be treated as having a single mass on each lens plane. Setting $g_i = 1$ $(i = 1, \dots, k)$ and $k = g$ in Theorem 12.8 yields (A.P. [Pet95a, pp. 4273, 4274]:

Corollary 12.9 *Consider the time delay function $T_{\mathbf{y}}^{pt}$ with $\mathbf{y}$ not a caustic point. Suppose that there are g lens planes with a single point mass on each plane. Then the total number of lensed images of a point light source at $\mathbf{y}$ is finite and satisfies:*

(1) *Weak inequalities:*

(i) $$M_0 \geq 1; \quad M_j \geq \binom{g}{j} \quad (1 \leq j \leq g);$$

$$M_j \geq 0 \;\; (g+1 \leq j \leq 2g-1); \qquad M_{2g} = 0;$$

(ii) $$M_+ \geq 2^{g-1}; \quad M_- \geq 2^{g-1}.$$

(2) *Strong inequalities:*

(i) $$M_j - M_{j-1} + \cdots (-1)^j M_0 \geq \binom{g-1}{j} \qquad (0 \leq j \leq g-1)$$

(ii) $$M_j - M_{j-1} + \cdots (-1)^j M_0 \geq 0, \qquad (g \leq j \leq 2g-1, \;\; g \geq 2)$$

(iii) $$M_+ = M_-.$$

(3) *Total number of lensed images:*

(i) $$N = 2M_+ = 2M_-$$

(ii) $$N \geq 2^g.$$

Let us compare multiple lensed-images by point masses on one lens plane with the case where the masses are spread over several lens planes, one on each plane. For g masses on the same lens plane (see Theorem 11.7, page 431): N is odd if and only if g is even. Moreover, if an extra point mass is added, then the total number of images increases at least by one (since $N \geq g+1$). If the g point masses are distributed in a multiplane configuration with one point mass on each lens plane, then N is always even and $N \geq 2^g$. Adding an extra point mass on a new lens plane will at least double the lower bound on N. Furthermore, if

$g \geq 5$, then N exceeds $g^2 + 1$, which is the maximum number of lensed images possible for g masses on a single lens plane (Theorem 11.11, page 439). The multiplane case has far more lensed images for g sufficiently large.

We shall show in the next section that two point masses on separate lens planes produce either *four* or *six* lensed images — see Figure 12.2. (Note that two masses on the same lens plane cannot produce four lensed images; they generate either three or five lensed images.) Let us bound the number of lensed images of a given type. If $N = 4$, then

$$M_0 = 1, \ M_1 = 2, \ M_2 = 1, \ M_3 = 0, \ M_4 = 0.$$

For $N = 6$, we have

$$1 \leq M_0 \leq 2, \ 2 \leq M_1 \leq 3, \ 1 \leq M_2 \leq 2, \ 0 \leq M_3 \leq 1, \ M_4 = 0,$$

where

$$M_0 + M_2 = M_1 + M_3 = 3.$$

12.3.2 Upper Bound on Number of Lensed Images

The total number $N^{(2D)}$ of lensed images of a light source, not on a caustic, that is lensed by g point masses on the same lens plane is bounded as follows (Theorem 11.11, page 439):

$$g + 1 \leq N^{(2D)} \leq g^2 + 1.$$

If the masses are distributed in a three-dimensional array with one on each lens plane, then the number $N^{(3D)}$ of lensed images obeys: $N^{(3D)} \geq 2^g$ (see Corollary 12.9(2,ii)). We shall present an upper bound on $N^{(3D)}$.

Requisite Mathematics: Section 11.5.1.

Consider g point masses $m_1, \ldots, m_g$ at respective positions $\boldsymbol{\xi}_1, \ldots, \boldsymbol{\xi}_g$ on distinct lens planes. Associated with this lens system is a *g-plane lensing map* $\boldsymbol{\eta} : P \to S$ given by

$$\boldsymbol{\eta}_{pt}(\mathbf{x}_1) = \mathbf{x}_1 - \sum_{\ell=1}^{g} m_\ell \frac{\boldsymbol{\eta}_\ell(\mathbf{x}_1) - \boldsymbol{\xi}_\ell}{|\boldsymbol{\eta}_\ell(\mathbf{x}_1) - \boldsymbol{\xi}_\ell|^2}, \tag{12.6}$$

with partial lensing maps (page 200)

$$\boldsymbol{\eta}_\ell(\mathbf{x}_1) = \mathbf{x}_1 - \sum_{r=1}^{\ell-1} \beta_{r,\ell}\, m_r \frac{\boldsymbol{\eta}_r(\mathbf{x}_1) - \boldsymbol{\xi}_r}{|\boldsymbol{\eta}_r(\mathbf{x}_1) - \boldsymbol{\xi}_r|^2}, \qquad (2 \leq \ell \leq g).$$

Here, $P = \mathbf{R}^2 - \mathsf{B}$, where B is the set of obstruction points of $\boldsymbol{\eta}_{pt}$ (cf. page 201). These points correspond to all $\mathbf{b} \in \mathbf{R}^2$ such that $|\boldsymbol{\eta}_\ell(\mathbf{x}_1) - \boldsymbol{\xi}_\ell|^{-1} \to \infty$ as $\mathbf{x}_1 \to \mathbf{b}$ for some $1 \leq \ell \leq g$. They are $\mathbf{x}_1 = \boldsymbol{\xi}_1$ and all solutions $\mathbf{x}_1$ of the equations

$$\boldsymbol{\eta}_\ell(\mathbf{x}_1) \ = \ \boldsymbol{\xi}_\ell \qquad (2 \leq \ell \leq g) \tag{12.7}$$

Note that no two of the above equations have a common solution (by construction of $\boldsymbol{\eta}_\ell$). *Henceforth, we shall assume (without loss of genericity) that all obstruction points are regular*, that is, $\boldsymbol{\xi}_i$ is not a critical value of the map $\boldsymbol{\eta}_i(\mathbf{x}_1)$ for $i = 2, \ldots, g$.

An upper bound on the total number of lensed images due to $\boldsymbol{\eta}_{pt}$ for a light source not on a caustic is given as follows (A.P. [Pet97, p. 1604]):

Theorem 12.10 *Suppose that* $\mathbf{y}$ *is not a caustic point of the g-plane point mass lensing map* $\boldsymbol{\eta}_{pt}$. *Then the total number N of lensed images, relative to* $\boldsymbol{\eta}_{pt}$, *of a light source at* $\mathbf{y}$ *is an even number that is bounded as follows:*

$$2^g \leq N \leq 2[2^{2(g-1)} - 1], \qquad \text{where } g \geq 2.$$

If $g = 1$, *then* $N = 2$; *we have* $N = 4$ *or* 6 *for* $g = 2$.

A double-plane binary lens ($g = 2$) achieves the upper bound of *six* images for light sources located inside caustics — see Figure 12.2.

Proof of Upper Bound in Theorem 12.10: View $\mathbf{y}$, $\mathbf{x}_\ell$, and $\boldsymbol{\xi}_\ell$ as complex variables. Then the lens equation $\mathbf{y} = \boldsymbol{\eta}_{pt}(\mathbf{x}_1)$ becomes

$$w = z_1 + \sum_{\ell=1}^{g} \frac{m_\ell}{\bar{\xi}_\ell - \bar{z}_\ell}, \tag{12.8}$$

where $\bar{z}_\ell$ is given recursively as a function of z_1 by

$$\bar{z}_\ell = \bar{z}_1 + \sum_{r=1}^{\ell-1} \frac{m_{r\ell}}{\xi_r - z_r}, \qquad (2 \leq \ell \leq g), \tag{12.9}$$

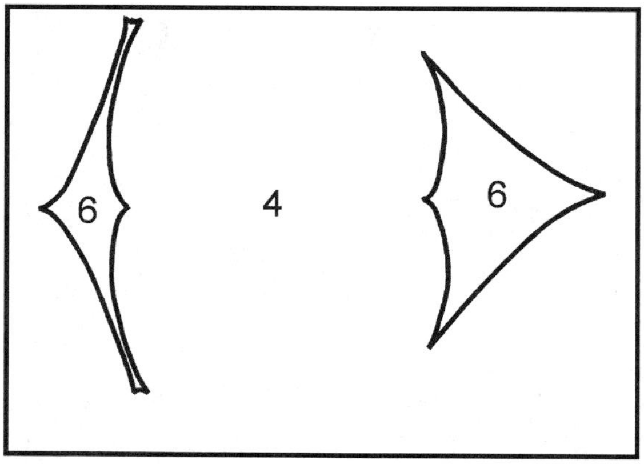

Figure 12.2. Caustics due to two point masses on distinct lens planes. The numbers indicate how many lensed images are seen of a point light source in the given region. (Courtesy of Wicklin.)

where, $m_{r\ell} = \beta_{r,\ell} m_r$. Clearing fractions in (12.8) yields

$$0 = (z_1 - w)\prod_{\ell=1}^{g}(\bar{\xi}_\ell - \bar{z}_\ell) + \sum_{\ell=1}^{g} m_\ell \prod_{\substack{r=1\\ r\neq\ell}}^{g}(\bar{\xi}_r - \bar{z}_r). \tag{12.10}$$

Notation:

(1) Let $q_\ell = \xi_\ell - z_1$ and set $Q_1 = q_1$.

(2) If $\ell = 1, \ldots, g$ and $P_\ell(z_1, \bar{z}_1)$ is a polynomial of degree s_ℓ and t_ℓ in z_1 and $\bar{z}_1$, respectively, then set $\mathrm{Deg}[P_\ell(z_1, \bar{z}_1)] = (s_\ell; t_\ell)$ and write

$$\sum_{\ell=1}^{g} \mathrm{Deg}[P_\ell(z_1, \bar{z}_1)] = \Big(\sum_{\ell=1}^{g} s_\ell; \sum_{\ell=1}^{g} t_\ell\Big).$$

Let the inequality $\mathrm{Deg}[P_\ell(z_1, \bar{z}_1)] \leq (\hat{s}_\ell; \hat{t}_\ell)$ mean that $s_\ell \leq \hat{s}_\ell$ and $t_\ell \leq \hat{t}_\ell$. Set

$$\max\{(s_\ell; t_\ell)\}_{\ell=1}^{g} = (\max\{s_\ell\}_{\ell=1}^{g}; \max\{t_\ell\}_{\ell=1}^{g}).$$

Next, we shall express the differences $\bar{\xi}_\ell - \bar{z}_\ell$ as rational functions in z_1 and $\bar{z}_1$, and determine upper bounds for the z_1 and $\bar{z}_1$ degrees of their numerators. By (12.9), we have $\bar{\xi}_2 - \bar{z}_2 = \bar{Q}_2 / Q_1$, where $\bar{Q}_2 = \bar{q}_2 Q_1 - m_{12}$ and $\mathrm{Deg}[\bar{Q}_2] \leq (1; 1)$.

Lemma 12.11 *(1) If $2 \leq \ell \leq g$, then*

$$\bar{\xi}_\ell - \bar{z}_\ell = \frac{\bar{Q}_\ell}{\prod_{r=1}^{\ell-1} Q_r},$$

where Q_ℓ is a polynomial in z_1 and $\bar{z}_1$ defined recursively by

$$\bar{Q}_\ell = (\bar{q}_\ell Q_1 - m_{1\ell})\prod_{r=2}^{\ell-1} Q_r - \sum_{r=2}^{\ell-2} m_{r\ell}\left[\prod_{s=1}^{r-1}|Q_s|^2\right]\left[\prod_{s=r+1}^{\ell-1} Q_s\right] - m_{\ell-1,\ell}\prod_{r=1}^{\ell-2}|Q_r|^2. \tag{12.11}$$

Note that $\bar{Q}_2 = \bar{q}_2 Q_1 - m_{12}$ and the summation term appears only for $4 \leq \ell \leq g$.
(2) If $2 \leq \ell \leq g$, then $\mathrm{Deg}[\bar{Q}_\ell] \leq \left(2^{\ell-2}; 2^{\ell-2}\right)$.

Proof of Lemma 12.11: By (12.9), we obtain

$$\bar{\xi}_3 - \bar{z}_3 = \bar{q}_3 - \frac{m_{13}}{Q_1} - \frac{m_{23}\bar{Q}_1}{Q_2} = \frac{\bar{q}_3 Q_1 Q_2 - m_{13} Q_2 - m_{23}|Q_1|^2}{Q_1 Q_2} = \frac{\bar{Q}_3}{Q_1 Q_2}.$$

Assume that for $1 \leq k \leq \ell - 1$ we have

$$\bar{\xi}_{k+2} - \bar{z}_{k+2} = \frac{\bar{Q}_{k+2}}{\prod_{r=1}^{k+1} Q_r}.$$

For $k = \ell$, equation (12.9) implies

$$\begin{aligned}\bar{\xi}_{\ell+2} - \bar{z}_{\ell+2} &= \bar{q}_{\ell+2} - \frac{m_{1,\ell+2}}{Q_1} - \frac{m_{2,\ell+2}\bar{Q}_1}{Q_2} - \frac{m_{3,\ell+2}\bar{Q}_1\bar{Q}_2}{Q_3} - \cdots \\ &\qquad - \frac{m_{\ell+1,\ell+2}\bar{Q}_1\bar{Q}_2\cdots\bar{Q}_\ell}{Q_{\ell+1}} \\ &= \frac{1}{\prod_{r=1}^{\ell+1} Q_r}\{\bar{q}_{\ell+2}Q_1Q_2\cdots Q_{\ell+1} - m_{1,\ell+2}Q_2\cdots Q_{\ell+1} \\ &\qquad - \sum_{r=2}^{\ell} m_{r,\ell+2}|Q_1|^2\cdots|Q_{r-1}|^2 Q_{r+1}\cdots Q_{\ell+1} \\ &\qquad - m_{\ell+1,\ell+2}|Q_1|^2\cdots|Q_\ell|^2\} \\ &= \frac{\bar{Q}_{\ell+2}}{\prod_{r=1}^{\ell+1} Q_r}.\end{aligned}$$

By complete induction on ℓ, we obtain the first part of the lemma.

Now, $\mathrm{Deg}[\bar{Q}_2] \leq (1; 1)$. Assume that $\mathrm{Deg}[\bar{Q}_k] \leq (2^{k-2}; 2^{k-2})$ for $k = 2, \ldots, \ell - 1$. To prove the case $k = \ell$, consider equation (12.11) and observe that

$$\mathrm{Deg}\left[\bar{q}_\ell Q_1 \prod_{r=2}^{\ell-1} Q_r\right] \leq \left(1 + \sum_{r=2}^{\ell-1} 2^{r-2}; 1 + \sum_{r=2}^{\ell-1} 2^{r-2}\right) = (2^{\ell-2}; 2^{\ell-2}),$$

$$\mathrm{Deg}\left[m_{1\ell} \prod_{r=2}^{\ell-1} Q_r\right] \leq \left(\sum_{r=2}^{\ell-1} 2^{r-2}; \sum_{r=2}^{\ell-1} 2^{r-2}\right) = (2^{\ell-2} - 1; 2^{\ell-2} - 1),$$

and

$$\begin{aligned}&\mathrm{Deg}\left[\sum_{r=2}^{\ell-2} m_{r\ell}\left(\prod_{s=1}^{r-1}|Q_s|^2\right)\left(\prod_{s=r+1}^{\ell-1} Q_s\right)\right] \\ &\leq \max\left\{\mathrm{Deg}\left[m_{r\ell}\left(\prod_{s=1}^{r-1}|Q_s|^2\right)\left(\prod_{s=r+1}^{\ell-1} Q_s\right)\right]\right\}_{r=2}^{\ell-2} \\ &\leq \max\left\{\mathrm{Deg}\left[\prod_{s=1}^{r-1}|Q_s|^2\right] + \mathrm{Deg}\left[\prod_{s=r+1}^{\ell-1} Q_s\right]\right\}_{r=2}^{\ell-2} \\ &\leq \max\left\{(2^{r-1} + (2^{\ell-2} - 2^{r-1}); 2^{r-1} + (2^{\ell-2} - 2^{r-1}))\right\}_{r=2}^{\ell-2}\end{aligned}$$

$= (2^{\ell-2}; 2^{\ell-2})$. In addition,

$$\mathrm{Deg}\left[m_{\ell-1,\ell} \prod_{r=1}^{\ell-2} |Q_r|^2\right] \leq (2^{\ell-2}; 2^{\ell-2}).$$

Hence, Lemma 12.11(1) implies: $\mathrm{Deg}[\bar{Q}_\ell] \leq (2^{\ell-2}; 2^{\ell-2})$. □ **(Lemma 12.11)**

We now convert equation (12.10) into polynomial form. By Lemma 12.11(1), equation (12.10) becomes

$$\begin{aligned} 0 &= (z_1 - w)\bar{Q}_1 \prod_{\ell=2}^{g} \left[\frac{\bar{Q}_\ell}{\prod_{r=1}^{\ell-1} Q_r} \right] + m_1 \prod_{\ell=2}^{g} \left[\frac{\bar{Q}_\ell}{\prod_{r=1}^{\ell-1} Q_r} \right] + \\ & \sum_{\ell=2}^{g} m_\ell \bar{Q}_1 \prod_{\substack{r=2 \\ r\neq\ell}}^{g} \left[\frac{\bar{Q}_r}{\prod_{s=1}^{r-1} Q_s} \right] \\ &\equiv p(z_1, \bar{z}_1), \end{aligned}$$

which is equivalent to the polynomial equation:

$$\begin{aligned} P(z_1, \bar{z_1}) &\equiv p(z_1, \bar{z}_1) \prod_{\ell=2}^{g} \prod_{r=1}^{\ell-1} Q_r \\ &= \left[(z_1 - w)\bar{Q}_1 + m_1\right] \prod_{\ell=2}^{g} \bar{Q}_\ell + \sum_{\ell=2}^{g} m_\ell \left[\prod_{r=1}^{\ell-1} Q_r \right] \left[\prod_{\substack{r=1 \\ r\neq\ell}}^{g} \bar{Q}_r \right] \\ &= 0. \end{aligned} \tag{12.12}$$

The corresponding conjugate equation is

$$\begin{aligned} \bar{P}(z_1, \bar{z_1}) &= \left[(\bar{z}_1 - \bar{w})Q_1 + m_1\right] \prod_{\ell=2}^{g} Q_\ell + \sum_{\ell=2}^{g} m_\ell \left[\prod_{r=1}^{\ell-1} \bar{Q}_r \right] \left[\prod_{\substack{r=1 \\ r\neq\ell}}^{g} Q_r \right] \\ &= 0. \end{aligned} \tag{12.13}$$

If z_1 is a lensed image of w, then $(z_1, \bar{z_1})$ simultaneously satisfies the polynomial equations $P(z_1, \bar{z_1}) = 0$ and $\bar{P}(z_1, \bar{z_1}) = 0$. Consequently, an upper bound on the number of common zeros of $P(z_1, \bar{z_1})$ and $\bar{P}(z_1, \bar{z_1})$ yields an upper bound on N. First, we determine the z_1- and $\bar{z}_1$-degrees of P and $\bar{P}$. Note that if $\text{Deg}[P(z_1, \bar{z}_1)] = (s; t)$, then $\text{Deg}[\bar{P}(z_1, \bar{z}_1)] = (t; s)$. Applying Lemma 12.11(2) to equation (12.12) yields

$$\begin{aligned} &\text{Deg}\left[\left((z_1 - w)\bar{Q}_1 + m_1\right) \prod_{\ell=2}^{g} \bar{Q}_\ell \right] \\ &\qquad \leq \left(1 + [2^{g-1} - 1]; 1 + [2^{g-1} - 1]\right) = \left(2^{g-1}; 2^{g-1}\right) \end{aligned} \tag{12.14}$$

and

$$\begin{aligned} &\text{Deg}\left[m_\ell \bar{Q}_1 Q_1 \left(\prod_{r=2}^{\ell-1} Q_r \right) \left(\prod_{\substack{r=2 \\ r\neq\ell}}^{g} \bar{Q}_r \right) \right] \\ &\leq \left(1 + (2^{\ell-2} - 1) + (2^{g-1} - 1 - 2^{\ell-2}); 1 + (2^{\ell-2} - 1) + (2^{g-1} - 1 - 2^{\ell-2})\right) \\ &= \left(2^{g-1} - 1; 2^{g-1} - 1\right). \end{aligned} \tag{12.15}$$

Hence,

$$\mathrm{Deg}[P(z_1,\bar{z}_1)] \le \left(2^{g-1};2^{g-1}\right) \qquad \text{and} \qquad \mathrm{Deg}[\bar{P}(z_1,\bar{z}_1)] \le \left(2^{g-1};2^{g-1}\right). \tag{12.16}$$

Next, write

$$P(z_1,\bar{z}_1) = a_0(\bar{z}_1) + a_1(\bar{z}_1)z_1 + \cdots + a_s(\bar{z}_1)z_1^s$$

and

$$\bar{P}(z_1,\bar{z}_1) = b_0(\bar{z}_1) + b_1(\bar{z}_1)z_1 + \cdots + b_t(\bar{z}_1)z_1^t.$$

Let $\mathrm{Res}(\bar{z}_1)$ be the z_1-Sylvester resultant of P and $\bar{P}$. Since we are considering conjugate pairs $(z_1,\bar{z}_1)$, the Resultant Theorem (Theorem 11.10, page 438) yields that the common zeros of P and $\bar{P}$ are in 1-1 correspondence with a subset of the zeros of $\mathrm{Res}(\bar{z}_1)$. This result and equation (11.5) (page 437) yield that

$$N \le \deg\left[\mathrm{Res}(\bar{z}_1)\right] \le t\left(\max\left\{\deg[a_\ell]_{\ell=0}^s\right\}\right) + s\left(\max\left\{\deg[b_\ell]_{\ell=0}^t\right\}\right). \tag{12.17}$$

Equation (12.16) implies that $t \le 2^{g-1}$, $\max\left\{\deg[a_\ell]_{\ell=0}^s\right\} \le 2^{g-1}$, $s \le 2^{g-1}$, and $\max\{\deg[b_\ell]_{\ell=0}^t\} \le 2^{g-1}$. Thus

$$N \le 2(2^{g-1})^2 = 2^{2g-1} \qquad (g \ge 1). \tag{12.18}$$

Note that since $N \ge 2^g$, we obtain $N = 2$ for $g = 1$.

We now show that the upper bound in equation (12.18) improves as follows:

$$N \le \deg\left[\mathrm{Res}(z_1)\right] \le 2^{2g-1} - 2, \qquad (g \ge 2).$$

By the Resultant Theorem, not all zeros of $\mathrm{Res}(\bar{z}_1)$ need to correspond to simultaneous solutions of P and $\bar{P}$, that is, the common zeros of $a_s(\bar{z}_1)$ and $b_t(\bar{z}_1)$. In fact, let $g \ge 2$ and consider

$$\begin{aligned} P(z_1,\bar{z}_1) &= (z_1 - w)\prod_{\ell=1}^{g}\bar{Q}_\ell + m_1\prod_{\ell=2}^{g}\bar{Q}_\ell + \sum_{\ell=2}^{g} m_\ell\left[\prod_{r=1}^{\ell-1}Q_r\right]\left[\prod_{\substack{r=1\\ r\neq\ell}}^{g}\bar{Q}_r\right] \\ &= 0. \end{aligned}$$

The second term on the right has z_1-degree less than the z_1-degree of the first term. Since $\bar{Q}_1$ is a factor of the first and third terms on the right, it is also a factor of the leading coefficient $a_s(\bar{z}_1)$. Similarly, the relation $Q_2 = q_2\bar{Q}_1 - m_{12}$ implies

$$
\begin{aligned}
\bar{P}(z_1, \bar{z_1}) &= (\bar{z}_1 - \bar{w})\prod_{\ell=1}^{g} Q_\ell + m_1\prod_{\ell=2}^{g} Q_\ell + \sum_{\ell=2}^{g} m_\ell\left[\prod_{r=1}^{\ell-1}\bar{Q}_r\right]\left[\prod_{\substack{r=1\\ r\neq\ell}}^{g} Q_r\right] \\
&= \bar{Q}_1\left[(\bar{z}_1 - \bar{w})Q_1 + m_1\right]q_2\prod_{\ell=3}^{g} Q_\ell \\
&\quad -m_{12}\left[(\bar{z}_1 - \bar{w})Q_1 + m_1\right]\prod_{\ell=3}^{g} Q_\ell \\
&\quad + m_2 Q_1\bar{Q}_1\prod_{\ell=3}^{g} Q_\ell + \sum_{\ell=3}^{g} m_\ell Q_1\bar{Q}_1\left[\prod_{r=2}^{\ell-1}\bar{Q}_r\right]\left[\prod_{\substack{r=2\\ r\neq\ell}}^{g} Q_r\right] \\
&= 0,
\end{aligned}
$$

The second term on the right has z_1-degree less than the z_1-degree of the first term. Consequently, $\bar{Q}_1$ is also a factor of $b_t(\bar{z}_1)$. By the Resultant Theorem, the zero $\bar{z}_1 = \boldsymbol{\xi}_1$ of $\bar{Q}_1$ is a zero of $\mathrm{Res}(\bar{z}_1)$. This zero is a point mass position and, so, is not a lensed image. Thus, $N \leq \deg[\mathrm{Res}(\bar{z}_1)] \leq 2^{2g-1} - 1$ for $g \geq 2$. However, the number N is even. Therefore, if $g \geq 2$, then: $N \leq \deg[\mathrm{Res}(\bar{z}_1)] \leq 2^{2g-1} - 2 = 2(2^{2(g-1)} - 1)$. □

Chapter 13

Total Magnification

Said the Moon to the Sun
"Is the daylight begun?"
Said the Sun to the Moon,
"Not a minute too soon."

"You're a Full Moon," said he.
She replied with a frown,
"Well! I never did see
So uncivil a clown!"

LEWIS CARROLL

This chapter investigates lower bounds on the total magnification of a lensed light source and presents a formula for magnification cross sections — that is, the areas of regions where light sources have a total magnification greater than some given amount (Problem **M2**).

The same Morse-theoretic tools employed for lensed-image counting can used to investigate lower bounds on the total magnification due to a general single-plane lens system (A.P. [Pet96a]). The lower bounds vary according to how "far inside" a caustic network the source is located, and is sensitive to the mass density of the lens in the vicinity of the images, as well as the number of singularities of the lens. These methods carry over naturally to the multiplane case. We shall also present a magnification cross section formula for gravitational lensing with any finite number of lens planes. This result will be base on the coarea formula. The asymptotic high-magnification limit of our magnification cross section formula will also be considered for the case of folds and cusps due to lensing by finitely many lens planes.

Outline: Section 13.1 presents lower bounds on the total magnification due to a general gravitational lens (Section 13.1.1) with applications to nonsingular isolated lenses (Section 13.1.2) and point mass lenses with continuous matter and shear from infinity (Section 13.1.3). The latter includes the cases of subcritical, strong shear, and supercritical lensing. Section 13.2 extends the main theorem of Section 13.1 to obtain lowers bounds on the magnification due to a general multiplane gravitational lens. Section 13.3 presents the coarea formula to obtain a general formula for magnification cross sections and the asymptotic limit of the magnification cross section near folds and cusps.

13.1 Lower Bounds on Total Magnification: Single-Plane Case

The total magnification due to a single-plane isolated gravitational lens system is bounded below by one. This is because such systems have at least one minimum lensed image (Theorem 11.4(1), page 424) and *minimum lensed images have magnification greater than or equal to one* (Schneider [Schn84, p. 123], [Schn-EF, p. 172, Thm2]); see equation (13.1) below. Now, consider a one point mass lens with continuous matter κ_c and shear γ from infinity. Set

$$\gamma_n = \frac{\gamma}{1-\kappa_c},$$

which is called *normalized shear*. If $\gamma_n < 1$, then an astroid caustic is produced about the origin, and the total magnification of a light source at the origin is given by (Mao [Mao, p. 64, Table 1]):

$$\mathrm{Mag}_{tot}(\mathbf{0}) = \frac{1}{\gamma_n(1-\gamma_n^2)}.$$

This attains a minimum total magnification at $\gamma_n = 1/\sqrt{3}$, which gives the following minimum value:

$$\mathrm{Mag}_{tot}(\mathbf{0}) = \frac{3\sqrt{3}}{2} \simeq 2.6.$$

For two point masses on a lens plane, the total magnification of a light source inside a caustic is at least three (Witt and Mao [Wit-M, p. L106, eq. (9)]). On the other hand, Rauch et al. [Rau-MWP, p. 34] found numerically that the total magnification due to many point masses on a plane with continuous matter (no shear) can be below three (in fact, they found total magnification values around 2.5).

We shall present lower bounds on the total magnification due to a *general* gravitational lens using the Morse inequalities. This is possible since the total magnification of a lensed source is affected by the number of minimum lensed images present. Morse inequalities allow us to express the number of minimum lensed images in terms of the total number of lensed images and singularities of the lens. Such a result yields a lower bound on the total magnification that can adjust according to the location of a light source inside a caustic network. Our treatment is based on [Pet96a].

13.1.1 General Single-Plane Gravitational Lens

The following result gives lower bounds on the total magnification for a general single-plane lens [Pet96a, p. 1480]:

Theorem 13.1 [1] *Let $T_{\mathbf{y}} : L \to \mathbf{R}$ be a single-plane time delay function induced at a non-caustic point* $\mathbf{y}$. *Suppose that $T_{\mathbf{y}}$ obeys Morse boundary conditions* B *relative to a regular domain $D \subset L$,*[2] *where D is punctured by g singularities of ψ and contains all light rays of $T_{\mathbf{y}}$. If the set of light rays of $T_{\mathbf{y}}$ consists of minima $\mathbf{x}^i_{min}$, where $i = 1, \ldots, M_{min}$, and at least one saddle or maximum, then the total magnification of a light source at* $\mathbf{y}$ *satisfies* **(the source is magnified)**:

$$1 \leq M_{min} \leq \sum_{i=1}^{M_{min}} \frac{1}{[1-\kappa(\mathbf{x}^i_{min})]^2} < \mathrm{Mag}_{tot}(\mathbf{y}) < \infty,$$

where

$$M_{min} = \frac{N - 2M_{max} - (M'_{min} - M'_{max}) - g + 1}{2}.$$

Proof: If $\mathbf{x}$ is a nondegenerate critical point, then $0 < \mathrm{Mag}(\mathbf{x};\mathbf{y}) < \infty$. Since there are finitely many lensed images, we obtain $\mathrm{Mag}_{tot}(\mathbf{y}) < \infty$. Now, suppose that $\mathbf{x}_{min}$ is a minimum lensed image. Since $\Delta T_{\mathbf{y}}(\mathbf{x}_{min}) = 2[1-\kappa(\mathbf{x}_{min})] > 0$, we have $0 \leq \kappa(\mathbf{x}_{min}) < 1$. Consequently,

$$0 < \det[\mathrm{Hess}\, T_{\mathbf{y}}](\mathbf{x}_{min}) = [1-\kappa(\mathbf{x}_{min})]^2 - \Gamma^2(\mathbf{x}) \leq [1-\kappa(\mathbf{x}_{min})]^2,$$

that is,

$$1 \leq \frac{1}{[1-\kappa(\mathbf{x}_{min})]^2} \leq \frac{1}{|\det[\mathrm{Hess}\, T_{\mathbf{y}}](\mathbf{x}_{min})|} = \mathrm{Mag}(\mathbf{x}_{min};\mathbf{y}). \tag{13.1}$$

Hence

$$1 \leq M_{min} \leq \sum_{i=1}^{M_{min}} \frac{1}{[1-\kappa(\mathbf{x}^i_{min})]^2} < \mathrm{Mag}_{tot}(\mathbf{y}) < \infty.$$

Note that by assumption, $M_{min} \geq 1$. Application of Theorem 11.1 on page 421 (cf. Theorem 12.4, page 452) and Theorem 11.2 on page 422 yields the desired expression for M_{min}. □

By (13.1), *minimum lensed images* $\mathbf{x}_{min}$ *are never de-magnified* [Schn84, p. 123]. In addition, we saw earlier (page 423) that minimum lensed images are located where the

[1] The proof of Theorem 13.1 uses the nonnegativity of κ. Also, recall that M'_{min} and M'_{max} are, respectively, the total number of minima and maxima of $T_{\mathbf{y}}|_{D_-}$ (see page 422).

[2] Morse boundary conditions B are readily satisfied through an arbitrarily small linear perturbations of $T_{\mathbf{y}}$ — Theorem 12.4, page 421.

surface mass density is subcritical ($\kappa(\mathbf{x}_{min}) < 1$), while maxima $\mathbf{x}_{max}$ are where the surface mass density is supercritical ($\kappa(\mathbf{x}_{max}) > 1$). Furthermore, since

$$\text{Mag}(\mathbf{x}_{max}) = \frac{1}{|[1-\kappa(\mathbf{x}_{max})]^2 - \Gamma^2(\mathbf{x}_{max})|},$$

if $\kappa(\mathbf{x}_{max})$ is sufficiently large, then $\mathbf{x}_{max}$ is de-magnified (e.g., if $\mathbf{x}_{max}$ lies in a sufficiently dense galactic nucleus).[3]

If a minimum lensed image is unmagnified, then it is at a point where there is no matter, while an unmagnified maximum lensed image is located where the density is at least two. To see this, notice that if $\mathbf{x}_0$ is either a maximum or minimum, then since the eigenvalues of $[\text{Hess}(T_\mathbf{y})](\mathbf{x}_0)$ are both positive or both negative, we get that $\det[\text{Hess}(T_\mathbf{y})](\mathbf{x}_0) = [1 - \kappa(\mathbf{x}_0)]^2 - \Gamma^2(\mathbf{x}_0) > 0$. The condition $\text{Mag}(\mathbf{x}_0) = 1$ then requires $\kappa(\mathbf{x}_0)[\kappa(\mathbf{x}_0) - 2] = \Gamma^2(\mathbf{x}_0) \geq 0$. Consequently, either $\kappa(\mathbf{x}_0) = 0$ (hence, $\Gamma(\mathbf{x}_0) = 0$) or $\kappa(\mathbf{x}_0) \geq 2$. Hence, $\kappa(\mathbf{x}_{min}) = 0$ if $\text{Mag}(\mathbf{x}_{min}) = 1$, while $2 \leq \kappa(\mathbf{x}_{max}) < \infty$ if $\text{Mag}(\mathbf{x}_{max}) = 1$.

Note that the magnitude of shear at a minimum lensed image is bounded as follows: $0 \leq \Gamma(\mathbf{x}_{min}) \leq [1 - \kappa(\mathbf{x}_{min})]^2 \leq 1$.

13.1.2 Nonsingular Isolated Lenses

The following theorem applies to any nonsingular isolated lens, for example, isothermal ellipsoids with softened core and elliptical potentials.

Theorem 13.2 *Let $T_\mathbf{y}$ be an isolated single-plane time delay function induced by a nonsingular gravitational lens potential, where $\mathbf{y}$ is not a caustic point. Then*

(1) *If at light source at $\mathbf{y}$ is multiply imaged, then*

$$1 \leq \frac{N - 2M_{max} + 1}{2} < \text{Mag}_{tot}(\mathbf{y}) < \infty$$

(2) *If a light source at $\mathbf{y}$ is multiply imaged and κ is everywhere subcritical,*[4] *then*

$$2 \leq \frac{N+1}{2} < \text{Mag}_{tot}(\mathbf{y}) < \infty.$$

Proof: (1) Since the light source is multiply imaged ($N > 1$) and $T_\mathbf{y}$ is a nonsingular isolated time delay function, we have by Theorem 11.5(1,2) (page 426) that $M_{min} \geq 1$ and $N = 2M_{sad} + 1 \geq 3$. The latter implies: $M_{sad} \geq 1$. Consequently, since $T_\mathbf{y}$ obeys Morse boundary conditions A (see parts (1-2) of proof of Theorem 11.4, page 425), Theorem 13.1

[3] This may explain why an even number of lensed images are observed in some lens systems for which modeling predicts an odd number (Theorem 11.5, page 426).

[4] That is, $\kappa(\mathbf{x}) < 1$ for all $\mathbf{x} \in \mathbf{R}^2$.

applies with all primed quantities zero. This yields $\text{Mag}_{tot}(\mathbf{y}) > M_{min} = (N - 2M_{max} + 1)/2$ (because $g = 0$).

(2) The subcritical condition on κ and the fact that $\kappa(\mathbf{x}_{max}) > 1$ (page 423) are incompatible, so no maximum lensed image can exist. Since the light source is multiply imaged and there is an odd number of lensed images (Theorem 11.5, page 426), we have $N \geq 3$. Now apply part (1). □

Let us illustrate Theorem 13.2 with an elliptical potential. This lens can produce a caustic network consisting of a single lips caustic as well as one with an astroid caustic inside a simple closed elliptical shaped caustic — see Figure 11.1 (page 426). From page 426, we know that a light source inside the lips caustic has three lensed images with $M_{max} = 0$; consequently, $\text{Mag}_{tot} > 2$. For the other caustic network, a source situated in the region between the elliptical shaped caustic and astroid has three images with $M_{max} = 1$. In this case, we have $\text{Mag}_{tot}(\mathbf{y}) > 1$, though the source is inside a caustic. Finally, a source inside the astroid has five lensed images with $M_{max} = 1$, which yields $\text{Mag}_{tot}(\mathbf{y}) > 2$ at such source positions.

13.1.3 Point Masses with Continuous Matter and Shear

For an isolated time delay function $T_{\mathbf{y}}$ induced by a gravitational lens potential with g singularities, Theorem 13.1 yields:

$$1 \leq \frac{N - 2M_{max} - g + 1}{2} \leq \text{Mag}_{tot}(\mathbf{y}) < \infty.$$

We shall obtain sharper lower bounds for the case of a time delay function $T_{\mathbf{y}}^{ptcs}$ due to g point masses with continuous matter κ_c and shear γ from infinity. This section will utilize results from Section 11.4 (page 429).

First, *if a lensed image of $T_{\mathbf{y}}^{ptcs}$ approaches a point mass, then the magnification of the lensed image tends to zero.*[5] Now, recall the notation $a = 1 - \kappa_c - \gamma$ and $b = 1 - \kappa_c + \gamma$ (page 431). We then have (see Theorem 11.7, page 431):

Theorem 13.3 (Subcritical) *Let $T_{\mathbf{y}}^{ptcs}$ be a single-plane time delay function $T_{\mathbf{y}}^{ptcs}$ that is subcritical* (i.e., $a > 0, b > 0$). *For almost all points $\mathbf{a}$ in $\mathbf{R}^2$ (except in a measure-zero set) with $|\mathbf{a}|$ sufficiently small, the linear perturbation, $[T_{\mathbf{y}}^{ptcs}]^{\mathbf{a}}(\mathbf{x}) = T_{\mathbf{y}}^{ptcs}(\mathbf{x}) - \mathbf{a} \cdot \mathbf{x}$, satisfies the following, where all quantities are defined relative to $[T_{\mathbf{y}}^{ptcs}]^{\mathbf{a}}$:*

(1) $$1 \leq M_{min} \leq \frac{N - g + 1}{2(1 - \kappa_c)^2} < \text{Mag}_{tot}(\mathbf{y}) < \infty.$$

[5] This holds since $|\det[\text{Hess}\, T_{\mathbf{y}}^{pt}](\mathbf{x})| \to \infty$ as $\mathbf{x}$ approaches a point mass.

(2) *If the associated lensing map* $\boldsymbol{\eta}_{ptcs}$ *is transverse stable,*[6] *then*

$$1 \leq M_{min} \leq \frac{1+N_p}{(1-\kappa_c)^2} < \mathrm{Mag}_{tot}(\mathbf{y}) < \infty,$$

where N_p *is the number of extra lensed image pairs — a minimum and saddle — beyond the minimum of* $g+1$ *lensed images.*

Proof of Theorem 13.3: (1) Theorem 11.1 (page 421) yields that Morse boundary conditions B hold for $T_{\mathbf{y}}^{\mathbf{a}}$, while Theorem 11.7 (page 431) gives $M_{min} \geq 1$, and $N = 2M_{min}+g-1$. Using the fact that $\kappa(\mathbf{x}) = \kappa_c$ for $\mathbf{x}$ not at a point mass position, Theorem 13.1 implies that $1 \leq M_{min} \leq M_{min}/[1-\kappa_c]^2 = (N-g+1)/(2[1-\kappa_c]^2) < \mathrm{Mag}_{tot}(\mathbf{y}) < \infty$.

(2) Since $\boldsymbol{\eta}$ is transverse stable, it is also locally stable (Theorem 7.56 (page 262). We can then apply Theorem 11.7(2) (page 431), which implies that $N = g+1$ with $M_{min} = 1$ and $M_{sad} = g$ for $|\mathbf{y}|$ sufficiently large. This is the minimum number of lensed images. Now, since $\boldsymbol{\eta}$ is transverse stable, Theorem 7.56 yields that each caustic is either a fold or cusp; all self-intersections of the fold caustic caustic curves occur at nonzero angles; no fold and cusp critical points are mapped by $\boldsymbol{\eta}$ to the same point; no two cusp critical points are mapped by $\boldsymbol{\eta}$ to the same point; and no more than two fold critical points are mapped by $\boldsymbol{\eta}$ to the same point. Consequently, if a light source with $g+1$ lensed images crosses a caustic curve in such a network, where the crossing does not occur at an intersection, then N increases by two — see Figure 9.12 (page 342). Hence, $N = g+1+2N_p$, where N_p is the total number of extra lensed image pairs beyond the minimum of $g+1$ images. This implies that either M_{min} or M_{sad} increases by at least one (since $M_{max} = 0$). Since $M_{min} = M_{sad} - g + 1$, we see that if M_{min} (resp., M_{sad}) increases by one, then M_{sad} (resp., M_{min}) must also increase by one. Thus, if N increases by exactly two lensed images, then one lensed image must be a minimum and the other a saddle. We also have $N = g+1+2N_p$, where N_p is the total number of extra lensed image pairs beyond the minimum of $g+1$ images. Inserting the previous expression for N into the former part (1) yields the desired lower bound. □

Theorem 13.4 (Strong Shear) *Let the time delay function* $T_{\mathbf{y}}^{ptcs}$ *be have strong shear at infinity* (i.e., $a > 0$, $b < 0$) *and suppose that* $0 \leq \kappa_c < 1$. *For almost all points* $\mathbf{a}$ *in* $\mathbf{R}^2$ *(except in a measure-zero set) with* $|\mathbf{a}|$ *sufficiently small, the linear perturbation,* $[T_{\mathbf{y}}^{ptcs}]^{\mathbf{a}}(\mathbf{x}) = T_{\mathbf{y}}^{ptcs}(\mathbf{x}) - \mathbf{a}\cdot\mathbf{x}$, *satisfies the following, where all quantities are defined relative to* $[T_{\mathbf{y}}^{ptcs}]^{\mathbf{a}}$*:*

(1) $$0 \leq M_{min} \leq \frac{N-(g+1)}{2(1-\kappa_c)^2} < \mathrm{Mag}_{tot}(\mathbf{y}) < \infty.$$

[6] See Theorem 8.6 (page 294) and Theorem 8.17 (page 311).

(2) *If the associated lensing map $\boldsymbol{\eta}_{ptcs}$ is transverse stable, then*

$$0 \leq M_{min} \leq \frac{1 + N_p}{(1 - \kappa_c)^2} < \mathrm{Mag}_{tot}(\mathbf{y}) < \infty.$$

Here N_p is the number of extra lensed image pairs (minimum and saddle) beyond the minimum of $g + 1$ lensed images.

Proof: We follow the proof of Theorem 13.3. For part (1), application of Theorems 11.1 (page 421) 11.8 (page 432) yields $M_{min} \geq 0$, and $N = 2(M_{min} + M_{max}) + g - 1$. Since $\kappa(\mathbf{x}) = \kappa_c$ for $\mathbf{x}$ not at a point mass position, Theorem 13.1 gives

$$0 \leq M_{min} \leq \frac{N - 2M_{max} - (g + 1)}{2[1 - \kappa_c]^2} < \mathrm{Mag}_{tot}(\mathbf{y}) < \infty$$

Since $0 \leq \kappa_c < 1$, Theorem 11.8 yields that $M_{max} = 0$. For part (2) of the theorem, note that for $|\mathbf{y}|$ sufficiently large, the minimum number of lensed images is achieved: $N = g + 1$ with $M_{min} = 0$ and $M_{sad} = g + 1$ (Theorem 11.8(3)). Arguing as in the proof of Theorem 13.3(2), we see that $N = g + 1 + 2N_p$, where each extra lensed image pair consists of a minimum and saddle. □

Supercritical Case: A single point mass supercritical lens with continuous matter κ_c and shear γ from infinity will produce a region with no lensed images ($N = 0$) if $\kappa_c > \sqrt{4/3}\,\gamma + 1$ — see Figure 11.3, page 436. For such parameter values, the lens generates a single simple closed caustic curve, which bounds the $N = 0$ region. Hence, a light source inside this region has $\mathrm{Mag}_{tot}(\mathbf{y}) = 0$. Thus, the inequality $\mathrm{Mag}_{tot}(\mathbf{y}) \geq 0$ is our best lower bound on the magnification due to a general supercritical single-plane lens system with point masses, continuous matter, and shear from infinity. Note that though minimum lensed images are never de-magnified (see (13.1), page 469), Theorem 11.9 (page 434) reminds us that no such lensed images occur for the given system.

13.2 Lower Bounds on Total Magnification: Multiplane Case

Consider a k-plane time delay function

$$T_{\mathbf{y}}(\mathbf{x}_1, \ldots, \mathbf{x}_k) = \sum_{i=1}^{k} \vartheta_i \left[\frac{|\mathbf{x}_{i+1} - \mathbf{x}_i|^2}{2} - \beta_i \psi_i^{\mathbf{a}_i}(\mathbf{x}_i) \right].$$

Set $v_1 = 0$ and $v_{j+1} = -\sqrt{\vartheta_j/\vartheta_{j+1}}$ for $j = 1, \ldots, k-1$. Let $\mathbf{I}$ be the 2×2 identity matrix and let

$$\mathrm{T}_\ell(\mathbf{x}_\ell) = (1 + v_\ell^2)\mathbf{I} - \beta_\ell(\mathrm{Hess}\,\psi_\ell)(\mathbf{x}_\ell), \qquad \ell = 1, \ldots, k.$$

Using the previous notation, we see that the next result generalizes (13.1) (page 469), showing that *minimum lensed images cannot be de-magnified in multiplane lensing* (Seitz and Schneider [Sei-S92, pp. 5, 6]).

Theorem 13.5 *Let $\mathbf{x}_{min}$ be the lensed image corresponding to a minimum light ray $\mathbf{X}_{min} = (\mathbf{m}_1, \ldots, \mathbf{m}_k)$ of the k-plane time delay function $T_{\mathbf{y}}$ (i.e., $\mathbf{x}_{min} = \mathbf{m}_1$). Then the magnification of $\mathbf{x}_{min}$ obeys the following:*

$$1 \le \prod_{\ell=1}^{k} \frac{1}{\left(\sqrt{\det \mathrm{T}_\ell(\mathbf{m}_\ell)} - v_\ell\right)^2} \le \mathrm{Mag}(\mathbf{x}_{min}; \mathbf{y}) < \infty$$

where $0 < \left(\sqrt{\det \mathrm{T}_\ell(\mathbf{m}_\ell)} - v_\ell^2\right) \le 1$ *for* $\ell = 1, \ldots, k$.

The proof of Theorem 13.5 involves extremely lengthy, tedious calculations whose methods will not be needed later. For this reason, we shall omit the proof and refer readers to [Sei-S92] for a detailed treatment.

We can now use Morse theory to present lowers bounds on the total magnification due to a general k-plane lens system (i.e., we generalize Theorem 13.1). Employing the notation of Theorem 12.1 and assuming Morse boundary conditions B, which hold after a slight linear perturbation of $T_{\mathbf{y}}$ (see Theorem 12.4), we obtain:[7]

Theorem 13.6 *Let $T_{\mathbf{y}} : L_{(k)} \to \mathbf{R}$ be a k-plane time delay function obeying Morse boundary conditions B relative to a regular domain $D \subset L_{(k)}$ that is punctured by $\mathbf{A}(k)$ and contains all light rays of $T_{\mathbf{y}}$. If the light rays of $T_{\mathbf{y}}$ consists of minima $\mathbf{X}^i_{min} = (\mathbf{m}^i_1, \ldots, \mathbf{m}^i_k)$, where $i = 1, \ldots, M_{min}$, and at least one non-minimum light ray, then* (**the source is magnified**)*:*

$$1 \le M_{min} \le \sum_{i=1}^{M_{min}} \left[\prod_{\ell=1}^{k} \frac{1}{\left(\sqrt{\det \mathrm{T}_\ell(\mathbf{m}^i_\ell)} - v_\ell\right)^2} \right] < \mathrm{Mag}_{tot}(\mathbf{y}) < \infty$$

with

$$M_{min} = \frac{N - 2M^*_+ - (M'_+ - M'_-) + \prod_{\ell=1}^{k}(1 - g_\ell)}{2},$$

where $M^*_+ = M_2 + M_4 + \cdots + M_{2k}$.

Proof: Since $\det[\mathrm{Hess}(T_{\mathbf{y}})](\mathbf{X})$ is finite and nonzero if $\mathbf{X} = (\mathbf{x}_1, \ldots, \mathbf{x}_k)$ is a nondegenerate critical point of $T_{\mathbf{y}}$, we get $0 < \mathrm{Mag}(\mathbf{x}_1; \mathbf{y}) < \infty$, where $\mathbf{x}_1$ is the lensed image associated with $\mathbf{X}$. Let $\mathrm{Mag}_{nmin}(\mathbf{y})$ be the total magnification due to all non-minimum lensed images (by hypothesis there is at least one). Then $0 < \mathrm{Mag}_{nmin}(\mathbf{y}) < \infty$. Consider the minimum lensed images $\mathbf{m}^1_1, \ldots, \mathbf{m}^{M_{min}}_1$. We have

$$\sum_{i=1}^{M_{min}} \mathrm{Mag}(\mathbf{m}^i_1; \mathbf{y}) < \mathrm{Mag}_{nmin}(\mathbf{y}) + \sum_{i=1}^{M_{min}} \mathrm{Mag}(\mathbf{m}^i_1; \mathbf{y}) = \mathrm{Mag}_{tot}(\mathbf{y}) < \infty.$$

[7]The proof of Theorems 13.5 and 13.6 uses the nonnegativity of the surface mass densities κ_i.

Theorem 13.5 gives lower bounds on $\mathrm{Mag}(\mathbf{m}_1^i; \mathbf{y})$, which yields

$$1 \leq M_{min} \leq \sum_{i=1}^{M_{min}} \left[\prod_{\ell=1}^{k} \frac{1}{\left(\sqrt{\det \mathrm{T}_\ell(\mathbf{m}_\ell^i)} - v_\ell\right)^2} \right] \leq \sum_{i=1}^{M_{min}} \mathrm{Mag}(\mathbf{m}_1^i; \mathbf{y}).$$

Theorem 12.1(3,i) (page 448) gives the formula for M_{min} (i.e., M_0). □

The following corollary gives lowers bounds on the total magnification due to multiplane nonsingular and point mass lenses:

Corollary 13.7

(1) *Let $T_\mathbf{y}$ be an isolated k-plane time delay function induced by nonsingular gravitational lens potentials. If* $\mathbf{y}$ *is not a caustic point and a light source at* $\mathbf{y}$ *is multiply imaged, then*

$$1 \leq \frac{N - 2M_+^* + 1}{2} < \mathrm{Mag}_{tot}(\mathbf{y}) < \infty.$$

(2) *If $T_\mathbf{y}^{pt}$ is a k-plane point mass time delay function with g_i point masses on the ith lens plane,*[8] *where* $\mathbf{y}$ *is not a caustic point and $i = 1, \ldots, k$, then*

$$1 \leq \frac{N - 2M_+^* + \prod_{\ell=1}^{k}(1 - g_\ell)}{2} < \mathrm{Mag}_{tot}(\mathbf{y}) < \infty.$$

Proof: (1) Since $N > 1$ and $T_\mathbf{y}$ is a nonsingular isolated k-plane time delay function, Theorem 12.7(1, 2i) that $M_0 \geq 1$ ($M_0 = M_{min}$) and $N = 2M_- + 1 \geq 3$. The latter yields $M_- \geq 1$. Since $T_\mathbf{y}$ obeys Morse boundary conditions A (cf. part (1–2) of proof of Theorem 11.4), Theorem 13.6 applies with all primed quantities zero. Because each g_ℓ is zero, we obtain $\mathrm{Mag}_{tot}(\mathbf{y}) > M_{min} = (N - 2M_+^* + 1)/2 \geq 1$.

(2) Theorem 12.8(1i) yields that $M_{min} \geq 1$ and at least one non-minimum lensed image exists. Because $T_\mathbf{y}$ obeys Morse boundary conditions A (see proof of Theorem 12.8), Theorem 13.6 applies with the primed quantities vanishing. □

13.3 Magnification Cross Section for Multiplane Lensing

A key problem in the theory of gravitational lensing is to determine the magnification cross section due to a given lens (Problem **M2**). In other words, determine the area of the locus of all points on the light source plane with magnification greater than some fixed value.

[8] See equation (12.5), page 457, for the explicit form of $T_\mathbf{y}^{pt}$.

Let us motivate this problem from the astrophysics of lensing. Since any gravitational lens changes the apparent brightness of a background source, one property that is of interest is the probability by which a source is magnified by more than a given magnification μ. For an isolated lens system (mass(es), position(s), distances given), one can determine the angular region in the sky that gets a magnification of at least μ. However, to translate these solid angle regions into probabilities, one has to define properly the "normalization region" $\mathcal{R}$ in the light source plane. For an isolated lens system, one could define such a probability as the ratio between the solid angle of the magnified region to the solid angle of the lens's Einstein disc (i.e., the disc bounded by the lens's Einstein ring[9]) projected into the light source plane. The normalization region $\mathcal{R}$ is the projected Einstein disc. In the case of a probability for a whole ensemble of isolated lenses randomly distributed over the sky, it is common to choose $\mathcal{R}$ to be the union of the projections into the light source plane of all the Einstein discs due to the lenses.

Now, since different cosmological models predict very different probabilities for highly magnified sources (i.e., giant luminous arcs — see [Bart98]), these "probabilities" or lensing cross sections are of high interest to astrophysicists in order to establish or narrow down the viable cosmological models from comparison between theory and observations. We note that different cosmological models also predict quite different probabilities for the fraction of sources that are multiply imaged [Koc96], [Wam-COT]. See Section 4.5 (page 137) for more on the statistical lensing properties of different cosmological models.

The coarea formula (Theorem 13.8 below) will now be applied to obtain a magnification cross-section formula for multiplane lensing.[10] This formula will be used to determine the asymptotic limit of the magnification cross section near fold and cusp caustics. The magnification cross section will then be used to present a formula for the probability density of having magnification greater than some given amount.

13.3.1 Coarea Formula for Magnification Cross Sections

We present the coarea formula along the lines of Chavel [Chav, p. 267] and Sakai [Sak, pp. 69, 70].

Let $\mathfrak{N}$ be a Riemannian manifold with $f : \mathfrak{N} \to \mathbf{R}$ a smooth proper function. The critical values of f form a set of measure zero in $\mathbf{R}$ (Sard's Theorem, page 176), while the regular values form a dense open set in $\mathbf{R}$ (see Corollary 6.4 and Proposition 6.5). For each regular value t of f in $\mathbf{R}$, the set

$$C_t = f^{-1}(t)$$

[9] If the lens is not a point mass, but an isolated (not necessarily circularly symmetric) extended lens with total mass m, then the lens's Einstein ring could be defined as the one produced by a point mass with mass m positioned at, the center-of-mass of the lens. However, for statistical considerations, the place where one positions the point mass m does not really matter.

[10] A.P. thanks Demetrios Christodoulou for pointing out the relevance of the coarea formula to magnification cross sections.

is a compact hypersurface in $\mathfrak{R}$ and the gradient vector $(\nabla f)(y)$, where $f(y) = t$, is orthogonal to $f^{-1}(t)$. Let g_t be the Riemannian metric induced on C_t by the Riemannian metric on $\mathfrak{R}$ and introduce the following volume function:

$$\mathsf{V}(t) = \mathrm{Vol}\{y \in \mathfrak{R} : f(y) > t\}.$$

Theorem 13.8 (Coarea Formula) *Let $f : \mathfrak{R} \to \mathbf{R}$ a smooth proper function on a Riemannian manifold $\mathfrak{R}$. Then*

$$\frac{d\mathsf{V}}{dt}(t) = -\int_{C_t} \frac{dv_{g_t}}{|\nabla f|}, \tag{13.2}$$

where dv_{g_t} is the Riemannian measure on C_t relative to the metric g_t. Moreover, the function $t \mapsto \mathsf{V}(t)$ is smooth at each regular value t of f such that $\mathsf{V}(t) < \infty$.

A detailed proof of the coarea formula is given by Sakai [Sak, pp. 69, 70]. Readers are referred to Federer [Fed, p. 248, Thm 3.2.11] for a generalization of the coarea formula to Lipschitz functions.

We now apply the coarea formula to magnification cross sections in gravitational lensing. Let $\boldsymbol{\eta} : P \longrightarrow S$ be a k-plane lensing map. Recall that $P \subseteq \mathbf{R}^2$ and $S = \mathbf{R}^2$, where both are assumed to have the Euclidean metric. Suppose that the set $\mathrm{Crit}(\boldsymbol{\eta})$ of critical points of $\boldsymbol{\eta}$ is bounded. Then the set $\mathrm{Caustic}(\boldsymbol{\eta})$ of caustics of $\boldsymbol{\eta}$ is compact (Theorem 8.4(1)). Let $\mathcal{R}$ be a bounded open subset of the open set $S - \mathrm{Caustic}(\boldsymbol{\eta})$ of regular values of $\boldsymbol{\eta}$. Suppose that the total magnification function Mag_{tot} of $\boldsymbol{\eta}$ is a smooth proper function on $\mathcal{R}$. Then the *magnification cross section* of $\boldsymbol{\eta}$ in $\mathcal{R}$ is defined by

$$\sigma_{\boldsymbol{\eta},\mathcal{R}}(\mu) = \mathrm{Area}\{\mathbf{y} \in \mathcal{R} : \mathrm{Mag}_{tot}(\mathbf{y}) > \mu\},$$

where $0 \leq \mu < \infty$. Note that $\sigma_{\boldsymbol{\eta},\mathcal{R}}(\mu) < \infty$ (since $\mathcal{R}$ is bounded) and $\sigma_{\boldsymbol{\eta},\mathcal{R}}(\mu)$ is nonincreasing with respect to μ. The function $\sigma_{\boldsymbol{\eta},\mathcal{R}}$ is actually the distribution function[11] of Mag_{tot} on $\mathcal{R}$.

Applying equation (13.2), we set: $t = \mu$; $f = \mathrm{Mag}_{tot}|\mathcal{R}$; $\mathsf{V}(\mu) = \sigma_{\boldsymbol{\eta},\mathcal{R}}(\mu)$; and C_μ is the level curve $(\mathrm{Mag}_{tot}|\mathcal{R})(\mathbf{y}) = \mu$. The volume element dv_{g_μ} now becomes the arc length element ds on C_μ. Hence:

$$\frac{d\sigma_{\boldsymbol{\eta},\mathcal{R}}}{d\mu}(\mu) = -\int_{C_\mu} \frac{ds}{|\nabla \mathrm{Mag}_{tot}(\mathbf{y})|}, \tag{13.3}$$

where ∇ is the gradient operator relative to the coordinates $\mathbf{y}$ on S. Note that $(\nabla \mathrm{Mag}_{tot})(\mathbf{y}) = (\nabla \mathrm{Mag}_{tot}|\mathcal{R})(\mathbf{y})$ for all $\mathbf{y}$ in $\mathcal{R}$. Integrating (13.3) from μ_1 to μ_2, where $0 \leq \mu_1 < \mu_2$, yields

[11] Readers are referred to Wheeden and Zygmund [Whe-Z, pp. 76–83] for an introduction to distribution functions.

the following:

$$\sigma_{\boldsymbol{\eta},\,\mathcal{R}}(\mu_2) = \sigma_{\boldsymbol{\eta},\,\mathcal{R}}(\mu_1) - \int_{\mu_1}^{\mu_2} \left[\int_{C_\mu} \frac{ds}{|\nabla \mathrm{Mag}_{tot}(\mathbf{y})|} \right] d\mu. \tag{13.4}$$

Since $\lim_{\mu\to\infty} \sigma_{\boldsymbol{\eta},\,\mathcal{R}}(\mu) = 0$, equation (13.4) immediately gives an integral formula for the magnification cross section:

Corollary 13.9 (Magnification Cross Section Formula) *Let* $\boldsymbol{\eta}$ *be a k-plane lensing map with* Crit($\boldsymbol{\eta}$) *bounded. Suppose that the associated total magnification* Mag_{tot} *is a smooth proper function on a bounded open subset* $\mathcal{R}$ *of the regular values of* $\boldsymbol{\eta}$. *Then*

$$\sigma_{\boldsymbol{\eta},\,\mathcal{R}}(\mu) = \int_{\mu}^{\infty} \left[\int_{C_\mu} \frac{ds}{|\nabla \mathrm{Mag}_{tot}(\mathbf{y})|} \right] d\mu. \tag{13.5}$$

Magnification Cross Section and Probability

The *probability* Pr that a light source will have magnification greater than μ, relative to $\mathcal{R}$, is defined by

$$\Pr\{\mathbf{y} \in \mathcal{R} : \mathrm{Mag}_{tot}(\mathbf{y}) > \mu\} \equiv \frac{\sigma_{\boldsymbol{\eta},\,\mathcal{R}}(\mu)}{\mathrm{Area}(\mathcal{R})},$$

where Mag_{tot} is treated as a random variable and Area($\mathcal{R}$) is the area of $\mathcal{R}$. Note that $\sigma_{\boldsymbol{\eta},\,\mathcal{R}}(\mu)$ is defined for negative values of μ; in fact, $\sigma_{\boldsymbol{\eta},\,\mathcal{R}}(\mu) = \sigma_{\boldsymbol{\eta},\,\mathcal{R}}(0)$ for $-\infty < \mu < 0$. Also, see the discussion at the start of Section 13.3 for more on the astrophysical choice of $\mathcal{R}$.

The *probability distribution function* of the random variable Mag_{tot} is

$$F_{\mathrm{Mag}_{tot}}(\mu) \equiv \Pr\{\mathbf{y} \in \mathcal{R} : \mathrm{Mag}_{tot}(\mathbf{y}) \le \mu\} = 1 - \frac{\sigma_{\boldsymbol{\eta},\,\mathcal{R}}(\mu)}{\mathrm{Area}(\mathcal{R})},$$

where we used the relation

$$\Pr\{\mathbf{y} \in \mathcal{R} : \mathrm{Mag}_{tot}(\mathbf{y}) > \mu\} = 1 - \Pr\{\mathbf{y} \in \mathcal{R} : \mathrm{Mag}_{tot}(\mathbf{y}) \le \mu\}.$$

Assume that Mag_{tot} has a *probability density* $f_{\mathrm{Mag}_{tot}}$. Then the following holds at points μ where $F_{\mathrm{Mag}_{tot}}$ is differentiable:

$$f_{\mathrm{Mag}_{tot}}(\mu) = \frac{dF_{\mathrm{Mag}_{tot}}}{d\mu}(\mu) = -\frac{1}{\mathrm{Area}(\mathcal{R})} \frac{d\sigma_{\boldsymbol{\eta},\,\mathcal{R}}}{d\mu}(\mu).$$

Note that $f_{\mathrm{Mag}_{tot}}$ integrates to one since

$$\int_{-\infty}^{\infty} \frac{d\sigma_{\boldsymbol{\eta},\,\mathcal{R}}}{d\mu} d\mu = \int_{\sigma_{-\infty}}^{\sigma_\infty} d\sigma_{\boldsymbol{\eta},\,\mathcal{R}} = \lim_{\mu\to\infty} \sigma_{\boldsymbol{\eta},\,\mathcal{R}}(\mu) - \lim_{\mu\to-\infty} \sigma_{\boldsymbol{\eta},\,\mathcal{R}}(\mu) = -\mathrm{Area}(\mathcal{R}).$$

Under the hypotheses of Corollary 13.9, equation (13.3) yields an integral formula for the probability density of Mag_{tot}:

$$f_{\text{Mag}_{tot}}(\mu) = \frac{1}{\text{Area}(\mathcal{R})} \int_{C_\mu} \frac{ds}{|\nabla \text{Mag}_{tot}(\mathbf{y})|}.$$

Finally, the probability of having magnification between μ and $\mu + \Delta\,\mu$ for sufficiently small increment $\Delta\,\mu$, is given approximately by

$$\Pr\{\mu \leq \text{Mag}_{tot} \leq \mu + \Delta\,\mu\} \simeq f_{\text{Mag}_{tot}}(\mu)\,\Delta\,\mu.$$

For more on magnification cross sections and their applications in gravitational lensing, see [Schn-EF, pp. 310–311, 313–320, 373–380] and [Kai92a]; also, consult [Tur-OG] and [Schn-EF] for more on statistical gravitational lensing.

Henceforth, the function Mag_{tot} *is not required to be a random variable.*

13.3.2 Magnification Cross Section Near Folds and Cusps

The qualitative behavior of a locally stable k-plane lensing map $\boldsymbol{\eta} : P \to S$ in a neighborhood of a critical point is described by one of the following:

Whitney fold:

$$\boldsymbol{\eta}_F : \mathbf{R}^2 \longrightarrow \mathbf{R}^2, \qquad \boldsymbol{\eta}_F(\mathrm{u}, \mathrm{v}) = (\mathrm{u}, a\mathrm{v}^2), \qquad a = \pm 1,$$

Whitney cusp:

$$\boldsymbol{\eta}_C : \mathbf{R}^2 \longrightarrow \mathbf{R}^2, \qquad \boldsymbol{\eta}_C(\mathrm{u}, \mathrm{v}) = (\mathrm{u}, a\mathrm{uv} + b\mathrm{v}^3), \qquad |a| = |b| = 1.$$

In other words, at each critical point $\mathbf{x}$ of $\boldsymbol{\eta}$, the map $\boldsymbol{\eta}$ is differentiably equivalent to either $\boldsymbol{\eta}_F$ or $\boldsymbol{\eta}_C$ — Theorem 7.49, page 251 (i.e., there are charts $\lambda : U \subseteq P \to \mathbf{R}^2$ and $\nu : V \subseteq S \to \mathbf{R}^2$ centered at $\mathbf{x}$ and $\boldsymbol{\eta}(\mathbf{x})$, respectively, such that on $\lambda(U)$ the map $\nu \circ \boldsymbol{\eta} \circ \lambda^{-1}$ equals either $\boldsymbol{\eta}_F$ or $\boldsymbol{\eta}_C$, depending on whether $\mathbf{x}$ is a fold or cusp critical point).[12]

We shall investigate the magnification cross section scalings for folds and cusps due to the k-plane lensing map $\boldsymbol{\eta}$. This will be done by first determining the scalings for Whitney folds $\boldsymbol{\eta}_F$ and Whitney cusps $\boldsymbol{\eta}_C$ (see Theorems 13.10 and 13.11 below). These results will then be used to establish (without assuming Taylor polynomial approximations) the magnification cross section scalings near caustics of $\boldsymbol{\eta}$ (Theorem 13.12).

[12]Consult Section 9.1 (page 328) for more on the lensing properties of $\boldsymbol{\eta}_F$ and $\boldsymbol{\eta}_C$.

Magnification Cross Section Near Whitney Folds

Theorem 13.10 *The magnification cross section of a Whitney fold $\boldsymbol{\eta}_F$ in an open disc D about a fold caustic point scales in μ as follows:*

$$\sigma_{\boldsymbol{\eta}_F,\mathcal{R}}(\mu) \sim \frac{\kappa_F}{\mu^2} \qquad \text{as } \mu \to \infty, \tag{13.6}$$

where the factor κ_F is independent of μ and $\mathcal{R} = D - \mathrm{Crit}(\boldsymbol{\eta}_F)$.

Proof of Theorem 13.10: Parametrize the domain and target of $\boldsymbol{\eta}_F$ by rectangular coordinates (u, v) and (s_1, s_2), respectively. We shall employ the results of Section 9.1.1 (page 329). It suffices to assume that $a = 1$ and use the origin as the fixed fold caustic point. We saw in Section 9.1.1 that the caustic curve C_∞ of $\boldsymbol{\eta}_F$ is the s_1-axis and a light source at (s_1, s_2) with $s_2 > 0$ has two lensed images:

$$r_\oplus = (s_1, \sqrt{s_2}), \qquad r_\ominus = (s_1, -\sqrt{s_2}).$$

Moreover, equation (9.1) on page 331 yields that the total magnification of a light source at $\mathbf{y} = (s_1, s_2)$ with $s_2 > 0$ is

$$\mathrm{Mag}_{tot}(\mathbf{y}) = \mathrm{Mag}(r_\oplus; \mathbf{y}) + \mathrm{Mag}(r_\ominus; \mathbf{y}) = \frac{1}{\sqrt{s_2}}.$$

The level curve $C_\mu : \mathrm{Mag}_{tot}(\mathbf{y}) = \mu$ is then given by $s_2 = \mu^{-2}$. As $\mu \to \infty$, the curve C_μ approximates the caustic $C_\infty : s_2 = 0$. In addition:

$$\frac{\partial \mathrm{Mag}_{tot}}{\partial s_1} = 0, \qquad \frac{\partial \mathrm{Mag}_{tot}}{\partial s_2} = \frac{1}{2}\frac{1}{s_2^{3/2}} = \frac{\mu^3}{2}.$$

In an open disc D about the origin, we see from the magnification cross section formula (13.5) that if $\mu \to \infty$, then

$$\sigma_{\boldsymbol{\eta}_F,\mathcal{R}}(\mu) = \int_\mu^\infty d\mu \int_{C_\mu} \frac{ds}{|\nabla \mathrm{Mag}_{tot}(\mathbf{y})|} \sim 2\left(\int_\mu^\infty \frac{d\mu}{\mu^3}\right)\int_{\tilde{C}_\infty} ds = \frac{1}{\mu^2}\int_{\tilde{C}_\infty} ds,$$

where $\tilde{C}_\infty$ is the portion of the line C_∞ in D and $\mathcal{R} = D - \mathrm{Crit}(\boldsymbol{\eta}_F)$. The integral on the right is the length of $\tilde{C}_\infty$, which is finite. □

Magnification Cross Section Near Whitney Cusps

Theorem 13.11 *The magnification cross section outside*[13] *the cusp caustic of a Whitney pleat $\boldsymbol{\eta}_C$ scales in μ as follows in an open disc D about the cusp:*

$$\sigma_{\boldsymbol{\eta}_C,\mathcal{R}}(\mu) \sim \frac{\kappa_C}{\mu^{5/2}}, \qquad \text{as } \mu \to \infty, \tag{13.7}$$

where the factor κ_C is independent of μ and $\mathcal{R} = D - \mathrm{Crit}(\boldsymbol{\eta}_C)$.

[13] i.e., in the singly covered region exterior to the cusped caustic curves (e.g., lobe in Figure 13.1).

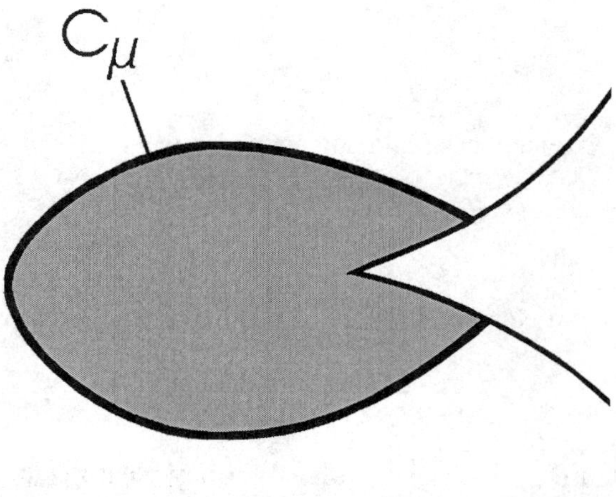

Figure 13.1. The bold curve C_μ is a contour of constant magnification μ. A light source inside the lobe (shaded region) has magnification at least μ.

Proof: Choose rectangular coordinates (u, v) and (s_1, s_2) in the domain and target, respectively, of $\boldsymbol{\eta}_C$. We shall draw on results from Section 9.1.2 (page 332). It suffices to assume that the origin is a positive cusp, say, $a = -1$ and $b = 1$. From Section 9.1.2, we know that there is a magnification lobe outside the cusp, which is shown in Figure 13.1.

A light source at $\mathbf{y} = (s_1, s_2)$ outside the cusp has one lensed image at (u, v), where

$$s_1 = \mathrm{u}, \qquad s_2 = -\mathrm{uv} + \mathrm{v}^3. \tag{13.8}$$

The total magnification is then (see (9.3), page 333):

$$\mathrm{Mag}_{tot}(\mathbf{y}) = \frac{1}{-\mathrm{u} + 3\mathrm{v}^2} > 0, \tag{13.9}$$

where (13.8) gives u and v in terms of the light source position $\mathbf{y} = (s_1, s_2)$. The magnification contour outside the cusp is the curve C_μ given by $\mathrm{Mag}_{tot}(\mathbf{y}) = \mu$ (Figure 13.1). Explicitly, the contour is determined by

$$\mathrm{u} = 3\mathrm{v}^2 - \mu^{-1}. \tag{13.10}$$

The u and v in (13.10) relate to the light source position $\mathbf{y}$ via (13.8). Plugging (13.10) into (13.8), we can express C_μ as a curve parametrized by v:

$$s_1 = 3\mathrm{v}^2 - \mu^{-1}, \qquad s_2 = -2\mathrm{v}^3 + \mathrm{v}\mu^{-1}. \tag{13.11}$$

Now, since a lensed image at (u, v) has magnification $1/(-\mathrm{u} + 3\mathrm{v}^2)$, where the light source lies on C_μ, we see that the equation $\mathrm{u} = 3\mathrm{v}^2 - \mu^{-1}$ also describes a curve c_μ of constant lensed-image magnification μ in the lens plane. Turning to Figure 9.11 on page 340, recall that the lensed-image dashed curve C_2 is given by $\mathrm{u} = 3\mathrm{v}^2/4$. Setting

$$\frac{3}{4}\mathrm{v}^2 = \mathrm{u} = 3\mathrm{v}^2 - \mu^{-1},$$

we see that c_μ and C_2 intersect at points determined by

$$\mathrm{v} = \pm\frac{2}{3}\mu^{-1}.$$

In other words, the parameter v for C_μ in (13.11) is restricted to the following range of values:

$$\frac{-2}{3}\mu^{-1} \le \mathrm{v} \le \frac{2}{3}\mu^{-1}.$$

This is because in Figure 9.11 (page 340) the Whitney pleat $\boldsymbol{\eta}_C$ maps region III, which is where c_μ is located, injectively onto the exterior of the cusped caustic curve in Figure 9.11, while regions I and II are mapped into the interior of the cusped curve (cf. [Schn-W92, p. 6]). As in [Schn-W92, p. 6], we also define a new parameter r by

$$\mathrm{v} = \frac{2}{3}\,r\mu^{-1/2}, \qquad -1 \le r \le 1.$$

In terms of the parameter r, the contour in (13.11) becomes $C_\mu(r) = (s_1(r), s_2(r))$, where

$$s_1(r) = \mu^{-1}\left(\frac{4}{3}r^2 - 1\right), \qquad s_2(r) = \frac{2}{3}\mu^{-3/2}\left(-\frac{8}{9}r^2 + 1\right).$$

Hence, if we let $\dot{s}_i = ds_i/dr$, then the following μ-scalings hold:

$$\mathrm{v} \sim \mu^{-1}, \quad s_1 \sim \mu^{-1}, \quad \dot{s}_1 \sim \mu^{-1}, \quad s_2 \sim \mu^{-3/2}, \quad \dot{s}_2 \sim \mu^{-3/2}. \tag{13.12}$$

We shall now employ (13.12) to determine the μ-scaling for $|\nabla \mathrm{Mag}_{tot}|^{-1}$. Expressing the total magnification (13.9) in terms of the light source position $\mathbf{y} = (s_1, s_2)$, we obtain (see (9.2) on page 333):

$$\mathrm{Mag}_{tot}(\mathbf{y}) = \frac{1}{-s_1 + 3\mathrm{v}^2},$$

where

$$\mathrm{v} = \left(\frac{s_2}{2} + \sqrt{C}\right)^{1/3} + \left(\frac{s_2}{2} - \sqrt{C}\right)^{1/3}$$

and

$$C = -\frac{s_1^3}{27} + \frac{s_2^2}{4}. \tag{13.13}$$

Note that points (s_1, s_2) outside the cusp are characterized by $C > 0$. Direct calculation yields

$$\begin{aligned}\frac{\partial \mathrm{Mag}_{tot}}{\partial s_1} &= -\left(-1 + 6\mathrm{v}\frac{\partial \mathrm{v}}{\partial s_1}\right)\left(-s_1 + 3\mathrm{v}^2\right)^{-2} \\ \frac{\partial \mathrm{Mag}_{tot}}{\partial s_2} &= -\left(6\mathrm{v}\frac{\partial \mathrm{v}}{\partial s_2}\right)\left(-s_1 + 3\mathrm{v}^2\right)^{-2}.\end{aligned} \tag{13.14}$$

Here

$$\begin{aligned}\frac{\partial \mathrm{v}}{\partial s_1} &= \frac{1}{2\cdot 3^3}\frac{s_1^2}{\sqrt{C}}\left[-\left(\frac{s_2}{2}+\sqrt{C}\right)^{-2/3} + \left(\frac{s_2}{2}-\sqrt{C}\right)^{-2/3}\right], \\ \frac{\partial \mathrm{v}}{\partial s_2} &= \frac{1}{2\cdot 3}\left[\frac{\left(1+\frac{s_2}{2\sqrt{C}}\right)}{\left(\frac{s_2}{2}+\sqrt{C}\right)^{2/3}} + \frac{\left(1-\frac{s_2}{2\sqrt{C}}\right)}{\left(\frac{s_2}{2}-\sqrt{C}\right)^{-2/3}}\right].\end{aligned} \tag{13.15}$$

Application of the scalings in (13.12) to equations (13.13) and (13.15) gives

$$C \sim \mu^{-3}, \qquad \sqrt{C} \sim \mu^{-3/2}, \qquad \frac{\partial \mathrm{v}}{\partial s_1} \sim \mu^{1/2}, \qquad \frac{\partial \mathrm{v}}{\partial s_2} \sim \mu,$$

which if applied to (13.14) yields

$$\frac{\partial \mathrm{Mag}_{tot}}{\partial s_1} \sim \mu^2, \qquad \frac{\partial \mathrm{Mag}_{tot}}{\partial s_2} \sim \mu^{5/2}.$$

Hence, as $\mu \to \infty$ we obtain

$$\frac{1}{|\nabla \mathrm{Mag}_{tot}(\mathbf{y})|} = \left[\left(\frac{\partial \mathrm{Mag}_{tot}}{\partial s_1}\right)^2 + \left(\frac{\partial \mathrm{Mag}_{tot}}{\partial s_2}\right)^2\right]^{-1/2} \sim \mu^{-5/2} \tag{13.16}$$

because $\partial \mathrm{Mag}_{tot}/\partial s_2$ becomes μ-dominated by $\partial \mathrm{Mag}_{tot}/\partial s_1$.

Equation (13.12) also shows that the arc length element ds along the contour $C_\mu(r) = (s_1(r), s_2(r))$, where $-1 \le r \le 1$, scales as

$$ds = \left[(\dot{s}_1(r))^2 + (\dot{s}_2(r))^2\right]^{1/2} dr \sim \mu^{-1} dr \tag{13.17}$$

in the limit $\mu \to \infty$ (since $(\dot{s}_1(r))^2$ becomes μ-dominated by $(\dot{s}_2(r))^2$). Employing the magnification cross section formula (13.5) on page 478, we see that equations (13.16) and (13.17) give the following in an open disc D about the cusp as $\mu \to \infty$:

$$\sigma_{\eta_C,\mathcal{R}}(\mu) = \int_\mu^\infty d\mu \left(\int_{-1}^{1} \frac{\left[(s_1(r))^2 + (s_2(r))^2\right]^{1/2} dr}{|\nabla \mathrm{Mag}_{tot}(s_1(r), s_2(r))|} \right) \sim K_* \int_\mu^\infty \frac{d\mu}{\mu^{7/2}} = \frac{\kappa_C}{\mu^{5/2}},$$

where $\mathcal{R} = D - \mathrm{Crit}(\eta_C)$ and the factor κ_C (and $\kappa_* = 5\kappa_C/2$) does not depend on μ. □

Remark: For the case of single-plane lensing, the scaling (13.6) was determined by Blandford and Narayan [Bla-N86, p. 577], while (13.7) was found by Mao [Mao, p. 389, eq. (18)] and Schneider and Weiss [Schn-W92, p. 6, eq. (2.39a)]. Their approach used Taylor polynomial approximations (see (9.18) and (9.19) on page 348) and did not employ the coarea formula.

Magnification Cross Section Near Caustics in k-Plane Lensing

Theorem 13.12 *Let η be a locally stable k-plane lensing map whose set of critical curves is bounded. Suppose that the total magnification* Mag_{tot} *due to η is a smooth proper function on a bounded open subset $\mathcal{R}$ of the regular values of η. Then the magnification cross section of η in $\mathcal{R}$ scales in μ as follows near caustics:*

$$\sigma_{\eta,\mathcal{R}}(\mu) \sim \frac{\mathsf{K}_F}{\mu^2} + \frac{\mathsf{K}_C}{\mu^{5/2}}, \qquad \text{as } \mu \to \infty, \tag{13.18}$$

where the factors K_F and K_C are independent of μ.

Proof: The first part of this proof establishes a "squeeze principle" (see (13.19) below) for the magnification cross sections due to differentiable equivalent maps between planes. This result is then applied to the lensing map η by using the differentiable equivalence of η to either a Whitney fold η_C or Whitney cusp η_C at critical points of η.

Let X and Y be open subsets of the plane. Consider smooth maps $f : X \to Y$ and $h : X \to Y$, where $f = \xi \circ h \circ \phi$ with ξ and ϕ diffeomorphisms of X and Y, respectively. Let J_f, J_h, J_ξ, and J_ϕ be the determinants of the Jacobian matrices of the indicated mappings. We shall work in the interior of a disc in X and one in Y. This implies that $|J_\xi|$ and $|J_\phi|$ are bounded above and below, say $c_0 < |J_\xi(\mathbf{y})| < c_1$ and $d_0 < |J_\phi(\mathbf{x})| < d_1$ for all $\mathbf{y}$ and $\mathbf{x}$ in these respective discs. Since $J_f = (J_\xi \circ (g \circ \phi))(J_h \circ \phi)(J_\phi)$, it follows that $c_0 d_0 |J_h(\phi(\mathbf{x}))| < |J_f(\mathbf{x})| < c_1 d_1 |J_h(\phi(\mathbf{x}))|$ for all $\mathbf{x}$ in the aforementioned discs. Consequently,

$$\frac{1}{c_1 d_1} \sum_{\mathbf{x} \in f^{-1}(\mathbf{y})} \frac{1}{|J_h(\phi(\mathbf{x}))|} < \sum_{\mathbf{x} \in f^{-1}(\mathbf{y})} \frac{1}{|J_f(\mathbf{x})|} < \frac{1}{c_0 d_0} \sum_{\mathbf{x} \in f^{-1}(\mathbf{y})} \frac{1}{|J_h(\phi(\mathbf{x}))|}.$$

Letting $\mathbf{y} = \xi(\mathbf{y}')$ and $\mathbf{x}' = \phi(\mathbf{x})$, we have that

$$\sum_{\mathbf{x}\in f^{-1}(\mathbf{y})} \frac{1}{|J_h(\phi(\mathbf{x}))|} = \sum_{\phi^{-1}(\mathbf{x}')\in f^{-1}(\mathbf{y})} \frac{1}{|J_h(\mathbf{x}')|} = \sum_{\mathbf{x}'\in\phi(f^{-1}(\mathbf{y}))} \frac{1}{|J_h(\mathbf{x}')|}$$
$$= \sum_{\mathbf{x}'\in h^{-1}(\xi^{-1}(\mathbf{y}))} \frac{1}{|J_h(\mathbf{x}')|} = \sum_{\mathbf{x}'\in h^{-1}(\mathbf{y}')} \frac{1}{|J_h(\mathbf{x}')|}.$$

Using the definition of total magnification, set

$$M_f(\mathbf{y}) = \sum_{\mathbf{x}\in f^{-1}(\mathbf{y})} \frac{1}{|J_f(\mathbf{x})|}, \qquad M_h(\mathbf{y}') = \sum_{\mathbf{x}'\in h^{-1}(\mathbf{y}')} \frac{1}{|J_h(\mathbf{x}')|}.$$

It follows that:

$$\frac{1}{c_1 d_1} M_h(\mathbf{y}') < M_f(\mathbf{y}) < \frac{1}{c_0 d_0} M_h(\mathbf{y}').$$

Now consider the set, $\{\mathbf{y}|M_f(\mathbf{y}) > \mu\}$, which by the preceding is contained in the following set:

$$\{\mathbf{y} = \xi(\mathbf{y}')|M_h(\mathbf{y}') > c_0 d_0 \mu\} = \xi[\{\mathbf{y}'|M_h(\mathbf{y}') > c_0 d_0 \mu\}].$$

We define

$$\sigma_f(\mu) = \text{Area}(\{\mathbf{y}|M_f(\mathbf{y}) > \mu\}), \qquad \sigma_h(\mu') = \text{Area}(\{\mathbf{y}'|M_h(\mathbf{y}') > \mu'\}).$$

Hence

$$\sigma_f(\mu) \leq \text{Area}(\xi[\{\mathbf{y}'|M_h(\mathbf{y}') > c_0 d_0 \mu\}]).$$

Since $c_0 < |J_\xi(\mathbf{y})| < c_1$ for all $\mathbf{y}$ in our disc in Y, we get the following for any subset A of that disc:

$$c_0 \text{Area}(A) < \text{Area}(\xi(A)) < c_1 \text{Area}(A).$$

Consequently, $\sigma_f(\mu) \leq c_1 \sigma_h(c_0 d_0 \mu)$; similarly, we obtain $c_0 \sigma_h(c_1 d_1 \mu) \leq \sigma_f(\mu)$. In other words, we can "squeeze" $\sigma_f(\mu)$ as follows:

$$c_0 \sigma_h(c_1 d_1 \mu) \leq \sigma_f(\mu) \leq c_1 \sigma_h(c_0 d_0 \mu). \tag{13.19}$$

Hence, if $\sigma_h(\mu) \sim K_h/\mu^\delta$ as $\mu \to \infty$, where $\delta > 0$ and K_h is a constant that is independent of μ, then by (13.19) the function $\sigma_f(\mu)$ also obeys this scaling.

Given that the k-plane lensing map $\boldsymbol{\eta}$ is locally stable, all its caustics are either folds or cusps, and $\boldsymbol{\eta}$ is differentiably equivalent to either a Whitney fold $\boldsymbol{\eta}_F$ or Whitney cusp $\boldsymbol{\eta}_C$ at every point of Crit($\boldsymbol{\eta}$) — see Theorem 7.49 (page 251) and Theorem 7.55 (page 261). Furthermore, since Crit($\boldsymbol{\eta}$) is bounded, it is also compact (Theorem 8.4(1), page 293). Consequently, there are finitely many caustic curves, each being a closed curve. In particular, there are finitely many cusps and fold arcs. By Theorems 13.10 and 13.11, we have the scalings $\sigma_{\boldsymbol{\eta}_F,\mathcal{R}}(\mu) \sim \kappa_F/\mu^2$ and $\sigma_{\boldsymbol{\eta}_C,\mathcal{R}}(\mu) \sim \kappa_C/\mu^{5/2}$ as $\mu \to \infty$. Applying (13.19) with $f = \boldsymbol{\eta}$, $h = \boldsymbol{\eta}_F$ and $f = \boldsymbol{\eta}$, $h = \boldsymbol{\eta}_C$, and summing the results over all fold arcs and cusps in $\mathcal{R}$, we obtain the desired scaling with new factors K_F and K_C. $\square$

Chapter 14

Computing the Euler Characteristic

A youth who had begun to read geometry with Euclid, when he had learnt the first proposition, inquired, 'What do I get by learning these things?' So Euclid called a slave and said, 'Give him threepence, since he must make a gain out of what he learns.'

STOBAEUS

One of the most important numerical invariants in mathematics is the *Euler characteristic*. We shall investigate the Euler characteristic of a surface in terms of the critical points of maps from the surface into the plane. This material will be applied later to the study of the global geometry of caustics (see Section 15.4.2).

Throughout the chapter M is a compact 2-manifold (i.e., a surface) and, except when stated to the contrary, we assume that M is without boundary.

Now, recall that a smooth function $f : M \to \mathbf{R}$ is a *Morse function* if f has only *nondegenerate critical points*. That is, $p \in M$ is a *critical point* of f if $df(p) = 0$, and a critical point p of f is *nondegenerate* if the Hessian Hess $f(p)$ of f at p is invertible. At a critical point p of f, *the index of p* is the index of Hess $f(p)$. If f is a Morse function and $N_i(f)$ is the number of critical points of index i, then the *Euler characteristic* $\chi(M)$ of M is given by (Corollary 10.8, page 406):

$$\chi(M) = N_0(f) - N_1(f) + N_2(f). \tag{14.1}$$

The above equation yields a computation of the topological invariant $\chi(M)$ via the critical points of any Morse function on M. This chapter will present a generalization due to H.L. [Lev66] of equation (14.1) to locally stable maps from M into the plane (see Theorem 14.6).

Outline: Section 14.1 reviews the characterization of locally stable maps from the surface M into the plane. Rotation numbers and projectivized rotation numbers are introduced in Section 14.2, while the orientation of critical curves is treated in Section 14.3. The chapter then concludes with a formula for computing the Euler characteristic in terms of the projectivized rotation numbers of the caustics of a locally stable map from M into the plane (Section 14.4).

14.1 Locally Stable Maps from Surfaces into the Plane

We shall review the definition and characterization of locally stable maps between two dimensional manifolds.

A smooth map $f : M \to \mathbf{R}^2$ is *stable at* $p \in M$ if the map-germ $[f]_p$ of f at p is stable, that is, if for each neighborhood V of p, there is a neighborhood $\mathfrak{F}$ of f in $C^\infty(M, \mathbf{R}^2)$ such that for every map $\tilde{f}$ in $\mathfrak{F}$ there is a point $\tilde{p}$ in V with $[\tilde{f}]_{\tilde{p}}$ differentiably equivalent to $[f]_p$ (see Section 7.2.1, page 226). We call f *locally stable* if f is stable at every point of M.

Locally stable maps $f : M \to \mathbf{R}^2$ are characterized by Theorem 7.49 (page 251) and Theorem 7.55 (page 261):

A smooth map $f : M \to \mathbf{R}^2$ *is locally stable if and only if:*

(1) *The rank of* df *is* 1 *or* 2. *We denote the set of all* $p \in M$ *such that* $\operatorname{rank}[df](p) = 1$ *by* $S_1(f)$.

(2) *If* $p \in S_1(f)$, *we can choose coordinates* $\{u, x\}$ *centered at* p *and* $\{U, X\}$ *centered at* $f(p)$ *such that* f *has one of the following Whitney forms:*

$$\text{(i)} \quad \begin{cases} U \circ f(u, x) = u \\ X \circ f(u, x) = a x^2 \end{cases}$$

$$\text{(ii)} \quad \begin{cases} U \circ f(u, x) = u \\ X \circ f(u, x) = axu + bx^3, \quad \text{where } |a| = |b| = 1. \end{cases}$$

Remark: A more concise, coordinate-free, equivalent description of locally stability is that f is 2-generic (Theorem 7.55, page 261). This means: (a) The 1-jet $j^1 f$ is transverse to the singularity manifolds $S_r(X, Y)$ for $r = 0, 1, 2$, and (b) the 2-jet $j^2 f$ is transverse to the singularity manifolds $S_{r,s}(X, Y)$ for $r = 0, 1, 2$ and $0 \leq s \leq r$ (see Definition 7.37, page 235, and Definition 7.40, page 236).

(3) *In case* M *has a nonempty boundary* ∂M, *we also assume that no point of* $S_1(f)$ *is on* ∂M.

For any locally stable map $f : M \to \mathbf{R}^2$, a point p is *regular* if $\operatorname{rank}(df)(p) = 2$: otherwise, p is a *critical point.* The set $\operatorname{Crit}(f)$ of critical points of f is given as follows when f is locally stable:

$$\operatorname{Crit}(f) = S_1(f).$$

Moreover, the set $\operatorname{Crit}(f)$ consists of a finite collection of nonintersecting, smoothly embedded circles in M (cf. page 245). Note that no component of $\operatorname{Crit}(f)$ is diffeomorphic to the real line since $\operatorname{Crit}(f)$ is compact. To see this, observe that f is a local diffeomorphism at each regular point p. Consequently, there is an open set of regular points of f about p. Hence, the set of regular points is open; that is, the set $\operatorname{Crit}(f)$ is closed. Since M is compact, the set $\operatorname{Crit}(f)$ is also compact.

The restriction of f to $\operatorname{Crit}(f)$ is an immersion except at a finite set of points, namely, the cusp critical points. This subset of $\operatorname{Crit}(f)$ is denoted by $\operatorname{Cusps}(f)$. The *components* of $S_1(f)$ are called the *critical curves of* f, and the points of $\operatorname{Folds}(f) = \operatorname{Crit}(f) - \operatorname{Cusps}(f)$ are called the *fold critical points* of f. By Theorems 7.49 (page 251), there are local coordinates such that at each point of $\operatorname{Folds}(f)$ the map f has one of the forms (2i), while at each point of $\operatorname{Cusps}(f)$, the map f has one of the forms (2ii).

14.2 Projectivized Rotation Numbers

We begin with a discussion of *rotation numbers* and then generalize them to *projectivized rotation numbers.*

In what follows, we always take the unit circle $\mathbf{S}^1$ and $\mathbf{R}^2$ to be oriented, where $\mathbf{S}^1$ is oriented as the boundary of the unit disk (namely, the positive sense on $\mathbf{S}^1$ is counterclockwise).

Rotation numbers

For any C^1 closed curve $h : \mathbf{S}^1 \to \mathbf{R}^2$ that has a nowhere-zero derivative $h' : \mathbf{S}^1 \to \mathbf{R}^2 - \{\mathbf{0}\}$, we define *the rotation number of* h denoted $R(h)$, by

$$R(h) = \frac{1}{2\pi}\left[\begin{array}{l}\textit{the net total angle traversed by } h'(s) \textit{ as } s\\ \textit{goes once around the domain } \mathbf{S}^1 \textit{ positively}\end{array}\right].$$

The net total angle traversed by $h'(s)$ is computed by taking the sense of traverse into account. That is, if $(h'/|h'|)(s)$ traces an arc traversed positively as s travels along one arc of the domain, $\mathbf{S}^1$, and traces the same arc negatively as s travels along some other arc of the domain, $\mathbf{S}^1$, then the contribution to $R(h)$ from those two arcs of $\mathbf{S}^1$ is zero. Here assume $\mathbf{S}^1$ and $\mathbf{R}^2$ are oriented as usual.

For such a closed curve, h, $R(h)$ is always an integer and, in particular, if h is a simple closed curve, $R(h) = \pm 1$. Furthermore, $R(h)$ does not change if we deform h through curves of the same type (i.e., through C^1 closed curves with nowhere-vanishing derivatives).

In other words, $R(h)$ is an invariant of regular homotopy for such immersions h (e.g., [Ber-G, p. 325]).

Notes:

(1) Another way of thinking about this is: If we consider the mapping $(h'/|h'|) : \mathbf{S}^1 \to \mathbf{S}^1$, then the rotation number $R(h)$ of h is the number of times —taking account of sign — that the map $\frac{h'}{|h'|}$ covers the circle. The integer $R(h)$ is also called the *Brouwer degree* of $\frac{h'}{|h'|}$ (e.g., [Mil, p. 27]). In a perhaps more familiar context, the rotation number of h is called the *winding number of* h' *around* $\mathbf{0}$ (see [doC, p. 36]).

(2) Recall that if $\jmath$: $\mathbf{S}^1 \to \mathbf{S}^1$ is a differentiable map, a point, $y \in \mathbf{S}^1$ is a *regular value of* $\jmath$ if $\jmath'(x) \neq 0$ for every $x \in \jmath^{-1}(y)$. If $h : \mathbf{S}^1 \to \mathbf{R}^2$ is twice differentiable and h' is nowhere-zero, to compute $R(h)$ or what is the same, the Brouwer degree of $\frac{h'}{|h'|}$, we choose any regular value y of $\frac{h'}{|h'|} = \jmath$: $\mathbf{S}^1 \to \mathbf{S}^1$. Let $x_1, \ldots, x_m$ be the points of $\jmath^{-1}(y)$ at which $\jmath$ is orientation preserving and $x_{m+1}, \ldots, x_{m+\ell}$ the points of $\jmath^{-1}(y)$ at which $\jmath$ is orientation reversing. Then $R(h) = m - \ell$.

Projectivized Rotation numbers

We extend the notion of rotation number to certain closed curves that may fail to have well-defined tangents at a finite set of points.

Let I be an open interval of the real line $\mathbf{R}$ or an open arc of the circle $\mathbf{S}^1$. Denote I by (a, b). We take I to be oriented by the orientation of $\mathbf{R}$ or the orientation of $\mathbf{S}^1$. For any C^1 curve $f : I \to \mathbf{R}^2$ that has a continuous nowhere-zero derivative f' on (a, b), we define the *rotation number of* f, denoted $R(f)$, by

$$R(f) = \frac{1}{2\pi}\left[\begin{array}{l}\textit{the net total angle traversed by } f'(s) \textit{ as} \\ s \textit{ travels along } I \textit{ in the positive sense.}\end{array}\right].$$

To make this definition explicit, note that we can write $\frac{f'(s)}{|f'(s)|} : I \to \mathbf{S}^1 : s \to (\cos\theta(s), \sin\theta(s))$ for some continuous function $\theta : I \to \mathbf{R}$. The function θ is not uniquely determined, but the difference of any two such functions is a constant multiple of 2π. In terms of any such θ:

$$R(f) = \frac{1}{2\pi}\left(\lim_{s\to b-}\theta(s) - \lim_{s\to a+}\theta(s)\right).$$

Here $R(f)$ need not exist and when it exists, $R(f)$ may not be an integer and does not have the same invariance under deformations through curves of the same type.

Now let $s_1 < s_2 < \ldots < s_k < s_1$ be points, in order, on $\mathbf{S}^1$ and let $I_j = (s_j, s_{j+1})$ be the positively oriented open arc on $\mathbf{S}^1$ with initial and final end points s_j and s_{j+1}, respectively. Here $j = 1, \ldots, k$ and $s_{k+1} = s_1$. Let $g : \mathbf{S}^1 \to \mathbf{R}^2$ be C^1, with $g' = 0$ precisely at the points s_j, where $j = 1, \ldots, k$. Set $\mathsf{g}_j = \mathsf{g}|_{I_j}$. We define the *rotation number of* g, denoted $R(\mathsf{g})$, by

$$R(\mathsf{g}) = \sum_{i=1}^{k} R(\mathsf{g}_j)\,.$$

Once again, such $R(\mathsf{g})$ may not exist, is not generally an integer, and does not enjoy the invariance under deformations.

We apply the preceding to the kinds of curves that arise as critical curves of locally stable maps of surfaces in $\mathbf{R}^2$. Suppose that $f : M \to \mathbf{R}^2$ is a locally stable map of a surface M into the plane. The critical set $\mathrm{Crit}(f)$ of f consists of a finite number of component curves c such that except for a finite set of points, $f|_c$ has a nonvanishing derivative. Those finite points are the cusp critical points of f on c and the other points are the fold critical points of f on c. By considering the local forms of f in a neighborhood of a cusp critical point p_0 (Section 14.1), we see that the limiting position of the tangent to the image of $f|_c$ near p_0, as p on c approaches p_0 from one side, is the negative of the limiting position of the tangent to the image of $f|_c$ as p approaches p_0 from the other side along c. That is,

$$\lim_{p\to p_0+} \frac{(f|_c)'}{|(f|_c)'|}(p) = -\lim_{p\to p_0-} \frac{(f|_c)'}{|(f|_c)'|}(p).$$

Here, we mean that both of the one sided limits exist and that they determine antipodal points on the unit circle. By $(f|_c)'$ we mean the derivative of f composed with any non-singular parametrization of c in a neighborhood of p_0. The statement above about the limiting positions of the tangents to $\mathrm{Crit}(f)$ at a cusp critical point is independent of the orientation of $\mathrm{Crit}(f)$ there. (We shall have more to say about the orientation of the $\mathrm{Crit}(f)$ curves in Section 14.3.) Moreover, the statement is pictorially clear: At a cusp caustic point the tangent line is well defined, but approaching and leaving the cusp give opposite senses to that line.

Proposition 14.1 *Let M be a surface and let $f : M \to \mathbf{R}^2$ be a locally stable map. Let c be a component of* $\mathrm{Crit}(f)$ *which is a closed curve on M. Then if c has k cusp critical points, then for any orientation of c the quantity $2R(f|_c)$ is an integer whose parity*[1] *is that of k.*

Proof: We consider the image of $f|_c$ near a cusp critical point p_0. Modify the curve by adding a teardrop curve falling from the cusp caustic point as in Figure 14.1. The resulting curve can be parametrized so that the tangents to the curve vary continuously. Such a teardrop curve contributes $\pm 1/2$ to the rotation number of the resulting curve.

Thus, if we modify the image of $f|_c$ in this way at each cusp critical point, then we know, calling the resulting curve h, that $R(h)$ is an integer. If there are j cusp critical points

[1] eveness or oddness

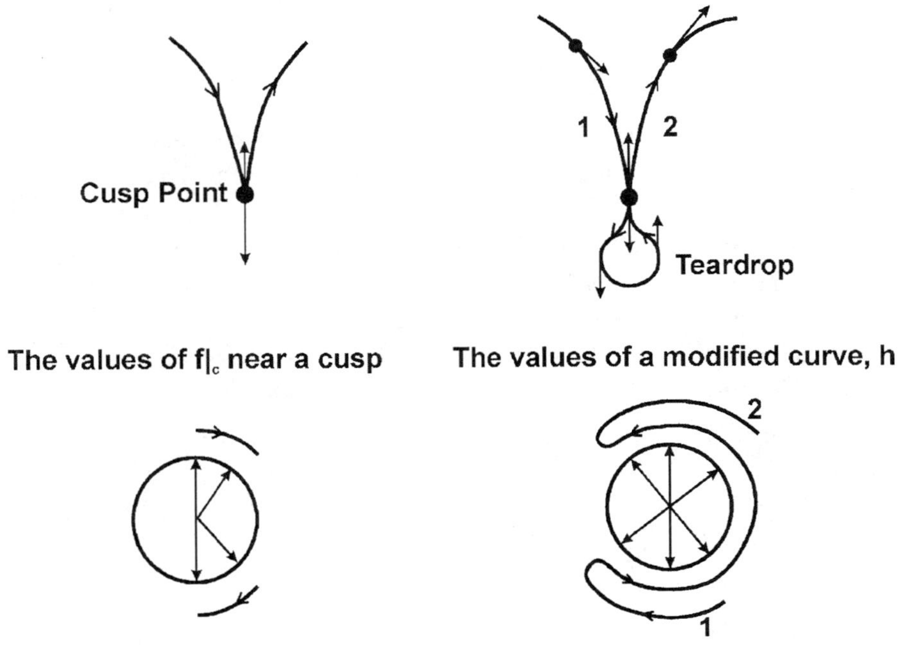

Figure 14.1. Motion of the unit tangent for the original curve (left) and the modified curve (right), which is obtained by adding a teardrop at each cusp.

that contribute $-\frac{1}{2}$ and $(k-j)$ cusp critical points that contribute $+\frac{1}{2}$ to $R(h)$, then

$$R(h) = R(f|_c) + \frac{1}{2}((k-j)-j) = R(f|_c) + \frac{1}{2}k - j. \tag{14.2}$$

Thus, $2R(f|_c) + k$ is even, so the parity of $2R(f|_c)$ is that of k. □

Motivated by Proposition 14.1, we make the following definition:

Definition 14.2 Given f and c as in Proposition 14.1, the *projectivized rotation number* of $f|_c$ is $2R(f|_c)$.

When it is clear which map f is meant, we shall denote the projectivized rotation number of $f|_c$ simply by $r(c)$.

Remark: The word "projectivized" is used since to compute a "projectivized rotation number," we regard a tangent vector and its negative as equal (see Remark at the end of this section).

We illustrate the previous rotation number notions in Figures 14.2 and 14.3. Notice that equation (14.2) applies.

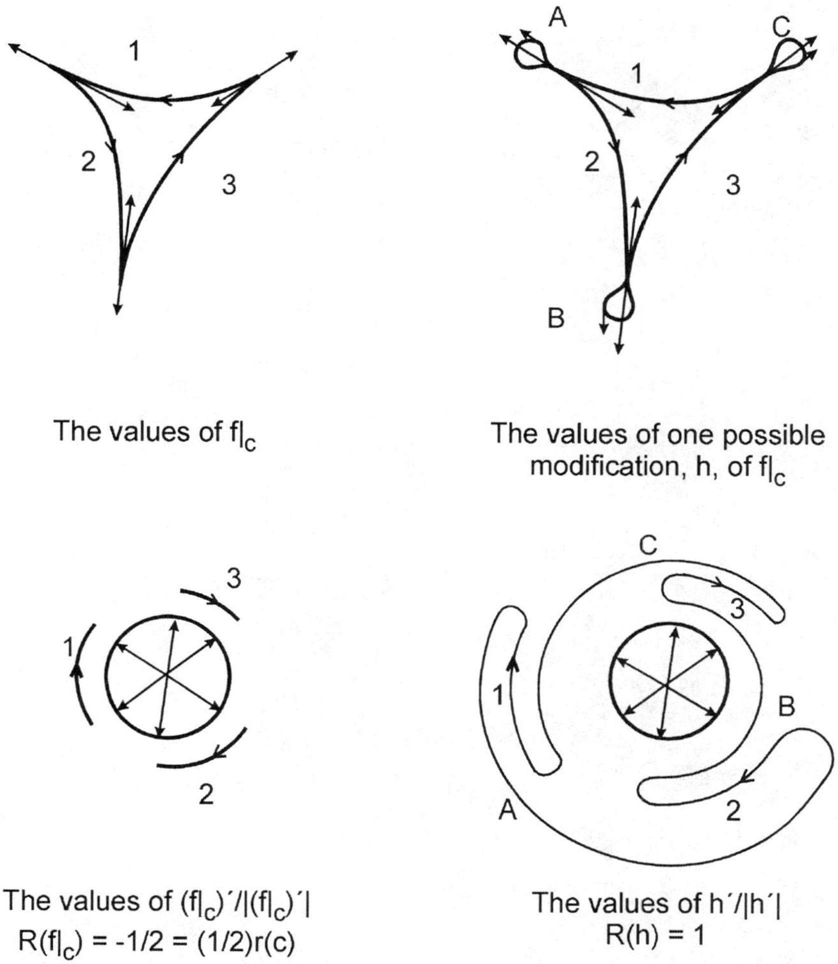

Figure 14.2. The hypocycloid caustic on the left has rotation number $-1/2$ and projectivized rotation number -1. The modified curve on the right has rotation number 1.

Definition 14.3 A C^1-mapping $g : \mathbf{S}^1 \to \mathbf{R}^2$ is called an *immersion with cusps* if g' vanishes only at a finite set of points, $s_1, \ldots, s_n$ at each of which the following holds:

$$\lim_{s \to s_i +} \frac{g'}{|g'|}(s) = - \lim_{s \to s_i -} \frac{g'}{|g'|}(s).$$

We now show that the projectivized rotation number $r(g) = 2R(g)$ of g is the Brouwer degree of a continuous map of $\mathbf{S}^1$ to itself. Let $h = g'/|g'| : \mathbf{S}^1 - \{s_1, \ldots, s_n\} \to \mathbf{S}^1$. Considering $\mathbf{S}^1$ as the unit circle in the complex plane $\mathbf{C}$, we let σ be the squaring map, which winds $\mathbf{S}^1$ twice around itself. The map $\sigma \circ h$ extends to a continuous map $k : \mathbf{S}^1 \to \mathbf{S}^1$.

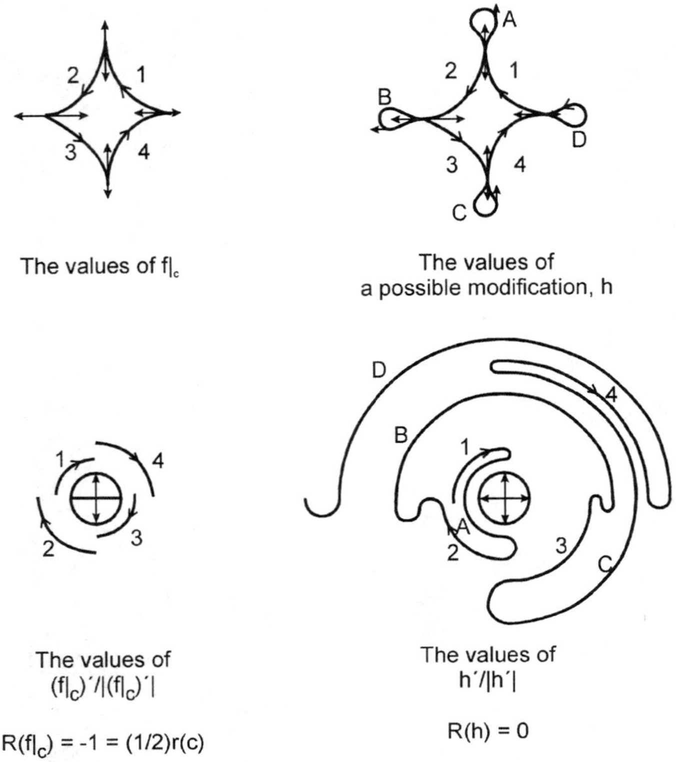

Figure 14.3. The astroid caustic on the left has rotation number -1 and projectivized rotation number -2, while the corresponding modified curve on the right has rotation number 0.

Proposition 14.4 *For* g *and* k *above, the projectivized rotation number* $r(\mathrm{g})$ *is the Brouwer degree of* k.

Proof: Let $s_1 < s_2 < \cdots < s_n < s_1$ be points, in order, on $\mathbf{S}^1$ and let $I_j = (s_j, s_{j+1})$ be the positively oriented, open arc on $\mathbf{S}^1$ with initial and final end points s_j and s_{j+1}, respectively. Here $j = 1, \ldots, n$ and $s_{n+1} = s_1$. With $\mathrm{g}_j = \mathrm{g}|_{I_j}$, the rotation number $R(\mathrm{g})$ of g is defined as

$$R(\mathrm{g}) = \sum_{j=1}^{n} R(\mathrm{g}_j)$$

and the projectivized rotation number is given by

$$r(\mathrm{g}) = 2R(\mathrm{g}) = \sum_{j=1}^{n} 2R(\mathrm{g}_j).$$

Notice that $2R(\mathrm{g}_j)$ is the total angle traversed by $k|_{I_j}$ and (since k is well defined and continuous on all of $\mathbf{S}^1$) $2R(\mathrm{g})$ is the total angle traversed by $k(s)$ as s goes once around $\mathbf{S}^1$ in the positive sense. Thus, $2R(\mathrm{g})$ is the Brouwer degree of k. □

According to Note (2) above, if k is differentiable, we can compute $r(\mathrm{g})$ by appropriately counting the pre-images of a regular value of k. However, suppose that for some point $z \in \mathbf{S}^1$, h is differentiable at every point $x \in k^{-1}(z)$ and $k'(x) \neq 0$. Then it is easy to see that, although k may not be differentiable on all of $\mathbf{S}^1$, we can still use this recipe for computing the degree of k. We continue to call such a point z a regular value of k. It is, however, most convenient to compute the degree of k in terms of the pre-images of $h = \mathrm{g}'/|\mathrm{g}'|$. Notice first that h is differentiable at $y \in \mathbf{S}^1$ if and only if k is differentiable and at such a y, $h'(y) \neq 0$ if and only if $k'(y) \neq 0$. Also, h is orientation preserving if and only if k is orientation preserving. For any point $z \in \mathbf{S}^1$, we have $\sigma^{-1}(z) = \pm y$, so $k^{-1}(z) = h^{-1}(\{y, -y\})$. If z is a regular value of k, both y and $-y$ are regular values of h.

To compute the Brouwer degree of k, we count all of the points in the h-preimage of $\{y, -y\}$, employing the usual sign convention: $(+1)$ for an orientation preserving point and (-1) for an orientation reversing point.

Remark: The projectivized rotation number may be cast in terms of the real projective line $P(\mathbf{R}^2)$. Define a map ϕ : Folds$(f) \to P(\mathbf{R}^2)$: $p \mapsto \phi(p)$, where $\phi(p)$ is the line through the origin of $\mathbf{R}^2$, parallel to the tangent to $f[\mathrm{Crit}(f)]$ at $f(p)$. Note that p is a fold critical point. The map ϕ extends to a smooth map Crit$(f) \to P(\mathbf{R}^2)$ [Lev66, p.359], which we also denote by ϕ. If c is an oriented component of Crit(f), then the Brouwer degree of $\phi|_c$ is precisely the *projectivized rotation number* $r(c)$ of c.

14.3 Orientation of Critical Curves

In what follows we make no assumption about the orientability of M, but always take $\mathbf{R}^2$ with a fixed orientation and assume that when coordinates (U, X) for $\mathbf{R}^2$ are given that $\frac{\partial}{\partial U} \wedge \frac{\partial}{\partial X}$ gives the orientation.

Consider the normal forms at a fold critical point:

$$\begin{cases} U \circ f(u,x) &= u \\ X \circ f(u,x) &= \pm x^2 . \end{cases}$$

By changing coordinates, if necessary, replacing u, x, U, X by their negatives, we may assume without loss of generality that the normal form at a fold critical point is

$$\begin{cases} U \circ f(u,x) &= u \\ X \circ f(u,x) &= x^2. \end{cases}$$

Near the origin of the domain of f in these coordinates the critical curve Crit(f) is given by $\{x = 0\}$. If we orient Crit(f) by $\frac{\partial}{\partial u}$, then $df(\partial/\partial u) = \partial/\partial U$ and orients the f-image of Crit(f) in such a way that the whole image of f locally lies to the left of $f(\mathrm{Crit}(f))$. This orientation makes sense and is coherent everywhere on Folds(f).

Claim: The above orientation on Folds(f) actually gives an orientation for all of Crit(f).

What we must show is that there is an orientation of Crit(f) at a cusp critical point that induces the above described orientation of the arcs of Folds(f) abutting the cusp critical point. We examine the normal forms of a cusp critical point with coordinates centered at the cusp point and at its f-image:

$$\begin{cases} U \circ f(u, x) = u \\ X \circ f(u, x) = axu + bx^3, \quad \text{where } |a| = |b| = 1. \end{cases}$$

N.B. Although these forms are only valid in a neighborhood of the cusp critical point and its image, for simplicity of language we shall discuss the map from $\mathbf{R}^2$ to $\mathbf{R}^2$ defined by these equations.

Again, by replacing all coordinates by their negatives, we lose no generality by considering just the two normal forms:

$$\begin{cases} U \circ f(u, x) = u \\ X \circ f(u, x) = -xu + bx^3 \quad \text{where } |b| = 1. \end{cases}$$

The equation for Crit(f) is $u = 3bx^2$, while the parametric equations for f(Crit(f)) are: $U = 3bx^2$, $X = -2bx^3$. Eliminating x we obtain that $U^3 = (27b/4)\,X^2$. The curve, $U^3 = (27b/4)\,X^2$, cuts the plane into two connected components. The component containing the half line, $\{X = 0,\ bU > 0\}$, is triply covered by f; the rest of the plane is singly covered by f (see page 333). Thus, if we orient Crit(f) so that its image is oriented as the boundary of the triply covered region, we know that locally, at each point of Folds(f), the image of f lies to the left of f(Crit(f)) as required. The required orientation of Crit(f) in both of these cases is given by $\partial/\partial x$ at the origin. We have thus proven:

Proposition 14.5 *For any locally stable map $f : M \to \mathbf{R}^2$, there is a unique orientation of* Crit(f) *such that at each point p of* Folds(f), *the function f has the normal form:*

$$\begin{cases} U \circ f(u, x) = u \\ X \circ f(u, x) = x^2 \end{cases}$$

where the orientation of Crit(f) *is given by $\frac{\partial}{\partial u}$ and $\frac{\partial}{\partial U} \wedge \frac{\partial}{\partial X}$ is the orientation of* $\mathbf{R}^2$.

14.4 A Formula for the Euler Characteristic

The following presents a formula due to H.L [Lev66] for computing the Euler characteristic in terms of projectivized rotation numbers:

Theorem 14.6 *Let M be any compact surface without boundary and let $f : M \to \mathbf{R}^2$ be a locally stable mapping. Orient the critical set* Crit(f) *of f so that, except for finite set of cusp critical points* Cusps(f), *locally the f-image of M lies to the left of the f-image of* Crit(f). *Then*

$$\chi(M) = \sum_C r(C),$$

where the sum runs over all components C of Crit(f) *and $r(C)$ is the projectivized rotation number of $f|_C$.*

A proof of Theorem 14.6 appears at the end of this section.

Later, we shall study the global geometry of gravitational lensing caustics (see page 554). This will require a generalization of Theorem 14.6 to surfaces with boundary [Lev-PW], [Lev95]:

Theorem 14.7 *Let M be a compact, two-dimensional manifold with boundary ∂M and let $f : M \to \mathbf{R}^2$ be a locally stable map with critical set* Crit(f) $= \bigcup_i c_i$, *where c_i are circles embedded in M disjoint from each other and ∂M. Let ∂M be a union of component circles b_j each of which is immersed by f. Then*

$$\chi(M) = \sum_i r(c_i) + \sum_j R(b_j),$$

where $r(c_i)$ is the projectivized rotation number of $f|_{c_i}$ relative to the natural orientation of $f(c_i)$ and $R(b_j)$ is the rotation number of $f|_{b_j}$ determined with $f(b_j)$ oriented as boundary, locally, of the image under f of its collar neighborhood.

Proof: To compute $\chi(M)$, we first compute $\chi(2M)$, where $2M$ is the *double* of M. To construct $2M$, consider the disjoint union

$$M_1 \dot{\bigcup} (\partial \mathrm{M}_1 \times [-1, 1]) \dot{\bigcup} \mathrm{M}_2,$$

where $\mathrm{M}_1 = M$ and $\mathrm{M}_2 = M$. Identify $x \in \partial \mathrm{M}_1$ and $x \in \partial \mathrm{M}_2$ with the points $(x, -1)$ and $(x, 1)$, resp., in $\partial \mathrm{M}_1 \times [-1, 1]$. The resulting quotient space, which we denote by $2M$, is a smooth, compact, 2-manifold without boundary. The manifold $2M$ is the quotient of $M \times \{-1, 1\}$ by the relation $(x, t) \sim (x', t')$ if and only if $x = x'$ and either $t = t'$ or $x \in \partial M$. Let us explicitly describe the differentiable structure assigned to $2M$. Define bijections $f_\pm : [M, \pm 1] \to M$ by $f_\pm(x, \pm) = x$. Next, recall that by the *collaring theorem* (e.g., [Broc-J, p. 132]), M has a *collar*, that is, there is a diffeomorphism k from $\partial M \times [0, 1)$ onto an open neighborhood V of ∂M in M with $k(x, 0) = x$ (which is the inclusion map on

∂M). Consider the set $2V$ and define a bijection $f_k : \partial M \times (-1, 1) \to 2V$ by $f_k(x, s) = [k(x, |s|), \frac{s}{|s|}]$ for $s \neq 0$ and $f_k(x, 0) = [x, \pm 1]$. Requiring that the maps f_+, f_-, and f_k be diffeomorphisms induces a differentiable structure on $2M$ that is compatible with the copies M_1 and M_2 of M as submanifolds of M.

Next, we compute $\chi(2M)$ using Theorem 14.6. Let $h : (-1, 1) \to [0, 1]$ be any smooth even function satisfying

$$h(s) = \begin{cases} s^2, & 0 \leq |s| < \epsilon < 1/2, \\ |s|, & 1 - \epsilon < |s|, \\ s\,h'(s) > 0, & s \neq 0. \end{cases}$$

Then define a smooth map $\hat{f} : 2M \to \mathbf{R}^2$ at points outside and inside, respectively, the double collar neighborhood of ∂M, by

$$\hat{f}([x, t]) = f(x), \qquad \text{for } x \notin k(\partial M \times [0, 1 - \epsilon]),$$

and

$$\hat{f}([k(x, s), t]) = f(k(x, h(s))), \qquad \text{for } (x, s) \in \partial M \times [0, 1 - \epsilon/2).$$

Outside the double collar neighborhood of ∂M, the map $\hat{f}$ is locally stable. On $\partial M \times (-\epsilon, \epsilon)$, we have

$$\hat{f}(f_k(x, s)) = f(k(x, s^2)).$$

A technical coordinate argument then shows that $\hat{f}$ can be transformed into the normal form for local stability (see [Lev95, pp. 2565–2567] for details). Applying Theorem 14.6 to $\hat{f}$ yields

$$\chi(2M) = 2\sum_i r(c_i) + \sum_j r(b_j),$$

where i and j run through the (finitely many) components of $\mathrm{Crit}(f)$ and ∂M, respectively. The reason that each $r(c_i)$ appears twice is that in M, in the notation representing $2M$ as a quotient of $M \times \{1, -1\}$, the curve c_i appears as $c_i \times \{-1\}$ and $c_i \times \{1\}$. Both copies of c_i are oriented in the same way, giving equal contributions to $\chi(2M)$. In addition, since each component of the boundary produces an immersed circle as a fold critical curve (i.e., no cusp critical points appear on b_j), $r(b_j) = 2R(b_j)$. Hence,

$$\chi(2M) = 2(\sum_i r(c_i) + \sum_i R(b_j)).$$

On the other hand,

$$\chi(2M) = 2\chi(M) - \chi(\partial M)$$

(e.g., [Gre-H, p. 240]). But $\chi(\partial M) = 0$ (since ∂M is odd dimensional). The theorem follows. □

Proof of Theorem 14.6: Let $f : M \to \mathbf{R}^2$ be locally stable. We now choose a linear function ς on $\mathbf{R}^2$ such that $\varsigma \circ f$ has only nondegenerate critical points all of which are in Folds(f). For any such f, almost all linear functions have those properties [Kui, p.79]. Let ξ be a unit vector in $\mathbf{R}^2$ in the kernel of ς and such that ς increases to the left of ξ. In other words, $\xi \wedge \frac{\partial}{\partial \varsigma}|_0$ is the orientation of $\mathbf{R}^2$. Here we identify $\mathbf{R}^2$ with its tangent space at the origin.

Henceforth, we take Crit(f) with the usual orientation described in Proposition 14.5. For any component c of Folds(f), we compute $(f|_c)'$ using any parametrization of c giving that orientation. This determines $(f|_c)'$ at a point of c only up to a positive factor. Such an ambiguity is eliminated when we compute $\frac{(f|_c)'}{|(f|_c)'|}$. For convenience denote $\frac{(f|_c)'}{|(f|_c)'|}$ by g_c.

For $\tau \in \{0, 1, 2\}$ and c a component of Folds(f), let $\mathfrak{N}_c^{\pm}(\tau)$ equal the number of points $p \in c$ at which $\varsigma \circ f$ is singular with index τ and $g_c(p) = \pm\xi$. Suppose $\mathbf{S}^1 \subseteq \mathbf{R}^2$ is oriented as usual. Let c be an oriented component of Folds(f) and $\beta : c \to \mathbf{S}^1$ be differentiable. For any regular value z of β, define $n(z; \beta)$ to be the signed number of β pre-images of z, counted as usual: $(+1)$ where β is orientation preserving and (-1) where β is orientation reversing.

Lemma 14.8 *Let c be an oriented component of* Folds(f) *Then:*

(1) *The points $\pm\xi$ are regular values of g_c.*

(2) $n(+\xi; g_c) = \mathfrak{N}_c^+(0) - \mathfrak{N}_c^+(1)$.

(3) $n(-\xi; g_c) = \mathfrak{N}_c^-(2) - \mathfrak{N}_c^-(1)$.

Proof: For simplicity let "iff" mean "if and only if."

(1) Note that $g_c(p) = \pm\xi$ iff $(f|_c)'(p)$ is a multiple of ξ iff $\varsigma \circ (f|_c)'(p) = 0$ iff $[(\varsigma \circ f)|_c]'(p) = 0$. Thus $g_c(p) = \pm\xi$ iff p is a singular point of $(\varsigma \circ f)|_c$. To check that $\pm\xi$ are regular values of g_c, we must check that $(g_c)'(p) \neq 0$. But since $\pm|g_c| \equiv 1$ on c, we know that if $g_c(p) = \pm\xi$, then $(g_c)'(p) \neq 0$ iff $\varsigma \circ (g_c)'(p) \neq 0$ iff $(\varsigma \circ g_c)'(p) \neq 0$. But $(\varsigma \circ g_c)'(p) \neq 0$ iff $[\varsigma \circ (f|_c)']'(p) \neq 0$ iff $[(\varsigma \circ f)|_c]''(p) \neq 0$. Thus we see that $\pm\xi$ is a regular value of g_c iff $(\varsigma \circ f)|_c$ has only nondegenerate singularities which is automatic by virtue of our choice of ς.

(2–3) Now suppose that $g_c(p) = \pm\xi$, which we know is equivalent to p being a nondegenerate singularity of $(\varsigma \circ f)|_c$. But since $p \in c \subseteq$ Folds(f) and all singularities of $\varsigma \circ f$ are contained in Folds(f), we know that this condition, in turn, is equivalent to p being a nondegenerate singularity of $\varsigma \circ f$. It only remains to determine the index of $\varsigma \circ f$ at such a point p. Working in local coordinates u, x, U, X centered at p and $f(p)$ which put f in normal form (2), we have $g_c(0, 0) = (1, 0)$. Suppose that

$$\varsigma(U, X) = aX + bU + eU^2 + h(U, X),$$

where h involves UX, X^2 and terms of order greater than two. Then

$$\varsigma \circ f(u,x) = \varsigma(u,x^2) = ax^2 + bu + eu^2 + h(u,x^2).$$

The order of $h \circ f(u,x)$ is greater than two in u and x. Since we are assuming that $(0,0)$ is a singular point of $\varsigma \circ f$ in these coordinates, we know that $b = 0$, and since the critical points of $\varsigma \circ f$ are nondegenerate, we know that $ae \neq 0$. Since $\varsigma \circ f$ and $(\varsigma \circ f)/|a|$ have the same critical points with the same indices we may assume that $a = \pm 1$ and so, $\varsigma \circ f(u,x) = \pm x^2 + eu^2 + h(u,x^2)$, and $\xi = (\pm 1, 0)$, as $a = \pm 1$. In these coordinates, the equation of $\mathrm{Crit}(f)$ is $x = 0$ and hence we have $\varsigma \circ f(u,0) = eu^2 + h(u,0)$, so $\varsigma \circ f|_c$ has a minimum if $e > 0$ and a maximum if $e < 0$. Hence, the points for which $g_c(0,0) = \xi$ are those for which $\varsigma \circ f(u,x) = x^2 + eu^2 + h(u,x^2)$ and g_c is orientation preserving (resp., reversing) when $e > 0$ (resp., $e < 0$). Thus the points of index 0 are counted positively and those of index 1, negatively. This gives

$$n(+\xi;\ g_c) = \mathfrak{N}_c^+(0) - \mathfrak{N}_c^+(1).$$

Similarly, the points for which $g_c(0,0) = -\xi$ are those for which

$$\varsigma \circ f(u,x) = -x^2 + eu^2 + h(u,x^2)$$

and g_c is orientation preserving (resp., reversing) when $e < 0$(resp. $e > 0$). Thus the points of index 2 are counted positively and those of index 1, negatively. This gives

$$n(-\xi;\ g_c) = \mathfrak{N}_c^-(2) - \mathfrak{N}_c^-(1).$$ □ **(Lemma 14.8)**

Remark. Points of index 0 or 2 arise only when $g_c = \xi$ or $-\xi$ respectively. So we can write simply $\mathfrak{N}_c(0)$ and $\mathfrak{N}_c(2)$ unambiguously.

By the paragraph after Proposition 14.4, the projectivized rotation number of any component C of $\mathrm{Crit}(f)$ is

$$r(C) = \sum_c (n(+\xi;\ g_c) + n(-\xi;\ g_c)),$$

where the sum runs over all of the components of $C - \mathrm{Cusps}(f)$.
Setting $\mathfrak{N}_c(1) = \mathfrak{N}_c^+(1) + \mathfrak{N}_c^-(1)$, Lemma 14.8 yields

$$r(C) = \sum_c (\mathfrak{N}_c(0) - \mathfrak{N}_c(1) + \mathfrak{N}_c(2)),$$

where the sum runs over all of the components of $C - \mathrm{Cusps}(f)$. Let $N(\tau)$ be the number of critical points of $\varsigma \circ f$ of index τ, where $\tau \in \{0,1,2\}$. Then:

$$N(0) = \sum_C \sum_c N_c(0),$$
$$N(1) = \sum_C \sum_c N_c(1),$$
$$N(2) = \sum_C \sum_c N_c(2).$$

Here, for each component C of $\mathrm{Crit}(f)$, we sum over all components c of $C - \mathrm{Cusps}(f)$ and then sum over all components C of $\mathrm{Crit}(f)$. By the Morse equality (14.1), we have that

$$\chi(M) = N(2) - N(1) + N(0) = \sum_C \sum_c [N_c(0) - N_c(1) + N_c(2)].$$

But that gives the desired result: $\chi(M) = \sum_C r(C)$. □ **(Theorem 14.6)**

Chapter 15

Global Geometry of Caustics

Geometry is to open my mind
so that I may see what has always been behind
the illusions that time and space construct.

DAVID HENDERSON

This chapter investigates the global geometric properties of caustics and critical curves due to single and multiple plane gravitational lensing, including links between a gravitational lens and the curvature of its caustics. In other words, the chapter addresses Problems **M3**, **M4**, and **M5**.

We shall use complex variables to investigate the number of caustics and geometric structures like cusps, beak-to-beaks, swallowtails, and elliptic umbilics due to g point masses with continuous matter κ_c and shear γ from infinity. For example, at most $3g - 3$ beak-to-beaks occur [Wit-P]. Moreover, increasing κ_c sufficiently large eliminates all non-fold, caustic points [Pet-Wit] and generates g oval, fold, caustic curves [Pet95b]. A global classification of caustics will also be given in certain cases. In addition, some of the global geometric features of caustics due to two point masses with continuous matter and shear from infinity will be contrasted for single and double-plane lensing. Furthermore, the light path obstruction points B of a general gravitational lens will also be studied and linked to the total signed-curvature K_f of its fold caustics [Pet95b, p. 4292]. Explicitly, we shall see that $K_f = -2\pi|\mathsf{B}|$, where $|\mathsf{B}|$ is the total number of obstructions. In addition, the signed-curvature K_{tot} of all the caustics is $K_{tot} = -2\pi|\mathsf{B}| + \pi N_{cusps}$, where N_{cusps} is the number of cusps. Bounds will also be given on $|\mathsf{B}|$ (hence K_f) and K_{tot}.

Outline: Section 15.1 formulates some of the basic equations for critical points of lensing maps in terms of complex variables. In Sections 15.2 and 15.3, these results are applied to the case of point masses on a lens plane with continuous matter and shear from infinity. Section 15.2 treats in some detail the number of caustics, cusps, and caustic metamorphoses due to a Chang–Refsdal lens (i.e., single point mass with continuous matter and shear) and binary point mass lenses (including continuous matter and shear) — see Sections 15.2.1 and 15.2.2. We also contrast the global properties of caustics for single and double-plane lensing by binary point masses (Section 15.2.2), and under certain limiting situations interpret single-plane binaries as a point mass plus Chang–Refsdal lens (Section 15.2.3). For any finite number of point masses on a lens plane with continuous matter and shear, Section 15.3 considers the elimination of non-fold caustics (Section 15.3.1) and upper bounds on the number of caustics and their metamorphoses (Section 15.3.2), including cusps (Section 15.3.3). Section 15.4 discusses light path obstruction points (Section 15.4.1) and their link to the signed-curvature of caustics (Section 15.4.2). This section applies the lensed-image counting results of Chapter 12 and the work of Chapter 14 on computing the Euler characteristic via projectivized rotation numbers.

15.1 Critical Points and Lens Equation in Complex Form

For the case of lensing due to point masses with continuous matter and shear from infinity, complex variables were used[1] to determine upper bounds on the total number of lensed images of a light source. The complex formulation will also provide a convenient useful tool for obtaining global information about caustics of such lens systems. This section casts in complex form some of the basic quantities used to study the critical points of a lensing map — see Sections 15.2 and 15.3 for applications.

General k-Plane Case

Consider the set of critical points of a k-plane lensing map $\boldsymbol{\eta}_t : P \to S$ that depends on n parameters $t \in \mathbf{R}^n$:

$$\mathrm{Crit}(\boldsymbol{\eta}_t) = \{\mathbf{x} \in P \subseteq \mathbf{R}^2 : J_t(\mathbf{x}) = 0\},$$

where

$$J_t(\mathbf{x}) = \det(\mathrm{Jac}\,\boldsymbol{\eta}_t)(\mathbf{x}).$$

For parameter values t and critical points $\mathbf{x}$ such that $(\mathrm{grad}\, J_t)(\mathbf{x}) \neq \mathbf{0}$, there is a neighborhood U of $\mathbf{x}$ where the critical points in U form a smooth simple arc passing through $\mathbf{x}$. However, if $(\mathrm{grad}\, J_t)(\mathbf{x})$ is initially nonzero and vanishes when t attains the value t_0, then $\mathrm{Crit}(\boldsymbol{\eta}_{t_0})$ fails to be a smooth curve at the point $\mathbf{x}$. Such parameter values then signal a change in the geometry of $\mathrm{Crit}(\boldsymbol{\eta}_t)$. In general, if the set of critical points forms a family of disjoint smooth simple critical curves, then a change in their geometry occurs locally at points $\mathbf{x}$ characterized by

$$J_t(\mathbf{x}) = 0, \qquad (\mathrm{grad}\, J_t)(\mathbf{x}) = \mathbf{0}. \tag{15.1}$$

[1] See Sections 11.5.2 (page 439) and 12.3.2 (page 459).

We shall see below that global changes in the topology of Crit($\boldsymbol{\eta}_t$) can arise when a simple closed critical curve suddenly becomes unbounded.

In this section, we are also interested in counting the number of cusp points of $\boldsymbol{\eta}_t$. These points are characterized via directional derivatives by the following equations (Table 9.1, page 381):

$$J_t(\mathbf{x}) = 0, \quad (\text{Jac}\,\boldsymbol{\eta}_t)(\mathbf{x}) \neq \mathbf{O}, \quad (\text{grad}\,J_t)(\mathbf{x}) \neq \mathbf{0},$$

$$(\nabla_{Z_t}\boldsymbol{\eta}_t)(\mathbf{x}) = \mathbf{0}, \quad (\nabla^{(2)}_{Z_t}\boldsymbol{\eta}_t)(\mathbf{x}) \neq \mathbf{0},$$

where

$$Z_t = -\frac{\partial J_t}{\partial \mathrm{v}}\frac{\partial}{\partial \mathrm{u}} + \frac{\partial J_t}{\partial \mathrm{u}}\frac{\partial}{\partial \mathrm{v}}$$

relative to rectangular coordinates (u, v) on the plane. At points $\mathbf{x}$ where $Z_t(\mathbf{x}) \neq \mathbf{0}$, the vector field Z_t is tangent to the critical curve $J_t = 0$. In addition, since

$$\nabla_{Z_t}\boldsymbol{\eta}_t = (\text{Jac}\,\boldsymbol{\eta}_t)(Z_t),$$

the vector field $\nabla_{Z_t}\boldsymbol{\eta}_t$ is tangent to the caustic curve at points $\boldsymbol{\eta}_t(\mathbf{x})$ with $Z_t(\mathbf{x}) \neq \mathbf{0}$.

It is difficult to find analytically the points $\mathbf{x}$ that obey all the equations characterizing cusp points. For this reason, we consider only the equations

$$J_t(\mathbf{x}) = 0, \qquad (\nabla_{Z_t}\boldsymbol{\eta}_t)(\mathbf{x}) = \mathbf{0}. \tag{15.2}$$

The solutions of (15.2) include not only cusp points, but also degenerate critical points like lips, beak-to-beaks, swallowtails, elliptic umbilics, and hyperbolic umbilics (Table 9.1, page 381).

Notation: For simplicity, we shall suppress the parameter t, writing $\boldsymbol{\eta}$, J, and Z in place of $\boldsymbol{\eta}_t$, J_t and Z_t, respectively.

Let us now express J, grad J, and $\nabla_Z\boldsymbol{\eta}$ in complex form. Using rectangular coordinates (u, v) on the complex plane $\mathbf{C}$, introduce the Wirtinger operators:

$$\frac{\partial}{\partial z} = \frac{1}{2}\left(\frac{\partial}{\partial \mathrm{u}} - i\,\frac{\partial}{\partial \mathrm{v}}\right), \quad \frac{\partial}{\partial \bar{z}} = \frac{1}{2}\left(\frac{\partial}{\partial \mathrm{u}} + i\,\frac{\partial}{\partial \mathrm{v}}\right).$$

These operators act as follows on a smooth function $F : U \subseteq \mathbf{C} \to \mathbf{C}$, where U is open and $F = A + iB$:

$$\frac{\partial F}{\partial z} = \frac{1}{2}\left(\frac{\partial F}{\partial \mathrm{u}} - i\,\frac{\partial F}{\partial \mathrm{v}}\right) = \frac{1}{2}\left[\left(\frac{\partial A}{\partial \mathrm{u}} + \frac{\partial B}{\partial \mathrm{v}}\right) + i\left(-\frac{\partial A}{\partial \mathrm{v}} + \frac{\partial B}{\partial \mathrm{u}}\right)\right] \tag{15.3}$$

and

$$\frac{\partial F}{\partial \bar{z}} = \frac{1}{2}\left(\frac{\partial F}{\partial \mathrm{u}} + i\,\frac{\partial F}{\partial \mathrm{v}}\right) = \frac{1}{2}\left[\left(\frac{\partial A}{\partial \mathrm{u}} - \frac{\partial B}{\partial \mathrm{v}}\right) + i\left(\frac{\partial A}{\partial \mathrm{v}} + \frac{\partial B}{\partial \mathrm{u}}\right)\right]. \tag{15.4}$$

Equations (15.3) and (15.4) yield

$$\frac{\partial \bar{F}}{\partial z} = \overline{\frac{\partial F}{\partial \bar{z}}}, \qquad \frac{\partial \bar{F}}{\partial \bar{z}} = \overline{\frac{\partial F}{\partial z}} \tag{15.5}$$

and

$$\det(\text{Jac}\, F) = \det \begin{bmatrix} \frac{\partial F}{\partial z} & \frac{\partial F}{\partial \bar{z}} \\ \frac{\partial \bar{F}}{\partial z} & \frac{\partial \bar{F}}{\partial \bar{z}} \end{bmatrix} = \left|\frac{\partial F}{\partial z}\right|^2 - \left|\frac{\partial F}{\partial \bar{z}}\right|^2. \tag{15.6}$$

When F is real-valued, that is, $F = A$, we have

$$\overline{\frac{\partial A}{\partial \bar{z}}} = \frac{\partial A}{\partial z}, \qquad \text{grad}\, A = \frac{\partial A}{\partial \mathrm{u}} + i\frac{\partial A}{\partial \mathrm{v}} = 2\frac{\partial A}{\partial \bar{z}}. \tag{15.7}$$

Viewing the real-valued function J and the lensing map $\boldsymbol{\eta}$ as functions of $(z, \bar{z})$, equations (15.5)–(15.7) yield

$$J = \det(\text{Jac}\, \boldsymbol{\eta}) = \left|\frac{\partial \boldsymbol{\eta}}{\partial z}\right|^2 - \left|\frac{\partial \boldsymbol{\eta}}{\partial \bar{z}}\right|^2, \qquad \text{grad}\, J = 2\frac{\partial J}{\partial \bar{z}}, \tag{15.8}$$

and

$$Z = -\frac{\partial J}{\partial \mathrm{v}} + i\frac{\partial J}{\partial \mathrm{u}} = 2i\frac{\partial J}{\partial \bar{z}}.$$

Setting $\boldsymbol{\eta} = a + ib$ and using the leftmost equality in (15.7), we see that the dot product $Z \cdot (\text{grad}\, a)$ can be written as follows:

$$Z \cdot (\text{grad}\, a) = 2\left(Z \cdot \frac{\partial a}{\partial \bar{z}}\right) = 2\text{Re}\left\{\bar{Z}\,\frac{\partial a}{\partial \bar{z}}\right\} = \overline{\frac{\partial a}{\partial \bar{z}}} Z + \frac{\partial a}{\partial \bar{z}}\bar{Z} = \frac{\partial a}{\partial z} Z + \frac{\partial a}{\partial \bar{z}}\bar{Z}.$$

A similar result holds for $Z \cdot (\text{grad}\, b)$. Hence

$$\nabla_Z \boldsymbol{\eta} = (Z \cdot \text{grad}\, a) + i(Z \cdot \text{grad}\, b) = \frac{\partial \boldsymbol{\eta}}{\partial z}\, Z + \frac{\partial \boldsymbol{\eta}}{\partial \bar{z}}\, \bar{Z}. \tag{15.9}$$

Single Lens Plane Case

We now restrict to single-plane lensing and connect with the complex variable formulation due to Witt [Wit90]. For a single-plane lensing situation with gravitational lens potential ψ, we have

$$\boldsymbol{\eta}(\mathbf{x}) = \mathbf{x} - \boldsymbol{\alpha}(\mathbf{x}), \qquad \text{Jac}\, \boldsymbol{\eta} = \begin{bmatrix} \frac{\partial a}{\partial \mathrm{u}} & \frac{\partial a}{\partial \mathrm{v}} \\ \frac{\partial b}{\partial \mathrm{u}} & \frac{\partial b}{\partial \mathrm{v}} \end{bmatrix},$$

where $\mathbf{x} = (\mathrm{u}, \mathrm{v})$, $\boldsymbol{\alpha}(\mathbf{x}) = \operatorname{grad} \psi(\mathbf{x})$, $a = \mathrm{u} - \partial\psi/\partial\mathrm{u}$, and $b = \mathrm{v} - \partial\psi/\partial\mathrm{v}$. It follows that

$$\frac{\partial a}{\partial \mathrm{v}} = \frac{\partial b}{\partial \mathrm{u}}. \tag{15.10}$$

In other words, the Jacobian matrix Jac $\boldsymbol{\eta}$ is symmetric, a property which can fail to hold for more than one lens plane.

Converting to the complex form $\boldsymbol{\eta} = a + ib$, equations (15.3) and (15.10) yield

$$\frac{\partial \boldsymbol{\eta}}{\partial z} = \frac{1}{2}\left(\frac{\partial a}{\partial \mathrm{u}} + \frac{\partial b}{\partial \mathrm{v}}\right) = \text{real function}. \tag{15.11}$$

The critical points of $\boldsymbol{\eta}$ are then determined by

$$0 = J = \left(\frac{\partial \boldsymbol{\eta}}{\partial z}\right)^2 - \frac{\partial \boldsymbol{\eta}}{\partial \bar{z}}\overline{\frac{\partial \boldsymbol{\eta}}{\partial \bar{z}}}. \tag{15.12}$$

Since $\partial\boldsymbol{\eta}/\partial z$ is real, equation (15.12) is solved by

$$\frac{\partial \boldsymbol{\eta}}{\partial \bar{z}} = \left|\frac{\partial \boldsymbol{\eta}}{\partial z}\right| e^{i\varphi}, \qquad \varphi \in [0, 2\pi). \tag{15.13}$$

Hence, the critical points of $\boldsymbol{\eta}$ are the points z such that (15.13) holds. Due to the phase angle φ, critical points obtained via (15.13) are said to have a *parametric representation*.

Taking the gradient of J in (15.12), we get

$$\operatorname{grad} J = 2\frac{\partial \boldsymbol{\eta}}{\partial z}\frac{\partial^2 \boldsymbol{\eta}}{\partial z \partial \bar{z}} - \frac{\partial \boldsymbol{\eta}}{\partial \bar{z}}\frac{\partial}{\partial \bar{z}}\left(\overline{\frac{\partial \boldsymbol{\eta}}{\partial \bar{z}}}\right) - \overline{\frac{\partial \boldsymbol{\eta}}{\partial \bar{z}}}\frac{\partial^2 \boldsymbol{\eta}}{\partial \bar{z}^2}.$$

By (15.3), (15.4), and (15.10), we get

$$\frac{\partial^2 \boldsymbol{\eta}}{\partial z \partial \bar{z}} = \overline{\frac{\partial^2 \boldsymbol{\eta}}{\partial z^2}}, \qquad \frac{\partial}{\partial \bar{z}}\left(\overline{\frac{\partial \boldsymbol{\eta}}{\partial \bar{z}}}\right) = \frac{\partial^2 \boldsymbol{\eta}}{\partial z^2}.$$

Hence

$$\operatorname{grad} J = 2\frac{\partial \boldsymbol{\eta}}{\partial z}\overline{\frac{\partial^2 \boldsymbol{\eta}}{\partial z^2}} - \frac{\partial \boldsymbol{\eta}}{\partial \bar{z}}\frac{\partial^2 \boldsymbol{\eta}}{\partial z^2} - \overline{\frac{\partial \boldsymbol{\eta}}{\partial \bar{z}}}\frac{\partial^2 \boldsymbol{\eta}}{\partial \bar{z}^2}. \tag{15.14}$$

Finally, the single lens plane version of the directional derivative of $\boldsymbol{\eta}$ with respect to Z is

$$\nabla_Z \boldsymbol{\eta} = \frac{\partial \boldsymbol{\eta}}{\partial z} Z + \frac{\partial \boldsymbol{\eta}}{\partial \bar{z}} \bar{Z}, \qquad \text{where } \frac{\partial \boldsymbol{\eta}}{\partial z} \text{ is real and } Z = 2i \operatorname{grad} J. \tag{15.15}$$

Here grad J is given by (15.14).

Remark: Equations (15.12), (15.14), and (15.15) agree with the (single-plane) complex formulas found by Witt [Wit90, p. 311–313; eqs. (7), (20), (21)].

Complex Lens Equation and Critical Points due to Point Masses

This subsection restricts to the special case of single-plane lensing by point masses with continuous matter and shear from infinity. It expresses in complex form the lens equation and some of the basic quantities needed to study the caustics and critical curves due to this lens system.

The following is the lens equation due to g point masses $m_1, \ldots, m_g$ at positions $\boldsymbol{\xi}_1, \ldots, \boldsymbol{\xi}_g$ on a lens plane with continuous matter κ_c and shear γ from infinity along the u-axis (see Section 3.3.1, page 101):

$$\mathbf{y} = \boldsymbol{\eta}_{ptcs}(\mathbf{x}) = \begin{bmatrix} 1-\kappa_c+\gamma & 0 \\ 0 & 1-\kappa_c-\gamma \end{bmatrix}\mathbf{x} - \sum_{\ell=1}^{g} m_\ell \frac{\mathbf{x}-\boldsymbol{\xi}_\ell}{|\mathbf{x}-\boldsymbol{\xi}_\ell|^2}, \tag{15.16}$$

where $\mathbf{x} = (\mathrm{u}, \mathrm{v})$. The parameters of this system are

$$t = (\kappa_c, \gamma, m_1, \ldots, m_g, \boldsymbol{\xi}_1, \ldots, \boldsymbol{\xi}_g),$$

where $\kappa_c \geq 0$, $\gamma \geq 0$, $m_\ell > 0$, and $\boldsymbol{\xi}_\ell \in \mathbf{R}^2$ for $\ell = 1, \ldots, g$. If $0 \leq \kappa_c < 1$, then $\boldsymbol{\eta}_{ptcs}$ is said to be *under-focused*, while for $\kappa_c > 1$ we call $\boldsymbol{\eta}_{ptcs}$ *over-focused.* It is sometimes convenient to express this in terms of

$$\varepsilon_c = \ \mathrm{sign}(1-\kappa_c),$$

where $\varepsilon_c = +1$ (resp., $\varepsilon_c = -1$) designates the under-focused (resp., over-focused) case.

Note. The lensing map $\boldsymbol{\eta}_{ptcs}$ can produce beak-to-beak, swallowtail, and elliptic umbilic metamorphoses, but not lips and hyperbolic umbilics (see Section 9.5.3, page 383).

Setting

$$\mathbf{y} = y_1 + iy_2 = \zeta, \qquad \mathbf{x} = \mathrm{u} + i\mathrm{v} = z, \qquad \boldsymbol{\xi}_\ell = \xi^1_\ell + i\xi^2_\ell = z_\ell,$$

the lens equation (15.16) converts into complex form as follows:

$$\zeta = (1-\kappa_c)z + \gamma\bar{z} + \sum_{\ell=1}^{g} \frac{m_\ell}{\bar{z}_\ell - \bar{z}}. \tag{15.17}$$

We have that

$$\frac{\partial\zeta}{\partial z} = 1-\kappa_c, \qquad \frac{\partial\zeta}{\partial\bar{z}} = \gamma + \sum_{\ell=1}^{g} \frac{m_\ell}{(\bar{z}_\ell - \bar{z})^2}, \qquad \frac{\partial^2\zeta}{\partial\bar{z}^2} = 2\sum_{\ell=1}^{g} \frac{m_\ell}{(\bar{z}_\ell - \bar{z})^3}. \tag{15.18}$$

Equation (15.13) then transforms into the following equation for the critical points of ζ:

$$\gamma + \sum_{\ell=1}^{g} \frac{m_\ell}{(\bar{z}_\ell - \bar{z})^2} = |1-\kappa_c|e^{i\varphi}, \qquad 0 \leq \varphi < 2\pi. \tag{15.19}$$

Due to the factor $|1-\kappa_c|$, the critical points are independent of ε_c. In addition, assuming the same position of the point masses, the under-focused case with $\kappa_c = t$, where $0 \le t < 1$, and the over-focused case with $\kappa_c = 2-t$, have identical critical curves — see Figure 15.1.

Now, since $\partial^2\zeta/\partial z^2 = 0$, equation (15.14) yields that the gradient reduces to

$$\operatorname{grad} J = -\overline{\frac{\partial\zeta}{\partial\bar{z}}}\,\frac{\partial^2\zeta}{\partial\bar{z}^2}.$$

By (15.18), we obtain the directional derivative:

$$\nabla_Z\zeta = (1-\kappa_c)\,Z + \left(\gamma + \sum_{\ell=1}^{g} \frac{m_\ell}{(\bar{z}_\ell - \bar{z})^2}\right)\bar{Z}, \tag{15.20}$$

where

$$Z = 2i \operatorname{grad} J = -4i\left(\gamma + \sum_{\ell=1}^{g} \frac{m_\ell}{(z_\ell - z)^2}\right)\left(\sum_{\ell=1}^{g} \frac{m_\ell}{(\bar{z}_\ell - \bar{z})^3}\right). \tag{15.21}$$

15.2 Caustics of One and Two Point Masses

We shall investigate the number of caustics and cusps due to single and binary point masses on a lens plane with continuous matter and shear. This section is based on Witt and A.P. [Wit-P] and A.P. and Wicklin [Pet-Wic95].

15.2.1 Chang–Refsdal Lens

We consider the critical curves, caustics, and cusp positions due to a point mass m at the origin of a lens plane with continuous matter κ_c and shear γ from infinity. This system is also known as a *Chang–Refsdal lens.* It was introduced to model the lensing action by a single star in a deflector galaxy on a "macro" lensed image created by the galaxy. The star splits the lensed image into sub-images and causes flux changes in the image [Chan-R79], [Chan-R84]. Our study of the critical curves and caustics due to this lens will follow [Wit-P].

The lens equation for the Chang–Refsdal lens is

$$\zeta = (1-\kappa_c)z + \gamma\bar{z} - \frac{m}{\bar{z}}, \tag{15.22}$$

The critical points of ζ are determined by (15.19):

$$\gamma + \frac{m}{\bar{z}^2} = |1-\kappa_c|e^{i\varphi}, \qquad 0 \le \varphi < 2\pi. \tag{15.23}$$

If $\kappa_c = 1$ and $\gamma = 0$, then $J = \det(\text{Jac}\,\zeta) = -m/|\bar{z}|^4 \neq 0$, so there are no critical curves (caustics). If $\kappa_c = 1$ and $\gamma > 0$, then the critical curves (and caustics) degenerate to two isolated points: $z = \pm i\sqrt{m/\gamma}$. We now assume that $\kappa_c \neq 1$ and set

$$\gamma_* = \frac{\gamma}{|1-\kappa_c|}, \qquad m_* = \frac{m}{|1-\kappa_c|}, \qquad \gamma_n = \frac{\gamma}{1-\kappa_c} = \varepsilon_c \gamma_*,$$

where γ_n is the normalized shear and $\varepsilon_c = \text{sign}(1-\kappa_c)$. Equation (15.23) now becomes

$$\gamma_* + \frac{m_*}{\bar{z}^2} = e^{i\varphi}, \qquad 0 \le \varphi < 2\pi. \tag{15.24}$$

This equation is solved by

$$z_\pm(\varphi) = \pm\sqrt{\frac{m_*}{e^{-i\varphi} - \gamma_*}}, \qquad 0 \le \varphi < 2\pi.$$

As φ varies over the interval $[0, 2\pi)$, the solutions $z_\pm(\varphi)$ trace out at most two critical curves. Furthermore, a change occurs in the number of critical curves during the transition from $|\gamma_n| < 1$ to $|\gamma_n| > 1$. For $|\gamma_n| < 1$ and $|\gamma_n| > 1$, we obtain one and two closed critical curves, respectively. In addition, since $|z_\pm(\varphi)| \to \infty$ as $\varphi \to \pi$, there are two critical curves extending to infinity for $|\gamma_n| = 1$; see Figure 15.1.

In order to count the number of cusp critical points of ζ, equation (15.2) shows that it suffices to determine the phases φ of $z_\pm(\varphi)$ for which

$$0 = \nabla_Z \zeta = (1-\kappa_c)Z + \left(\gamma + \frac{m}{\bar{z}^2}\right)\bar{Z}, \tag{15.25}$$

where

$$Z = -4i\left(\gamma + \frac{m}{z^2}\right)\frac{m}{\bar{z}^3}.$$

Equation (15.25) is equivalent to

$$0 = Z_* + \varepsilon_c\left(\gamma_* + \frac{m_*}{\bar{z}^2}\right)\bar{Z}_*, \tag{15.26}$$

where

$$Z_* = -4i\left(\gamma_* + \frac{m_*}{z^2}\right)\frac{m_*}{\bar{z}^3}.$$

Inserting $z_\pm(\varphi)$ into (15.26), equation (15.26) transforms into the following after squaring:

$$0 = (e^{-i\varphi} - \gamma_*)^3 e^{6i\varphi} - (e^{i\varphi} - \gamma_*)^3.$$

This equation is equivalent to a sixth-degree polynomial in $e^{i\varphi}$:

$$\begin{aligned} 0 &= [e^{2i\varphi} - 1][\gamma_*^2 e^{4i\varphi} - 3\gamma_* e^{3i\varphi} + (\gamma_*^2 + 3)e^{2i\varphi} - 3\gamma_* e^{i\varphi} + \gamma_*^2] \\ &= [e^{2i\varphi} - 1]p_4(e^{i\varphi}). \end{aligned} \tag{15.27}$$

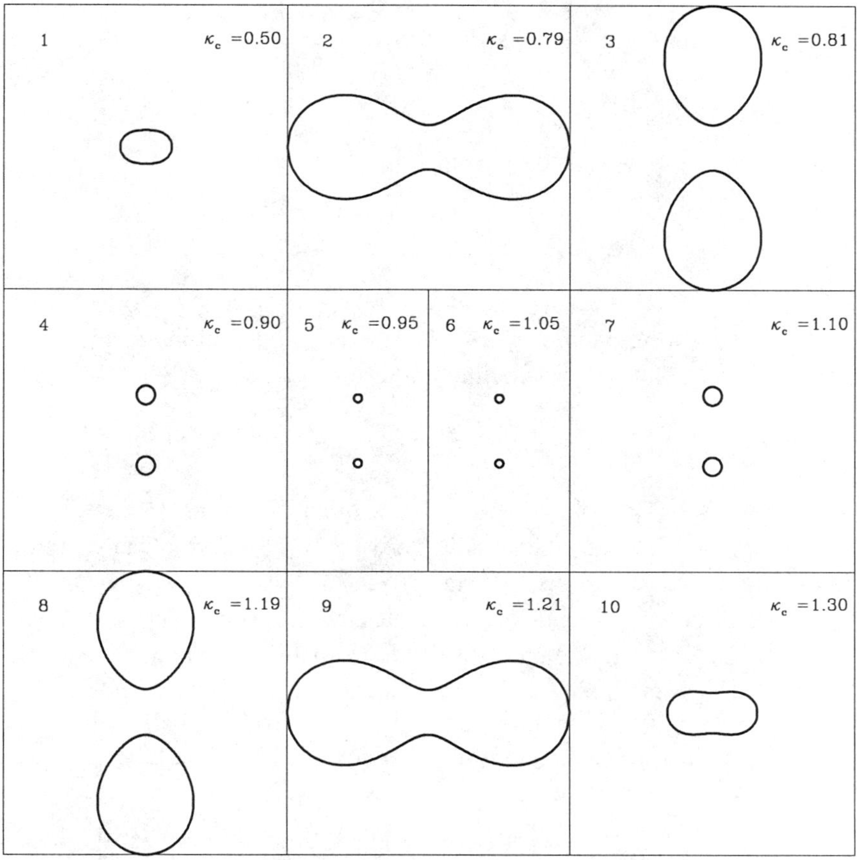

Figure 15.1. Critical curves due to a point mass with shear $\gamma = 0.2$ from infinity and continuous matter with increasing density κ_c. The critical curves extend to infinity as κ_c passes through the values $\kappa_c = 0.8$ (i.e., $\gamma_n = 1$) and $\kappa_c = 1.2$ (i.e., $\gamma_n = -1$). Note that critical curves for the under-focused case $\kappa_c = t$, where $0 \leq t < 1$, and over-focused one $\kappa_c = 2 - t$ have identical critical curves (e.g., panels 2 and 9). The caustics associated with the critical curves in this figure are shown in Figure 15.3.

Note that the solutions of (15.26) are a subset of the solutions of (15.27) (due to the squaring). Two immediate solutions of (15.27) are

$$\varphi_1 = 0, \qquad \varphi_2 = \pi.$$

Using $e^{i\varphi} = \cos\varphi + i\sin\varphi$, the fourth degree polynomial in (15.27) can be expressed as:

$$p_4(e^{i\varphi}) = e^{2i\varphi}[4\gamma_*^2 \cos^2\varphi - 6\gamma_* \cos\varphi + 3 - \gamma_*^2].$$

The equation

$$4\gamma_*^2 \cos^2\varphi - 6\gamma_* \cos\varphi + 3 - \gamma_*^2 = 0$$

has two solutions, say, φ_3 and φ_4, determined by

$$\cos\varphi_{3,4} = \frac{3 \pm \sqrt{4\gamma_n^2 - 3}}{4\gamma_n}, \qquad \gamma_n^2 \geq \frac{3}{4}.$$

Note that $\gamma_n^2 = \gamma_*^2$ and replacing γ_n by γ_* in the above denominator yields the same two solutions. Two more solutions are obtained in terms of φ_3 and φ_4:

$$\varphi_5 = 2\pi - \varphi_3, \qquad \varphi_6 = 2\pi - \varphi_4.$$

Observe that for $0 \leq \gamma_n^2 < 3/4$ and $\varepsilon_c = -1$, the quantities $z_\pm(\varphi_1)$ and $z_\pm(\varphi_2)$ do not obey (15.26), that is, $\nabla_Z \zeta \neq 0$. In other words, for such a regime the Chang–Refsdal lens has only fold caustic points (cf. Table 9.1, page 381).

The number of cusps and caustics associated with the solutions φ_i $(i = 1, \ldots, 6)$ are shown in Table 15.1. Note that the positions of the cusp caustics are of the form $\zeta(z_\pm(\varphi_i))$, where $i = 1, 2, 3, 4, 5$ and some φ_i do not exist for certain ranges of γ_n.

	$\varepsilon_c = 1$		$\varepsilon_c = -1$		
Shear	φ	N_{cusps}	φ	N_{cusps}	N_{caustics}
$0 \leq \gamma_n^2 < 3/4$	φ_1, φ_2	4	$\cdots$	0	1
$3/4 \leq \gamma_n^2 < 1$	φ_1, φ_2	4	$\varphi_3, \varphi_4, \varphi_5, \varphi_6$	8	1
$\gamma_n^2 > 1$	$\varphi_2, \varphi_3, \varphi_5$	6	$\varphi_1, \varphi_4, \varphi_6$	6	2

Table 15.1. Number of cusps N_{cusps} and caustics $N_{caustics}$ due a one point mass lens with continuous matter κ_c and shear γ from infinity. The normalized shear is given by $\gamma_n = \varepsilon_c\gamma/|1 - \kappa_c|$, where $\varepsilon_c = \operatorname{sign}(1 - \kappa_c)$.

The results of Table 15.1 are depicted in Figure 15.2 in terms of the continuous matter κ_c and shear γ from infinity.

The caustics corresponding to parameter values (γ, κ_c) in regions 4, 6, 8, and 0 of Figure 15.2 have the forms given, respectively, in panels 1, 3, 9, and 10 of Figure 15.3 (see Figure 15.1 for the critical curves). For parameter values (γ, κ_c) in Figure 15.2 that lie on the solid line below the dashed line as well as on the boundary line between the 8- and 6-cusp regions, the caustics extend to infinity. These caustics correspond in Figure 15.3 to the cases $\kappa_c = 0.8$ and $\kappa_c = 1.2$. Values of (γ, κ_c) on the line separating the 0- and 8-cusp regions in Figure 15.2 have caustics with four swallowtails (see panels 9 and 10 in Figure 15.3), while two elliptic umbilics occur (cf. panels 4–7) for the dashed line. Consequently, there are no merging of critical curves — hence, no beak-to-beak metamorphoses.

If $\gamma = 0$, then there is one critical curve — a circle of radius $\sqrt{m/|1-\kappa_c|}$ centered at the origin. The associated caustic is the origin of the light source plane if $0 \leq \kappa_c < 1$, while it is a circle of radius $2\sqrt{|1-\kappa_c|m}$ for $\kappa_c > 1$.

When $\kappa_c = 0$ and $\gamma \neq 1$, the caustics have the diamond and triangular forms in panels

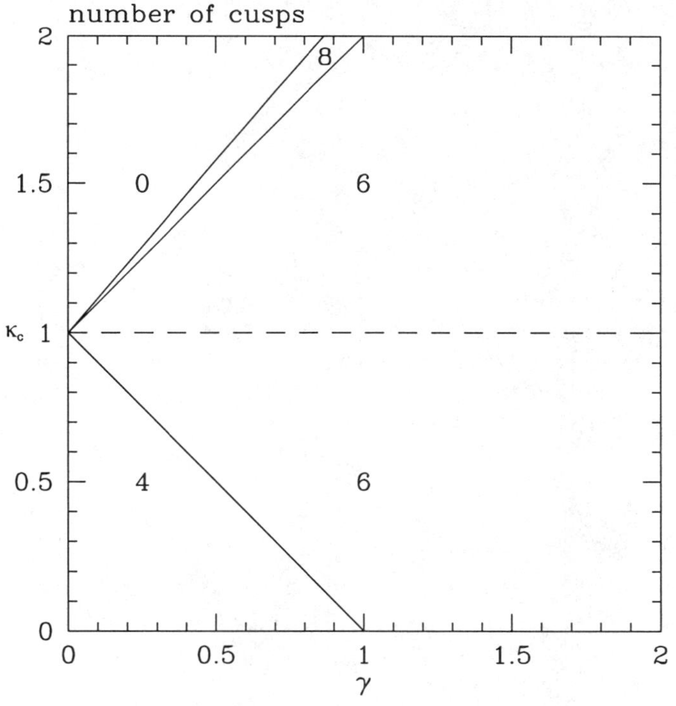

Figure 15.2. Parameter space for the number of cusps for the one point mass gravitational lens with shear γ from infinity and continuous matter κ_c. The dashed line divides the under- and over-focused regions. The solid line below the dashed one is given by $\kappa_c = 1 - \gamma$, while the line separating 8 and 6-cusp regions is $\kappa_c = 1 + \gamma$. The boundary between the 0 and 8-cusp regions is $\kappa_c = 1 + \sqrt{4/3}\,\gamma$. (After [Wit-P, p. 4099].)

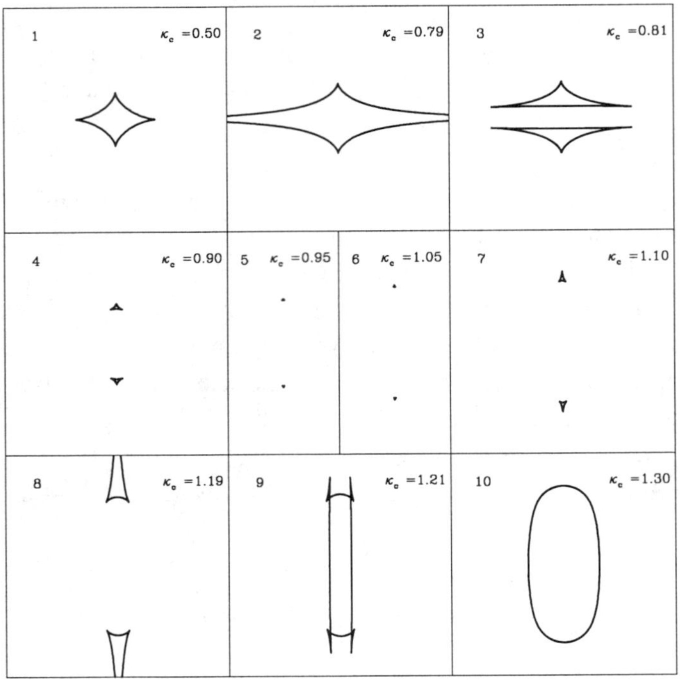

Figure 15.3. Caustics due to a point mass with shear $\gamma = 0.2$ from infinity and continuous matter with increasing density κ. The caustics extend to infinity as κ_c varies through the values $\kappa_c = 0.8$ (i.e., $\gamma_n = 1$) and $\kappa_c = 1.2$ (i.e., $\gamma_n = -1$). The caustics in panel 8 are two triangular caustics with each having a cusp extending out of the panel. Note that the caustic in panel 10 has no cusp. A light source inside this caustic has no lensed image, while outside the source has two lensed images. Elliptic umbilics and swallowtails occur at $\kappa_c = 1$ and for $\kappa_c = 1 + \sqrt{3/4}\,\gamma$, respectively (see the metamorphoses for panels 5–7 and 9–10).

1 and 3 of Figure 15.3 for $0 < \gamma < 1$ and $\gamma > 1$, respectively. In the limit $\gamma \to 1$, the critical curves increase without bound. Explicitly, for $0 < \gamma < 1$ the u-intercepts of the critical curve scales like $|1 - \gamma|^{-1/2}$; a similar scaling holds for the v-intercepts of the two critical curves when $\gamma > 1$ (e.g., [Dom, p. 111]). As $\gamma \to 1$, these critical curves are mapped via (15.22) to caustics that extend to infinity in a direction parallel to the y_1-axis (cf. [Chan-R84, pp. 170, 171]).

15.2.2 Binary Point Masses

The current subsection treats the number of critical curves and cusps due to single-plane two point mass lenses (pure binaries) as well as two point masses with continuous matter and shear from infinity. We also contrast the caustics due to single and double-plane binary point masses with continuous matter and shear. The main references for this section are [Wit-P] and [Pet-Wic95].

Unless explicitly stated, we shall not assume that $\mathbf{m_1 + m_2 = 1}$.

Pure Binary Point Mass Lens: No Continuous Matter and Shear

The lens equation for two point masses m_1 and m_2 on a lens plane is

$$\zeta = z + \frac{m_1}{\bar{z}_1 - \bar{z}} + \frac{m_2}{\bar{z}_2 - \bar{z}},$$

which is a *pure binary point mass* lens system. It models lensing by binary stars as well as a star with a planetary companion. Readers are referred to Section 5.5 (page 159) for astrophysical applications of this model.

The parametric equation for the critical curve of ζ is given by (15.19):

$$\frac{m_1}{(\bar{z}_1 - \bar{z})^2} + \frac{m_2}{(\bar{z}_2 - \bar{z})^2} = e^{i\varphi}, \qquad 0 \leq \varphi < 2\pi. \tag{15.28}$$

Without loss of generality, we shall assume that the masses lie on the x-axis symmetrically about the origin, that is, $z_2 = -z_1$ with z_1 real. However, sometimes it will be convenient to select the origin at one of the masses' positions (e.g., when studying star-planet systems — $m_1 \gg m_2$). For $z_2 = -z_1$, equation (15.28) is equivalent to

$$0 = m_1(z_1 + z)^2 + m_2(z_1 - z)^2 - e^{-i\varphi}(z_1 + z)^2(z_1 - z)^2 \equiv p(z). \tag{15.29}$$

Now, observe that if different parameters $\varphi_1, \varphi_2 \in [0, 2\pi)$ correspond to the same solution z of (15.19) (page 508), then we obtain the same value on the left-hand side of (15.19), but different values on the right-hand side. This is a contradiction. Hence, distinct values of the parameter $\varphi \in [0, 2\pi)$ correspond to distinct positions of the solutions of (15.19). Thus, the points where the critical curves merge are determined by the double or higher order zeros of the polynomial equivalent of (15.19). For our case (namely, equation (15.29)), this occurs when

$$p(z) = 0 = \frac{dp}{dz}(z). \tag{15.30}$$

Using

$$p(z) = (z_1 + z)^2(z_1 - z)^2 \left(\frac{\partial \bar{\zeta}}{\partial z} - e^{-i\varphi} \right) = 0,$$

we see that $p(z) = 0$ if and only if $\partial \bar{\zeta} / \partial z = e^{-i\varphi}$. Consequently, equation (15.30) holds if and only if we have the following:

$$\frac{\partial^2 \bar{\zeta}}{\partial z^2} = 0, \qquad \text{on critical curves,}$$

that is,

$$m_1(z_1 + z)^3 + m_2(z_1 - z)^3 = 0. \tag{15.31}$$

Notice that since

$$\frac{\partial \bar{\zeta}}{\partial z} = \overline{\frac{\partial \zeta}{\partial \bar{z}}}, \qquad \frac{\partial^2 \bar{\zeta}}{\partial z^2} = \overline{\frac{\partial^2 \zeta}{\partial \bar{z}^2}},$$

equation (15.30) is equivalent to the vanishing of the tangent,

$$Z = -2i \overline{\frac{\partial \zeta}{\partial \bar{z}}} \frac{\partial^2 \zeta}{\partial \bar{z}^2},$$

to the critical curves (since $\partial \zeta / \partial \bar{z} = e^{i\varphi} \neq 0$ along each critical curve).

The previous discussion shows that a change in the number of critical curves occurs when (15.29) and (15.31) (which are now viewed as polynomial equations in the variable z) have a common zero. Observe that the coefficients of the highest order z-terms in (15.29) and (15.31) do not have a common zero. Consequently, the Resultant Theorem (Theorem 11.10, page 438) yields that (15.29) and (15.31) have a common zero if and only if their resultant vanishes. This resultant is a third degree polynomial equation in $e^{i\varphi}$ (cf. [Wit-P, p. 4098, eq. (17)]):

$$0 = [e^{i\varphi}(m_1 + m_2) - d^2]^3 + 27 e^{2i\varphi} d^2 m_1 m_2, \tag{15.32}$$

where d is the distance between the two masses ($d = 2z_1$).

Now take the complex conjugate of (15.32) and multiply it by $e^{3i\varphi}$. The resulting new equation is also of third degree in $e^{i\varphi}$ and it holds if and only if (15.32) does. The conditions on d, m_1, and m_2 for which both of the third degree equations are solved by $e^{i\varphi}$ is determined by the vanishing of the resultant of the two equations. In this case, the resultant factors into three polynomials, one of which has no real zeros and the following two with real zeros [Wit-P, pp. 4098, 4099]:

$$0 = [(m_1 + m_2)^2 - d^4]^3 - 27 m_1 m_2 d^8 \tag{15.33}$$

$$0 = (m_1 + m_2 - d^2)^3 + 27 m_1 m_2 d^2. \tag{15.34}$$

The above two equations describe surfaces that divide the parameter space (d, m_1, m_2) into regions corresponding to different numbers of critical curves.

Viewing equations (15.33) and (15.34) as polynomials in d, we see that each has only one real solution. Denote the d-solution of (15.33) by

$$d_c = d_c(m_1, m_2) = \left(\frac{(m_1 + m_2)^2 - d_c^4}{3 m_1^{1/3} m_2^{1/3}} \right)^{3/8}$$

and note that (15.34) is solved by

$$d_w = d_w(m_1, m_2) = \left(m_1^{1/3} + m_2^{1/3}\right)^{3/2}.$$

A single-plane two point mass lens system is called a *close* (resp., *intermediate*, *wide*) *binary* if the point masses have separation $0 < d < d_c$ (resp., $d_c < d < d_w$, $d_w < d < \infty$). In Figure 15.4, we depict the curves $d_c(1-m_2, m_2)$ and $d_w(1-m_2, m_2)$, where $m_1+m_2 = 1$.

Close, intermediate, and wide binaries produce three, one, and two caustics, respectively. Furthermore, the caustics have no transverse crossings, and only beak-to-beak metamorphoses are produced (see page 390). There are ten, six, and eight cusps, respectively, when the binary is close, intermediate, and wide [Wit-P, p. 4106]. Parameter values (d_c, m_2) and (d_w, m_2) yield caustics with two and one beak-to-beak metamorphoses, respectively. Figure 15.5 illustrates the caustics and critical curves due to close, intermediate, and wide binaries for $m_1 + m_2 = 1$ with $m_1 = m_2$. Note that in this case we have $d_c = 1/\sqrt{2}$ and $d_w = 2$.

In modeling physical lens systems using pure binaries, an ambiguity can arise due to the similar morphology of the diamond-shaped central caustic of a close binary and the

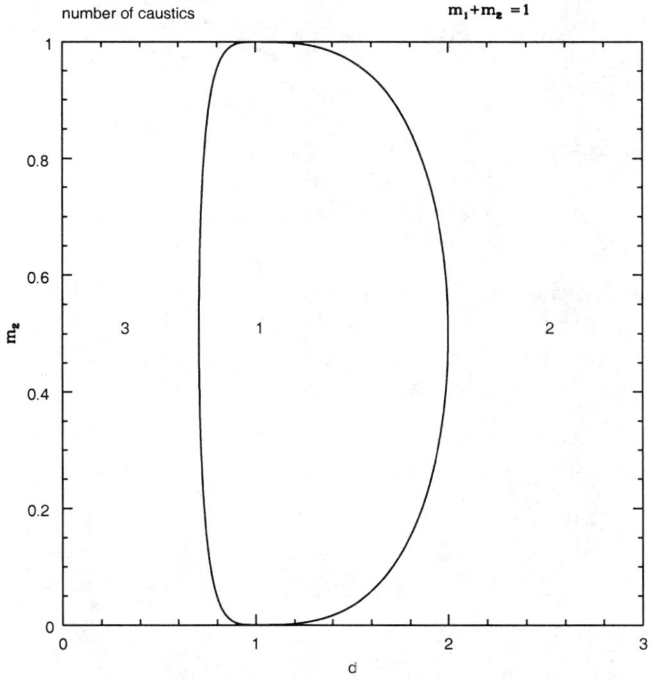

Figure 15.4. Parameter space for a binary point mass lens with $m_1 + m_2 = 1$. The left and right curves are plots of the functions $d_c(1 - m_2, m_2)$ and $d_w(1 - m_2, m_2)$. The numbers indicate how many critical curves (caustics) occur for parameter values (d, m_2) in the given region.

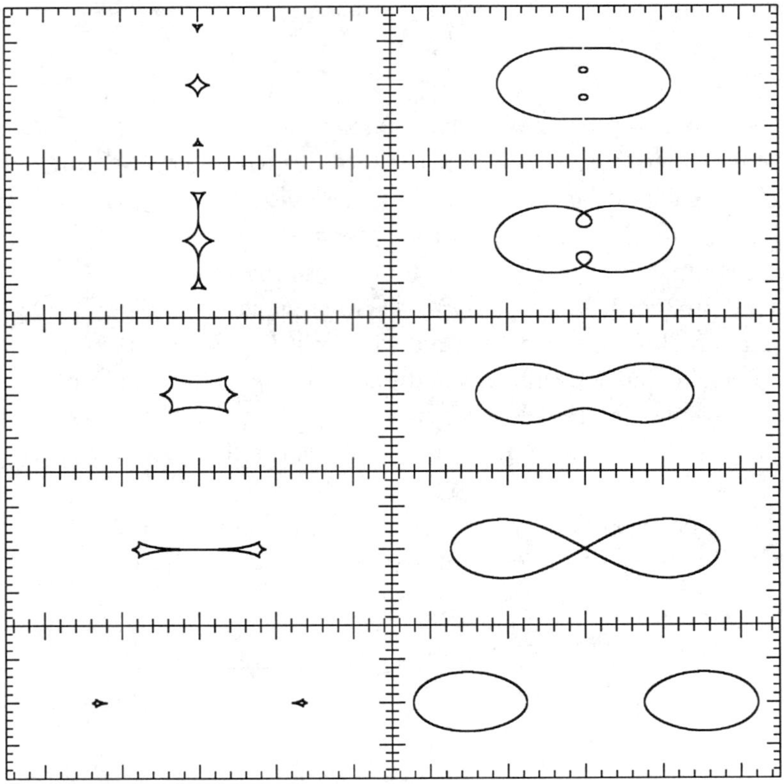

Figure 15.5. Caustics (left) and critical curves (right) due to close (top row of panels), intermediate (third row of panels), and wide binaries (bottom row of panels) for $m_1 + m_2 = 1$ and $m_1 = m_2$. The second and fourth rows correspond to the masses having separations $d_c = 1/\sqrt{2}$ and $d_w = 2$, respectively.

two diamond-shaped caustics of a wide binary [Dom, pp. 118–122]. For example, this is relevant when trying to fit real light curve data with the theoretical light curve of a source traversing a diamond-shaped caustic due to a pure binary. In some cases, however, the ambiguity between close and wide binaries can be broken (e.g., [Alb00b, p. 8]).

Binary Point Masses with Continuous Matter and Shear

We shall consider the number of critical curves and cusps due to two point masses with continuous matter and shear. Our development will parallel the treatment in Section 15.2.1. However, since the equations will become much more complicated, especially when considering cusps, most of the results will simply be quoted from [Wit-P].

The lens equation is given by

$$\zeta = (1 - \kappa_c)z + \gamma\bar{z} + \frac{m_1}{\bar{z}_1 - \bar{z}} + \frac{m_2}{\bar{z}_2 - \bar{z}}.$$

It is assumed that the shear γ from infinity acts along the u-axis, where $z = \mathrm{u} + i\mathrm{v}$. This is relative to ray tracing from the observer to the light source plane.[2] For negative values of γ, the shear from infinity is along the v-axis. Unless stated to the contrary, we always assume that $\gamma \geq 0$.

The critical curves of ζ are determined by the following, where (without loss of generality) the mass m_2 is assumed to lie on the u-axis at $z_2 = -z_1$:

$$\gamma + \frac{m_1}{(\bar{z}_1 - \bar{z})^2} + \frac{m_2}{(\bar{z}_1 + \bar{z})^2} = |1 - \kappa_c|e^{i\varphi}, \qquad 0 \leq \varphi < 2\pi,$$

or, equivalently,

$$0 = m_1(z_1 + z)^2 + m_2(z_1 - z)^2 - (\gamma - |1 - \kappa_c|e^{-i\varphi})(z_1 + z)^2(z_1 - z)^2. \qquad (15.35)$$

This is a polynomial of degree four in $\bar{z}$, so *there are at most four critical curves.* The critical curves (possibly less than four) are traced out by the previous four solutions as φ varies over the interval $[0, 2\pi)$. Furthermore, the merger of the critical curves requires (15.35) to have a double or higher order zero (see page 515), which forces $\partial^2\bar{\zeta}/\partial z^2$ to vanish on the critical curves, that is,

$$m_1(z_1 - z)^3 - m_2(z_1 + z)^3 = 0.$$

It follows that *there are at most three merging points — hence, at most three beak-to-beak metamorphoses.*

The previous upper bounds on the number of critical curves and beak-to-beak metamorphoses are achievable. For example, consider the case of equal masses $m_1 = m_2 = m$ and set $z_1 = d(\cos\theta + i\sin\theta)/2$, where θ is the angle between the u-axis and the line between the masses. The condition for a change in the number of critical curves is [Wit-P, p. 4096]:

$$\begin{aligned} 0 &= (\gamma_n^2 - 1)d^4 - 2\gamma_n m d^2\cos(2\theta) + m^2 \\ 0 &= (\gamma_n^2 - 1)d^4 + 16\gamma_n m d^2\cos(2\theta) + 64m^2. \end{aligned} \qquad (15.36)$$

Analogous to (15.33) and (15.34), the two equations in (15.36) define three-dimensional surfaces that divide the parameter space (d, m, γ_n, θ) into regions associated with different numbers of critical curves. Part of this parameter space is depicted in Figure 15.6 for $m = 1$ and $\theta = 0$. Observe that four critical curves do occur.

In Figure 15.6, parameter values on the thick line in the portion of the parameter space with $-1 < \gamma_n \leq -7/9$ and $3/\sqrt{2} \leq d$ determine a single critical curve having three beak-to-beak metamorphoses. Figure 15.7 depicts this for parameter values close to the thick

[2]See "Shear From Infinity," page 103.

line. Readers are referred to Grieger et al. [Gri-KRS] for numerous illustrations of the critical curves and caustics due to under- and over-focused binaries for different values of γ, θ, and d.

Equal masses with continuous matter and shear from infinity: Let us consider the number of cusps due to a single-plane binary point mass lens with, say $m_1 = m = m_2$, continuous matter κ_c, and shear γ from infinity. We shall proceed as in the case of the Chang–Refsdal lens (Section 15.2.1).

Consider solutions $z = z(\varphi)$ of (15.35), where the phase angle φ is such that

$$0 = \nabla_Z \zeta = (1 - \kappa_c)\, Z + \left(\gamma + \frac{m}{(\bar{z}_1 - \bar{z})^2} + \frac{m}{(\bar{z}_1 + \bar{z})^2}\right) \bar{Z}. \tag{15.37}$$

Here

$$Z = -4i\left(\gamma + \frac{m}{(z_1 - z)^2} + \frac{m}{(z_1 + z)^2}\right)\left(\frac{m}{(\bar{z}_1 - \bar{z})^3} - \frac{m}{(\bar{z}_1 + \bar{z})^3}\right).$$

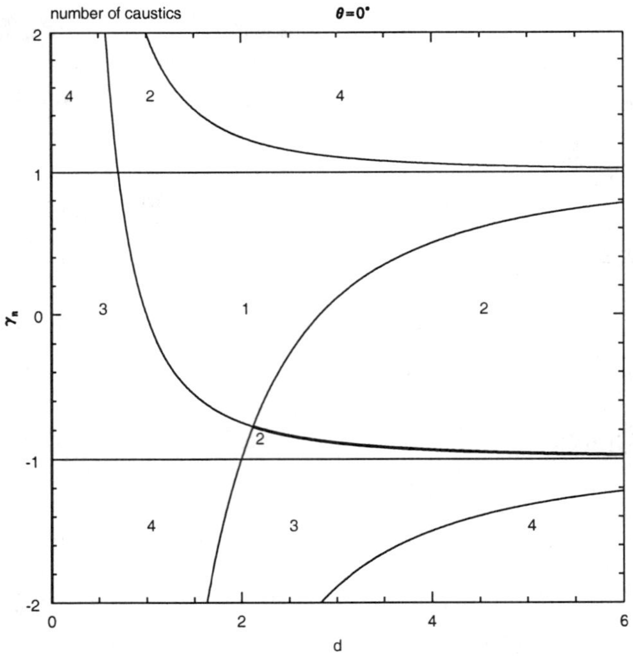

Figure 15.6. Parameter space for two point masses with continuous matter and shear from infinity. Here, $m_1 = m_2 = 1$ and the masses lie at the points $(\pm d/2, 0)$. The quantity γ_n is the normalized shear. Negative values of γ_n are included and used to designate the case when the shear from infinity is along the v-axis. The numbers indicate how many critical curves (caustics) occur for parameters in the given region. Parameters on the thick line yield a single caustic that has three beak-to-beak metamorphoses — see Figure 15.7. (After [Wit-P, p. 4097].)

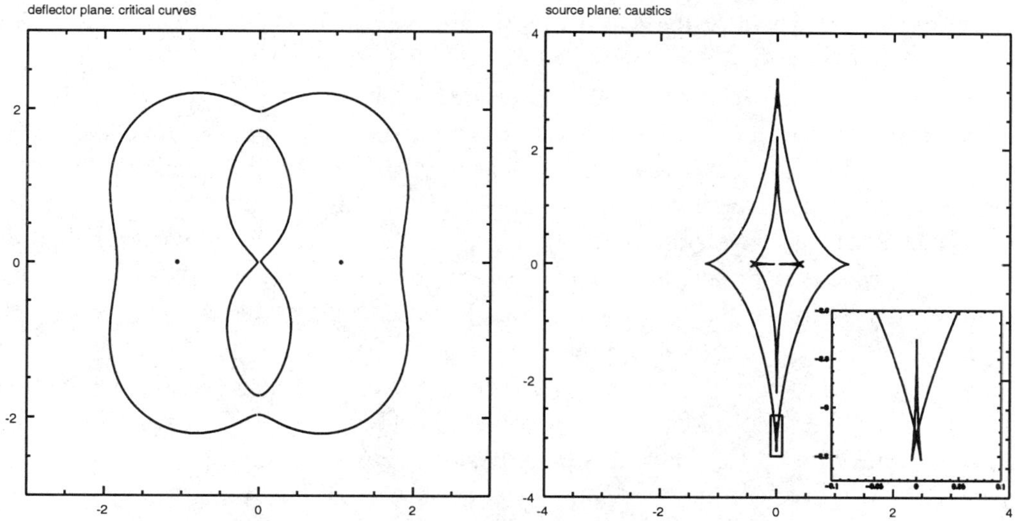

Figure 15.7. Critical curves (left) and caustics (right) due to two point masses $m_1 = m_2 = 1$ with under-focused continuous matter ($\varepsilon = 1$, separation $d = 3/\sqrt{2}$, and normalized shear $\gamma_n = -0.78$ along the v-axis. The masses are depicted by the dots in the left panel. There are four butterflies and 16 cusps. Three beak-to-beaks are about to form. (After [Wit-P, p. 4098].)

Furthermore, we assume that $z_2 = -z_1$ and $z_1 = d(\cos\theta + i\sin\theta)/2$, where θ is the angle between the u-axis and straight line through the masses. Due to the complexity of the equations involved, we shall simply state the results pertaining to the number of cusps for this lens system. Readers should consult [Wit-P] for more detail.

Equation (15.37) is equivalent to a polynomial equation of degree 24 into $e^{i\varphi}$:

$$p_{24}(e^{i\varphi}) = 0. \tag{15.38}$$

When the two masses lie on the u- or v-axis (i.e., $\theta = 0$ or $\pi/2$), equation (15.38) has solutions $\varphi_1 = 0$ and $\varphi_2 = \pi$. Additional solutions can be found by inserting $e^{i\varphi} = \cos\varphi + i\sin\varphi$ in (15.38). This reduces the problem to obtaining solutions of a tenth degree polynomial equation in $\cos\varphi$:

$$p_{10}(\cos\varphi) = 0.$$

As in the case of the Chang–Refsdal lens, and since $\theta = 0$ or $\pi/2$, the solutions φ_i determine the cusp critical points $z(\varphi_i)$ for given ranges of the parameters (γ_n, m, d).

For the case $m = 1$, $\theta = 0$, and $0 \le \kappa_c < 1$ (under-focused), the top panel in Figure 15.8 depicts a portion of the parameter space (d, γ_n) along with the number of cusps for different regions of the space. The figure has a region with 16 cusps. An example of the corresponding caustics is shown in Figure 15.7, which also has four butterflies and three beak-to-beaks.

The parameter space for the over-focused case is shown in the bottom panel of Figure 15.8. Note that there is a region with 18 cusps. An example of the caustics appears in Figure 15.9, which also displays eight swallowtails.

Different masses with continuous matter and no shear: For the under-focused case, there are either one, two, or three caustics, none of which has self-intersections. Six, eight, or ten cusps occur if there are one, two, or three caustics, respectively. In the over-focused case, the caustics can have self-intersections and there are parameter-space regions with zero, two, four, six, eight, and ten cusps. Readers are referred to [Wit-P, pp. 4103–4108] for further discussion of the cusps due to this lens system, as well as the case of *equal masses with continuous matter and no shear*.

In Figures 15.10 and 15.11, we show some examples of the caustics for the over-focused situation with $m_2 \gg m_1$ (e.g., star-planet system) and $\gamma = 0$. The former figure has two caustics — one with no cusp and an extremely small caustic having four swallowtail self-intersections.

An evolution showing the emergence of the tiny caustic in Figure 15.10 is depicted in Figure 15.11. The tiny caustic evolves into one with a Chang–Refsdal geometry for certain ranges of κ_c (cf. panels 9 and 10 of Figure 15.3, page 514).

Single versus Double Plane Two Point Mass Lens

We shall briefly contrast the global geometry of caustics due to single and double plane binary point masses with continuous matter and shear from infinity (cf. Section 9.5.4, page 388, and Figure 15.19, page 556).

Consider gravitational lensing by two point masses on distinct lens planes, where each lens plane L_i $(i = 1, 2)$ has continuous matter with constant-density κ_i and shear γ_i from infinity. Suppose that the first mass m_1 is at the origin of L_1 and the second m_2 at $\mathbf{d}$ on L_2. Then the lensing map of this system can be expressed as (see (9.54), page 389):

$$\begin{aligned} \boldsymbol{\eta}^{(3D)}_{ptcs} &= \mathbf{x}_1 - \left(m_1 \frac{\mathbf{x}_1}{|\mathbf{x}_1|^2} + \kappa_1 \mathbf{x}_1 + \gamma_1(-\mathrm{u}_1, \mathrm{v}_1) \right) \\ &\quad - \left(m_2 \frac{\mathbf{x}_2 - \mathbf{d}}{|\mathbf{x}_2 - \mathbf{d}|^2} + \kappa_2 \mathbf{x}_2 + \gamma_2(\mathrm{u}_2, -\mathrm{v}_2) \right), \end{aligned}$$

where

$$\mathbf{x}_2 = \mathbf{x}_1 - \beta \left(m_1 \frac{\mathbf{x}_1}{|\mathbf{x}_1|^2} + \kappa_1 \mathbf{x}_1 + \gamma_1(-\mathrm{u}_1, \mathrm{v}_1) \right)$$

with $\mathbf{x}_i = (\mathrm{u}_i, \mathrm{v}_i)$ and $\beta = d_{1,2} d_S / d_2 d_{1,S}$.

The caustics due to $\boldsymbol{\eta}^{(3D)}_{ptcs}$ display some striking global geometric features that cannot be realized in the single plane two point mass case. For example, the lensing map $\boldsymbol{\eta}^{(3D)}_{ptcs}$ can produce caustics with one cusp: teardrops and cardioids. Due to the violation of local convexity (Section 9.3, page 353), teardrops cannot be produced by point masses on a single

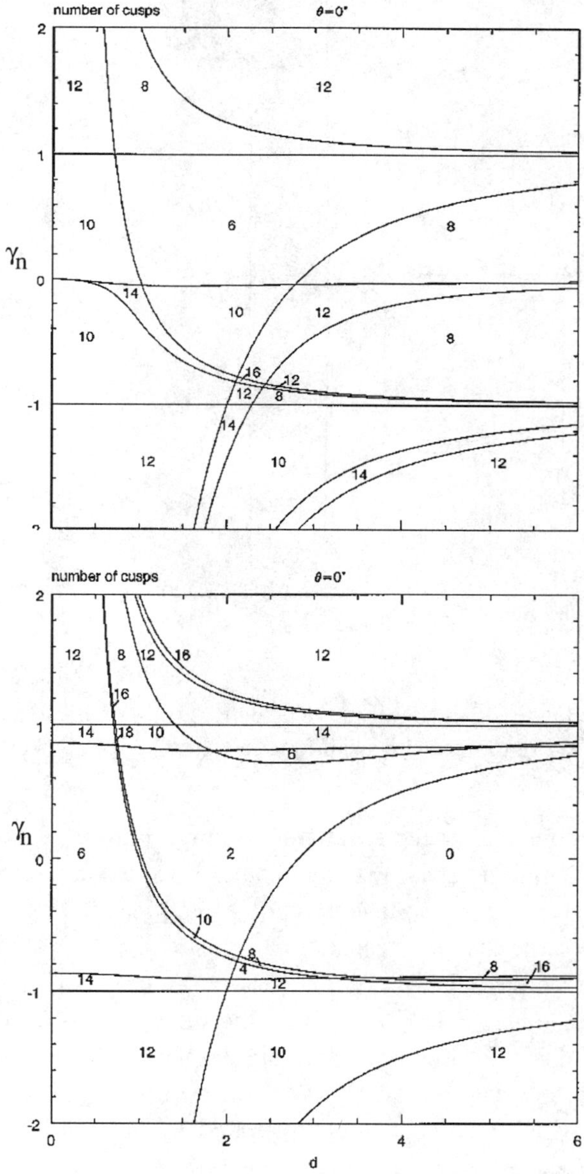

Figure 15.8. Part of the parameter space for the number of cusps of an under-focused two point mass lens with normalized shear γ_n and $m_1 = m_2 = 1$. Here d denotes the distance between the two lenses. The lenses are aligned on the u-axis. The top and lower panels correspond to the under- and over-focused cases, respectively. For the under-focused case, there is a region with 16 cusps — see Figure 15.7 for an example. In the over-focused situation, Figure 15.9 depicts a caustic associated with the 18-cusp region. (After [Wit-P, pp. 4108, 4110].)

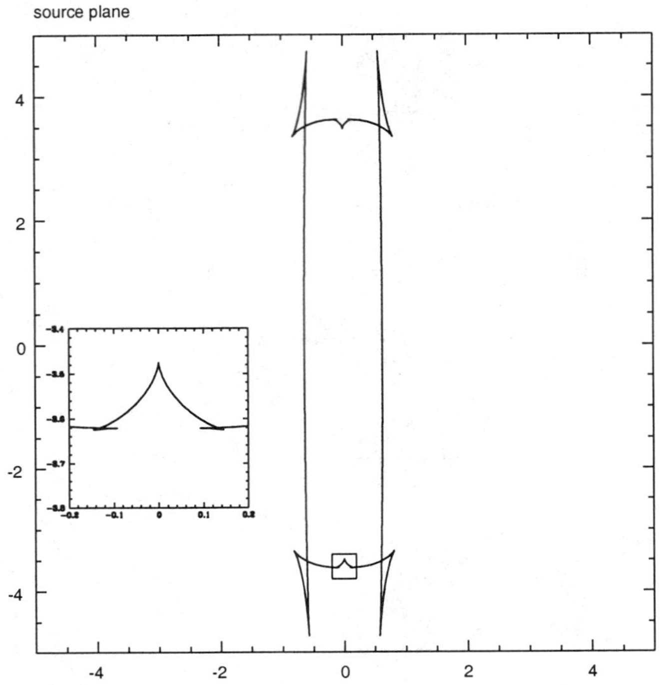

Figure 15.9. A caustic due to a two point mass lens with $\varepsilon_c = -1$, $m_1 = m_2 = 1$, $d = 0.72$ and $\gamma_n = 0.95$. The caustic has 18 cusps and 8 swallowtails. (After [Wit-P, p. 4110].)

lens plane with continuous matter and shear from infinity. Figure 15.12(a) depicts teardrop and cardioid caustics, including swallowtail and butterfly substructures.

Earlier we saw that for a single-plane two point mass lens with continuous matter and shear from infinity, no more than *four* caustics can exist and at most 18 cusps are known to occur for the system (see page 519 and Figure 15.9). Figure 15.12(b) displays 9 caustics and 28 cusps due to the lens system $\boldsymbol{\eta}_{ptcs}^{(3D)}$. Notice the simultaneous occurrences of four lips caustics in the figure. Lips are impossible for single-plane point masses with continuous matter and shear (due to local convexity). We note that the lensing map $\boldsymbol{\eta}_{ptcs}^{(3D)}$ produces hyperbolic umbilics, which also cannot be generated by single-plane point masses with continuous matter and shear (see Figure 9.25, page 390).

15.2.3 Binaries as a Point Mass Plus Chang–Refsdal Lens

The relationship between single-plane two point mass lenses and a Chang–Refsdal lens with pure shear (no continuous matter) was indicated for certain limiting cases by Schneider and Weiss [Schn-W86, pp. 245,246] and Gould and Loeb [Gou-L, pp.105,106]. Their results were later incorporated into a systematic, comprehensive treatment by Dominik [Dom] of

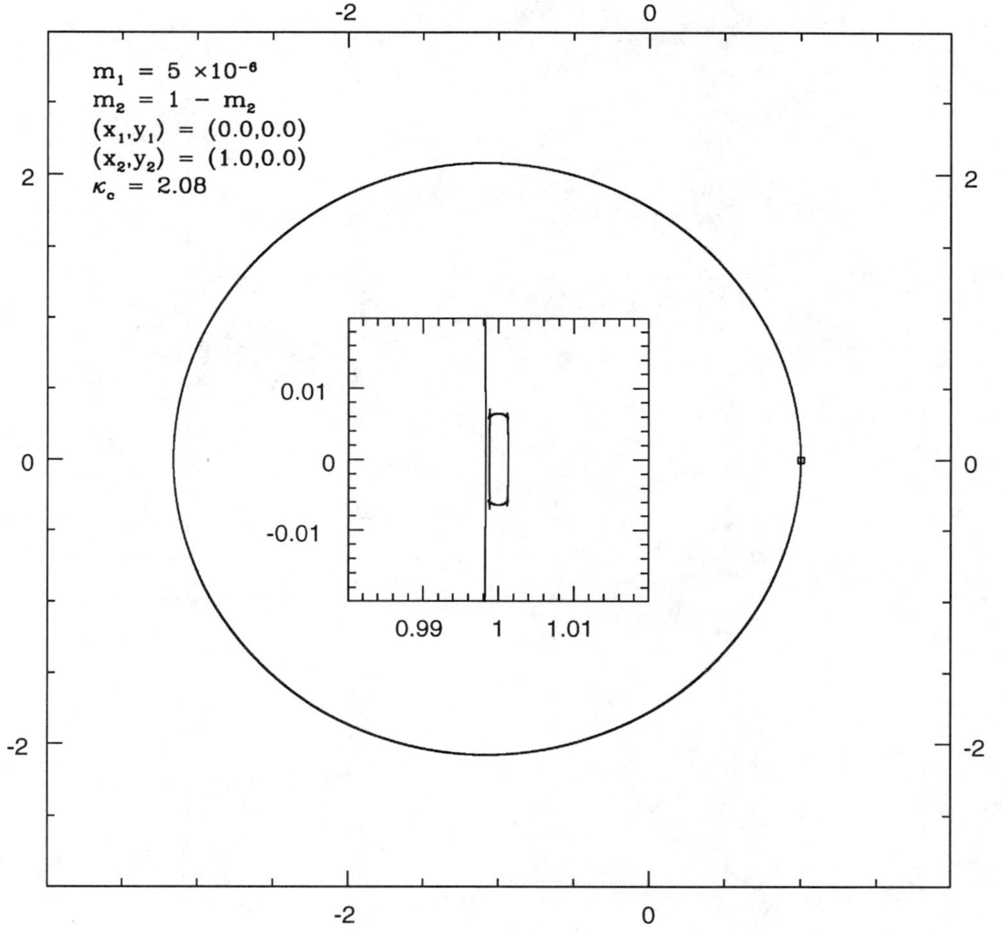

Figure 15.10. An example of caustics due to two point masses $m_2 \gg m_1$ on a lens plane with continuous matter κ_c and no shear from infinity. A caustic is revealed in the close up (middle panel) for the tiny box on the right.

pure binaries in the limit of extremely small mass ratios. For the single-plane case of two point masses with continuous matter and shear from infinity, the link to a Chang–Refsdal lens was also observed by Witt and A.P. [Wit-P, p. 4108]. A good illustration appears in Figure 15.11 (page 526). In the situation of any finite number of point masses on a lens plane with continuous matter and shear from infinity, the link was shown and utilized by A.P. and Witt [Pet-Wit, pp. 2922, 2923] (see Section 15.3.1, page 531).

The current subsection shows that a single-plane pure binary lens with point masses m_1 and m_2 having separation d such that $d/\sqrt{m_1} \neq 1$ and sufficiently small mass ratio,

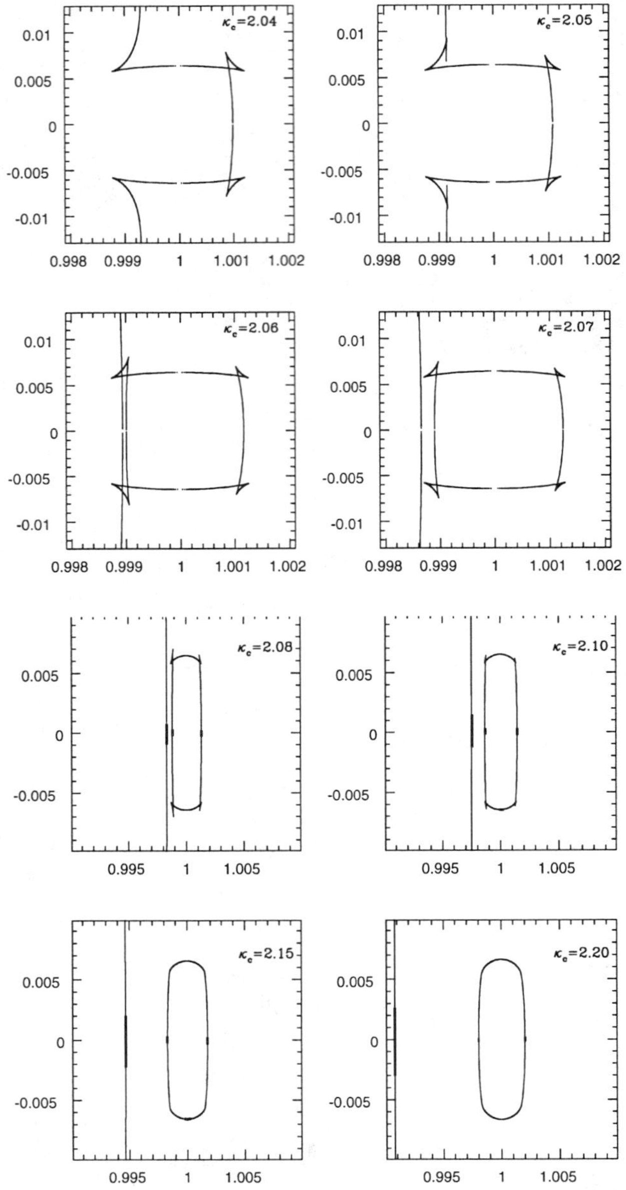

Figure 15.11. Emergence of the tiny caustic in Figure 15.10. There is a small change in the scale of the top four panels to reveal more of the curves' topology. The transition from $\kappa_c = 2.04$ to $\kappa_c = 2.05$ involves two swallowtail metamorphoses, resulting in four swallowtails in the $\kappa_c = 2.05$ panel. Note that the transition from $\kappa_c = 2.05$ to $\kappa_c = 2.06$ then involves a beak-to-beak metamorphosis. The evolution ends with all cusps being eliminated.

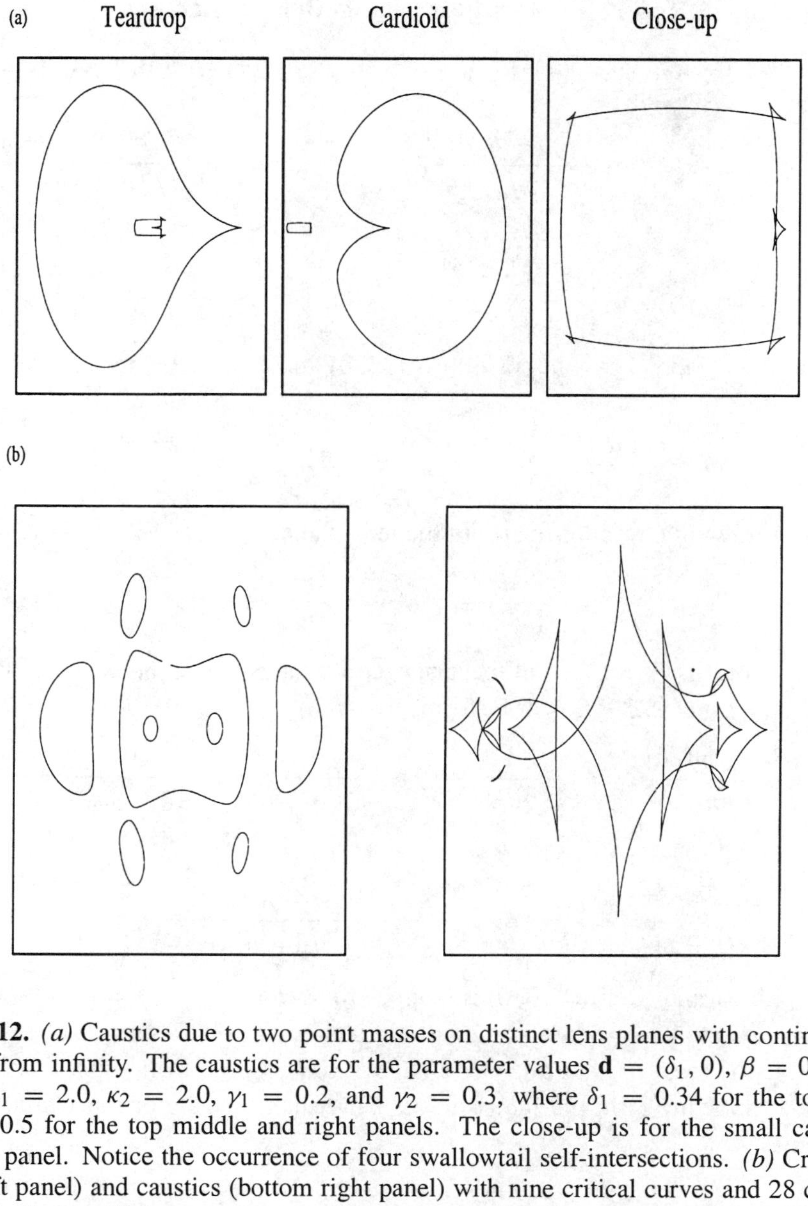

Figure 15.12. *(a)* Caustics due to two point masses on distinct lens planes with continuous matter and shear from infinity. The caustics are for the parameter values $\mathbf{d} = (\delta_1, 0)$, $\beta = 0.6$, $m_1 = 1$, $m_2 = 1$, $\kappa_1 = 2.0$, $\kappa_2 = 2.0$, $\gamma_1 = 0.2$, and $\gamma_2 = 0.3$, where $\delta_1 = 0.34$ for the top left panel and $\delta_1 = 0.5$ for the top middle and right panels. The close-up is for the small caustic in the top middle panel. Notice the occurrence of four swallowtail self-intersections. *(b)* Critical curves (bottom left panel) and caustics (bottom right panel) with nine critical curves and 28 cusps occurs at $\mathbf{d} = (0.1, 0)$, $\beta = 0.6$, $m_1 = 1$, $m_2 = 1$, $\kappa_1 = 0$, $\kappa_2 = 0$, $\kappa_2 = 0$, $\gamma_1 = 1$, and $\gamma_2 = 2$. (After [Pet-Wic95, pp. 1402, 1403].)

say $m_1 \gg m_2$, can be interpreted as a primary point mass lens m_1 with one of its two lensed images perturbed by the secondary mass m_2, where the perturbation is described by a Chang–Refsdal lens in a neighborhood of m_2. Our presentation will parallel the discussion in [Dom].

Interpreting Binaries as a Point Mass Plus Chang–Refsdal Lens

The lens equation for a binary point mass lens with $m_1 + m_2 = 1$ is given in component form as follows:

$$y_1 = \mathrm{u} - m_1 \frac{\mathrm{u} - d/2}{(\mathrm{u} - d/2)^2 + \mathrm{v}^2} - (1 - m_1)\frac{\mathrm{u} + d/2}{(\mathrm{u} + d/2)^2 + \mathrm{v}^2}$$

$$y_2 = \mathrm{v} - m_1 \frac{\mathrm{v}}{(\mathrm{u} - d/2)^2 + \mathrm{v}^2} - (1 - m_1)\frac{\mathrm{v}}{(\mathrm{u} + d/2)^2 + \mathrm{v}^2}, \qquad (15.39)$$

where the masses m_1 and m_2 are at positions $(d/2, 0)$ and $(-d/2, 0)$, respectively.

Let

$$q = \frac{m_2}{m_1}, \quad d_{[1]} = \frac{d}{\sqrt{m_1}} = d\sqrt{1+q}, \quad d_{[2]} = \frac{d}{\sqrt{m_2}} = d\sqrt{1+q^{-1}} = \frac{d_{[1]}}{\sqrt{q}}.$$

Introduce the following new coordinates in the lens plane:

$$\hat{\mathrm{u}} = \frac{\mathrm{u} + d/2}{\sqrt{1 - m_1}}, \qquad \hat{\mathrm{v}} = \frac{\mathrm{v}}{\sqrt{1 - m_1}}.$$

Note that the origin $(\hat{\mathrm{u}}, \hat{\mathrm{v}}) = (0, 0)$ of these new coordinates lies at the position of the mass m_2. The lens equation (15.39) now transforms to

$$\frac{y_1 + d/2 - m_1/d}{\sqrt{1 - m_1}} = -\frac{1}{q d_{[2]}} + \hat{\mathrm{u}} - \frac{\hat{\mathrm{u}}}{\hat{\mathrm{u}}^2 + \hat{\mathrm{v}}^2} - \frac{1}{q}\frac{\hat{\mathrm{u}} - d_{[2]}}{(\hat{\mathrm{u}} - d_{[2]})^2 + \hat{\mathrm{v}}^2},$$

$$\frac{y_2}{\sqrt{1 - m_1}} = \hat{\mathrm{v}} - \frac{\hat{\mathrm{v}}}{\hat{\mathrm{u}}^2 + \hat{\mathrm{v}}^2} - \frac{1}{q}\frac{\hat{\mathrm{v}}}{(\hat{\mathrm{u}} - d_{[2]})^2 + \hat{\mathrm{v}}^2}.$$

Consequently, the origin $(\hat{\mathrm{u}}, \hat{\mathrm{v}}) = (0, 0)$ is mapped to position $(y_1, y_2) = (-d/2 + m_1/d, 0)$ in the light source plane. In other words, a light source at $(-d/2 + m_1/d, 0)$ undergoing lensing only by the mass m_1 has a lensed image at the position of m_2. Motivated by this, we define new coordinates in the light source plane such that the origin lies at $(-d/2 + m_1/d, 0)$, namely,

$$\hat{y}_1 = \frac{y_1 + d/2 - m_1/d}{\sqrt{1 - m_1}}, \qquad \hat{y}_2 = \frac{y_2}{\sqrt{1 - m_1}}.$$

The lens equation now becomes

$$\hat{y}_1 = -\frac{1}{q\, d_{[2]}} + \hat{\mathrm{u}} - \frac{\hat{\mathrm{u}}}{\hat{\mathrm{u}}^2 + \hat{\mathrm{v}}^2} - \frac{1}{q}\frac{\hat{\mathrm{u}} - d_{[2]}}{(\hat{\mathrm{u}} - d_{[2]})^2 + \hat{\mathrm{v}}^2},$$

$$\hat{y}_2 = \hat{\mathrm{v}} - \frac{\hat{\mathrm{v}}}{\hat{\mathrm{u}}^2 + \hat{\mathrm{v}}^2} - \frac{1}{q}\frac{\hat{\mathrm{v}}}{(\hat{\mathrm{u}} - d_{[2]})^2 + \hat{\mathrm{v}}^2}. \qquad (15.40)$$

Hence, a light source at the origin of $(\hat{y}_1, \hat{y}_2)$ that is lensed by m_1 has a lensed image at the origin of $(\hat{\mathrm{u}}, \hat{\mathrm{v}})$ (i.e., the position of m_2).

The Taylor expansion about $(\hat{\mathrm{u}}, \hat{\mathrm{v}}) = (0, 0)$ of the rightmost terms in (15.40) converges for

$$\sqrt{\hat{\mathrm{u}}^2 + \hat{\mathrm{v}}^2} < d_{[2]} = \sqrt{q}\, d_{[1]}$$

and yields the following expansion of (15.40) up to first order [Dom, pp. 116, 124, 125]:

$$\hat{y}_1 = (1 + \gamma_{[1]})\,\hat{\mathrm{u}} - \frac{\hat{\mathrm{u}}}{\hat{\mathrm{u}}^2 + \hat{\mathrm{v}}^2} + O\left(\frac{\sqrt{q}}{d_{[1]}}\right),$$

$$\hat{y}_2 = (1 - \gamma_{[1]})\,\hat{\mathrm{u}} - \frac{\hat{\mathrm{v}}}{\hat{\mathrm{u}}^2 + \hat{\mathrm{v}}^2} + O\left(\frac{\sqrt{q}}{d_{[1]}}\right), \qquad (15.41)$$

where

$$\gamma_{[1]} = \frac{1}{(d_{[1]})^2}.$$

Now, the Chang–Refsdal lens equation is given by (15.22), which in component form becomes:

$$y_1 = (1 - \kappa_c + \gamma)\mathrm{u} - m\,\frac{\mathrm{u}}{\mathrm{u}^2 + \mathrm{v}^2},$$

$$y_2 = (1 - \kappa_c - \gamma)\mathrm{u} - m\,\frac{\mathrm{v}}{\mathrm{u}^2 + \mathrm{v}^2}. \qquad (15.42)$$

Since

$$\frac{\sqrt{q}}{d_{[1]}} = \frac{1}{d\sqrt{1 + q^{-1}}},$$

taking the limit $q \to 0$ in equation (15.41) for fixed d such that $\gamma_{[1]} \neq 1$, we obtain a pure shear unit mass Chang–Refsdal lens at the origin, that is, equation (15.42) with $m = 1$, $\kappa_c = 0$, and $\gamma = \gamma_{[1]}$. This limit fails for $\gamma_{[1]} = 1$ because the caustic(s) for the associated Chang–Refsdal lens would extend to infinity (see page 514), contradicting the fact that caustics due to binary point masses are bounded.

The previous discussion shows that for $\gamma_{[1]} \neq 1$ and sufficiently small mass ratios q (e.g., star-planet systems), our binary point mass lens can be approximated by a point mass m_1 plus a Chang–Refsdal lens at m_2 with $\kappa_c = 0$ and $\gamma = \gamma_{[1]}$. The shear $\gamma_{[1]}$ from infinity arises from the tidal action of m_1 on m_2. Treating m_1 as a point mass lens that creates two lens images (minimum and saddle) of a light source at the origin of $(\hat{y}_1, \hat{y}_2)$, one of the lensed images is located at the position of m_2. The Chang–Refsdal lensing action in a neighborhood of m_2 splits the latter lensed image into two or four lensed images, depending on whether the light source is inside or outside the caustic created by the Chang–Refsdal

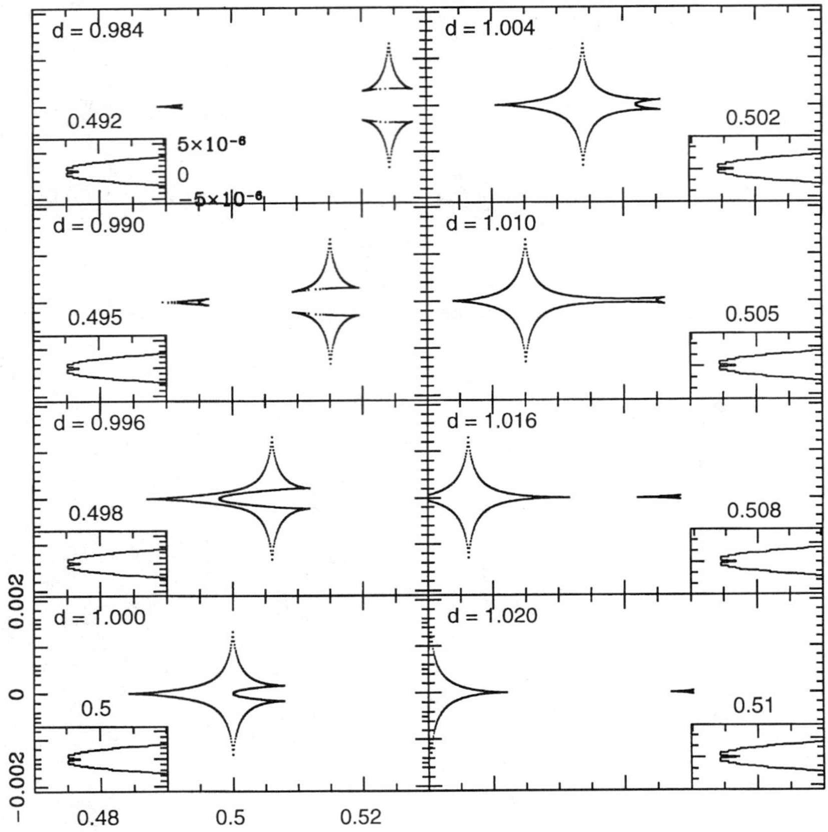

Figure 15.13. Caustics due to two point masses with $m_1 = 1 - 10^{-6}$, $m_2 = 10^{-6}$, and varying separation d. The masses m_1 and m_2 are at $(d/2, 0)$ and $(-d/2, 0)$, respectively. Some of the caustics have the triangular and diamond-shaped forms characteristic of caustics due to a Chang–Refsdal lens with pure shear — see the panels for $d = 0.964, 0.990$ and $d = 1.016, 1.020$). The Chang–Refsdal approximation fails for the caustics in the remaining panels. Note that each close up sub-panel reveals a cusp.

lens (cf. Figure 3.15, page 105). This results in the usual three or five lensed images due to a pure binary point mass lens (Theorem 11.11, page 439). We add that for $\gamma_{[1]} < 1$ the perturbed lensed image is a minimum and the Chang–Refsdal caustic is diamond shaped, while for $\gamma_{[1]} > 1$ the lensed image is a saddle and the caustic has two triangular shaped components [Dom, p. 118]. Note that the case $\gamma_{[1]} = 1$, where the approximation breaks down, the mass m_2 lies on the Einstein ring of m_1 (cf. [Dom, p.119]).

Figure 15.13 displays the caustics due to two point masses $m_1 = 1 - 10^{-6}$ and $m_2 = 10^{-6}$ with varying separation d. The Chang–Refsdal limit becomes apparent from the caustics for certain ranges of d. In fact, for q sufficiently small, we have [Dom, p.115]:

$$d_c \simeq 1 - \frac{3}{4}\, q^{1/3}, \qquad d_w \simeq 1 + \frac{3}{2}\, q^{1/3}.$$

Consequently, for Figure 15.13 we have:

$$d_c \simeq 0.9925, \qquad d_w \simeq 1.015, \qquad d_{[1]} \simeq d.$$

The ranges $d < d_c$ (close binary) and $d > d_w$ (wide binary), display Chang–Refsdal type caustics, namely, the two triangular caustics and diamond-shaped caustic (cf. panels 2 and 3 of Figure 15.3, page 514). The approximation breaks down for $d_c < d < d_w$ (intermediate binary).

15.3 Caustics of Finitely Many Point Masses

This section establishes the elimination of non-fold caustics and determines upper bounds on the number of caustics, caustic metamorphoses, and cusps due to finitely many point masses on a lens plane with continuous matter and shear from infinity.

15.3.1 Elimination of Non-Fold Caustics: Ovals

Consider a gravitational lens consisting of point masses $m_1, \ldots, m_g$ at positions $z_1, \ldots, z_g$ on a lens plane with continuous matter κ_c and shear γ from infinity. The associated lens equation is

$$\zeta = (1 - \kappa_c)z + \gamma \bar{z} + \sum_{\ell=1}^{g} \frac{m_\ell}{\bar{z}_\ell - \bar{z}}.$$

Of the five typical local planar caustic metamorphoses — namely, lips, beak-to-beaks, swallowtails, elliptic umbilics, and hyperbolic umbilics (Section 9.5.1, page 375) — the lensing map ζ can generate only beak-to-beaks, swallowtails, and elliptic umbilics (Section 9.5.3, page 383). We shall show that all these caustics, along with cusps, can be eliminated.

The fact that certain types of caustics can be removed was shown in Figure 15.3, where all non-fold caustics of a Chang–Refsdal lens were eliminated by increasing the continuous matter κ_c sufficiently high (cf. Figure 15.11, page 526). In Figure 15.14, we see that for several point masses on a lens plane with continuous matter and shear from infinity, all the cusps can be eliminated by increasing κ_c. The next theorem due to A.P. and Witt [Pet-Wit, p. 2921] shows that for such lens systems, all the non-fold caustics (not only cusps) can be removed.

Theorem 15.1 *Let ζ be the lensing map due to g point masses on a lens plane with continuous matter κ_c and shear γ from infinity. If κ_c is sufficiently large, then the caustics consist of g simple, closed fold curves.*

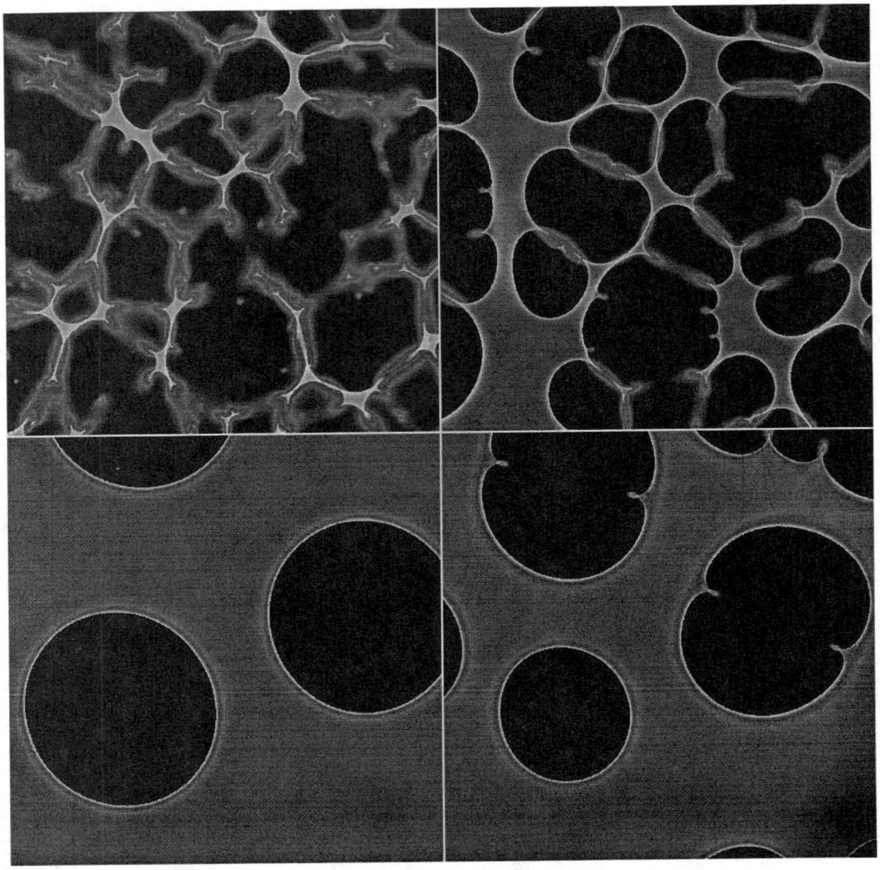

Figure 15.14. Magnification pattern of caustics (white curves) due to many point masses on a lens plane with fixed shear from infinity and increasing continuous matter (clockwise through panels, starting from top left). Lighter regions correspond to higher magnification. All cusps are eliminated in the bottom left panel.

Before proving Theorem 15.1, we present one of its corollaries. Recall that an *oval* is a simple, closed, regular, planar curve that bounds a convex set. Also, a *vertex* of a regular, planar curve is a point where the derivative of its signed-curvature vanishes (e.g., the points where the axes cross a standard ellipse are its vertices). Then the following result of A.P. [Pet95b, p. 4289] classifies the caustics of ζ in the regime of high-density continuous matter:

Corollary 15.2 *If* κ_c *is sufficiently large, then the caustics of* ζ *consist of* g *clockwise, oval, fold curves. The signed-curvature function of each caustic is then strictly negative and* (by the Four-Vertex Theorem[3]) *either constant or has at least four vertices — at least two maxima and two minima.*

[3] See [doC, pp. 37–41] and [Ber-G, pp. 334–337] for a discussion and proof.

Proof: Since by Theorem 15.1 the lensing map ζ has only fold critical points for κ_c sufficiently large, the map ζ is locally stable (Theorem 8.5(1), page 294). Moreover, the associated simple, closed, (smooth) fold caustic curves are locally convex (Corollary 9.5, page 360), so the signed-curvature function does not change sign along each such curve (page 358). But a simple, closed, regular, planar curve is an oval if and only if its signed-curvature has constant sign [Ber-G, p. 331, Thm. 9.6.2]. Consequently, the caustics of Theorem 15.1 are ovals. In addition, they are all clockwise since if an oval caustic c is counterclockwise, then a point source inside c will have two more lensed images than one outside and sufficiently close to c. This would violate local convexity. Hence, the signed-curvature function is strictly negative along the caustics. □

Proof of Theorem 15.1: The critical points of ζ are determined by equation (15.19) (page 508), which is equivalent to:

$$p(\bar{z}) \equiv p_\infty(\bar{z}) + \sum_{\ell=1}^{g} \left(\hat{m}_\ell \prod_{\substack{r=1 \\ r\neq\ell}}^{g} (\bar{z}_r - \bar{z})^2 \right) = 0,$$

where

$$p_\infty(\bar{z}) = \prod_{\ell=1}^{g} (\bar{z}_\ell - \bar{z})^2, \qquad \hat{m}_\ell = \frac{m_\ell}{|1-\kappa_c|e^{i\varphi} - \gamma}, \qquad 0 \leq \varphi < 2\pi.$$

Since

$$\lim_{\kappa_c \to \infty} p(\bar{z}) = p_\infty(\bar{z}),$$

the coefficients of $p(\bar{z})$ approach those of $p_\infty(\bar{z})$ as $\kappa_c \to \infty$. Hence, in the latter limit the zeros of $p(\bar{z})$ are in one-to-one correspondence with the zeros of $p_\infty(\bar{z})$, where zeros are counted with multiplicity, and approach the double zeros $z_1, \ldots, z_g$ of $p_\infty(\bar{z})$. This employs the fact that the zeros of a complex polynomial are continuous functions of the coefficients of the polynomial [Cool].

It follows that for κ_c sufficiently large, the critical curves lie in small discs about each of the point mass positions $z_1, \ldots, z_g$ — see Figure 15.15. Now consider any such disc, say, $D(z_j; r_j)$ with radius r_j and center at z_j $(1 \leq j \leq g)$. The larger the value of κ_c, the smaller one can choose r_j. Write the lensing map as

$$\zeta = (1-\kappa_c)z + \gamma\bar{z} + \frac{m_j}{\bar{z}_j - \bar{z}} + R(\bar{z}),$$

where

$$R(\bar{z}) = \sum_{\substack{\ell=1 \\ \ell\neq j}}^{g} \frac{m_\ell}{\bar{z}_\ell - \bar{z}}.$$

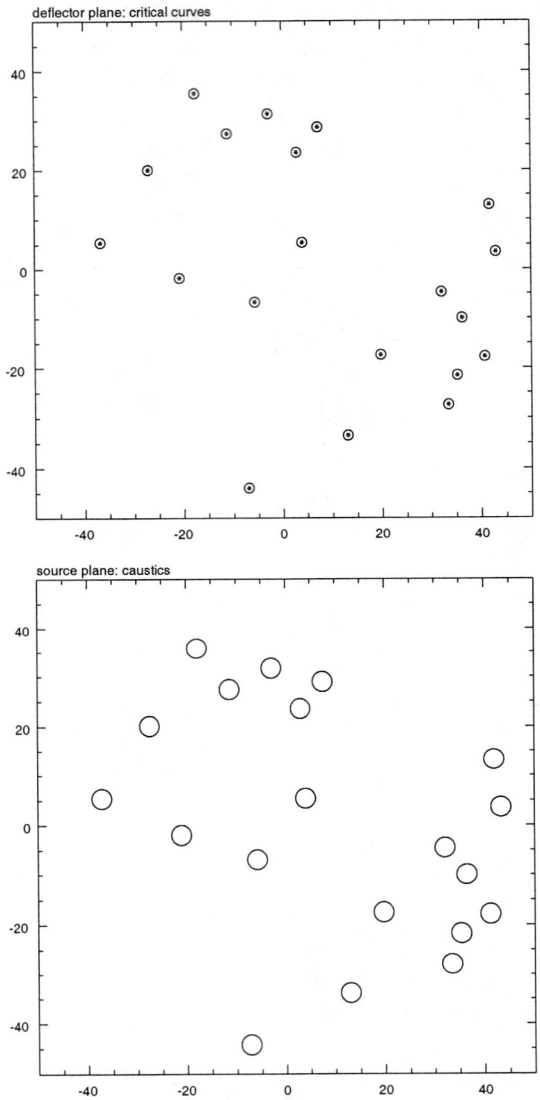

Figure 15.15. Critical curves and caustics due to several point masses with a very high density continuous matter. Each critical curve surrounds a single point mass and there are as many critical curves as point masses (top panel). The caustics (bottom panel) are oval, fold curves.

For $j = 1, \ldots, g$, we can restrict ζ to the disc $D(z_j; r_j)$ and make r_j sufficiently small so as to approximate ζ arbitrarily close by

$$\zeta_j = (1 - \kappa_c)z + \gamma\bar{z} + \frac{m_j}{\bar{z}_j - \bar{z}},$$

which is a Chang–Refsdal lens of mass m, at position z_j. We saw in Section 15.2.1 that for

$\kappa_c > \sqrt{4/3}\,\gamma + 1$ the Chang–Refsdal lens has a single caustic and it is a simple, closed, fold curve — see Table 15.1 (page 512) and Figure 15.3 (page 514). Thus, for large enough κ_c the caustics of ζ consist of g simple, closed, fold curves. □

Remark: The theory of cusp elimination for differentiable maps was introduced in singularity theory by H.L. [Lev63] in 1965 for purely mathematical reasons. It was further investigated by Eliashberg [Eli], culminating in the Levine-Eliashberg theorem: A locally stable map from a compact, oriented, boundary-less n-manifold M into the plane is homotopic to a locally stable map with *zero* or *one* cusp depending on whether the Euler characteristic of M is even or not. Though this theorem cannot be used directly as a basis for cusp elimination in gravitational lensing (since if one starts with a locally stable lensing map between celestial spheres, there is no guarantee that each stage of the theorem's homotopy is a lensing map), it is gratifying to see that cusp elimination was studied independently by singularity theorists.

15.3.2 Bounds on Number of Caustics and Metamorphoses

For a single-plane lensing map ζ due to a g-point mass lens with continuous matter κ_c and shear γ from infinity, Corollary 15.2 (page 532) yields that for κ_c sufficiently large, all the caustics are oval, fold curves. In other words, all non-fold caustics of ζ — in particular, *all beak-to-beaks, swallowtails, and elliptic umbilics — can be eliminated,* which gives zero as an achievable lower bound on the number of such caustic metamorphoses.

The critical points of a single-plane lensing map ζ due to g-point mass lens with continuous matter κ_c and shear γ from infinity are given parametrically by the solutions $z = z(\varphi)$ of equation (15.19) (page 508), which is equivalent to the following $2g$-degree polynomial:

$$(\gamma - |1-\kappa_c|e^{i\varphi}) \prod_{\ell=1}^{g} (\bar{z}_\ell - \bar{z})^2 + \sum_{\ell=1}^{g} \left(m_\ell \prod_{\substack{r=1 \\ r\neq \ell}}^{g} (\bar{z}_r - \bar{z})^2 \right) = 0, \tag{15.43}$$

where $0 \le \varphi < 2\pi$. As φ varies over the interval $[0, 2\pi)$, the solutions $z_1(\varphi), \ldots, z_{2g}(\varphi)$ of (15.43) trace out the critical curves. We note that one or more solutions may trace out a single critical curve. Consequently, *there are at least one and at most* $2g$ *critical curves (caustics). If* $\gamma = |1-\kappa_c|$ *and* $\varphi = 0$*, then there are at most* $2g-2$ *critical curves.*

We now determine an upper bound on the number of beak-to-beaks. The critical curves of ζ merge at points determined by the double or higher order zeros of (15.43), which requires the following to hold on the critical curves (see page 515):

$$\frac{\partial^2 \zeta}{\partial \bar{z}^2} = 0.$$

This is equivalent to a $(3g-3)$-degree polynomial equation:

$$\sum_{\ell=1}^{g} m_\ell \left(\prod_{\substack{r=1 \\ r\neq \ell}}^{g} (\bar{z}_r - \bar{z})^3 \right) = 0. \tag{15.44}$$

In other words, as φ varies over the interval $[0, 2\pi)$ only those solutions of (15.44) that lie on the critical curves will determine the merger points. Consequently, no more than $3g-3$ such points occur, so *there are at most* $3g-3$ *beak-to-beak metamorphoses.*

For the case of elliptic umbilics, Table 9.1 (page 381) shows that at such points the Jacobian matrix of ζ is identically zero. This implies that the potential ψ of the lens obeys

$$1 - \psi_{\mathrm{uu}} = 0 = 1 - \psi_{\mathrm{vv}},$$

which yields

$$1 = \kappa(\mathbf{x}) = \kappa_c.$$

If $\kappa_c = 1$, then (15.43) implies that the critical curves and caustics consist of $2g$ isolated points, which are counted with multiplicity. Hence, *there are at most* $2g$ *elliptic umbilics for* $\gamma > 0$ *and* $2g-2$ *for* $\gamma = 0$. Note that *no elliptic umbilics occur for the case with only point masses* (i.e., $\kappa = 0 = \gamma$).

In the situation of swallowtails, note from Table 9.1 (page 381) that the characterizing equations are complicated and include equation (15.43) and $\nabla_Z \zeta = 0$. The latter two equations are also the ones we shall study in Section 15.3.3 to obtain an upper bound on the number of cusps. Consequently, an upper bound on the number of swallowtails will not be pursued directly. Instead, *we shall simply note that bounds on the number of cusps yield bounds on the number of swallowtails, though these bounds will be exorbitant.*

15.3.3 Bounds on Number of Cusps

Consider the single-plane lensing map ζ due to g point masses with continuous matter κ_c and shear γ from infinity. We shall determine bounds on the number of cusps of ζ in terms of g for $\gamma > 0$ and $\gamma = 0$.

Requisite Mathematics: Section 11.5.1

Our main result is due to A.P. and Witt [Pet-Wit, p.2921]:

Theorem 15.3 *The total number N_{cusps} of cusps of ζ obeys:*

$$0 \leq N_{cusps} \leq \begin{cases} 12g^2 & \text{if } \gamma > 0 \\ 12g(g-1) & \text{if } \gamma = 0. \end{cases} \tag{15.45}$$

If κ_c is sufficiently large, then $N_{cusps} = 0$.

The upper bounds in Theorem 15.3 on the number of cusps may seem much higher than the actual maximum number of cusps for certain special configurations. The reason for this lies in the parametric representation of the critical curves — see (15.19), page 508. Two configurations of caustics always occur, namely, the under-focusing case ($0 \leq \kappa_c < 1$) and over-focusing one ($\kappa_c > 1$). These two configurations have the same critical curves for $\kappa_c = t$ and $\kappa_c = 2 - t$, where $0 \leq t < 1$. Since we only set upper bounds on the number of cusp critical points, equation (15.45) applies simultaneously to the under- and over-focused cases. It is not known *a priori* whether a point on a critical curve is mapped to a cusp for the under- or over-focused situation. Hence, *equation (15.45) actually yields upper bounds on the sum of the maximum number of cusps for the under and over-focused cases.*

Example (Chang–Refsdal lens): If $g = 1$ and $\gamma > 0$, then consulting Figure 15.2 (page 513) we see that there are *four* cusps for the under-focused case with $0 \leq \kappa_c < 1-\gamma$, and a maximum of *eight* cusps for the over-focused case with $1 + \gamma < \kappa_c$. For $1 - \gamma < \kappa_c < 1 + \gamma$ with $\kappa_c \neq 1$, we get *six* cusps in the under- and over-focused cases. The total sums agree with the upper bound of 12 given by (15.45): $4 + 8 = 12$ For $0 \leq \kappa_c < 1 - \gamma$ and $\kappa_c > 1+\gamma$ we have $4+8 = 12$, while for $1-\gamma < \kappa_c < 1+\gamma$ (with $\kappa_c \neq 1$) the upper bound is $6 + 6 = 12$.

We believe that it is extremely difficult to separate cusps according to the under- and over-focused cases, and, hence, to derive an upper bound for each such case. Furthermore, the equations we use to investigate cusps are given by (15.2), which are not the complete equations for cusps (see Table 9.1, page 381) and so allow for non-cusp, degenerate critical points. Hence, it is not surprising that there are situations for which the upper bounds of Theorem 15.3 cannot be achieved. For instance, if $g = 2$ and $\gamma = 0$, then a maximum of 10 cusps occur for the separate cases of under- and over-focused lensing [Wit-P, pp. 4105, 4108]. This yields a combined maximum of 20 cusps, but Theorem 15.3 predicts a combined maximum of 24.

Proof of Theorem 15.3: The cusp critical points of the lensing map

$$\zeta = (1 - \kappa_c)z + \gamma\bar{z} + \sum_{\ell=1}^{g} \frac{m_\ell}{\bar{z}_\ell - \bar{z}}.$$

are a subset of the solutions z of the following equations (cf. (15.2), page 505):

$$J = 0 = \nabla_Z \zeta,$$

where (see (15.12) on page 507 and (15.20), page 509):

$$J = \det(\text{Jac}\,\zeta) = \left(\frac{\partial \zeta}{\partial z}\right)^2 - \frac{\partial \zeta}{\partial \bar{z}}\,\overline{\frac{\partial \zeta}{\partial \bar{z}}}$$

and

$$\nabla_Z \zeta = (1 - \kappa_c)\; Z + \left(\gamma + \sum_{\ell=1}^{g} \frac{m_\ell}{(\bar{z}_\ell - \bar{z})^2}\right)\; \bar{Z}$$

with

$$Z = -2i\overline{\frac{\partial \zeta}{\partial \bar{z}}}\,\frac{\partial^2 \zeta}{\partial \bar{z}^2} = -4i\left(\gamma + \sum_{\ell=1}^{g} \frac{m_\ell}{(z_\ell - z)^2}\right)\left(\sum_{\ell=1}^{g} \frac{m_\ell}{(\bar{z}_\ell - \bar{z})^3}\right).$$

Notice that if $\kappa_c = 1$, then (15.43) on page 535 yields that the critical curves and caustics of ζ are isolated points, so no cusps occur. *Without loss of generality, we shall then assume throughout the proof that* $\kappa_c \neq 1$.

We have

$$\frac{\partial \zeta}{\partial z} = 1 - \kappa_c, \qquad \frac{\partial \zeta}{\partial \bar{z}} = \gamma + \sum_{\ell=1}^{g} \frac{m_\ell}{(\bar{z}_\ell - \bar{z})^2}, \qquad \frac{\partial^2 \zeta}{\partial \bar{z}^2} = 2\sum_{\ell=1}^{g} \frac{m_\ell}{(\bar{z}_\ell - \bar{z})^3}.$$

Set

$$p_{\zeta,1}(\bar{z}) = \frac{\partial \zeta}{\partial \bar{z}} \prod_{\ell=1}^{g} (\bar{z}_\ell - \bar{z})^2 = \gamma \prod_{\ell=1}^{g} (\bar{z}_\ell - \bar{z})^2 + \sum_{\ell=1}^{g} m_\ell \prod_{\substack{r=1\\ r\neq \ell}}^{g} (\bar{z}_r - \bar{z})^2$$

and

$$p_{\zeta,2}(\bar{z}) = \frac{1}{2}\frac{\partial^2 \zeta}{\partial \bar{z}^2} \prod_{\ell=1}^{g} (\bar{z}_\ell - \bar{z})^4 = \left(\prod_{\ell=1}^{g} (\bar{z}_\ell - \bar{z})\right)\left(\sum_{\ell=1}^{g} m_\ell \prod_{\substack{r=1\\ r\neq \ell}}^{g} (\bar{z}_r - \bar{z})^3\right)$$

Direct computation shows that the equations $\det(\mathrm{Jac}\,\zeta) = 0$ and $\nabla_Z \zeta = 0$ are equivalent, respectively, to

$$\begin{aligned} P_J(z,\bar{z}) &\equiv (J)\prod_{\ell=1}^{g} (\bar{z}_\ell - \bar{z})^2 (z_\ell - z)^2 \\ &= \left\{(1-\kappa_c)^2 \prod_{\ell=1}^{g} (\bar{z}_\ell - \bar{z})^2\right\} \prod_{\ell=1}^{g} (z_\ell - z)^2 \\ &\quad - p_{\zeta,1}(\bar{z})\left\{\gamma \prod_{\ell=1}^{g} (z_\ell - z)^2 + \sum_{\ell=1}^{g} m_\ell \prod_{\substack{r=1\\ r\neq \ell}}^{g} (z_r - z)^2\right\} = 0, \end{aligned} \tag{15.46}$$

and

$$\begin{aligned} P_Z(z,\bar{z}) &\equiv \frac{i}{4}(\nabla_Z \zeta) \prod_{\ell=1}^{g} (z_\ell - z)^3 (\bar{z}_\ell - \bar{z})^4 \\ &= (1-\kappa_c) p_{\zeta,2}(\bar{z}) \left\{\gamma \prod_{\ell=1}^{g} (z_\ell - z)^3 + \prod_{\ell=1}^{g} (z_\ell - z) \sum_{\ell=1}^{g} m_\ell \prod_{\substack{r=1\\ r\neq \ell}}^{g} (z_r - z)^2\right\} \\ &\quad - p_{\zeta,1}^2(\bar{z}) \sum_{\ell=1}^{g} m_\ell \prod_{\substack{r=1\\ r\neq \ell}}^{g} (z_r - z)^3 = 0. \end{aligned} \tag{15.47}$$

Using equations (15.46) and (15.47), we can express $P_J(z, \bar{z})$ and $P_Z(z, \bar{z})$ as polynomials in z with coefficients that are polynomials in $\bar{z}$:

$$P_J(z, \bar{z}) = a_0(\bar{z}) + a_1(\bar{z})z + \cdots + a_{2g}(\bar{z})z^{2g}$$

and

$$P_Z(z, \bar{z}) = b_0(\bar{z}) + b_1(\bar{z})z + \cdots + b_{3g}(\bar{z})z^{3g}.$$

Note that if $\gamma = 0$, then $b_{3g} = b_{3g-1} = 0$.

Notation: In order to capture the cases $\gamma = 0$ and $\gamma > 0$ simultaneously, let $n(\gamma) = 3g - 2$ if $\gamma = 0$ and $n(\gamma) = 3g$ if $\gamma > 0$.

Rewrite P_Z as:

$$P_Z(z, \bar{z}) = b_0(\bar{z}) + b_1(\bar{z})z + \cdots + b_{n(\gamma)}(\bar{z})z^{n(\gamma)}.$$

Now if z is a cusp critical point of ζ, then $(z, \bar{z})$ satisfies

$$P_J(z, \bar{z}) = 0 = P_Z(z, \bar{z}).$$

Consequently, an upper bound on the number of common zeros of P_J and P_Z yields an upper bound on the total number N_{cusps}.

Let $\mathrm{Res}(\bar{z})$ be the z-Sylvester resultant of P_J and P_Z. Since we are considering conjugate pairs $(z, \bar{z})$, the Resultant Theorem (Theorem 11.10, page 438) implies that the common zeros of P_J and P_Z are in 1-1 correspondence with a subset of the zeros of $\mathrm{Res}(\bar{z})$. This result and equation (11.5) (page 437) yield that:

$$N_{cusps} \leq \deg\,[\mathrm{Res}(\bar{z})] \leq [n(\gamma)]r + (2g)s, \tag{15.48}$$

where

$$r = \max\left\{\deg[a_\ell]_{\ell=0}^{2g}\right\}, \qquad s = \max\left\{\deg[b_\ell]_{\ell=0}^{n(\gamma)}\right\}.$$

Upper bounds on the degree of $\mathrm{Res}(\bar{\mathbf{z}})$**:** In this part of the proof, we shall first determine r and s to obtain an upper bound on the degree of $\mathrm{Res}(\bar{z})$ via (15.48). The resulting upper bound will then be improved further.

The $\bar{z}$-coefficients in equation (15.46) have a maximum degree of $2g$ for $\gamma \geq 0$. In equation (15.47), the maximum degree of the $\bar{z}$-coefficients is $4g - 3$ for $\gamma = 0$ and $4g$ for $\gamma > 0$. Consequently, if $\gamma = 0$, then $r = 2g$ and $s = 4g - 3$, while $r = 2g$ and $s = 4g$ for $\gamma > 0$. By (15.48), we obtain:

$$N_{cusps} \leq \deg\,[\mathrm{Res}(\bar{z})] = \begin{cases} 14g^2 & \text{if } \gamma > 0 \\ 14g^2 - 10g & \text{if } \gamma = 0 \end{cases} \tag{15.49}$$

We now show that the upper bound in (15.49) improves as follows when there is no shear from infinity:

$$N_{cusps} \leq \deg\,[\mathrm{Res}(\bar{z})] \leq 14g^2 - 12g, \qquad \text{if } \gamma = 0. \tag{15.50}$$

Closer inspection of the coefficients a_ℓ and b_ℓ of the polynomials P_J and P_Z, respectively, reveals the following structure:

$$a_\ell(\bar{z}) = a_{\ell,1}(\bar{z}) + p_{\zeta,1}(\bar{z})a_{\ell,2} \quad \text{and} \quad b_r(\bar{z}) = b_{r,1}(\bar{z}) + p_{\zeta,1}^2(\bar{z})b_{r,2},$$

where $\ell = 0, \dots, 2g$ and $r = 0, \dots, n(\gamma)$. Note that $a_{\ell,2}$ and $b_{r,2}$ are complex constants. Equations (15.46) and (15.47) imply that

$$P_J(z_j, \bar{z}) = \sum_{\ell=0}^{2g} a_\ell(\bar{z})z_j^\ell = -p_{\zeta,1}(\bar{z})m_j \prod_{\substack{\ell=1\\ \ell\neq j}}^{g}(z_\ell - z_j)^2$$

and

$$P_Z(z_j, \bar{z}) = \sum_{\ell=0}^{n(\gamma)} b_\ell(\bar{z})z_j^\ell = -p_{\zeta,1}^2(\bar{z})m_j \prod_{\substack{\ell=1\\ \ell\neq j}}^{g}(z_\ell - z_j)^3.$$

Since $p_{\zeta,1}(\bar{z})$ (resp., $p_{\zeta,1}^2(\bar{z})$) is not a factor of

$$\sum_{\ell=0}^{2g} a_{\ell,1}(\bar{z})z_j^\ell \quad \left(\text{resp.,} \quad \sum_{\ell=0}^{n(\gamma)} b_{\ell,1}(\bar{z})z_j^\ell\right),$$

we get

$$\sum_{\ell=0}^{2g} a_{\ell,1}(\bar{z})z_j^\ell = 0 \quad \text{and} \quad \sum_{\ell=0}^{n(\gamma)} b_{\ell,1}(\bar{z})z_j^\ell = 0.$$

By the Resultant Theorem, if two single-variable polynomials have a common root, then their resultant vanishes. Consequently,

$$\mathrm{Res}(a_{0,1}(\bar{z}), \dots, a_{2g,1}(\bar{z}); b_{0,1}(\bar{z}), \dots, b_{n(\gamma),1}(\bar{z})) = 0.$$

Now equation (11.6) on page 437 implies:

$$\begin{aligned}
&\mathrm{Res}(a_0(\bar{z}), \dots, a_{2g}(\bar{z}); b_0(\bar{z}), \dots, b_{n(\gamma)}(\bar{z}))\\
&\quad = \mathrm{Res}(a_{0,1}(\bar{z}), \dots, a_{2g,1}(\bar{z}); b_{0,1}(\bar{z}), \dots, b_{n(\gamma),1}(\bar{z}))\\
&\qquad + [\text{sum of terms each containing at least one}\\
&\qquad\qquad \text{factor with a } (*, 2)\text{–index}]\\
&\quad = [\text{sum of terms each containing at least one}\\
&\qquad\qquad \text{factor with a } (*, 2)\text{–index}].
\end{aligned} \tag{15.51}$$

Furthermore,

$$\max[\deg\{a_{\ell,1}(\bar{z}) : \ell = 0, \dots, 2g\}] = 2g,$$

$$\max[\deg\{p^2_{\zeta,1}(\bar{z})b_{\ell,2} : \ell = 0, \dots, n(\gamma)\}] = 4g - 4 \text{ for } \gamma = 0,$$

$$\max[\deg\{p_{\zeta,1}(\bar{z})a_{\ell,2} : \ell = 0, \dots, 2g\}] = 2g - 2 \text{ for } \gamma = 0,$$

and

$$\max[\deg\{b_{\ell,1}(\bar{z}) : \ell = 0, \dots, n(\gamma)\}] = 4g - 3.$$

Hence, the maximum of the degrees of the terms in (15.51) coincides with the degree of one of the following terms:

$$a^{i_0}_{0,1} \cdots a^{i_{2g}}_{2g,1}(p^2_{\zeta,1}b_{0,2})^{j_0} \cdots (p^2_{\zeta,1}b_{n(\gamma),2})^{j_{n(\gamma)}}, \tag{15.52}$$

$$(p_{\zeta,1}a_{0,2})^{i_0} \cdots (p_{\zeta,1}a_{2g,2})^{i_{2g}} b^{j_0}_{0,1} \cdots b^{j_{n(\gamma)}}_{n(\gamma),1},$$

and

$$(p_{\zeta,1}a_{0,2})^{i_0} \cdots (p_{\zeta,1}a_{2g,2})^{i_{2g}} (p^2_{\zeta,1}b_{0,2})^{j_0} \cdots (p^2_{\zeta,1}b_{n(\gamma),2})^{j_{n(\gamma)}}.$$

By equation (11.7), we get $i_0 + \cdots + i_{2g} = n(\gamma)$ and $j_0 + \cdots + j_{n(\gamma)} = 2g$, where $n(\gamma) = 3g - 2$ (since $\gamma = 0$). Of the above listed terms, the one in (15.52) has the largest degree:

$$(2g)(3g - 2) + (4g - 4)(2g) = 14g^2 - 12g.$$

This establishes (15.50).

The polynomial $\mathbf{p}^{\mathbf{g}}_{\zeta,1}(\bar{\mathbf{z}})$ is a factor of $\mathrm{Res}(\bar{\mathbf{z}})$**:** This result will be established and then used in the next subsection to further improve the upper bounds in (15.49) and (15.50). We begin with the following:

Lemma. *Let $p(z) = a_0 + a_1 z + \cdots + a_m z^m$ and $q(z) = b_0 + b_1 z + \cdots + b_n z^n$ be polynomials over* $\mathbf{C}$. *If $p(\alpha) = C_0C_1 \neq 0$ and $q(\alpha) = C_0C_2 \neq 0$, then*

$$\mathrm{Res}(a_0, \dots, a_{2g}; b_0, \dots, b_{n(\gamma)}) = C_0\mathrm{Res}(C_1, a_1, \dots, a_{2g}; C_2, b_1, \dots, b_{n(\gamma)}).$$

Proof (Lemma): First, notice that

$$\begin{aligned}&\mathrm{Res}(a_0, a_1, \dots, a_m; b_0, b_1, \dots, b_n)\\ &= \mathrm{Res}(C_0C_1 - \sum_{\ell=1}^{m} a_\ell\alpha^\ell, a_1, \dots, a_m; C_0C_2 - \sum_{\ell=1}^{n} b_\ell\alpha^\ell, b_1, \dots, b_n).\end{aligned}$$

Since the last column of the resultant matrix of p and q contains a_0 and b_0 with zeros everywhere else (see equation (11.5.1), we can expand the resultant in the coefficients a_0 and b_0. This yields

$$\begin{aligned}&\mathrm{Res}(a_0, a_1, \dots, a_m; b_0, b_1, \dots, b_n)\\ &= \mathrm{Res}(C_0C_1, a_1, \dots, a_m; C_0C_2, b_1, \dots, b_n)\\ &+\mathrm{Res}(-\sum_{\ell=1}^{m} a_\ell\alpha^\ell, a_1, \dots, a_m; -\sum_{\ell=1}^{n} b_\ell\alpha^\ell, b_1, \dots, b_n).\end{aligned}$$

The last resultant vanishes because the polynomials

$$p_1(z) = \left(-\sum_{\ell=1}^{m} a_\ell \alpha^\ell\right) + a_1 z + \cdots + a_m z^m$$

and

$$q_1(z) = \left(-\sum_{\ell=1}^{n} b_\ell \alpha^\ell\right) + b_1 z + \cdots + b_n z^n$$

satisfy $p_1(\alpha) = q_1(\alpha) = 0$. The lemma now follows since

$$\mathrm{Res}(C_0 C_1, a_1, \ldots, a_m; C_0 C_2, b_1, \ldots, b_n) = C_0 \mathrm{Res}(C_0, a_1, \ldots, a_m; C_2, b_1, \ldots, b_n).$$

□ **(Lemma)**

Applying the previous lemma to P_J and P_Z with $C_1 = 1$ yields that

$$C_0 = -p_{\zeta,1}(\bar{z}) m_j \prod_{\substack{\ell=1 \\ \ell \neq j}}^{g} (z_\ell - z_j)^2$$

factors out of $\mathrm{Res}(a_0(\bar{z}), \ldots, a_{2g}(\bar{z}); b_0(\bar{z}), \ldots, b_{n(\gamma)}(\bar{z}))$ for $j = 1, \ldots, g$. Hence

$$(-1)^g p_{\zeta,1}^g(\bar{z}) \prod_{j=1}^{g} m_j \prod_{\substack{\ell=1 \\ \ell \neq j}}^{g} (z_\ell - z_j)^2$$

is a factor of the resultant of P_J and P_Z.

Eliminating zeros of $\mathrm{Res}(\bar{\mathbf{z}})$ **that do not map to cusps:** On a critical curve with $\kappa_c \neq 1$, we have $|\partial\zeta/\partial\bar{z}| = |1 - \kappa_c| \neq 0$. Since

$$p_{\zeta,1}(\bar{z}) = \frac{\partial \zeta}{\partial \bar{z}} \prod_{\ell=1}^{g} (\bar{z}_\ell - \bar{z})^2,$$

it follows that the zeros of the polynomial $p_{\zeta,1}(\bar{z})$ are the conjugated point mass positions $\bar{z}_\ell$. Consequently, no zero of $p_{\zeta,1}(\bar{z})$ (or its conjugate) can lie on a critical curve with $\kappa_c \neq 1$. But $p_{\zeta,1}^g(\bar{z})$ is a factor of the resultant $\mathrm{Res}(\bar{z})$. Thus, we can improve the upper bound on the number of cusps as follows:

$$N_{cusps} \leq \deg\left[\mathrm{Res}(\bar{z})\right] \leq \begin{cases} 14g^2 - (2g)g = 12g^2 & \text{if } \gamma > 0 \\ 14g^2 - 12g - (2g-2)g \;\; = 12g^2 - 10g & \text{if } \gamma = 0. \end{cases} \tag{15.53}$$

By the Resultant Theorem not all the zeros of

$$\mathrm{Res}(\bar{z}) = \mathrm{Res}(a_0(\bar{z}), \ldots, a_{2g}(\bar{z}); b_0(\bar{z}), \ldots, b_{n(\gamma)}(\bar{z}))$$

necessarily correspond to cusps. There are common zeros of $a_{2g}(\bar{z})$ and $b_{n(\gamma)}(\bar{z})$ that are not cusp points. In fact, if $\gamma = 0$, then

$$a_{2g}(\bar{z}) = (1-\kappa_c)^2 \prod_{\ell=1}^{g} (\bar{z}_\ell - \bar{z})^2,$$

$$a_{2g-1}(\bar{z}) = (1-\kappa_c)^2 \left(-2\sum_{\ell=1}^{g} z_\ell\right) \prod_{\ell=1}^{g} (\bar{z}_\ell - \bar{z})^2,$$

and

$$b_{3g-2}(\bar{z}) = (1-\kappa_c) \left(\prod_{\ell=1}^{g} (\bar{z}_\ell - \bar{z})\right) \left(\sum_{\ell=1}^{g} m_\ell \prod_{\substack{r=1 \\ r\neq\ell}}^{g} (\bar{z}_r - \bar{z})^3\right).$$

The coefficients a_{2g}, a_{2g-1} and b_{3g-2} have a common factor.

Now, denote the $\bar{z}$-resultant matrix of P_J and P_Z by $\mathbf{S}_{2g+3g-2}$ (cf. equation (11.5.1). Let $\mathbf{S}_{i,j}$ be the submatrix of $\mathbf{S}_{2g+3g-2}$ obtained by removing the ith row and jth column, and let $\mathbf{S}_{i,j}(r,s)$ be the submatrix of $\mathbf{S}_{i,j}$ obtained by removing the rth row and sth column. If we expand the determinant of $\mathbf{S}_{2g+3g-2}$ by the first two columns, then

$$\begin{aligned} \det[\mathbf{S}_{2g+3g-2}] &= a_{2g} \det[\mathbf{S}_{1,1}] + (-1)^{3g-1} b_{3g-2} \det[\mathbf{S}_{3g-1,1}] \\ &= a_2 \det[\mathbf{S}_{1,1}] + (-1)^{3g-1} b_{3g-2} \left[a_{2g-1} \det[\mathbf{S}_{3g-1,1}(1,1)]\right. \\ &\quad \left. - a_{2g} \det[\mathbf{S}_{3g-1,1}(2,1)] + (-1)^{3g-1} b_{3g-2} \det[\mathbf{S}_{3g-1,1}(3g-1,1)]\right]. \end{aligned} \tag{15.54}$$

We observe that in the expansion of equation (15.54) the coefficient

$$(1-\kappa_c)^2 \prod_{\ell=1}^{g} (\bar{z}_\ell - \bar{z})^2$$

must factor out of the resultant for the case $\gamma = 0$. Hence, the points $\bar{z}_1, \ldots, \bar{z}_g$ are double zeros of the resultant. But no $\bar{z}_\ell$ or z_ℓ ($\ell = 1, \ldots, g$) can be mapped to a cusp point because as $\bar{z} \to \bar{z}_\ell$ the lensing map satisfies $|\zeta(\bar{z})| \to \infty$. Thus, for $\gamma = 0$ we can exclude an additional $2g$ zeros from $\mathrm{Res}(\bar{z})$. The latter and equation (15.53) imply that

$$0 \le N_{cusps} \le \begin{cases} 12g^2 & \text{if } \gamma > 0 \\ 12g(g-1) & \text{if } \gamma = 0. \end{cases}$$

By Theorem 15.1 (page 531), we achieve $N_{cusps} = 0$ for κ_c sufficiently large. This completes the proof of Theorem 15.3. □ **(Theorem 15.3)**

15.4 Curvature of Caustics

The current section connects the light path obstruction points (or "opacity") of a gravitational lens to the signed-curvature of its caustics. Bounds will also be given on these quantities.

15.4.1 Light Path Obstruction Points

The notion of "light path obstruction points" was introduced in lensing by A.P. [Pet95b] to study the global geometry of caustics. Lower bounds on the total number $|\mathsf{B}|$ of such points were obtained in [Pet95b, 4277-4280] through the application of Morse-theoretic, lensed-image counting techniques (Chapters 11 and 12). We shall discuss these results and present upper bounds on $|\mathsf{B}|$ for point mass lenses. Applications to bounds on the total signed-curvature of caustics will be given in Section 15.4.2.

We begin by recalling the concept of a "light path obstruction point." Consider gravitational lensing by potentials $\psi_1, \ldots, \psi_k$ on distinct lens planes $L_1, \ldots, L_k$, respectively, with L_1 the first lens plane in the direction from observer to the light source plane S. Let T and $\boldsymbol{\eta}$ be, respectively, the k-plane time-family and lensing map induced by the potentials $\psi_1, \ldots, \psi_k$. Suppose that light rays are traced backwards from the observer to S. and let $\mathbf{R}^2$ be a plane containing L_1. The points of $\mathbf{R}^2 \supseteq L_1$ at which a light ray either terminates by impacting a singularity of ψ_1, or, passes through and terminates at a singularitiy of one of the potentials $\psi_2, \ldots, \psi_k$ (so never reaching S), are called *light path obstruction points* (or simply, *obstruction points*) of $\boldsymbol{\eta}$ (Definition 6.14, page 201). The set of all such points is denoted by B and the total number by $|\mathsf{B}|$. Loosely speaking, obstruction points determine the "opacity" of a gravitational lens.

Let us characterize obstruction points more precisely. The lensing map $\boldsymbol{\eta}$ is given explicitly by (Definition 6.15, page 201):

$$\boldsymbol{\eta}(\mathbf{x}_1) = \mathbf{x}_1 - \sum_{i=1}^{k} (\boldsymbol{\alpha}_i \circ \boldsymbol{\eta}_i)(\mathbf{x}_1),$$

where $\boldsymbol{\alpha}_i$ is the bending angle vector field on L_i and $\boldsymbol{\eta}_i$ is the ith partial lensing map relative to $\boldsymbol{\eta}$. The map $\boldsymbol{\eta}_i$ traces a light ray from the observer to $\mathbf{x}_1 \in L_1$ and from $\mathbf{x}_1$ to L_i, with deflections at $L_1, \ldots, L_{i-1}$. These maps are defined recursively:

$$\boldsymbol{\eta}_i(\mathbf{x}_1) = \mathbf{x}_1 - \sum_{j=1}^{i-1} \beta_{j,i} \boldsymbol{\alpha}_j(\boldsymbol{\eta}_j(\mathbf{x}_1)), \quad (2 \le i \le k), \tag{15.55}$$

where $\boldsymbol{\eta}_1(\mathbf{x}_1) = \mathbf{x}_1$ and $\boldsymbol{\eta}_{k+1} = \boldsymbol{\eta}$. We note that the domain of $\boldsymbol{\eta}_i$ is defined such that it excludes all obstruction points through which a light ray terminates on one of the lens planes $L_1, \ldots, L_{i-1}$. In other words, for $i = 1, \ldots, k+1$ the domain of each $\boldsymbol{\eta}_i$ does not include the obstruction points of $\boldsymbol{\eta}_1, \ldots, \boldsymbol{\eta}_{i-1}$ — see page 200.

Denote the singularities of the potential ψ_j by $\mathbf{a}^j_\ell$, where $j = 1, \ldots, k$ and $\ell = 1, \ldots, g_j$. The obstruction points of $\boldsymbol{\eta}$ are then given by the solutions $\mathbf{x}_1$ to the following $g_1 + g_2 + \cdots + g_k$ equations:

$$\boldsymbol{\eta}_1(\mathbf{x}_1) = \mathbf{a}^1_\ell \qquad (\ell = 1, \ldots, g_1) \tag{15.56}$$

$$\boldsymbol{\eta}_j(\mathbf{x}_1) = \mathbf{a}^j_\ell \qquad (j = 2, \ldots, k;\ \ell = 1, \ldots, g_j). \tag{15.57}$$

Note that no two of the equations in (15.56) and (15.57) have a common solution $\mathbf{x}_1$ (since by contruction of $\boldsymbol{\eta}_j$, the obstruction points of $\boldsymbol{\eta}_j$ does not include those of $\boldsymbol{\eta}_1, \ldots, \boldsymbol{\eta}_{j-1}$). We shall say that the obstruction points of $\boldsymbol{\eta}$ are *regular obstruction points* if none of the singularities $\mathbf{a}^j_\ell$ is a critical value of the map $\boldsymbol{\eta}_j$ for $j = 1, \ldots, k$ and $\ell = 1, \ldots, g_j$

Obstruction points are difficult to count for an arbitrary lens system. In order to get a handle on the problem, we enforce two physically reasonable assumptions.

Throughout the remainder of this chapter, we shall suppose that $\boldsymbol{\eta}$ has finitely many obstruction points, all of which are regular obstruction points.

Equation (15.56) shows that the set of obstruction points B contains at least the g_1 singularities of ψ_1. For $j = 2, \ldots, k$, treat each singularity $\mathbf{a}^j_\ell$ of the potential ψ_j as a light source. Equation (15.57) allows us to interpret the lensed images of $\mathbf{a}^j_\ell$ (under the action of $\boldsymbol{\eta}_j$) as obstruction points. We can then obtain further counting information on $|\mathsf{B}|$ by associating the lensed images of $\mathbf{a}^j_\ell$ with critical point of a time delay function and then apply the Morse-theoretic results of Chapters 11 and 12.

Let us link obstruction points to critical points of a time delay function. The k-plane time delay family T is given by

$$T(\mathbf{x}_1, \ldots, \mathbf{x}_k, \mathbf{y}) = T_\mathbf{y}(\mathbf{x}_1, \ldots, \mathbf{x}_k, \mathbf{y}),$$

where $T_\mathbf{y}$ is the time delay function at $\mathbf{y}$:

$$T_\mathbf{y}(\mathbf{x}_1, \ldots, \mathbf{x}_k, \mathbf{y}) = \sum_{i=1}^{k} \vartheta_i \left[\frac{1}{2}|\mathbf{x}_i - \mathbf{x}_{i+1}|^2 - \beta_i \psi_i(\mathbf{x}_i)\right]$$

with $\mathbf{x}_{k+1} = \mathbf{y}$. Define a *partial time delay function* relative to T by

$$T_{\mathbf{x}_j}(\mathbf{x}_1, \ldots, \mathbf{x}_{j-1}) = \sum_{i=1}^{j-1} \vartheta_i \left[\frac{1}{2}|\mathbf{x}_i - \mathbf{x}_{i+1}|^2 - \beta_{i,i+1} \psi_i(\mathbf{x}_i)\right],$$

where $2 \leq j \leq k$. The critical points of $T_{\mathbf{x}_j}$ are the solutions $(\mathbf{x}_1, \ldots, \mathbf{x}_{j-1})$ of the following $(j-1)$ vector equations

$$\mathrm{grad}_{\mathbf{x}_i}[T_{\mathbf{x}_j}(\mathbf{x}_1, \ldots, \mathbf{x}_{j-1})] = \mathbf{0}, \qquad (i = 1, \ldots, j-1),$$

or, equivalently,

$$\begin{aligned} \mathbf{x}_2 &= \mathbf{x}_1 - \beta_1 \boldsymbol{\alpha}_1(\mathbf{x}_1) \\ \mathbf{x}_{i+1} &= \mathbf{x}_i + \frac{\vartheta_{i-1}}{\vartheta_i}(\mathbf{x}_i - \mathbf{x}_{i-1}) - \beta_i \boldsymbol{\alpha}_i(\mathbf{x}_i), \qquad (2 \le i \le j-1). \end{aligned} \tag{15.58}$$

By Theorem 6.13 (page 199), equation (15.58) is equivalent to (15.55). Hence, a point $(\mathbf{x}_1, \ldots, \mathbf{x}_{j-1})$ is a critical point of the partial time delay function $T_{\mathbf{x}_j}$ if and only if $\boldsymbol{\eta}_i(\mathbf{x}_1) = \mathbf{x}_i$ for $i = 1, \ldots, j-1$. Therefore, for $j = 2, \ldots, k$ and $\ell = 1, \ldots, g_j$, the critical points of $T_{\mathbf{a}_\ell^j}$ are in 1-1 correspondence with the obstruction points defined by equation (15.57).

With the previous correspondence available, we can obtain a lower bound on the number of solutions $\mathbf{x}_1$ of (15.57) via lower bounds on the number of critical points of all the partial time delay functions $T_{\mathbf{a}_\ell^j}$, where $j = 2, \ldots, k$ and $\ell = 1, \ldots, g_j$. We shall now use some of the results of Section 12.1.1 (page 446). Let

$$D_\ell^{j-1} = D_{g_1}^{j-1} \times \cdots \times D_{g_{j-1}}^{j-1},$$

where $D_{g_e}^{j-1}$ $(1 \le e \le j-1)$ is a closed disc punctured by g_e disjoint, open discs centered at the singularities of the potential ψ_e. Suppose that $D_\ell^{j-1} \subset \mathbf{R}^{2(j-1)}$ contains all the critical points of $T_{\mathbf{a}_\ell^j}$. Note that since we assume that there are finitely many obstruction points, each function $T_{\mathbf{a}_\ell^j}$ has finitely many critical points. Consequently, the space D_ℓ^{j-1} can always be made to contain all critical points of $T_{\mathbf{a}_\ell^j}$ (by choosing the radius and punctures of each disc $D_{g_e}^{j-1}$ $(1 \le e \le j-1)$ sufficiently large and small, respectively).

Let $\tilde{D}_\ell^{j-1}$ be a smoothing of D_ℓ^{j-1} that contains all the critical points of $T_{\mathbf{a}_\ell^j}$ (see Section 12.1.1, page 446, for smoothing corners in a product of manifolds with boundary). Without loss of genericity, we can suppose that $T_{\mathbf{a}_\ell^j}$ satisfies Morse boundary conditions B relative to $\tilde{D}_\ell^{j-1}$ (Theorem 12.1.2, page 451). Let $(\tilde{D}_\ell^{j-1})_-$ be the union of all open subsets of the boundary $\partial \tilde{D}_\ell^{j-1}$ on which grad $T_{\mathbf{a}_\ell^j}$ is directed into the interior of $\tilde{D}_\ell^{j-1} \subset \mathbf{R}^{2(j-1)}$. Set

$$\begin{aligned} N_\ell'^{j-1} &= \text{number of critical points of } T_{\mathbf{a}_\ell^j}|_{(\tilde{D}_\ell^{j-1})_-} \\ N_\ell^{j-1} &= \text{number of critical points of } T_{\mathbf{a}_\ell^j}. \end{aligned}$$

The critical points of $T_{\mathbf{a}_\ell^j}$ obey the Morse-theoretic, lensed-image-counting relations in Theorem 12.1 (page 448) with $k = j-1$, $\mathbf{y} = \mathbf{a}_\ell^j$, and $D = \tilde{D}_\ell^{j-1}$. In particular, Theorem 12.1(3, ii) implies that

$$N_\ell^{j-1} \ge \prod_{e=1}^{j-1}(1 + g_e) - N_\ell'^{j-1}.$$

Applying this to (15.57), we see that for each j, where $2 \leq j \leq k$, the g_j equations $\boldsymbol{\eta}_j(\mathbf{x}_1) = \mathbf{a}^j_\ell$ ($\ell = 1, \ldots, g_j$) have at least the following number of solutions:

$$g_j \left[\left(\prod_{e=1}^{j-1} (1 + g_e) \right) - \frac{N'^{j-1}}{g_j} \right], \qquad \text{where} \quad N'^{j-1} = \sum_{\ell=1}^{g_j} N'^{j-1}_\ell .$$

Summing over all lens planes, we obtain a lower bound on the number of solutions to (15.57).

The results of the previous work are summarized in the following [Pet95b, p. 4279]:

Theorem 15.4 *Let T and $\boldsymbol{\eta}$ be the k-plane time delay family and lensing map, respectively, due to the potentials $\psi_1, \ldots, \psi_k$. Denote the set of singularities of ψ_i by $A_i = \{\mathbf{a}^i_1, \ldots, \mathbf{a}^i_{g_i}\}$, where $i = 1, \ldots, k$. Suppose that each partial time delay function $T_{\mathbf{a}^i_\ell}$ with respect to T obeys Morse boundary conditions B relative to $\tilde{D}^{i-1}_\ell$, where $i = 2, \ldots, k$ and $\ell = 1, \ldots, g_i$. Assume that the obstruction points B of $\boldsymbol{\eta}$ are regular obstruction points and $|\mathsf{B}|$ is finite. Then:*

(1) *If $k = 1$, then $\mathsf{B} = A_1$ and $|\mathsf{B}| = g_1$. For $k > 1$, then B contains A_1 and $\mathsf{B} - A_1$ is in 1-1 correspondence with the critical points of all the functions $T_{\mathbf{a}^i_\ell}$, where $i = 2, \ldots, k$ and $\ell = 1, \ldots, g_i$. Consequently,*

$$|\mathsf{B}| = g_1 + \sum_{i=2}^{k} N^{i-1},$$

where $N^{i-1} = \sum_{\ell=1}^{g_i} N^{i-1}_\ell$ with N^{i-1}_ℓ the number of critical points of $T_{\mathbf{a}^i_\ell}$.

(2) *If $k > 1$, then*

$$|\mathsf{B}| \geq g_1 + \sum_{i=2}^{k} g_i \left[\prod_{j=1}^{i-1} (1 + g_j) - \frac{N'^{i-1}}{g_i} \right].$$

The quantity N'^{i-1} vanishes if each $T_{\mathbf{a}^i_\ell}$ obeys Morse boundary conditions A relative to $\tilde{D}^{i-1}_\ell$.

Let us illustrate how Theorem 15.4(1) can be used to count obstruction points. Consider two point masses on separate planes with continuous matter and shear from infinity. If the continuous matter on each lens plane is sufficiently over-focused, then the caustics have the form in Figure 15.16. The deflector on the first lens plane L_1 is an over-focused Chang–Refsdal lens that produces an oval, fold caustic c on the second lens plane L_2. If the mass m_2 on L_2 is inside c, then it will have no lensed images on L_1. Hence, such a lens configuration has exactly *one* obstruction point — Figure 15.16(a). This is less than the total number of point masses. Thus, it is not true in general that $|\mathsf{B}| > g_1 + \cdots + g_k$ (total number of

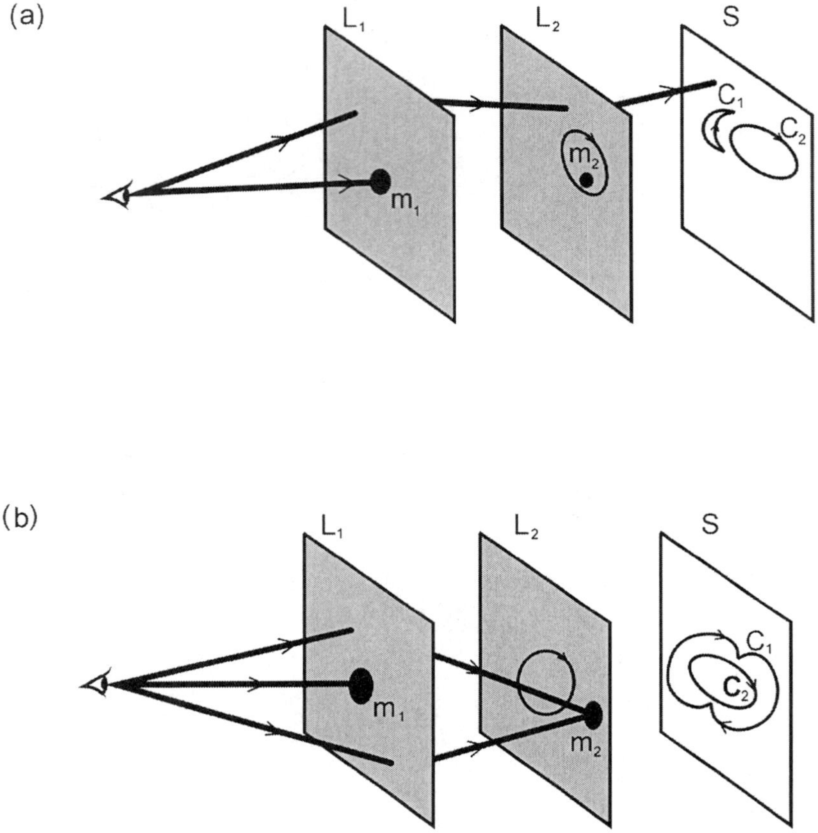

Figure 15.16. Gravitational lensing by masses m_1 and m_2 on lens planes with shear from infinity and over-focused continuous matter. In (a), there is one obstruction point corresponding to the position of m_1. In (b), the obstruction points consist of the position of m_1 and the impact points on L_1 of the two light rays converging at m_2 (i.e., the two lensed images of m_2 under the lensing action of m_1).

singularities). If m_2 is outside c, then there are two lensed images of m_2 on L_1. For such a configuration, there are *three* obstruction points — Figure 15.16(b). These results show that in the multiplane case $k > 1$, the total number of obstruction points is *not* independent of the positions of the singularities of the potentials. This is unlike the case for single plane lensing, where the number of obstructions is the number of singularities of the potential (Theorem 15.4(1)).

Theorem 15.4(2) shows that if each partial time delay function $T_{\mathbf{a}_\ell^i}$ obeys Morse boundary conditions A relative to $\tilde{D}_\ell^{i-1}$ and every potential ψ_i has only one singularity, then

$$|\mathsf{B}| \geq 2^k - 1, \qquad |\mathsf{B}| = \text{odd number} \qquad (k \geq 2). \tag{15.59}$$

The parity of $|\mathsf{B}|$ follows from Theorem 15.4(1) Since $g_1 = 1$ and N^{i-1} is even (see Theorem 12.1(3,ii), page 448 with $M'' = 0$ and $g_i = 1$), the parity of $|\mathsf{B}|$ follows from Theorem 15.4(1). This situation applies to point mass lensing. For example, suppose that two point masses m_1 and m_2 are on separate lens planes L_1 and L_2, respectively. If m_2 is not located at the point caustic created by m_1 on L_2, then m_1 creates on L_1 two lensed images of m_2. These two lensed images and the position of m_1 are the *three* obstruction points due to the two masses.

For g point masses in a three-dimensional array with one mass on each lens plane, we can apply Theorem 12.10 (page 460) to obtain an upper bound on the number of obstruction points. Combined with (15.59), we get

$$2^g - 1 \leq |\mathsf{B}| \leq 3 + N_g, \qquad (g \geq 2), \tag{15.60}$$

where $|\mathsf{B}|$ is an odd number, $N_2 = 0$, and $N_g = 2\sum_{i=2}^{g-1}[2^{2(i-1)} - 1]$ for $g \geq 3$. Equation (15.60) implies that $7 \leq |\mathsf{B}| \leq 9$ for $g = 3$; hence, $|\mathsf{B}| = 7$ or 9 (since $|\mathsf{B}|$ is odd). We can also show this result directly. Theorem 12.10 (page 460) yields that the first two point masses would produce either *four* or *six* lensed images of the third mass, depending on whether the third mass is inside or outside one of the caustic curves due to the first two masses. Note that these caustic curves have no self-intersections. Hence, the set B consists of the position of the first point mass, the two lensed images of the second point mass due to the first, and the four or six lensed images of the third mass due to the lensing action of the first two masses: $|\mathsf{B}| = [1 + 2 + (4 \text{ or } 6)] = 7$ or 9.

15.4.2 Linking Curvature and Obstruction Points

For a k-plane gravitational lens system, consider the number $|\mathsf{B}|$ of light path obstruction points and total signed-curvature K_f of its fold caustics. By utilizing the fact that $|\mathsf{B}|$ coincides with the first Betti number of the domain of the associated lensing map and explicitly calculating K_f for transverse stable caustics, A.P. [Pet95b, p. 4292] realized that H.L.'s work on computing the Euler characteristic via projectivized rotation numbers (Theorem 14.6, page 497) can be applied to obtain a formula connecting K_f to $|\mathsf{B}|$. The latter also yields a formula linking the total signed-curvature K_{tot} of the caustics $|\mathsf{B}|$ and the number N_{cusps} of cusps. The proof and applications of these results will be the focus of the current section.

Preliminaries

The following concepts, notation, and assumptions are necessary:

Let a be a regular planar curve parametrized by arc length s. If $a(0)$ is the initial point of a, then $a(0)$ is an *outside starting point* if there is a *line of support* at $a(0)$, that is, a straight line L through $a(0)$ such that either L coincides with the path traced out by a or the path lies on one side of L. *Henceforth, assume each caustic c is (piecewise) parametrized according to arc length and has a specified outside starting point.* Note that the latter is defined even if the initial point of c is a cusp.

Let $\boldsymbol{\eta} : P \to S$ be a transverse stable k-plane lensing map $\boldsymbol{\eta}$. The transverse stability of $\boldsymbol{\eta}$ is equivalent to: Each caustic point of $\boldsymbol{\eta}$ is either a fold or cusp; all fold caustics curves intersect at nonzero angles; no more than two fold caustic curves meet at any point; no two cusp caustics meet at a point; and a fold caustic curve never passes through a cusp (Theorem 8.6, page 294). In addition, transverse stability is a generic property of lensing maps (Theorem 8.17, page 311).

Suppose that c is a closed, caustic curve of $\boldsymbol{\eta}$ that is parametrized by arc length s in the direction of the orientation of c. Recall that caustics are oriented so that, locally, point sources to the left of c have two more lensed images than those to the right (see page 341). We assume that the initial point of c corresponds to $s = 0$. Write c as a union of regular arcs ordered with respect to the orientation of c, say

$$c = v_1 v_2 \cup v_2 v_3 \cup \cdots \cup v_{N_{cusps}(c)} v_1 .$$

Here the point v_i is a cusp, $v_{i-1} v_i$ is the (open) fold arc from v_{i-1} to v_i, $N_{cusps}(c)$ is the number of cusps on c, and the initial point of c lies on the arc $v_1 v_2$. Let k_c^i be the signed-curvature function along the oriented arc $v_i v_{i+1}$, where $v_{N_{cusps}(c)+1} = v_1$. Explicitly,

$$k_c^i(s) =< \frac{d\mathbf{t}}{ds}(s), \mathbf{n}(s) > .$$

The symbols $<, >$ denote the $\mathbf{R}^2$ dot product and $\mathbf{t}(s)$, $\mathbf{n}(s)$ are the unit tangent and normal, respectively, at $(v_i v_{i+1})(s)$. The *total signed fold curvature* of a caustic $c = v_1 v_2 \cup v_2 v_3 \cup \cdots \cup v_{N_{cusps}(c)} v_1$ is then defined by

$$K_f(c) = \sum_{i=1}^{N_{cusps}(c)} \int_{s_{i-1}}^{s_i} k_c^i(s) ds,$$

where $0 = s_0 < s_1 < \cdots < s_{N_{cusps}(c)}$ with $s_{N_{cusps}(c)}$ the length of c, and $k_c^i(s)$ is the curvature function along $v_i v_{i+1}$. The quantity K_f measures the total bending of the fold arcs of c. Since each cusp contributes an external angle of $+\pi$, the *total signed-curvature* of c is defined by

$$K_{tot}(c) = K_f(c) + \pi N_{cusps}(c). \tag{15.61}$$

Set

$$K_{tot} = \sum_c K_{tot}(c), \qquad N_{cusps} = \sum_c N_{cusps}(c),$$

where the sum runs over all caustics of $\boldsymbol{\eta}$.

Next, consider the obstruction points B of the transverse stable lensing map $\boldsymbol{\eta} : P \to S$. Define a closed, $|\mathsf{B}|$-holed disc $\mathcal{D} \subset P$ to be *punctured by* B if each hole of $\mathcal{D}$ contains exactly one element of B. Suppose that all critical curves of $\boldsymbol{\eta}$ are both contained in $\mathcal{D}$ and disjoint from the boundary of $\partial\mathcal{D}$. Then critical curves of $\boldsymbol{\eta}$ consist of a finite collection of nonintersecting, smoothly imbedded circles in $\mathcal{D}$ (see page 489). In addition, each component b_i of $\partial\mathcal{D}$ is a circle that is immersed by $\boldsymbol{\eta}$. We orient each of these immersed circles $\boldsymbol{\eta}(b_i)$ as boundary, locally, of the $\boldsymbol{\eta}$-image of its collar neighborhood. The rotation number $R(b_i)$ of $\boldsymbol{\eta}|_{b_i}$ is determined with $\boldsymbol{\eta}(b_i)$ having the previous orientation. For example, suppose that

$$|\boldsymbol{\eta}(\mathbf{x}_1)| \to \infty \text{ as } |\mathbf{x}_1| \to \infty \text{ and } \mathbf{x}_1 \to \mathsf{B}$$

(as in the case of point mass lensing). For $\mathcal{D}$ with sufficiently large radius and sufficiently small holes, the $\boldsymbol{\eta}$-image of the outer and inner boundary circles will be counterclockwise oriented. In fact, if b^{out} is the outer boundary circle of $\mathcal{D}$, then a collar neighborhood of b^{out} is mapped to a neighborhood to the left of $\boldsymbol{\eta}(b^{out})$ (because $|\boldsymbol{\eta}(\mathbf{x}_1)| \to \infty$ as $|\mathbf{x}_1| \to \infty$). Since points closer to an obstruction point get mapped further away than those not as close ($|\boldsymbol{\eta}(\mathbf{x}_1)| \to \infty$ as $\mathbf{x}_1 \to \mathsf{B}$), a collar neighborhood of an inner boundary curve b_i^{inn} of $\partial\mathcal{D}$ is mapped to a neighborhood to the left of $\boldsymbol{\eta}(b_i^{inn})$. Hence, the curves $\boldsymbol{\eta}(b^{out})$ and $\boldsymbol{\eta}(b_i^{inn})$ are counterclockwise. Thus, $R(b^{out}) = +1 = R(b_i^{inn})$.

Main Theorem

With the previous concepts and notation available, we state this section's central theorem [Pet95b, p. 4292]:

Theorem 15.5 *Let $\boldsymbol{\eta} : P \to S$ be a transverse stable k-plane lensing map with obstruction points* B*. Assume that there is a closed, $|\mathsf{B}|$-holed disc $\mathcal{D} \subset P$ that is punctured by* B *and contains all critical curves of $\boldsymbol{\eta}$. The critical curves of $\boldsymbol{\eta}$ are nonintersecting, smoothly embedded circles in $\mathcal{D}$ and each is assumed to be disjoint from the boundary $\mathcal{D}$. Suppose that for every component b_i of $\partial\mathcal{D}$, the rotation number of $\boldsymbol{\eta}|_{b_i}$ is $R(b_i) = +1$.*[4] *Then the total signed fold curvature of the caustics of $\boldsymbol{\eta}$ is*

$$K_f = -2\pi|\mathsf{B}|.$$

Proof: Let $c_i^{\pm}$ be the counterclockwise and clockwise Jordan curves (i.e., simple, closed curves) in the Seifert decomposition of a caustic c of $\boldsymbol{\eta}$. An illustration of this decomposition appears in Figure 15.17.

Translate the curves $c_i^{\pm}$ in the Seifert decomposition of c so that none intersects the interior of any other. Observe that if the oriented curves $c_i^{\pm}$ are considered relative to the usual orientation as boundary of their (disjoint) simply connected interiors, then the orientation of

[4] This condition is satisfied by lensing maps obeying: $|\boldsymbol{\eta}(\mathbf{x}_1)| \to \infty$ as $|\mathbf{x}_1| \to \infty$ and $\mathbf{x}_1 \to \mathsf{B}$.

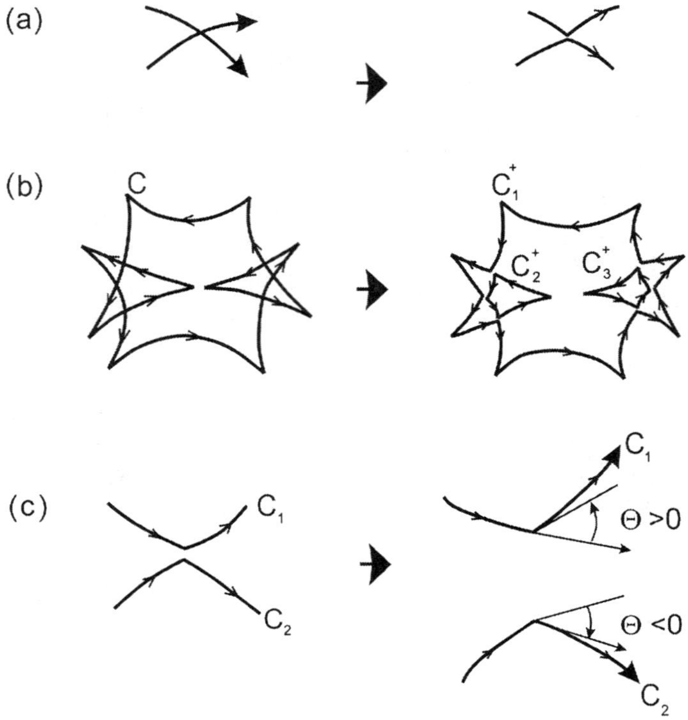

Figure 15.17. (a) Seifert decomposition of a transverse crossing of two curves. (b) Application of a Seifert decomposition at each self-intersection of a transverse stable caustic curve results in a collection of simple closed curves. (c) All exterior angles Θ at corners due to a Seifert decomposition, occur twice with opposite sign.

c_i^+ coincides with that induced as a boundary curve, while c_j^- has the opposite orientation. The Theorem of Turning Tangents (e.g., [doC, p. 267]) yields

$$K_{tot}(c_i^{\pm}) = \pm 2\pi,$$

where $+2\pi$ (resp., -2π) is associated with c_i^+ (resp., c_i^-). Using the fact that (relative to our orientation of caustics) cusps contribute an external angle of $+\pi$, we have

$$K_f(c_i^{\pm}) = \pm 2\pi \quad - \sum_{j=1}^{N_{crn}(c_i^{\pm})} \theta_j^{\pm} - \pi N_{cusps}(c_i^{\pm}).$$

Here $N_{crn}(c_i^{\pm})$ is the number of non-cusped corners on $c_i^{\pm}$, and $\theta_j^{\pm}$ is the external angle at the jth corner of $c_i^{\pm}$. In the previous equation, a choice of $+$ (resp., $-$) on the left is to be matched with choices of $+$ (resp., $-$) on the right, and a corner of $c_i^{\pm}$ is a point where $c_i^{\pm}$ is not regular (e.g., Figure 15.17(c)).

Let N_c^+ and N_c^- be, respectively, the number of counterclockwise and clockwise Jordan curves in the Seifert decomposition of c. By construction of Seifert decompositions, all exterior angles at corners occur twice with opposite orientation (see Figure 15.17(c)). Consequently,

$$\sum_{i=1}^{N_c^+} \sum_{j=1}^{N_{crn}(c_i^+)} \theta_j^+ + \sum_{i=1}^{N_c^-} \sum_{j=1}^{N_{crn}(c_i^-)} \theta_j^- = 0.$$

Hence,

$$K_f(c) = \sum_{i=1}^{N_c^+} K_f(c_i^+) + \sum_{i=1}^{N_c^-} K_f(c_i^-) = 2\pi(N_c^+ - N_c^-) - \pi N_{cusps}(c). \tag{15.62}$$

Every cusp of c can be rounded off, leaving $N_{cusps}(c)$ hypocycloid curves c_i^Δ (one for each cusp) and a regular curve $\tilde{c}$ — see Figure 15.18.

Since $\tilde{c}$ is regular, we have

$$K_f(\tilde{c}) = K_{tot}(\tilde{c}) = 2\pi R(\tilde{c}). \tag{15.63}$$

Let $\tilde{c}_i^\pm$ be the counterclockwise and clockwise Jordan curves in the Seifert decomposition of $\tilde{c}$. Then the Theorem of Turning Tangents implies that

$$K_f(\tilde{c}_i^\pm) = \pm 2\pi \;-\; \sum_{j=1}^{N_{crn}(c_i^\pm)} \theta_j^\pm.$$

Note that $N_{crn}(\tilde{c}_i^\pm) = N_{crn}(c_i^\pm)$. Consequently,

$$K_f(\tilde{c}) = \sum_{i=1}^{N_c^+} K_f(\tilde{c}_i^+) + \sum_{i=1}^{N_c^-} K_f(\tilde{c}_i^-) = 2\pi(N_c^+ - N_c^-). \tag{15.64}$$

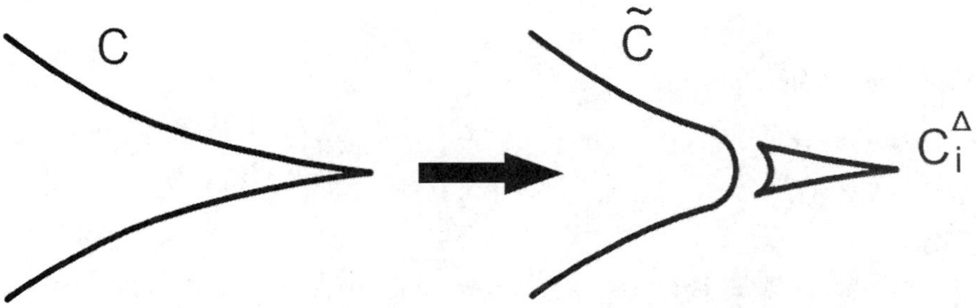

Figure 15.18. Rounding off each cusp on c yields a regular curve $\tilde{c}$ and hypocycloid caustics c_i^Δ.

By (15.62) and (15.64), we get

$$K_f(c) = K_f(\tilde{c}) - \pi N_{cusps}(c).$$

Note that $K_{tot}(c) = K_f(\tilde{c}) = K_{tot}(\tilde{c})$. Equation (15.63) then yields

$$K_f(c) = 2\pi R(\tilde{c}) - \pi N_{cusps}(c). \tag{15.65}$$

Now, each hypocycloid curve $c_i^{\triangle}$ obtained from rounding off a cusp of c has projectivized rotation number -1 (see Figure 14.2, page 493). Consequently, the projectivized rotation number of c is

$$r(c) = r(\tilde{c}) + \sum_{i=1}^{N_{cusps}(c)} r(c_i^{\triangle}) = 2\pi R(\tilde{c}) - \pi N_{cusps}(c). \tag{15.66}$$

Equations (15.65) and (15.66) then imply that

$$K_f(c) = \pi r(c). \tag{15.67}$$

By Theorem 14.7 (page 497), we can express the Euler characteristic of $\mathcal{D}$ in terms of projectivized rotation numbers:

$$\chi(\mathcal{D}) \;=\; \sum_{i=1}^{N_{caustics}} r(c_i) + \sum_{j=1}^{N_{bd}} R(b_j). \tag{15.68}$$

Here $N_{caustics}$ is the total number of caustics (or critical curves) of $\boldsymbol{\eta}$ and N_{bd} is the number of components b_j of $\partial\mathcal{D}$. Since $\chi(\mathcal{D}) = 1 - |\mathsf{B}|$, $N_{bd} = 1 + |\mathsf{B}|$, and $R(b_j) = +1$, equations (15.67) and (15.68) imply that

$$K_f = \sum_{i=1}^{N_{caustics}} K_f(c_i) = \pi \sum_{i=1}^{N_{caustics}} r(c_i) = -2\pi |\mathsf{B}|. \square$$

Applications of Main Theorem

Theorem 15.5 links the geometry of caustics to the "opacity" of a gravitational lens:

$$K_f = -2\pi |\mathsf{B}|.$$

Note that the "opacity" $|\mathsf{B}|$ is a topological invariant (since $|\mathsf{B}| = B_1(\mathcal{D})$ — the first Betti number of $\mathcal{D}$). For the case of nonsingular, k-plane gravitational lens system, the deflector on each lens plane is transparent (no "opacity"), so

$$K_f = 0.$$

By using equation (15.65) to calculate K_f, the reader can immediately verify the result directly for an isothermal ellipsoid with softened core or an elliptical potential (e.g., Figure 9.9 on page 337, where all the caustics are counterclockwise).

In the situation of singular lenses, Theorem 15.4 (page 547) yields a counting formula and bound for $|\mathsf{B}|$, hence, K_f. For example, application of Theorem 15.4(1) shows that a single-plane lens with g point masses, continuous matter, and shear from infinity has:

$$K_f = -2\pi g.$$

This can be seen directly via (15.65) for the caustics of a Chang–Refsdal lens in Figure 15.3 (the oval in panel 10 is clockwise) and binary point masses in Figure 15.5 (all the caustics are counterclockwise).

Let us use the lower bound in equation (15.60) (page 549), which is a consequence of Theorem 15.4(2), to contrast the geometry of caustics due to single and multiple plane lensing. If $K_f^{(2D)}$ is the total signed fold curvature when g point masses are on the same lens plane, and if $K_f^{(3D)}$ is the total signed fold curvature when these point masses are spread over g lens planes with one on each lens plane, then

$$K_f^{(3D)} \leq -2\pi(2^g - 1) < K_f^{(2D)} = -2\pi g.$$

The caustics are more negatively curved for the three-dimensional point mass configuration. For example, the caustics of two point masses on the same plane have total fold curvature $K_f^{(2D)} = -4\pi$. If the masses are put (generically) on separate planes, then there are three obstruction points by (15.60) and, hence, $K_f^{(3D)} = -2\pi(3) = -6\pi$. These results are shown in Figure 15.19.

Theorem 15.5, along with equation (15.61), also shows that the total signed-curvature of the caustics of $\boldsymbol{\eta}$ is given by

$$K_{tot} = -2\pi|\mathsf{B}| + \pi N_{cusps}.$$

We see that K_{tot} varies with smooth parameter evolutions of $\boldsymbol{\eta}$ that preserve $|\mathsf{B}|$, but change N_{cusps}. This is unlike the case for K_f.

Nonsingular lenses have

$$K_{tot} = \pi N_{cusps}.$$

In the singular situation of lensing by g point masses on a plane with continuous matter κ_c and shear γ from infinity, we have $|\mathsf{B}| = g$ (by Theorem 15.4(1)). Consequently, the bounds on N_{cusps} in Theorem 15.3 (page 536) immediately yield:

$$-2\pi g \leq K_{tot} \leq \begin{cases} 2\pi g(6g-1) & \text{if } \gamma > 0 \\ 2\pi g(6g-7) & \text{if } \gamma = 0. \end{cases}$$

Moreover, if κ_c is sufficiently large, then

$$K_{tot} = -2\pi g.$$

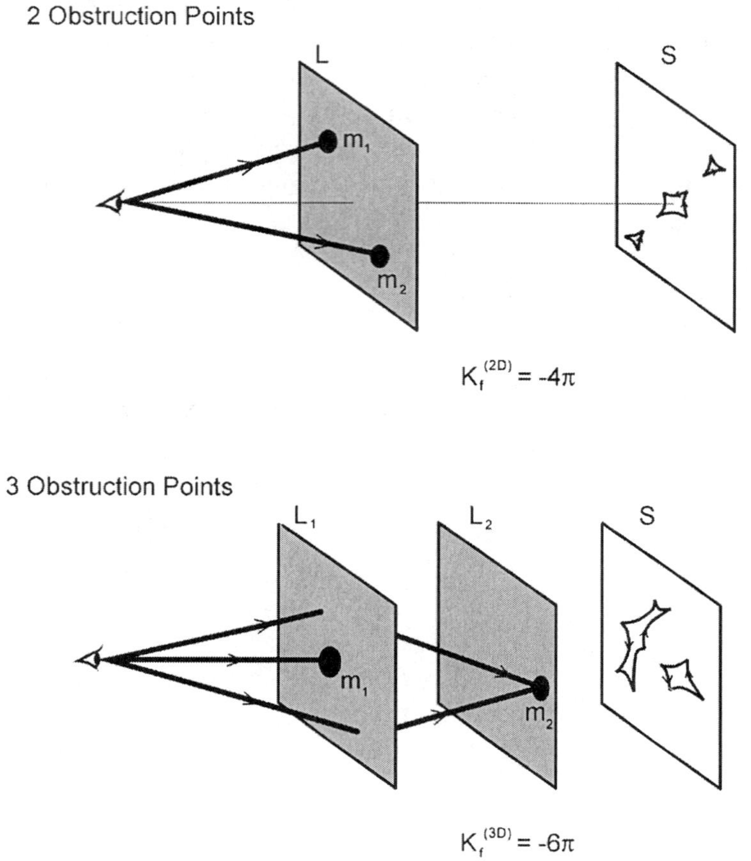

Figure 15.19. Gravitational lensing by two points masses in single (top figure) and double-plane (bottom figure) lensing configurations. The total signed-curvature of the fold caustics is given by $K_f = -2\pi|\mathsf{B}|$, where $|\mathsf{B}|$ is the number of obstruction points. The single plane case has two obstruction points (position of m_1 and m_2), while three occur for the two-plane case (position of m_1 plus the two impact points on L_1 of the rays converging at m_2). The fold caustics in the double-plane case are more negatively curved than those in the single plane case: $-6\pi = K_f^{(3D)} < K_f^{(2D)} = -4\pi$.

This agrees with the fact that for κ_c sufficiently large, the caustics consist of g clockwise, oval fold curves (Corollary 15.2).

The remainder of this section will focus on a counting formula for the total number of cusps. The following concepts and notation will be needed: If by a rigid motion the path traced out by a regular curve $a(s)$ (parameterized by arc length s) is positioned such that the x-axis is a line of support at $a(0)$ and the path lies on the side of the positive y-direction, then a is called *positive* (resp., *negative*) if the initial tangent vector $\dot{a}(0)$ is in the positive (resp., negative) x-direction. *In the case of a caustic c, if the initial point $c(0)$ is a cusp, then* *c is assumed positive.* Figure 15.20 shows a positive caustic produced by a Chang–Refsdal lens. Let

$$N^{\pm}_{caustics} = \text{number of positive/negative caustics of } \boldsymbol{\eta}.$$

If $a(s_1) = a(s_2)$, where $s_1 < s_2$, is a normal self-intersection of a, and if the ordered pair $\{\dot{a}(s_1), \dot{a}(s_2)\}$, where $\dot{a}(s_i)$ is the tangent vector at $a(s_i)$, is opposite (resp., identical) to the xy-orientation of the plane, then the self-intersection is called *positive* (resp., *negative*). Let

$$N^{\pm}_{self}(c) = \text{number of positive/negative self-intersections of caustic } c.$$

For example, the caustic c in Figure 15.20 has $N^{+}_{self}(c) = 3$ and $N^{-}_{self}(c) = 1$. Set

$$N^{\pm}_{self} = \sum_{c} N^{\pm}_{self}(c),$$

where the sum runs over all caustics c of $\boldsymbol{\eta}$.

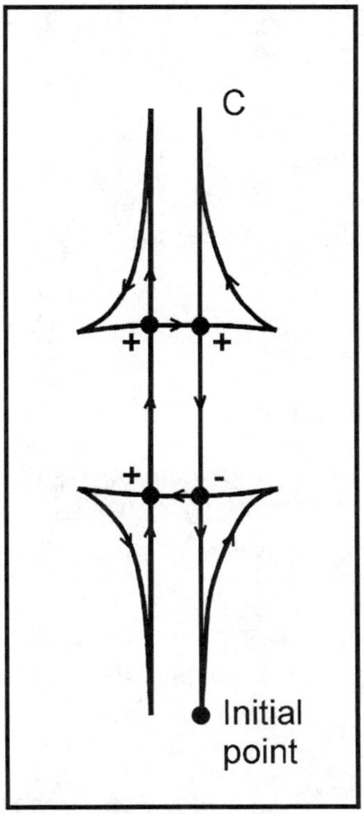

Figure 15.20. Positive and negative self-intersections of a positive caustic due to an over-focused Chang–Refsdal lens.

We now present a formula observed by A.P. [Pet95b, p. 4282] that expresses the total number of cusps in terms of the number of obstructions points and the algebraic number of caustics and self-intersections:

Theorem 15.6 (Cusp-Counting Formula) *Assume the hypotheses of Theorem 15.5. Then the total number of cusps of the k-plane lensing map η is even and given by*

$$N_{cusps} = 2\Big[|\mathsf{B}| + (N^{+}_{caustics} - N^{-}_{caustics}) + (N^{+}_{self} - N^{-}_{self})\Big].$$

Proof: Let c be a caustic curve of η and consider the regular curve $\tilde{c}$ obtained by rounding off the cusps of c. Then $\tilde{c}$ is a positive (resp., negative) curve if and only if c is positive (resp., negative). Also, since $\tilde{c}$ and c have the same self-intersections, we have $N^{\pm}_{self}(\tilde{c}) = N^{\pm}_{cusps}(c)$. Application of Whitney's Rotation Number Formula [Whi37, p. 281] then yields

$$R(\tilde{c}) = \pm 1 + N^{+}_{self}(c) - N^{-}_{self}(c), \tag{15.69}$$

where a plus or minus is chosen in the first term on the right depending on whether c is positive or negative. Equations (15.65) and (15.69) now imply that

$$\begin{aligned} N_{cusps} &= \sum_{i=1}^{N_{caustics}} N_{cusps}(c_i) \\ &= \frac{1}{\pi} \sum_{i=1}^{N_{caustics}} \big[-K_f(c_i) + 2\pi R(\tilde{c}_i)\big] \\ &= -\frac{K_f}{\pi} + 2\Big[(N^{+}_{caustics} - N^{-}_{caustics}) + (N^{+}_{self} - N^{-}_{self})\Big]. \end{aligned}$$

The result now follows from Theorem 15.5. □

Let us illustrate Theorem 15.6 with the caustic in Figure 15.20. This figure shows a caustic due to a Chang–Refsdal lens. We have $|\mathsf{B}| = 1$, $N^{+}_{caustics} = 1$, $N^{-}_{caustics} = 0$, $N^{+}_{self} = 3$, and $N^{-}_{self} = 1$. Theorem 15.6 then gives $N_{cusps} = 2[1 + (1-0) + (3-1)] = 8$, which agrees with the figure.

If the caustics of η have no self-intersections (though different caustics may cross), then $N^{\pm}_{caustics}$ are, respectively, the number of counterclockwise and clockwise caustics. In this case, Theorem 15.6 implies that

$$N_{cusps} = 2(|\mathsf{B}| + N^{+}_{caustics} - N^{-}_{caustics}).$$

For example, a single-plane binary lens has one, two, or three caustics (Figure 15.4), none of which has self-intersections (see page 390). All the caustics are counterclockwise (because five lensed images occur of light sources inside a caustic, while sources outside all the caustics have three). Since $|\mathsf{B}| = 2$, we then have

$$N_{cusps} = 4 + 2N_{caustics}.$$

This yields *six*, *eight*, or *ten* cusps depending on whether there is one, two, or three caustics, respectively (cf. Figure 15.5). If the two point masses are on separate lens planes (with the more distant mass not at the point caustic due to the closer mass), then equation (15.60) yields $|\mathsf{B}| = 3$. Also, the caustics have no transverse self-intersections (see page 391) and are all counterclockwise oriented (since sources inside the caustics have six lensed images, while those outside all the caustics have four). It follows that

$$N_{cusps} = 6 + 2N_{caustics}.$$

The lens system produces two, three, or four caustics (see [Erd-S, p. 463]), which then have 10, 12, or 14 cusps, respectively.

Bibliography

[Ab-M] Abraham, R., and Marsden, J.E., *Foundations of Mechanics* (Benjamin/Cummings, Reading, 1978).

[Ada] Adams, C., *The Knot Book*, W.H. Freeman, New York, 1994.

[Adl] Adler, R., *The Geometry of Random Fields*, John Wiley and Sons, New York, 1981.

[Ala-MG] Alard, C., Mao, S., Guibert, J., Object DUO2: a new binary lens candidate?, *Astron. Astrophys.* **300**, L17 (1995).

[Alb96] Albrow, M., et al., "The PLANET collaboration," in: *IAU Symposium 173*, eds. C.S. Kochanek and J.N. Hewitt, p. 227, Kluwer, Dordrecht, 1996.

[Alb99] Albrow, M., et al. (The PLANET collaboration), Limb darkening of a K giant in the galactic bulge: Planet photometry of Macho 97-BLG-28, *Astrophys. J.*, **522**, 1011 (1999).

[Alb00a] Albrow, M., et al. (The PLANET collaboration), Limits on stellar and planetary companions in microlensing event OGLE-1998-BUL-14, *Astrophys. J.*, **535**, 176 (2000).

[Alb00b] Albrow, M., et al. (The PLANET collaboration), PLANET observations of microlensing event OGLE-1999-BUL-23: limb darkening measurement of the source star, http://arXiv.org/abs/astro-ph/0004243, (2000).

[Alc93] Alcock, C., Akerlof, C.W., Allsman, R.A., Axelrod, T. S., Bennett, D.P., et al., Possible gravitational microlensing of a star in the Large Magellanic Cloud, *Nature* **365**, 621 (1993).

[Alc95] Alcock, C., Allsman, R.A., Alves, D., Axelrod, T.S., Bennett, D.P., Cook, K.H., et al. (The MACHO Collaboration), First observation of parallax in a gravitational microlensing event, *Astrophys. J.*, **454**, L125 (1995).

[And] Anderson, M.R., "Gravitational Lensing by Curved Cosmic Strings," *in: IAU Symposium 173*, eds. C.S. Kochanek and J.N. Hewitt, p. 377, Kluwer, Dordrecht, 1996.

[Ans] Ansari, R., Aurière, M., Baillon, P., Bouquet, A., Coupinot, G., et al., AGAPE: a search for dark matter towards M31 by microlensing effects on unresolved stars, *Astron. Astrophys.* **324**, 843 (1997).

[Arf-W] Arfken, G., and Weber, H., *Mathematical Methods for Physicists*, 4th Ed., Academic Press, San Diego, 1995.

[Arn73] Arnold, V.I., Normal forms for functions near degenerate critical points, the Weyl groups of A_k, D_k, E_k and Lagrangian singularities, *Funct. Anal. Appl.* **6**, 254 (1973).

[Arn86] Arnold, V.I., Evolution of singularities of potential flows in collision-free media and the metamorphoses of caustics in three-dimensionl space, *J. Sov. Math.* **32**, 229 (1986).

[Arn89a] Arnold, V.I. (Ed.), *Dynamical Systems V: Bifurcation Theory and Catastrophes*, Springer-Verlag, Berlin, 1989.

[Arn89b] Arnold, V.I., *Mathematical Methods of Classical Mechanics*, 2nd Ed., Springer-Verlag, Berlin, 1989.

[Arn90] Arnold, V.I. (Ed.), *Dynamical Systems IV: Symplectic Geometry and Applications*, Springer-Verlag, Berlin, 1990.

[Arn91] Arnold, V.I., *Singularities of Caustics and Wave Fronts*, Kluwer Academic Publishers, Dordrecht, 1991.

[Arn92] Arnold, V.I., *Catastrophe Theory*, 3rd Ed., Springer-Verlag, Berlin, 1992.

[Arn93a] Arnold, V.I. (Ed.), *Dynamical Systems VI: Singularity Theory I*, Springer-Verlag, Berlin, 1993.

[Arn93b] Arnold, V.I.(Ed.), *Dynamical Systems VIII: Singularity Theory II*, Springer-Verlag, Berlin, 1993.

[Arn-GVI] Arnold, V.I., Gusein-Zade, S.M., and Varchenko, A.N., *Singularities of Differentiable Maps. Volume I*, Birkhäuser, Boston, 1985.

[Arn-GVII] Arnold, V.I., Gusein-Zade, S.M., and Varchenko, A.N., *Singularities of Differentiable Maps. Volume II*, Birkhäuser, Boston, 1985.

[Art] Artin, M., *Algebra*, Prentice Hall, Englewood Cliffs, 1991.

[Aub] Aubourg, E., Bareyre, P., Bréhin, S., Gros, M., Lachlèze-Rey, M., et al., Evidence for gravitational microlensing by dark objects in the Galactic halo, *Nature* **365**, 623 (1993).

[Bac] Bacon, B., Detection of weak gravitational lensing by large-scale structure, *Mon. Not. R. Astron. Soc.* **318**, 625 (2000).

[Bah-O] Bahcall, J.N., and Ostriker, J.P., eds., *Unsolved Problems in Astrophysics*, Princeton University Press, Princeton, 1997.

[Ban-GM] Banchoff, T., Gaffney, T., and McCrory, C., *Cusps of Gauss Mappings*, Pitman Research Notes in Mathematics, Vol. 55, Pitman Publishing Inc., London, 1982.

[Bar] Bar-Kana, R., Effect of Large-Scale Structure on Multiply Imaged Sources, *Astrophys. J.* **468**, 17 (1996).

[Barn] Barnothy, J.M., Quasars and the gravitational image intensifier, *Astron. J.* **70**, 666 (1965).

[Barn-B] Barnothy, J.M., Barnothy, M.F., Expected density of gravitational-lens quasars, *Astrophys. J.* **174**, 477 (1972).

[Bart95] Bartelmann, M., Cluster estimates from weak lensing, *Astron. Astrophys.* **303**, 643 (1995).

[Bart98] Bartelmann, M., Huss, A., Colberg, J.M., Jenkins, A., and Pearce, F.R. Arc statistics with realistic cluster potentials. IV. Clusters in different cosmologies, *Astron. Astrophys.* **330**, 1 (1998).

[Bart-N] Bartelmann, M., and Narayan, R., The lens parallas method: determining redshifts of faint blue galaxies through gravitational lensing, *Astrophys. J.* **451**, 60 (1995).

[Bart-NSS] Bartelmann, M., Narayan, R., Seitz, S., Schneider, P., Maximum likelihood cluster reconstruction, *Astrophys. J. Lett.* **464**, L115-L118 (1996).

[Bart-S] Bartelmann, M., and Schneider, P., Weak gravitational lensing, *Phys. Reports* **340**, 291 (2001).

[Bee-E] Beem, J., and Ehrlich, P., *Global Lorentzian Geometry*, Marcel Dekker, New York, 1981.

[Ber-G] Berger, M., and Gostiaux, B., *Differential Geometry: Manifolds, Curves, and Surfaces*, Springer-Verlag, Berlin, 1988.

[Berr76] Berry, M.V., Waves and Thom's theorem, *Adv. Phys.* **25**, 1 (1976).

[Berr81] Berry, M.V, "Singularities in waves and rays," in: *Physics of Defects*, Les Houches Lectures XXXIV, eds. R.D. Balain, M. Kleman, and J-P Poirier, p. 453, North-Holland, Amsterdam, 1981.

[Berr-U] Berry, M.V, and Upstill, C., "Catastrophe Optics: morphologies of caustics and their diffraction patterns," in: *Progress in Optics XVIII*, ed. E. Wolf, p. 257, North-Holland, Amsterdam, 1980.

[Bir] Birkhoff, D., *Collected Mathematical Papers*, vol. II, Amer. Math. Soc., New York, 1950.

[Bla90] Blandford, R., Gravitational lenses, *Q. Jl. R. Astron. Soc.* **31**, 305 (1990).

[Bla-K87a] Blandford, R., and Kochanek, C., "Gravitational lenses," in: *Dark Matter in the Universe*, eds. J. Bahcall, T. Piran, and S. Weinberg, p. 131, World Scientific, Singapore, 1987.

[Bla-K87b] Blandford, R., and Kochanek, C., Gravitational imaging by isolated elliptical potential wells. I. Cross sections, *Astrophys. J.* **321**, 658 (1987).

[Bla-N86] Blandford, R., and Narayan, R., Fermat's principle, caustics, and the classification of gravitational lens images, *Astrophys. J.* **310**, 568 (1986).

[Bla-N92] Blandford, R., and Narayan, R., Cosmological applications of gravitational lensing *Ann. Rev. Astr. Ap.* **30**, 311 (1992).

[Bla-SBV] Blandford, R., Saust, A.B., Brainerd, T.G., and Villumsen, J.V., The distortion of distant galaxy images by large-scale structure, *Mon. Not. R. Astron. Soc.* **251**, 600 (1991).

[Bli-M] Bliokh, P., and Minakov, A., *Gravitational Lenses*, Naukova Dumka, Kiev, 1989.

[Boa] Boardman, J. M., Singularities of differentiable maps, *Publ. Math. I.H.E.S.* **33**, 21 (1967).

[Bor-W] Born, M., and and Wolf, E., *Principles of Optics*, 6th Ed., Cambridge University Press, Cambridge, U.K., 1980.

[Bot] Bott, R., Marston Morse and his mathematical works, *Bull. Am. Math. Soc.* **3**, 907 (1980).

[Bou-KN] Bourassa, R.R., Kantowski, R., Norton, T.D., The spheroidal gravitational lens, *Astrophys. J.* **185**, 747 (1973).

[Bou-K] Bourassa, R.R., Kantowski, R., The theory of transparent gravitational lenses, *Astrophys. J.* **195**, 13 (1975).

[Bre] Brecher, K., "Newton, Einstein, and Gravity," in: *Revealing the Universe*, eds. J. Cornell and A.P. Lightman, p. 17, MIT Press, Cambridge 1982.

[Bri] Brill, D., Horowitz, G., Kastor, D., and Trafchen, J., Testing cosmic censorship censorship with black hole collisions, *Phys. Rev. D***49**, 840 (1994).

[Broa-TP] Broadhurst, T.J., Taylor, A.N., Peacock, J.A., Mapping cluster mass distributions via gravitational of background galaxies, *Astrophys. J.* **438**, 49 (1995).

[Broc-J] Bröcker, TH., and Jänich, K., *Introduction to Differential Topology*, Cambridge University Press, Cambridge, U.K., 1982.

[Broc-L] Bröcker, TH., and Lander, L., *Differentiable Germs and Catastrophes*, Cambridge University Press, Cambridge, U.K., 1975.

[Bru-G] Bruce, J.W., and Giblin, P.J., *Curves and Singularities*, Cambridge U. Press, Cambridge, U.K., 1992.

[Bru-M] Bruce, J.W., and Mond, D., Eds., *Singularity Theory*, Cambridge University Press, Cambrige, 1999.

[Bur] Burke, W., Multiple gravitational imaging by distributed masses, *Astrophys. J. Lett.* **244**, L1 (1981).

[Can] Canizares, C.R., Manifestations of a cosmological density of compact objects in quasar light, *Astrophys. J.* **263**, 508 (1982).

[Car-PT] Carroll, S.M., Press, W.H., and Turner, E.L., The cosmological constant, *Ann. Rev. Astr. Ap.* **30**, 499 (1992).

[Cas-H] Castrigiano, D., and Hayes, S., *Catastrophe Theory*, Addison-Wesley, Reading, 1993.

[Cha] Chandrasekhar, S., *The Mathematical Theory of Black Holes*, Oxford University Press, Oxford, 1992.

[Chan-R79] Chang, K., Refsdal, S., Flux variations of QSO 0957+561 A, B and image splitting by stars near the light path, *Nature* **282**, 561 (1979).

[Chan-R84] Chang, K., Refsdal, S., Star disturbances in gravitational lens galaxies, *Astron. Astrophys.* **132**, 168 (1984).

[Chav] Chavel, I., *Riemannian Geometry: A Modern Introduction*, Cambridge University Press, Cambrige, U.K., 1993.

[Chr84] Christodoulou, D., Violation of cosmic censorship in the gravitational collapse of a dust cloud, *Commun. Math. Phys.* **93**, 171 (1984).

[Chr94] Christodoulou, D., Examples of naked singularity formation in the gravitational collapse of a scalar field, *Annals of Math.* **140**, 607 (1994).

[Chr99] Christodoulou, D., The instability of naked singularities in the gravitational collapse of a scalar field, *Annals of Math.* **149**, 189 (1999).

[Chw] Chwolson, O., Über eine mögliche Form fiktiver Doppelsterne, *Astron. Nachr.* **221**, 329 (1924).

[Cla] Clark, E.E., The uniform transparent gravitational lens, *Mon. Not. R. Astron. Soc.* **158**, 233 (1972).

[Coh] Cohen, N., *Gravity's Lens* (John Wiley and Sons, New York, 1988).

[Col-L] Coles, P., and Lucchin, F., *Cosmology*, John Wiley and Sons, New York, 1995.

[Col] Colley, W.N., Tyson, J.A., and Turner, E.L., Unlensing multiple arcs in 0024+1654: reconstruction of the source image, *Astrophys. J.* **461**, L83 (1996).

[Cook-K] Cooke, J.H., and Kantowski, R., Time delays for multiply imaged quasars, *Astrophys. J.* **195**, L11 (1975).

[Cool] Coolidge, J., The continuity of the roots of an algebraic equation, *Ann. Math.* **9**, 116 (1908).

[Cor-SS] Cornish, N., Spergel, D., and Starkman, G., Circles in the sky: Finding topology with the microwave background radiation, *Class. Quant. Grav.* **15**, 2657 (1998).

[Cor-W] Cornish, N., and Weeks, J., Measuring the shape of the universe, *Notices of the Amer. Math. Soc.* **45**, 1463 (1998).

[Dal] Dalcanton, J.J., Canizares, C.R., Granados, A., Steidel, C.C., Stocke, J.T., Observational limits on Ω in stars, brown dwarfs, and stellar remnants from gravitational microlensing, *Astrophys. J.* **424**, 550 (1994).

[Del-T] Dell'Antonio, I.P., Tyson, J.A., Galaxy dark matter: Galaxy-galaxy lensing in the Hubble Deep Field *Astrophys. J.* **473**, L17 (1996).

[deS] de Silva, L.N.K., On gravitational-lens quasars, *Astrophys. J.* **189**, 177 (1974).

[Die] Dieudonné, J., *A History of Algebraic and Differential Topology 1900—1960*, Birkhäuser, Boston, 1989.

[dIn] d'Inverno, R., *Introducing Einstein's Relativity*, Clarendon Press, Oxford, 1998.

[doC] do Carmo, M., *Differential Geometry of Curves and Surfaces*, Prentice-Hall, Englewood Cliffs, 197).

[Dom] Dominik, M., The binary gravitational lens and its extreme cases, *Astron. Astrophys.* **349**, 108 (1999).

[Dub-FN] Dubrovin, B., Fomenko, A., and Novikov, S., *Modern Geometry — Methods and Applications. Part I.*, Springer-Verlag, Berlin, 1984.

[Dug] Dugundji, J., *Topology*, Allyn and Bacon, Boston, 1966.

[Dye-R72] Dyer, R., and Roeder, R., The distance-redshift relation for universes with no intergalactic medium, *Astrophys. J.* **174**, L115 (1972).

[Dye-R73] Dyer, R., and Roeder, R., Distance-redshift relations for universes with some intergalactic medium, *Astrophys. J.* **180**, L31 (1973).

[Edd] Eddington, A., *Space, Time and Gravitation*, Cambridge U. Press, Cambridge, U.K., 1920.

[Ehl-N] Ehlers, J., and Newman, E., The theory of caustics and wave front singularities with physical applications, *J. Math. Phys.* **41**, 3344 (2000).

[Eil-S] Eilenberg, S., and Steenrod, N., *Foundations of Algebraic Topology*, Princeton University Press, Princeton, 1952.

[Ein11] Einstein, A., Über den Einfluß der Schwerkraft auf die Ausbreitung des Lichtes, *Annalen der Physik* **35**, 898 (1911).

[Ein16] Einstein, A., Die Grundlage der allgemeinen Relativitätstheorie, *Annalen der Physik* **49**, 769 (1916).

[Ein22] Einstein, A., *Vier Vorlesungen über Relativitätstheorie*, Vieweg, Braunschweig 1922.

[Ein36] Einstein, A., Lens-like action of a star by the deviation of light in the gravitational field, *Science* **84**, 506 (1936).

[Eli] Eliashberg, J., On the singularities of folding type, *Math. USSR Izvestija* **4**, 1119 (1970).

[Erd-S] Erdl, H., and Schneider, P., Classification of the multiple deflection two point-mass gravitational lens models and applications of catastrophe theory in lensing, *Astron. Astrophys.* **268**, 453 (1993).

[Fah-KSW] Fahlman, G., Kaiser, N., Squires, G., Woods, D., Dark matter in MS1224 from distortion of background galaxies, *Astrophys. J.* **437**, 56 (1994).

[Fal-GS] Falco, E.E., Gorenstein, M.V., and Shapiro, I.I., On model-dependent bounds on H_0 from gravitational images: Application to Q0957+561, *Astrophys. J.* **289**, L1 (1985).

[Fal-KM] Falco, E.E., Kochanek, C.S., and Munoz, J.A., Limits on cosmological models from radio-selected gravitational lenses *Astrophys. J.* **494**, 47, (1998)

[Fal-SMD] Falco, E.E., Shapiro, I.I., Moustakas, L.A., Davis, M., An estimate of H_0 from Keck spectroscopy of the gravitational lens system 0957+561, *Astrophys. J.* **484**, 70 (1997).

[Fed] Federer, H., *Geometric Measure Theory*, Springer-Verlag, New York, 1969.

[Fer] Fermat, P., *Oeuvres de Fermat*, vol. 2, p. 354, Paris, 1894.

[Fey] Feynman, R., *The Character of Physical Law*, MIT Press, Cambridge, U.K., 1965.

[Fil-R] Filippenko, A., and Riess, A., Type Ia supernovae and their cosmological implications, http://arXiv.org/abs/astro-ph/9905049 (1999).

[Fis-BRT] Fischer, P., Bernstein, G., Rhee, G., and Tyson, J.A., The mass distribution of the cluster 0957+561 from gravitational lensing, *Astron. J.* **113**, 521 (1997).

[Fol] Folland, G., *Introduction to Partial Differential Equations*, Princeton University Press, Princeton, 1976.

[For-M] Fort, B., and Mellier, Y., Arc(let)s in clusters of galaxies, *Astron. Astrophys. Rev.* **5**, 239 (1994).

[Franc] Francis, G., *A Topological Picturebook*, Springer-Verlag, New York, 1987.

[Frank] Frankel, T., *Gravitational Curvature*, W. H. Freeman and Co., San Francisco, 1979.

[Franx] Franx, M., Illingworth, G.D., Kelson, D., van Dokkum, P.G., and Tran, K.-V., A pair of lensed galaxies at z=4.92 in the field of CL1358+62, *Astrophys. J.* **486**, L75 (1997).

[Fri-S] Friedrich, H., and Stewart, J., Characteristic initial data and wavefront singularities in general relativity, *Proc. Roy. Soc. Lond.* **A385**, 345 (1983).

[Fri-KN] Frittelli, S., Kling, T., and Newman, Ezra, Spacetime perspective of Schwarzschild lensing, *Phys. Rev. D* **61**, 064021 (2000).

[Fuk-FK] Fukugita, M., Futamase, T., and Kasai, M., A possible test for the cosmological constant with gravitational lenses, *Mon. Not. R. Astron. Soc.* **246**, 24p (1990).

[Fuk-FKT] Fukugita, M., Futamase, T., Kasai, M., and Turner, E.L., Statistical properties of gravitational lenses with a non-zero cosmological constant, *Astrophys. J.* **393**, 3 (1992).

[Fut-S] Futamase, T., and Sasaki, M., Light propagation and the distance-redshift relation in a realistic inhomogeneous universe, *Phys. Rev. D* **40**, 2502 (1989).

[Gau-S] Gaudi, S., and Sackett, P., Detection efficiencies of microlensing data sets to stellar and planetary companions, *Astrophys. J.* **528**, 56 (2000).

[Ger] Geroch, R., *Mathematical Physics*, The University of Chicago Press, Chicago, 1985.

[Ger-J] Geroch, R., and Jang, P., Motion of a body in general relativity, *J. Math. Phys.* **16**, 65 (1975).

[Gia-MP] Giannoni, F., Masiello, A., and Piccione, P., A variational theory for light rays in stably causal Lorentzian manifolds: regularity and multiplicity results, *Commun. Math. Phys.* **187**, 375 (1997).

[Gib] Gibson, C.G., *Singular Points of Smooth Mappings*, Pitman, London, 1979.

[Gil] Gilmore, R., *Catastrophe Theory for Scientists and Engineers*, Dover, New York, 1981.

[Gio-L] Gioia, I., and Luppino, G., The EMSS catalog of X-ray-selected clusters of galaxies. I. An atlas of CCD images of 41 distant clusters *Astrophys. J. Suppl.* **94**, 583 (1994).

[Gol-G] Golubitsky, M., and Guillemin, V., *Stable Mappings and Their Singularities*, Springer-Verlag, Berlin, 1973.

[Goren-FS] Gorenstein, M.V., Falco, E.E., and Shapiro, I.I., Degeneracies in parameter estimates for models of gravitational lens systems, *Astrophys. J.* **327**, 693 (1988).

[Gores-M] Goresky, M., and MacPherson, R., *Stratified Morse Theory*, Springer-Verlag, Berlin, 1980.

[Got81] Gott III, J., Are heavy halos made of low mass stars? A gravitational lens test, *Astrophys. J.* **243**, 140 (1981).

[Got85] Gott III, J., Gravitational lensing effects of vacuum strings: exact solutions, *Astrophys. J.* **288**, 422 (1985).

[Got-G] Gott III, J., and Gunn, J., The double quasar 1548 + 115 a, b as a gravitational lens, *Astrophys. J.* **190**, L105 (1974).

[Got-PL] Gott III, J.., Park, M., and Lee, H., Microlensing light curves: a new and efficient numerical method, *Astrophys. J.* **338**, 1 (1989).

[Gou] Gould, A., Femtolensing of gamma-ray bursters, *Astrophys. J.*, **386**, L5 (1992).

[Gou-L] Gould, A., and Loeb, A., Discovering planetary systems through gravitational microlenses, *Astrophys. J.*, **396**, 104 (1992).

[Gre-H] Greenberg, M., and Harper, J., *Algebraic Topology: A First Course*, Benjamin/Cummings, Menlo Park, 1981.

[Gri-KRS] Grieger, B., Kayser, R., Refsdal, S., and Stabell, R., The two point mass gravitational lens, *Abhandlungen aus der Hamburger Sternwarte* **X**, **4**, Hamburg-Bergedorf (1989).

[Gri-KS] Grieger, B., Kayser, R., and Schramm, R., The deconvolution of the quasar structure from microlensing light curves, *Astron. Astrophys.* **252**, 508 (1991)

[Grie] Griest, K., Galactic microlensing as a method of detecting massive compact halo objects, *Astrophys. J.* **366**, 412 (1991).

[Guc73] Guckenheimer, J., Catastrophes and partial differential equations, *Ann. Inst. Fourier. Grenoble* **23**, 4 (1973).

[Guc74] Guckenheimer, J., Caustics and non-degenerate Hamiltonians, *Topology* **13**, 127 (1974).

[Gui-P] Guillemin, V., and Pollack, A., *Differential Topology*, Prentice-Hall, Englewood Cliff, 1974.

[Gun67a] Gunn, J.E., A fundamental limitation on the accuracy of angular measurements in observational cosmology, *Astrophys. J.* **147**, 61 (1967).

[Gun67b] Gunn, J.E., On the propagation of light in inhomogeneous cosmologies. I - Mean effects, *Astrophys. J.* **150**, 737 (1967).

[Haw-E] Hawking, S., and Ellis, G., *The Large Scale Structure of Space-Time*, Cambridge U. Press, Cambridge, U.K., 1973.

[Hew] Hewitt, J.N., Turner, E.L., Schneider, D.P., Burke, B.F., Langston, G.I., and Lawrence, C.R., 'The unusual radio source MG 1131+0456: a possible Einstein ring, *Nature* **333**, 537 (1988).

[Hir] Hirsch, M. W., *Differential Topology*, Springer-Verlag, Berlin, 1976.

[Hoe-FKS] Hoekstra, H., Franx, M., Kuijken, K., and Squires, G., Weak lensing analysis of CL 1358+62 using HST observations, *Astrophys. J.* **999**, 999 (1998).

[Irw] Irwin, M.J., Webster, R.L., Hewett, P.C., Corrigan, R.T., and Jedrzejewski, R.I., Photometric variations in the Q 2237+0305 system: first detection of a microlensing event, *Astron. J.* **98**, 1989 (1989).

[Jain-Sel] Jain, B., Seljak, U. Cosmological Model Predictions for Weak Lensing: Linear and Nonlinear Regimes *Astrophys. J.* **484**, 560 (1997).

[Jak] Jaki, S.L., Johann Georg von Soldner and the gravitational bending of light, with an English translation of his essay on it published in 1801, *Foundations of Physics* **8**, 927 (1978).

[Jar] Jaroszyński, M., Multi-plane gravitational lensing by isothermal spheres, *Acta. Astr.* **39**, 301 (1989).

[Jar-P] Jaroszyński, M., and Paczyński, B., Diffraction effects in microlensing of Q2237+0305, *Astrophys. J.*, **455**, 443 (1995).

[Kai92] Kaiser, N., Weak gravitational lensing of distant galaxies, *Astrophys. J.* **388**, 272 (1992).

[Kai92a] Kaiser, N., "Statistics of gravitational lensing 1: Strong lenses" in: *New Insights into the Universe*, eds. V.J. Martinez, M. Portilla, and S. Saez, p. 247, Springer-Verlag, Berlin, 1992.

[Kai95] Kaiser, N., Nonlinear cluster lens reconstruction, *Astrophys. J. Lett.* **439**, L1 (1995).

[Kai98] Kaiser, N., Weak lensing and cosmology, *Astrophys. J.* **498**, 26 (1998).

[Kai-S] Kaiser, N., and Squires, G., Mapping the dark matter with weak gravitational lensing, *Astrophys. J.* **404**, 441 (1993).

[Kai-WL] Kaiser, N., Wilson, G., and Luppino, G.A., Large-scale cosmic shear measurements, http://arXiv.org/abs/astro-ph/0003338 (2000).

[Kas-K] Kassiola, A., and Kovner, I., Elliptical mass distributions versus elliptical potentials in gravitational lenses, *Astrophys. J.* **417**, 450 (1993).

[Kas-KB] Kassiola, A., Kovner, I., and Blandford, R., On the straight arc in Abell 2390, *Astrophys. J.* **396**, 10 (1992).

[Kee-00] Keeton, C. R., Falco, E. E., Impey, C. D., Kochanek, C. S., Lehár, J., McLeod, B. A., Rix, H.-W., Munoz, J. A., and Peng, C. Y. The Host Galaxy of the Lensed Quasar Q0957+561, *Astrophys. J.* **542**, 74 (2000).

[Kee-K96] Keeton, C., and Kochanek, C.S., "Summary of multiply imaged systems," in: *IAU Symposium 173*, eds. C.S. Kochanek and J.N. Hewitt, p. 419, Kluwer, Dordrecht, 1996.

[Kee-K97] Keeton, C., and Kochanek, C.S., Determining the Hubble Constant from the Gravitational Lens PG 1115+080, *Astrophys. J.* **487**, 42 (1992).

[Kee-K98] Keeton, C., and Kochanek, C.S., Gravitational lensing by spiral galaxies, *Astrophys. J.* **495**, 157 (1998).

[Kin] King L.J., Browne I.W.A., Muxlow, T.W.B., Narasimha, D., Patnaik, A.R., et al., Multifrequency radio observations of the gravitational lens system 1938+666, *Mon. Not. R. Astron. Soc.* **289**, 450 (1997).

[Kli] Klimov, Y.G., The deflection of light rays in the gravitational fields of galaxies, *Soviet Phys. Doklady* **8**, 119 (1963).

[Kne] Kneib, J.-P., Ellis, R.S., Smail, I., Couch, W.J., Sharples, R.M., Hubble Space Telescope Observations of the Lensing Cluster Abell 2218 *Astrophys. J.,* 471, 643 (1996).

[Kno] Knorr, W.R., *The Ancient Tradition of Geometric Problems*, Dover, Mineola, 1986.

[Koc91] Kochanek, C.S., Selection effects in optical surveys for gravitational lenses, *Astrophys. J.* **379**, 517 (1991).

[Koc92] Kochanek, C.S., Cluster Gravitational Lenses in *Gravitational Lenses*, eds. R. Kayser, T. Schramm, and L. Nieser (Lecture Notes in Physics **406**, Springer-Verlag, Berlin, 1992), p. 279.

[Koc93] Kochanek, C.S., Analytic results for the gravitational lens statistics of singular isothermal spheres in general cosmologies *Mon. Not. R. Astron. Soc.* **261**, 453 (1993).

[Koc95] Kochanek, C.S., Evidence for dark matter in MG1654+134, *Astrophys. J.* **445**, 559 (1995).

[Koc96] Kochanek, C.S., Is there a cosmological constant?" *Astrophys. J.* **466**, 638 (1996).

[Koc-A] Kochanek, C.S., and Apostolakis, J., The two-screen gravitational lens, *Mon. Not. R. Astron. Soc.* **235**, 1073 (1988).

[Koc-KM] Kochanek, C.S., Keeton, C.R., and McLeod, B.A. The importance of Einstein Rings, http://arXiv.org/abs/astro-ph/0006116 (2000).

[Kol-Tk] Kolb, E.W., Tkachev, I.I., Femtolensing and picolensing by axion miniclusters *Astrophys. J.,* **460**, L25 (1996).

[Kol-Tu] Kolb, E.W., and Turner, M.S., *The Early Universe,* Addison-Wesley, Redwood City, 1990.

[Kov87] Kovner, I., The thick gravitational lens: a lens composed of many elements at different distances, *Astrophys. J.* **316**, 52 (1987).

[Kov90] Kovner, I., Fermat principle in arbitrary gravitational fields, *Astrophys. J.* **351**, 114 (1990).

[Kui] Kuiper, N., "Immersions with minimal total curvature," in: *Colloque de Géométrie Différentielle Globale*, p. 75 (Centre Belge des Recherches Mathematiques, Louvain, 1959).

[Kun95] Kundić, T., Colley, W., Gott, III, J.R., Malhotra, S., Pen, U., et al., An event in the light curve of 0957+561A and prediction of the 1996 image B light curve, *Astrophys. J.* **455**, L5 (1995).

[Kun97] Kundić, T., Turner, E., Colley, W., Gott, III, J., Rhoads, J., et al., A robust determination of the time delay in 0957+561A, B and a measurement of the global value of Hubble's constant, *Astrophys. J.* **482**, 75 (1997).

[Lap] Laplace, P.S., *Exposition du systéme du monde*, vol. II, p. 305, De 1'Imprimerie du Circle Social, Paris, An IV, 1796.

[Leb] Lebach D. E., Corey B.E., Shapiro I.I., Ratner M.I., Weber J.C., Rogers A.E.E., Davis J.L., Herring T.A., Measurement of the solar gravitational deflection of radio waves Very-Long-Baseline Interferometry, *Phys. Rev. Lett.* **75**, 1439 (1995).

[Lev59] Levine, H., Singularities of Differentiable Mappings, Bonn Notes (1959); reprinted in: *Lectures Notes in Math* **191**, 1 (1971).

[Lev63] Levine, H., Elimination of Cusps, *Topology* **3**, 263 (1963).

[Lev64] Levine, H., The singularities, S_1^q, *Illinois J. of Math.* **8**, 152 (1964).

[Lev66] Levine, H., Mappings of manifolds into the plane, *Am. J. Math.* **88**, 357 (1966).

[Lev95] Levine, H., Computing the Euler characteristic of a manifold with boundary, *Proc. Am. Math. Soc.* **123**, 2563 (1995).

[Lev-P] Levine, H., and Petters, A., "Singularités et lentilles gravitationnelles," in: *Formes et dynamiques, renaissance d'un paradigme. Hommage á René Thom*, ed. M. Porte, p. 439, E.N.S. Éditions, Fontenay-St Cloud, 1994.

[Lev-PW] Levine, H., Petters, A. O., and Wambsganss, J. Applications of singularity theory to gravitational lensing. I. Multiple lens planes, *J. Math. Phys.* **34**, 4781 (1993).

[Lew-MRW] Lewis, G., Miralda-Escudé, J., Richardson, D., and Wambsganss, J., Microlensing light curves: a new and efficient numerical method, *Mon. Not. R. Astron. Soc.* **261**, 647 (1993).

[Lic] Lichtenberg, G.C., *Schriften und Briefe (Band 1),* Herausgegeben von W. Promies, München, 1968.

[Lie] Liebes, S., Gravitational lenses, *Phys. Rev. B* **133**, 835 (1964).

[Lip] Lipschutz, M., *Differential Geometry*, McGraw-Hill, New York, 1969.

[Lod] Lodge, O., Gravitation and light, *Nature* **104**, 354 (1919).

[Lor-EMW] Lorentz, H., Einstein, A., Minkowski, H., and Weyl, H., *The Principle of Relativity: A Collection of Original Memoirs on the Special and General Theory of Relativity*, Dover, New York, 1952.

[Lov] Lovelock, D., The four-dimensionality of spacetime and the Einstein tensor, *J. Math. Phys.* **13**, 874 (1972).

[Lu] Lu, Y., *Singularity Theory and an Introduction to Catastrophe Theory*, Springer-Verlag, Berlin, 1976.

[Lyn-P] Lynds, R., Petrosian, V., Giant luminous arcs in galaxy clusters, *Bull. Am. Astr. Soc.* **18**, 1014 (1986).

[Maj] Majthay, A., *Foundations of Catastrophe Theory*, Pitman, Boston, 1985.

[Mao] Mao, S., Gravitational microlensing by a single star plus external shear, *Astrophys. J.* **389**, 63 (1992).

[Mao-P] Mao, S., Paczyński, B., Gravitational microlensing by double stars and planetary systems, *Astrophys. J.* **374**, L37 (1991).

[Maoz-R] Maoz, D., Rix, H.-W., Early-type galaxies, dark halos, and gravitational lensing statistics, *Astrophys. J.*, **416**, 425 (1993).

[Mars-T] Marsden, J., Tromba, A., *Vector Calculus*, W. H. Freeman and Co., New York, 1988.

[Mart] Martinet, J., *Singularities of Smooth Functions*, Cambridge U. Press, Cambridge, U.K., 1982.

[Mas] Massey, W., *Homology and Cohomology Theory*, Marcel Dekker, Inc., New York, 1978.

[MatI] Mather, J., Stability of C^∞ mappings. I: The division theorem, *Ann. Math.* **87**, 89 (1968).

[MatII] Mather, J., Stability of C^∞ mappings. II: Infinitesimal stability implies stability, *Ann. Math.* **89**, 254 (1969).

[MatIII] Mather, J., Stability of C^∞ mappings. III: Finitely determined map germs, *Publ. Math. I.H.E.S.* **35**, 127 (1968).

[MatIV] Mather, J., Stability of C^∞ mappings. IV: Classification of stable germs by R-algebras, *Publ. Math. I.H.E.S.* **37**, 223 (1969).

[MatV] Mather, J., Stability of C^∞ mappings. V: Transversality, *Advances in Math.* **4**, 301 (1970).

[MatVI] Mather, J., Stability of C^∞ mappings. VI: The nice dimensions *Lecture Notes in Math* **192**, 207 (1971).

[Mel] Mellier, Y., Probing the Universe with Weak Lensing, *Ann. Rev. Astr. Ap.* **37**, 127 (1999).

[Met] Metzner, A.W.K., Observable properties of large relativistic masses, *J. Math. Phys.* **4**, 1194 (1963).

[Miche] Michell, J., On the means of discovering the distance, magnitude, &c. of the fixed stars, ..., *Philo. Trans. R. Soc. Lon.* **74**, 35 (1784). Reprinted in: Misner, C., Thorne, K., and Wheeler, J., *Black Holes: Selected Reprints*, ed. S. Detweiler, Am. Assoc. of Physics Teachers, Stony Brook, 1982.

[Micho] Michor, P. W., *Manifolds of Differentiable Mappings*, Shiva Pub. Ltd., Orpington, 1980.

[Mil] Milnor, J., *Morse Theory*, Princeton University Press, Princeton, 1963.

[Mir] Miralda-Escudé, J., Correlation function of galaxy ellipticities, *Astrophys. J.* **380**, 1 (1991).

[Mis-TW] Misner, C., Thorne, K., and Wheeler, J., *Gravitation*, W.H. Freeman and Co., San Francisco, 1973.

[Mor] Morse, M., Relations between the critical points of a real function of n variables, *Trans. Am. Math. Soc.* **27**, 345 (1925).

[Mor-C] Morse, M., and Cairns, S., *Critical Point Theory in Global Analysis and Differential Topology*, Academic Press, New York, 1969.

[Mun74] Munkres, J., *Topology*, Prentice-Hall, Englewood Cliffs, 1974.

[Mun84] Munkres, J., *Elements of Algebraic Topology*, Benjamin/Cummings, Menlo Park, 1984.

[Nar-B] Narayan, R., and Bartelmann, M. "Gravitational Lensing," in: *Formation of Structure in the Universe*, eds. A. Dekel and J.P. Ostriker, p. 360, Cambridge University Press, Cambridge, U.K., 1999.

[Nar-G] Narayan, R., and Grossman, S., The versatile elliptical gravitational lens, in *Gravitational Lenses*, eds. J. Moran, J. Hewitt, and K. Lo, p. 31, Lecture Notes in Physics **330**, Springer-Verlag, Berlin, 1989.

[Nar-W] Narayan, R., and Wallington, S., "Introduction to concepts of gravitational lensing" in: *Gravitational Lenses*, eds. R. Kayser, T. Schramm, and L. Nieser, Lecture Notes in Physics **406**, Springer-Verlag, Berlin, 1992, p. 12.

[Nem] Nemiroff, R.J., Constraining compact dark matter with gravitational lensing, *Comments on Astrophysics* **15**, 139 (1991).

[New] Newton, I., *Opticks* 1730, 4th edition, reprint Dover, New York, 1952.

[Nit] Nityananda, R., Gravitational lenses, *Current Science* **59**, 1044 (1990).

[Nit-S] Nityananda, R., Samuel, J., Fermat's principle in general relativity, *Phys. Rev. D* / **45**, 3862 (1992).

[Noo] Noonan, T.W., The effect of the Einstein light deflection on observed properties of clusters of galaxies, *Astron. J.* **76**, 765 (1971).

[Nye] Nye, J., *Natural Focusing and Fine Structure of Light*, Inst. of Physics Pub., Bristol, 1999.

[OLe] O'Leary, A.J. Redshift and Deflection of Photons bt Gravitation: A Comparison of Relativistic and Newtonian Treatments, *Am. J. Phys.* **32**, 52 (1964).

[ONe] O'Neill, B., *The Geometry of Kerr Black Holes*, A K Peters, Wellesley, 1995.

[Ost-V] Ostriker, J.P., Vietri, M., The statistics of gravitational lensing. III - Astrophysical consequences of quasar lensing, *Astrophys. J* . **300**, 68 (1986).

[Pac86a] Paczyński, B., Gravitational microlensing at large optical depth, *Astrophys. J.* **301**, 503 (1986).

[Pac86b] Paczyński, B., Gravitational microlensing by the galactic halo, *Astrophys. J.* **304**, 1 (1986).

[Pac87] Paczyński, B., Giant luminous arcs discovered in two clusters of galaxies, *Nature* **325**, 572 (1987).

[Pac96] Paczyński, B., Gravitational microlensing in the Local Group, *Ann. Rev. Astr. Ap.* **34**, 419 (1996).

[Pad] Padmanabhan, T., *Structure Formation in the Universe*, Cambridge University Press, Cambridge, U.K., 1993.

[Pai] Pais, A., *Subtle is the Lord*, Oxford U. Press, Oxford, 1982.

[Pee] Peebles, P.J.E., *Principles of Physical Cosmology*, Princeton University Press, Princeton, 1993.

[Pen66] Penrose, R., "General-relativistic energy flux and elementary optics," in: *Perspectives in Geometry and Relativity* ed. B. Hoffman, p. 259, Indiana U. Press, Bloomington, 1966.

[Pen98] Penrose, R., "The question of cosmic censorship," in: *Black Holes and Relativistic Stars*, ed. R. W. Wald, p. 103, The University of Chicago Press, Chicago, 1998.

[Per90a] Perlic, V., On Fermat's principle in general relativity: I. The general case, *Class. Quantum Grav.* **7**, 1319 (1990).

[Per90b] Perlic, V., On Fermat's principle in general relativity: II. The conformally stationary case, *Class. Quantum Grav.* **7**, 1849 (1990).

[Per95] Perlic, V., Infinite dimensional Morse theory and Fermat's principle in general relativity. I, *J. Math. Phys.* **36**, 6915 (1995).

[Per00a] Perlick, V., "Gravitational lensing from a geometric viewpoint," in: *Einstein's Field Equations and Their Physical Implications: Essays in Honour of Juergen Ehlers*, ed. B. Schmidt, Lecture Notes in Physics **540**, Springer-Verlag, Berlin, 2000.

[Per00b] Perlick, V., *Ray Optics, Fermat's Principle, and Applications to General Relativity*, Springer-Verlag, Berlin, 2000.

[Pet91] Petters, A., *Singularities in Gravitational Microlensing*, Ph.D. thesis, MIT, Cambridge, U.K., 1991.

[Pet92] Petters, A. O., Morse theory and gravitational microlensing, *J. Math. Phys.* **33**, 1915 (1992).

[Pet93] Petters, A. O., Arnold's singularity theory and gravitational lensing, *J. Math. Phys.* **34**, 3555 (1993).

[Pet95a] Petters, A. O., Multiplane gravitational lensing. I. Morse theory and image counting, *J. Math. Phys.* **36**, 4263 (1995).

[Pet95b] Petters, A. O., Multiplane gravitational lensing. II. Global Geometry of Caustics, *J. Math. Phys.* **36**, 4276 (1995).

[Pet96a] Petters, A. O., Lower bounds on image magnification in gravitational lensing, *Proc. R. Soc. Lond. A* **452**, 1475 (1996).

[Pet96b] Petters, A., "Mathematical aspects of gravitational lensing," in: *Proceedings of the Seventh Marcel Grossmann Meeting on General Relativity*, eds. R. T. Jantzen and G. Mac Keiser, p. 1117, World Scientific, Singapore, 1996.

[Pet97] Petters, A. O., Multiplane gravitational lensing III: Upper bound on number of images, *J. Math. Phys.* **38**, 1605 (1997).

[Pet-Wic95] Petters, A. O., and Wicklin, F. J., Caustics of the double-plane two-point-mass gravitational lens with continuous matter and shear, *Mon. Not. R. Astron. Soc.* **277**, 1399 (1995).

[Pet-Wic98] Petters, A. O., and Wicklin, F. J., Fixed points due to gravitational lenses, *J. Math. Phys.* **39**, 1011 (1998).

[Pet-Wit] Petters, A. O., and Witt, H., Bounds on number of cusps due to point mass gravitational lenses, *J. Math. Phys.* **37**, 2920 (1996).

[Poi] Poincaré, H., *Oeuvres*, vol. I, Gauthier-Villars, Paris, 1928.

[Pol] Polya, G., *How to Solve it*, Princeton University Press, Princeton, 1973.

[Pos-S] Poston, T., and Stewart, I., *Catastrophe Theory and its Applications*, Dover, New York, 1978.

[Pra] Pratt, M.R., et al., "Real-time detection of gravitational microlensing," in: *IAU Symposium 173*, eds. C.S. Kochanek and J.N. Hewitt, p. 221, Kluwer, Dordrecht, 1996.

[Pre] Press, W.H., Prognosticating the future of gravitational lenses," in: *Astrophysical Applications of Gravitational Lensing*, eds. C.S. Kochanek and J.N. Hewitt, p. 407, Kluwer, Dordrecht, 1996.

[Pre-G] Press, W.H., and Gunn, J.E., Method for detecting a cosmological density of condensed objects, *Astrophys. J.* **185**, 397 (1973).

[Rau] Rauch, K., Gravitational microlensing of high redshift supernovae by compact objects, *Astrophys. J.* **374**, 83 (1991).

[Rau-B] Rauch, K., and Blandford, R., Optical caustics in a Kerr spactime and the origin of rapid X-ray variability in active galactic nuclei, *Astrophys. J.* **421**, 46 (1994).

[Rau-MWP] Rauch, K., Mao, S., Wambganss, J., and Paczynski, B., Caustic-induced features in microlensing magnification probability distributions, *Astrophys. J.* **386**, 30 (1992).

[Ray-BB] Raychaudhuri, A., Banerji, S., and Banerjee, A., *General Relativity, Astrophysics, and Cosmology*, Springer-Verlag, Berlin, 1992.

[Ree-S] Reed, M., and Simon, B., *Methods of Modern Mathematical Physics I*, Academic Press, New York, 1980.

[Ref64a] Refsdal, S., The gravitational lens effect, *Mon. Not. R. Astron. Soc.* **128**, 295 (1964).

[Ref64b] Refsdal, S., On the possibility of determining Hubble's parameter and the masses of galaxies from the gravitational lens effect, *Mon. Not. R. Astron. Soc.* **128**, 307 (1964).

[Ref66] Refsdal, S., On the possibility of testing cosmological theories from the gravitational lens effect, *Mon. Not. R. Astron. Soc.* **132**, 101 (1966).

[Ref70] Refsdal, S., On the Propagation of light in universes with inhomogeneous mass distribution, *Mon. Not. R. Astron. Soc.* **159**, 357 (1970).

[Ref94] Refsdal, S., Gravitational lenses, *Rep. Prog. Phys.* **56**, 117 (1994).

[Ren-SS] Renn, J., Sauer, T., and Stachel, J. The Origin of Gravitational Lensing: A Postscript to Einstein's 1936 *Science* paper, *Science* **275**, 184 (1997).

[Rie] Riess, A.G., et al., Observational evidence from supernovae for an accelerating universe and a cosmological constant," *Astronom. J.* **116**, 1009 (1998).

[Rin] Rindler, W., *Essential Relativity*, Springer-Verlag, Berlin, 1977.

[Rix-MTF] Rix, H.-W., Maoz, D., Turner, E.L., Fukugita, M., Galaxy mergers and gravitational lens statistics, *Astrophys. J.* **435**, 49 (1994).

[Rix] Rix, H.-W., Schneider, D.P., Bahcall, J.N., Hubble Space Telescope wide field camera imaging of the gravitational lens 2237+0305, *Astron. J.* **104**, 959 (1992).

[Rus] Russell, H.N., A relativistic eclipse, *Scientific American* **156**, 76 (1937).

[Sach] Sachs, R., Gravitational waves in general relativiy VI. The outgoing radiation condition, *Proc. Roy. S. Lond.* **264A**, 309 (1961).

[Sach-W] Sachs, R., and Wu, H., *General Relativity for Mathematician*, Springer, Berlin, 1977.

[Sack] Sackett, P., "The 24-hour night shift: astronomy from microlensing monitoring networks," in: *Gravitational Lensing: Recent Progress and Future Goals*, eds. T. Brainerd and C. Kochanek, Astronomical Society of the Pacific Conference Series, 2000.

[Sak] Sakai, T., *Riemannian Geometry*, American Mathematical Society, Providence, 1996.

[San] Sanitt, N., Quasi-stellar objects and gravitational lenses, *Nature* **234**, 199 (1971).

[Sas] Sasaki, M., Cosmological gravitational lens equations — its validity and limitations, *Progress of Theo. Phys.* **90**, 753 (1993).

[Sche] Schechter, P.L., Bailyn, C.D., Barr, R., Barvainis, R., Becker, C.M., et al., The quadruple gravitational lens PG 1115+080: time delays and models, *Astrophys. J.* **475**, L85 (1997).

[Schi90] Schild, R.E., The time delay in the twin QSO Q 0957+561, *Astron. J.* **100**, 1771 (1990).

[Schi96] Schild, R., Microlensing variability of the gravitationally lensed quasar Q0957+561A, B, *Astrophys. J.* **464**, 125 (1996).

[Schm-W] Schmidt, R., and Wambsganss, J., Limits on MACHOs from microlensing in the double quasar," *Astron. Astrophys.* **335**, 379 (1998).

[Schn84] Schneider, P., The amplification caused by gravitational bending of light, *Astron. Astrophys.* **140**, 119 (1984).

[Schn85] Schneider, P., A new formulation of gravitational lens theory, time-delay, and Fermat's principle, *Astron. Astrophys.* **143**, 413 (1985).

[Schn95] Schneider, P., Cluster lens reconstruction using only observed local data, *Astron. Astrophys.* **302**, 639 (1995).

[Schn97] Schneider, P., The cosmological lens equation and equivalent single-plane gravitational lens. *Mon. Not. R. Astron. Soc.* **292**, 673 (1997).

[Schn-EF] Schneider, P., Ehlers, J., and Falco, E.E., *Gravitational Lenses*, Springer, Berlin, 1992.

[Schn-Sc] Schneider, P., and Schmid-Burgk, J., Mutual coherence of gravitationally lensed images, *Astron. Astrophys.* **148**, 369, (1985)

[Schn-Se] Schneider, P., and Seitz, C., Steps towards nonlinear cluster inversion through gravitational distortions. I.Basic considerations and circular clusters, *Astron. Astrophys.* **294**, 411 (1995).

[Schn-W86] Schneider, P., and Weiss, A., The two-point-mass lens: detailed investigation of a special asymmetric gravitational lens, *Astron. Astrophys.* **164**, 237 (1986).

[Schn-W88] Schneider, P., and Weiss, A., Light propagation in inhomogeneous universes: the ray-shooting method, *Astrophys. J.* **330**, 1 (1988).

[Schn-W92] Schneider, P., and Weiss, A., The gravitational lens equation near cusps, *Astron. Astrophys.* **260**, 1 (1992).

[Schr-K] Schramm, T., and Kayser, R., The complex theory of gravitational lensing, *Astron. Astrophys.* **299**, 1 (1995).

[Schu] Schutz, B., *A First Course in General Relativity*, Cambridge University Press, Cambridge, U.K., 1985.

[Sei-KSS] Seitz, C., Kneib, J.-P., Schneider, P., Seitz, S., The mass distribution of CL0939+4713 obtained from a weak lensing analysis of a WFPC2 image, *Astron. Astrophys.* **314**, 707 (1996).

[Sei-S92] Seitz, S., and Schneider, P., Two theorems in multiple deflection gravitational lens theory, *Astron. Astrophys.* **265**, 1 (1992).

[Sei-S94] Seitz, S., and Schneider, P., Some remarks on multiple deflection gravitational lensing, *Astron. Astrophys.* **287**, 349 (1994).

[Sei-S95] Seitz, S., and Schneider, P., Steps towards nonlinear cluster inversion through gravitational distortions. II. Generalization of the Kaiser and Squires method, *Astron. Astrophys.* **302**, 9 (1995).

[Sei-S96] Seitz, S., and Schneider, P., Cluster lens reconstruction using only observed data: an improved finite-field inversion technique, *Astron. Astrophys.* **305**, 383 (1996).

[Sei-SE] Seitz, S., Schneider, P., and Ehlers, J., Light propagation in arbitrary spacetimes and the gravitational lens approximation, *Class. Quant. Grav.* **11**, 2345 (1994).

[Sel94] Seljak, U., Large-scale structure effects on the gravitational lens image positions and time delay, *Astrophys. J.* **436**, 509 (1994).

[Sel96] Seljak, U., Gravitational lensing effect on cosmic microwave background anisotropies: a Power spectrum approach, *Astrophys. J.* **463**, 1 (1996).

[Sha] Shapiro, I., Fourth test of general relativity, *Phys. Rev. Letters*, **13**, 789 (1964).

[Sol] Soldner, J., Uber die Ablenkung eines Lichtstrals von seiner geradlinigen Bewegung durch die Attraktion eines Weltkörpers, an welchem er nahe vorbeigeht, *Berliner Astron. Jahrb. 1804*, p. 161 (1801).

[Sou-FMP] Soucail, G., Fort, B., Mellier, Y., Picat, J.P., A blue ring-like structure in the center of the A370 cluster of galaxies, *Astron. Astrophys.* **172**, L14 (1987).

[Spa] Spanier, E., *Algebraic Topology*, McGraw-Hill, New York, 1966; reprinted by Springer-Verlag, Berlin, 1982.

[Squ-K] Squires, G., and Kaiser, N., Unbiased cluster lens reconstruction, *Astrophys. J.* **473**, 65 (1996).

[Step-H76] Stephani, H., and Herlt, E., Wave optics of the spherical gravitational lens -I: Diffraction of a plane electromagnetic wave, *Intern. J. Theoret. Physics* **15**, 45 (1976).

[Step-H78] Stephani, H., and Herlt, E., Wave optics of the spherical gravitational lens -II: Diffraction of a plane electromagnetic wave, *Intern. J. Theoret. Physics* **17**, 189 (1978).

[Stew] Stewart, J., *Advanced General Relativity*, Cambridge University Press, Cambridge, U.K., 1990.

[Str] Strang, G., *Linear Algebra and its Applications*, 3d ed., Harcourt Brace Jovanovich, Fort Wort, 1988.

[Sur] Surdej, J. 1996 and Pospieszalska-Surdej, A.: http://www.stsci.edu/ftp/stsci/library/ grav_lens/grav_lens.html.

[Tho] Thom, R., Les singularités des applications differentiables, *Annales de l'Institut Fourier* **6**, 43 (1956).

[Thu] Thurston, W., *Three-Dimensional Geometry and Topology,* Princeton University Press, Princeton, 1997.

[Tik] Tikhov, G.A., Sur le Deviation des Rayons Lumineux dans le Champ de Gravitation des Etoiles, *Dokl. Akad. Nauk. S.S.S.R.* **16**, 199 (1937).

[Too] Toomer, G.J., *Diocles: On Burning Mirrors*, Springer-Verlag, Berlin, 1976.

[Tri] Trimble, V., "The first lenses," in: *Gravitational Lensing: Recent Progress and Future Goals*, eds. T. Brainerd and C. Kochanek, Astronomical Society of the Pacific Conference Series, 2000.

[Tur] Turner, E.L., Gravitational lensing limits on the cosmological constant in a flat universe, *Astrophys. J.* **365**, L43 (1990).

[Tur-OG] Turner, E., Ostriker, J., and Gott III, J., The statistics of gravitational lenses: the distribution of image angular separations and lens redshift, *Astrophys. J.* **284**, 1 (1984).

[Tur-T] Turner, M., and Tyson, J.A., Cosmology at the millennium, *Rev. Mod. Phys.* **71**, S145 (1999).

[Tys-KD] Tyson, J.A., Kochanski, G., and Dell'Antonio, I., Detailed mass map of CL0024+1654 from strong lensing, *Astrophys. J. Lett.* **498**, L101 (1998).

[Tys-VW] Tyson, J.A., Valdes, F., Wenk, R.A., Detection of systematic gravitational lens image alignments: mapping dark matter in galaxy clusters, *Astrophys. J. Lett.* **349**, L1 (1990).

[Uda93] Udalski, A., Szymański, M., Kałużny, J., Kubiak, M., Krzemiński, W., et al., The optical gravitational lensing experiment. Discovery of the first candidate microlensing event in the direction of the Galactic Bulge, *Acta Astron.* **43**, 289 (1993).

[Uda94] Udalski, A., Szymański, M., Mao, S., Di Stefano, R., Kaluzny, J., et al., The optical gravitational lensing experiment. Ogle No 7: binary microlens or a new unusual variable, *Astrophys. J.* **436**, L103 (1994).

[Ulm-G] Ulmer, A., and Goodman, J., Femtolensing: beyond the semiclassical approximation, *Astrophys. J.,* **442**, 67 (1995).

[Val-TJ] Valdes, F., Tyson, J.A., Jarvis, J., Alignment of faint galaxy images: Cosmological distortion and rotation, *Astrophys. J.* **271**, 431 (1983).

[van00] van Waerbecke, et al., Detection of correlated galaxy ellipticities from CFHT data: first evidence for gravitational lensing by large-scale structures, *Astron. Astrophys. J.,* **358**, 30 (2000).

[Vic] Vick, J., *Homology Theory*, Academic Press, New York, 1973.

[Vil] Vilenkin, A., Cosmic Strings as Gravitational Lenses, *Astrophys. J.,* **282**, L51 (1984).

[Vill] Villumsen, J., Weak lensing by large scale structure in open, flat, and closed universes, lens system 1938+666, *Mon. Not. R. Astron. Soc.* **281**, 369 (1996).

[Vir-E] Virbhadra, K., and Ellis, G., Schwarzchild black hole lensing, *Phys. Rev. D* **62**, 084003 (2000).

[Vir-NC] Virbhadra, K., Narasimha, D., and Chitre, S., Role of the scalar field in gravitational lensing, *Astron. Astrophys.* **337**, 1 (1998).

[Wald84] Wald, R., *General Relativity*, U. Chicago Press, Chicago, 1984.

[Wald97] Wald, R., Gravitational collapse and cosmic censorship, *http://xxx.lanl.gov/list/gr-qc/9710:* gr-qc/9710068 (1997).

[Wall] Wallace, A., *Algebraic Topology: Homology and Cohomology*, W. A. Benjamin, Inc., New York, 1970.

[Wals] Walsh, D., "0957+561: The unpublished Story," in: *Lecture Notes in Physics* **330**, eds. Moran, J.M., et al., p. 11, Springer-Verlag, Berlin, 1989.

[Wals-CW] Walsh, D., Carswell, R.F., and Weymann, R.J, 0957+561 A, B: twin quasistellar objects or gravitational lens?, *Nature* **279**, 381 (1979).

[Wam90] Wambsganss, J., *Gravitational Microlensing*, Max Planck Report MPA 550, Garching, 1990.

[Wam93] Wambsganss, J., "Gravitational Microlensing: Simulations and Observations," in: *Gravitational Lenses in the Universe*, eds. J. Surdej et al., p. 369, Universite de Liege, Liege, 1993.

[Wam97] Wambsganss, J., Discovering Galactic planets by gravitational microlensing: magnification patterns and light curves, *Mon. Not. R. Astron. Soc.* **284**, 172 (1997).

[Wam-CO] Wambsganss, J., Cen, R., and Ostriker, J.P., Testing cosmogonic models by gravitational lensing: method and first applications, *Astrophys. J.* **494**, 29 (1998).

[Wam-COT] Wambsganss, J., Cen, R., Ostriker, J.P., and Turner, E.L., Testing cosmogonic models with gravitational lensing, *Science* **268**, 274 (1995).

[Wam-COX] Wambsganss, J., Cen, R., Ostriker, J.P., and Xu, G., Effects of weak lensing on the determination of q_0, *Astrophys. J.* **475**, L81 (1997).

[Wam-K] Wambsganss, J., and Kundić, T., Gravitational microlensing by random motion of stars: analysis of light curves, *Astrophys. J.* **450**, 19 (1995).

[Wam-P] Wambsganss, J., and Paczyński, B., Parameter degeneracy in models of the quadruple lens system Q2237+0305, *Astron. J.* **108**, 1156 (1994).

[Wam-PS] Wambsganss, J., Paczyński, B., and Schneider, P., Interpretation of the microlensing event in QSO 2237+0305, *Astrophys. J.* **358**, L33 (1990).

[Wam-WS] Wambsganss, J., Witt, H., and Schneider, P., Gravitational microlensing: powerful combination of ray-shooting and parametric representation of caustics, *Astron. Astrophys.* **258**, 591 (1992).

[War] Warner, F., *Foundations of Differentiable Manifolds and Lie Groups*, Springer-Verlag, Berlin, 1983.

[Was] Wasserman, G., Stability of Unfoldings, *Lecture Notes in Mathematics* **393** (1971).

[Wee85] Weeks, J., *The Shape of Space*, Marcel Dekker, New York, 1985.

[Wee98] Weeks, J., Reconstructing the global topology of the universe from the cosmic microwave background radiation, *Class. Quant. Grav.* **15**, 2599 (1998).

[Wei72] Weinberg, S., *Gravitation and Cosmology*, John Wiley and Sons, New York, 1972.

[Wei97] Weinberg, S. " Theories of the cosmological constant," in: *Critical Dialogues in Cosmology*, ed. Neil Turok, World Scientific, p. 195, Singapore 1997.

[Wei00] Weinberg, S. The cosmological constant problems, http://arXiv.org/abs/astro-ph/0005265 (2000).

[Whe-Z] Wheeden, R., and Zygmund, A., *Measure and Integral*, (Marcel Dekker, New York, 1977).

[Whi37] Whitney, H., On regular closed curves in the plane, *Compositio Math.* **4**, 276 (1937).

[Whi55] Whitney, H., On singularities of mappings of Euclidean spaces. I. Mappings of the plane into the plane, *Annals of Math.* **62**, 374 (1955).

[Whi58] Whitney, H., "Singularities of mappings of Euclidean spaces," in: *Symposium International de Topologia Algebraica*, p. 258, Mexico City, 1958.

[Wil] Will, C., Henry Cavendish, Johann von Soldner, and the deflection of light, *Am. J. of Phys.* **56**, 413 (1988).

[Wit90] Witt, H., Investigation of high amplification events in light curves of gravitationally lensed quasars, *Astron. Astrophys.* **236**, 311 (1990).

[Wit93] Witt, H., An efficient method to compute microlensed light curves for point sources, *Astrophys. J.* **403**, 530 (1993).

[Wit95] Witt, H., The effect of the stellar size on microlensing at the Baade window, *Astrophys. J.* **449**, 42 (1995).

[Wit-M] Witt, H.J., and Mao, S., Can lensed stars be regarded as pointlike for microlensing by MACHOs?, *Astrophys. J.* **430**, 505 (1994).

[Wit-P] Witt, H.J., and Petters, A. O., Singularities of the one- and two-point mass gravitational lens, *J. Math. Phys.* **34**, 4093 (1993).

[Wittm00] Wittman, D.M., et al., Detection of weak gravitational lensing distortions of distant galaxies by cosmic dark matter at large scales, *Nature* **405**, 143 (2000).

[Wu-M] Wu, X., and Mao, S., The cosmological constant and statistical lensing of giant arcs, *Astrophys. J.* **463**, 404 (1996).

[Zaka76] Zakalyukin, V. M., Lagrangain and Legendrian singularities, *Funct. Anal. Appl.* **10**, 23 (1976).

[Zaka84] Zakalyukin, V. M., Reconstructions of fronts and caustics depending on a parameter ans versality of mappings, *J. Sov. Math.* **27**, 2713 (1984).

[Zaka92] Zakalyukin, V. M., "Nice dimensions and their generalizations in singularity theory," in: *Singularity Theory and Some Problems of Functional Analysis*, ed. S. G. Gindikin, p. 137, American Mathematical Society, Providence, 1992.

[Zakh97] Zakharov, A., *Gravitational Lenses and Microlenses*, Yanus, Moscow, 1997.

[Zakh99] Zakharov, A., "Gravitational lens equation near cusps," in: *Proceedings of the Eighth Marcel Grossmann Meeting on General Relativity*, eds. T. Piran and R. Ruffini, World Scientific, Singapore, 1999.

[Zee-T] Zeeman, E.C. and Trotman, D.J.A., *The classification of elementary catastrophes of co-dimension* $\leq$ *5*, *Lecture Notes in Math* **525**, 263 (1976).

[Zwi33] Zwicky, F., Die Rotverschiebung von extragalaktischen Nebeln, *Helv. Phys. Acta* **6**, 110 (1933).

[Zwi37a] Zwicky, F., Nebulae as gravitational lenses, *Phys. Rev.* **51**, 290 (1937).

[Zwi37b] Zwicky, F., On the probability of detecting nebulae which act as gravitational lenses *Phys. Rev.* **51**, 679 (1937).

[Zwi57] Zwicky, F., *Morphological Astronomy*, Springer-Verlag, Berlin, 1957.

Index of Notation

Index

Bold page number locates mathematical definition of entry